SALMONELLOSIS

MONOGRAPHIAE BIOLOGICAE

EDITORES

W. W. WEISBACH
The Hague

P. VAN OYE
Ghent

VOLUMEN XIII

SPRINGER-SCIENCE+BUSINESS MEDIA, B.V. 1964

THE WORLD PROBLEM
OF
SALMONELLOSIS

EDITED BY

E: VAN OYE
Brussels

SPRINGER-SCIENCE+BUSINESS MEDIA, B.V. 1964

ISBN 978-94-011-9844-8   ISBN 978-94-011-9842-4 (eBook)
DOI 10.1007/978-94-011-9842-4

Zuid-Nederlandsche Drukkerij N.V. — 's-Hertogenbosch

CONTENTS

INTRODUCTION

PAR

A. LAFONTAINE
Directeur de l'Institut d'Hygiène et d'Epidémiologie, Bruxelles

L'histoire des fièvres typhoïdes et paratyphoïdes présente une étonnante opposition entre le caractère définitif des descriptions magistrales des cliniciens du XIX° siècle et la fluctuation des découvertes biologiques qui, à chacune de leurs étapes, ont à la fois ouvert un chapitre pathogénique nouveau et servi de départ à des techniques nouvelles.

C'est en 1820 que BRETONNEAU décrit la dothienentérie; neuf ans plus tard, LOUIS réalise la synthèse anatomo-clinique de la maladie et, en 1834, CHAUVEL lui donne le nom de fièvre typhoïde: leurs descriptions restèrent inattaquables jusqu'au moment où les fièvres typhoïdes et paratyphoïdes virent leur pronostic modifié et leur évolution accélérée par l'usage du chloramphénicol de WOOD-WAARD et SMADEL.

La nature épidémique et l'allure contagieuse furent démontrées par TROUSSEAU, par PETTENKOFER et par BUDD, mais ce n'est qu'en 1880 qu' EBERTH décela le bacille typhique dont les caractères furent mis en évidence par GAFFKI: c'était la première *Salmonella*, début de la longue série que nous continuons à identifier.

WIDAL et SCHOTTMÜLLER ouvrirent deux nouvelles étapes biologiques de la maladie en ajoutant aux possibilités de diagnostic l'appoint considérable de l'agglutination spécifique et de l'hémoculture.

Les épidémiologistes purent mieux connaître les réservoirs de virus et les modes de transmission et en arrivèrent à modifier la réceptivité par la vaccination. Après des succès généralisés, la diminution progressive de l'activité de celle-ci demanda une explication: une connaissance approfondie du comportement biochimique des souches avec comme corollaires la découverte des milieux sélectifs et des milieux d'enrichissement, l'étude détaillée de la structure antigénique et l'application de la bactériophagie spécifique permit de comprendre pourquoi la protection apportée par les vaccins n'était pas universelle.

En 1939 l'école uruguayenne avec HORMAECHE attirait par ailleurs l'attention sur le rôle de plus en plus large que jouaient les souches de *Salmonella* d'origine animale dans la pathologie humaine et en particulier chez les enfants.

En même temps qu'une meilleure compréhension de la structure antigénique, les travaux de pathologie expérimentale, les recherches biochimiques et les études génétiques éclairaient d'autres aspects passionnants du germe *Salmonella*.

La fièvre typhoïde décrite voici 140 ans a ouvert un chapitre de la pathologie et de la recherche médicale qui est loin d'être clos.

L'universalité de l'agent causal et ses relations étroites avec certains organismes fort répandus dans la nature a ouvert des aspects nouveaux de l'apparition des souches pathogènes.

Le polymorphisme antigénique des *Salmonella* et le damier de leur sensibilité bactériophagique sollicitent des études nouvelles en génétique microbienne en même temps que l'adoption à l'échelle mondiale d'un système de classification unique. Leur dispersion chez tous les vertébrés, leur adaptabilité éventuelle à de nouveaux hôtes en font un sujet de choix pour l'étude de l'évolution étiologique des affections qu'elles provoquent.

La persistance des *Salmonella* dans le sol, les eaux, les denrées alimentaires en font un des principaux soucis des hygiénistes de tous les pays d'autant plus que les échanges multiples de personnes et de produits de consommation font apparaître des souches nouvelles en des endroits inattendus: les dégâts qu'elles causent au cheptel animal, les frais qu'entraînent le traitement des aliments susceptibles d'en contenir, les complications humaines qu'elles provoquent constituent en outre un problème économique important pour toutes les nations.

Ce livre vient en son temps et la collaboration d'auteurs de toutes les parties du monde permet de le considérer comme une somme de nos connaissances actuelles: les jeunes chercheurs ne le considéreront point comme une conclusion mais comme un chapitre inachevé sur un des sujets les plus captivants de la microbiologie pure et de la bactériologie au service de l'épidémiologie, de la pathogénie et de l'hygiène.

LES SALMONELLOSES: 80 ANS D'HISTOIRE

PAR

GUIDO GHYSELS
Institut d'Hygiène et d'Epidémiologie, Bruxelles

L'histoire des salmonelloses se confond en partie, surtout à ses débuts, avec celle de la fièvre typhoïde, la première des infections à *Salmonella* qui fut identifiée bactériologiquement, il y a 80 ans de cela.

La fièvre typhoïde est une de ces maladies qui semblent avoir existé de tous temps et certains veulent en retrouver une description dans les écrits d'HIPPOCRATE. Mais le terme "typhus" employé depuis l'antiquité ne désignait pas uniquement la vraie fièvre typhoïde, maladie infectieuse bien délimitée: il couvrait également un groupe disparate d'affections morbides caractérisées par un état plus ou moins prononcé de stupeur, de prostration et d'adynamie qui frappait les cliniciens. Chose curieuse, les médecins du XVII° siècle savaient probablement reconnaître encore la maladie typhoïdique, mais par après de nombreux cliniciens s'attachèrent à distinguer les "typhus" les uns des autres en se basant uniquement sur la symptomatologie. Leurs efforts n'aboutirent qu'à créer des classifications artificielles et erronées. Il fallut attendre le début du XIX° siècle pour que, de ce chaos de fièvres, la typhoïde soit à nouveau dégagée en tant qu'entité clinique. A partir de cette époque l'étude clinique fut progressivement complétée par des recherches de laboratoire.

Au cours d'une première étape un rapprochement fut fait entre la symptomatologie et les lésions anatomo-pathologiques. L'honneur de ces progrès revient à des cliniciens français. PROST, en 1804, est frappé par l'aspect des altérations de l'intestin grêle au cours de certaines fièvres, à cette époque dénommées ,,fièvres muqueuses, adynamiques et ataxiques", sans toutefois reconnaître la spécificité de la maladie. BRETONNEAU, en 1818, s'attache à l'étude de cette "fièvre entéro-mésentérique" qu'il appelle *dothiénentérite*, nom changé par son élève TROUSSEAU en *dothienentérie*. Mais c'est à LOUIS que revient le mérite d'avoir fait la synthèse: dans un chef d'oeuvre de la littérature médicale intitulé:" Recherches anatomiques, pathologiques et thérapeutiques sur la maladie connue sous les noms de gastroentérite, fièvre putride, adynamique" paru en 1829, il démontre la spécificité des lésions anatomo-pathologiques qu'il retrouvait toujours dans les nombreuses fièvres que ses prédécesseurs avaient individualisées. Il regroupe toutes ces affections et en fait une seule maladie bien déterminée qu'il appelle "fièvre typhoïde".

Longtemps encore après le travail fondamental de Louis, l'étiologie et la transmission de la maladie ont été l'objet d'hypothèses les plus variées et qui furent défendues avec acharnement par leurs protagonistes. En 1828, Leuret emploie pour la première fois le mot *contagion* à l'occasion d'une épidémie de fièvre typhoïde. Bretonneau et d'autres soutiennent cette thèse mais les premiers "contagionistes" pour lesquels la maladie était due à un germe d'origine exogène n'ont pas pu ébranler l'ancienne croyance aux miasmes. En 1858, Murchinson défend encore une théorie "pythogénique" selon laquelle l'affection serait due à une intoxication par des gaz provenant de la fermentation des matières organiques et fécales et cela indépendamment d'un cas antérieur. Cette théorie, si fausse qu'elle fût, a eu au moins le mérite de provoquer des travaux d'assainissement dans les agglomérations de Grande Bretagne.

Pettenkofer, en 1868, essaie d'expliquer l'épidémiologie par une théorie qui attache une grande importance au sol et aux variations du niveau des nappes d'eau après qu'en 1856 Budd avait avancé une hypothèse qui reste presque entièrement valable: elle est basée sur le postulat que chaque cas de fièvre typhoïde provient d'un cas antérieur et que la toxine spécifique est répandue par les matières fécales d'un malade.

Tous ces travaux ont en fait préparé la voie à l'étape bactériologique au cours de laquelle des recherches devaient permettre de préciser l'origine et le mode de transmission de l'infection. Dès le moment où la maladie fut définie sous ses aspects étiologique, pathologique et symptomatologique, elle devint l'objet de recherches approfondies et attira l'attention de nombreux bactériologues dont les noms restent inscrits dans l'histoire de la microbiologie. La fièvre typhoïde devient le type même de la maladie infectieuse. Les techniques et les connaissances nouvelles acquises grâce à son étude furent étendues aux recherches sur les autres grandes infections humaines et animales.

L'ère bactériologique de l'étude des salmonelloses débute en 1880 avec la découverte de l'agent causal de la fièvre typhoïde: à un groupe de médecins de Zürich Eberth communique qu'il a pu déceler le bacille typhique dans la rate et dans les ganglions mésentériques d'une personne décédée à la suite de cette maladie.

En 1872 déjà, Coze & Feltz avaient décrit un bâtonnet mobile dans le sang d'un malade typhoïdique mais il ne fut pas prouvé qu'il s'agissait du bacille typhique. La découverte d'Eberth, par contre, fut presque immédiatement confirmée par Robert Koch qui réussit en 1881 à colorer et à photographier le bacille typhique dans les organes d'un cadavre. La même année encore, Meyer publie les résultats de ses recherches microscopiques tendant à prouver que la fièvre typhoïde se développe à partir d'une infection primaire intestinale. En 1884, Gaffky, qui était à ce moment assistant de Koch, réussit pour la première fois à isoler le bacille en culture pure sur un milieu solide, ce qui permit une étude approfondie du germe. GAFFKY fut également le premier à décrire la répartition du bacille dans l'organisme du malade et à prouver qu'il est toujours décelable dans les cas de fièvre typhoïde alors qu'il ne l'est jamais dans une autre maladie. Les recherches de GAFFKY forment le point de départ de tous les travaux ultérieurs sur le diagnostic, la pathologie et la propagation de la fièvre typhoïde.

Depuis les travaux de GAFFKY, le bacille d'EBERTH fut reconnu comme l'agent spécifique de la fièvre typhoïde abdominale et un grand nombre de bactériologistes s'attachèrent à l'isoler et à l'étudier. Mais les moyens techniques dont les laboratoires disposaient à l'époque ne permettaient pas toujours d'arriver à des résultats concordants. Le plus souvent il fut impossible de séparer le bacille typhique de la flore saprophyte. L'identification du germe isolé restait la plupart du temps douteuse. Et tandis qu'un nombre toujours plus grand de caractères soit-disant spécifiques du bacille typhique furent décrits, le grand groupe de bacilles que certains considéraient comme étant des bacilles coli et que d'autres classaient parmi les "pseudo-typhiques" se révéla de plus en plus variable. Dans beaucoup de cas furent considérés comme bacilles typhiques des germes qui n'avaient en commun que quelques-uns des caractères du bacille d'EBERTH. Ces résultats divergents amenèrent une confusion dans les esprits et consécutivement provoquèrent du scepticisme envers les concepts de GAFFKY. En 1890 encore KOCH conclut prudemment qu'un critère absolu permettant de distinguer le bacille typhique n'existe pas et qu'il faut tenir compte de l'origine de la souche examinée. D'autres auteurs, et non des moindres, arrivaient à des conclusions moins nuancées. SANARELLI (1894) par exemple, trouvant ses coprocultures toujours négatives nie l'origine et la localisation intestinale de l'infection. Pour ROUX et d'autres le bacille d'EBERTH n'est pas un germe bien déterminé mais bien appartenant, avec de nombreuses variantes et formes intermédiaires, au même groupe que le *Bacillus coli communis* que ESCHERICH décrit entretemps (1885).

Fort heureusement, la bactériologie fit de rapides progrès. L'emploi de nouveaux milieux rendit possible une étude plus complète des caractères culturels et facilita le diagnostic différentiel. CASTELLANI (1899) et SCHOTTMÜLLER (1900) introduisent des techniques nouvelles pour l'hémoculture qui devint la méthode la plus rapide et la plus sûre pour l'isolement du bacille typhique chez le malade.

L'identification du bacille d'EBERTH est rendue beaucoup plus aisée par l'emploi de techniques sérologiques. En 1896, PFEIFFER & KOLLE reprennent avec des bacilles typhiques les recherches immunologiques qu'ils avaient faites sur le vibrion du choléra, et ils trouvent que ces germes sont tués par un immun-sérum spécifique quand ils sont introduits dans la cavité péritonéale d'un cobaye. La même année, GRUBER & DURHAM décrivent un phénomène qui avait été constaté également par PFEIFFER & KOLLE: en mélangeant dans un tube une suspension de bacilles d'EBERTH et le sérum d'un animal immunisé vis-à-vis de ces germes, on observe la formation d'amas de microbes. Le phénomène fut appelé *agglutination* et les anticorps responsables de sa production *agglutinines*. Ces divers auteurs reconnurent la spécificité de la réaction et apportèrent par l'introduction de la technique d'agglutination un moyen précieux pour le diagnostic différentiel entre le bacille typhique et le bacille coli. Ces mêmes auteurs constatent que le sérum des convalescents possède un pouvoir agglutinant identique à celui du sérum d'un animal immunisé. Pour le diagnostic direct pourtant la découverte annoncée simultanément à Paris par WIDAL et à Londres par GRÜNBAUM est beaucoup plus importante: en 1896, WIDAL et GRÜNBAUM avaient observé que le sérum d'un malade possède déjà le même pouvoir agglutinant que le sérum d'un convalescent et que la réaction d'agglutination permet donc assez tôt au cours de l'évolution de la maladie d'établir un *séro-diagnostic* de l'infection.

L'année 1896 vit donc paraître, présentés par différentes écoles de bactériologie, un certain nombre de travaux importants sur la sérologie de la fièvre typhoïde. Attribuer à chaque auteur la priorité à laquelle il a droit n'est guère facile, à cause surtout des discussions âpres, virulentes et peu objectives qui enveniment leurs rapports. La plupart d'entre eux, WIDAL & SICARD (1897) notamment, reconnaissent néanmoins le rôle déterminant joué par JULES BORDET dont les recherches immunologiques avaient jeté la base théorique qui permettra d'aboutir au séro-diagnostic connu universellement sous le nom de ''réaction de Widal''.

En 1902, ALDO CASTELLANI, qui à ce moment travaille à l'Institut d'Hygiène de Bonn, publie une modification de la méthode sérologique qui devait rendre plus aisée l'interprétation du séro-diagnostic et éclairer le problème des ''co-agglutinines'': il s'agit de la technique de saturation des anticorps qui se montrera plus tard indispensable pour l'étude des antigènes des *Salmonella* et qui rendit possible le sérotypage des souches microbiennes.

Les recherches bactériologiques devaient aboutir non seulement à un diagnostic précis mais aussi expliquer la pathogénie de la fièvre typhoïde. C'est ainsi que la technique de l'hémoculture a mis en évidence le caractère septicémique de l'infection: en trouvant le bacille dans le sang au début de la maladie, CASTELLANI (1899) et SCHOTTMÜLLER (1900, 1901) ont prouvé que la fièvre typhoïde n'est pas une maladie localisée de l'intestin mais bien une affection généralisée de nature septicémique.

La découverte de GAFFKY qui lia d'une manière indiscutable bacille typhique et fièvre typhoïde ne permit pour autant pas d'éclairer par elle même l'épidémiologie de l'infection. La théorie des miasmes avait bien été ébranlée mais les concepts de PETTEN-KOFER continuèrent pendant longtemps encore à influencer les épidémiologistes et ceux ci essayaient d'accomoder les nouvelles données aux vieilles théories, et cela malgré les travaux de l'école de KOCH qui avaient pourtant démontré le rôle important joué par l'eau dans la propagation des épidémies. Le bacille fut surtout recherché dans les milieux extérieurs où il était sensé subir une évolution ou se maintenir à l'état de vie latente. Partant, la prophylaxie se limitait à une désinfection des locaux et des objets mais elle oubliait le malade et ses excreta. La possibilité d'une transmission directe d'homme à homme ne fut acceptée qu'au début du XX° siècle. En 1902, KOCH établit que dans certaines circonstances la maladie se transmet par contact direct plutôt que par la contamination du milieu extérieur, et il souligne la nécessité de rechercher le germe chez les contacts restés sains. C'est à ce moment précisément que des auteurs signalent l'existence de porteurs (et excréteurs) de germes non-malades responsables des "maisons à typhus" et des foyers d'infection qui entretiennent les endémies et qui peuvent être à l'origine de poussées épidémiques de fièvre typhoïde. Les travaux de KOCH et de ses élèves établissent l'importance respective des différents mécanismes de propagation de la maladie et rendent possible une prophylaxie efficace. La prévention passive fut bientôt complétée par l'immunisation active: la vaccination.

Longtemps avant les découvertes d'EBERTH et de GAFFKY il était connu qu'un même sujet n'est pour ainsi dire jamais atteint deux fois de fièvre typhoïde. En 1903, FROSCH parle même déjà de "l'immunité régionale" qui semble protéger une ville après qu'une épidémie en avait touché de nombreux habitants. Les premiers essais d'immunisation active contre la fièvre typhoïde furent faits avec succès sur des animaux par BEUMER & PEIPER en 1886. Tandis que ces auteurs utilisaient des petites doses de bacilles vivants, CHANTEMESSE & WIDAL employaient des bacilles tués et obtenaient des résultats comparables (1888). Encouragé par l'exemple de HAFFKINE, qui avait introduit aux Indes la vaccination anti-cholérique, WRIGHT utilise en 1896 pour la première fois un vaccin anti-typhique chez l'homme. PFEIFFER & KOLLE, indépendamment de WRIGHT, publient au cours de la même année les résultats de l'application du vaccin chez des sujets humains; ces derniers auteurs avaient trouvé dans leurs recherches sérologiques antérieures la base scientifique de l'immunisation prophylactique.

D'autre part, les techniques bactériologiques utilisées pour le diagnostic de la fièvre typhoïde permirent la découverte des infections à Bacilles paratyphiques: ACHARD & BENSAUDE communiquent en 1896 leurs observations sur deux cas de "fièvre typhoïde" chez lesquels ils avaient isolé un germe ayant certains caractères culturels du bacille d'EBERTH mais chez lesquels la réaction de Widal était restée négative. ACHARD & BENSAUDE donnent à la maladie le nom de *infection paratyphoïdique* et à son agent celui de *bacille paratyphique*. Ces termes ne furent pas acceptés à l'époque et les bacilles furent classés parmi les coli-bacilles. Plusieurs auteurs, aussi bien avant qu'après ACHARD & BENSAUDE ont décrit des germes analogues. En 1890 déjà, BABES avait isolé une bactérie intermédiaire entre le bacille typhique et le bacille coli et qui, peut-être, était un bacille paratyphique. En 1898, GWYN isola du sang d'un malade un germe apparenté au bacille d'EBERTH et l'appella "paracoli bacillus".

Ce n'est que plus tard, après les travaux de SCHOTTMÜLLER (1901) que les bacilles d'ACHARD & BENSAUDE furent reconnus comme étant des bacilles paratyphiques du type B et que le bacille de GWYN fut identifié comme paratyphique A. Le terme "bacilles paratyphiques" devait par ailleurs être admis après la publication par SCHOTTMÜLLER (1900, 1901) des résultats d'hémocultures faites chez des malades suspects d'être atteints de fièvre typhoïde. SCHOTTMÜLLER donna lui aussi aux germes qu'il avait isolés, et qui étaient différents du bacille typhique, le nom de "bacilles paratyphiques". L'auteur se rendait compte que toutes ses souches n'étaient pas tout à fait identiques et qu'elles appartenaient à deux types qui se distinguaient par leurs caractères biochimiques et sérologiques. BRION & KAYSER (1902) confirmèrent les résultats de SCHOTTMÜLLER et imposèrent les noms de Paratyphus A et B.

Ces bacilles furent immédiatement l'objet de nombreuses études qui permirent de les identifier avec des bactéries décrites antérieurement dans des cas de soit-disant fièvre typhoïde, tels que les bacilles de GWYN et d'ACHARD & BENSAUDE. C'est surtout le *Bacillus paratyphus B* qui attira l'attention des chercheurs, les infections par le type A se montrant beaucoup moins fréquentes. On s'aperçut rapidement que le B. paratyphique B pouvait provoquer non seulement un syndrome de fièvre typhoïde mais également des infections pouvant évoluer sous une forme clinique complètement différente, se présentant notamment comme une gastro-entérite aiguë due aux toxines élaborées par les germes. En outre, on se rendit compte de ce qu'on connaissait déjà avant l'isolement du B. paratyphique B de SCHOTTMÜLLER un certain nombre de germes identiques, ou du moins très fortement apparentés, qui avaient provoqué tant chez l'homme que chez l'animal des infections évoluant avec une symptomatologie différente de celle de la fièvre typhoïde.

En 1885, GAFFKY & PAAK isolèrent lors d'une intoxication due à de la viande de cheval un germe qui se révéla pathogène pour les animaux de laboratoire. En 1888, GÄRTNER, à l'occasion également d'une intoxication carnée, isola un germe toxigène à la fois de la viande incriminée et de la rate d'une personne décédée à la suite de l'ingestion de cette viande. GÄRTNER établit le rôle pathogène du bacille qu'il avait isolé et confirma l'étiologie bactérienne de certaines toxi-infections d'origine alimentaire. Des bacilles analogues furent isolés dans d'autres cas et ils furent tous groupés sous le même nom de *Bacillus enteritidis* GÄRTNER. DE NOBELE (1901), après de minutieuses études sérologiques poursuivies durant trois ans conclut qu'il fallait classer ces germes dans deux groupes distincts, l'un représenté par le *B. enteritidis* et l'autre par le *B. Aertrijcke* qu'il avait isolé en 1898 lors d'une épidémie de toxi-infection alimentaire due à de la viande de veau survenue dans le village d'Aertrijcke. Ce dernier germe n'est autre que le *B. typhimurium* que LÖFFLER avait découvert en 1892.

Des bactéries ressemblant aux bacilles paratyphiques avaient été trouvées également lors d'épizooties. La première découverte est celle du *Bacillus cholerae-suis* faite en 1885 par SALMON & SMITH chez des porcs atteints du "hog cholera", maladie qu'on attribua erronément à ce germe.

Le Bacille paratyphique de SCHOTTMÜLLER et les bacilles isolés de viandes avariées et capables de provoquer des épidémies et des épizooties furent réunis dans un même groupe, le groupe des *Salmonella*, nom générique proposé par LIGNIÈRES en 1900 pour honorer la découverte, par SALMON, du premier bacille paratyphique. Bien que tout l'honneur de cette découverte ait été ainsi attribué à SALMON, il semble bien que le mérite en revient plutôt à son assistant, THEOBALD SMITH. Justice a par ailleurs été partiellement rendue puisque le microbe qu'ils ont découvert porte maintenant le nom officiel de *Salmonella cholerae-suis* SMITH.

Le groupe des *Salmonella* fut d'abord nettement séparé du bacille d'EBERTH, ou *Eberthella*, en raison des images cliniques différentes, les infections paratyphiques évoluant le plus souvent sous la forme d'une gastro-entérite. Certaines *Salmonella* ont même été considérées à un certain moment comme de simples saprophytes devenues accidentellement pathogènes. Aux *Salmonella* vinrent s'ajouter par après les autres bacilles paratyphiques tels que le *B. paratyphus C* qui fut décrit au cours de la première guerre mondiale, e.a. par HIRSCHFELD (1919). Peu à peu régna une grande confusion qui ne fit que croître au fur et à mesure que le nombre de souches isolées augmenta: le problème put paraître à certain moment insoluble.

Seule l'étude détaillée des antigènes rendit possible une identification et une classification valables des souches. Cette étude eut pour base quelques travaux fondamentaux: En 1903, SMITH & REAGH démontrent la différence qui existe entre les antigènes somatiques et flagellaires des *Salmonella* mais leur observation tomba dans l'oubli. En 1917, MAURICE NICOLLE définit les microbes comme "une mosaïque d'antigènes et de caractères biologiques" et fait e.a. la distinction entre le pouvoir agglutinogène et l'agglutinabilité. En 1918, WEIL & FELIX décrivent à nouveau les antigènes O et H, mais chez *Proteus X* cette fois. L'application de leur découverte aux *Salmonella* permettra l'étude sérologique précise de celles-ci. La démonstration faite par ANDREWES (1923) des deux phases de l'antigène H marque une étape importante dans l'histoire des salmonelloses. En 1934, FELIX & PITT découvrent l'antigène Vi des bacilles typhiques et fournissent l'explication de la non-agglutinabilité de certaines souches.

Grâce à ces travaux précurseurs BRUCE WHITE a pu présenter en 1926 une première esquisse de la structure antigénique des espèces de *Salmonella* connues à ce moment. Les recherches de BRUCE WHITE furent reprises et développées par KAUFFMANN (1941) qui soumit à l'approbation de la Société Internationale de Microbiologie, lors de son congrès de 1933, le Schéma de KAUFFMANN-WHITE, base de toutes les études dans le domaine des salmonelloses. Ce Schéma a maintenant 30 ans et, bien que rien n'ait été modifié aux principes de base sur lesquels il a été construit, il a au cours des années suivi une évolution constante due à l'opposition qui se fait sentir continuellement entre le désir légitime de précision et les exigences impérieuses de la pratique. Ce Schéma donne a chacune des centaines d'espèces de *Salmonella* connues à ce jour un visage personnel et il a rallié l'accord général.

De tous les groupes de bactéries, le groupe des *Salmonella* est probablement celui qui est le mieux étudié. Les maladies parfois très graves que ces germes provoquent chez les humains et chez les animaux, leur présence si souvent observée dans des denrées alimentaires les plus variées et l'incidence économique de cette présence tant sur les plans nationaux que sur celui des échanges internationaux obligent les médecins, les vétérinaires, les hygiénistes et même les économistes à leur consacrer une attention qui va toujours grandissante.

Il nous parait bon, en exergue de cette publication, de souligner que l'historique des salmonelloses présente un exemple classique de l'évolution des connaissances microbiologiques et de l'impossibilité qu'il y a à dissocier dans le domaine des maladies infectieuses la recherche pure et la recherche appliquée, l'une ouvrant la voie à l'autre et réciproquement, avec pour but commun la santé des individus.

BIBLIOGRAPHIE

ACHARD, CH. & BENSAUDE, R. (1896): Infections paratyphoïdiques. *Bull. et Mém. Soc. méd. Hôp.*, 27 novembre 1896. Cité par WIDAL, LEMIERRE, & ABRAMI (1921), p. *48*.

BABES (1890). Cité par UHLENHUTH & HÜBENER (1913), p. *1005*.

BEUMER & PEIPER (1886). Cité par FORNET (1913), p. *839*.

BRETONNEAU, P. (1818). Cité par THOINET & RIBBIERE (1912), p. *6*.

BRION, A. & KAYSER, H. (1902): Über eine Erkrankung mit dem Befund eines typhusähnlichen Bakteriums im Blute (Paratyphus). *Münch. med. Wschr.*, **49**: *611—615*.

BUDD (1856): On intestinal fever, etc., *Lancet*. Cité par NEUFELD (1903), p. *291*.

CASTELLANI, A. (1899). Cité par NEUFELD (1912), p. *249*.

CASTELLANI, A. (1902): Die Agglutination bei gemischter Infektion und die Diagnose der letzteren. *Z. Hyg. Infektkr.*, **40**: *1—20*.

CHANTEMESSE, A. & WIDAL. F. (1888): De l'immunité contre le virus de la fièvre typhoïde conférée par des substances solubles. *Ann. Inst. Pasteur*, **2**: *55—59*.

COZE & FELTZ (1872): Recherches cliniques et expérimentales sur les maladies infectieuses. Cité par MACÉ: Traité pratique de Bactériologie, Paris, 1901.

DE NOBELE, J. (1898): Du sérodiagnostic dans les affections gastrointestinales d'origine alimentaire. *Ann. Soc. Méd. de Gand*, **77**: *281—306*.

DE NOBELE, J. (1901): Le séro-diagnostic dans les affections gastrointestinales d'origine alimentaire (2° Mémoire): *Ann. Soc. Méd. de Gand*, **80**: *74—98*.

EBERTH C. J. (1880): Die Mikro-organismen in den Organen bei Typhus abdominalis. *Arch. path. Anat. u. Physiol. u.f. klin. Med. (Virchow's Arch.)*, **81**: *58—74*.

FELIX, A. & PITT, R. M. (1934). Cité par KAUFFMANN (1941), p. *85*.

FORNET, W. (1913): Immunität bei Typhus. In: Handbuch der pathogenen Mikro-organismen, von KOLLE & VON WASSERMANN, **3**: *837—898*, 2° Edit., Jena (1913).

FROSCH (1903). Cité par DUNSCHMANN, H.: Etudes sur la fièvre typhoïde. *Ann. Inst. Pasteur*, **23**: *29—69* (1909).

GAFFKY (1884). Cité par NEUFELD (1903), p. *268*.

GAFFKY & PAAK (1885). Cité par UHLENHUTH & HÜBENER (1913), p. *1037*.

GÄRTNER, A. (1888): Über die Fleischvergiftung in Frankenhausen und den Erreger derselben. *Correspondenzbl. d. allg. ärtztl. Verein v. Thüringen*. Cité par WIDAL, LEMIERRE & ABRAMI (1921), p. *51*.

GRUBER, M. & DURHAM, H. E. (1896): Eine neue Methode zur raschen Erkennung des Choleravibrio und des Typhusbacillus. *Münch. med. Wschr.*, **43**: *285—286*.

GRÜNBAUM, A. S. (1896): Preliminary note on the use of the agglutinative action of human serum for the diagnosis of enteric fever. *Lancet*, *806—807*.

GWYN (1898). Cité par POTTEVIN: Contribution à la bactériologie des gastro-entérites infectieuses. *Ann. Inst. Pasteur*, **19**: *426—448* (1905).

HIRSCHFELD, L. (1919): A new germ of paratyphoïd. *Lancet*, *296—297*.

KAUFFMANN, F. (1941): Die Bakteriologie der Salmonella-Gruppe. Edit. Copenhague.

KOCH, R. (1881). Cité par NEUFELD (1903), p. *268*.

KOCH, R. (1890). Cité par NEUFELD (1903), p. *206*.

KOCH, R. (1902): Bekämpfung des Typhus. Cité par BRAUN: La recherche du bacille d'Eberth. *Ann. Inst. Pasteur*, **19**: *578—591*.

LEURET (1828). Cité par THOINET & RIBBIERRE (1912), p. *7*.

LIGNIÈRES (1900). Cité par WIDAL, LEMIERRE & ABRAMI (1921), p. *51*.

LÖFFLER, F. (1892): Über Epidemien unter den im Hygienischen Institute zu Greifswald gehaltenen Mäusen und über die Bekämpfung der Feldmausplage. *Zbl. Bakt.*, 21: *129—141.*

LOUIS (1829). Cité par THOINOT & RIBBIERRE (1912), p. *6.*

MEYER, W. (1881). Cité par NEUFELD (1903), p. *268.*

MURCHINSON (1858). Cité par THOINOT & RIBBIERRE (1912), p. *8.*

NEUFELD, F. (1903): Typhus. In: Handbuch der pathogenen Mikroorganismen, von KOLLE & VON WASSERMANN, 2: *204—308,* 1° Edit., Jena.

NICOLLE, M., RAPHAEL, A. & DEBAINS, E. (1917): Etudes sur le bacille d'Eberth et les bacilles paratyphiques. *Ann. Inst. Pasteur*, 31: *373—402.*

PETTENKOFER (1868). Cité par NEUFELD (1903), p. *291.*

PFEIFFER, R. & KOLLE, W. (1896): Zur Differentialdiagnose der Typhusbacillen vermittels Serum der gegen Typhus immunisierten Thiere. *Dtsch. med. Wschr.*, 22: *185—186.*

PFEIFFER, R. & KOLLE, W. (1896b): Experimentelle Untersuchungen zur Frage der Schutzimpfung des Menschen gegen Typhus abdominalis. *Dtsch. med. Wschr.*, 22: *735—737.*

PROST (1804). Cité par SERGENT, E., RIBADEAU, L. & BABONEIX, L.: Traité de pathologie médicale, Paris (1921).

SALMON, D. E. & SMITH, TH. (1885). Cité par KAUFFMANN (1941), p. *223.*

SANARELLI (1894). Cité par WATHELET, A.: Recherches bactériologiques sur les déjections dans la fièvre typhoïde. *Ann. Inst. Pasteur*, 9: *252—257.*

SCHOTTMÜLLER, H. (1900): Über eine das Bild des Typhus bietende Erkrankung hervorgerufen durch Typhusähnliche Bacillen. *Dtsch. med. Wschr.*, 26: *511—512.*

SCHOTTMÜLLER, H. (1901): Weitere Mitteilungen über mehrere das Bild des Typhus bietende Krankheitsfälle, hervorgerufen durch typhusähnliche Bacillen (Paratyphus). *Z. Hyg. Infektkr.*, 36: *366—396.*

SMITH, TH. & REAGH, A. L. (1903). Cité par HAUDUROY, P.: Microbiologie générale et Techniques microbiologiques, Paris (1947).

THOINOT, L. & RIBBIERRE, P. (1912) In: Nouveau Traité de Médecine et de Thérapeutique, par GILBERT A. & THOINOT L., Paris.

TROUSSEAU. Cité par WIDAL, LEMIERRE & ABRAMI (1912), p. *2.*

UHLENHUTH, P. & HÜBENER, E. (1913): Infektiöse Darmbakterien der Paratyphus- und Gärtnergruppe. In: Handbuch der pathogenen Mikroorganismen, von KOLLE & VON WASSERMANN, 3: *1005—1156,* 2° Edit., Jena.

WIDAL, F. (1896): Le sérodiagnostic de la fièvre typhoïde. *Bull. Soc. méd. des Hôp. de Paris*, 26 juin 1896. Traduction dans: *Münch. med. Wschr.*, 43: *690* (1896).

WIDAL, F., LEMIERRE, A. & ABRAMI, P. (1921): Fièvres typhoïdes et paratyphoïdes. In: ROGER, G. H., WIDAL, F. & TEISSIER, P. J.: Nouveau Traité de Médecine, fasc. 3, Paris.

WIDAL, F. & SICARD, A. (1897): Etude sur le sérodiagnostic et sur la réaction agglutinante chez les typhiques. *Ann. Inst. Pasteur*, 11: *353—432.*

WHITE, P. BRUCE (1926). Cité par KAUFFMANN (1941), p. *128.*

WRIGHT, A. E. (1896). Cité par FORNET (1913), p. *868.*

PART I
THE SALMONELLA

DAS KAUFFMANN-WHITE-SCHEMA

VON

F. KAUFFMANN *)

Leiter der Internationalen Salmonella-Centrale, Statens Seruminstitut, Kopenhagen

Einleitung

Unter den *Salmonella*-Problemen nimmt das diagnostische Antigenschema, das KAUFFMANN-WHITE-SCHEMA (K-W-Schema) eine zentrale Stellung ein. Es ist nämlich nicht nur für diagnostische und klassifikatorische, sondern auch für epidemiologische, immun-chemische und genetische Probleme von grundlegender Bedeutung. Die grossen Fortschritte, die im Laufe der letzten Jahre auf allen diesen Gebieten erzielt wurden, sind in erster Linie diesem Antigenschema zu verdanken. Epidemiologische Zusammenhänge lassen sich ohne exakte Typendiagnose überhaupt nicht erfassen, für immun-chemische Analysen ist die serologische Antigenanalyse die unerlässliche Voraussetzung (KAUFFMANN, LÜDERITZ, STIERLIN & WESTPHAL) und eine erfolgreiche Durchführung genetischer Versuche unter Berücksichtigung der Antigenstruktur ist ohne das K-W-Schema unmöglich. Diese Tatsachen beweisen klar die überragende Bedeutung der Serologie, nicht nur für das *genus Salmonella*, sondern auch für die übrigen *genera* der *Enterobacteriaceae* und anderer Familien.

Bevor wir aber näher auf das K-W-Schema eingehen, wollen wir zunächst definieren, was wir unter dem *genus Salmonella* und seinen *species* verstehen und einige, allgemeine Bemerkungen über dieses Problem vorausschicken.

Da es in der Natur kein scharf abgegrenztes *genus Salmonella* gibt, kann man auch keine scharfe Definition geben. Um den natürlichen Verhältnissen gerecht zu werden, muss daher unsere Definition unscharf sein. Je schärfer wir eine Definition fassen, um so unbrauchbarer wird sie, da sie um so grössere Abweichungen und Schwierigkeiten ergibt.

Das entsprechende gilt auch für die Definition der Antigene im K-W-Schema. Je weiter wir die einzelnen Antigen-Komplexe, wie zum Beispiel den G-Komplex, in Partialantigene aufteilen, um so grössere Schwierigkeiten erhalten wir und um so grösser werden die Differenzen zwischen den Resultaten der verschiedenen Laboratorien. Vereinfachen wir aber die Diagnose des G-Komplexes, so er-

---

*) Adresse: Professor Dr. F. Kauffmann, Statens Seruminstitut, Kopenhagen S., Dänemark.

halten wir sichere und übereinstimmende Diagnosen, aber nur eine geringe Differenzierungs-Möglichkeit. Es ist unmöglich, beide Vorteile: Einfachheit der Diagnose und grosse Differenzierungs-Möglichkeit mit einander zu vereinen.

Die Erfahrung hat gezeigt, dass die im originalen K-W-Schema durchgeführte Aufteilung des G-Komplexes in zahlreiche Partialantigene die Grenze der praktischen Zuverlässigkeit bereits überschritten hat. Es ist daher vorzuziehen, auf die so erzielte, hohe Differenzierungs-Möglichkeit zu verzichten und diesen Komplex, so wie es im vereinfachten K-W-Schema geschah, nur mit G zu bezeichnen. Man darf hierbei auch nicht vergessen, dass dieses ein konsequentes Vorgehen ist, da alle anderen H-Antigen-Komplexe wie a, b, c, d etc. in der entsprechenden Weise bereits im originalen K-W-Schema vereinfacht sind, sodass hierauf die Sicherheit ihrer Diagnose beruht.

Aus diesem Grunde haben KAUFFMANN & ROHDE (1) vorgeschlagen, alle *neuen species* des *Salmonella sub-genus* I mit den H-Antigen-Komplexen G und L nach dem vereinfachten K-W-Schema zu diagnostizieren. Ferner schlugen sie vor, alle *neuen species* des *sub-genus* II ebenfalls nach dem vereinfachten K-W-Schema zu bestimmen, um die ständig anwachsende Zahl der *species* zu verringern. Es ist natürlich möglich, in besonderen, epidemiologisch wichtigen Fällen neue *species* mit Hilfe des originalen K-W-Schemas, das seine volle Gültigkeit behält, zu diagnostizieren. KAUFFMANN & ROHDE stellten die Frage zur Diskussion, ob nicht jetzt der Zeitpunkt gekommen sei, alle *neuen species* des *sub-genus* I nach dem vereinfachten K-W-Schema zu bestimmen. Wird z.B. eine neue *species* nur einmal aus einem Salamander oder einem anderen Tiere isoliert, so erscheint es uns nicht als notwendig, diese *species* in das originale K-W-Schema einzufügen.

Das K-W-Schema ist der Versuch, aus der sehr komplizierten Antigenstruktur der Bakterien ein praktisch brauchbares Schema zu abstrahieren, sodass es nicht unnötig kompliziert, sondern so weit wie möglich vereinfacht werden soll.

Vor allem sollte man sich aber darüber im klaren sein, dass nur unter Berücksichtigung der serologischen Variation ein diagnostisches Antigenschema aufgestellt und benutzt werden kann. Sowohl in der Serologie der Geissel-Antigene, der H-Antigene, beim Phasen-Wechsel, als auch in der Serologie der somatischen Antigene, beim Formen-Wechsel, handelt es sich nicht um allmähliche Übergänge zwischen den einzelnen Phasen oder Formen, sondern um sprunghafte Vorgänge, die durch entsprechende "quantenartige" Veränderungen im genetischen Apparat bedingt sein müssen. Die serologische Variation verläuft also nicht kontinuierlich, sondern diskontinuierlich. Verschiedene stationäre Zustände gehen mutationsartig in einander über, sodass wir es mit

einem nicht genau, sondern nur statistisch berechenbaren, dynamischen Geschehen zu tun haben. Deshalb ist auch eine Standardi-
sierung serologischer Reaktionen ohne Berücksichtigung dieser
sprunghaften Vorgänge nicht möglich.

Das Genus Salmonella

Das *genus Salmonella* besteht aus einer Gruppe verwandter *species*.
Die Namen und Antigen-Formeln der zum *sub-genus* I und II ge-
hörenden *species* sind im originalen K-W-Schema enthalten,
während die *species* des *sub-genus* III (= *Salmonella arizonae*) nur im
vereinfachten K-W-Schema erscheinen, z.B. als *S. arizonae* $51:z_4:-$.
In Zukunft werden auch alle *species* des *sub-genus* II sowie diejenigen des *sub-genus* I, welche die G- und L-Komplexe enthalten, nur
nach dem vereinfachten K-W-Schema diagnostiziert werden (siehe
bei KAUFFMANN & ROHDE[1]).

Beim *genus Salmonella* handelt es sich um Gram-negative, aerobe,
nicht sporenbildende Stäbchen, die auf den üblichen Nährböden
wachsen und Nitrat zu Nitrit reduzieren. Sie sind in der Regel peri-
trich begeisselt, doch kommen auch geissellose, unbewegliche sowie
geisselhaltige, aber unbewegliche Formen vor. Sie spalten nicht
Adonit und Saccharose, bilden kein Indol, spalten nicht Urea und
geben eine negative VOGES-PROSKAUER-Reaktion, aber eine positive
Methylrot-Reaktion. Die *species* der *sub-genera* I und II greifen in
der Regel Lactose nicht an, während die Arizona-Kulturen Lactose
prompt, verzögert oder gar nicht spalten. Dulcit wird von den
sub-genera I und II meist gespalten, während die Stämme des *sub-
genus* III (Arizona) Dulcit nicht spalten. Die Mehrzahl der *species*
des *sub-genus* I verflüssigt nicht Gelatine, während alle Angehörigen
der *sub-genera* II und III Gelatine verflüssigen. Alle Kulturen spalten
Glukose, meist mit Gasbildung. Salicin wird nicht prompt gespalten,
in einigen Fällen aber verzögert.

Auf Grund internationaler Vereinbarungen werden bestimmte
species mit *Salmonella* O- und H-Antigenen zum *genus Salmonella*
gerechnet, auch wenn ihr biochemisches Verhalten von der Regel
abweicht.

Alle *species* sind für Menschen, Tiere oder beide pathogen.

Nachdem die ersten *Salmonella-species* vom Salmonella-Subcommittee in seiner Publikation "The genus Salmonella Lignières, 1900"
(*J. of Hygiene* 34, *333*, 1934) publiziert waren, hat das Subcommittee
in seinen Berichten an die Internationalen Kongresse in New York
1939 und in Kopenhagen 1947 weitere neue *Salmonella-species* unter
der Bezeichnung *"species"* publiziert, sodass im ganzen über 140
Salmonella-species anerkannt wurden. Das Internationale Nomen-
klatur-Committee und die Vollversammlungen der Internationalen
Mikrobiologen-Kongresse in New York und Kopenhagen haben diese

Berichte gebilligt und in den Proceedings der Kongresse publiziert.

Da jedoch in den folgenden Jahren das Enterobacteriaceae-Sub-committee nur vulgäre Namen wie "group", "type" oder "sero-type" benutzte, hat KAUFFMANN (1) in seinem Buche "Die Bakteriologie der Salmonella-Species" alle bis Ende 1960 publizierten Typen als species veröffentlicht und in seiner Arbeit "The species-definition in the family Enterobacteriaceae" (2) die *species* wie folgt definiert:

"Eine species ist eine Gruppe verwandter, serofermentativer Phag-Typen".

Als Beispiel hierfür gilt die *species S. typhi-murium*, eine der grössten *species*, die aus zahlreichen sero-fermentativen Phag-Typen besteht.

Da diese *species*-Definition: "A species is a group of related sero-fermentative phage-types" trotz ihrer Einfachheit und Klarheit miss-verstanden wurde, so sei noch folgendes gesagt:

Die kleinste Einheit ist der s e r o - f e r m e n t a t i v e P h a g - T y p, der also serologisch-vergärungsmässig und phagmässig definiert ist. Eine grössere Anzahl derartiger, verwandter Elementareinheiten = sero-fermentative Phag-Typen bildet eine Gruppe, welche *species* genannt ist, z.B. *Salmonella typhi-murium*.

Alle Kulturen von *Salmonella typhi-murium* gehören zu dem Serotyp 1, 4, 5, 12:i:1, 2 des K-W-Schemas, liegen aber in minde-stens 4 verschiedenen serologischen Subtypen vor.

$$1, 4, 5, 12:i:1, 2$$
$$4, 5, 12:i:1, 2$$
$$1, 4, 12:i:1, 2$$
$$4, 12:i:1, 2$$

Diese serologischen Subtypen (1, 4, 5, 12 — 4, 5, 12 — 1, 4, 12 — 4, 12) sind im KAUFFMANN-WHITE-Schema zwecks Vereinfachung zum Serotyp 1, 4, 5, 12:i:1, 2 zusammengefasst.

Bei jedem dieser serologischen Subtypen treten zahlreiche, ver-schiedene Vergärungstypen auf, z.B. Inosit-positive und Inosit-negative, d-Tartrat-positive und d-Tartrat-negative u.s.w. Ferner können die einzelnen Kulturen in zahlreiche Phagtypen eingeteilt werden. So gehört z.B. eine bestimmte Elementareinheit dieser *species S. typhi-murium* zu einem ganz bestimmten sero-fermentati-ven Phagtyp, der die Antigene 4, 5, 12:i:1, 2 besitzt, der Inosit und d-Tartrat spaltet und zum Phagtyp Nr. 1 gehört. Eine solche Kultur ist die kleinste Einheit, die wir in unserer Routine-Diagnostik be-stimmen, sie ist also ein sero-fermentativer Phagtyp, der mit Sicherheit diagnostiziert werden kann.

Wenn wir nun 20 derartige Einheiten, die alle zum Serotyp 1, 4, 5, 12:i:1, 2 gehören, zu einer Gruppe vereinen, so ist dieses Vorgehen absolut berechtigt und im Interesse einer Klassifikation auch not-wendig. Ebenso ist es aus Gründen der Nomenklatur berechtigt, eine derartige Gruppe *"species"* zu nennen. Da die offiziell acceptierte "*type-species*" des *genus Salmonella* die *species S. cholerae-suis* ist, ist es korrekt und logisch, eine entsprechende Einheit wie die Gruppe von sero-fermentativen Phag-Typen, die zu *S. typhi-murium* gehören, eine *species* zu nennen. Sie wurde daher auch als solche im 1. Report des Salmonella Subcommittee's als *species* aufgeführt. Eine jede andere Bezeichnung als *species* würde die Regeln des Bacteriological Code verletzen.

Schematisch
A. *Genus.*
B. *Sub-genus.*
C. *Species.*
D. *Sero-fermentativer Phag-Typ.*

Da es nur ein einziges genus *Salmonella* gibt, so ist der Plural "*Salmonellae*" unzulässig.

Da dieselben Vergärungstypen auch bei anderen Serotypen vorkommen können und da die Phagtypen nicht immer bestimmt werden, ist die Bestimmung des *Serotyps* das Wichtigste und die Basis der ganzen *Salmonella*-Klassifikation. Deshalb hat man auch abgekürzt die *species* als "Serotyp" bezeichnet und kann es auch der Einfachheit wegen in Zukunft tun, doch darf man nicht vergessen, dass die *species* in Wirklichkeit eine *Gruppe verwandter, sero-fermentativer Phag-Typen* ist.

Um Missverständnisse zu vermeiden, sei betont, dass die frühere Definition der *species* als biochemische (vergärungsmässige) Unterteilung des *genus* aufgegeben wurde. Derartige Unterteilungen werden jetzt als "*sub-genera*" bezeichnet.

Falls man aber die *species* nur mit Hilfe von Vergärungsreaktionen oder bestimmten, anderen biochemischen Proben definieren will, so müssten die 3 *sub-genera* I, II und III als *species* betrachtet werden.

Im folgenden Schema ist die orthodoxe, biochemische Klassifikation der modernen Klassifikation, die biochemisch und serologisch definiert ist, gegenüber gestellt.

Orthodoxe Klassifikation (Biochemische Einteilung)	Moderne Klassifikation (Biochemische und serologische Einteilung)
Genus: Salmonella *Species* I. *Species* II. *Species* III. (*Species* = *Sub-genera*, nur biochemisch definiert).	*Genus: Salmonella* *Sub-genus* I. *Sub-genus* II. *Sub-genus* III. *Species* = Gruppen verwandter sero-fermentativer Phag-Typen

In der orthodoxen Klassifikation, die durch BERGEY's Manual vertreten ist, werden die *species* nur biochemisch definiert und haben deshalb nur eine untergeordnete und begrenzte Bedeutung. In der modernen Klassifikation, die mit dem 1. Bericht des Salmonella-Subcommittee's 1934 beginnt, werden die *species* sowohl biochemisch als auch serologisch definiert und umfassen daher die entscheidenden Elementar-Einheiten, d.h. die sero-fermentativen Phag-Typen. Während die *genera* und *sub-genera* biochemisch definiert sind, beruht die moderne Klassifikation der *species* in erster Linie auf serologischen Merkmalen.

Es handelt sich hier nicht um einen Streit um Worte und um die Frage, welche Kategorie man als *species* betrachten will, sondern um eine fundamentale Entscheidung, nämlich um die zu Grunde liegende experimentelle Methode. Um eine brauchbare Klassifikation zu erhalten, muss diejenige Kategorie, durch welche die *species* sowohl biochemisch als auch serologisch definiert ist, gewählt werden.

KAUFFMANN (3) hat das *genus Salmonella* mit Hilfe biochemischer Methoden in 2 verschiedene *sub-genera* eingeteilt und ferner (1) die Arizona-Gruppe als *sub-genus* III des *genus Salmonella* betrachtet. KAUFFMANN & ROHDE (2) haben dann eine vereinfachte serologische Diagnose der Arizona-Kulturen mit Hilfe des vereinfachten K-W-Schemas vorgeschlagen.

Wir unterscheiden also im *genus Salmonella* 3 biochemisch definierte *sub-genera*:

Sub-genus I oder typische *Salmonella*-Bakterien,

sub-genus II oder atypische *Salmonella*-Bakterien und

sub-genus III oder Arizona-Bakterien = *Salmonella arizonae*.

Diese 3 *sub-genera* werden mit Hilfe der Serologie in *species* = Gruppen verwandter, sero-fermentativer Phag-Typen eingeteilt und können in serologischer Hinsicht alle mit Hilfe des K.W-Schemas diagnostiziert werden.

Die "*type-species*" des *sub-genus* I ist *S. cholerae-suis*, diejenige des *sub-genus* II ist *S. dar-es-salaam*, während die "*type-species*" des *sub-genus* III die Arizona-*species* 1, 2:1, 2, 5 (= $51:z_4$) ist.

Die Antigen-Formel dieser *type-species*, des originalen Arizona-Stammes, war von KAUFFMANN (4) mit $33:z_4$, z_{23}, z_{26} angegeben worden. Die O-Gruppe 33 wurde aber aus dem KAUFFMANN-WHITE-Schema ausgeschlossen, als eine selbständige Arizona-Gruppe von EDWARDS, WEST & BRUNER aufgestellt wurde. Die heutige *Salmonella* O-Gruppe 51 ist mit der früheren O-Gruppe 33 identisch. Die früher mit z_4, z_{23}, z_{26} angegebenen H-Antigene (= 1, 2, 5) des originalen Arizona-Stammes werden im vereinfachten K-W-Schema und damit auch im vereinfachten Arizona-Schema mit z_4 bezeichnet.

Betreffs der Serologie der Arizona-Gruppe sei auf die letzte zusammenfassende Arbeit von EDWARDS, FIFE & RAMSEY sowie auf die dort angegebene Literatur verwiesen.

Die nahen Antigen-Beziehungen zwischen *Salmonella-* und Arizona-Kulturen waren von Anfang an bekannt und sind in der Arbeit von EDWARDS, FIFE & RAMSEY zusammengestellt. Inzwischen sind weitere Antigen-Beziehungen festgestellt worden, sodass man sagen kann, dass die meisten Arizona O- und H-Antigene im K-W-Schema vorhanden sind und deshalb mit *Salmonella*-Antigenformeln wiedergegeben werden können.

Obwohl es also möglich ist, die Arizona-*species* in das originale K-W-Schema einzufügen, wollen wir dieses nicht tun, sondern stellen ein besonderes Arizona-Schema, entsprechend dem vereinfachten K-W-Schema, auf. In diesem Schema werden so weit wie möglich die Symbole des vereinfachten K-W-Schemas benutzt und nur in denjenigen Fällen, in denen noch keine Antigengemeinschaften zwischen Arizona und *Salmonella* nachgewiesen sind, werden bis auf weiteres Arizona-Symbole angewandt.

Mit Rücksicht auf die bakteriologische Praxis ist es ausreichend, Arizona-Kulturen mit Hilfe des vereinfachten K-W-Schemas zu diagnostizieren, d.h. die Diagnose mit Hilfe der diagnostischen *Salmonella*-Seren zu stellen. Es sei betont, dass das originale Arizona-Schema nach EDWARDS und Mitarbeitern seine volle Gültigkeit behält und dass es jedem Untersucher freisteht, dieses Schema anzuwenden.

In einer im Druck befindlichen Arbeit "Zur Differentialdiagnose der Salmonella-Sub-Genera I, II und III" hat KAUFFMANN ausser den bisher üblichen Methoden den von LE MINOR & BEN HAMIDA empfohlenen β-Galaktosidase-Test angewandt und konnte hiermit deutliche Unterschiede zwischen den 3 *sub-genera* nachweisen. In der folgenden Übersicht ist die Diagnose der 3 *sub-genera* und ihr typisches Verhalten wiedergegeben:

	I	II	III
Lactose	—	—	$+^1$ oder $\times$
β-Galaktosidase	—	— oder $\times$	$+$
Dulcit	$+^1$	$+^1$	—
Gelatine	—	$+$	$+$
d-Tartrat	$+^1$	— oder $\times$	— oder $\times$
l-Weinsäure	d	—	—
i-Weinsäure	d	—	—
Natrium-Citrat	$+^1$	$+^2$	$+^{2-4}$
Mukat	$+^1$	$+^{1-2}$	d
Malonat	—	$+^{1-2}$	$+^{1-2}$

Zeichenerklärung: — = negativ, + = positiv, $+^1$ = positiv nach 1 Tage, $\times$ = spät und unregelmässig positiv oder negativ, bei β-Galaktosidase nach 1—4 Tagen positiv, d = differente, biochemische Typen.

28

Um von der Stellung des *genus Salmonella* in der Familie der *Enterobacteriaceae* ein Bild zu geben, ist in der folgenden Tabelle I die Einteilung der *Enterobacteriaceae* in 3 *tribus* und 12 *genera* (nach KAUFFMANN) gegeben.

Tabelle I.

Enterobacteriaceae

Tribus	Genera
Eschericheae	1. *Escherichia* 2. *Shigella* 3. *Salmonella* (incl. Arizona) 4. *Citrobacter*
Klebsielleae	1. *Klebsiella* 2. *Enterobacter (= Cloaca)* 3. *Hafnia* 4. *Serratia*
Proteae	1. *Proteus* 2. *Morganella* 3. *Rettgerella* 4. *Providencia*

In der Tabelle I sind die *genera Escherichia, Shigella, Salmonella* und *Citrobacter* in einem einzigen *tribus Eschericheae* zusammengefasst, da sie sehr nahe mit einander verwandt sind, und die Zahl der aufgestellten *tribus* möglichst verringert werden sollte. Ferner sei erwähnt, dass das *genus Shigella* als *sub-genus* des *genus Escherichia* aufgefasst werden kann. Wir unterscheiden also zwischen 3 verschiedenen *tribus: Eschericheae, Klebsielleae* und *Proteae*. Abschliessend sei hervorgehoben, dass wir die *Proteae* in 4 *genera: Proteus, Morganella, Rettgerella* und *Providencia* einteilen.

Das Kauffmann-White-Schema

Das KAUFFMANN-WHITE-Schema ist eine diagnostische Antigen-Tabelle, in der die verschiedenen Körper- und Geissel-Antigene (besser: Antigen-Komponenten) der *Salmonella-species* gekürzt und formelmässig wiedergegeben sind. Es handelt sich also um eine serologische Bestimmungstabelle und nicht um ein Verzeichnis aller nachweisbaren Antigene. Es gibt erheblich mehr Antigen-Faktoren als die Tabelle aufweist, da sie absichtlich vereinfacht wurde. Diese Vereinfachung ist jedoch ein schwieriges Problem, da man einerseits Gefahr läuft, wichtige Dinge ausser Acht zu lassen und andererseits die Tabelle für die Praxis unnötig zu komplizieren. Es bleibt ferner zu berücksichtigen, dass die Antigen-Formeln relative Formeln sind. Sie entsprechen nur dem jeweiligen Stande unserer Kenntnisse und sind daher bei der Auffindung neuer *species* mit bisher unbekannten Antigenbeziehungen zu überprüfen und eventuell abzuändern. Die systematische Anordnung der einzelnen Antigene und ihre typische Kombination gestatten die Voraussage, welche *species* noch fehlen und daher zu erwarten sind. Wenn wir aber auch theoretisch mit einer sehr grossen Zahl von *Salmonella-species* — mit vielen Tausenden — zu rechnen haben, so kann diese Mannigfaltigkeit der Formen nicht mehr zur Verwirrung führen, da die Grundzüge des gesetzmässigen Antigen-Aufbaues und speziell des serologischen Formenwechsels im Prinzip erkannt und im Kauffmann-White-Schema niedergelegt sind.

Wir sind uns bewusst, dass dieses Schema nur den Ausschnitt eines

Tabelle II.

O-Gruppen-Einteilung.

Gruppe	Antigene	Gruppe	Antigene
A	1, 2, 12	K	18
		L	21
B	1, 4, 5, 12	M	28
	1, 4, 12, 27	N	30
C 1	6, 7	O	35
C 2	6, 8	P	38
C 3	(8), 20	Q	39
C 4	6, (7), (14)	R	1,40
		S	41
D 1	1, 9, 12	T	1,42
D 2	(9), 46	U	43
		V	44
E 1	3, 10	W	45
E 2	3, 15	X	1,47
E 3	(3), (15), 34	Y	48
E 4	1, 3, 19	Z	50
			1,51
F	11		52
			53
G 1	1, 13, 22		54
G 2	1, 13, 23		55
			56
H	1, 6, 14, 24		57
	1, 6, 14, 25		58
			59
I	16		60
J	17		

in Wirklichkeit viel grösseren Mosaikbildes darstellt und dass es nicht scharf abgegrenzt werden kann. In dem unerschöpflichen Formenreichtum der Natur ist aber ein fester Standpunkt, ein System, notwendig, um nicht jede Orientierung zu verlieren.

Die Fundamente des KAUFFMANN-WHITE-Schemas sind:

1. Die Differenzierung zwischen Körper- und Geissel-Antigenen.
2. Die Einteilung des *genus Salmonella* in O-Gruppen.
3. Die Berücksichtigung der serologischen Variation:
   a. S-T-R-Formen-Wechsel,
   b. O-Formen-Wechsel,
   c. Phasen-Wechsel und
   d. S-R-Phasen-Wechsel.

Die Einteilung des *genus Salmonella* in O-Gruppen und O-Untergruppen nach KAUFFMANN ist in Tabelle II wiedergegeben, während die serologische Variation weiter unten besprochen wird.

Als Ergänzung zur Tabelle II sei noch erwähnt, dass mehrere der höheren O-Gruppen komplex gebaut sind und in folgende O-Untergruppen eingeteilt werden können:

$$O\ 28 = 28_1, 28_2 \qquad = S.\ tel\text{-}aviv$$
$$28_1, 28_3 \qquad = S.\ dakar$$
$$O\ 30 = 30_1, 30_2 \qquad = S.\ urbana$$
$$30_1 \qquad = S.\ soerenga$$
$$O\ 40 = 40_1, 40_2 \qquad = S.\ riogrande$$
$$1,40_1, 40_3 \qquad = S.\ bulawayo$$
$$O\ 42 = 42_1 \qquad = S.\ weslaco$$
$$42_1, 42_2 \qquad = S.\ loenga$$
$$O\ 43 = 43_1, 43_2, 43_3 \qquad = S.\ milwaukee$$
$$43_1, 43_2 \qquad = S.\ kingabwa$$
$$43_1, (43_3), 43_4 \qquad = S.\ bunnik$$
$$O\ 45 = 45_1, 45_2 \qquad = S.\ deversoir$$
$$45_1, 45_3 \qquad = S.\ dugbe$$
$$O\ 47 = 47_1, 47_2 \qquad = S.\ bergen$$
$$47_1, 47_3 \qquad = S.\ kaolack$$
$$O\ 48 = 48_1, 48_2 \qquad = S.\ dahlem$$
$$48_1, 48_2, 48_3 \qquad = S.\ djakarta$$
$$48_1, 48_3, 48_4 \qquad = Citrobacter\ 2624/36.$$
$$O\ 50 = 50_1, 50_2, 50_4 \qquad = S.\ greenside$$
$$50_1, 50_2, 50_3 \qquad = S.\ wassenaar$$
$$50_1, 50_3 \qquad = S.\ hooggraven$$

Betreffs der notwendigen Faktor-Seren und weiterer Einzelheiten sei auf die Publikationen von KAUFFMANN & PETERSEN verwiesen.

Die weitere Einteilung der O-Gruppen in *species* erfolgt auf Grund verschiedener H-Antigene und ermöglicht die Unterscheidung von monophasischen und diphasischen *species*. Bei den diphasischen *species* kommen über 20 verschiedene H-Antigenkomplexe der 1. Phase mit über 10 verschiedenen H-Antigenkomplexen der 2. Phase in allen möglichen Kombinationen vor. Die H-Antigenkomplexe der 1. Phase sind wie folgt bezeichnet: a - b - c - d - e, h - i - k - l, v - l, w - l, z_{13} - l, z_{13}, z_{28} - l, z_{28} - r - r(i) - y - z - z_{29} - z_{35} - z_{41} etc. Bei den meist monophasischen *species* treten folgende Phasen auf: Der G-Komplex wie f, g - f, g, t - g, m - g, p etc. und z_4, z_{23} - z_4, z_{24} - z_4, z_{32} - z_{36} - z_{38} etc.

Als 2. Phasen treten folgende Antigenkomplexe auf: 1,2 - 1,5 - 1,6 - 1,7 - z_6 - e, n, x - e, n, z_{15} - l, w - l, z_{13}, z_{28} etc.

Betrachtet man das KAUFFMANN-WHITE-Schema, so sieht man, dass fast alle möglichen Kombinationen zwischen den Antigenen der 1. und 2. Phase verwirklicht sind, wenn wir alle O-Gruppen zusammenrechnen. Betrachtet man aber nur eine einzige O-Gruppe, wie z.B. die häufige C 1-Gruppe, sieht man, dass bisher weniger als die Hälfte aller möglichen Kombinationen nachgewiesen ist. Da es sich um etwa 20 verschiedene H-Komplexe der 1. Phase, die mit etwa 10 verschiedenen H-Komplexen der 2. Phase kombiniert sind, handelt, so sind innerhalb einer jeden O-Gruppe ca. 200 verschiedene *species* möglich. Bei einer Zahl von 50 O-Gruppen würden wir also auf ca. 10.000 mögliche *Salmonella-species*, die mit Hilfe des KAUFFMANN-WHITE-Schemas diagnostiziert werden können, kommen.

Von diesen theoretisch möglichen *species* sind bisher nur ca. 830 festgestellt worden, sodass es kein Wunder ist, wenn im Laufe der letzten Jahre jährlich ca. 50–70 neue *species* gefunden wurden.

Hierbei ist zu beachten, dass die Zahl von 10.000 *Salmonella-species* als Minimalzahl betrachtet werden muss, weil das KAUFFMANN-WHITE-Schema aus praktischen Gründen vereinfacht ist. Falls wir alle nachweisbaren H-Partialantigene berücksichtigen, können wir ca. 40 H-Antigenkomplexe der 1. Phase und ca. 20 der 2. Phase unterscheiden. Wir gelangen so zu 800 verschiedenen H-Antigen-Kombinationen, die innerhalb von 50 O-Gruppen auftreten können, sodass wir zu einer Zahl von ca. 40.000 *species* gelangen.

Um eine Vereinfachung zu erreichen, wurde das KAUFFMANN-WHITE-Schema stark verkürzt und kann noch weiter vereinfacht werden, wie es der Vorschlag von EDWARDS & KAUFFMANN zeigt. Man kann auf diese Weise die Zahl der H-Komplexe der 1.Phase auf etwa 15 und diejenige der 2. Phase auf 2 H-Komplexe beschränken. Mit Hilfe dieses vereinfachten Schemas kann man über 1000 verschiedene *Salmonella*-Typen diagnostizieren, während mit Hilfe des originalen KAUFFMANN-WHITE-Schemas ca. 10.000 verschiedene Serotypen bestimmt werden können. Man kann also das diagnostische Antigenschema vereinfachen oder komplizieren, je nach den Bedürfnissen des Untersuchers. Das KAUFFMANN-WHITE-Schema steht zwischen dem "vereinfachten" und dem "kompletten" Schema und ist mit Rücksicht auf die diagnostische Praxis stark vereinfacht.

Allgemein-gültige Regeln für die Vereinfachung der serologischen Bestimmung können nicht gegeben werden, da diese Frage von jedem einzelnen Laboratorium selbst entschieden werden muss. Doch sei betont, dass die Bestimmung aller bekannten *Salmonella-species* prinzipiell nur in den hierzu eingerichteten Salmonella-Centralen ausgeführt werden soll, also nicht in den gewöhnlichen, bakteriologischen Laboratorien, die sich mit der Diagnose pathogener Darmbakterien befassen. Diese Laboratorien sollen nur die *Salmonella*-Diagnose stellen und die wichtigsten *species* diagnostizieren, wie *S. paratyphi-A*, B und C, *S. typhi*, *S. typhi-murium*, *S. cholerae-suis* und *S. enteritidis*.

Zu diesem Zwecke genügen 15 nichtabsorbierte Seren: O - S e r e n : A = 1, 2, 12 — B = 4, 5, 12 — C = 6, 7, 8 — D = 9, 12 — E = 3, 10, 15 und ein polyvalentes O-Serum A + B + C + D + E. Ferner ein V i - S e r u m und die H - S e r e n : a - b - c - d - e, n, x - g, p - i und 1, 2, 5.

Absorbierte Seren, sogenannte Faktorseren, sollen nur in Salmonella-Centralen für eigenen Gebrauch hergestellt werden. Die fabrikmässige Herstellung derartiger Faktorseren und ihre Lieferung an bakteriologische Laboratorien werden dringend widerraten.

Die serologische Typen-Bestimmung, bei der die K ö r p e r - und G e i s s e l - Antigene getrennt berücksichtigt werden, erfolgt mit Hilfe der Agglutination, in der Regel mit Hilfe der Objektglas-Agglutination. Es sei betont, dass nicht alle im Kauffmann-White-Schema angegebenen O-Antigene in chemisch-physikalischer Hinsicht den klassischen O-Antigen entsprechen, sondern dass verschiedene Antigene (oder besser: Antigen-Komponenten oder Faktoren) eine Sonderstellung einnehmen. Um einen neutralen Namen zu haben, sind daher die somatischen Antigene, einschliesslich des Vi-Antigens, als "K ö r p e r - A n t i g e n e" (somatic antigens) oder als "O-Antigene im weiteren Sinne" bezeichnet. Wenn man das Vi-Antigen als abweichendes O-Antigen auffasst und das M-Antigen (Schleim-Antigen) gesondert aufführt, so kann man die Bezeichnung "K - A n t i g e n e" vermeiden und so die schematische Darstellung vereinfachen.

Man muss sich darüber im klaren sein, dass alle Körper- oder O-Antigene sehr komplex gebaut sind, dass nicht alle Partial-Faktoren im K-W-Schema angegeben sind, und dass wir über die Verschiedenheiten innerhalb der Körper-Antigene noch sehr wenig wissen, sodass eingehende serologische und chemische Untersuchungen dringend erwünscht sind.

Die G e i s s e l - A n t i g e n e oder H-Antigene liegen bei zahlreichen *species* in 2 verschiedenen, serologischen Erscheinungformen, die P h a s e n genannt werden, vor. Auch diese Antigene sind sehr komplex gebaut und im Kauffmann-White-Schema stark vereinfacht.

Serologische Variation

Die Erfolge der *Salmonella*-Serologie sind in erster Linie der Aufklärung der serologischen Variation, speziell des Formen-Wechsels bei den somatischen Antigenen und des Phasen-Wechsels bei den Geissel-Antigenen, zu verdanken. Ohne Kenntnis und Berücksichtigung dieser dynamischen, serologischen Vorgänge kann keine erfolgreiche Serologie betrieben werden. Es muss aber betont werden, dass wir von einer völligen Klarstellung dieser serologischen Variation noch weit entfernt sind, besonders hinsichtlich des S-T-R-Formen-Wechsels und des S-R-Phasen-Wechsels. Wie kürzliche Untersuchungen von KAUFFMANN & ROHDE über den O-Formen-Wechsel zeigen, gibt es noch immer auf dem *Salmonella*-Gebiete die Möglichkeit, durch neue Befunde die serologische Diagnostik zu verbessern und zu vereinfachen.

Um einen kurzen Überblick über dieses umfangreiche Gebiet zu geben, sind im folgenden die verschiedenen Arten der Variation aufgezählt.

I. H-O-Variation = Verlust der H-Antigene, Übergang der OH-Form zur O-Form.

II. S-T-R-Variation = Qualitative Veränderungen der O-Antigene, die bis zum Verlust der O-Antigene führen. Übergang der S-Form zur T-Form (KAUFFMANN) oder zur R-Form.

Die Ausdrücke S, T und R werden rein serologisch gebraucht. Die morphologisch glatt oder rauh wachsenden Kulturen werden "Glatt-Formen" (smooth forms) oder "Rauh-Formen" (rough forms) genannt. Die T-Formen sind morphologisch Glatt-Formen.

III. Form-Variation (KAUFFMANN) = Quantitative Veränderungen der somatischen Antigene:

1. O-Variation, innerhalb bestimmter O-Antigene.

2. Vi-Variation oder V-W-Variation, innerhalb des Vi-Antigens.

IV. Phasen-Variation (phase-variation) oder Phasen-Wechsel = Qualitative Veränderungen der H-Antigene, d.h. sprunghafter Übergang von einer Phase in die andere.

1. S-Phasen-Wechsel, innerhalb der S-Phasen.

(a) Spezifisch-unspezifischer Phasen-Wechsel (ANDREWES) z.B. b:1, 2.

(b) α-β-Phasen-Wechsel (KAUFFMANN & MITSUI), z.B. b:e, n, x.

(c) 1, w-Phasen-Wechsel (EDWARDS & BRUNER), z.B. b:l, w.

Um einen neutralen Ausdruck zu gebrauchen, werden die üblichen S-Phasen mit 1. und 2. Phase bezeichnet.

2. S-R-Phasen-Wechsel (KAUFFMANN).

Wechsel zwischen den S- und R-Phasen. Die R-Phasen wurden früher auch als "künstliche" oder "induzierte" Phasen bezeichnet. So ist zum Beispiel die j-Phase von *S. typhi* eine typische R-Phase.

(Näheres siehe in KAUFFMANN: "Die Bakteriologie der Salmonella-Species" (1961)).

An dieser Stelle sei nur hervorgehoben, dass der Verfasser alle derartigen Phasen wie j - z_5 - z_{11} - z_{33} - z_{34} - z_{37} - z_{43} - z_{45} - z_{46} - z_{47} - etc. als R-Phasen betrachtet, d.h. als spontan entstandene Varianten, die als Degenerationsprodukte aufzufassen sind. Es handelt sich hierbei also nicht um normale Phasen oder um einen normalen Phasenwechsel, sondern um pathologische Vorgänge, die dem S-R-Formenwechsel bei den somatischen Antigenen entsprechen. Aus diesem Grunde sind alle derartigen R-Phasen aus dem KAUFFMANN-WHITE-Schema ausgeschlossen, da dieses Schema nur eine diagnostische Bestimmungstabelle der normalen *Salmonella*-Antigene und Phasen ist.

Es ist möglich, dass einige der R-Phasen wie z_{27}, z_{43}, z_{45} und z_{46} Hybriden sind, so dass die R-Phasen in Mutanten und Hybriden eingeteilt werden können.

So wurde z.B. *S. simsbury* = 1, 3, 19:z_{27}:- aus dem K-W-Schema gestrichen, da es sich hierbei um eine R-Phase handelt. Ferner liegt bei der von LE MINOR & EDWARDS beschriebenen *Salmonella*-Kultur mit 3 Phasen nicht eine neue *species*, sondern *S. goerlitz* = 3, 15: e, h:1, 2 mit einer R-Phase z_{27} vor. Wir haben es hierbei und in allen ähnlichen Fällen mit 2 normalen S-Phasen und 1 R-Phase derselben *species* zu tun. In der gleichen Weise, in der beim S-R-Formenwechsel Zwischenformen auftreten, können auch beim S-R-Phasenwechsel intermediäre Phasen, die gleichzeitig S- und R-Antigene enthalten, vorkommen, z.B. bei *S. meleagridis* = e, h, z_{45} und l, w, z_{45}. Hier besitzen beide S-Phasen (e, h und l, w) das R-Antigen z_{45}, können aber nach DOUGLAS & EDWARDS durch Züchtung in z_{45}-Serum rein gewonnen werden. Eine entsprechende *S. infantis*-Kultur, deren beide Phasen r-1,5 das z_{49}-Antigen, ein R-Antigen, enthielten, wurde von EDWARDS, McWHORTER & DOUGLAS beschrieben.

Mit diesem z_{49}-Antigen ist das H-Antigen z_{48} verwandt, das nach McWHORTER, DOUGLAS & EDWARDS in einem neuen Serotyp "S. cook" = 39:z_{48}:1,5 vorkommt. Der Verfasser hat jedoch diesen Typ nicht in das K-W-Schema eingefügt, da es sich beim z_{48}-Antigen um eine R-Phase von *S. champaign* handeln könnte.

Aus diesem Grunde wurde auch *S. rutgers* = 3,10:l, z_{40}:1,7 aus dem K-W-Schema gestrichen, da l, z_{40} wahrscheinlich eine R-Phase ist.

Da wir bis heute bereits über 830 *Salmonella-species* kennen, so liegt kein Grund vor, fragwürdige Typen in das K-W-Schema einzuordnen.

Während es sich bei der H-O-Variation und der S-T-R-Variation um eine Verlust-Mutation, um das Verschwinden der normalen H- oder O-Antigene handelt, treten beim S-Phasen-Wechsel und bei der O- und Vi-Variation sprunghafte Veränderungen diskreter

Zustände auf, sodass es zu einem Wechsel zwischen 2 verschiedenen Erscheinungsformen kommt. Beim S-Phasen-Wechsel handelt es sich um qualitative Antigen-Veränderungen, während es sich bei der O- und Vi-Variation um quantitative Antigen-Veränderungen handelt. Das serologische Geschehen in einer Kultur verläuft also in einem bestimmten Rhythmus, es ist nicht stationär, sondern dynamisch.

Die O-Variation (Kauffmann).

Bei der O-Variation oder Form-Variation kommt es zu quantitativen Veränderungen innerhalb bestimmter O-Antigene:

1-Variation (Kauffmann),

innerhalb des O 1-Antigens. Bei der Objektglas-Agglutination kann man bei zahlreichen *species* Kolonien mit stark oder schwach entwickeltem O 1-Antigen, die einander abspalten, feststellen. Durch Infektion mit bestimmten Phagen (Conversion) kann das O 1-Antigen bei verschiedenen *species* der A, B, D und anderen Gruppen auftreten.

6_1-Variation (Edwards),

innerhalb des 6_1-Antigens. Es treten Kolonien mit stark oder schwach entwickeltem 6_1-Antigen auf und spalten einander ab.

12-Variation (Kauffmann),

innerhalb des 12-Antigens. Die verschiedenen *species* enthalten entweder die 12_1, 12_2-Antigene, die 12_1, 12_3-Antigene oder die 12_1, 12_2, 12_3-Antigene. Der 12-Formenwechsel spielt sich am 12_2-Antigen ab und führt zu Kolonien, die das 12_2-Antigen stark oder schwach entwickelt haben und einander abspalten.

An ein- und derselben Kultur (z.B. bei *S. paratyphi-B*) kann der O 1-und O 12_2-Formen-Wechsel gleichzeitig auftreten.

Der *S. typhi*-Stamm T 2 A.S. (A.S. = ALMON & STOVALL) enthält nicht das 12_2-Antigen.

Während die *S. paratyphi-A*-Kulturen meist in der 12_1, 12_3-Form vorliegen, kommen nach SCHMID & KAUFFMANN gelegentlich auch Stämme mit 12_2-Antigen und dem 12_2-Formen-Wechsel vor.

Kürzlich konnte der Verfasser einen Teil des 12_2-Antigens auch in der E 3-Gruppe = (3), (15), 34 nachweisen. Schon früher (1942) hatte er auf O-Antigenbeziehungen zwischen der E 3-Gruppe (*S. illinois*) und der D-Gruppe (*S. typhi*) hingewiesen und gesagt, dass das O 34-Serum durch Absorption eines (3), (15), 34-Serum mit *S. newington* + *S. typhi* hergestellt werden solle. Kürzlich konnte er zeigen, dass diese Antigenbeziehungen auf einem Teile des 12_2-Antigens beruhen. Bisher konnte aber kein 12_2-Formen-Wechsel innerhalb der E 3-Gruppe nachgewiesen werden.

Weitere O-Formen-Wechsel wurden von KAUFFMANN & ROHDE

(3) beschrieben, und zwar bei den Antigenen 19, 24, 25, 27, 36 und 37. Man muss beachten, dass auch bei anderen Körper-Antigenen mit einem Formen-Wechsel zu rechnen ist, da derartige Untersuchungen bisher nur an einem ausgewählten Material vorgenommen sind. Wahrscheinlich haben wir es bei der O-Variation mit einer ganz allgemeinen Erscheinung in der Serologie der *Enterobacteriaceae* zu tun.

Betreffs Serologie der R-Formen und R-Phasen sei auf das Buch von KAUFFMANN: „Die Bakteriologie der Salmonella-Species" verwiesen, ebenso auch betreffs der Definition der einzelnen Körper- und Geissel-Antigene.

Dagegen sei hier auf das vereinfachte KAUFFMANN-WHITE-Schema kurz eingegangen, da es auf Grund der Arbeit von KAUFFMANN & ROHDE (2) für die Diagnose von Arizona-Kulturen von Bedeutung ist. Der Zweck dieser Vereinfachungen war, die Typendiagnose für kleinere Laboratorien zu erleichtern und absorbierte Seren möglichst zu vermeiden. Man muss hierbei aber in Kauf nehmen, dass eine grössere Anzahl der *species* nicht exakt bestimmt und daher auch nicht benannt werden kann. Im folgenden sind die zur vereinfachten Bestimmung nötigen Seren angegeben.

O-Seren

A	=	*S. paratyphi-A*	42	=	*S. weslaco*
B	=	*S. paratyphi-B + S. schleissheim*	43	=	*S. milwaukee*
C	=	*S. thompson + S. newport +*	44	=	*S. niarembe*
		S. carrau + S. onderstepoort	45	=	*S. deversoir + S. dugbe*
D	=	*S. gallinarum + S. strasbourg*	47	=	*S. bergen + S. kaolack*
E	=	*S. anatum + S. newington*	48	=	*S. dahlem + S. djakarta*
11	=	*S. aberdeen*	50	=	*S. greenside + S. wassenaar*
13	=	*S. poona + S. grumpensis*	51	=	*S. treforest*
16	=	*S. gaminara*	52	=	*S. utrecht*
17	=	*S. kirkee*	53	=	*S. humber*
18	=	*S. cerro*	54	=	*S. uccle*
21	=	*S. minnesota*	55	=	*S. tranoroa*
28	=	*S. tel-aviv + S. dakar*	56	=	*S. artis*
30	=	*S. urbana*	57	=	*S. locarno*
35	=	*S. adelaide*	58	=	*S. basel*
38	=	*S. inverness*	59	=	*S. betioky*
39	=	*S. champaign*	60	=	*S. luton*
40	=	*S. riogrande + S. bulawayo*	Vi	=	*Ballerup* original.
41	=	*S. waycross*			

H-Seren

a	=	*S. paratyphi-A*	r	=	*S. rubislaw*
b	=	*S. java* oder *S. hvittingfoss*	y	=	*S. madelia*
c	=	*S. cholerae-suis*	z	=	*S. poona*
d	=	*S. typhi*	z_4	=	*S. cerro + S. duesseldorf +*
e, h	=	4, 5, 12: e, h: —			*S. tallahassee*
G	=	*S. derby + S. oranienburg +*	z_{10}	=	*S. illinois*
		S. dublin	z_{29}	=	*S. tennessee*
i	=	*S. typhi-murium* oder *S. bonariensis*	l	=	*S. newport + S. thompson +*
					S. kentucky (2. Phase)
k	=	*S. thompson*	e, n	=	*S. abortus-equi.*
L	=	*S. worthington* (l, w)			

Alle übrigen H-Seren können in kleineren Laboratorien ausgelassen werden, da diese H-Antigene relativ selten vorkommen.

Unter der Bezeichnung "G" sind alle diejenigen H-Antigene, die zum g . . . -Komplex gehören, zusammengefasst:

f, g	g, m, s	g, q
f, g, t	g, m, t	g, s, t
g, m	g, p	m, t
g, m, q	g, p, u	u.s.w.

In der entsprechenden Weise sind die l, v - l, w - l, z_{13} - l, z_{28} - l, z_{13}, z_{28}-Phasen unter der Bezeichnung "L" zusammengefasst.

Die mit einander verwandten H-Antigene z_4, z_{23} - z_4, z_{24} und z_4, z_{32} sind mit "z_4" bezeichnet.

Die H-Antigene der 2. Phase mit dem 1 . . .-Antigen-Komplex sind mit "1" angegeben und umfassen 1,2 - 1, 5 - 1, 6 - 1, 7 und z_6.

Die H-Antigene der 2. Phase mit dem e, n . . .-Antigen-Komplex sind mit "e, n" bezeichnet und umfassen die Phasen e, n, x - e, n, z_{15} etc.

In der Tabelle III ist die vereinfachte, serologische Diagnose von *Salmonella arizonae*-species nach KAUFFMANN & ROHDE (2) angegeben.

Tabelle III.

Antigenstruktur von Salmonella arizonae-Species
(Vereinfacht nach KAUFFMANN & ROHDE)

O	H 1. Phase	H 2. Phase	O	H 1. Phase	H 2. Phase
C*	L	Ar. 25	45	a	e, n
C	z_4	—	45	z_{29}	—
C + 18	z_{10}	Ar. 38	47	i	Ar. 25
11	k	Ar. 25	47	L	Ar. 25
13	z_4	—	48	z_{36}	—
16	z_{10}	e, n	50	G	—
17	z_4	—	50	i	z
18	z_4	—	51	z_4	—
21	z_4	—	52	L	Ar. 25
35	L	1	53	z_4	—
38	k	Ar. 37	56	z_4	—
38	L	Ar. 34	57	i	e, n
38	r	z	58	z_{52}**	z
38	Ar. 39	Ar. 25	59	z_4	—
40	G	—	60	k	Ar. 25
40	z_4	—	Ar. 6	G	—
41	G	—	Ar. 8	z_4	—
42	G	—	Ar. 26	c	z_{35}
43	z_4	—	Ar. 29	i	Ar. 40
44	z_4	—	Ar. 30	L	z

* C = 6,7 + 6,8 + (8),20 + 6,(7),(14) + 6,14,24 + 1,6,14,25.
** *Salmonella* z_{52} = Ar. H 26.
O 59 ist verwandt mit Ar. O 19
O 60 ist identisch mit Ar. O 24.

Um einen Überblick über die bisher festgestellten Arizona-Antigene zu geben, sind in der Tabelle IV die vereinfachten Antigen-Komponenten, die serologischen "Elemente", aus denen sich die Antigen-"Verbindungen" der Arizona-*species* aufbauen, dargestellt. So ist z.B. die *type-species* aus den "Elementen" O 1, 2 und H 1, 2, 5 oder vereinfacht aus $51:z_4$ aufgebaut. Die O-Antigen-Komponenten 1 und 2 entsprechen dem O-Antigen-Komplex 51 des K-W-Schemas, während die H-Antigen-Komponenten 1, 2 und 5 dem z_4-Komplex entsprechen. Während das K-W-Schema eine Liste aller bisher festgestellten Antigen-"Verbindungen" ist, gibt die Tabelle IV nur die fundamentalen Antigen-Komponenten oder die serologischen "Elemente" der Arizona-*species* wieder.

Tabelle IV.

Die Antigen-Komponenten der Salmonella arizonae-Species
(vereinfacht)

Körper-Antigene		Geissel-Antigene	
		1. Phase	2. Phase
C	47	a	1
C + 18	48	c	e, n
11	50	G	z
13	51	i	z_{35}
16	52	k	
17	53	L	Ar. 25
18	56	r	Ar. 34
21	57	z_4	Ar. 37
35	58	z_{10}	Ar. 38
38	59	z_{29}	Ar. 40
40	60	z_{36}	
41	Ar. 6	z_{52}	
42	Ar. 8		
43	Ar. 26	Ar. 39*	
44	Ar. 29		
45	Ar. 30		

* Nach EDWARDS & EWING ("Identification of Enterobacteriaceae" 2. edition) ist das Ar. 39 H-Antigen mit dem *Salmonella* H-Antigen z_{47} verwandt. Da es sich nach Ansicht des Verfassers hierbei um eine R-Phase handeln dürfte, so kann dieses Antigen aus dem K-W-Schema ausgelassen werden.

Die mit Buchstaben oder mit Zahlen bezeichneten Antigene sind *Salmonella*-Antigene. Nur die allein in der Arizona-Gruppe vorkommenden Antigene sind mit Ar. 6 etc. angegeben.

In der entsprechenden Weise sind in der Tabelle V die Antigen-Komponenten oder serologischen "Elemente" der *Salmonella-species* der *sub-genera* I und II wiedergegeben.

Tabelle V.

Die Antigen-Komponenten der Salmonella-Species
(Sub-genera I und II, vereinfacht)

Körper-Antigene		Geissel-Antigene	
		1. Phase	2. Phase
A	43	a	1
B	44	b	e, n
C	45	c	L
D	46	d	.
E	47	e, h	.
11	48	G	.
13	50	i	
16	51	k	
17	52	L	
18	53	r	
21	54	y	
28	55	z	
30	56	z_4	
35	57	z_{10}	
38	58	z_{29}	
39	59	.	
40	60	.	
41	Vi		
42			

Aus der Tabelle V kann man ohne weiteres ablesen, welche Antigen-Verbindungen oder *species* möglich sind. So entspricht z.B. die Antigen-Verbindung 11:r:1 der *Salmonella-species S. aberdeen*, deren vollständigere Formel 11:r:1, 2 lautet.

Von den in Tabelle III angegebenen *species* können 6 auf Grund ihres biochemischen Verhaltens zum *Salmonella sub-genus* II gerechnet werden:

$$C:z_4:- \qquad = \text{Ar. } 32:1, 2, 6$$
$$21:z_4:- \qquad = \text{Ar. } 22:1, 2, 5, 6$$
$$40:G:- \qquad = \text{Ar. } 10a, c:13, 15$$
$$43:z_4:- \qquad = \text{Ar. } 21:1, 2, 6$$
$$45:a:e, n = \text{Ar. } 11:35:28$$
$$50:G:- \qquad = \text{Ar. } 9a, b:13, 15$$

Diese 6 *species* wurden früher, vor der Aufstellung des *sub-genus* II, als Arizona-Kulturen betrachtet. Sie nehmen aber in biochemischer Hinsicht eine Zwischenstellung ein und gehören zu den sogenannten "atypischen *sub-genus* II" Kulturen.

Das Kauffmann-White-Schema

Species	Körper-Antigene	Geissel-Antigene 1. Phase	Geissel-Antigene 2. Phase
Gruppe A			
S. paratyphi-A	1, 2, 12	a	—
S. paratyphi-A var. durazzo	2, 12	a	—
S. kiel	1, 2, 12	g, p	—
Gruppe B			
S. kisangani	1, 4, 5, 12	a	1, 2
S. hessarek	4, 12	a	1, 5
S. hessarek var. 27	4, 12, 27	a	1, 5
S. fulica	4, 5, 12	a	1, 5
S. arechavaleta	4, 5, 12	a	1, 7
S. bispebjerg	1, 4, 5, 12	a	e, n, x
S. abortus-equi	4, 12	—	e, n, x
S. tinda	1, 4, 12, 27	a	e, n, z_{15}
S. paratyphi-B	1, 4, 5, 12	b	1, 2
S. paratyphi-B var. odense	1, 4, 12	b	1, 2
S. java	1, 4, 5, 12	b	[1, 2]
* S. sofia	4, 12	b	—
* S. sofia var. 27	4, 12, 27	b	—
S. limete	1, 4, 12, 27	b	1, 5
S. canada	4, 12	b	1, 6
S. uppsala	4, 12, 27	b	1, 7
S. abony	1, 4, 5, 12	b	e, n, x
S. abony var. haifa	4, 12	b	e, n, x
S. abortus-bovis	1, 4, 12, 27	b	e, n, x
S. wagenia	1, 4, 12, 27	b	e, n, z_{15}
S. schleissheim	4, 12, 27	b	—
S. wien	1, 4, 12	b	l, w
S. legon	4, 12	c	1, 5
S. abortus ovis	4, 12	c	1, 6
S. altendorf	4, 12	c	1, 7
S. womba	4, 12, 27	c	1, 7
S. jericho	1, 4, 12, 27	c	e, n, z_{15}
S. bury	4, 12, 27	c	z_6
S. stanley	4, 5, 12	d	1, 2
S. cairo	1, 4, 12, 27	d	1, 2
S. eppendorf	1, 4, 12, 27	d	1, 5
S. schwarzengrund	1, 4, 12, 27	d	1, 7
S. sarajane	4, 12, 27	d	e, n, x
S. duisburg	4, 12	d	e, n, z_{15}
S. duisburg var. 27	1, 4, 12, 27	d	e, n, z_{15}
S. salinatis	4, 12	d, e, h	d, e, n, z_{15}
S. mons	1, 4, 12	d	l, w
S. ayinde	4, 12, 27	d	z_6
S. saint-paul	1, 4, 5, 12	e, h	1, 2
S. reading	4, 5, 12	e, h	1, 5

Species	Körper-Antigene	Geissel-Antigene	
		1. Phase	2. Phase
S. kaapstad	4, 12	e, h	1, 7
S. chester	4, 5, 12	e, h	e, n, x
S. san-diego	4, 5, 12	e, h	e, n, z_{15}
* S. makumira	4, 12	e, n, x	1, 7
S. derby	1, 4, 5, 12	f, g	—
S. agona	4, 12	f, g, s	—
S. essen	4, 12	g, m	—
* S. caledon	4, 12	g, m	e, n, x
S. hato	4, 5, 12	g, m, s	—
S. travis	4, 5, 12	g, (p), z_{51}	1, 7
S. kingston	1, 4, 12, 27	g, s, t	—
S. kingston var. copenhagen	4, 12	g, s, t	—
S. budapest	1, 4, 12	g, t	—
* S. bechuana	4, 12, 27	g, t	—
S. california	4, 5, 12	m, t	—
S. typhi-murium	1, 4, 5, 12	i	1, 2
S. typhi-murium var. copenhagen	1, 4, 12	i	1, 2
S. agama	4, 12	i	1, 6
S. gloucester	1, 4, 12, (27)	i	l, w
S. neumuenster	1, 4, 12, 27	k	1,6
S. texas	4, 5, 12	k	e, n, z_{15}
S. azteca	4, 5, 12	l, v	1, 5
S. bredeney	1, 4, 12, 27	l, v	1, 7
S. kimuenza	1, 4, 12, 27	l, v	e, n, x
S. brandenburg	4, 12	l, v	e, n, z_{15}
* S. kilwa	4, 12	l, w	e, n, x
S. vom	4, 12, 27	l, z_{13}, z_{28}	e, n, z_{15}
S. kunduchi	1, 4, 12, 27	l, z_{28}	1, 2
S. heidelberg	1, 4, 5, 12	r	1, 2
S. bradford	4, 12, 27	r	1, 5
S. remo	1, 4, 12, 27	r	1, 7
S. africana	4, 12	r (i)	l, w
S. coeln	4, 5, 12	y	1, 2
S. teddington	4, 12, 27	y	1, 7
S. ruki	4, 5, 12	y	e, n, x
S. ball	1, 4 ,12, 27	y	e, n, x
S. jos	1, 4, 12, 27	y	e, n, z_{15}
S. kamoru	4, 12, 27	y	z_6
S. shubra	4, 5, 12	z	1, 2
S. kiambu	4, 12	z	1, 5
S. indiana	1, 4, 12	z	1, 7
* S. nordenham	1, 4, 12, 27	z	e, n, x
S. preston	1, 4, 12	z	l, w
S. entebbe	1, 4, 12, 27	z	z_6
S. stanleyville	1, 4, 5, 12	z_4, z_{23}	[1, 2]

Species	Körper-Antigene	Geissel-Antigene	
		1. Phase	2. Phase
S. jaja	4, 12, 27	z_4, z_{23}	—
S. kalamu	4, 12	z_4, z_{24}	—
S. haifa	1, 4, 5, 12	z_{10}	1, 2
S. ituri	1, 4, 12	z_{10}	1, 5
S. tudu	4, 12	z_{10}	1, 6
S. albert	4, 12	z_{10}	e, n, x
S. fortune	4, 12, 27	z_{10}	z_6
S. brancaster	1, 4, 12, 27	z_{29}	—
S. tafo	1, 4, 12, 27	z_{35}	1, 7
S. tejas	4, 12	z_{36}	—
S. wilhelmsburg	4, 12, 27	z_{38}	—
* S. durbanville	4, 12	z_{39}	1, 5, 7
Gruppe C 1			
S. san-juan	6, 7	a	1, 5
S. umhlali	6, 7	a	1, 6
S. austin	6, 7	a	1, 7
S. oslo	6, 7	a	e, n, x
S. denver	6, 7	a	e, n, z_{15}
S. coleypark	6, 7	a	l, w
* S. calvinia	6, 7	a	z_{42}
S. brazzaville	6, 7	b	1, 2
S. edinburg	6, 7	b	1, 5
S. georgia	6, 7	b	e, n, z_{15}
S. ohio	6, 7	b	l, w
S. leopoldville	6, 7	b	z_6
S. kotte	6, 7	b	z_{35}
* S. bloemfontein	6, 7	b	z_{42}
S. paratyphi-C	6, 7, Vi	c	1, 5
S. cholerae-suis	6, 7	c	1, 5
S. typhi-suis	6, 7	c	1, 5
S. decatur	6, 7	c	1, 5
S. birkenhead	6, 7	c	1, 6
S. mission	6, 7	d	1, 5
S. isangi	6, 7	d	1, 5
S. kivu	6, 7	d	1, 6
S. amersfoort	6, 7	d	e, n, x
S. gombe	6, 7	d	e, n, z_{15}
S. livingstone	6, 7	d	l, w
S. wil	6, 7	d	l, z_{13}, z_{28}

Species	Körper-Antigene	Geissel-Antigene	
		1. Phase	2. Phase
S. larochelle	6, 7	e, h	1, 2
S. lomita	6, 7	e, h	1, 5
S. norwich	6, 7	e, h	1, 6
S. braenderup	6, 7	e, h	e, n, z_{15}
S. rissen	6, 7	f, g	—
S. montevideo	6, 7	g, m, s	—
S. othmarschen	6, 7	g, m, t	—
S. alamo	6, 7	g, (p), z_{51}	1, 5
S. menston	6, 7	g, s, t	—
S. riggil	6, 7	g, t	—
S. haelsingborg	6, 7	m, p, t, u	—
S. oranienburg	6, 7	m, t	—
S. oritamerin	6, 7	i	1, 5
S. garoli	6, 7	i	1, 6
S. norton	6, 7	i	l, w
S. galiema	6, 7	k	1, 2
S. thompson	6, 7	k	1, 5
S. daytona	6, 7	k	1, 6
S. singapore	6, 7	k	e, n, x
S. escanaba	6, 7	k	e, n, z_{15}
S. concord	6, 7	l, v	1, 2
S. irumu	6, 7	l, v	1, 5
S. bonn	6, 7	l, v	e, n, x
S. potsdam	6, 7	l, v	e, n, z_{15}
S. gdansk	6, 7	l, v	z_6
S. colorado	6, 7	l, w	1, 5
S. ness-ziona	6, 7	l, z_{13}	1, 5
S. kenya	6, 7	l, z_{13}	e, n, x
S. neukoelln	6, 7	l, z_{13}, z_{28}	e, n, z_{15}
S. makiso	6, 7	l, z_{13}, z_{28}	z_6
* S. heilbron	6, 7	l, z_{28}	1, 5
S. virchow	6, 7	r	1, 2
S. infantis	6, 7	r	1, 5
S. nigeria	6, 7	r	1, 6
S. colindale	6, 7	r	1, 7
S. papuana	6, 7	r	e, n, z_{15}
S. richmond	6, 7	y	1, 2
S. bareilly	6, 7	y	1, 5
S. gatow	6, 7	y	1, 7
S. hartford	6, 7	y	e, n, x
S. mikawasima	6, 7	y	e, n, z_{15}
* S. tosamanga	6, 7	z	1, 5
S. businga	6, 7	z	e, n, z_{15}
S. aequatoria	6, 7	z_4, z_{23}	e, n, z_{15}
S. inganda	6, 7	z_{10}	1, 5
S. eschweiler	6, 7	z_{10}	1, 6

Species	Körper-Antigene	Geissel-Antigene	
		1. Phase	2. Phase
S. djugu	6, 7	z_{10}	e, n, x
S. mbandaka	6, 7	z_{10}	e, n, z_{15}
S. tennessee	6, 7	z_{29}	—
* S. bacongo	6, 7	z_{36}	z_{42}
S. lille	6, 7	z_{38}	—
* S. gilbert	6, 7	z_{39}	1, 7
S. hillsborough	6, 7	z_{41}	1, w
* S. sullivan	6, 7	z_{42}	1, 7

Gruppe C 2

Species	Körper-Antigene	Geissel-Antigene	
		1. Phase	2. Phase
S. curacao	6, 8	a	1, 6
S. narashino	6, 8	a	e, n, x
S. leith	6, 8	a	e, n, z_{15}
* S. tulear	6, 8	a	z_{52}
S. nagoya	6, 8	b	1, 5
S. stourbridge	6, 8	b	1, 6
S. gatuni	6, 8	b	e, n, x
S. bukuru	6, 8	b	1, w
S. banalia	6, 8	b	z_6
S. wingrove	6, 8	c	1, 2
S. utah	6, 8	c	1, 5
S. bronx	6, 8	c	1, 6
S. belem	6, 8	c	e, n, x
S. quiniela	6, 8	c	e, n, z_{15}
S. muenchen	6, 8	d	1, 2
S. manhattan	6, 8	d	1, 5
S. sterrenbos	6, 8	d	e, n, x
S. labadi	6, 8	d	z_6
S. newport	6, 8	e, h	1, 2
S. kottbus	6, 8	e, h	1, 5
S. tshiongwe	6, 8	e, h	e, n, z_{15}
S. chincol	6, 8	g, m, s	e, n, x
* S. baragwanath	6, 8	m, t	1, 5
* S. germiston	6, 8	m, t	e, n, x
S. lindenburg	6, 8	i	1, 2
S. takoradi	6, 8	i	1, 5
S. bonariensis	6, 8	i	e, n, x
S. aba	6, 8	i	e, n, z_{15}

Species	Körper-Antigene	Geissel-Antigene	
		1. Phase	2. Phase
S. blockley	6, 8	k	1, 5
S. litchfield	6, 8	l, v	1, 2
S. manchester	6, 8	l, v	1, 7
S. holcomb	6, 8	l, v	e, n, x
S. edmonton	6, 8	l, v	e, n, z_{15}
S. fayed	6, 8	l, w	1, 2
S. bovis-morbificans	6, 8	r	1, 5
S. akanji	6, 8	r	1, 7
S. hidalgo	6, 8	r	e, n, z_{15}
S. gold-coast	6, 8	r	l, w
S. tananarive	6, 8	y	1, 5
S. alagbon	6, 8	y	1, 7
S. praha	6, 8	y	e, n, z_{15}
S. kuru	6, 8	z	l, w
S. chailey	6, 8	z_4, z_{23}	e, n, z_{15}
S. duesseldorf	6, 8	z_4, z_{24}	—
S. tallahassee	6, 8	z_4, z_{32}	—
S. mapo	6, 8	z_{10}	1, 5
S. hadar	6, 8	z_{10}	e, n, x
S. glostrup	6, 8	z_{10}	e, n, z_{15}
S. wippra	6, 8	z_{10}	z_6
S. uno	6, 8	z_{29}	—

Gruppe C 3

Species	Körper-Antigene	Geissel-Antigene	
		1. Phase	2. Phase
S. sanga	(8)	b	1, 7
S. shipley	(8), 20	b	e, n, z_{15}
S. virginia	(8)	d	1, 2
S. emek	(8), 20	g, m, s	—
S. kentucky	(8), 20	i	z_6
S. kentucky var.	(8)	i	z_6
S. amherstiana	(8)	l, v	1, 6
S. hindmarsh	(8)	r	1, 5
S. pikine	(8), 20	r	z_6
S. altona	(8), 20	r (i)	z_6
S. brunei	(8), 20	y	
S. kralingen	(8), 20	y	
S. corvallis	(8), 20)	z_4, z_{23}	—

Species	Körper-Antigene	Geissel-Antigene	
		1. Phase	2. Phase
S. albany	(8), 20	z_4, z_{24}	—
S. molade	(8), 20	z_{10}	z_6
S. tamale	(8), 20	z_{29}	—
Gruppe C 4			
S. kaduna	6, (7), (14)	c	e, n, z_{15}
S. eimsbuettel	6, (7), (14)	d	l, w
S. ardwick	6, (7), (14)	f, g	—
S. gelsenkirchen	6, (7), (14)	l, v	z_6
S. jerusalem	6, (7), (14)	z_{10}	l, w
S. bornum	6, (7), (14)	z_{38}	—
Gruppe D 1			
S. sendai	1, 9, 12	a	1, 5
S. miami	1, 9, 12	a	1, 5
S. (wuerzburg, S. bambesa)			
S. os	9, 12	a	1, 6
S. saarbruecken	1, 9, 12	a	1, 7
S. loma-linda	9, 12	a	e, n, x
S. durban	9, 12	a	e, n, z_{15}
S. onarimon	1, 9, 12	b	1, 2
S. frintrop	1, 9, 12	b	1, 5
* S. mjimwema	1, 9, 12	b	e, n, x
* S. blankenese	1, 9, 12	b	z_6
S. goeteborg	9, 12	c	1, 5
S. ipeko	9, 12	c	1, 6
S. alabama	9, 12	c	e, n, z_{15}
S. ridge	9, 12	c	z_6
* S. zuerich	1, 9, 12	c	z_{39}
S. typhi	9, 12, Vi	d	—
S. ndolo	9, 12	d	1, 5
S. tarshyne	9, 12	d	1, 6
S. zega	9, 12	d	z_6
S. jaffna	1, 9, 12	d	z_{35}
S. eastbourne	1, 9, 12	e, h	1, 5
S. israel	9, 12	e, h	e, n, z_{15}
* S. lindrick	9, 12	e, n, x	1, 5, 7
* S. lindrick var. 1, 7	9, 12	e, n, x	1, 7

Species	Körper-Antigene	Geissel-Antigene	
		1. Phase	2. Phase
S. berta	9, 12	f, g, t	—
S. enteritidis	1, 9, 12	g, m	—
S. blegdam	9, 12	g, m, q	—
* S. kuilsrivier	9, 12	g, m, s, t	e, n, x
* S. manica	1, 9, 12	g, m, s, t	z_{42}
S. dublin	1, 9, 12	g, p	—
S. naestved	1, 9, 12	g, p, s	—
S. rostock	1, 9, 12	g, p, u	—
S. new-mexico	9, 12	g, (p), z_{51}	1, 5
S. moscow	9, 12	g, q	—
* S. neasden	9, 12	g, s, t	e, n, x
* S. hamburg	1, 9, 12	g, t	—
S. pensacola	9, 12	m, t	—
S. seremban	9, 12	i	1, 5
S. claibornei	1, 9, 12	k	1, 5
S. mendoza	9, 12	l, v	1, 2
S. panama	1, 9, 12	l, v	1, 5
S. kapemba	9, 12	l, v	1, 7
S. goettingen	9, 12	l, v	e, n, z_{15}
S. victoria	1, 9, 12	l, w	1, 5
* S. dar-es-salaam	1, 9, 12	l, w	e, n, x
S. miyazaki	9, 12	l, z_{13}	1, 7
S. napoli	1, 9, 12	l, z_{13}	e, n, x
S. javiana	1, 9, 12	l, z_{28}	1, 5
S. jamaica	9, 12	r	1, 5
S. lawndale	1, 9, 12	z	1, 5
* S. stellenbosch	1, 9, 12	z	1, 7
* S. angola	1, 9, 12	z	z_6
* S. hueningen	9, 12	z	z_{39}
S. wangata	1, 9, 12	z_4, z_{23}	[1, 7]
S. portland	9, 12	z_{10}	1, 5
* S. canastel	9, 12	z_{29}	1, 5
S. penarth	9, 12	z_{35}	z_6
* S. wynberg	1, 9, 12	z_{39}	1, 7
S. gallinarum-pullorum	1, 9 ,12	—	—

Gruppe D 2

S. baildon	(9), 46	a	e, n, x
S. zadar	(9), 46	b	1, 6

Species	Körper-Antigene	Geissel-Antigene 1. Phase	2. Phase
* S. lundby	(9), 46	b	e, n, x
S. bamboye	(9), 46	b	l, w
S. itutaba	(9), 46	c	z_6
S. strasbourg	(9), 46	d	1, 7
S. plymouth	(9), 46	d	z_6
S. bergedorf	(9), 46	e, h	1, 2
S. wernigerode	(9), 46	f, g	—
S. gateshead	(9), 46	g, s, t	—
S. mathura	(9), 46	i	e, n, z_{15}
S. marylebone	(9), 46	k	1, 2
S. ceyco	(9), 46	k	z_{35}
S. shoreditch	(9), 46	r	e, n, z_{15}
* S. haarlem	(9), 46	z	e, n, x
S. ekotedo	(9), 46	z_4, z_{23}	—
S. lishabi	(9), 46	z_{10}	1, 7
S. ouakam	(9), 46	z_{29}	—
S. fresno	(9), 12, 46	z_{38}	—

Gruppe E 1

Species	Körper-Antigene	Geissel-Antigene 1. Phase	2. Phase
S. aminatu	3, 10	a	1, 2
S. oxford	3, 10	a	1, 7
S. kalina	3, 10	b	1, 2
S. butantan	3, 10	b	1, 5
S. huvudsta	3, 10	b	1, 7
S. benfica	3, 10	b	e, n, x
S. yaba	3, 10	b	e, n, z_{15}
S. epicrates	3, 10	b	1, w
S. pramiso	3, 10	c	1, 7
S. anderlecht	3, 10	c	l, w
S. okefoko	3, 10	c	z_6
S. shangani	3, 10	d	1, 5
S. onireke	3, 10	d	1, 7
S. souza	3, 10	d	e, n, x
S. birmingham	3, 10	d	l, w
S. weybridge	3, 10	d	z_6
S. maron	3, 10	d	z_{35}

Species	Körper-Antigene	Geissel-Antigene	
		1. Phase	2. Phase
S. vejle	3, 10	e, h	1, 2
S. muenster	3, 10	e, h	1, 5
S. anatum	3, 10	e, h	1, 6
S. nyborg	3, 10	e, h	1, 7
S. newlands	3, 10	e, h	e, n, x
S. meleagridis	3, 10	e, h	l, w
S. sekondi	3, 10	e, h	z_6
* S. chudleigh	3, 10	e, n, x	1, 7
S. suberu	3, 10	g, m	—
S. amsterdam	3, 10	g, m, s	—
S. westhampton	3, 10	g, s, t	—
* S. islington	3, 10	g, t	—
S. falkensee	3, 10	i	e, n, z_{15}
S. zanzibar	3, 10	k	1, 5
S. new-rochelle	3, 10	k	l, w
S. nchanga	3, 10	l, v	1, 2
S. sinstorf	3, 10	l, v	1, 5
S. london	3, 10	l, v	1, 6
S. give	3, 10	l, v	1, 7
S. ruzizi	3, 10	l, v	e, n, z_{15}
* S. fuhlsbuettel	3, 10	l, v	z_6
S. uganda	3, 10	l, z_{13}	1, 5
S. fallowfield	3, 10	l, z_{13}, z_{28}	e, n, z_{15}
* S. westpark	3, 10	l, z_{28}	e, n, x
S. seegefeld	3, 10	r (i)	1, 2
S. elisabethville	3, 10	r	1, 7
S. simi	3, 10	r	e, n, z_{15}
S. weltevreden	3, 10	r	z_6
S. amager	3, 10	y	1, 2
S. orion	3, 10	y	1, 5
S. mokola	3, 10	y	1, 7
S. ohlstedt	3, 10	y	e, n, x
S. bolton	3, 10	y	e, n, z_{15}
S. stockholm	3, 10	y	z_6
* S. alexander	3, 10	z	1, 5
* S. finchley	3, 10	z	e, n, x
S. clerkenwell	3, 10	z	l, w
S. adabraka	3, 10	z_4, z_{23}	[1, 7]
S. okerara	3, 10	z_{10}	1, 2
S. lexington	3, 10	z_{10}	1, 5
S. coquilhatville	3, 10	z_{10}	1, 7
S. kristianstad	3, 10	z_{10}	e, n, z_{15}
S. jedburgh	3, 10	z_{29}	—

Species	Körper-Antigene	Geissel-Antigene	
		1. Phase	2. Phase
S. cairina	3, 10	z_{35}	z_6
S. macallen	3, 10	z_{36}	—
S. bolombo	3, 10	z_{38}	—
* S. mpila	3, 10	z_{38}	z_{42}
* S. winchester	3, 10	z_{39}	1, 7

Gruppe E 2

Species	Körper-Antigene	1. Phase	2. Phase
S. rosenthal	3, 15	b	1, 5
S. pankow	3, 15	d	1, 5
S. goerlitz	3, 15	e, h	1, 2
S. new-haw	3, 15	e, h	1, 5
S. newington	3, 15	e, h	1, 6
S. selandia	3, 15	e, h	1, 7
S. cambridge	3, 15	e, h	l, w
S. drypool	3, 15	g, m, s	—
* S. parow	3, 15	g, m, s, t	—
S. halmstad	3, 15	g, s, t	—
S. portsmouth	3, 15	l, v	1, 6
S. new-brunswick	3, 15	l, v	1, 7
S. kinshasa	3, 15	l, z_{13}	1, 5
S. lanka	3, 15	r	z_6
S. tuebingen	3, 15	y	1, 2
S. binza	3, 15	y	1, 5
S. tournai	3, 15	y	z_6
S. manila	3, 15	z_{10}	1, 5

Gruppe E 3

Species	Körper-Antigene	1. Phase	2. Phase
S. arkansas	(3), (15), 34	e, h	1, 5
S. minneapolis	(3), (15), 34	e, h	1, 6
S. wildwood	(3), (15), 34	e, h	l, w
S. canoga	(3), (15), 34	g, s, t	—
S. menhaden	(3), (15), 34	l, v	1, 7
S. thomasville	(3), (15), 34	y	1, 5
S. illinois	(3), (15), 34	z_{10}	1, 5
S. harrisonburg	(3), (15), 34	z_{10}	1, 5

Species	Körper-Antigene	Geissel-Antigene	
		1. Phase	2. Phase

Gruppe E 4

Species	Körper-Antigene	1. Phase	2. Phase
S. gwoza	1, 3, 19	a	e, n, z_{15}
S. broughton	1, 3, 19	b	l, w
S. accra	1, 3, 19	b	z_6
S. ahmadi	1, 3, 19	d	1, 5
S. liverpool	1, 3, 19	d	e, n, z_{15}
S. tilburg	1, 3, 19	d	l, w
S. niloese	1, 3, 19	d	z_6
S. calabar	1, 3, 19	e, h	l, w
S. maiduguri	1, 3, 19	f, g, t	e, n, z_{15}
S. senftenberg	1, 3, 19	g, s, t	—
S. taksony	1, 3, 19	i	z_6
S. ngor	1, 3, 19	l, v	1, 5
S. westerstede	1, 3, 19	l, z_{13}	—
S. lokstedt	1, 3, 19	l, z_{13}, z_{28}	1, 2
S. bedford	1, 3, 19	l, z_{13}, z_{28}	e, n, z_{15}
S. yalding	1, 3, 19	r	e, n, z_{15}
S. krefeld	1, 3, 19	y	l, w
S. korlebu	1, 3, 19	z	1, 5
S. schoeneberg	1, 3, 19	z	e, n, z_{15}
S. carno	1, 3, 19	z	l, w
S. llandoff	1, 3, 19	z_{29}	—
S. chittagong	(1), 3, 10, (19)	b	z_{35}
S. ilugun	(1), 3, 10, (19)	z_4, z_{23}	z_6
S. dessau	(1), 3, 15, (19)	g, s, t	—

Gruppe F

Species	Körper-Antigene	1. Phase	2. Phase
S. marseille	11	a	1, 5
S. luciana	11	a	e, n, z_{15}
S. leeuwarden	11	b	1, 5
S. pharr	11	b	e, n, z_{15}
S. chandans	11	d	e, n, x
* S. montgomery	11	d (a)	d, e, n, z_{15}

Species	Körper-Antigene	Geissel-Antigene	
		1. Phase	2. Phase
S. chingola	11	e, h	1, 2
S. adamstua	11	e, h	1, 6
S. redhill	11	e, h	1, z_{13}, z_{28}
* S. grabouw	11	g, m, s, t	z_{39}
* S. lincoln	11	m, t	e, n, x
S. aberdeen	11	i	1, 2
S. brijbhumi	11	i	1, 5
S. heerlen	11	i	1, 6
S. veneziana	11	i	e, n, x
S. pretoria	11	k	1, 2
S. abaetetuba	11	k	1, 5
S. kisarawe	11	k	e, n, x
S. stendal	11	l, v	1, 2
S. maracaibo	11	l, v	1, 5
S. fann	11	l, v	e, n, x
S. osnabrueck	11	l, z_{13}, z_{28}	e, n, x
* S. huila	11	l, z_{28}	e, n, x
S. senegal	11	r	1, 5
S. rubislaw	11	r	e, n, x
S. volta	11	r	1, z_{13}, z_{28}
S. solt	11	y	1, 5
S. nyanza	11	z	z_6
* S. parera	11	z_4, z_{23}	—
S. wentworth	11	z_{10}	1, 2
S. straengnaes	11	z_{10}	1, 5
S. tel-hashomer	11	z_{10}	e, n, x
S. maastricht	11	z_{41}	1, 2

Gruppe G 1

Species	Körper-Antigene	Geissel-Antigene	
		1. Phase	2. Phase
S. mim	13, 22	a	1, 6
S. ibadan	13, 22	b	1, 5
S. bahati	13, 22	b	e, n, z_{15}
S. friedenau	13, 22	d	1, 6
S. willemstad	1, 13, 22	e, h	1, 6
S. bron	13, 22	g, m	e, n, z_{15}
* S. limbe	13, 22	g, m, t	—

Species	Körper-Antigene	Geissel-Antigene 1. Phase	2. Phase
* S. rotterdam	1, 13, 22	g, t	1, 5
S. tanger	1, 13, 22	y	1, 6
S. poona	13, 22	z	1, 6
S. poona var. 37	1, 13, 22	z	1, 6
S. bristol	13, 22	z	1, 7
S. roodepoort	13, 22	z_{10}	1, 5
* S. clifton	13, 22	z_{29}	1, 5
* S. goodwood	13, 22	z_{29}	e, n, x
S. mampong	13, 22	z_{35}	1, 6

Gruppe G 2

Species	Körper-Antigene	Geissel-Antigene 1. Phase	2. Phase
S. chagoua	1, 13, 23	a	1, 5
S. mississippi	1, 13, 23	b	1, 5
S. atlanta	13, 23	b	—
S. bracknell	13, 23	b	1, 6
S. durham	13, 23	b	e, n, z_{15}
* S. acres	1, 13, 23	b	z_{42}
S. mishmar-haemek	1, 13, 23	d	1, 5
S. grumpensis	13, 23	d	1, 7
S. tel-el-kebir	13, 23	d	e, n, z_{15}
S. wichita	1, 13, 23	d	—
S. putten	13, 23	d	l, w
S. isuge	13, 23	d	z_6
S. havana	1, 13, 23	f, g	—
S. agbeni	13, 23	g, m	—
S. okatie	13, 23	g, s, t	—
* S. luanshya	13, 23	g, s, (t)	—
S. congo	13, 23	g, t	—
* S. gojenberg	1, 13, 23	g, t	1, 5
* S. katesgrove	1, 13, 23	m, t	1, 5
* S. worcester	1, 13, 23	m, t	e, n, x
S. kintambo	13, 23	m, t	—
S. idikan	13, 23	i	1, 5
S. jukestown	13, 23	i	e, n, z_{15}
S. linton	13, 23	r	e, n, z_{15}
S. tunis	1, 13, 23	y	z_6
* S. nachshonim	1, 13, 23	z	1, 5
S. worthington	1, 13, 23	z	l, w

Species	Körper-Antigene	Geissel-Antigene	
		1. Phase	2. Phase
S. ajiobo	13, 23	z_4, z_{23}	—
S. demerara	13, 23	z_{10}	l, w
S. cubana	1, 13, 23	z_{29}	—
S. fanti	13, 23	z_{38}	—
* S. stevenage	1, 13, 23	z_{42}	1, 7

Gruppe H

Species	Körper-Antigene	Geissel 1. Phase	Geissel 2. Phase
S. garba	1, 6 , 14, 25	a	1, 5
S. ferlac	1, 6, 14, 25	a	e, n, x
S. tucson	1, 6, 14, 25	b	—
S. blijdorp	1, 6, 14, 25	c	1, 5
S. heves	6, 14, 24	d	1, 5
S. finkenwerder	1, 6, 14, 25	d	1, 5
S. florida	1, 6, 14, 25	d	1, 7
S. lindern	6, 14, 24	d	e, n, x
S. charity	1, 6, 14, 25	d	e, n, x
S. teko	1, 6, 14, 25	d	e, n, z_{15}
S. albuquerque	1, 6, 14, 24	d	z_6
S. bahrenfeld	6, 14, 24	e, h	1, 5
S. onderstepoort	1, 6, 14, 25	e, h	1, 5
S. magumeri	1, 6, 14, 25	e, h	1, 6
S. warragul	1, 6, 14, 25	g, m	—
S. caracas	1, 6, 14, 25	g, m, s	—
S. kaitaan	1, 6, 14, 25	m, t	—
S. mampeza	1, 6 ,14, 25	i	1, 5
S. buzu	1, 6, 14, 25	i	1, 7
S. schalkwijk	(6), 14, (24)	i	e, n..
S. harburg	1, 6, 14, 25	k	1,5
S. boecker	6, 14	l, v	1, 7
S. horsham	1, 6, 14, 25	l, v	e, n, x
S. surat	1, 6, 14, 25	r (i)	e, n, z_{15}
S. carrau	6, 14, 24	y	1, 7
S. madelia	1, 6, 14, 25	y	1, 7
S. fischerkietz	1, 6, 14, 25	y	e, n, x

55

Species	Körper-Antigene	Geissel-Antigene	
		1. Phase	2. Phase
S. homosassa	1, 6, 14, 25	z	1, 5
S. soahanina	6, 14, 24	z	e, n, x
S. sundsvall	1, 6, 14, 25	z	e, n, x
S. uzaramo	1, 6, 14, 25	z_4, z_{24}	—

Gruppe I

Species	Körper-Antigene	Geissel-Antigene	
S. hannover	16	a	1, 2
S. brazil	16	a	1, 5
S. amunigun	16	a	1, 6
S. fischerhuette	16	a	e, n, z_{15}
S. heron	16	a	z_6
S. hull	16	b	1, 2
S. wa	16	b	1, 5
S. glasgow	16	b	1, 6
S. hvittingfoss	16	b	e, n, x
S. malstatt	16	b	z_6
S. vancouver	16	c	1, 5
S. shamba	16	c	e, n, x
S. oldenburg	16	d	1, 2
S. gaminara	16	d	1, 7
S. nottingham	16	d	e, n, z_{15}
S. barmbek	16	d	z_6
S. weston	16	e, h	z_6
* S. bellville	16	e, n, x	1, 7
S. adeoyo	16	g, m	—
* S. mobeni	16	g, m, s, t	—
* S. merseyside	16	g, t	1, 5
* S. rowbarton	16	m, t	—
S. amina	16	i	1, 5
S. frankfurt	16	i	e, n, z_{15}
S. szentes	16	k	1, 2
S. orientalis	16	k	e, n, z_{15}
S. shanghai	16	l, v	1, 6
S. welikade	16	l, v	1, 7
S. salford	16	l, v	e, n, x
S. burgas	16	l, v	e, n, z_{15}
S. saphra	16	y	1,5
S. akuafo	16	y	1, 6
S. kikoma	16	y	e, n, x
* S. haddon	16	z_4, z_{23}	—

Species	Körper-Antigene	Geissel-Antigene 1. Phase	Geissel-Antigene 2. Phase
S. lisboa	16	z_{10}	1, 6
S. redlands	16	z_{10}	e, n, z_{15}
* S. jacksonville	16	z_{29}	—
* S. woodstock	16	z_{42}	1, (5), 7
* S. elsiesrivier	16	z_{42}	1, 6

Gruppe J

Species	Körper-Antigene	Geissel-Antigene 1. Phase	Geissel-Antigene 2. Phase
S. jangwani	17	a	1, 5
S. kinondoni	17	a	e, n, x
S. kirkee	17	b	1, 2
* S. hillbrow	17	b	e, n, x, z_{15}
S. victoriaborg	17	c	1, 6
S. berlin	17	d	1, 5
S. niamey	17	d	l, w
* S. verity	17	e, n, x, z_{15}	1, 6
* S. bleadon	17	f, g, t	[e, n, x, z_{15}]
S. irenea	17	k	1, 5
S. matadi	17	k	e, n, x
S. morotai	17	l, v	1, 2
S. michigan	17	l, v	1, 5
S. carmel	17	l, v	e, n, x
S. kandla	17	z_{29}	—

Gruppe K

Species	Körper-Antigene	Geissel-Antigene 1. Phase	Geissel-Antigene 2. Phase
S. usumbura	18	d	1, 7
S. langenhorn	18	m, t	—
S. memphis	18	k	1, 5
S. cerro	18	z_4, z_{23}	—
S. siegburg	6, 14, 18	z_4, z_{23}	—
S. blukwa	18	z_4, z_{24}	—
* S. zeist	18	z_{10}	z_6
* S. beloha	18	z_{36}	—

Species	Körper-Antigene	Geissel-Antigene	
		1. Phase	2. Phase

Gruppe L

Species	Körper-Antigene	1. Phase	2. Phase
S. assen	21	a	—
S. ghana	21	b	1, 6
S. minnesota	21	b	e, n, x
S. rhone	21	c	e, n, x
S. spartel	21	d	1, 5
S. magwa	21	d	e, n, x
S. ruiru	21	y	e, n, x
* S. soesterberg	21	z_4, z_{23}	—
* S. gwaai	21	z_4, z_{24}	—
* S. wandsbek	21	z_{10}	z_6
S. gambaga	21	z_{35}	e, n, z_{15}

Gruppe M

Species	Körper-Antigene	1. Phase	2. Phase
S. solna	28	a	1, 5
S. dakar	28	a	1, 6
S. seattle	28	a	e, n, x
S. moëro	28	b	1, 5
S. ashanti	28	b	1, 6
S. bokanjac	28	b	1, 7
S. langford	28	b	e, n, z_{15}
* S. kaltenhausen	28	b	z_6
S. hermannswerder	28	c	1, 5
S. halle	28	c	1, 7
S. wedding	28	c	e, n, z_{15}
S. techimani	28	c	z_6
S. mundonobo	28	d	1, 7
S. mocamedes	28	d	e, n, x
S. patience	28	d	e, n, z_{15}
S. friedrichsfelde	28	f, g	—
S. abadina	28	g, m	e, n, z_{15}
S. ona	28	g, s, t	—
S. vinohrady	28	m, t	—
S. kuessel	28	i	e, n, z_{15}

Species	Körper-Antigene	Geissel-Antigene	
		1. Phase	2. Phase
S. guilford	28	k	1, 2
S. ilala	28	k	1, 5
S. adamstown	28	k	1, 6
S. taunton	28	k	e, n, x
S. ank	28	k	e, n, z_{15}
S. leoben	28	l, v	1, 5
S. vitkin	28	l, v	e, n, x
S. nashua	28	l, v	e, n, z_{15}
S. chicago	28	r	1, 5
S. kibusi	28	r	e, n, x
S. nima	28	y	1, 5
S. pomona	28	y	1, 7
S. tel-aviv	28	y	e, n, z_{15}
S. ezra	28	z	1, 7
S. brisbane	28	z	e, n, z_{15}
* S. ceres	28	z	z_{39}
S. babelsberg	28	z_4, z_{23}	e, n, z_{15}
S. umbilo	28	z_{10}	e, n, x
S. luckenwalde	28	z_{10}	e, n, z_{15}
S. moroto	28	z_{10}	l, w
S. aderike	28	z_{38}	—

Gruppe N

Species	Körper-Antigene	Geissel-Antigene	
		1. Phase	2. Phase
S. zehlendorf	30	a	1, 5
S. urbana	30	b	e, n, x
S. godesberg	30	g, m	—
S. wayne	30	g, (p), z_{51}	—
S. landau	30	i	1, 2
S. morehead	30	i	1, 5
S. soerenga	30	i	l, w
S. ramat-gan	30	k	1, 5
S. aqua	30	k	1, 6
S. angoda	30	k	e, n, x
S. odozi	30	k	e, n, x, z_{15}
S. donna	30	l, v	1, 5
S. morocco	30	l, z_{13}, z_{28}	e, n, z_{15}
S. gege	30	r	1, 5

Species	Körper-Antigene	Geissel-Antigene	
		1. Phase	2. Phase
S. matopeni	30	y	1, 2
S. bodjonegoro	30	z_4, z_{24}	—
S. kumasi	30	z_{10}	e, n, z_{15}

Gruppe O

S. umhlatazana	35	a	e, n, z_{15}
S. tchad	35	b	—
S. yolo	35	c	—
S. adelaide	35	f, g	—
S. ealing	35	g, m, s	—
S. agodi	35	g, t	—
S. monschaui	35	m, t	—
S. gambia	35	i	e, n, z_{15}
S. massakory	35	r	l, w
S. alachua	35	z_4, z_{23}	—
S. westphalia	35	z_4, z_{24}	—
S. camberene	35	z_{10}	1, 5
S. ligna	35	z_{10}	z_6

Gruppe P

S. sheffield	38	c	1, 5
* S. carletonville	38	d	—
S. thiaroye	38	e, h	1, 2
S. kasenyi	38	e, h	1, 5
S. korovi	38	g, m, s	—
* S. foulpointe	38	g, t	—
S. mgulani	38	i	1, 2
S. lansing	38	i	1, 5
S. inverness	38	k	1, 6
S. roan	38	l, v	e, n, x
S. lindi	38	r	1, 5
S. emmastad	38	r	1, 6

Species	Körper-Antigene	Geissel-Antigene	
		1. Phase	2. Phase
S. freetown	38	y	1, 5
S. colombo	38	y	1, 6
S. perth	38	y	e, n, x
S. yoff	38	z_4, z_{23}	1, 2

Gruppe Q

Species	Körper-Antigene	1. Phase	2. Phase
S. wandsworth	39	b	1, 2
S. mara	39	e, h	—
S. hofit	39	i	1, 5
S. champaign	39	k	1, 5
S. kokomlemle	39	l, v	e, n, x
* S. mondeor	39	l, z_{28}	e, n, x
S. anfo	39	y	1, 2

Gruppe R

Species	Körper-Antigene	1. Phase	2. Phase
S. shikmonah	40	a	1, 5
S. greiz	40	a	z_6
* S. springs	40	a	z_{39}
S. riogrande	40	b	1, 5
S. johannesburg	1, 40	b	e, n, x
S. duval	1, 40	b	e, n, z_{15}
S. benguella	40	b	z_6
* S. suarez	1, 40	c	e, n, x, z_{15}
S. driffield	1, 40	d	1, 5
S. tilene	1, 40	e, h	1, 2
* S. alsterdorf	1, 40	g, m, t	—
* S. boksburg	40	g, s	e, n, x, z_{15}
S. allandale	1, 40	k	1, 6
S. millesi	1, 40	l, v	1, 2
S. bukavu	1, 40	l, z_{28}	1, 5
S. santhiaba	40	l, z_{28}	1, 6
* S. bulawayo	1, 40	z	1, 5
S. nowawes	40	z	z_6

Species	Körper-Antigene	Geissel-Antigene	
		1. Phase	2. Phase
* S. sachsenwald	1, 40	z_4, z_{23}	—
* S. degania	40	z_4, z_{24}	—
* S. bern	1, 40	z_4, z_{32}	—
S. omifisan	40	z_{29}	—
* S. fandran	1, 40	z_{35}	e, n, x, z_{15}
* S. grunty	1, 40	z_{39}	1, 6
S. karamoja	1, 40	z_{41}	1, 2

Gruppe S

Species	Körper-Antigene	Geissel-Antigene	
		1. Phase	2. Phase
S. vietnam	41	b	—
S. egusi	41	d	—
* S. lethe	41	g, t	—
S. waycross	41	z_4, z_{23}	—
S. ipswich	41	z_4, z_{24}	—
* S. negev	41	z_{10}	1, 2
S. leipzig	41	z_{10}	1, 5
S. inpraw	41	z_{10}	e, n, x
* S. lichtenberg	41	z_{10}	z_6
S. offa	41	z_{38}	—

Gruppe T

Species	Körper-Antigene	Geissel-Antigene	
		1. Phase	2. Phase
* S. chinovum	42	b	1, 5
* S. uphill	42	b	e, n, x, z_{15}
S. egusitoo	1, 42	b	z_6
S. kampala	1, 42	c	z_6
* S. fremantle	42	(f), g, t	—
S. maricopa	1, 42	g, (p), z_{51}	1, 5
S. kaneshie	1, 42	i	l, w
* S. portbech	42	l, v	e, n, x, z_{15}
* S. nairobi	42	r	—
S. harvestehude	1, 42	y	z_6
* S. detroit	42	z	1, 5
* S. rand	42	z	e, n, x, z_{15}

Species	Körper-Antigene	Geissel-Antigene 1. Phase	Geissel-Antigene 2. Phase
S. loenga	1, 42	z_{10}	z_6
S. weslaco	42	z_{36}	—

Gruppe U

Species	Körper-Antigene	1. Phase	2. Phase
S. berkeley	43	a	1, 5
S. milwaukee	43	f, g	—
S. mbao	43	i	1, 2
S. ahuza	43	k	1, 5
S. farcha	43	y	1, 2
S. kingabwa	43	y	1, 5
* S. houten	43	z_4, z_{23}	—
* S. tuindorp	43	z_4, z_{32}	—
S. irigny	43	z_{38}	—
* S. bunnik	43	z_{42}	[1, 5, 7]

Gruppe V

Species	Körper-Antigene	1. Phase	2. Phase
S. niarembe	44	a	l, w
S. fischerstrasse	44	d	e, n, z_{15}
S. vleuten	44	f, g	—
S. gamaba	44	g, m, s	—
S. uhlenhorst	44	z	l, w
S. christiansborg	44	z_4, z_{24}	—
S. guinea	44	z_{10}	—

Gruppe W

Species	Körper-Antigene	1. Phase	2. Phase
* S. vrindaban	45	a	e, n, x
* S. ejeda	45	a	z_{10}
S. deversoir	45	c	e, n, x
S. dugbe	45	d	1, 6
S. karachi	45	d	e, n, x

Species	Körper-Antigene	Geissel-Antigene	
		1. Phase	2. Phase
S. suelldorf	45	f, g	—
S. tornow	45	g, m,	—
* S. bremen	45	g, m, s, t	e, n, x
* S. windhoek	45	g, t	1, 5
S. apapa	45	m, t	—
* S. perinet	45	m, t	e, n, x, z_{15}
* S. klapmuts	45	z	z_{39}
S. jodhpur	45	z_{29}	—

Gruppe X

Species	Körper-Antigene	Geissel-Antigene	
* S. bilthoven	47	a	—
S. saka	47	b	—
* S. phoenix	47	b	1, 5
S. stellingen	47	d	e, n, x
* S. quimbamba	47	d	z_{39}
S. luke	1, 47	g, m	—
S. bergen	47	i	e, n, z_{15}
S. bootle	47	k	1, 5
S. lyon	47	k	e, n, z_{15}
S. teshie	1, 47	l, z_{13}, z_{28}	e, n, z_{15}
S. moualine	47	y	1, 6
S. mount-pleasant	47	z	1, 5
S. kaolack	47	z	1, 6
* S. chersina	47	z	z_6
S. bere	47	z_4, z_{23}	z_6
S. quinhon	47	z_{44}	—

Gruppe Y

Species	Körper-Antigene	Geissel-Antigene	
S. hisingen	48	a	1, 5, 7
* S. hammonia	48	e, n, x, z_{15}	z_6
S. dahlem	48	k	e, n, z_{15}
* S. sakaraha	48	k	z_{39}

Species	Körper-Antigene	Geissel-Antigene 1. Phase	Geissel-Antigene 2. Phase
S. djakarta	48	z_4, z_{24}	—
* S. ngozi	48	z_{10}	1, 5

Gruppe Z

Species	Körper-Antigene	1. Phase	2. Phase
* S. krugersdorp	50	e, n, x	1, 7
* S. wassenaar	50	g, (p), z_{51}	—
* S. greenside	50	z	e, n, x
* S. bonaire	50	z_4, z_{32}	—
* S. hooggraven	50	z_{10}	$z_6 : z_{42}$

Gruppe 51

Species	Körper-Antigene	1. Phase	2. Phase
S. meskin	51	e, h	1, 2
S. overschie	51	l, v	1, 5
S. antsalova	51	z	1, 5
S. treforest	51	z	1, 6
* S. roggeveld	51	—	1, 7

Gruppe 52

Species	Körper-Antigene	1. Phase	2. Phase
S. flottbek	52	b	—
S. utrecht	52	d	1, 5
S. sainte-marie	52	g, t	—
* S. lobatsi	52	—	1, 5, 7

Gruppe 53

Species	Körper-Antigene	1. Phase	2. Phase
* S. midhurst	53	1, z_{28}	z_{39}
* S. humber	53	z_4, z_{24}	—

Gruppe 54

Species	Körper-Antigene	1. Phase	2. Phase
S. uccle	54	g, s, t	—

Species	Körper-Antigene	Geissel-Antigene	
		1. Phase	2. Phase
Gruppe 55			
* S. tranoroa	55	k	z_{39}
Gruppe 56			
* S. artis	56	b	—
Gruppe 57			
* S. locarno	57	z_{29}	z_{42}
* S. manombo	57	z_{39}	e, n, x, z_{15}
Gruppe 58			
* S. basel	58	1, z_{13}, z_{28}	1, 5
Gruppe 59			
* S. betioky	59	k	(z)
Gruppe 60			
* S. luton	60	z	e, n, x

* = sub-genus II.
[ ] = kann fehlen.
( ) = nicht vollständig vorhanden.

66

LITERATUR

DOUGLAS, G. W. & EDWARDS, P. R.: "Complex flagellar phases in Salmonella". *J. gen. Microbiol.* **29**, *367—372*, 1962.

EDWARDS, P. R., FIFE, M. A. & RAMSEY, C. H., "Studies on the Arizona group of Enterobacteriaceae". *Bact. Rev.* **23**, *155—174*, 1959 sowie die dort angegebene Literatur.

EDWARDS, P. R., McWHORTER, A. C. & DOUGLAS, G. W.: "A culture of Salmonella infantis of complex antigenic constitution". *J. Bact.* **84**, *95—98*, 1962.

KAUFFMANN, F. (1): "Die Bakteriologie der Salmonella-Species", bei E. Munksgaard, Copenhagen, 256 Seiten, 1961 sowie die dort angegebene Literatur.

KAUFFMANN, F. (2): "The species-definition in the family Enterobacteriaceae" *Int. Bull. bact. Nomencl.* **11**, *5—6*, 1961.

KAUFFMANN, F. (3): "Two biochemical sub-divisions of the genus Salmonella". *Acta path. microbiol. scand.* **49**, *393—396*, 1960.

KAUFFMANN, F. (4): "Über mehrere neue Salmonella-Typen". *Acta path. microbiol. scand.* **18**, *351—366*, 1941.

KAUFFMANN, F. & PETERSEN, A.: "Zur Serologie der Salmonella O-Gruppen 28, 40, 45 und 47". *Acta path. microbiol. scand.* **56**, *343—351*, 1962.

KAUFFMANN, F. & PETERSEN, A.: "Zur Serologie der Salmonella O-Gruppen 30, 42, 43, 48 und 50". *Acta path. microbiol. scand.* **58**, *99—108*, 1963.

KAUFFMANN, F., LÜDERITZ, O., STIERLIN, H. & WESTPHAL, O.: "Zur Immunchemie der O-Antigene von Enterobacteriaceae. I. Analyse der Zuckerbausteine von Salmonella O-Antigenen". *Zbl. Bakt.* I. Abt. Orig. **178**, *442—458*, 1960.

KAUFFMANN, F. & ROHDE, R. (1): "Eine Vereinfachung der serologischen Salmonella-Diagnose". *Acta path. microbiol. scand.* **56**, *341—342*, 1962.

KAUFFMANN, F. & ROHDE, R. (2): "Eine Vereinfachung der serologischen Arizona-Diagnose". *Acta path. microbiol. scand.* **54**, *473—478*, 1962.

KAUFFMANN, F. & ROHDE, R. (3): "Neue Befunde beim O-Formen-Wechsel der Salmonella-Species". *Acta path. microbiol. scand.* **52**, *211—216*, 1961.

McWHORTER, A. C., DOUGLAS, G. W. & EDWARDS, P. R.: "A new salmonella serotype, Salmonella cook $(39:z_{48}:1,5)$ containing an undescribed flagellar antigen". *Int. Bull. bact. Nomencl.* **12**, *181—183*, 1962.

MINOR, L. LE & BEN HAMIDA, F.: "Avantages de la recherche de la β-Galactosidase sur celle de la fermentation du lactose en milieu complexe dans le diagnostic bactériologique, en particulier des Enterobacteriaceae". *Ann. Inst. Pasteur* **102**, *267—277*, 1962.

MINOR, L. LE & EDWARDS, P. R.: "Présence de trois phases de l'antigène flagellaire chez des bactéries du groupe Salmonella". *Ann. Inst. Pasteur* **99**, *469—474*, 1960.

LA LYSOTYPIE DE SALMONELLA TYPHI. SON PRINCIPE, SA TECHNIQUE, SON APPLICATION À L'ÉPIDÉMIOLOGIE DE LA FIÈVRE TYPHOÏDE

PAR

PIERRE NICOLLE

Chef de Service à l' Institut Pasteur

Introduction

Les espèces homogènes et les espèces hétérogènes sous le rapport de leur sensibilité aux bactériophages.

L'idée d'utiliser la spécificité des bactériophages pour rendre plus rapide et plus sûr le diagnostic bactériologique de certaines espèces bactériennes (lysodiagnose) et éventuellement pour mettre en évidence des variétés à l'intérieur de quelques-unes d'entre elles (lysotypie) découle de faits entrevus par F.W. Twort, 1915[55] et surtout par F. d'Hérelle, 1919[30].

Un peu plus tard, précisant cette notion, ce dernier auteur, 1921, 1926[31] divisa les espèces bactériennes sensibles à la bactériophagie en deux groupes d'après leur comportement vis-à-vis d'un bactériophage: le premier de ces groupes comprenait les espèces "homogènes", c'est-à-dire les espèces dont toutes les souches éprouvées se montraient sensibles à ce phage, avec seulement quelques différences dans le degré de cette sensibilité et le second, les espèces "hétérogènes" qui réunissaient des souches sensibles et des souches réfractaires.

Le groupe homogène, d'après le même auteur, ne comptait qu'un très petit nombre d'espèces, parmi lesquelles les bacilles dysentériques Shiga, Hiss, Flexner, le bacille de la Peste et celui du Barbone du Buffle.

Le groupe hétérogène était formé de toutes les autres espèces sensibles à la bactériophagie, en particulier les *Salmonella*, les *Escherichia*, le vibrion cholérique, le bacille diphtérique, le staphylocoque, le streptocoque, etc.

Premières tentatives pour établir une lysodiagnose et une lysotypie de certaines *Salmonella*, en particulier *S. typhi*.

Sonnenschein, 1925, 1928, 1929[62, 63, 64], affirma qu'il avait isolé un bactériophage spécifique pour le bacille typhique et un autre spécifique pour le bacille paratyphique B. Il proposa d'utiliser ces phages spécifiques comme moyens de différenciation rapide entre les deux espèces. Le phage actif sur *S. typhi* présentait notamment une spécificité remarquable: sur 109 souches de bacille typhique soumises à son action, 104 subirent la lyse (92%) tandis que, sur

512 souches étrangères à l'espèce S. *typhi*, 403 furent complètement résistantes.

Par la suite, le même auteur a constaté que son phage typhique acquérait par passages sur le bacille typhique un pouvoir lytique plus considérable et surtout étendu à un plus grand nombre de souches de ce germe.

Les conclusions de SONNENSCHEIN furent confirmées par plusieurs auteurs parmi lesquels MARCUSE, 1931[40] et discutées par d'autres en particulier MASSA, 1931[43], qui exprima la crainte que la lysodiagnose ne conduisit souvent à des erreurs.

MARCUSE en 1934[41, 42], utilisa un phage spécifique pour S. *typhi* qui lysait 331 souches de cette espèce sur 469 et se montrait dépourvu d'activité sur 138 autres. Par des essais d'adaptation de ce phage à des souches sensibles et à des souches résistantes, il obtint 4 nouveaux phages qui agissaient sur 133 souches parmi celles qui étaient réfractaires au premier. Au moyen de ces 5 phages adaptés, il réussit à répartir sa collection de souches de S. *typhi* en 5 groupes. En outre, il constata que les bacilles isolés à plusieurs reprises chez une même personne appartenaient toujours à un même groupe. Il a donc été le premier inventeur de la méthode de la lysotypie pour le bacille typhique.

Cependant l'application de ces phages à l'épidémiologie se révéla décevante. Il fallut la découverte des phages Vi et les travaux de CRAIGIE sur la remarquable adaptabilité du phage Vi II pour que la lysotypie du bacille typhique prît son essor.

Les dangers d'invasion de la fièvre typhoïde en Europe et en Amérique du Nord

La fièvre typhoïde tend à disparaître de presque tous les pays d'Europe et d'Amérique du Nord (Mexique excepté). Mais elle demeure toujours pour ces régions une sérieuse menace.

Premièrement, parce qu'elle sévit encore gravement en Afrique, en Asie et en Amérique latine et que ces pays à endémicité typhoïdique élevée, dont certains sont à nos portes, peuvent à tout moment être le point de départ de nouvelles invasions épidémiques. La situation deviendrait particulièrement grave en cas de bouleversements politiques, sociaux, militaires et de cataclysmes naturels, qui rendraient inefficace la fragile surveillance sanitaire actuelle.

Deuxièmement, sans qu'il soit nécessaire d'imaginer de pareilles catastrophes, parce que si, depuis la fin de la dernière guerre mondiale, la fièvre typhoïde est en nette régression en Europe et en Amérique du Nord, si l'on n'observe plus les grandes flambées épidémiques qui causaient encore tant de ravages parmi les populations urbaines à la fin du siècle dernier, elle persiste sous forme de nombreux petits foyers disséminés aussi bien dans les villes qu'à la campagne, qui doivent généralement leur apparition à des porteurs de germes. Chacun de ces porteurs de germes, chacun de ces petits foyers risquent d'être à leur tour la source de nouveaux cas. Eventuellement même, si les sujets atteints, si les porteurs de germes exercent un métier qui les amènent à manipuler les denrées alimentaires, ou bien si la malchance veut que leurs bacilles, excrétés dans la nature ou véhiculés par un cours d'eau ou un égout s'en aillent, à la faveur de mauvaises connexions, contaminer un puits, une nappe d'eau ou une conduite, ils pourront se trouver à l'origine de véritables épidémies. C'est ce qu'on a observé à Zermatt en 1963.

Tant que des cas sporadiques et des porteurs de germes demeurent nombreux dans un pays, les menaces d'épidémies continuent donc de peser sur la population.

Lutte contre l'endémicité typhoïdique.

De quels moyens les Services d'Hygiène disposent-ils pour écarter ces dangers? Comment pourraient-ils même parvenir à l'éradication totale de l'infection?

On devrait d'abord obtenir d'importants résultats par l'application de mesures générales telles que l'amélioration du régime des eaux de consommation dans les petites villes, dans les villages et à la campagne, le renforcement du contrôle bactériologique de certaines denrées alimentaires, l'interdiction de l'arrosage des jardins maraîchers avec de l'eau d'égout insuffisamment épurée, l'extension de la vaccination dans les régions où sévit une endémicité élevée, le dépistage systématique et la surveillance des porteurs de germes, etc.

Mais surtout, comme la fièvre typhoïde vraie est une infection exclusivement humaine, les mesures à prendre doivent viser essentiellement à empêcher ou du moins à limiter les contages interhumains. Pour y arriver, on peut beaucoup attendre de l'étude épidémiologique attentive des foyers sporadiques, qui seule permet de trouver la solution appropriée à chaque cas.

Certes, il faut reconnaître que, s'il est relativement aisé de trouver l'origine d'une épidémie importante en raison du nombre des faits convergents qui peuvent être recueillis, la tâche des épidémiologistes, en présence d'une poussière de minuscules foyers, est autrement difficile, fastidieuse et ingrate. Cependant, s'ils veulent se donner la peine de demander l'aide des laboratoires, leurs enquêtes pourront se trouver orientées et souvent merveilleusement éclairées par les résultats que ceux-ci leur communiqueront.

L'isolement du germe responsable doit être obtenu.

Dans l'étude d'un foyer de fièvre typhoïde, il faudrait en premier lieu obtenir du médecin traitant qu'il fasse, dans tous les cas, avant l'administration d'antibiotiques, les prélèvements biologiques indispensables (sang pour l'hémoculture et le sérodiagnostic chez les malades et pour la recherche des agglutinines Vi chez les porteurs de germes; matières fécales pour la coproculture; urine, bile, liquide céphalo-rachidien, pus d'abcès pour l'isolement du germe dans ces produits). Ces précautions n'apporteraient qu'un retard insignifiant à la mise en application de la thérapeutique. Faute de les prendre, comme on l'observe trop souvent, le médecin traitant prive l'épidémiologiste de très précieux renseignements pour ses enquêtes, en se privant du reste lui-même d'informations utiles, non seulement au diagnostic, mais encore au traitement et au pronostic de la maladie, qu'il prétend soigner ainsi à l'aveuglette. Il nous paraît indispensable d'insister en particulier sur la nécessité de l'isolement du bacille responsable. Pour l'épidémiologiste, les sérodiagnostics ne sont pas suffisants. Il faut qu'il dispose du germe en culture. Il faut que les caractères de ce germe soient soigneusement étudiés et comparés aux caractères des germes isolés chez les autres malades ou les porteurs de germes du voisinage.

Utilité du diagnostic bactériologique.

Il ne viendrait à l'idée de personne de nier l'utilité du diagnostic bactériologique en épidémiologie. La première question qui se pose lorsqu'on signale une poussée de fièvre typhoïde, c'est d'établir si cette infection est due à un bacille typhique vrai ou à une autre *Salmonella*.

Dans l'épidémiologie d'une région, on doit donc soigneusement séparer ce qui revient à l'un de ces germes de ce qui revient à chacun des autres. Cette discrimination est le point de départ de toute enquête épidémiologique, car il est bien évident que la filiation entre des cas à bacille typhique et des cas à bacille paratyphique B par exemple doit être exclue. On sera en droit au contraire d'envisager la possibilité d'une relation entre tous les cas de fièvre à bacille typhique survenus dans une même contrée; il en sera de même, entre tous les cas de fièvre à bacille paratyphique B.

Imaginons que l'espèce *Salmonella typhi* puisse être divisée en plusieurs variétés faciles à reconnaître et stables. Le diagnostic de la variété ou des variétés en cause offrira pour l'épidémiologiste un intérêt aussi grand que le diagnostic des espèces. On pourra, par exemple, marquer sur une carte à l'aide de crayons de couleurs, les différentes variétés rencontrées dans une région. Un simple coup d'oeil permettra de comprendre qu'il y a dans cette région au moins autant de sources de contamination distinctes qu'il y a de couleurs sur la carte, c'est-à-dire de variétés en présence. En outre, on comprendra immédiatement quels sont les cas entre lesquels on peut envisager une filiation sinon certaine, du moins possible (les cas appartenant à une même variété) et ceux pour lesquels toute relation est absolument improbable (les cas appartenant à des variétés différentes).

Or, de telles variétés existent. L'espèce *Salmonella typhi* a pu être divisée, par des méthodes de laboratoire, en variétés morphologiques, culturales, biochimiques, sérologiques et lysotypiques.

Homogénéité apparente de l'espèce *Salmonella typhi*.

Parmi les diverses Entérobactériacées qui peuvent déterminer le syndrome typhoïdique, est-il une espèce en apparence plus homogène dans ses propriétés que celle du classique bacille d'Eberth?

Il n'y a pas longtemps, si la clinique pouvait à la rigueur distinguer, dans certains foyers de fièvre typhoïde, des bacilles typhiques de virulence élevée, et dans d'autres foyers des bacilles typhiques de faible virulence, le laboratoire s'avouait incapable de reconnaître des caractères bactériologiques nettement différents entre les échantillons qui lui étaient soumis. Depuis une vingtaine d'années, le bloc jusqu'alors sans fissures de l'espèce *Salmonella typhi*, sous l'effet de recherches antigéniques, biochimiques et bactériologiques dont il a été l'objet, a pu être fragmenté en un certain nombre de variétés. On doit donc admettre aujourd'hui qu'il n'y a pas un bacille typhique unique, mais plusieurs sortes de bacilles typhiques facilement et sûrement reconnaissables.

Variétés antigéniques du bacille typhique

Le bacille typhique (bacille d'Eberth, *Salmonella typhi*), lorsque son équipement antigénique est complet, ce qui est la règle générale, possède les antigènes somatiques 9 et 12 (facteurs O, lipido-glucido-protéiniques, thermorésistants et doués d'une toxicité élevée), l'antigène d'enveloppe Vi (facteur capsulaire ou K, un peu moins toxique, mais protégeant les bacilles contre les moyens de défense de l'organisme, en particulier la phagocytose, donc vraisemblablement un des facteurs de la virulence) et l'antigène flagellaire d, également thermolabile, facteur de mobilité des germes (facteur H).

Mais on rencontre parfois des bacilles qui sont plus ou moins déficients en l'un ou l'autre de ces trois types d'antigènes. Il y a des souches qui n'ont pas d'antigène Vi ou qui n'en ont que très peu (forme de virulence atténuée). Il y en a qui ne sont pas mobiles. Il y en a enfin dont les antigènes O n'existent qu'à l'état de traces. Ces différences antigéniques ne sont pas stables. La souche 0 901 (Vi négative, H négative) a été obtenue par FELIX en 1930[22] à partir de colonies isolées de la souche H 901. En gélose molle, elle récupère facilement sa mobilité et, par passage sur la souris, KAUFFMANN, 1936[25] lui a redonné de l'antigène Vi. Il est remarquable que cette culture rendue Vi positive ait donné à la lysotypie les mêmes réactions que la souche Ty 2 dont elle dérive, celles du lysotype E1 identifié par CRAIGIE & YEN en 1938[19].

On peut conclure de ces faits que la recherche de la composition antigénique des bacilles typhiques ne peut pas être d'une grande utilité en épidémiologie.

Variétés culturales.

Dans un autre groupe, on peut ranger les variétés de bacille typhique d'après l'aspect de leurs colonies sur gélose nutritive: dimensions, structure de leur bord, opacité à la lumière (GIOVANARDI, 1938 [29]), irisation en transillumination oblique (NICOLLE, JUDE & LE MINOR, 1950[47]; NICOLLE, JUDE & LE MINOR, 1955[49]; JUDE, LE MINOR & NICOLLE, 1955[34]). Ces variétés correspondent du reste assez souvent à des différences dans la structure antigénique, déjà signalées plus haut.

Variétés biochimiques ou chimiotypes.

Un troisième groupe est constitué par les différences biochimiques étudiées par KRISTENSEN, 1926, 1938[37], [36].

Les quatre cinquièmes des souches de bacille typhique isolées dans la plupart des pays d'Europe font fermenter le xylose en moins de trois jours, mais non l'arabinose (chimiotype I).

La grande majorité du cinquième restant n'attaquent ni l'un, ni l'autre de ces sucres (chimiotype II). KRISTENSEN admettait l'existence d'une troisième variété les décomposant tous les deux (chimiotype III). Enfin, DE BLASI & BUOGO, 1952[11] ont décrit une quatrième variété qui, à l'instar du bacille paratyphique A, respecte le xylose et utilise l'arabinose (chimiotype IV). Notre expérience portant sur plusieurs milliers de souches nous incite à penser que ces deux variétés n'existent pas ou bien qu'elles sont rarissimes. Nous ne les avons pour notre part jamais rencontrées avec certitude NICOLLE, NICOLLE & DIVERNEAU, 1961[52].

D'après les observations de KRISTENSEN, celles D'OLITZKI et collab., 1944 et 1948 [55], [56], celles des auteurs polonais LACHOWICZ & BUCZOWSKI, 1950[38], CHOMICZEWSKI, 1960[14] et les nôtres (JUDE & NICOLLE, 1949[32], PAVLATOU & NICOLLE, 1953[57]), les chimiotypes ont sur les variétés antigéniques l'avantage d'être stables. Ils offrent par là-même un très grand intérêt épidémiologique: dans un même foyer de contagion, le type fermentatif est le même pour tous les bacilles isolés des divers cas. Si l'on constate au contraire, dans une région où sévit la fièvre typhoïde, la présence de deux chimiotypes, il faut nécessairement en conclure qu'il n'y a pas une source unique de contamination, mais au moins deux. Malheureusement, la très grande fréquence du chimiotype I et la rareté extrême, pour ne pas dire l'inexistence des chimiotypes III et IV, réduisent beaucoup l'utilité de cette méthode de différenciation biochimique en épidémiologie. Nous verrons plus loin qu'elle apporte néanmoins une aide appréciable à la méthode lysotypique en permettant la subdivision binaire des lysotypes qui sont biochimiquement hétérogènes.

La lysotypie du bacille typhique

Variétés lysotypiques: la lysotypie du bacille typhique par la méthode de CRAIGIE.

Enfin un quatrième groupe de variétés de bacille typhique comprend les types bactériophagiques ou lysotypes: la différenciation la plus variée, la plus fidèle et la plus utile entre les bacilles typhiques est celle que l'on obtient au moyen des bactériophages.

Les phages spécifiques de l'antigène Vi; l'adaptation remarquable du phage Vi II.

Deux ans après la découverte de l'antigène Vi par FELIX & PITT, 1934[26, 27], en trois pays différents, au Canada CRAIGIE & BRANDON, 1936[15, 16], en Hollande SCHOLTENS[60] et en France SERTIC & BOULGAKOV[61] constatent que certains bactériophages actifs sur la plupart des cultures Vi positives du bacille typhique ne produisent jamais d'effet lytique sur les cultures Vi négatives. Ils prouvent que la présence de l'antigène Vi est indispensable pour que les corpuscules bactériophages puissent se fixer sur les bactéries. L'antigène Vi est donc le récepteur bactérien de ces phages particuliers. Puisqu'ils sont doués d'affinité spécifique pour l'antigène Vi, on les appelle les phages Vi.

Un peu plus tard, en 1937 et 1938, CRAIGIE & YEN[18, 19] découvrent que l'un de leurs phages Vi, le phage Vi II, est doué d'une très remarquable plasticité d'adaptation.

Si l'on régénère, par exemple, le phage Vi II sur dix cultures Vi positives de *Salmonella typhi* provenant de régions variées, on obtient des préparations bactériophagiques adaptées qui ne se comporteront pas toutes de la même manière. Ainsi, par exemple, les préparations 1, 5, 6 et 9 donneront, à leur dilution limite, la lyse confluente sur les souches 1, 5, 6 et 9 et rien ou seulement quelques plages isolées sur les autres souches. Les préparations 2, 3 et 4, à leur dilution limite, agiront sur les souches 2, 3 et 4 et seront très faiblement actives ou totalement inactives sur les autres souches. Les préparations 7, 8 et 10 lyseront totalement les souches correspondantes à la dilution limite et ne produiront aucun effet sur les autres souches. Ainsi, nous avons obtenu trois sortes de préparations au moyen desquelles nous pourrons répartir nos 10 souches de bacille typhique en 3 variétés bactériophagiques ou lysotypes.

En généralisant le procédé sur un grand nombre de cultures de *Salmonella typhi* provenant de régions variées, CRAIGIE & YEN, 1938[20], obtinrent, après élimination de celles qui faisaient double emploi, 18 préparations adaptées du phage Vi II, spécifiques de 18 types bactériophagiques ou lysotypes du bacille typhique.

Par la suite, le nombre des lysotypes de *S. typhi* fut porté successivement à 24 (CRAIGIE & FELIX, 1947[17]), puis à 33 (FELIX,

1955[23]), puis à 44 (ANDERSON, 1962[5]). Le nombre des lysotypes Vi de *Salmonella typhi* officiellement reconnus par le Comité International de la Lysotypie Entérique lors de ses réunions pendant le Congrès International de Microbiologie (Montréal, 1962) est de 72 (ANDERSON, sous presse).

Ce sont les lysotypes et sous-types suivants: lysotype A, les 3 sous-types du groupe B (B1, B2, B3), les 9 sous-types du groupe C (de C1 à C9), les 10 sous-types du groupe D (de D1 à D11, le sous-type D3 manque), les 10 sous-types du groupe E (de E1 à E10), les 5 sous-types du groupe F, les lysotypes G et H, les 3 sous-types du groupe J (de J1 à J3), les 2 sous-types du groupe K, les 2 sous-types du groupe L (L1 et L2), les 3 sous-types du groupe M (M1, M2, M3), les lysotypes N, O, T, 25, 26, 27, 28, 29, 32, 34, 35, 36, 37, 38, 39, 40, 41 et 42 plus 4 lysotypes nouvellement reconnus par le Comité International de la Lysotypie Entérique, dont le dernier porte la dénomination 46 bien qu'il soit le 72ème lysotype.

Les souches non lysotypables du bacille typhique.

Une certaine proportion de souches de *S. typhi*, variable suivant les régions, échappe toujours à l'identification lysotypique par la méthode de CRAIGIE & FELIX. Ces souches non lysotypables peuvent appartenir aux trois groupes suivants:

a) Souches Vi négatives – En l'absence de l'antigène Vi qui est le récepteur bactérien spécifique des phages Vi, ceux-ci ne peuvent se fixer sur les bactéries, et comme la fixation préalable est indispensable pour que la bactériophagie se produise, on n'observe aucune action lytique.

Parfois la perte de l'antigène Vi n'est qu'un accident de l'isolement du germe lorsque, par exemple, un hasard malheureux a fait prélever une colonie Vi négative parmi beaucoup d'autres qui étaient Vi-positives. Pour éviter ce risque, FELIX conseillait de prélever toujours au moins 6 colonies au hasard.

Parfois, la souche est Vi négative dans la majorité de ses bactéries. Dans ce cas, avec un peu de chance, on peut souvent la sauver pour la lysotypie: un examen attentif des boîtes de gélose d'isolement montrera, parmi un grand nombre de colonies ternes, à irisation bleutée, quelques colonies opaques, lumineuses et présentant une irisation rose, jaune ou jaune orangé. Ce sont souvent des colonies Vi positives. Enfin, il arrive parfois que des souches ne contiennent aucun élément Vi positif, soit qu'elles aient perdu leur antigène Vi au cours des repiquages, soit qu'elles en fussent dépourvues, même au sortir de l'organisme humain. Ces souches ne peuvent pas être lysotypées par la méthode de CRAIGIE & FELIX. Nous verrons plus loin qu'elles peuvent l'être indirectement par le moyen d'une lysotypie complémentaire avec des phages non Vi.

b) Le groupe des souches Vi positives non lysotypables (groupe

I + IV) — Certaines souches Vi positives ne réagissent à aucune des préparations adaptées du phage Vi II. Cependant, elles sont généralement sensibles à d'autres phages Vi, au phage Vi I ou au phage Vi IV notamment, ou aux deux. Pour cette raison, on a proposé de classer de telles souches Vi positives non lysotypables dans un groupe provisoire, le groupe I + IV, qui contient probablement quelques lysotypes en puissance, auxquels on parviendra peut-être un jour à adapter le phage Vi II. En épidémiologie, ce groupe disparate rend les mêmes services qu'un lysotype véritable.

c) Enfin, dans une troisième catégorie de souches non lysotypables, on classe des souches dont les réactions aux préparations adaptées du phage Vi II sont fortes, nombreuses, en quelque sorte anarchiques et souvent même inconstantes. Il est donc impossible d'identifier de telles souches que nous désignons comme "aliénosensibles", parce qu'elles sont sensibles à des phages étrangers à leur lysotype.

Technique de la lysotypie du bacille typhique.

Technique de CRAIGIE: on ensemence une plaque de gélose en déposant des gouttelettes de culture en milieu liquide au moyen d'une anse de platine calibrée et en étalant chaque gouttelette de manière à dessiner un disque d'un centimètre de diamètre environ. On formera ainsi avec la même culture autant de disques sur la gélose qu'il doit y avoir de phages à essayer.

Ensuite, on déposera, toujours avec l'anse de platine, au centre de chaque disque ensemencé, une gouttelette de phage (un phage différent pour chaque disque) et on l'étalera en un second disque légèrement plus petit que le premier, de manière à ménager entre les deux disques une marge au niveau de laquelle la culture n'aura pas été mélangée avec le phage.

Après une nuit d'incubation, certains disques de culture ne montreront aucune trace de lyse (réaction négative), d'autres au contraire présenteront une lyse confluente entourée d'un cercle de culture, d'autres enfin seront parsemés de plages plus ou moins nombreuses.

Autre technique: on peut remplacer les ensemencements par disques, qui prennent beaucoup de temps, par un ensemencement unique de toute la surface de la plaque de gélose, soit en déposant quelques gouttes d'une culture jeune (moins de 24h) en bouillon, de la souche à lysotyper et en les étalant avec un étaleur de verre préalablement flambé et refroidi, soit en inondant toute la surface de la gélose avec plusieurs centimètres cubes de culture en milieu liquide, puis en enlevant l'excès de liquide au moyen d'une aspiration mécanique (pipette effilée reliée à une fiole de Kitasato, connectée elle-même avec une trompe à eau ou une pompe à vide). Puisqu'on manipule des germes pathogènes, il conviendra de placer dans la fiole un certain volume (20 à 50 cm³) d'une solution antiseptique (eau de Javel).

Les boîtes de gélose ont été préalablement divisées sur leur face inférieure par un quadrillage dessiné à l'encre à marquer le verre. Les gouttes des différents phages sont déposées dans les carrés de ce quadrillage au moyen de pipettes effilées ou de seringues armées d'aiguilles hypodermiques ou encore de flacons compte-gouttes dont une tubulure latérale a été munie également d'une aiguille hypodermique.

Analyses des diverses images lysotypiques du bacille typhique.

Si nous analysons le tableau de la lysotypie de *Salmonella typhi* par la méthode de CRAIGIE & FELIX, complétée par FELIX, puis par divers membres du Comité International de la Lysotypie Entérique*, nous voyons que certains lysotypes subissent la lyse confluente avec le phage homonyme et qu'ils ne réagissent pas, ou qu'ils réagissent peu en formant quelques plages isolées avec les autres phages. C'est le cas du lysotype C4 ou du lysotype G par exemple. Ces lysotypes présentent donc une sensibilité très étroitement spécifique.

Dans une autre catégorie de lysotypes, on classera ceux qui en plus de la lyse confluente avec le phage homonyme, donnent des réactions très fortes avec plusieurs autres phages. Les uns sont sensibles à tous les phages appartenant au même groupe que le phage homonyme. C'est le cas des lysotypes chefs de file d'un groupe ou d'une famille, par exemple le lysotype C1, qui est sensible aux phages spécifiques des 9 sous-types du groupe C, du lysotype E1, qui est sensible aux phages spécifiques des 10 sous-types du groupe E, du lysotype M1, qui est sensible aux phages spécifiques des 3 sous-types du groupe M.

Les autres sont sensibles, non seulement à leur phage homonyme, mais aussi à des phages n'appartenant pas à un groupe défini. Par exemple, le lysotype 29 donne la lyse confluente avec le phage 29 et aussi avec les phages C2, C3, D6, D11, E7, E9, F2, K2, 36, 40 et 46.

Ce sont des lysotypes de sensibilité moins spécifique. Pour leur diagnostic, on devra tenir compte non seulement de la réaction spécifique avec le phage homonyme, mais aussi des réactions avec les phages hétéronymes. Autrement dit, ce n'est pas seulement la réaction spécifique qui définit le lysotype dans ce cas, c'est toute l'image lysotypique.

Une troisième catégorie comprend les lysotypes du groupe B. Ceux-ci subissent bien la lyse confluente avec leur phage homonyme et ils donnent en outre des réactions très importantes (nombreuses

---

* ANDERSON, BERCOVICI, BORMAN, CLARK, DESRANLEAU, FELIX & FRASER, FUKUMI, LIE KIAN JOE, NICOLLE & BRAULT, NICOLLE & DIVERNEAU, NICOLLE, VIDAL-TORT & BRAULT, SCHOLTENS, SECHTER, E. M. J. WILSON & ANDERSON et V. WILSON & EDWARDS.

plages ou lyse presque confluente) avec un grand nombre de phages hétéronymes. Le diagnostic en est rendu presque toujours assez difficile.

De telles cultures sont très voisines de celles que nous avons mentionnés plus haut sous le nom de souches "aliénosensibles" (ou souches Vi dégradées de FELIX).

Enfin, dans une dernière catégorie, nous placerons le lysotype A, qui est sensible à toutes les préparations adaptées du phage Vi II. Lorsqu'on a réussi à adapter le phage Vi II à un nouveau lysotype, on peut être à peu près sûr qu'à la dilution limite (concentration d'épreuve pour la routine de la lysotypie), la préparation adaptée donnera la lyse confluente sur le lysotype A.

Certaines cultures aliénosensibles sont très voisines du lysotype A. Si tous les phages, à l'exception de deux ou trois, produisent une lyse confluente ou presque confluente, nous les appelons: cultures A imparfaites. Signalons aussi que certains lysotypes, en particulier les lysotypes B1, B2, B3, C1, D1 et E1 notamment, peuvent se transformer assez rapidement, sur mileu de conservation, en cultures aliénosensibles, puis en cultures A imparfaites, enfin en lysotype A normal. Il semble que la même transformation puisse se faire aussi par mutation, au moins dans certains cas (D1 transformé en A) (JUDE, NICOLLE & DUCREST, 1951[33]).

Stabilité des lysotypes.

La valeur d'une lysotypie dépend avant tout, on le comprend, de la stabilité des lysotypes *in vivo* et, d'une manière moins impérieuse, de leur stabilité *in vitro*, c'est-à-dire dans les milieux de culture et dans les milieux de conservation.

Pour *Salmonella typhi*, la stabilité *in vivo* est généralement assez satisfaisante: chez un même malade, pendant toute la durée de son infection, de sa convalescence, au cours de ses rechutes s'il en fait, et pendant tout le temps qu'il reste porteur de germes s'il le devient, c'est le même type que l'on observe pour les bacilles qu'on isole par hémoculture, coproculture, biliculture, culture de l'urine, du pus d'abcès, du liquide céphalorachidien, etc. Le traitement par le chloramphénicol ne l'altère pas non plus.

Voici un exemple emprunté à notre observation personnelle: un travailleur du laboratoire s'est contaminé accidentellement avec une souche de *Salmonella typhi*, type C1, provenant d'un aumônier militaire en traitement au Val-de-Grâce. Au cours des trois épisodes fébriles (première infection et deux rechutes) et d'un court portage de germes entre les deux rechutes, ce patient, qui a ingéré au total plus de 100 g de chloramphénicol a fourni 8 cultures (2 hémocultures, 5 coprocultures et 1 biliculture par tubage duodénal) qui toutes ont donné les réactions du lysotype C1. Le traitement par l'antibiotique n'a pas exercé de dégradation sur le lysotype.

Les bacilles isolés des divers malades d'un même foyer épidémique, qu'il s'agisse de quelques cas familiaux ou d'une épidémie groupant plusieurs centaines ou même quelques milliers d'individus appartiennent nécessairement à un lysotype unique si la source de contamination est la même pour tous.

La stabilité *in vitro* est également importante pour la valeur de la lysotypie. Mais elle n'est pas aussi primordiale que la stabilité *in vivo* et, de fait, elle est bonne, mais non absolue pour les bacilles typhiques à condition, toutefois, qu'ils soient maintenus dans la forme Vi positive. Pour obtenir une bonne conservation de l'antigène Vi et de l'image lysotypique, on utilisera le milieu de Dorset, la gélose en piqûre profonde ou mieux encore on lyophilisera les cultures aussitôt que possible après leur isolement (Vieu & Nicolle, 1962 [67]).

Cependant, même en prenant les précautions requises, on observe parfois deux sortes de dégradations dans les réactions des bacilles typhiques aux bactériophages.

C'est d'abord une diminution de la sensiblité au phage spécifique du lysotype. Ainsi, nous avons observé plusieurs fois que, parmi des cultures du type F1 provenant d'un même foyer, certaines ne donnaient plus, comme les autres, la lyse confluente avec les phages F1 et F2, mais seulement des plages en plus ou moins grand nombre. Certaines subcultures de colonies isolées de ces cultures ont donné des réactions nettes avec les phages F1 et F2, mais la plupart ne réagissaient que faiblement. On se trouvait, dans ce cas, en présence de cultures qui avaient perdu une grande partie de leur sensibilité aux phages spécifiques. Elles tendaient vers la forme Vi positive non caractérisable, c'est-à-dire vers le groupe I + IV.

On observe aussi parfois la dégradation inverse: une augmentation de la sensibilité aux phages hétéronymes. Les souches acquièrent ainsi l'image lysotypique du groupe aliénosensible. Les cultures des lysotypes B1, B2, B3, C1, E1, ou D1 passent souvent par cette forme avant d'aboutir au lysotype A imparfait, puis au lysotype A.

Intérêt de la lysotypie

L'intérêt de la lysotypie est multiple. Avant tout, c'est une méthode qui peut rendre de grands services en épidémiologie. Mais, même dans cet ordre d'idées, on peut lui reconnaître un intérêt en épidémiologie régionale ou appliquée et un intérêt en épidémiologie mondiale. En outre, elle présente un intérêt théorique considérable.

En épidémiologie régionale ou appliquée.

Dans les enquêtes régionales, cette méthode aide souvent à établir la filiation des cas chez des malades et porteurs de germes vivant dans une même localité, lorsque leurs bacilles appartiennent à un même lysotype ou, au contraire, elle permet d'éliminer avec certitude toute éventualité de contagion entre des sujets dont les bacilles donnent les réactions de lysotypes différents. De nombreux auteurs, pratiquement tous ceux qui les ont tant soit peu appliquées, ont reconnu à la fois la valeur et l'utilité de cette méthode.

On ne peut prétendre aujourd'hui avoir réuni tous les éléments d'une enquête épidémiologique complète si l'on ne connaît pas les types bactériophagiques des bacilles typhiques en cause chez des personnes que l'on soupçonne de s'être contaminées à une même source ou entre eux.

Quelques exemples empruntés à notre expérience personnelle illustreront les notions qui viennent d'être développées sur l'intérêt que présente la connaissance des lysotypes du bacille typhique en épidémiologie régionale.

Dans un établissement hospitalier de la banlieue parisienne groupant une centaine d'enfants, tous les ans des cas de fièvre typhoïde apparaissaient. En 1950, les cas furent plus nombreux que d'habitude. Le personnel commençait à s'inquiéter. Il suspectait l'eau de canalisation et réclamait des mesures. L'un de nous, averti par hasard, parvint à se procurer des cultures isolées chez plusieurs malades par différents laboratoires hospitaliers et privés. Ces cultures appartenaient toutes au lysotype D1, qui n'est pas très fréquent en France. Cette communauté de lysotype chez les malades de l'établissement en question et l'allure intermittente des apparitions de cas permettaient déjà de rejeter l'origine hydrique de l'infection et de penser à des contaminations par porteurs de germes. La recherche des agglutinines Vi, telle qu'elle a été préconisée par FELIX, chez les personnes chargées de la cuisine et des soins aux enfants, et les coprocultures pratiquées chez celles qui présentèrent un sérodiagnostic Vi positif, révélèrent la présence, non pas d'un seul porteur de germes, mais de plusieurs. Les bacilles isolés chez toutes ces personnes appartenaient sans exception au lysotype D1. Finalement, l'histoire de ce foyer épidémique a pu être reconstituée: une cuisinière, porteuse de germes, avait probablement contaminé quelques années auparavant plusieurs membres du personnel. Certains d'entre eux étaient devenus à leur tour porteurs de germes, d'où la communauté de type bactériophagique de leur bacille. De temps en temps, l'un ou l'autre de ces porteurs infectait les petits pensionnaires de l'établissement (ALIMANESTIANU-BUTAS, BONNEFOI & NICOLLE, 1951[1]).

Des mesures simples: porteurs de germes éloignés de la cuisine, obligation pour tout le personnel de passer les mains dans un bain d'eau de Javel diluée chaque fois qu'il se disposait à toucher des aliments ou à donner des soins aux enfants, suffirent à empêcher l'éclosion de nouveaux cas dans l'établissement en question.

Voici enfin un dernier exemple: à la fin de janvier et au début de février 1954, une épidémie de fièvre typhoïde survint dans plusieurs casernes de Lyon. Au total, 150 soldats et civils prenant leurs repas dans ces casernes furent atteints. Les 200 cultures, isolées par hémocultures et coprocultures au cours de l'épidémie ou à l'occasion de rechutes appartenaient toutes au lysotype C1 et au chimiotype I (xylose positives, arabinose négatives). Les épidémiologistes furent amenés à penser que cette épidémie avait été provoquée par le rinçage dans les bacs à vaisselle, dont l'eau n'était pas changée souvent, de salades provenant d'un jardin maraîcher des environs de Lyon. Dans une des salades prélevées dans une resserre attenant au jardin en question, on a isolé un bacille typhique qui a donné lui-aussi les réactions du lysotype C1 et du chimiotype I. La preuve étant acquise de l'origine de l'épidémie, le maraîcher finit par avouer que, suivant une habitude commune dans cette profession, il arrosait les salades entassées dans la resserre avec de l'eau de latrines voisines pour leur conserver leur fraîcheur (GAUBERT et collab., 1955)[28].

Intérêt de la lysotypie en épidémiologie mondiale.

En dehors de l'utilisation de la lysotypie dans l'étude régionale des foyers épidémiques ou endémiques, cette méthode fournit de très intéressantes informations concernant l'épidémiologie générale de la fièvre typhoïde.

Elle a permis de constater déjà que la distribution des lysotypes variait souvent considérablement d'une région du globe à l'autre. Si certains types sont cosmopolites (A, B2, C1, D1, E1, F1, N, O, T, 28, le groupe I + IV et le groupe aliénosensible), d'autres ne se rencontrent guère que dans certaines régions: le lysotype G, absent ou très rare en Europe et en Amérique du Nord, est relativement fréquent dans les pays proches du pourtour de l'Océan Indien (région orientale du Congo (Léopoldville), Madagascar, Iran, Inde, Indonésie, Vietnam). Les types du groupe M (M1, M2, M3, M4) également très rares ou absents en Europe et en Amérique du Nord ne se rencontrent guère qu'en Extrême-Orient, surtout le lysotype M1. C'est le type dominant à Hanoï; il est encore abondant à Saïgon. On le trouve à Java et au Japon. Nous avons des raisons de penser qu'il est fréquent en Chine du Sud. Il est encore assez fréquent de l'autre côté du Pacifique, sur la rive occidentale de l'Amérique du Sud.* Les lysotypes D2, D4, D6, E10, J1 n'ont été trouvés en France qu'exceptionnellement. Ils sont communs en Extrême-Orient (Vietnam, Cambodge).

La variété Centre Africaine du lysotype C1 (C1 (C.A.)) (NICOLLE, VAN OYE, CROCKER & BRAULT, 1955[54]; PRUNET & NICOLLE, 1962[58]), n'a été identifiée jusqu'à présent qu'en Afrique Centrale (République

---

* Où il a été vraisemblablement implanté par l'immigration chinoise et japonaise.

Centre Africaine, les deux Congo et Madagascar). Cependant, nous l'avons identifié une fois dans un lot de souches envoyées par le Laboratoire Départemental de Strasbourg (DR. A. LUTZ). Renseignements pris, cette souche avait été isolée à Strasbourg du sang d'un enfant qui, atteint de fièvre typhoïde, venait d'arriver de Bangui (République Centre Africaine) pour être soigné dans sa famille.

Les types L1 et L2 n'ont jamais été rencontré en France. Ils l'ont été très rarement en Grande-Bretagne, en Pologne, aux Etats-Unis. Au contraire, ils ne sont pas très rares en Afrique du Nord et spécialement au Maroc (région de Fez) (5,02% sur 399 cultures).

Le lysotype 42 (SPANO, 1956[65]; NICOLLE, HUET & NINARD, 1958[51]) est fréquent en Afrique du Nord et en Sicile, très rare ailleurs. Les lysotypes 26 et 35 ne sont guère rencontrés qu'au Mexique, le lysotype 46, qu'en Espagne et en Amérique Latine (NICOLLE, 1962[44]).

Tout se passe comme si la fièvre typhoïde, initialement homogène, s'était diversifiée, peut-être sous l'influence de facteurs écologiques, en évoluant dans différentes parties du globe, et que certaines de leurs variétés, débordant des limites de leur domaine primitif d'apparition, sans doute en raison de l'intensité des déplacements humains, avaient été propagées au loin, tandis que d'autres variétés au contraire, moins favorisées sous ce rapport, ne s'étaient guère étendues au delà des limites des territoires où elles s'étaient formées.

La lysotypie, par conséquent, peut nous renseigner sur la distribution actuelle des lysotypes dans les différentes parties du monde et sur les changements qui surviennent dans cette distribution.

Les notions ainsi acquises, en dehors de l'intérêt théorique qui s'y attache et qui est certainement d'une importance capitale pour la connaissance de ce que CHARLES NICOLLE appelait le "Destin des maladies infectieuses", permettraient de tirer un meilleur parti de la lysotypie en épidémiologie régionale. L'interprétation des résultats dépend, en effet, dans une certaine mesure de la connaissance que l'on a de la fréquence ou de la rareté des types rencontrés dans un foyer de fièvre typhoïde. Si les bacilles de plusieurs cas contigus appartiennent à un type fréquent dans la région, il serait imprudent d'affirmer qu'il y a communauté de contagion. La lysotypie doit être considérée plutôt en de telles circonstances comme une méthode d'exclusion: on peut admettre en toute certitude que deux bacilles de types différents n'ont pas la même origine.

Au contraire, lorsque les bacilles trouvés chez des malades et des porteurs de germes voisins appartiennent à un type rare dans la région, cette similitude constitue un argument très important en faveur d'une contagion commune.

Enfin, en dehors de son intérêt épidémiologique, la lysotypie peut rendre de grands services dans une infinité de recherches d'ordre théorique et pratique en bactériologie. On comprend quels précieux renseignements peuvent être obtenus grâce à la possibilité qu'offre cette méthode de suivre, dans une foule d'expériences, le sort des bacilles en quelque sorte "marqués" par leur lysotype.

Les limites de la lysotypie. Ses améliorations possibles

Lorsque, dans un pays, le nombre des lysotypes habituellement présents est très petit, si l'on n'en trouve qu'un seul ou deux par exemple, ou si l'un d'entre eux ou bien deux sont très fortement prédominants, il est bien évident que la lysotypie ne peut rendre que des services médiocres, puisque presque tous les foyers reconnaîtront pour cause un seul ou bien deux seuls lysotypes. Au contraire, dans les pays où l'on observe une grande variété de lysotypes, surtout si aucun n'atteint un pourcentage élevé, les renseignements fournis par la méthode auront une plus grande valeur.

Dans les pays d'Europe, la distribution des lysotypes est en général moyennement bonne. Cependant, certains d'entre eux sont fréquents ou très fréquents. Ce sont toujours les mêmes: les lysotypes E1, A, C1, D1, F1, le groupe I + IV et le groupe aliénosensible. Dans ces conditions, on s'est demandé s'il ne serait pas possible, pour augmenter l'efficacité de la méthode, de subdiviser les lysotypes les plus fréquents.

Une première tentative de subdivision a été proposée par OLITZKI, 1944, 1948[55, 56], puis reprise par JUDE & NICOLLE, 1949[32], par PAVLATOU & NICOLLE, 1953[57] et NICOLLE, NICOLLE & DIVERNEAU, 1961[52]. Il s'agissait d'associer la méthode biochimique à la méthode lysotypique. Nous avons vu plus haut que, d'après KRISTENSEN, 1926, 1938[36, 37], l'espèce *Salmonella typhi* comprend trois variétés biochimiques: le chimiotype I, qui fait fermenter le xylose, mais non l'arabinose; le chimiotype II, qui les respecte l'un et l'autre; le chimiotype III, qui les attaque tous les deux. La plupart des lysotypes sont homogènes ou presque homogènes dans leurs propriétés biochimiques. Pour eux, la méthode de KRISTENSEN ne sera donc d'aucune utilité: parmi eux, certains sont du chimiotype I, ce sont notamment les lysotypes C1, D1, D2, D4, E1, E3, E4, F1, G, J, L1, 26, 35, 37 et d'autres du chimiotype II, en particulier les lysotypes M1, 40 et 42.

Un petit nombre de lysotypes sont hétérogènes. Leurs cultures peuvent être xylose positives ou xylose négatives. Ce sont les lysotypes A, B2, C3, D6, L2, N, O, T, 28, 29, le groupe I + IV et le groupe aliénosensible.

Comme on le voit, la combinaison de la méthode de KRISTENSEN avec la lysotypie ne présente d'intérêt que pour un nombre restreint de lysotypes. De plus, dans les cas les plus favorables, elle ne va guère au-delà d'une division binaire.

Certains auteurs ont proposé diverses variétés de lysotypie complémentaires d'après les différences de sensibilité des cultures d'un même lysotype à d'autres bactériophages. Ce sont les variétés Aφ et Aψ du lysotype A (Desranleau, 1947[21]) qui se distinguent du lysotype A normal, sensible aux phages Vi I, III, IV, V et VI non adaptés, par la résistance aux phages Vi I et V pour la variété Aφ et aux phages Vi III et IV pour la variété Aψ; la variété E1 (b) (Brandis, 1955[12]) qui, au contraire de ce que l'on observe pour la variété commune E1 (a), n'est pas lysée par le phage Vi VII. Enfin, Nicolle, Pavlatou & Diverneau, 1953[48], puis Nicolle, Diverneau & Brault, 1958[50] ont mis au point une lysotypie complémentaire au moyen de 2 jeux de phages; le premier jeu comprend 6 phages non Vi et 2 phages Vi; le second, 6 phages non Vi extraits des sous-types lysogènes identifiés par le 1er jeu de phages. Ce second jeu de phages représente donc une lysotypie rationnelle qui a pour rôle de confirmer la lysotypie empirique du premier jeu. Certains sous-types du lysotype A sont cosmopolites (Tananarive, Montréal, Chamblee); d'autres ont une répartition géographique particulière: Coquilhatville, Douala et Léopoldville sont fréquents en Afrique Noire. Maracaïbo est presque le seul sous-type du lysotype A que l'on rencontre dans la région des Caraïbes (Venezuela, Guyane, Antilles); d'autres enfin, n'ont été rencontrés qu'en Angleterre et en Extrême-Orient (Welshpool, Oswestry).

Les bases fondamentales de la lysotypie du bacille typhique

La lysotypie du bacille typhique doit son existence et sa précision aux circonstances suivantes qui concernent d'une part le bactériophage Vi II et ses propriétés adaptatives et d'autre part le bacille typhique et ses divers états lysogènes.

1 — Le bactériophage Vi II et ses propriétés adaptatives — Le phage Vi II de Craigie est doué d'une très exceptionnelle plasticité adaptative que le rend spécifique, à sa dilution limite, du lysotype du bacille auquel il a été adapté. Cette adaptation peut être la conséquence, comme Anderson en 1955, 1956[2, 8] et 1962[5] l'a bien mis en lumière, de deux types de variations que subit le phage ou d'une combinaison des deux. Le premier type est une modification phénotypique du spectre d'activité du phage, du genre de celle qui a été décrite par Luria & Human, en 1952[39] et Bertani & Weigle, en 1953[10]. Cette modification est provoquée par la bactérie et elle est entièrement réversible par un simple passage du phage sur une autre souche, en particulier sur le lysotype A, qui est le lysotype ancestral du bacille typhique. Le phage désadapté reprend les caractères du phage A qu'il avait avant son adaptation. Par exemple, la préparation M1 obtenue par adaptation du phage A qui est le type sauvage ou ancestral du phage Vi II au lysotype M1, lorsqu'elle est repassée sur le lysotype A perd son adaptation au lysotype M1 pour reprendre les caractères du phage A.

La seconde modification est une modification génotypique indépendante des conditions extérieures et de son contact avec le lysotype. Elle est irréversible et héréditaire. C'est une véritable mutation. Par exemple, la préparation adaptée D1 est obtenue par multiplication du phage A sur le lysotype D1. Lorsqu'on la repasse sur le lysotype A, elle garde intégralement son adaptation au lysotype D1.

Dans certains cas, on observe une combinaison de ces deux types de modifications. Par exemple, pour obtenir la préparation adaptée au lysotype D6, il faut d'abord adapter le phage A au lysotype D1 (modification génotypique), puis adapter le phage D1 au lysotype D6 (modification phénotypique).

Parfois, on observe des modifications encore plus surprenantes, comme c'est le cas pour la préparation adaptée au lysotype M2. Pour parvenir à l'adaptation du phage Vi II au lysotype M2, il faut d'abord adapter le phage A au lysotype M1, car le phage A n'agit pas directement sur le lysotype M2. A partir de la préparation M1 ainsi adaptée (modification phénotypique), on peut en régénérant le phage M1 sur le lysotype M2, l'adapter à ce dernier. La préparation M2 ainsi obtenue, si on la repasse sur le lysotype A perd son expression phénotypique pour M2, mais après passage sur le lysotype M1 cette nouvelle préparation acquiert en même temps l'adaptation au lysotype M1 et l'adaptation au lysotype M2 (NICOLLE, HAMON & DIVERNEAU, 1962[53]; NICOLLE & VIEU, 1962[46]).

La première adaptation du phage M1 au lysotype M2 n'était donc pas exclusivement phénotypique comme l'adaptation du phage A au lysotype M1: elle avait laissé au phage adapté au lysotype M1, puis désadapté par passage sur le lysotype A, le souvenir de son adaptation antérieure au lysotype M2, souvenir qui lui a donné la possibilité de réacquérir, par passage sur le lysotype M1, l'adaptation perdue au lysotype M2.

L'adaptation au lysotype M2 est donc à la fois phénotypique dans sa spécificité (phénotypique puisqu'elle s'efface par passage sur le lysotype A) et génotypique, par l'aptitude du phage adapté au lysotype M2, puis phénotypiquement désadapté (par passage sur le lysotype A), à récupérer en une seule étape (par passage sur le lysotype M1), l'activité lysotypique qu'il avait acquise la première fois en deux étapes (par passage sur M1, puis sur M2) (NICOLLE & VIEU, 1962[46]).

2 — Le bacille typhique et ses divers états lysogènes — NICOLLE & HAMON, dès 1951[45], ont établi, par l'analyse de la lysogénie et par des expériences de lysogénisation, que la diversité de ces états lysogènes des bacilles paratyphiques B étaient l'une des causes, certainement même la cause principale, de la diversité de leurs lysotypes, tels qu'ils peuvent être identifiés par la méthode de FELIX & CALLOW, 1951[25]. Nous avons donné le nom de "phage déterminant pour un

lysotype donné" à tout phage de lysogénie dont la présence imposait à la souche bactérienne qui le portait son image lysotypique caractéristique, donc son lysotype, et celui de "phage indifférent" à tout phage de lysogénie qui ne joue aucun rôle dans la formation des lysotypes. Certains lysotypes doivent leurs caractères lysotypiques à la présence de deux phages. Chacun de ceux-ci est dit "semi-déterminant pour un lysotype donné".

Un peu plus tard, FELIX & ANDERSON, 1951[24], puis ANDERSON & FELIX 1953[7] ont formulé des conclusions analogues pour *S. typhi*. Les expressions de "phage déterminant pour un lysotype donné" et de "phage non déterminant" sont aujourd'hui passées dans le langage courant.

La plupart des adaptations du phage Vi II à des lysotypes non lysogènes, ou plus exactement à des lysotypes qui n'élaborent pas un phage déterminant pour le lysotype, résultent de modifications phénotypiques seulement, tandis que celles qui sont obtenues en utilisant les lysotypes à phage déterminant pour le lysotype, sont dues à de véritables mutations. Elles ont donc une stabilité génotypique.

Le fait que l'on puisse obtenir artificiellement des lysotypes semblables à ceux que l'on rencontre dans la nature, par lysogénisation des lysotypes dépourvus de phages déterminants pour le lysotype, permet d'assigner une "formule structurale" à la plupart des lysotypes Vi naturels connus. Cette formule est composée du symbole du lysotype précurseur (non lysogène ou plutôt ne portant pas de phage déterminant) suivi entre parenthèses, de celui du phage tempéré déterminant pour le lysotype. Ainsi le lysotype F2 n'est autre chose que le lysotype F1 porteur du prophage f2. Sa formule structurale est donc: F1 (f2).

Dans diverses publications de 1956[9], 1957[3] et 1959[4], ANDERSON a complété la notion des formules structurales des lysotypes à phages déterminants de *S. typhi*. On en trouvera la liste dans le tableau suivant.

READ & FERGUSON, 1961[59], ont confirmé pour *S. typhi* les observations de NICOLLE & HAMON pour *S. paratyphi B* au sujet des phages semi-déterminants. Ils montrent que, dans certains cas, il faut deux phages déterminants pour former un lysotype donné. Par exemple, le lysotype D8 est doublement lysogène; ses phages sont d8 alpha et d8 beta. Le phage d8 alpha, donnant des plages semi-voilées, transforme le lysotype A en lysotype D1. Il est donc identique au phage de lysogénie d1. Le phage d8 beta, donnant des plages voilées est sans action transformatrice sur le lysotype A, mais il transforme le lysotype D1 en lysotype D8. La formule structurale du lysotype D8 est donc D1 (d8 alpha ou d1 + d8 beta). Il en est de même pour les lysotypes D9 et D10 dont les formules structurales sont: A (d9 alpha ou d1 + d9 beta) et ? (d10 alpha + d10 beta); le

Lysotypes à phages déterminants	Phages déterminants	Lysotypes précurseurs	Formules structurales
C9	c9		
D1	d1	A	A (d1)
D6	d6	A	A (d6)
C2	d6	C1	C1 (d6)
E9	d6	E1	E1(d6)
F2	f2 ou 30′	F1	F1 (d6)
C3	f2 ou 30′	C1	C1 (f2)
E7	f2 ou 30′	E1	E1(f2)
29	f2 ou 30′	A	A (f2)
J2	j2		
T	t		
25	25′		
26	26′	A	A (26′)
E8	26′	E1	E1(26′)
C8	26′	C1	C1(26′)

point d'interrogation signifie que le précurseur du lysotype D10 est encore inconnu.

Des résultats analogues ont été obtenus par CEFALU & FICHERA en 1961[13] avec les sous-types du lysotype A tels qu'ils peuvent être identifiés par la lysotypie complémentaire de NICOLLE, PAVLATOU & DIVERNEAU. Les auteurs siciliens ont confirmé, par l'obtention artificielle de sous-types du lysotype A au moyen de la lysogénisation du sous-type Tananarive par les phages de lysogénie extraits de certains de ces sous-types, leur existence réelle. Ils ont confirmé également le rôle, affirmé par NICOLLE, PAVLATOU & DIVERNEAU, des divers états lysogènes dans la différenciation du lysotype A en sous-types. Par exemple, en lysogénisant le sous-type Tananarive (non lysogène) par les phages de lysogénie des sous-types Montréal, Chamblee et Welshpool, ils ont obtenu respectivement les sous-types Montréal, Chamblee et Welshpool.

Le Comité International de la Lysotypie Entérique

Pour assurer la meilleure constance dans les résultats et pour conserver à la méthode toute sa valeur et toute son efficacité, les auteurs (CRAIGIE & FELIX) ont codifié la technique et standardisé les préparations bactériophagiques adaptées du phage Vi II. Ces préparations sont distribuées par le Laboratoire International de Référence (Enteric Phage Typing Reference Laboratory, Public Health Laboratory Service, Colindale Avenue, London) à un certain nombre de Centres de lysotypie, en principe un seul centre par pays (exceptionnellement, on compte plusieurs centres en Italie, en

Allemagne, en Pologne et aux Etats-Unis). La découverte de nouveaux lysotypes ou de nouveaux sous-types de *Salmonella typhi* et leur reconnaissance officielle par le Comité International de la Lysotypie Entérique qui groupe tous les directeurs des Centres nationaux est donc le résultat d'une collaboration internationale qui s'est montrée très fructueuse (ANDERSON, 1962[5]; NICOLLE, 1962[44]).

BIBLIOGRAPHIE

1. ALIMANESTIANU-BUTAS, C., BONNEFOI, A. & NICOLLE, P., *Presse méd.* 1951, **59**, *1613.*
2. ANDERSON, E. S., *Nature, Lond.* 1955, **175**, *171.*
3. ANDERSON, E. S., *Zbl. Bakt.*, I.O., 1957, **168**, *489.*
4. ANDERSON, E. S., *Nature, Lond.* 1959, **184**, *1822.*
5. ANDERSON, E. S., *Ann. Inst. Pasteur*, 1962, **102**, *379.*
6. ANDERSON, E. S., *Brit. med. Bull.*, 1962, **18**, *64.*
7. ANDERSON, E. S. & FELIX, A., *J. gen. Microbiol.*, 1953, **9**, 65.
8. ANDERSON, E. S. & FRASER, A., *J. gen. Microbiol.*, 1956, **15**, *225.*
9. ANDERSON, E. S. & WILLIAMS, R. E. O., *J. clin. Path.*, 1956, **15**, *225.*
10. BERTANI, G. & WEIGLE, J. J., *J. Bact.*, 1953, **65**, 113.
11. BLASI, R. DE & BUOGO, H. *Riv. Ital. Ig.*, 1952, **12**, *15.*
12. BRANDIS, H., *Zbl. Bakt.*, I.O., 1955, **162**, *223.*
13. CEFALU, M. & FICHERA, G., *Zbl. Bakt.*, I.O., 1961, **181**, *373.*
14. CHOMICZEWSKI, J., *Med. Dosw. Mikrobiol.*, 1960, **12**, *195.*
15. CRAIGIE, J. & BRANDON, K. F., *J. Path. Bact.*, 1936, **43**, *233.*
16. CRAIGIE, J. & BRANDON, K. F., *J. Path. Bact.*, 1936, **43**, *249.*
17. CRAIGIE, J. & FELIX, A., *Lancet*, 1947, **252**, *823.*
18. CRAIGIE, J. & YEN, C. H., *Trans. Roy. Soc. Canada*, 1937, **5**, *49.*
19. CRAIGIE, J. & YEN, C. H., *Canad. Publ. Hlth.*, 1938, **29**, *484.*
20. CRAIGIE, J. & YEN, C. H., *Canad. Publ. Hlth*, 1938, **29**, *448.*
21. DESRANLEAU, J. M., *Canad. J. Publ. Hlth.*, 1947, **38**, *343* et 1950, **41**, *128.*
22. FELIX, A., *Lancet*, 1930, **I**, *505.*
23. FELIX, A., *Bull. Org. Mond. Santé*, 1955, **13**, *109.*
24. FELIX, A. & ANDERSON, E. S., *Nature, Lond.*, 1951, **167**, *603.*
25. FELIX, A. & CALLOW, B. R., *Lancet*, 1951, **261**, *10.*
26. FELIX, A. & PITT, R. M., *J. Path. Bact.*, 1934, **38**, *409.*
27. FELIX, A. & PITT, R. M., *Lancet*, 1934, **2**, *186.*
28. GAUBERT, Y., BENAZET, F., MIGEON, A. & SOHIER, R., *Soc. Med. Hôp. Paris*, 1953, **71**, *431.*
29. GIOVANARDI, A., *Zbl. Bakt.*, I.O., 1938, **141**, *341.*
30. HÉRELLE, F. D', *C.R. Acad. Sci.*, 1919, **168**, *631.*
31. HÉRELLE, F. D', Le Bactériophage, son rôle dans l'immunité. Masson et Cie, Paris, 1921.
    Le Bactériophage et son comportement. Masson et Cie, Paris, 1926.
32. JUDE, A. & NICOLLE, P., *Ann. Inst. Pasteur*, 1949, **77**, *550.*
33. JUDE, A., NICOLLE, P. & DUCREST, P., *Ann. Inst. Pasteur*, 1951, **81**, *245.*
34. JUDE, A., LE MINOR, L. & NICOLLE, P., *C.R. Acad. Sci.*, 1955, **240**, *822.*
35. KAUFFMANN, F., *Z. Hyg.*, 1936, **117**, *778.*
36. KRISTENSEN, M., *J. Hyg. Camb.*, 1938, **38**, *688.*
37. KRISTENSEN, M. & HENRIKSEN, H. C. D., *Acta path. microbiol. scand.*, 1926, 3, *531.*
38. LACHOWICZ, K. & BUCZOWSKI, Z., *Med. Dosw. Mikrobiol.*, 1950, **2**, *390.*
39. LURIA, E. S. & HUMAN, M. L., *J. Bact.*, 1952, **64**, *557.*
40. MARCUSE, K., *Klin. Wschr.*, 1931, à, *732.*

41. MARCUSE, K., *Zbl. Bakt.*, I.O., 1934, **131**, *49*.
42. MARCUSE, K., *Zbl. Bakt.*, I.O., 1934, **131**, *206*.
43. MASSA, M., *Z. Immun Forsch.*, 1931, **70**, *525*.
44. NICOLLE, P., *Ann. Inst. Pasteur*, 1962, **102**, *389* et *580*.
45. NICOLLE, P. & HAMON, Y., *C.R. Acad. Sci.*, 1951, **232**, *898*.
46. NICOLLE, P. & VIEU, J. F., *C.R. Acad. Sci.*, 1962, **254**, *4527*.
47. NICOLLE, P., JUDE, A. & LE MINOR, L., *Ann. Inst. Pasteur*, 1950, **78**, *572*.
48. NICOLLE, P., PAVLATOU, M. & DIVERNEAU, G., *C.R. Acad. Sci.*, 1953, **236**, *2453*.
49. NICOLLE, P., JUDE, A. & LE MINOR, L., *C.R. Acad. Sci.*, 1955, **240**, *694*.
50. NICOLLE, P., DIVERNEAU, G. & BRAULT, J., *Bull. Res. Council Israel*, 1958, **7E**, *89*.
51. NICOLLE, P., HUET, M. & NINARD, B., *Arch. Inst. Pasteur, Tunis*, 1958, **35**, *153*.
52. NICOLLE, P., NICOLLE, J. & DIVERNEAU, G., *Ann. Inst. Pasteur*, 1961. **101**, *211*.
53. NICOLLE, P., HAMON, Y. & DIVERNEAU, G., *Arch. Roum. Path. exp. Microbiol.*, 1962, **21**, *315*.
54. NICOLLE, P., VAN OYE, E., CROCKER, C. G. & BRAULT, J. *Bull. Soc. Path. exot.* 1955, **48**, *492*.
55. OLITZKI, A. L. & SHELUBSKY, M., *Res. Council Israel*, Jérusalem, 1944.
56. OLITZKI, A. L., SHELUBSKY, M. & STRAUSS, W., *Harefuah*, 1948, *107*.
57. PAVLATOU, M. & NICOLLE, P., *Ann. Inst. Pasteur*, 1953 **85**, *185*.
58. PRUNET, J. & NICOLLE, P., *Ann. Inst. Pasteur*, 1962, **103**, *536*.
59. READ, K. S. & FERGUSON, W. W., *Ann. Inst. Pasteur*, 1961, **100**, *120*.
60. SCHOLTENS, R. TH., *J. Hyg. Camb.*, 1936, **36**, *452*.
61. SERTIC, V. & BOULGAKOV, N. A., *C.R. Soc. Biol.*, 1937, **126**, *737*.
62. SONNENSCHEIN, C., *Münch. med. Wschr.*, 1925, *1443*.
63. SONNENSCHEIN, C., *Dtsch. med. Wschr.*, 1928, *1034*.
64. SONNENSCHEIN, C., *Münch. med. Wschr.*, 1929, *355*.
65. SPANO, C., *Riv. Ital. Sieroter. Ital.*, 1956, **31**, *104*.
66. TWORT, F. W., *Lancet*, 1915, **II**, *1241*.
67. VIEU, J. F. & NICOLLE, P., *C.R. Acad. Sci.*, 1962, **255**, *209*.

THE PHAGE TYPING OF SALMONELLAE OTHER THAN S. TYPHI

BY

E. S. ANDERSON

Director, Enteric Reference Laboratory, Central Public Health Laboratory, Colindale, London, N.W. 9.
(with 1 fig.)

The emergence of salmonellosis as a world problem has been dealt with elsewhere in this Monograph. The complexity of the problem and the certainty that there is strain diversity within each serotype have necessitated the development of special methods of strain identification for epidemiological purposes. Among such methods, that of bacteriophage typing has emerged as the one giving at the same time the most reliable results and the maximum amount of strain differentiation. It has also the advantage of employing a relatively simple technique which is easy to standardize for widespread routine use. Most of the phages used are stable and can be relied upon to yield reproducible lytic spectra on defined organismal types. The schemes employed have been exhaustively tested to establish their value in epidemiological investigations, and the stability of the organismal phage types they distinguish has been proved over long periods. Thus, provided that a salmonella serotype can be divided into an adequate number of phage types, and provided that the frequency distribution of these types is not such that there is an overwhelming preponderance of a single type, a great deal of valuable information can be contained by using the phage-typing method. The method may be used on a local basis to identify the types responsible for given outbreaks of salmonellosis and to help to trace their sources. It may be employed regionally, to define the type distribution in larger areas of a country, or nationally in attempts to discover the possible channels of transfer of certain local types to other parts of a country. Its use on an international scale is helpful in several ways. In the first place, because of the speed of modern travel, persons may arrive in a country while incubating a disease such as typhoid fever. The phage types with which they are infected are occasionally characteristic of the region they have come from, and any secondary cases they cause may be rapidly identified by means of phage typing. Infections imported in this way may even form part of an outbreak which originated at an international holiday centre, and which may therefore be distributed between a number of countries. Phage typing will rapidly establish the homogeneity of such an outbreak. An excellent example of this was provided by the typhoid outbreak due to Vi-phage type E1, originating in Zermatt, Switzerland, in February-March 1963. Patients in the early stages of the disease, or still in the incubation period, returned to their home countries where the typhoid infection was diagnosed. This epidemic affected at least six countries.

Another valuable international use of phage typing is in the identification of the sources and countries of origin of salmonellae imported in foodstuffs and other organic materials such as animal feeding stuffs and fertilizers.

The great majority of salmonella phage-typing schemes in use today have been modelled on the pattern established by the Vi-phage typing scheme of CRAIGIE & YEN (1938) for *Salmonella typhi*. This, which is now in wide international use, is described in detail elsewhere in this Monograph. We shall limit ourselves here to mentioning that it has the unique feature of depending on the versatility of host range of a single phage, Vi-phage II, which has been successively adapted to 78 different types of the typhoid bacillus. None of the schemes that have subsequently been developed for other salmonella serotypes possess this quality. All employ a diversity of phages for typing, though some of these phages have a limited capacity for adaptation, and use is made of this property where possible. In the development of a typing scheme, however, the aim is to distinguish the maximum possible number of different phage-sensitivity spectra in the serotype under scrutiny and, provided that such spectra are easily recognizable and constant for a given phage type, the fact that phages of diverse origin are employed for their recognition is of secondary importance.

The technique of phage typing

The technique followed in the Enteric Reference Laboratory (E.R.L.) is basically that described by CRAIGIE & FELIX (1947). Most other countries using the method follow a closely similar technique, which, with only minor variations, is used for all salmonella phage typing.

The nutrient medium ("Difco broth") routinely employed has the following composition:

Bacto dehydrated nutrient broth (Difco Laboratories)	20	g
NaCl	8.5	g
Distilled water	1000	ml

The medium is autoclaved at 15 lb. for 15 minutes. The final pH is 6.8. For the preparation of solid medium ("Difco agar"), 1.3% Davis New Zealand powdered agar is added.

Cultures to be phage typed are inoculated into 2 ml of Difco broth to give a barely visible turbidity and are incubated with agitation at 37 °C until they reach an opacity of about 4×10^8 cells/ml. The culture is inoculated on to a suitably marked Difco agar plate, either as a series of discrete spots or by flooding, and is allowed to dry. The typing phages are then spotted on the culture in a predetermined order corresponding to the marking of the plate. The plate is incubated at 38.5 °C.

Phage-typing plates are usually read after $4\frac{1}{2}$ to 5 hr. and 24 hr. incubation. Readings are carried out with a X10 aplanat hand lens through the bottom of the plates, using transmitted oblique illumination. All-glass or transparent plastic Petri dishes are used so that the lids need not be removed for reading. The various degrees of lysis are recorded as shown in Table I.

Table I.

Method of recording degrees of lysis on salmonella phage typing plates

Plaque Sizes	Plaque Numbers	Lysis
L = large	0 to 5 plaques	SCL = semi-confluent lysis
N = normal	± = 6— 20 plaques	CL = confluent lysis
S = small	+ = 21— 40 plaques	<SCL⎱ = intermediate
		< CL⎰ degrees of lysis
m = minute, visible only	+± = 41— 60 plaques	OL = confluent "opaque" lysis
with hand	++ = 61— 80 plaques	(opacity due to
lens.	++± = 81—120 plaques	heavy secondary
μ = micro	+++ = >120 plaques	growth)

At a given phage concentration, the amount of culture destroyed by a phage is obviously dependent on the size of the plaques; relatively few large plaques (often under 100) are needed to produce confluent lysis, whereas thousands of plaques of a minute-plaque phage may be necessary to produce lysis of a similar degree. The plaque sizes of typing phages may vary considerably, and this is one of the factors to be considered when deciding on the phage dilution to be used for routine typing, which is known as the routine test dilution (R.T.D.).

The R.T.D. is mainly important in this work, however, because if the typing phages for any salmonella serotype are used undiluted, they will usually lyse indiscriminately all phage types of the organism, so that no lytic patterns can be distinguished. This can be avoided by using the R.T.D., which is the highest phage dilution that will produce confluent lysis, or a reaction approaching that order, on the type the phage is primarily intended to distinguish. The use of this dilution minimizes cross reactions on other types and enables us to narrow the host specificity of the phages as far as is practicable. Thus, although the ideal of having a single phage specific for each type can rarely be attained, the use of the R.T.D. helps to make the patterns on which type diagnosis depends distinct.

The reader is referred to ANDERSON & WILLIAMS (1956) for fuller technical details on phage typing.

Fig. 1 shows a typical phage-typing plate, the organism in this case being type 1 of *S. paratyphi B*.

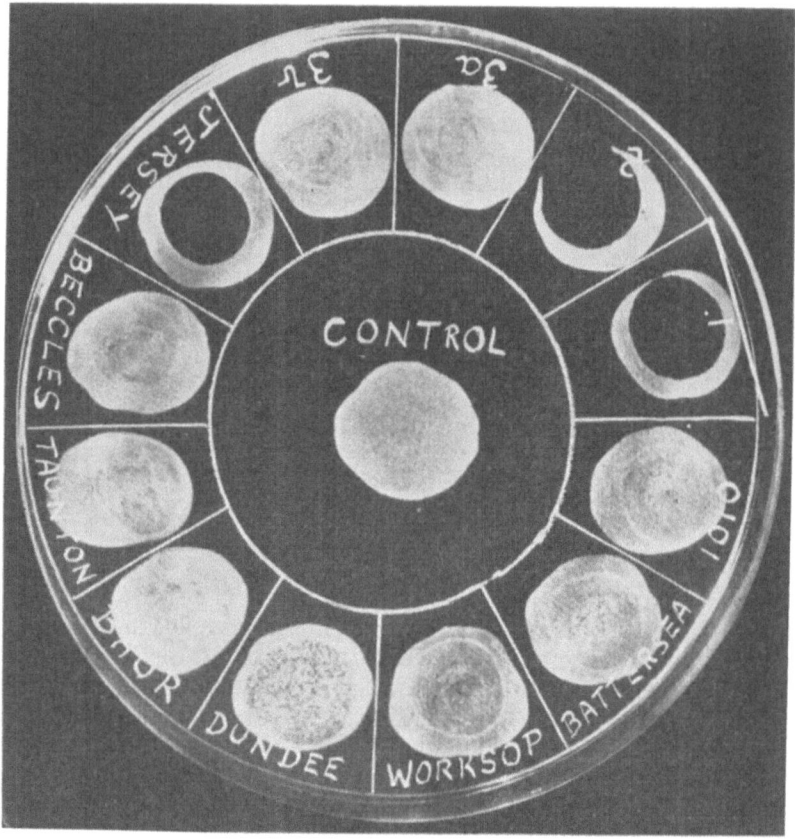

Fig. 1.

The phage typing of Salmonella paratyphi B

The typing scheme for this organism was developed by FELIX & CALLOW (1943, 1951). It originally distinguished four types, a number which soon increased to 10 (see Table II). The number of types, subtypes and "variations" now recognized in the E.R.L. is 48, and their reaction patterns are shown in Table III.

The limitation of the scheme lies in the fact that certain types preponderate in frequency over wide areas, and this restricts the

Table II.

Reactions of type strains of *Salmonella paratyphi B* with routine test dilutions of the typing phages

Type Strain	Typing Phages									
	1	2	3a	3b	Jersey	Beccles	Taunton	B.A.O.R.	Dundee	1010
1	CL	CL	++	—	CL	—	—	—	<SCL	—
2	—	CL		—	—	—	—	—	SCL	—
3a	—	—	<CL	OL	—	<CL	<CL	OL	<CL	CL
3aI	—	—	<CL	—	—	—	—	—	<CL	CL
3b	—	—	—	OL	—	<CL	<CL	OL	<CL	CL
Jersey	—	—	—	—	<CL	<CL	<CL	<CL	<CL	CL
Beccles	—	—	—	—	—	<CL	<CL	—	<CL	CL
Taunton	—	—	—	—	—	—	<CL	—	<CL	CL
B.A.O.R.	—	—	—	—	—	—	—	OL	—	SCL
Dundee	—	—	—	—	—	—	—	—	<CL	CL

Table III.

Reactions of types, subtypes and variations of *S. paratyphi B* with typing phages in routine test dilution

Type strains	Typing Phages											
	1	2	3a	3b	Jersey	Beccles	Taunton	B.A.O.R.	Dundee	Battersea	Worksop	1010
1	CL	CL	+++	++	CL	+++	+++	−	+++	−	−	−
1 var.1	CL	CL	++	++	CL	SCL	SCL	∓	SCL	−	−	CL
1 var.2	CL	CL	CL	CL	CL	CL	CL	OL	+++	SCL	++	CL
1 var.3	CL	CL	CL	CL	CL	++	−	OL	SCL	++	++	CL
1 var.4	CL	CL	CL	CL	CL	++	∓	OL	SCL	++	++	−
1 var.5	SCL	CL	SCL	−	SCL	−	−	±	−	OL	−	SCL
1 var.6	+++	+++	−	−	CL	−	−	−	−	OL	−	SCL
1 var.7	SCL	SCL	−	OL	−	+++	SCL	−	SCL	−	++	CL
1 var.8	++	+++	−	−	−	OL	−	−	−	−	−	−
1 var.9	OL	OL	∓	−	OL	−	CL	OL	+++	−	−	OL
1 var.10	++	++	−	OL	++	CL	CL	−	SCL	++	++	OL
1 var.11	CL	CL	CL	−	CL	CL	OL	OL	∓	+++	−	SCL
1 var.12	OL	CL	−	−	CL	−	−	OL	−	OL	SCL	OL
2	CL	CL	−	−	−	−	−	−	+++	−	−	−
2 var.1	CL	CL	−	−	−	SCL	CL	−	SCL	−	−	OL
3a	−	−	CL	CL	−	SCL	OL	OL	SCL	−	++	OL
3a var.2	−	−	CL	CL	−	−	−	−	+++	−	++	SCL
3a var.4	−	−	OL	+++	−	SCL	OL	SCL	SCL	−	±	OL
3a var.6	−	−	OL	OL	−	CL	OL	OL	+++	−	++	−
3a var.7	−	−	CL	CL	−	−	SCL	OL	<SCL	CL	CL	CL
3aI	−	−	OL	−	−	−	−	∓+	+++	−	−	++
3aI var.1	−	−	OL	−	−	OL	OL	++	SCL	−	−	OL
3aI var.4	−	−	OL	−	−	∓+	∓+	−	+++	−	−	OL

3b	OL	++	—	SCL	OL	SCL	—	OL	—	—
3b var.1	OL	—	—	—	OL	SCL	—	++	—	—
3b var.2	+++	+++	—	+	—	—	—	+++	—	—
3b var.3	SCL	+++	—	OL	OL	—	OL	OL	—	OL
3b var.6	CL	+++	—	+++	OL	SCL	—	OL	—	OL
3b var.7	CL	<CL	SCL	+++	—	—	—	SCL	—	SCL
Jersey	CL	—	—	+++	—	SCL	—	CL	CL	—
Jersey var.1	CL	—	—	+++	—	—	OL	OL	OL	—
Jersey var.2	—	—	—	+++	—	SCL	SCL	SCL	SCL	—
Beccles	OL	—	—	+++	—	OL	SCL	SCL	—	—
Beccles var.1	+++	—	—	+++	—	+	SCL	SCL	—	++
Beccles var.2	—	—	—	OL	—	CL	CL	—	—	∓
Beccles var.3	OL	—	—	—	+	SCL	SCL	—	—	—
Beccles var.4	OL	—	++	—	+	OL	SCL	—	—	—
Beccles var.5	OL	—	CL	—	—	OL	SCL	—	—	—
Beccles var.6	CL	—	—	<OL	—	SCL	SCL	∓	—	∓
Scarborough	OL	—	—	SCL	—	+++	+++	+	—	—
						SCL	SCL			
Taunton	OL	—	—	SCL	—	SCL	—	—	—	—
Taunton var.1	—	—	—	SCL	—	SCL	—	—	—	—
B.A.O.R.	OL	—	—	—	OL	—	—	—	—	—
Dundee	SCL	—	—	SCL	—	—	—	—	—	—
Dundee var.1	—	—	—	+++	—	—	—	∓	—	∓
Dundee var.2	CL	—	<SCL	SCL	—	—	—	—	—	—
			<OL							
Battersea	∓	—	OL	—	—	∓	∓	—	—	—
Worksop	+++	OL	∓	—	—	—	—	+	—	—

For interpretation of degrees of lysis see Table I.
Where a range of reactions may be found with a particular phage, the highest expected reaction is shown beneath the lowest: thus: +∓.

value of the method unless such types can be still further subdivided. For this reason attempts have been made (SCHOLTENS 1959) to subdivide type Taunton, which is commonest in continental Europe, and types Beccles and Dundee.

The practical results of phage typing of Salmonella paratyphi B

The value of this scheme has been recognized for many years and is indicated in numerous publications in the United Kingdom and other countries. It was given international recognition in 1947 and is now widely used throughout the world.

The phage-type distribution of S. *paratyphi B* infections in the United Kingdom during 1962 is shown in Table IV.

Table IV.

Frequency distribution of phage types of *Salmonella paratyphi B* in the United Kingdom during the year 1962

Phage types	No. of foci	%	Total number of cases	%
1	19	18.1	41	20.8
1 var.1	2	1.9	4	2.0
1 var.6	3	2.9	11	5.6
1 var.9	2	1.9	3	1.5
2	1	0.9	1	0.5
3a	5	4.8	8	4.1
3a var.4	2	1.9	3	1.5
3aI var.1	5	4.8	7	3.6
3b	3	2.9	4	2.0
3b var.2	2	1.9	3	1.5
3b var.7	1	0.9	1	0.5
B.A.O.R.	1	0.9	5	2.5
Battersea	5	4.8	17	8.6
Beccles	8	7.6	12	6.1
Beccles var.4	1	0.9	1	0.5
Beccles var.5	1	0.9	1	0.5
Dundee	3	2.9	5	2.5
(from abroad)	5	4.8	8	4.1
Dundee var.1	1	0.9	1	0.5
Jersey (from abroad)	1	0.9	1	0.5
Scarborough	5	4.8	8	4.1
Taunton	15	14.3	28	14.2
(from abroad)	6	5.7	12	6.1
Untypable	7	6.7	11	5.6
(from abroad)	1	0.9	1	0.5
Totals	105	99.9	197	99.9

The year 1962 happened to be one of unusually low incidence of this disease, 197 cases being recorded. The total number of cases during the preceding year, as judged by cultures typed in the E.R.L., was 423.

Type 1 is usually commonest, although this position is sometimes usurped by type Taunton. However, a number of infections due to the latter type are contracted abroad, whereas those due to type 1 are usually autochthonous.

The use of phage typing as an epidemiological guide in outbreaks of paratyphoid B fever is satisfactory in that it shows that the patients and symptomless excreters concerned are infected with the same type, which indicates that their infections originate from a single source. As far as the United Kingdom is concerned, however, the ultimate sources of such outbreaks long remained obscure. Epidemiological investigations revealed that many of the outbreaks of paratyphoid fever that occurred after World War II were connected with bakeries, and it was suggested by COCKBURN, JAMESON & FENTON (1951) and THOMSON (1953) that the paratyphoid B bacilli entered the bakeries in some commonly used product. It was not till 1955, however, that the material probably responsible for most of such outbreaks was identified. It proved to be egg imported from China (NEWELL 1955a, b; NEWELL, HOBBS & WALLACE, 1955; HOBBS & SMITH, 1955; SMITH & HOBBS, 1955). This egg is imported into England in large quantities, mostly as broken-out whole egg or albumen frozen in large cans, or in the form of dried albumen. Long term investigations of this material, which is widely used by bakeries, have shown that it is infected with a multiplicity of salmonella serotypes, and at least 13 phage types of S. *paratyphi* B have been isolated from it, (see Table V).

Table V.

Phage types of S. *paratyphi* B isolated from Chinese egg.

1	3aI var.1
1 var.5	3b
1 var.6	Taunton
2	Dundee
3a	Battersea
3a var.2	Untypable
3aI	

Most of these phage types are commonly found in human infections in the United Kingdom (Report 1958), and there is little doubt that a proportion, undetermined as yet, of these infections originate from contaminated Chinese egg. The difficulties in the way of final proof of this aetiology in any given incident stem from the fact that the incubation period of the disease is sufficiently long for the supply, not only of the food which acted as the vehicle of infection to be exhausted, but for the tins of egg from which it was prepared or contaminated to be used up by the time the victims fall ill. However, as the batches in which the egg is prepared are large, it is sometimes possible to isolate the offending phage type — and often others — of *S. paratyphi B* from other tins of the same batch. Such tins may be found at the port of importation or among the stocks held by other consumers.

The danger of imported Chinese egg has also been pointed out by workers in Germany (ROHDE & ADAM, 1956). The mode of contamination of the egg with *S. paratyphi B* during its manufacture has not yet been determined, but the organisms are clearly of human origin. It is evident that more control is needed, both of manufacture and importation of this product, for outbreaks of paratyphoid fever attributable to this source continue to occur. More effective guarantees are necessary of the safety of the egg at the time of its export from China. It goes without saying that thorough bacteriological examination of all batches should be carried out, using isolation techniques of maximum efficiency. As far as decontamination is concerned, pasteurization of the material either before it is exported from China or after it reaches its country of destination would probably render it safe, though it seems strange that a foodstuff known to present such a hazard should be imported at all*.

It is apparent from the foregoing account that a factor has been introduced that calls for rethinking in the epidemiology of paratyphoid fever. While we can accept that the ultimate source of the paratyphoid B bacilli is the human excreter, it emerges that, with the wide distribution of potentially infected substances such as Chinese egg, excreters may be separated from the patients they infect by thousands of miles. Provided that the organisms are transported in a medium such as egg at a low temperature, they will survive almost indefinitely, to cause outbreaks the true source of which may long remain obscure.

Another foodstuff which may be contaminated with enteric pathogens is desiccated coconut. WILSON & MACKENZIE (1955) reported an outbreak of typhoid fever due to infected Papuan coconut. KOVACS (1959) and GALBRAITH, HOBBS, SMITH & TOMLINSON (1960) found Ceylonese desiccated coconut to contain many salmonella serotypes including *S. paratyphi B*. The first incrimination of this material in human infection in the United Kingdom was described by ANDERSON (1960). An expectant mother with an appetite for raw desiccated coconut was suffering from severe vomiting

---

* Since this was written, regulations have been introduced requiring the pasteurization of liquid whole egg intended for human consumption in England and Wales.

and diarrhoea when she was admitted to the maternity home for her confinement. Ultimately, 12 persons, among them 5 infants and 6 hospital staff were infected, but the only seriously ill patient was the original case. Indeed, 7 of the patients were symptomless excreters. The phage type concerned was 1 var. 9, which was new at that time.

Between December 1959 and May 1960, no less than 8 phage types were found in desiccated coconut. The investigation was pursued, and up to the time of writing 21 types of S. *paratyphi B* have been identified in this material. Again, the actual mode of contamination has not yet been determined. It must be pointed out that the Ceylonese Government has now instituted control of coconut manufacture and we can probably be optimistic about the outcome. However, until this material can be regarded as safe, some form of sterilization or pasteurization treatment is advisable if it is to be imported.

All the phage types of S. *paratyphi B* found in coconut have also been isolated from man. Many of the types have not been isolated from any foodstuff other than coconut. The types concerned are largely D-tartrate positive, and slime-wall negative. On this basis they would be classified by some workers as belonging to S. *paratyphi B* var. *java*, and would therefore be regarded as potential causes of food poisoning rather than true paratyphoid fever. However, experience with some of these types has indicated that they may occupy a position intermediate between classical S. *paratyphi B* and the *java* variety, in that they have caused infections diagnosed in some instances as paratyphoid fever and in others as food poisoning.

The phage types isolated from Ceylonese coconut are listed in Table VI.

Table VI.

Phage types of S. *paratyphi B* isolated from desiccated coconut

1 var.9	Beccles var.5
1 var.10	Beccles var.6
3b	Dundee
3b var.2	Dundee var.1
3b var.3	Dundee var.2
Battersea	Jersey
Beccles	Jersey var.2
Beccles var.1	Scarborough
Beccles var.2	Taunton
Beccles var.3	Untypable
Beccles var.4	

With the exception of the Scandinavian countries, the incidence of paratyphoid fever in continental Europe is much higher than in the United Kingdom. A number of cases of the disease occurring in the United Kingdom are infected while abroad. Although the frequency distribution of phage types in England differs from that in Europe generally, type 1 usually being commonest in the former and type Taunton in the latter, there are no types that are peculiar to any country in the European group. It is therefore impossible to inculpate the country of origin by identification of the phage-type of the infecting strain in returning travellers.

The phage typing of Salmonella typhi-murium

S. typhi-murium is the commonest of all salmonellae in the United Kingdom. In 1959, for example it caused 3,241 out of 5,132 incidents (63%) of salmonella food poisoning in England and Wales (Report, 1960), the remaining 1,891 incidents being distributed between at least 76 other salmonella serotypes. Only a proportion of cases of salmonellosis are bacteriologically examined, and we believe that the total number of cases of S. typhi-murium infection occurring in man per annum in England and Wales may be in the neighbourhood of 50,000.

The problem of S. typhi-murium infection is thus one of a totally different order of magnitude from that of S. paratyphi B. Man is not the host responsible for the initiation of infection, though he may of course help to disseminate it once he is infected. The organism is the most ubiquitous of all salmonellae, and is responsible for much disease of animals used by man as food, such as chickens, calves and adult cattle, turkeys, pigs and ducks. There are therefore multiple indigenous reservoirs of the organism, all of which are important sources of human infection. In addition, there are imported foods from which S. typhi-murium has been isolated on a number of occasions. Examples of such foods are boneless beef from the Argentine, boneless veal from New Zealand (HOBBS & WILSON, 1959), and frozen egg from Australia, and there is good evidence that such materials may convey infection to man (see KALLINGS, LAURELL & ZETTERBERG, 1959). The picture is still further complicated by the fact that it has been shown that animal feeding stuffs and fertilizers may contain S. typhi-murium in addition to a multiplicity of other serotypes (see ROHDE & BISCHOFF, 1956; WALKER 1957; Report, 1959) and so constitute an added potential source of infection of livestock, and ultimately of man, with this organism.

With such a multiplicity of sources of human infection it would be expected that the tracing of the origins of outbreaks would be a task of great complexity. Infected food may precipitate the disease wherever it is distributed in a country. Outbreaks of independent origin may occur simultaneously in a single area, and there may be little to indicate the existence of this multiplicity or the extent of each outbreak.

The most useful tool so far evolved for unravelling this intricate network of S. *typhi-murium* infection is that of phage typing. We owe the systems in use in the United Kingdom at the present day largely to the patient and dedicated work of the late Miss BESSIE R. CALLOW. The first scheme evolved was that of FELIX & CALLOW (1943, 1951) (see also FELIX 1956). It employed 10 phages to identify 12 types of S. *typhi-murium*. The scheme is shown in slightly modified form in Table VII. Since it was first devised two types, 5 (PARKHOUSE unpublished) and 2e have been added, and a number of variations of the main lytic patterns have been recognized. This has enabled us to extend the scheme to that shown in Table VIII which distinguishes 33 types, subtypes and variations.

With a comprehension of the magnitude of the S. *typhi-murium* problem, Miss CALLOW set herself the task of devising a new scheme of greater sensitivity and precision. This resulted in that shown in Table IX (CALLOW 1959). It initially distinguished 34 phage types of S. *typhi-murium*, but this number has now greatly increased (ANDERSON, CALLOW & GUNNELL, to be published).

Table IX shows the new type designations together with the types to which they correspond on the earlier scheme. It can be seen that some of the old types are subdivided by the new scheme. For example, type 2 has been broken into 5 new types and type 2b into 4. Moreover, 9 types that were untypable with the older method can now be typed (types 21 to 25, type 27, and types 29 to 31). Many new types have been recognized that are not shown in Table IX. These have not yet been given final designations; some are shown in Table VIII as numbers prefaced by the letter U.

For some years we have used both methods in parallel and have accumulated much valuable information on the type distribution of S. *typhi-murium* in the United Kingdom. A widespread investigation of strains of the organism isolated from man, from livestock and from foodstuffs has shed light on most of the probable sources of human infection. An example of this was shown in an analysis of the frequency distribution of phage types of S. *typhi-murium* in man and livestock during the last quarter of 1958 (ANDERSON, 1962). Of the six most frequent types found in each group, five were common to both, which suggested that the livestock were probably the source of the human infections. Moreover, of 314 cultures of animal origin, 304 were isolated from chicks and calves. Of these, 73 cultures came from chicks and 231 from calves. The commonest type in chicks was 14*; the commonest in calves was 20a. These two types were also the most frequent in man during the same period. There was a long "tail" of types in man that had not been found in native

---

* Only new type designations will be given.

Table VII.

Reactions of type strains of *Salmonella typhi-murium* with routine test dilutions of the typing phages.

Type strains	Typing Phages											
	1	1a	1a var.1	1b	2	2a	2b	2c	2d	3	3a	4
1	OL	OL	OL	<CL	CL	CL	<CL	CL	OL	OL	OL	±
1a	—	OL	OL	<CL	CL	<CL	OL	<CL	OL	OL	OL	—
1a var.1	—	OL	<CL	<CL	—	±	OL	—	OL	OL	<CL	—
1b	—	—	<CL	<CL	CL	CL	OL	<CL	OL	<CL	<CL	—
2	—	—	—	—	CL	CL	—	CL	—	—	—	—
2a	—	—	—	—	—	CL	<CL	CL	<CL	—	—	—
2b	—	—	—	—	—	—	OL	—	OL	—	—	—
2c	—	—	—	—	<CL	<CL	<CL	<CL	<CL	—	—	—
2d	—	—	—	+++	±	±	±	±	<CL	—	—	—
3	—	—	±	—	—	—	—	—	—	OL	++	—
3a	—	—	±	—	—	—	—	—	SCL	+	<CL	—
4	—	—	—	—	—	±	—	±	±	—	—	<CL

livestock, and it was suggested that some of these were possibly of imported origin.

The similarity in type distribution between man and livestock, found in 1958, has been maintained to the present day. Furthermore, the types commonest in chicks and calves have remained common in man, which suggests that much human infection originates, directly or indirectly, from these two sources. It may be added that type 14 is still the most frequent in chicks and is also found in English frozen egg. It has caused a great deal of human infection. The type distribution in calves is somewhat broader than that in fowls, but all the types found in calves are common in man.

ANDERSON, GALBRAITH & TAYLOR (1961) published an account of an outbreak due to type 20a. This occurred as a widely scattered series of small incidents, 55 in all, between July and October, 1958. The total number of cases was only 90 and the outbreak was spread over the southeastern quadrant of England. During the course of routine phage typing it was discovered that type 20a was causing infection in calves at the same time. The calves concerned had been bought from a collecting centre near Oxford and it was found that most of the patients were supplied with meat by butchers who drew on abattoirs handling calves purchased from the collecting centre concerned. Type 20a was isolated from 12.7% of rectal swabs of apparently healthy calves at one of the abattoirs connected with the outbreak, and from 9.5% of calves' palates at a butcher's shop. An important result of this investigation was the finding that, although the initial rate of infection of calves with $S.$ $typhi$-$murium$ type 20a was only 0.5%, this rate rose to 12.7% in animals that had been herded together in the collecting centre for 4 or 5 days before being sent to the abattoir. The source of infection of the calves was not identified. ANDERSON et al. emphasized in this paper that the risk of human infection did not necessarily arise directly from the veal alone, but also from contamination of other foodstuffs, of utensils and of the environment in abattoirs, in butchers' shops and in kitchens. This is, of course, a common feature of the spread of all salmonella infection to man. It was also pointed out that so widely dispersed was this quite small outbreak that it would have been impossible to establish its unity without the help of phage typing, and that it was the routine phage typing of cultures isolated from calves, initially without reference to the outbreak, that identified the infective source and potentiated the entire animal investigation.

A comparison of the phage types of $S.$ $typhi$-$murium$ isolated from diseased calves, from abattoir swabs and from man was described by HARVEY & PHILLIPS (1961). Of 30 typable human strains of the organism, 23 (76.7%) belonged to phage types found in the abattoirs or in local farm animals. 25 out of 30 cultures were isolated from apparently sporadic cases of human infection, and the types concerned

Table

Reactions of *S. typhi-murium* types, subtypes and

Type strains	Typing phages						
	1	1a	1a var.1	1b	2	2a	2b
1	CL	CL	CL	CL	CL	CL	CL
1 var.1	CL	CL	—	—	—	—	SCL
1 var.2	CL	CL	CL	CL	—	—	CL
1 var.5	SCL	SCL	—	—	+∓	—	SCL
1 var.6	CL	CL	CL	CL	—	CL	CL
1 var.7	+++	+++	—	—	CL	CL	CL
1a	—	CL	CL	CL	CL	CL	CL
{ 1a var.1	—	CL	CL	CL	—	—	CL
{ 1a var.2	—	CL	CL	CL	—	—	CL
Derby	—	CL	—	—	—	—	—
1b	—	—	CL	CL	SCL	SCL	CL
1b var.1	—	—	CL	CL	SCL	+∓	—
1b var.2	—	—	+++	+++	—	—	CL
2	—	—	—	—	CL	CL	—
2 (4 + ve)	—	—	—	—	CL	CL	—
2 (5 + ve)	—	—	—	—	CL	CL	—
2 (poor)	—	—	—	—	+++	+++	—
2a	—	—	—	—	—	CL	CL
2a (poor)	—	—	—	—	—	+++	CL
2b	—	—	—	—	—	—	CL
2b (poor)	—	—	—	—	—	—	+++
2c (common)	—	—	—	—	+++	+++	+∓+
2c (rare)	—	—	—	—	CL	CL	CL
2c/2b	—	—	—	—	+++	+++	OL
2c (4 + ve)	—	—	—	—	SCL	SCL	SCL
2c (5 + ve)	—	—	—	—	SCL	+++	+++
2d	—	—	—	—	—	—	—
2e	—	—	—	—	CL	CL	—
3	—	—	± / SCL	SCL	—	—	—
3a	—	—	+∓+	—	—	—	CL
3a var.1	—	—	SCL	+∓+	—	—	—
4	—	—	—	—	—	—	—
5	—	—	—	—	—	—	—
	1	1a	1a var.1	1b	2	2a	2b

For interpretation of degrees of lysis see Table I.

Where a range of reactions may be found with a particular phage, the highest expected reaction is shown beneath the lowest: thus: +∓.

variations with typing phages in routine test dilution.

Typing phages						Provisional designations on new typing scheme of CALLOW (1959)
2c	2d	3	3a	4	5	
CL	CL	CL	+++	—	CL	1
—	—	—	—	—	CL	U35
—	CL	CL	CL	—	CL	U41
—	SCL	—	—	—	CL	U9
CL	CL	CL	CL	+++	SCL	U86
CL	CL	—	—	—	—	U56
CL	CL	CL	CL	—	CL	2
—	CL	CL	CL	—	+++	3 anaerogenic
—	CL	CL	CL	—	+++	3 aerogenic
—	—	—	—	—	—	24
SCL	CL	CL	SCL	—	—	4
+∓	—	CL	SCL	—	—	4a
—	CL	CL	+∓+	+∓+	—	U59
CL	—	—	—	—	±	12 (and others)
CL	—	—	—	+++	—	9
CL	—	—	—	—	CL	U105
+++	—	—	—	—	—	U101 (and others)
CL	CL	—	—	—	±	13
+++	CL	—	—	—	+++	U55
—	CL	—	—	—	—	18 (and others)
—	+++	—	—	—	+∓+	16 (and others)
+++	+∓+	—	—	—	—	14
CL	CL	—	—	—	—	15a (and others)
+++	OL	—	—	—	—	U79
SCL	SCL	—	—	+++	—	28
+++	+++	—	—	—	OL	U93
—	OL	—	—	—	—	20a (and others)
CL	CL	—	—	+∓+	++	U75
—	—	CL	SCL	—	—	5
—	SCL	—	OL	+∓+	—	6
—	—	—	CL	—	—	U108
—	—	—	—	SCL	—	8 (and others)
—	—	—	—	—	OL	26 (and others)
2c	2d	3	3a	4	5	

were again those found in abattoir swabs and the faeces of local live-stock. These authors pointed out that such comparative surveys may provide valuable clues to the source of so-called sporadic infections in man, as well as indicating the sources of current epidemics. There is evidently a strong case for the large-scale routine phage typing of cultures of S. typhi-murium isolated from livestock, from home-produced foodstuffs and from animal feeding stuffs and fertilizers. The tendency of the frequency distribution of phage types in these materials to be characteristic of the sources from which they are isolated enables comparison to be made with the distribution of types in man during the same period, and may help to identify the origins of the human infection. Some of the phage types encountered in man have never been isolated from livestock in the United King-dom. One such type — 24 — has caused a number of outbreaks of food poisoning in man in recent years. The only materials from which it has been isolated are imported beef and egg, and it can be reasona-bly assumed that these are the sources from which the human cases were infected.

It is frequently impossible to examine the suspected vehicle of infection in outbreaks of salmonella food poisoning, because the food concerned is either completely consumed or its remains discarded before the onset of symptoms in the patients. However successful the general epidemiology may be in indicating the probable infecting foodstuff, therefore, it is often argued that final proof is lacking because the suspected vehicle is no longer available for examination, and thus the infecting organism cannot be isolated from it. Further-more, the cause of infection of the food may not be directly through use of the original contaminated material. Environmental contami-nation of the type mentioned earlier may result in the organism reaching and multiplying in a food of a totally different sort. With the use of phage typing, investigations are much more precise than formerly, as the outbreak reported by ANDERSON et al. (1961) shows. The construction of such a network of circumstantial evidence as they described, when supported by phage typing, may indicate with reasonable certainty the source of an outbreak of S. typhi-murium infection, and may be the only way in which the epidemio-logical study of this disease can be effective. If this evidence is not accepted, it seems to the writer that much of the object of the epi-demiological investigation is defeated.

The phage typing of S. typhi-murium by the method of Lilleengen.

LILLEENGEN (1948) devised a typing scheme independently of FELIX & CALLOW. Using 12 phages he was able to subdivide S. typhi-murium into 24 types. This scheme is used mainly in Scandi-navia, and has been employed uccessfully in investigations of outbreaks of human infection (LUNDBECK, PLAZIKOWSKI & SILVER-STOLPE (1955); KALLINGS, LAURELL & ZETTERBERG (1959). KALLINGS & LAURELL (1957) have indicated the value of identifying the fermentation types of phage types of *S. typhi-murium*. Assuming that the fermentation reactions of a given strain are constant and that a diversity of fermentation types can be defined, this may enable strain identification to be carried beyond the point reached by phage typing. Their work referred to LILLEENGEN's phage types but the argument is equally applicable to any phage-typing scheme.

The phage typing of Salmonella paratyphi A

FELIX & BANKER (see FELIX, 1955) divided *S. paratyphi A* into two types, using a phage isolated from sewage. BANKER (1955) added two more types. Since then four further types have been added to the scheme; its development has been slow, however, and it is in only limited use. It is hoped that further work will be done in this field, because paratyphoid A fever is common in a number of parts of the world, and there is no reason to suppose that the *S. paratyphi A* phage-typing scheme cannot be developed to a level of precision similar to that attained with other salmonellae.

The phage typing of other salmonellae

Phage typing is primarily an epidemiological tool and, although the method could probably be applied to all salmonellae, there is clearly little point in evolving typing schemes for serotypes that are uncommon. Among the commoner serotypes for which schemes have been devised are *S. dublin* (LILLEENGEN, 1950; SMITH 1951a): *S. enteritidis* (LILLEENGEN, 1950): *S. gallinarum* and *S. pullorum* (LILLEENGEN, 1952) and *S. thompson* (SMITH, 1951b & c). These schemes are not in general use, however, and will not be described here. The reader is referred to the original articles for further information on them. Schemes have been evolved independently in the E.R.L., where they are in routine use, for *S. thompson, S. enteritidis* and *S. pullorum*. Details of these will be published later.

It is hoped that the foregoing account has shown how much more accuracy is brought to the epidemiological study of certain common salmonelloses by the phage-typing method. There are clearly two ways in which the method may be employed. One is in the investigation of particular outbreaks in which it is important to show that organisms isolated from the patients, vehicles of infection, carriers and so on, all belong to the same phage type. The second is the routine phage typing of as many cultures as possible of the serotype under scrutiny, isolated from whatever source, in the country con-

108

cerned. In the long term this is perhaps the more important application of phage typing, because it delineates the pattern of type distribution in the country and enables us to identify sources of human (and animal) infection that might not otherwise be suspected. For the prosecution of the latter type of programme, however, close collaboration is necessary between medical and veterinary clinicians and bacteriologists, public health and veterinary authorities, epidemiologists and, of course, phage-typing centres. It is only by such concerted effort that the problem can be effectively studied and its extent defined. Once this is done, we are a step nearer to its control.

REFERENCES

ANDERSON, E. S. (1960). The occurrence of Salmonella paratyphi B in desiccated coconut from Ceylon. *Mon. Bull. Minist. Hlth. Lab. Serv.* **19**, *172—175.*

ANDERSON, E. S. (1962). The use of bacteriophage typing in the investigation of outbreaks of salmonella food poisoning. In *Food Poisoning.* London Roy. Soc. Hlth.

ANDERSON, E. S., GALBRAITH, N. S. & TAYLOR, C. E. D. (1961). An outbreak of human infection due to Salmonella typhi-murium phage-type 20a associated with infection in calves. *Lancet,* **i**, *854—858.*

ANDERSON, E. S. & WILLIAMS, R. E. O. (1956). Bacteriophage typing of enteric pathogens and staphylococci and its use in epidemiology. *J. clin. Path.* **9**, *94—127.*

BANKER, D. D. (1955) Paratyphoid-A Phage Typing. *Nature, Lond.* **175**, *309—310.*

CALLOW, BESSIE R. (1959). A new phage-typing scheme for Salmonella typhi-murium. *J. Hyg., Camb.* **57**, *346—359.*

CRAIGIE, J. & FELIX, A. (1947) Typing of typhoid bacilli with Vi bacteriophage. *Lancet,* **i**, *823—827.*

CRAIGIE, J. & YEN, C. H. (1938a). The demonstration of types of B. typhosus by means of preparations of Type II Vi-phage. 1. *Canad. publ. Hlth. J.* **29**, *448—463.*

CRAIGIE, J. & YEN, C. H. (1938b). The demonstration of types of B. typhosus by means of preparations of Type II Vi-phage. 2. *Canad. publ. Hlth. J.* **29**, *484—496.*

FELIX, A. (1955). World survey of typhoid and paratyphoid B phage types. *Bull. Wld. Hlth. Org.* **13**, *109—170.*

FELIX, A. (1956). Phage typing of Salmonella typhi-murium, its place in epidemiological and epizootiological investigations. *J. gen. Microbiol.* **14**, *208—222.*

FELIX, A. & CALLOW, BESSIE R. (1943) Typing of paratyphoid B bacilli by means of Vi bacteriophage. *Brit. med. J.* **ii**, *127—130.*

FELIX, A. & CALLOW, BESSIE R. (1951) Paratyphoid B Vi-phage typing. *Lancet.* **ii**, *10—14.*

GALBRAITH, N. S., HOBBS, BETTY C., SMITH, MURIEL E., & TOMLINSON, A. J. H. (1960) Salmonellae in desiccated coconut. *Mon. Bull. Minist. Hlth. Lab. Serv.* **19**, *99—106.*

HARVEY, R. W. S. & PHILLIPS, W. P. (1961). An environmental survey of bakehouses and abattoirs for salmonellae. *J. Hyg., Camb.* **59**, *93—103.*

Table IX.

Reactions of type strains of *S. typhi-murium* with routine test dilutions of the new typing phages (CALLOW, 1959)

Type strains		Typing Phages																												
Old type	New type	1	2	3	4	5	6	7	8	9	10	11	12	13	14	15	16	17	18	19	20	21	22	23	24	25	26	27	28	29
1	1	CL	CL	CL	CL	CL	CL	CL				CL	CL	CL	CL	CL	CL	CL	CL	CL	CL	CL	<CL	+	CL	SCL	CL	CL	+	CL
1a	2		CL	CL	CL	CL	CL			+	CL	CL	CL	CL	+	CL	CL	CL		CL	CL	CL	<CL	CL	CL	CL	+	CL	+	CL
1a var.1	3		+	CL	CL	CL	CL			+	CL	CL	+	+	+	+	CL	SCL		SCL	CL	+	+CL	CL	CL	CL	+	CL	+	CL
1b	4		+		+	CL	+CL			+	CL	+	+		+	+	CL	+		+	CL	+	+CL	CL				+	+	
3	5						+					+					+	+					+	+		+			+	
3a	6																+									+			+	
2d	7																+									+			+	
4	8									+																	+		+	SCL
	9								CL	CL	CL	CL	CL	CL					CL											OL
2	10								CL	CL	CL	SCL	SCL	CL				SCL	CL	SCL	SCL		<CL	CL			SCL	<CL		
	11								CL	+CL	CL	OL	CL	CL				OL		CL	CL		<CL	CL		+CL	SCL	SCL		OL
12	12								CL				SCL	SCL				+OL		CL	CL		<CL	CL		+CL	CL	+CL		
	12a											OL	CL	CL	OL	OL	OL	OL		OL	OL		<CL			OL	SCL	<CL		
2a	13														OL	OL		SCL	<CL	OL	OL		+				SCL			SCL
	14																			+	CL						SCL			OL
2c	15														CL	+CL		CL	CL	OL	<CL					OL	SCL	CL		
	15a												+CL	CL		SCL		SCL		+	CL	+				OL	SCL	<CL		OL
16	16																		CL		CL			+		CL	SCL			
2b	17																				SCL				+	CL				
	18																								+	CL				
	19	+	+	+																+						+	+	+	++	+
2d	20																									OL	OL	OL	+	+
	20a	+	+	+	+							+														+	+	+	+	+
Untypable	21																					+	+			+	+	+	+	+
Untypable	22																				+					OL	OL			OL
Untypable	23																								CL		OL	OL		OL
Untypable	24																								CL	OL	OL			++
Untypable	25														SCL										CL	OL				+
5	26																									++	SCL			
Untypable	27																									+	OL	OL		OL
2c	28								CL	CL			SCL	SCL		<CL		<CL	<CL	<CL	<CL									++
Untypable	29								SCL			+CL								<CL	<CL					CL				+++
Untypable	30																				+									
Untypable	31																										+			+++
		1	2	3	4	5	6	7	8	9	10	11	12	13	14	15	16	17	18	19	20	21	22	23	24	25	26	27	28	29

HOBBS, BETTY C. & SMITH, MURIEL E. (1955). Outbreaks of paratypoid B fever associated with imported frozen egg. II Bacteriology. *J. appl. Bact.* **18**, *471—477*.

HOBBS, BETTY C. & WILSON, J. G. (1959) Contamination of wholesale meat supplies with salmonellae and heat-resistant Clostridium welchii. *Mon. Bull. Minist. Hlth Lab. Serv.* **18**, *198—206*.

KALLINGS, L. O. & LAURELL, ANNA-BRIT (1957). Relation between phage types and fermentation types of Salmonella typhi-murium. *Acta path. microbiol. scand.* **40**, *328—342*.

KALLINGS, L. O., LAURELL, ANNA-BRIT, & ZETTERBERG, B. (1959) An outbreak due to Salmonella typhi-murium in veal with special reference to phage and fermentation typing. *Acta. path. microbiol. scand.* **45**, *347—356*.

KOVACS, N. (1959) Salmonellae in desiccated coconut, egg pulp, fertilizer, meat meal and mesenteric glands. *Med. J. Aust.* **1**, *557—559*.

LILLEENGEN, K. (1948). Typing of Salmonella typhi-murium by means of bacteriophage. *Acta path. microbiol. scand.* Supplement 77.

LILLEENGEN, K. (1950). Typing of Salmonella dublin and Salmonella enteritidis by means of bacteriophage. *Acta path. microbiol. scand.* **27**, *625—640*.

LILLEENGEN, K. (1952). Typing of Salmonella gallinarum and Salmonella pullorum by means of bacteriophage. *Acta path. microbiol. scand.* **30**, *194—202*.

LUNDBECK, H., PLAZIKOWSKI, U. & SILVERSTOLPE, L. (1955). The Swedish Salmonella outbreak of 1953. *J. appl. Bact.* **18**, *535—548*.

NEWELL, K. W. (1955a) Outbreaks of paratyphoid B fever associated with frozen egg. 1. Epidemiology. *J. appl. Bact.* **18**, *462—470*.

NEWELL, K. W. (1955b) Paratypoid B fever possibly associated with Chinese frozen egg. *Mon. Bull. Minist. Hlth Lab. Serv.* **14**, *146—154*.

NEWELL, K. W., HOBBS, BETTY C. & WALLACE, E. J. G. (1955). Paratyphoid fever associated with Chinese frozen whole egg. *Brit. med. J.* **ii**, *1296—1298*.

NEWELL, K. W., McCLARIN, R., MURDOCK, C. R., MacDONALD, W. N. & HUTCHINSON, H. L. (1959). Salmonellosis in Northern Ireland, with special reference to pigs and salmonella-contaminated pig meal. *J. Hyg. Camb.* **57**, *92—105*.

Report (1958). The contamination of egg products with salmonellae, with particular reference to S. paratyphi B. *Mon. Bull. Minist. Hlth Lab. Serv.* **17**, *36—51*.

Report (1959) Salmonella organisms in animal feeding stuffs and fertilizers. *Mon. Bull. Minist. Hlth Lab. Serv.* **18**, *26—35*.

ROHDE, R., & ADAM, W. (1956) Über die Gefahren bakteriologisch nicht kontrollierter Ei-bzw. Eikonservenimporte. *Zbl. Bakt.* Abt. 1/Orig. **166**, *329-334*.

ROHDE, R. & BISCHOFF, J. (1956) Die epidemiologische Bedeutung salmonella infizierter Tierfuttermittel (insbesondere Knochenschrot und Fischmehl) als Quelle verschiedener Lebensmittelvergiftungen. *Zbl. Bakt.* Abt. 1. (Ref.), **159**, *145—164*.

SCHOLTENS, R. T. (1959) Comparison of the Felix & Callow system of phage typing of Salmonella paratyphi B with the natural system used in the Netherlands. *Ant. v. Leeuwenhoek.* **25**, *403—421*.

SMITH, H. W. (1951a) The typing of Salmonella dublin by means of bacteriophage. *J. gen. Microbiol.* **5**, *919—925*.

SMITH, H. W. (1951b) Some observations on lysogenic strains of Salmonella. *J. gen. Microbiol.* **5**, *458—471*.

SMITH, H. W. (1951c) The typing of Salmonella thomson by means of bacteriophage. *J. gen. Microbiol.* **5**, *472—479*.

SMITH, MURIEL E. & HOBBS, BETTY C. (1955) Salmonella in Chinese frozen egg. *Mon. Bull. Minist. Hlth Lab. Serv.* **14**, *154—160*.

THOMPSON, S. (1953) Paratyphoid fever & bakers' confectionery. *Mon. Bull. Minist. Hlth Lab. Serv.* **12**, *187—199*.

WILSON, M. M. & MACKENZIE, E. F. (1955). Typhoid fever and salmonellosis due to the consumption of infected desiccated coconut. *J. appl. Bact.* **18**, *510—521*.

GENETICS OF SALMONELLA

BY

TETSUO IINO*

Department of Microbial Genetics, National Institute of Genetics. Misima, Shizuoka-ken, Japan

AND

JOSHUA LEDERBERG**

Department of Genetics, School of Medicine, Stanford University, Palo Alto, California, U.S.A.

(with 2 figs.)

I. OUTLOOK

The genetics of *Salmonella* can be traced back to the discoveries and descriptions of several remarkable phenomena of antigenic variation in this genus, namely O-H variation, S-R variation, form variation and phase variation (reviewed by KAUFFMANN, 1954).

The discovery of phage-mediated transduction in 1952 (ZINDER & LEDERBERG, 1952) introduced the possibility of the detailed genetic analysis of these intriguing phenomena. Antigenicities and antigenic variations of *Salmonella* were re-examined with transductional technique, and detailed schemes on the genetic determination of antigenicity (Section IV & V) and on the cellular regulatory system of gene expression (Section VII) have been constructed. Further, an evolutionary pattern of *Salmonella* serotypes has been postulated, based on their genotypic constitutions (Section VI).

Transductional analysis was not confined only to the antigenic characters, but extended also to many other bacterial characters. In relation to the genetics of flagellar antigen, the genetic study of motility has been undertaken with paralyzed mutants (STOCKER *et al.*, 1953; ENOMOTO, 1962). Various nutritional mutants including auxotrophs for amino acids or nucleic acid bases, and mutants deficient in sugar fermentations were isolated by the application of the selection techniques established on *Escherichia coli* (LEDER-

---

* Whose studies have been supported by a research grant from the National Institute of Allergy and Infectious Diseases, (E-2872), U.S. Public Health Service.

** Whose studies have been supported in part by National Institutes of Health Training Grant 26-295 and Research Grant C-4496, and by the National Science Foundation.

BERG, 1950). Detailed fine structure analyses have been carried out with these mutants (Section III). The mutants well analyzed for their mutational sites have been used for the investigation on the mutagenesis of base analogues (RUDNER, 1961a, b; MARGOLIN & MUKAI, 1961). A mutator gene which influences the frequency of various auxotrophic mutations has been discovered in a strain of *S. typhi-murium* (MIYAKE, 1960).

The second epoch of *Salmonella* genetics came with the success in transfer of the fertility factor from *E. coli* to *Salmonella* and the hybrid formation between them (Section III). It has led to the establishment of the sexual recombination system in *Salmonella* and has been used for the construction of the chromosome map of *Salmonella*. It also opened a way to investigate phylogenetic relations of *Salmonella* with other genera in Enterobacteriae.

The studies on the episomal multi-drug resistant factor *R*, which confers on a bacterial cell simultaneous resistance to sulfanilamide, streptomycin, chloramphenicol, and tetracycline, are under active investigation, with *Salmonella* as well as other Enterobacteria. (Section II). Among the chromosomal genes which control drug resistance, a streptomycin-resistant mutant showing pleiotropism has been studied (WATANABE, 1960). This requires thiamine and nicotinic acid, as well as being streptomycin indifferent. Transductional analysis indicated that all of these changes are caused by one step recessive mutation of a single gene. The pleiotropism of drug-resistant mutants was observed also on a chloramphenicol resistant mutant (WEINER & SWANSON, 1960). The mutant differs from the sensitive parent in morphological and cultural characteristics and in somatic antigens.

Together with the problem of drug-resistance, the problems of pathogenicity and host specificity are important subjects that *Salmonella* genetics may contribute to epidemiology as well as basic bacteriology. It has been known that nutritional factors influence the virulence of the pathogen (GOWEN, 1951) and that certain auxotrophic mutations of *S. typhi-murium* accompany the loss of virulence (BARON, 1953). The abundant accumulations of auxotrophic mutants in *Salmonella* provide a rich source for systematic study in this direction. The well analyzed antigen types in this group also favor the genetic investigation of the host-parasitic relationships. The genetic approach has already been started (FURNESS & ROWLEY, 1956). The development of this field in future is expected.

It is not the purpose of this article to review the entire field of *Salmonella* genetics comprehensively. Rather, the subjects will be focused on a few topics, especially on the immunogenetic aspects, which are specific to *Salmonella* and have not been systematically reviewed elsewhere.

II. GENETIC TRANSFER SYSTEM

Transduction

Salmonella is the prototype of phage-mediated transduction (ZINDER & LEDERBERG, 1952). At first, the combination of the competent bacteria was restricted to those in group A, B, and D, because phage *PLT22*, a type of A1-group phage, can attack only those bacteria which have somatic antigen *12*. Later, transduction was demonstrated with other *Salmonella* phages having host ranges different from A1-group, and the range of the competent group is being extended (BARON *et al.*, 1953; SAKAI & ISEKI, 1954; EDWARDS *et al.*, 1955). Several techniques have been invented to use virulent phages for transduction: namely the use of lysogenic recipient (ZINDER, 1955), lowering of multiplicity of infection (STARLINGER, 1958) or pretreatment of the lysate with ultraviolet light (GOLD-SCHMIDT & LANDMAN, 1962). The last method is based on the information that the plaque-forming ability of a lysate is more sensitive to ultraviolet light than its transducing ability. These methods have been successfully extended to the heterogenetic transduction of nutritional markers between *Salmonella* and *E. coli* with the virulent phage "chi" which can infect certain motile strains of both genera (TSUJI & IINO, unpublished data).

The transductions in *Salmonella* so far studied are of the generalized type. Each gene of the donor bacteria has approximately an equal chance, about one to ten per 10^6 phage particles, to be incorporated into transducing phage and transferred to the recipient bacteria. The size of genetic fragment transduced by a phage particle at one time corresponds to a small fraction of the bacterial chromosome; chemically it is around 10^4 nucleotide pairs, less than one hundredth of the whole chromosome length. Therefore, only very closely linked genes are transduced simultaneously. The transduced fragments may have predeterminate broken ends (OZEKI, 1959).

The incorporation of a donor chromosome fragment in a recipient cell results in the formation of a hybrid cell which is heterozygous for a small fraction of the chromosome. Such cell is termed "heterogenote", and the donor chromosome fragment is called "exogenote", in contrast to "endogenote", which is a part of the intact recipient chromosome homologous to a given exogenote (MORSE et al., 1956). Recombination between exogenote and endogenote occurs in a fraction of heterogenotes, and consequently stable transductional recombinants are produced from time to time. The rest of the transductional clones remains as persistent heterogenotes. In generalized transduction as seen in *Salmonella*, the exogenotes of the transductional cells cannot multiply and the heterogenotic state is maintained only in a single cell in each cell generation of a transductional

clone until the exogenote is either lost or integrated into recipient
chromosomes. The genetic characters expressed by such heterogenotes are inherited linearly through cell generations. The linear
inheritance of motility appears as a trail of compact colonies on a
semi-solid medium (LEDERBERG, 1956; STOCKER, 1956a) and that
of nutritional markers as a minute colony on a minimal medium
(OZEKI, 1956). The phenomenon of linear inheritance in transductional heterogenote has been used as a main criterion of "dominance" and "gene complementation". For example, the production of trails on transduction from a flagellated strain to a non-
flagellated one indicated that the flagellation is dominant to non-
flagellation (STOCKER et al., 1953). The production of trails in combination between two non-flagellated mutants indicates that the
flagellation alleles of these two mutants are complementary to each
other and presumably belong to different genetic functional units.

There is another related phenomenon of the hereditary change
caused by infection with bacteriophage: lysogenic conversion. In
transduction a phage particle carries a genetic fragment of the host
bacteria. Contrasting with it, in *conversion*, phage genes function as
part of the bacterial organism. Such modification occurs soon after
phage infection and lasts until lysis occurs. When a bacterium is
lysogenized by a conversion phage, the modified character is inherited through cell generations to the offspring as long as the lysogenic state persists. The conversion of somatic antigen type in
Salmonella is one of the most intensively studied examples (Section
IV).

Conjugation

As well as phage-mediated genetic transfer system, sexual mating
system has been recently established in *Salmonella* for the genetic
recombination analysis. The sex factor F is transferred from *E. coli*
to various *Salmonella* strains (BARON et al., 1959a; ZINDER, 1960;
MIYAKE, 1962) and from the latter to other *Salmonella* (MÄKELA
et al., 1962). Both the ability to be infected by F and the degree
of fertility of male $(F+)$ *Salmonella* differ between different combinations of the strains. Generally homologous strains show the
highest efficiency. The conjugation was also demonstrated between
E. coli Hfr and certain strains of *Salmonella* (MIYAKE & DEMEREC,
1959; ZINDER, 1960; FALKOW et al., 1962). They yield hybrids
between *E. coli* and *Salmonella*.

There are three other systems of conjugation in *Salmonella:*
namely F-duction (or sex-duction), colicinoduction, and R-duction.
They are mediated by different episomal factors: sex factor F,
colicinogenic factor C, and multidrug resistant factor R respectively.
These systems are common with the phage-mediated transduction
to the point that an episome is a vector of the transmission of a piece

of genetic fragment, and also akin to the sexual conjugation in that both require cell contact for the transfer.

In "F-duction", a chromosome fragment is attached to F, replicates synchronously with F, and can be transferred to competent cells in high frequency, one per 10^2 to 10^4 cells in 1: 1 mixture of donor and recipient cells in homologous combinations. Such an autonomously replicating complex of F and a chromosome fragment, designated *F-prime (F')*, was first obtained from *E. coli* and introduced to *Salmonella* (BARON et al., 1959a; ØRSKOV et al., 1961). Later, an F' which is combined with lactose fermentation gene *(lac)* was demonstrated in a lactose positive *Salmonella* isolated from nature (BARON et al., 1959b). The F' introduced to *Salmonella* can be retransferred to the more distantly related bacteria such as *Shigella* (BARON *et al.*, 1959b), *Vibrio* (BARON & FALKOW, 1961), *Serratia* (FALKOW *et al.*, 1961) or *Klebsiella* (MÄKELA *et al.*, 1962) at lower frequencies, as well as to other *Salmonella* or *Escherichia*. The genes introduced to other genera by F-duction are usually not integrated into the chromosome of recipient cell; the cell multiplies as persistent heterogenote.

In "cin-duction", a bacterial chromosome fragment is transferred along with the transmission of certain colicinogenic factors (OZEKI & HOWARTH, 1961; SMITH & STOCKER, 1962). The sizes of chromosome fragments transferred by F-duction or cin-duction vary in different systems, but they are usually far longer than those in phage-mediated transduction.

Like F', "multi-drug resistant factor R," originally detected from *Shigella*, has a diverse range of transferability among *Salmonella*, *Escherichia*, *Shigella* and even *Serratia* (reviewed by WATANABE, 1963), R-factor interferes with the genetic transfer system by F; however, the joint transfer of a chromosomal fragment with it has not been demonstrated.

Owing to the establishment of the two major recombination systems, i.e. transduction and conjugation, it is now possible to study both micro- and macro-topology of the *Salmonella* chromosome. The former is applied for the fine structure analysis of a gene or genes closely linked to each other, but it is not adequate for the mapping of the various genes distributed through the entire chromosome. Sexual conjugation and cin-duction cover the gaps of transduction analysis, and they have started to be used for the mapping of *Salmonella* chromosome (Section III). Because of its wide range and high efficiency of transfer of a particular genetic fragment, F-duction promises to be a unique system for genetic and chemical investigations of the localized chromosomal regions. It will supplement the absence of specialized transduction in *Salmonella*.

Several consequences of possible DNA-transformation in *Sal-*

monella have been reported by some investigators, but they are still ambiguous.

The detail on these genetic transfer processes, as well as the nature of the mediatory factors, will not be discussed here, as they were covered in a recently published book by JACOB & WOOLMAN (1961) and in two review articles by LURIA (1962) and by CLARK et al. (1962).

III. CHROMOSOME MAP

Microtopology

In *Salmonella*, recombination analysis has been focused mainly on the fine structure analysis in a gene or a short segment of its chromosome by means of transduction. Mapping of the various genes on a chromosome had been left behind until recently, as the conjugation system was established on the organism.

The genetic fine structure analysis on *Salmonella* raised several interesting problems on the gene structure in relation to its function. For example, (1) clustering of the genes with phenotypically similar effect, (2) coincidence of the order of the genes in a gene cluster with their sequence in respect to biochemical blocks in the chain of reaction leading to the synthesis of a certain final product, (3) hetero-geneity among mutant sites in a genetic functional unit in respect to mutation rate, phenotypic expression or competence to suppressor. We will not discuss these problems here; they have been reviewed in detail by DEMEREC & HARTMAN (1959), and since then work with *Salmonella* has mainly elaborated on the analysis of individual genes. It may be worth noting, however, that the concept of "operon" as a unit of gene expression is being extended for the understanding of the phenomenon of gene clustering in *E. coli* and several other organisms, and it is interesting to examine to what extent this concept can be applied to the gene clusters found in *Salmonella* (Section VII).

Macrotopology

Mapping of the various genes on *Salmonella typhi-murium* strains LT2 and LT7 are summarized in Fig. 1 (DEMEREC & HARTMAN, 1959; FALKOW et al., 1962; MIYAKE, 1962; SMITH & STOCKER, 1962, personal comm.; MÄKELÄ, personal comm.; IINO, unpublished data). It is represented by a ring as revealed in *E. coli* K-12. The sequence and map distance of each marker on the chromosome is roughly the same as in *E. coli* (c.f. chromosome map of *E. coli* illustrated by FALKOW et al., 1962). Together with the successful hybridization between these two genera, the parallelism of their linkage maps may reflect their phylogenetic intimacy. On detailed structure, however, their chromosomes might differ in many respects, as already suggested by preferential transduction of *Salmonella* markers from *E. coli-Salmonella* hybrids by *Salmonella* phage (ZINDER, 1960).

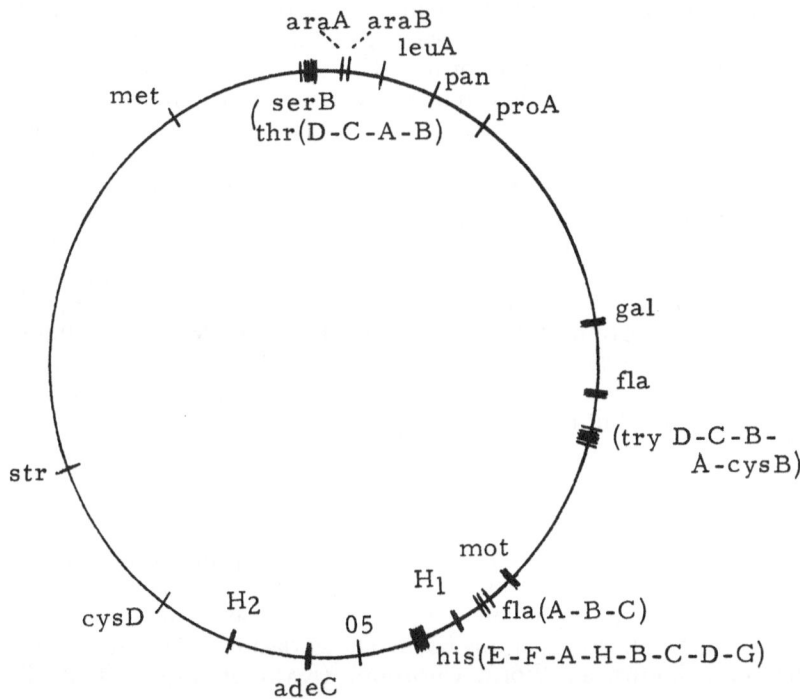

Fig. 1. Linkage map of *Salmonella typhi-murium*.
met: methionine, cys: cysteine, ade: adenine, pro: proline, pan: pantothenic acid, leu: leucine, ser: serine, thr: threonine, gal: galactose, ara: arabinose, str: streptomycin, O5: O-antigen 5, H_1: phase-1 flagellin, H_2: phase-2 flagellin, fla: flagellation, mot: motility.
A capital letter after a gene symbol represents a cistron.

Information obtained from molecular biology endorses these situations: the overall base composition of chromosome DNA is similar between *Escherichia* and *Salmonella*, but DNA molecules from each genus are imperfectly complementary as tested by their capacity for annealing with one another (FALKOW et al., 1962). Further comparative studies on the chromosome structures of various *Salmonella* species and the related bacteria from both genetic and molecular standpoints will throw light on the evolutionary history of *Salmonella*.

A distinguished difference between *Salmonella* and *Escherichia* is that the former cannot ferment lactose. In *E. coli*, a gene *lac*, which is an operon controlling lactose fermentation (Section VII; PARDEE et al., 1959), is present near the *pro* gene. These two loci are three units apart. By introduction of a chromosome fragment of *E. coli* of this region to *Salmonella* by F-duction, the recipient *Salmonella* cell acquires the ability to ferment lactose. In *E. coli* Hfr-*Salmonella* conjugation, stable hybrids carrying *lac* are obtained by transfer of *pro-lac* region from the former. The recombination between *pro* and *lac* genes is never detected among the hybrids, and the recombination value between *lac* and such markers as *asa* or *gla*, which are more distal than *pro* from *lac*, is also very low compared with homospecific conjugation in *E. coli* (MIYAKE, 1962). These results indicate that the incapability of lactose fermentation in *Salmonella* is caused by deletion of *lac* region on its chromosome. As well as the *lac*-locus, the *P*-locus, which controls the production and antigenic specificity of fimbriae, is present in *E. coli* K-12 strain but absent in a certain strain of *S. typhi* (BRINTON & BARON, 1960). Conversely, *Salmonella* has a duplication of H-antigen type determinants, only one of which is present in *Escherichia*. This will be discussed in detail later.

IV. GENETICS OF SOMATIC ANTIGENS

O-antigen is a heat stable antigen of microcapsule on the surface of bacterial cell wall. Its specificity is determined by sugar terminals of lipo-polysaccharide chains. Chemistry of the antigens is described by STAUB et al. (1964) in this book. The variation of O-antigen type has been known as "Form variation" (KAUFFMANN, 1940, 1954).

Phage conversion

The genetic determination of certain O-antigen types is an excellent example of conversion by bacteriophage. Since the conversion of *3, 10* type *Salmonella* (E_1-group) to *3, 15* type (E_2-group) by infection of phage ε was demonstrated (ISEKI & SAKAI, 1953), several antigen types have been shown to be converted by infection of certain bacteriophages (Table I). A clear-cut scheme on the correlations between antigen type, its chemical structure and infecting phage was constructed on the antigenic conversions among E-group *Salmonella* (Fig. 2, UCHIDA et al., 1963). E_1-group *(3, 10)* *Salmonella* converts to E_2 type *(3, 15)* by infection of phage ε^{15}. The new antigen type is detected in a few minutes (UETAKE et al., 1958). Naturally occurring E_2-group *Salmonella* are lysogenic for ε^{15} or the phages relating to it. It is possible to isolate E_1 *(3, 10)* type *Salmonella* from E_2 type through anti-15 serum selection. The E_1 type cells thus obtained had lost the ε^{15} type phage and be sensitive to it. The antigenic change caused by the infection of

Table I.

Antigenic conversion caused by phage infection in Salmonella.

conversion phage	host bacteria	changes in O-antigen	reference
ε, ε^{15}	E_1 group	$3,10 \rightarrow 3,15$	Iseki & Sakai (1953)
ε^{34}	E_2 group	$3,15 \rightarrow (3,15)34$	Uetake & Hagiwara (1960)
$\varepsilon^{15} + \varepsilon^{34}$	E_1 group	$3,10 \rightarrow (3,15)34$	Uetake & Hagiwara (1960)
ι, P22 Al-grop phage	4, 12 type group B Salmonella	$4,12 \rightarrow 1,4,12$	Iseki & Kashiwagi (1957)
phage 27	1,4,12 type group B Salmonella	$1,4,12 \rightarrow 1,4,12,27$	LeMinor et al. (1961)

ε^{15} to E_1 *Salmonella* is expressed biochemically as the replacement of the terminal sugar acetyl-galactose by galactose. The change of E_2 *(3, 15)* to E_3 *(3, 15, 34)* by infection of ε^{34} is the addition of a sugar, glucose, to the terminal. The addition of glucose occurs when the terminal sugar is galactose but not acetyl-galactose. Consequently, E_1 *(3, 10)* *Salmonella* does not produce antigen-34 after infection of ε^{34}, but E_1 cells lysogenic for ε^{34} produce *34* antigen after double lysogenization with ε^{15}. The production of antigen-1 by Al-group phage in A, B, D, group *Salmonella* is also an additive reaction of glucose to a terminal of the antigenic chain by 1:6 glucosidic linkage (Stocker et al., 1961). Thus, several O-antigen factors were found to be determined each by a specific bacteriophage genome. It may control directly or indirectly the production of an enzyme which catalyzes, modifies or inhibits the addition of a specific sugar to the terminal of polysaccharide chains. Phage mutants which differ in their conversion abilities have been obtained (Terada et al., 1956; Uetake & Uchida, 1959).

It is interesting to note that an episome, *F*, also confers to the carrier cell a new surface substance involved in male fertility, possibly a surface polysaccharide (Sneath & Lederberg, 1961). Immunologically, the f^+ antigen was demonstrated on F^+ and Hfr clones of *E. coli* (Ørskov & Ørskov, 1960) and *S. abony* (Mäkelä et al., 1962).

The determinant of O-antigen type in bacterial chromosome is not disclosed yet, except the approximate position of the antigen-*5* determinant in *S. typhi-murium* (Fig. 1). We also have no direct information on the mode of gene control of the backbone structure of O-antigen and type specific polysaccharides which are not determined by conversion phage.

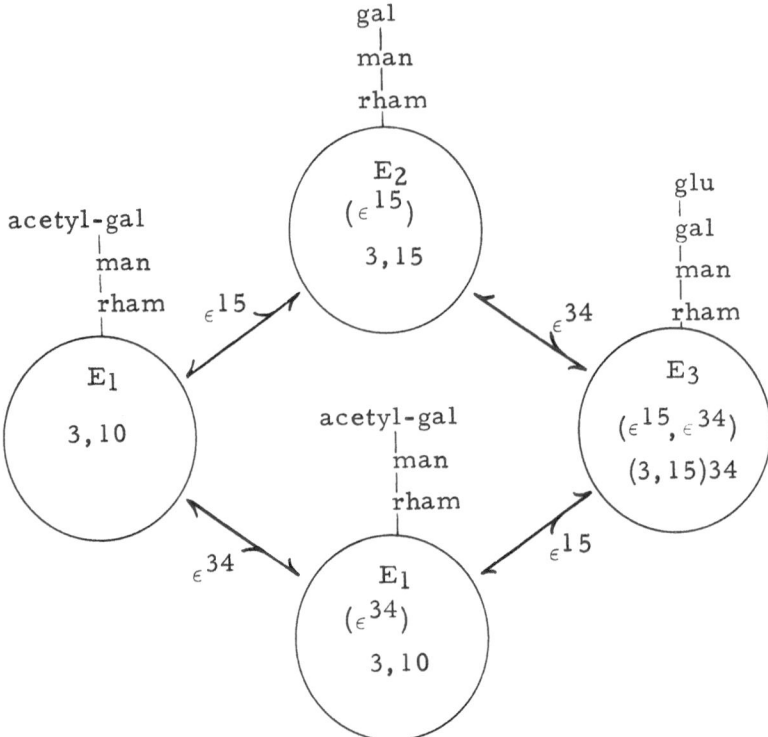

Fig. 2. Antigenic conversion by phage ϵ^{15} and ϵ^{34} in E-group Salmonella (constructed from UCHIDA et al., 1963). rham: rhamnose, man: mannose, gal: galactose, acetyl-gal: acetyl galactose.

S-R Variation

The variant cells which had lost O-antigens manifest another type of antigen called R-antigen. R refers to the roughness of the colonies in contrast with the smoothness (S) of those colonies which have O-antigens. Their type specificity has not been worked out as extensively as the O and H antigens. The changes from S to R type, called S-R variation, were first noted by ARKWRIGHT (1921). Since then the phenomenon has been reported by many investigators. S and R types are not uniquely distinct types, for there are many intermediate types. The frequency of S-R variation has been studied by several workers (ZELLE, 1942; PAGE et al., 1951). ZELLE found an extraordinarily high value in one unstable strain of *S. typhi-murium*, amounting to one R mutation per 14 cell divisions in an S culture.

A suggestive result concerning the genetic basis of S-R variation and the gene control of polysaccharide structure on cell surface was obtained with a galactose sensitive mutant of *S. typhi-murium* (FUKASAWA & NIKAIDO, 1961a). The mutant, called M-mutant, has a single defect in the uridine-diphospho (UDP)-galactose-epimerase gene and was not only unable to ferment galactose but also showed galactose-induced bacteriolysis. On ordinary media, devoid of galactose, they form rough colonies, and the sugar composition of their cell walls is only glucose; while the wild type is smooth and contains galactose, mannose, rhamnose and 3,6-didesoxyhexose in addition to glucose. The mutant also acquired the resistance to phage *PLT22*. These pleiotropic effects of the epimerase mutant are explained as follows. The polysaccharide structure at the surface of the wild type cell is composed of three layers. The inner core consists of polyglucose. Galactose links to the glucose and forms the second core. The sugars of the outer layer, which is responsible for O-antigenicity, connect to these galactose molecules of the second core. The accumulated chemical evidence (STAUB et al., 1964) has shown that nucleotide-phosphate-sugars, for example uridine-diphospho-(UDP)-galactose for galactose, are generally used as the donors of the sugar component in polysaccharide synthesis. The defect of UDP-galactose epimerase in the mutant blocks the conversion of UDP-glucose to UDP-galactose. Then, the reaction of the polysaccharide synthesis must stop at the step of the second core formation. The loss of outer layer results in the simultaneous loss of O-antigenicity, including antigen-12, which is the receptor of phage *PLT22*, and the change of colonial type occurs from smooth to rough. On the mechanism of bacteriolysis in galactose-containing media, several possibilities were discussed in relation to the enormous accumulation of UDP-galactose (FUKASAWA & NIKAIDO, 1961a). The most plausible one is that the compound interferes with the incorporation of other nucleotide-phosphate-sugar compounds into the cell surface structure. These results and the accumulated information on the chemical changes which are accompanied with the S-R variation of various bacteria (reviewed by BEALE & WILKERSON, 1961) together with the antigen conversion by phage, lead to the speculation that the biosynthesis of polysaccharide in the cell surface consists of stepwise reactions: a series of genes function by bringing about an orderly conversion of the precursor, by the addition of a new terminal sugar together with adjacent groups.

V. GENETICS OF H-ANTIGENS

Flagella and flagellin as H-antigen

Flagella are long filamentous appendages, having the length of several times that of the bacterial cell. They originate in the cytoplasm and project through the cell membrane. The number and distribution of flagella differ in different bacterial strains and cultural conditions. In *Salmonella*, like other flagellated enterobacteria, generally 5 to 10 flagella are observed around a cell. Their diameter, measured on electron-micrographs, is 10 to 15 mμ. As can be seen on dark-field microscope under very powerful illumination, the shape of each flagellum is helical, with a pitch and diameter characteristic of the strain. Flagella can be isolated easily by mechanical vibration of the cell suspension, and purified by high-speed centrifugation and chromatography (WEIBULL, 1950; KOBAYASHI et al., 1959; ENOMOTO & IINO, 1962). A purified flagellum is a fibrous polymer of a unit protein. X-ray diffraction analysis indicated that it belongs to the keratin-myosine-epidermin-fibrinogen type. In acid solution below pH 3.5, a flagellum dissociates into monomers called "flagellin". Molecular weight of a *Salmonella* flagellin is about 38,000. The particle size is 45 Å in diameter. From the electron microscopy, a three stranded complex helical structure has been proposed for the arrangement of flagellin in a flagellum (KERRIDGE et al., 1962). Amino acid composition of flagellin has been studied with several *Salmonella* serotypes (AMBLER & REES, 1959; AMBLER & STOCKER, personal comm.). They contain only 15 to 17 kinds of amino acids, 373 in number (Table II). Their N-terminal is lysine. In certain *Salmonella* flagellin, an unusual amino acid "N-methyl lysine" is detected (AMBLER & REES, 1959). The details and references on the studies of general structure and function of flagella are found in a review by STOCKER (1956b).

There is abundant evidence that flagella are responsible for the H-antigenicity among the cell surface structures. The non-flagellated mutants or natural isolations, called O-type, are without exception H-agglutination negative. The reversion of these O-type strain to flagellated type (H-type) always restores the H-agglutinability. The parallelism between flagellation and H-agglutination is observed when flagellated cells lose flagella by growing in a medium containing certain concentration of phenol or at the higher temperature. Although the specificity of H-antigen is usually examined by H-agglutination of the flagellated bacterial cells for diagnostic purposes, H-antigenicity is demonstrated with flagella which have been isolated from the cells or with purified flagellins. For the latter, precipitation test, such as gel-diffusion technique, has been successfully applied (KERRIDGE et al., 1962). Both isolated flagella and flagellins are also able to induce corresponding H-antibody by

Table II.

Amino Acid Content of Bacterial Flagellins.
moles amino acid/mole flagellin

amino acid	*S. typhi-murium* SW 1061, 1.2 type[1]	*Proteus vulgaris* P[2]
glycine.	22	14
alanine	42	18
valine	17	11
leucine.	21	14
isoleucine.	14	9
glutamic acid	27	16
aspartic acid	44	27
serine	17	11
threonine	29	14
tyrosine	6	2
phenylalanine.	4	4
tryptophan	0	0
proline.	4	0
cystine/2	0	1
methionine	1	1
lysine	10	8
ε-N-methyl-lysine	7	0
arginine	1	6
histidine	7	0
total.	373	156
molecular weight	38,000	17,300

1. AMBLER & REES, 1959 & personal comm.
2. KOBAYASHI et al., 1959.

immunization of rabbit. Their H-antigen type specificity is not qualitatively distinguishable from that of the whole cell. Though the possibility exists that the antigenicities of a flagellum and its component flagellin are quantitatively different (KOBAYASHI et al., 1959), these results strongly support the idea that a flagellin molecule is the unit of H-antigens of a *Salmonella* cell. In other words, H-antigen specificity is a mirror of the tertiary structure of a flagellin molecule.

H-genes as the structural determinants of flagellin

Before starting the discussion on the genetics of H-antigen determinants, it is convenient to introduce a unique phenomenon of antigenic variation. A culture of many *Salmonella* strains manifests two alternative types of H-antigen. The one is called the "specific phase", or "phase-1", and another is the "non-specific phase", or "phase-2". Each *Salmonella* serotype has its own specific type of phase-1 and phase-2 H-antigens manifested by flagellar protein.

When a mass culture of such strain is plated it dissociates into the colonies of phase-1 and phase-2. During successive cultures of the two types, the population of each type give rise to cells of an alternative phase. The phenomenon was first studied by ANDREWS (1922) and has been known as antigenic "phase variation". The frequency of phase variation differ in different strains ranging from 10^{-5} to 10^{-3} per bacterial division (STOCKER, 1949). The *Salmonella* strains expressing phase-1 and phase-2 alternatively are called diphasic ones.

H-antigen determinants were disclosed by transductional analysis (LEDERBERG & EDWARDS, 1953). The transduction between different diphasic serotypes, for example from *S. typhi-murium i : 1.2* to *S. abony b : enx* gave the recombinants *i : enx* and *b : 1.2* but none of the *b : i, enx : 1.2* types (an exceptional type will be described in Section VI). This means that the antigenic specificities of phase-1 and phase-2 are determined by independent loci in each, which are symbolized by H_1 and H_2. They are on the same bacterial chromosome but located far enough from each other not to be transduced simultaneously by a phage particle (Fig. 1). Further transductional experiments in reverse direction and between different serotypes proved the generality of this conclusion and led to the establishment of two series of multiple alleles of H_1 and H_2. The genotype of H-antigens of *S. typhi-murium* is described by $H_1{}^i H_2{}^{1.2}$, and that of *S. abony* is $H_1{}^b H_2{}^{enx}$.

It must be emphasized that the antigen of each phase is a complex of at least several subunits, for example *e*, *n*, and *x* in phase-2 of *S. abony*, though they are often listed by a simplified symbol for descriptive convenience, such as *b* in phase-1 of *S. abony*. Many subunits, if not all, can mutate independently from the other (IINO, 1959). In genetic recombination, however, they are transferred as a unit by one of the *H*-genes.

H_1 and H_2 have been defined first as phase-1 and phase-2 antigen type determinants. Later, certain flagellar shapes were found to be controlled by the *H*-genes (IINO, 1962a). The most common mutant type in flagellar shape of *Salmonella* is "curly". Curly flagellum is characterized by a tighter pitch, half that found in the normal type (LEIFSON & HUGH, 1953). In serum agglutination test, H-antigens of normal and its curly mutant flagella behave identically. The curly character is phase specific and its expression accompanies phase variation. For example, a curly mutant of *S. typhi-murium* is curly in phase-1 while normal in phase-2. In transduction from a normal flagella strain to such a curly phase-1 strain, transductional clones with normal flagella were isolated. The transductional clones thus produced showed the antigen of the donor in phase-1 and that of the recipient in phase-2. The parallel results were obtained on curly phase-2 mutants. The replacement of the phase-2 curly deter-

125

minant by normal one was always accompanied by the replacement
of the phase-2 antigen type by that of the donor. From these
results it was shown that genetic determinant of curly phase-1 is
at H_1 and curly phase-2 is at H_2.

The curly flagella are formed not only by curly mutants but also
by normal cells when flagella are synthesized in a medium containing para-fluorophenylalanine in place of phenylalanine. This fact
was demonstrated by flagellar regeneration experiment (KERRIDGE,
1959). Para-fluorophenylalanine is an amino acid analogue which
is incorporated into proteins, replacing phenylalanine (reviewed by
RICHMOND, 1962). When the proteins have certain biological roles,
their function may be altered or lost by the replacement. In the
synthesis of flagella, replacement of phenylalanine by the analogue
in flagellin molecules may cause alteration of their molecular configuration, and consequently the change in the mode of their polymerization, resulting in the production of flagella with changed
curvature. The biochemical process of flagellar formation in curly
mutants might be analogous with that described above: the replacement of phenylalanine in flagellin by certain other amino acid
might occur by curly mutation.

Just like the curly flagellar character, one type of resistance to a
motility-specific phage has been demonstrated to be phase-specific
(MEYNELL, 1961). Motility-specific phage attacks motile, not non-
motile, bacteria. Flagellins or part of them are presumably the receptor of this phage. Among *Salmonella* serotypes, those which have
g . . . antigens are excluded from this rule. Even if motile they are
resistant to a motility-phage, *chi*. When g . . . antigen is introduced
into diphasic strains by transduction of $H_1{}^g$ . . ., the resulting transductional clones are resistant to *chi*-phage in phase-1 while sensitive
in phase-2. The antigenic phase specificity of the resistance is also
observed on a motile *chi*-resistant mutant of *S. typhi-murium*. The
mutant is resistant in phase-2, *1.2*-type, while sensitive in phase-1,
i-type (SASAKI, 1961). The determinant of the phage resistance and
1.2-type determinant, H_2, are always co-transduced.

The primacies of H_1 and H_2 as shown in the determination of the
specificity of flagella strongly support the idea that H_1 and H_2 are
the primary structural determinants, structure genes, of flagellin
in phase-1 and phase-2 respectively. A mutation in one of these
genes produces an altered configuration of the corresponding flagellin, resulting in a change in antigen type, a modification of the
flagella shape or an alteration of the receptor site to certain bacteriophage.

As can be seen the excellent examples in the genetic and chemical
analysis of human hemoglobin (INGRAM, 1957), and certain bacterial
enzymes (YANOFSKY et al., 1961), it is the current view that the
structural gene carries the genetic code of amino acid sequence in

the form of base sequence in its DNA chain, and that the amino acid sequence in polypeptide primarily determines the tertiary configuration of the protein. An alternative mechanism was proposed on the genetic determination of immunological specificity of insulin (BERSON & YALOW, 1961), and enzymatic and immunological specificities of tyrosinase (FOX & BURNETT, 1962). That is, certain genetic determinants of the protein specificity may control the type of folding of polypeptide; thus they may provide the chance of producing the proteins of different specificity with the same amino acid sequence under different genetic background (KARUSH, 1960). However, there is no definite evidence so far for the second mechanism, and the folding of a protein may be completely predetermined by its amino acid sequence.

Though the amino acid sequence is not known yet, the comparative amino acid analysis and the analysis of tryptic peptides of *Salmonella* flagellin (AMBLER & STOCKER, personal comm.; ENOMOTO & IINO, unpublished data) strongly suggest that the *H*-gene determines the actual amino acid sequence in the polypeptide of flagellin. The relative amount of each component amino acid differed among the flagellins of different antigen type. Differences are noticed even between two antigen types which represent phase-1 and phase-2 of the same strain. The finger prints of tryptic digests of *i*-flagellin and *1.2* flagellin, which represent the phase-1 and phase-2 antigen of *S. typhi-murium*, respectively, differ in 15 among 35 spots (STOCKER & AMBLER, personal comm.). The difference of certain amino acid composition between flagellins of normal and curly mutant has been suggested by preliminary experiment but is not conclusive yet. For the clear understanding of the determination of flagellin structure by the *H*-genes, we need the detailed fine structure analysis of the *H*-genes, especially the localization of the antigenic subunit determinants in *H*, and in parallel the analysis of the amino acid sequence of flagellin. Some hopeful steps have been taken in this direction. The comparative studies on the subunits of $g \ldots$ groups antigens (IINO, 1959) and on the mutants of *i*-antigens (JOYS, 1961) have shown that in H_1 there are sequential mutational subunits which correspond to the antigenic subunits. The intra-H_2 recombination was obtained between $H_2{}^{1.2}$ and $H_2{}^{enx}$ (IINO, 1960, unpublished data).

Genetic modification of H-antigen

Besides the specificity in the relative amount of each amino acid species, *Salmonella* flagellin has a remarkable chemical feature in that some of them contain N-methyl-lysin (NML), an amino acid that has not been previously found to occur naturally (AMBLER & REES, 1959). The amount of NML is about equal with that of lysine. Among the serotypes tested, *S. typhi-murium (i:1.2.3)* were NML-positive in both antigenic phases. In certain other serotypes, for example *S. derby (gp: -)*, both NML-positive and -negative strains were found.

The analysis with transduction has shown that a gene which determines the presence or absence of NML is linked to H_1 (STOCKER et al., 1961). Unlike H_1 and H_2, this one gene determines the presence or absence of NML in both phase-1 and phase-2 flagellin. The immunological study of the recombinants between NML$^+$ and NML$^-$ strain disclosed an interesting correlation between H-antigen type and the amino acid: when *1.2* type NML-negative cells are changed to NML-positive by introduction of NML$^+$-gene, antigen *3* as well as *1* and *2* appear, and vice versa (STOCKER, personal comm.). The correlation between NML-positive and antigen-*3* positive is observed, only when the antigen type of the flagella is *1.2.(3)*. The H-antigen of *S. abony (b:e.n.x.)* is not altered by the presence or absence of NML.

In the foregoing chapter we came to the conclusion that H_1 and H_2 genes determine respectively the complete amino acid sequences of the phase-1 and phase-2 flagellins. Consequently, the NML-gene, which is phase non-specific and at a different locus from H_1 and H_2, is thought to modify the product of the H_1 gene, by engendering an enzyme which methylates some lysine radicals of already formed flagellin. The methylation of lysine, either in the form of free lysine or S-RNA bound lysine before its incorporation into flagellin is less likely, under our present understanding of protein synthesis.

Genetics of antigenic phases

In diphasic strains, two antigenic phases are determined at the separate loci. Therefore phase variation cannot be construed as the mutation of an antigen type determinant from one specific allele to another. Instead, it must be the *alternative manifestation* of each of the two antigenic specificities already inherent in the genotype. The transduction between single phase cultures of diphasic strains in all possible combinations showed that H_1 can be expressed only when transduced into phase-1 cells regardless of the phase of the donor, whereas H_2 can be expressed in any phases of the recipient but only when donor is in phase-2. This result, together with other supporting data (LEDERBERG & IINO, 1956), indicates that the H_2 gene plays a decisive role in the expression of the antigenic phases. The process of phase variation is thus explained as follows: "H_2 takes two different states, active and inactive. Active H_2 is epistatic to H_1 and inhibits the production of the phase-1 antigen, while it carries out the production of phase-2 antigen. When H_2 changes to the inactive state, corresponding to the change from phase-2 to phase-1, the production of phase-2 antigen stops and alternatively the production of phase-1, specified by H_1, proceeds."

Occasionally serotypic variants which express three or four anti-genic phases have been isolated from nature (EDWARDS et al., 1962). As far as has been analyzed genetically, such tri- or tetra-phasic strains are duplications of H_1 and/or H_2 (LEDERBERG, 1961). They might have been produced by either translocation or by heterozy-gosis. The persistent heterozygotes of H_1 and H_2 are obtained either by transduction (Spicer and Datta, 1959) or by sexual re-combination (HIROKAWA & IINO, 1961). In every case only one antigen type among three or four is expressed at a time.

As well as *Salmonella* having more than two antigenic phases, there are *Salmonella* serotypes and certain mutants of diphasic strains which express only one antigenic phase, either phase-1 or phase-2. They are called monophasic-1 or monophasic-2 type respectively. The serotypes carrying antigen g and the related antigens, such as *S. enteritidis gm:-*, are almost always monophasic-1 types, and H_2 could not be transferred from any diphasic strains to them, though the diphasic combinations of these factors are readily synthesized by transduction of the H_1 to any other diphasic strain. The genotype of monophasic-1 *(g)* type is represented by $H_1{}^g$ H_2-deficient, for no homologous locus with H_2 of other strains can be detected by genetic recombination. *S. typhi d: -*, and some monophasic mutants of *S. paratyphi B b: -*, appear to belong to the same category.

Another group of monophasic-1 types, which have been isolated as variants of some diphasic serotypes, especially of *S. typhi-murium* and *S. paratypi B*, is the H_2 inactive type. They can become di-phasic type by reversion or by transduction of H_2 from a diphasic donor strain. In some mutants, the phase-2 antigen type recovered by the transduction is consistently the same as that of the donor, indicating that the inactivation of H_2 is caused by mutation of a site not separable from the antigen type determinants; while in other mutants, diphasic transductional types with concealed phase-2 antigen of the recipient appear as well as the donor phase-2 type. Linkage analysis of such mutants disclosed a factor termed ah_2, which is adjoined to the phase-2 antigen type determinant, H_2, and controls the activity of H_2 (IINO, 1958a, 1962b): that is, the mutation of $ah_2{}^+$ to an allele $ah_2{}^-$ causes the inactivation of H_2 and consequently the change from diphasic type to monophasic-1 type.

In contrast to monophasic-1 types, H_1 deficient type has not been detected among monophasic-2 strains either from nature or as laboratory mutants. All the monophasic-2 mutants so far examin-ed were found to be the mutants in a region, termed ah_1, adjoining to H_1 (IINO, 1961a). The function of ah_1 parallels that of ah_2 in relation to H_2. The mutation of $ah_1{}^-$ causes the inactivation of H_1, but it does not concern the activity of H_2. In such mutants H_2

changes its activity as in diphasic strains. Consequently, in phase-1 both H_1 and H_2 are inactive and the production of flagella is entirely stopped, while in phase-2 normal flagella are produced: the mutants undergo oscillatory changes between flagellated phase-2 and non-flagellated phase-1.

The last group of monophasic types include those which actually carry both H_1 and H_2 genes but have a greatly reduced rate of variation. A representative of this group is S. *abortus-equi* (EDWARDS & BRUNER, 1939). S. *abortus-equi* isolated from nature is generally in phase-2, *enx*-type. The phase-2 culture of this serotype is very stable and the alternative phase, *a*-type, is occasionally isolated after anti-*enx* serum selection. The phase-1 clone thus obtained is also stable: reversion to phase-2 can be detected only with strenuous selection with anti-a serum. Though there is no detailed quantitative measurement of phase variation on these serotypes, it is estimated to be less than 10^7 per bacterial division: this value is less than one thousandth of the frequency of phase variation in ordinary diphasic strains. Thus S. *abortus-equi* carries both H_1 and H_2 and undergoes phase variation, but the frequency is so low that a culture behave as if it is either (usually) a monophasic-1 or (rarely) monophasic-2 type. The transductions between such stable phase strain and a diphasic strain demonstrated that a major factor which regulates the stability of antigenic phases is transduced simultaneously with H_2 at the frequency of 30% (IINO, 1961b). The factor is designated vh_2. An allele vh_2^- is in S. *abortus-equi* and stabilizes H_2 in its existing state, whether inactive or active, and produces monophasic-1 or monophasic-2 types respectively. The ah_1^- monophasic-2 and the vh_2^- monophasic-2 type are identified by plating the culture on semi-solid media; the former dissociates non-flagellated phase-1 cells while the latter is stable in flagellated phase-2. The distinction between ah_2^- monophasic-1 type and vh_2^- monophasic-1 type is possible only after the stability of the phase of their phase-2 derivatives is examined: the phase-2 culture originated from the former by reversion of ah_1^- to ah_1^+ is diphasic, whereas those from the latter is monophasic-2. The genetic change from vh_2^- to vh_2^+ or the opposite direction has not been reported since. A similar type of monophasic behavior as that of S. *abortus-equi* has been reported in S. *paratyphi* A (BRUNER & EDWARDS, 1941; EDWARDS et al., 1950). The monophasic property of this serotype may be caused by stabilization of H_2 as in S. *abortus-equi*. Whether the genetic factor which stabilizes the antigenic phase in S. *paratyphi* A is identical to that of S. *abortus-equi* or not awaits further genetic analysis.

An extreme type of mutation in flagellar antigenic phase is that from flagellated to non-flagellated (O) type. The change has been called O-H variation. The mapping of flagellation genes has been carried out in parallel with transduction (IINO, 1958b and unpublished data) and with cin-duction (SMITH & STOCKER, 1962 and unpublished data). The results are consistent with each other, and two chromosomal regions responsible for the flagellation both in phase-1 and phase-2 are assigned. The one is adjacent to H_1 and composed from at least three cistrons, fla-A, fla-B and fla-C. The other is located between try and gal loci, and the mutants in this region so far examined belong to one cistron. Fla^- mutation of any one of these genes causes the loss of flagella.

Among the *Salmonella* serotypes, *S. gallinarum* and *S. pullorum* in group-D do not produce flagella and are H-negative. The spontaneous mutation to H-type of these serotypes has not been observed. Though the detailed genetic analysis of these serotypes has not yet been carried out, the preliminary transduction experiment suggests that they carry $H_1^{g\,m}$ gene, but non-flagellated because of the deletion or multi-site mutation of a *Fla* region (LEDERBERG & EDWARDS, 1953).

The genotypes of several representative H-antigen types are summarized in Table III.

VI. EVOLUTIONARY ASPECTS OF SEROTYPES

As it can be presumed from the earlier discussion on the genetic determination of O-antigen, the differentiation of the O-antigen types may proceed by the loss, gain or modification of the ability to add a specific sugar to the polysaccharide skeleton of the microcapsule. This may be attained stepwise by either phage conversion or mutation of antigen type determinants. Actually, the differences of O-antigen types between E_1, E_2, E_3 and E_4 groups are explained by the presence or absence of symbiotic conversion phage(s) (Fig. 2). The differences of O-antigen type in group B are also explained by two conversion phages for antigens *1*, and *12*, and a mutation in a particular chromosomal gene for antigen *5*. As we have only scarce information on the chromosomal genes controlling the O-antigen production, it is not certain whether the differences in O-antigens between different groups are solely explained by accumulation of stepwise mutations or not.

The H-antigen situation is somewhat more complicated. If you look through the Kauffmann-White scheme (KAUFFMANN, 1964) you may notice that many H antigen types appear only in phase-1 while others in phase-2. However, there are several antigen types, like *e*, *l*, and *w*, which appear both in phase-1 and phase-2, indicating that the antigen type is not specific to antigen phase. Indeed, there are occasions that the same type of antigenic subunit appears in both phase-1 and phase-2 of the same strain as, in *S. salinatis deh*, $denz_{15}$. The same antigen *lw* is determined by H_2 in *S. wien b: lw*

Table III.

Genotypes of the H-antigen determinants in the representative Salmonella serotypes.

serotype	antigenic phase	antigen type	genotype						
			ah_1	H_1	ah_2	H_2	vh_2	ph_2	fla
S. typhi-murium	diphasic	i: 1.2	+	i	+	1.2	+	v	+
S. typhi-murium mutant SW1061	monophasic-1	—(i): 1.2	—	i	+	1.2	+	?	+
S. typhi-murium mutant SW1166	monophasic-2	i: (1.2)	+	i	—	1.2	+	?	+
S. abony	diphasic	b: enx	+	b	+	enx	+	v	+
S. paratyphi A	monophasic-1	a: (1.5)	+	a	+	1.5	—	—	+
S. abortus-equi	monophasic-2	(a): enx	+	a	+	enx	—	+	+
S. essen	monophasic-1	gm:	+	gm	0	0	?	0	+
S. gallinarum	nulliphasic	(gm):	+	gm	0	0	?	0	0

( ): unexpressed antigen type.   0: deletion or multisite mutant.   v: unstable state. ph_2: antigenic phase determinant.

while by H_1 in S. *dar-es-salaam lw: enz$_{18}$* (EDWARDS et al., 1955).
In some instances the antigenic phases classified by antigen type
do not agree with the phases defined by genetic analysis. The phase-1
and phase-2 of S. *worthington* have been registered as *lw* and *z*
respectively; in transduction, *lw* is found to be determined by H_2
and *z* by H_1 (EDWARDS et al., 1955). There is an isolation of S.
paratyphi B java, which expresses *b* and *1.2* alternatively but at a
lower frequency than the ordinary phase variation. Not only the *b*,
but also the *1.2* factor of this strain was found to be determined by
an allele of H_1 (LEDERBERG, 1961). The strain also carries a null-
allele of H_2, allowing the synthesis of such types as $(H_1{}^b) H_1{}^{1.2} H_2{}^{enx}$.

Regardless of the seeming confusion in the correlation between
antigen type and antigenic phases, there is actually a regular strict
separation of H_1 and H_2 homologies in all strains above. Assign-
ments of alleles to H_1 versus H_2 can be made unambiguously by
genetic analysis, and the further behavior of each strain is perfectly
consistent. They still precisely fit the rule of phase variation:
Frequent state change of H_2 and epistasis of H_2 to H_1.

An unique anomaly in the homology of H-genes was, however,
found on one occasion of transduction from S. *abony b: enx* to a
monophasic-1 type of S. *typhi-murium-: 1.2* (IINO, 1961c). The
recombinant expresses non-flagellated phase and *b*-phase, *b:-*, alter-
nately, instead of *b: 1.2* or *-: enx*. By further transduction or reversion, *i:b* type was obtained from the recombinant. H^b of the strain
behaves as an allelic locus of phase-2 antigen type determinant,
$H_2{}^b$ rather than the original $H_1{}^b$. For example, the transduction
from the recombinant to *a:1.5* type gave recombinants *a:b* and
i:1.5. Phase variation in these strains occurs as frequently as the
original recipient strain. Thus the anomalous recombinant is presumed to have originated by an exceptional recombination: H_2 locus is
replaced by H_1 of the donor in the transduction, indicating residual
structural homology between H_1 and H_2. The very rare chance of
the event and the exclusive affinity of the translocated $H_2{}^b$ to
other H_2 genes indicate that the barriers of the synapsis between
H_1 and H_2 are not so much their own structural differences but
may also be the differences of the other genes involved in a transduction fragment together with H_1 or H_2.

The genetic homology between H_1 and H_2 revealed from the
studies of unequal recombination leads us to the speculation on the
phylogenic relations of phase-1 and phase-2. We may infer that one
of the H genes originated by duplication of the other. Then, which
of H_1 and H_2 is the original locus? Like many other questions on
evolution, one may never get the conclusive evidence on it. However, there are several indications which favor the idea that H_2
arose by duplication of H_1. For instance, there are numbers of
monophasic-1 types deficient in H_2, but no H_1 deficient types have

been detected among monophasic-2 types. Clusters of the genes which control the sequence of a reaction process have been demonstrated for many nutritional characters. A similar gene cluster of flagellar formation is present at the H_1 region. On the contrary, the genetic factors known to be closely linked to H_2 are only the H_2 activity controler, ah_2 and its stability controller, vh_2. The interaction between H_2 and vh_2 is analogous to the variegated type position effect observed by translocation of certain genes to heterochromatic region (reviewed by LEWIS, 1950). Furthermore, in *E. coli* which is monophasic, only one H gene has been found and linked to *his*, to which H_1 of *Salmonella* is also linked (Fig. 1). Though allelism test of *E. coli H* and *Salmonella H* has not been led to a definite conclusion, it suggests that the H locus of *E. coli* is allelic to H_1 of *Salmonella*.

Based on these speculations, we may trace the evolutional pathway of antigenic phases as follows: The original type is primary monophasic-1 type having H_1 but not H_2. Duplication and translocation of H_1 happened in the monophasic-1 type. The synaptic homology between H_1 and duplicated H_1 was blocked by the disparity of the loci closely linked to each of H_1. The translocated H is now identified as a new locus, H_2, different from the original H_1. The H_2-gene was unstabilized by an influence of the adjoining chromosomal region. Then, the structural differentiation might have occurred between H_1 and H_2 and now the strain is recognized as a diphasic strain. Subsequently, a variety of further types has evolved through mutations and deletions in the diphasic system (of EDWARDS & BRUN, 1936), affecting the H_1 and H_2 genes themselves, their controllers ah_1 and ah_2, the variation factor vh_2 and the *fla* genes, respectively.

Next we will look into the combination of H-antigen and O-antigen of a serotype. The various H-antigen types are almost equally distributed to each group and it is difficult to find any correlations in the combination of O- and H-antigens. It might be a possibility that same types of mutations of H (or O) antigen have occurred in parallel in two independent strains having different O (or H) antigens. The more important mechanism might be, however, the formation of the new combination through genetic recombination. Several facts have been known to support this idea: Namely wide distribution of a specific genotype such as $H_1{}^g H_2{}^- \text{-x}$ in various O-groups, *in vivo* occurrence of phage mediated transduction and conversion on antigenic characters (VELAUDAPILLAI, 1960) and the isolation of *Salmonella-Escherichia* hybrid (BARON et al., 1959b).

VII. REGULATION OF GENE ACTION

The mechanism of regulation of gene expression is a highlight of modern genetics which aims to bridge between heredity and differentiation. The progress of biochemical and molecular genetics in the past decade succeeded in the establishment of the fundamental scheme of protein synthesis. The primary structure of a protein is determined by the nucleotide sequence of a chromosome region called structural gene. Messenger-RNA, which is complementary to the structural gene, is synthesized copying the later as the primer. The messenger-RNA (m-RNA) moves from nucleus to cytoplasm and fixes on a ribosome system. Transfer-RNA (t-RNA), each molecule of which carries an activated amino acid, comes to the ribosome and arranges itself along the m-RNA. A t-RNA has both amino acid specificity and pairing specificity to a coding unit of m-RNA. The activated amino acids, which arranged together with t-RNAs on the m-RNA, combine with polypeptide linkage side by side and leave from the ribosome system. The sequential polypeptide formation and stripping from the ribosome allow the synthesized polypeptide to fold in proper order and to form a protein with a specific tertiary structure. The mechanism of genetic regulation is being discussed in connection with these biochemical steps of gene expression (FRISCH, 1961).

A prevailing concept on the regulation of gene function is the regulator-operator theory formalized by JACOB & MONOD (1961). There is a unit segment of chromosome called operon. An operon is composed from an operator-gene and one or more functionally related structural genes linked to it. The operator-gene switches on and off the expression of the structural gene(s) directly, without mediation of cytoplasmic factor. Its function is presumably the initiation of m-RNA synthesis on the template of the following segment of DNA. The function of the operator-gene is managed by the product, termed repressor, of another gene, regulator-gene. The attachment of the repressor to the operator-gene prevents m-RNA initiation in that region and consequently inactivates the structural genes in an operon. The regulator gene may or may not be linked to the operon. In enzyme synthesis, the "inducer" is explained by its combination with the repressor, to inhibit the latter from attaching to the operator-gene. The repressor has been presumed to be non-protein substance; for example RNA. The possibility still remains, however, that certain kinds of protein might be a repressor (HORI-UCHI & NOVIK, 1961), e.g. histone, which combines with chromosomal DNA and suppresses the synthesis of chromosomal RNA (HUANG & BONNER, 1962).

In *Salmonella*, a cluster of genes controlling the related function has been found on various auxotrophic mutants and fermentation mutants (DEMEREC & HARTMAN, 1959). In two groups among them, results have been reported which seem to suggest that the cluster fit to the concept of operon. They are histidine mutants (AMES et al.,

1960) and galactose mutants (FUKASAWA & NIKAIDO, 1961b). In each of these groups, a type of mutants show pleiotropic effect to block the function of all structure genes in the cluster (LEDERBERG, E.M., 1961). The mutants are not complemented by any of the structural gene mutants. Their mutant sites map at one extremity of the cluster. They were explained as the mutants in the operator gene.

The genetic system controlling the flagellar formation provides an interesting model of gene interaction. In the flagellar forming system, the genes H_1 and H_2 are assigned as the structural genes. The expression of the H-genes are regulated by several other genes: namely ah_1, ah_2, vh_2 and $flas$, as well as the interactions between H_1 and H_2 (section V). The ah-genes are phase specific regulators. An allele ah^+ switches on the production of flagellin controlled by adjoining structural gene, ah_1 for H_1 and ah_2 for H_2 respectively. A mutant allele, ah_1^-, or ah_2^- switches off the reaction in the corresponding phase. Ah^+ is dominant to ah^-, but the function of ah appears only in cis-position with the adjacent H (IINO, 1962b). It means that H_1 (or H_2) and ah_1 (or ah_2) behave as two component parts of a genetic functional unit. In chemical terms, an ah-H complex may be a unit of transcription of genetic code carried by a chromosomal DNA. It must determine the production and structure of a m-RNA. The ah-region in the ah-H complex may be a terminal segment which codes a part of flagellin polypeptide not responsible for the antigenic specificity but important for the folding of flagellin polypeptide; or it may be a region where the synthesis of the complementary m-RNA starts; or it may correspond to a terminal of m-RNA where polypeptidation of the orderly arranged amino acid starts. Another possibility is that ah carries the code of a region of m-RNA which is not responsible for the amino acid sequence in flagellin but for the specific affinity of the m-RNA to ribosome. The comparative analysis of the genetic fine structure of ah-H region and the amino acid sequence in flagellin will give the final conclusion on these aspects in the future.

You may notice that the ah-H system is in many respects analogous to the operon. The ah_1 (or ah_2) may correspond to the operator of H_1 (or H_2), and ah_1^- and ah_2^- to the operator negative mutants (O°), which is inactive regardless the presence or absence of repressor. Then, is the phase variation explained as the repression and derepression of the ah_2-H_2 operon? If it is the case, we may assign ah_2^+ as the repressible operator (O^+); in phase-1, O^+ is repressed and H_2 is inactive, while in phase 2, O^+ is unrepressed and H_2 is active. The phase controller in diphasic strains is functionally indistinguishable from ah_2 except for the high frequency of state change in the former. Moreover, the recombination of the phase controller has not been demonstrated with neither H_2 nor ah_2.

Therefore, it is an acceptable hypothesis that both the phase controller and the ah_2^+ gene are identical. So far we have not detected any agent which works as exogenous inducer or repressor of phase variation. Consequently, it is premature to speculate on the nature of the endogenous factors. We can infer a phase-2 specific product which accumulates in the phase-2 cell and at a certain level of concentration represses the function of ah, perhaps the same agent that maintains the H_2 activity. The possibility that either flagella or flagellin in phase-2 is the repressor is excluded because of the fact that the state change of H_2 occurs even in the mutant which has lost the ability to synthesize flagellin (IINO, unpublished data). It is also unlikely that vh_2 is equivalent to the regulator gene which produces a repressor because in vh_2^- cell both phase-1 and phase-2 are stable.

"Formation and dissociation of the ah_2-repressor complex" is an attractive hypothesis to explain the state change in phase variation but we must keep in mind that it is not the sole possible explanation. It must be recalled that the H_2 locus had been originated by translocation of H_1 gene (Section VI). Thus the translocation may result in a certain structural anomaly at the junction of the chromosome and the translocated segment of the H_2 region. The vh_2 factor would then represent such a structural anomaly which causes the instability of the adjacent region where ah_2 is located; consequently ah_2 frequently changes its activity independently of any repressor substance. Both the constancy of the frequency and the randomness of the occurrence of the phase variation in different ages of a single phase colony under various cultural conditions (STOCKER, 1949, IINO, unpublished data) favors this hypothesis rather than the hypothesis of self-reinforced repressor production in phase-2, at least for the H_2 effect.

On the expression of the H-genes, the interaction between the two homologous genes, H_1 and H_2, superimposes on the ah-H system. In di-, tri or tetra-phase strains of various origin, we have noticed that only flagella of one phase are produced at a time by a cell. It is the general rule on all of these strains that H_2 is epistatic to H_1 regardless of the antigen type. How is the epistasis of a gene, namely H_2, to another homologous one, H_1, established? Several working hypotheses may be proposed for this question. It might be the interaction at the chromosomal level; for example, the ordered transcription of the genom in m-RNA synthesis favors the preferential production of phase-2 m-RNA, or either phase-2 m-RNA or (less likely) phase-2 flagellin acts as a cross-repressor of H_1, presumably by complexing at ah_1. Still other possibilities remain that the functional competition occurs at a localized site in cytoplasm, between phase-1 and phase-2 m-RNAs on ribosomes or between phase-1 and phase-2 flagellins at the polymerization step on a flagella-forming apparatus. For the present, we can not exclude the possibility of any one of them.

In contrast to the above regulation systems, the regulations by the *fla*-genes are non-specific with respect to antigenic phase. The *fla*-genes are in functional units distinct from both of H_1 and H_2 genes. The dominance of *fla⁺* to *fla⁻* excludes the possibility that *fla⁻* produces repressor substance of the *H*-genes by itself. We may assume three other mechanisms on the function of the *fla*-genes. (1) *Fla⁺* supports the polymerisation of flagellin monomers. (2) Flagellin molecules are synthesized on specialized "flagellosomes" and *fla⁺* controls the production of such specific ribosomes. (3) *Fla⁺* produces internal inducer of flagellin. The test on the production of immunologically cross-reacting-material (CRM) of flagellin with twenty-five *fla⁻* mutants of *S. typhi-murium* and *S. abortus-equi* showed that, with one exception, they do not produce flagellin CRM at all (IINO & ENOMOTO, 1952). Presumably, they belong to the category (2) or (3). The remaining one, obtained from a strain of *S. abortus-equi*, produces a flagellin antigenically indistinguishable from the parental flagellin (IINO & HARUNA, 1960); the mutant can produce flagellin molecules but cannot polymerize them into a flagellum, hypothesis (1). The mutant gene is in a region different from either any other *fla*-genes or the *H*-genes.

As for hypothesis (2), the presence of specialized flagellar-forming apparatus in the flagellated *Salmonella* cells has been predicted from the flagellar regeneration experiment (KERRIDGE, 1960) and the deflagellation experiment with phenol (IINO & MITANI, 1962). The CRM⁺ *fla⁻* mutant may be an excellent material for the studies on the genetic control of such a flagellar forming apparatus. Hypotheses of type (3) are always available when other evidence is not.

The central problem of development in higher organisms is the quasi-stable switching of gene activity — for example the transition from fetal to adult hemoglobin synthesis, the differentiation of isozymes, or the formation of specialized nerve-endings in the brain. Phase variation in *Salmonella* is a prototype of endogenous switching, in a simple organism amenable to genetic analysis. (Mainly lacking is a direct criterion of the chemical state of DNA from H_2-active and -inactive states.) It has already justified continued study by the specialist for its significance in the epidermology and evolution of *Salmonella*. Even more important it embodies the contemporary tradition of bacteriological research: Its unification with general biochemistry, genetics and cell biology.

138

REFERENCES

AMBLER, R. P. & REES, M. W., 1959. ε-N-Methyl-lysine in bacterial flagella protein. *Nature, Lond.* **184,** *56—57.*

AMES, B. N., GARRY, B. & HERZENBERG, L. A., 1960. The genetic control of the enzymes of histidine biosynthesis in *Salmonella typhimurium.* *J. gen. Microbiol.,* **22,** *369—378.*

ANDREWS, F. W., 1922. Studies in group-agglutination. I. The Salmonella group and its antigenic structure. *J. Path. Bact.* **25,** *515—521.*

BARON, L. S., 1953. Genetic transfer by means of Vi phage lysates. *Cold Spring Harbor Symp. quant. Biol.* **18,** *271—272.*

BARON, L. S., SPILMAN, W. M. & CAREY, W. F. 1959a. Hybridization of *Salmonella* species by mating with *E. coli. Science.* **130,** *566—567.*

BARON, L. S., CAREY, W. F. & SPILMAN, W. M. 1959b. Characteristics of a high frequency of recombination (Hfr) strain of *Salmonella typhosa* compatible with Salmonella, Shigella and Escherichia species. *Proc. Nat. Acad. Sci. U.S.* **45,** *1752—1757.*

BARON, L. S. & FALKOW, S., 1961. Genetic transfer of episomes from *Salmonella typhosa* to *Vibrio cholerae. Rec. Genet. Soc. Amer.* **30,** *59.*

BARON, L. S., FORMAL, S. B. & SPILMAN, W., 1953. Uses of Vi phage lysates in genetic transfer. *Proc. Soc. exp. Biol. Med.* **83,** *292—295.*

BEALE, G.H. & WILKINSON, J. F., 1961. Antigenic variation in unicellular organisms. *Ann. Rev. Microbiol.* **15,** *263—296.*

BERSON, S. A. & YALOW, R. S., 1961. Immunochemical distinction between insulins with identical amino-acid sequences. *Nature, Lond.* **19,** *1392—1393.*

BRINTON, C. C. & BARON, L. S., 1960. Transfer of piliation from *Escherichia coli* to *Salmonella typhosa* by genetic recombination. *Biochim. biophys. Acta* **42,** *298—311.*

BRUNER, D. W. & EDWARDS, P. R., 1941. The demonstration of non-specific components in *Salmonella paratyphi A* by induced variation. *J. Bact.* **42,** *467—478.*

CLARK, A. J. & ADELBERG, E. A., 1962. Bacterial Conjugation. *Ann. Rev. Microbiol.* **16,** *289—319.*

DEMEREC, M. & HARTMAN, P. E., 1959. Complex loci in microorganisms. *Ann. Rev. Microbiol.* **13,** *377—406.*

EDWARDS, P. R. & BRUNER, D. W., 1939. The demonstration of phase variation in *Salmonella abortus-equi. J. Bact.* **58,** *63—72.*

EDWARDS, P. R., BARNES, L. A. & BABCOCK, M. C., 1950. The natural occurrence of phase 2 of *Salmonella paratyphi A. J. Bact.* **59,** *135—136.*

EDWARDS, P. R., DAVIS, B. R. & CHERRY, W. B., 1955. Transfer of antigens by phage lysates with particular reference to the *l. w* antigens of *Salmonella. J. Bact.* **70,** *279—284.*

EDWARDS, P. R., SAKAZAKI, R. & KATO, I., 1962. Natural occurrence of four reversible flagellar phases in cultures of *Salmonella Mikawashima. J. Bact.* **84,** *99—103.*

ENOMOTO, M., 1962. Grouping of paralyzed mutants in *Salmonella. Ann. Rept. Nat. Inst. Genetics (Japan).* **13,** *75.*

ENOMOTO, M. & IINO, T., 1962. Purification, chromatography and electrophoresis of Salmonella flagellin. *Ann. Rept. Nat. Inst. Genetics. (Japan).* **12.**

FALKOW, S., MARMUR, J., CAREY, W. F., SPILMAN, W. M. & BARON, L. S., 1961. Episomic transfer between *Salmonella typhosa* and *Serratia marcescens. Genetics* **46,** *703—706.*

FALKOW, S., ROWND, R. & BARON, L. S., 1962. Genetic homology between *Escherichia coli*-K12 and *Salmonella. J. Bact.* **34,** *1300—1312.*

Fox, A. S. & Burnett, J. B., 1961. Tyrosinases of diverse thermostabilities and their interconversion in *Neurospora crassa*. *Biochem. biophys. Acta* **61**, *108—120*.

Frisch, L. (ed.), 1961. Cellular regulatory mechanisms. *Cold Spring Harbor Symp. quant. Biol.* **26**, pp. *408*.

Fukasawa, T. & Nikaido, H., 1961a. Galactose-sensitive mutants of *Salmonella* II. Bacteriolysis induced by galactose. *Biochim. biophys. Acta.* **48**, *470—483*.

Fukasawa, T. & Nikaido, H., 1961b. Galactose mutants of *Salmonella typhimurium*. *Genetics*. **46**, *1295—1303*.

Furness, G. & Rowley, D., 1956. Transduction of virulence within the species *Salm. typhimurium*. *J. gen. Microbiol.* **15**, *140—145*.

Goldshmidt, E. P. & Landman, O. E., 1962. Transduction by ultraviolet irradiated virulent bacteriophage. *J. Bact.* **83**, *690—691*.

Gowen, J. W., 1951. Genetics and disease resistance. Genetics in the 20th Century, (L. C. Dunn ed.). *401—429*.

Hirokawa, H. & Iino, T., 1961. H-antigen of heterozygous hybrids between *Salmonella abony* and *Salmonella typhimurium*. *Ann. Rept. Nat. Inst. Genetics (Japan)*, **12**, *81—82*.

Horiuchi, T. & Novick, A., 1961. A thermolabile repression system. *Cold Spring Harbor Symp. quant. Biol.* **26**, *247—248*.

Huang, R. C. & Bonner, J., 1962. Histone, a suppressor of chromosomal RNA synthesis. *Proc. Nat. Acad. Sci., U.S.* **48**, *1217—1222*.

Iino, T., 1958a. *Immunogenetics of Salmonella*. Thesis. University of Wisconsin.

Iino, T., 1958b. Cistron test of motility genes in *Salmonella*. *Ann. Rept. Nat. Inst. Genetics (Japan)*. **9**, *96*.

Iino, T., 1959. Subunits of H_1 gene in *Salmonella*. *Ann. Rept. Nat. Inst. Genetics (Japan)* **10**, *111—113*.

Iino, T., 1960. Transductions between curly flagellar mutants in *Salmonella*. *Ann. Rept. Nat. Inst. Genetics (Japan)* **11**, *73—74*.

Iino, T., 1961a. Genetic analysis of O-H variation in *Salmonella*. *Japan. J. Genet.* **36**, *268—275*.

Iino, T., 1961b. A stabilizer of antigenic phase in *Salmonella abortus-equi*. *Genetics* **46**, *1465—1469*.

Iino, T., 1961c. Anomalous homology of flagellar phases in *Salmonella*. *Genetics* **46**, *1471—1474*.

Iino, T., 1962a. Curly flagellar mutants in *Salmonella*. *J. gen. Microbiol.* **27**, *167—175*.

Iino, T., 1962b. Phase specific regulation of the flagellin genes (H_1 and H_2) in *Salmonella*. *Ann. Rept. Nat. Inst. Genetics (Japan)* **13**, *72—73*.

Iino, T. & Enomoto, M., 1962. Further studies on the non-flagellated mutants of *Salmonella*. *Ann. Rept. Nat. Inst. Genetics (Japan)* **13**, *73—74*.

Iino, T. & Haruna, I., 1960. A non-flagellated mutant which produces flagellar protein in *Salmonella*. *Ann. Rept. Nat. Inst. Genetics (Japan)* **11**, *74*.

Iino, T. & Mitani, M., 1962. Effect of phenol on the flagellation of Salmonella cell. *Ann. Rept. Nat. Inst. Genetics (Japan)* **13**, *74*.

Iseki, S. & Kashiwagi, K., 1957. Induction of somatic antigen 1 by bacteriophage in *Salmonella B* group. *Proc. Japan Acad.* **33**, *481—485*.

Iseki, S. & Sakai, T., 1953. Artificial transformation of O antigens in *Salmonella E* group. II. Antigen-transforming factor in bacilli of subgroup E_2. *Proc. Jap. Acad.* **29**, *127—131*.

Ingram, V. M., 1957. Gene mutaton in human hemoglobin: the chemical differences between normal and sickle-cell hemoglobin. *Nature, Lond.* **180**, *326—328*.

Jacob, F. & Monod, J., 1961. On the regulation of gene activity. *Cold Spring Harbor Symp. quant. Biol.* **26**, *193—211*.

140

JACOB, F. & WOLLMAN, E. L., 1961. Sexuality and the genetics of bacteria. Academic Press. New York & London. pp. 374.

JOYS, T. M., 1961. Mutation in flagellar antigen *i* of *Salmonella typhimurium*. Thesis, London University.

KARUSH, F., 1960. Role of disulfide pairing in the biosynthesis of antibody. *Science* **132**, *1494*.

KAUFFMANN, F., 1940. Zur Serologie des I-antigens in der Salmonella Gruppe. *Acta path. microbiol. scand.* **17**, *135—144*.

KAUFFMANN, F., 1954. *Enterobacteriaceae*. 2nd ed. E. Munksgroard, Copenhagen.

KAUFFMANN, F., 1964. Das Kauffmann-White Scheme, in *"The World Problem of Salmonellosis"*. Dr. W. Junk-Publishers, The Hague. p. *21—66*.

KARRIDGE, D., 1959. The effect of amino acids analogues on the synthesis of bacterial flagella. *Biochim. biophys. Acta* **31**, *579—581*.

KERRIDGE, D., 1960. The effect of inhibitors on the formation of flagella by *Salmonella typhimurium*. *J. gen. Microbiol.* **33**, *519—538*.

KERRIDGE, D., HORNE, R. W. & GLAUERT, A. M., 1962. Structural components of flagella from *Salmonella typhimurium*. *J. Mol. Biol.* **4**, *227—238*.

KOBAYASHI, T., RINKER, J. N. & KOFFLER, H., 1959. Purification and chemical properties of flagellin. *Arch. Biochem. Biophys.* **84**, *342—361*.

LEDERBERG, J., 1950. Isolation and characterization of biochemical mutants of bacteria. *Methods in med. Res.* **3**, *5—22*.

LEDERBERG, J., 1956. Linear inheritance in transductional clones. *Genetics* **41**, *845—871*.

LEDERBERG, J., 1961. A duplication of the H_1 (flagellar antigen) locus in *Salmonella*. *Genetics* **46**, *1475—1481*.

LEDERBERG, J. & EDWARDS, P. R., 1953. Serotypic recombination in *Salmonella*. *J. Immunol.* **71**, *232—240*.

LEDERBERG, J. & IINO, T., 1956. Phase variation in *Salmonella*. *Genetics* **41**, *744—757*.

LEIFSON, E. & HUGH, R., 1953. Variation in shape and arrangement of bacterial flagella. *J. Bact.* **65**, *263—271*.

LE MINOR, L., LE MINOR, S. & NICOLLE, P., 1961. Conversion de cultures de *Salmonella schwarzengrund* et *Salmonella bredeney*, dépourvues de l'antigène 27, en cultures 27 positives par la lysogénisation. *Ann. Inst. Pasteur* **101**, *571—589*.

LEWIS, E. B., 1950. The phenomenon of position effect. *Advances in Genetics* **3**, *73—115*.

LURIA, S. E., 1962. Bacteriophage genes and bacterial functions. *Science* **136**, *685—692*.

MARGOLIN, P. & MUKAI, F. H., 1961. The pattern of mutagen-induced back mutations in *Salmonella typhimurium*. *Z. Vererbungsl.*, **92**, *330—335*.

MÄKELÄ, P. H., LEDERBERG, J. & LEDERBERG, E. M., 1962. Patterns of sexual recombination in enteric bacteria. *Genetics* **47**, *1427—1439*.

MARMUR, J., FALKOW, S. & MANDEL, M., 1963. New approaches to bacterial taxonomy. *Ann. Rev. Microbiol.* **17**, *329—372*.

MEYNELL, E. W., 1961. A phage, øx, which attacks motile bacteria. *J. gen. Microbiol.* **25**, *253—290*.

MIYAKE, T., 1960. Mutator factor in *Salmonella typhimurium*. *Genetics* **45**, *11—14*.

MIYAKE, T., 1962. Exchange of genetic material between *Salmonella typhimurium* and *Escherichia coli* K-12. *Genetics* **47**, *1043—1052*.

MIYAKE, T. & DEMEREC, M., 1959. Salmonella-Escherichia hybrids. *Nature, Lond.* **183**, *1586*.

MORSE, M. L., LEDERBERG, E. M. & LEDERBERG, J., 1956. Transductional heterogenotes in *Escherichia coli*. *Genetics* **41**, *758—779*.

ØRSKOV, S. & ØRSKOV, E., 1960. An antigen termed f+ occurring in F+ *E. coli* strains. *Acta path. microbiol. scand.* **48**, *37—46.*

ØRSKOV, F., ØRSKOV, I. & KAUFFMANN, F., 1961. The fertility of Salmonella strains determined in mating experiments with Escherichia strains. *Acta path. microbiol. scand.* **51**, *291—296.*

OZEKI, H., 1956. Abortive transduction in purine-requiring mutants of *Salmonella typhimurium*. Genetic Studies with Bacteria. *Carnegie Inst. Wash. Publ.* **612**, *97—106.*

OZEKI, H., 1959. Chromosome fragments participating in transduction in *Salmonella typhimurium*. *Genetics* **44**, *454—470.*

OZEKI, H. & HOWARTH, S., 1961. Colicine factors as fertility factors in bacteria. *Nature, Lond.*, **190**, *986—989.*

PAGE, L. A., GROODLOW, R. J. & BRAWN, K., 1951. The effects of threonine of population changes and virulence of *Salmonella typhimurium*. *J. Bact.* **62**, *639—647.*

PARDEE, A. B., JACOB, F. & MONOD, J., 1959. The genetic control and cytoplasmic expression of "inducibility" in the synthesis of β-galactosidase by *E. coli*. *J. Mol. Biol.* **1**, *165—178.*

RICHMOND, M. H., 1962. The effect of amino acid analogues on growth and protein synthesis in microorganisms. *Bact. Rev.* **26**, *398—420.*

RUDNER, R., 1961a. Mutation as an error in base pairing. I. The mutagenicity of base analogues and their incorporation into the DNA of *Salmonella typhimurium*. *Z. Vererbungsl.* **92**, *336—360.*

RUDNER, R., 1961b. Mutation as an error in base pairing. II. Kinetics of 5-bromodeoxyuridine and 2-aminopurine-induced mutagenesis. *Z. Vererbungsl.* **92**, *361—379.*

SAKAI, T. & ISEKI, S., 1954. Transduction of flagella antigen in *Salmonella* E. Group. *Geumma J. med. Sci.* **3**, *195—199.*

SASAKI, I., 1961. Chi-phage resistance of the Salmonella serotypes having g-antigen. *Ann. Rept. Nat. Inst. Genetics. (Japan)* **12**, *82—83.*

SMITH, S. M. & STOCKER, B. A. D., 1962. Colicinogeny and recombination. *Brit. med. Bull.* **18**, *46—51.*

SNEATH, P. H. A. & LEDERBERG, J., 1961. Inhibition by periolate mating in *Escherichia coli* K. 12. *Proc. Nat. Acad. Sci., U.S.* **47**, 86.

SPICER, C. C. & DATTA, N., 1959. Reversion of transduced antigenic characters in *Salmonella typhimurium*. *J. gen. Microbiol.* **20**, *136—143.*

STARLINGER, P., 1958. Über einen Defect des transduzierenden Salmonella-Phagen. *Z. Naturforsch.* **136**, *489—493.*

STAUB, A. M. & RAYNAUD, M., 1964. Connaissances actuelles sur la nature chimique des antigènes présents dans les *Salmonella*, in "The World Problem of Salmonellosis". Dr. W. Junk Publishers, The Hague, p. *143—170.*

STOCKER, B. A. D., 1949. Measurement of rate of mutation of flageller antigenic phase in *Salmonella typhimurium*. *J. Hyg.* **47**, *308—413.*

STOCKER, B. A. D., 1956a. Abortive transduction of motility in *Salmonella*, a non-replicated gene transmitted through many generations to single descendant. *J. gen. Microbiol.* **15**, *575—598.*

STOCKER, B. A. D., 1956b. Bacterial flagella: morphology, constitution and inheritance. *Symp. Soc. gen. Microbiol.* **6**, *19—40.*

STOCKER, B. A. D., STAUB, A. M., TINELLI, R. & KOPACKA, B., 1960. Étude immunochimique sur les *Salmonella*. VI. Étude de l'antigène 1 présent sur deux *Salmonella* B et E_4. *Ann. Inst. Pasteur* **98**, *505—523.*

STOCKER, A. D., ZINDER, N. D. & LEDERBERG, J., 1953. Transduction of flagellar characters in *Salmonella*. *J. gen. Microbiol.* **9**, *410—433.*

TERADA, M., TOMII, T. & KUROSAKA, K., 1956. A doubly lysogenic strain of *S. typhimurium*. *Virus* **6**, *274—281.*

142

UETAKE, H., LURIA, S. E. & BURROUS, J. W., 1958. Conversion of somatic antigens in Salmonella by phage infection leading to lysis or lysogeny. *Virology* **5**, *68—91.*

UCHIDA, T., ROBBINS, P. W. & LURIA, S. E., 1963. *Biochemistry* (in press).

UETAKE, H. & HAGIWARA, S., 1960. Somatic antigen 15 as a precursor of antigen 34 in Salmonella. *Nature, Lond.*, **186**, *261—262.*

UETAKE, H. & UCHIDA, T., 1959. Mutants of Salmonella phage 15 with abnormal conversion properties. *Virology* **9**, *495—505.*

VELAUDAPILLAI, T., 1960. Transduction *in vivo. Z. f. Hyg.* **146**, *470—480.*

WATANABE, T., 1960. Transductional studies of thiamine and nicotinic acid requiring streptomycin resistant mutants of *Salmonella typhimurium. J. gen. Microbiol.* **22**, *102—112.*

WATANABE, T., 1963, Infective heredity of multiple drug resistance in bacteria. *Bact. Rev.* **27**, *87—115.*

WEIBULL, C., 1950. Investigations on bacterial flagella. *Acta chem. scand.* **4**, *268—276.*

WEINER, L. M. & SWANSON, R. E., 1960. Chloramphenicol-resistant strains of *Salmonella typhosa. J. Bact.* **79**, *863—868.*

YANOFSKY, C., HELINSKI, D. R. & MALING, B. D., 1961. The effects of mutation on the composition and properties of the A protein of *Escherichia coli* tryptophan synthetase. *Cold Spring Harbor Symp. quant. Biol.* **26**, *11—24.*

ZELLE, M., 1942. Genetic constitution of host and pathogen in mouse typhoid. *J. infect. Dis.* **71**, *131—152.*

ZINDER, N. D., 1955. Bacterial transduction *J. Cell. comp. Physiol.* **45** (Suppl. 2), *23—49.*

ZINDER, N. D., 1960. Hybrids of *Escherichia and Salmonella. Science* **131**, *813—815.*

ZINDER, N. D. & LEDERBERG, J., 1952. Genetic exchange in *Salmonella. J. Bact.* **64**, *679—699.*

CONNAISSANCES ACTUELLES SUR LA NATURE CHIMIQUE DES ANTIGÈNES PRÉSENTS DANS LES SALMONELLA

PAR

ANNE MARIE STAUB & MARCEL RAYNAUD

(Institut Pasteur — Paris et Garches)

(avec 5 figs.)

La précision avec laquelle la conjonction de la biochimie, de la sérologie et de la lysotypie, permet de déterminer une *Salmonella* donnée, est certes d'une grande aide pour l'épidémiologiste mais la multiplicité des *Salmonella* ainsi individualisées n'est pas sans poser un problème aussi bien au taxonomiste qu'au généticien. Faut-il considérer chaque nouvelle *Salmonella* comme une nouvelle espèce, ne s'agit-il seulement que de multiples mutants d'une même souche mère, ou existe-t-il quelques espèces comportant un grand nombre de mutants?

Les études génétiques et en particulier la conversion par les phages sont un moyen d'aborder le problème. L'étude chimique des antigènes présents dans les *Salmonella* est un autre moyen d'aborder le même problème.

Les antigènes susceptibles d'être décelés par un sérum précipitant sont multiples. De récentes études effectuées sur les souches Smooth (O+, H+, Vi+; O+ H+, O+)[1, 2] et Rough (R)[3] ont en effet montré que certains extraits contenaient non seulement les antigènes O, Vi, H et R classiques mais encore d'autres antigènes communs non seulement aux *Salmonella* (smooth ou rough) mais aussi à d'autres Enterobacteriaceae comme les *Shigella*.

Jusqu'à présent peu ou pas d'études chimiques ont été faites sur les antigènes autres que les antigènes O, R, H et Vi.

Antigène O

Depuis les travaux de BOIVIN[4] et de RAISTRICK[5], on sait que l'on peut extraire des Salmonella un antigène purifié qui, injecté aux lapins, provoque la formation d'agglutinines comme le germe entier, qui élimine totalement les agglutinines d'un sérum antimicrobien[6] et qui est capable de protéger la souris contre l'infection expérimentale vraie par un germe pathogène pour cet animal[7]. De plus cet antigène est toxique comme les germes tués: c'est l'antigène O. Les petites quantités d'azote qu'il contient (2 à 4% suivant la Salmonella) étaient considérées comme des impuretés par BOIVIN, d'où le nom d'antigène "glucido-lipidique" que lui donna cet auteur. Les tra-

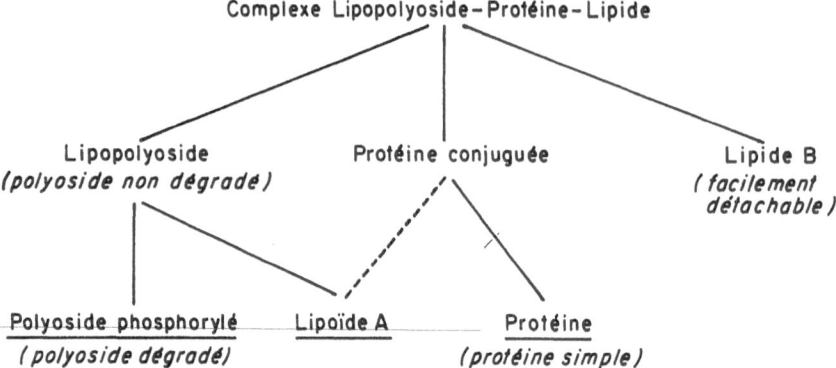

Fig. 1. Composants des endotoxines O des bactéries à Gram négatif d'après
Westphal & Lüderitz[10].

vaux de Morgan[8] montrèrent que cet azote était partie intégrante
de l'antigène qui est en réalité un complexe: glucido-lipido-protéi-
nique. Il est possible d'obtenir un lipopolyoside sans protéine en
extrayant les germes non plus par l'acide trichloracétique[4], la tryp-
sine[5] ou l'éthylène glycol[8], mais par le phénol à chaud, comme l'ont
montré Westphal et ses collaborateurs[9, 10]. On obtient alors un
produit de haut poids moléculaire comme le complexe complet qui
est aussi toxique que ce dernier mais peu (ou pas) antigénique chez
le lapin et qui n'élimine toutes les agglutinines d'un sérum antimicrobien qu'à de très fortes concentrations[11]. Il contient encore
quelques acides aminés, mais il ne s'agit plus d'une protéine.

Il est donc possible que le rôle de la protéine dans l'antigène O
complet se réduise à lui donner son caractère immunogène, encore
faut-il souligner que des préparations très appauvries en protéine[12]
sont encore antigéniques. La spécificité des antigènes O du tableau
de Kauffman-White dépend-elle en partie de sites présents sur cette
protéine? Il est impossible, à l'heure actuelle, de répondre d'une
façon certaine dans un sens ou dans l'autre. Les préparations de
polyosides dont nous disposons actuellement sont en effet incapables
d'éliminer toutes les agglutinines (cf. plus bas) et les quantités énor-
mes de lipopolyoside nécessaires pour y parvenir (4 à 5 mg pour 1 ml
de sérum 1/100) peuvent s'expliquer aussi bien par une contamination
avec quelques restes de protéines dans ces préparations que l'on
sait hétérogènes que par la rareté de certains sites propres au lipopolyoside à la surface de ces très grosses molécules. Les protéines
que Morgan[8] a séparées du complexe ne provoquent pas, chez le
lapin, la formation d'agglutinines spécifiques du germe dont elles
proviennent, mais il est possible qu'elles aient été partiellement
dénaturées. En dehors de Morgan, personne ne s'est d'ailleurs

occupé de cette fraction protéinique*, alors que le lipide et le polyoside ont fait l'objet de nombreuses recherches.

Bien que le lipide ait un rôle certain dans la toxicité de l'antigène complet et ait retenu ainsi l'attention d'un grand nombre de chercheurs ([23]), on ne sait rien de son rôle dans la spécificité des antigènes O qui nous intéresse ici. Aussi passerons nous tout de suite à l'étude des polyosides que l'on a pu extraire de l'antigène complet ou des germes entiers.

Il convient d'ajouter que l'on ignore tout du rôle de l'antigène O en général et de ses composants dans la protection de l'homme et des animaux contre les Salmonelloses. Certes, les anticorps O sont opsonisants et dans ce domaine les anticorps antipolyosidiques sont aussi actifs que les anticorps obtenus avec l'antigène complet [14, 15, 16]. C'est à cette activité opsonisante qu'il faut sans aucun doute rattacher le pouvoir protecteur relatif des anticorps O[17, 7] et des antipolyosides[18] contre une infection massive de la souris par S. enteritidis ou S. typhi-murium. Mais ces mêmes anticorps sont incapables de protéger la souris contre une infection provoquée par quelques germes avec ces 2 Salmonella[19]. Enfin s'il est certain qu'on peut vacciner la souris contre l'infection vraie par des injections répétées d'antigène complet comme l'a montré BOIVIN[8], les résultats positifs obtenus par le même auteur avec le polyoside dégradé, obtenu par hydrolyse acétique, adsorbé sur de l'alumine[20] doivent être interprétés avec une certaine réserve et devraient être reproduits avec une préparation très purifiée ne contenant certainement plus d'antigène complet. On sait en effet que de très petites quantités de ce dernier suffisent pour provoquer la formation d'anticorps[21]. Enfin, on ne sait rien du rôle des anticorps O dans la protection de l'homme par les vaccins tués[22].

Polyoside

a. Présence des facteurs O sur les polyosides

L'hydrolyse acétique du complexe entier ou du lipopolyoside fournit un polyoside qui précipite les sérums agglutinants (à condition qu'ils soient obtenus après une immunisation assez prolongée) et reflète la spécificité de l'antigène dont ils proviennent[24, 4, 8]. Par exemple le polyoside extrait de l'antigène O de S. *typhi-murium* précipite comme cet antigène les sérums anti S. *typhi-murium*, anti S. *paratyphi B*, anti S. *typhi* et anti S. *senftenberg*. On retrouve donc dans ce polyoside la spécificité des facteurs O dénombrés dans le tableau de KAUFFMANN-WHITE. Néanmoins, il est impossible de précipiter avec eux la totalité des agglutinines du sérum. Lorsqu'il s'agit d'un sérum obtenu après une faible immunisation, possédant toutefois un bon titre d'agglutinines, celui-ci peut à peine baisser. Un excès de polyoside n'inhibe pas l'agglutination.[6]

---

* Les fractions "protéiniques" de MESROBEANU et coll.[13] ne contiennent en effet que 10% d'azote: aussi est-il difficile de savoir si les antigènes qu'elles contiennent sont vraiment des holo-protéines. Il s'agit en fait, d'après les données analytiques des auteurs eux-mêmes[13], de complexes glucido-lipido-protéiques à teneur en protéine plus élevée que celle des complexes préparés par les méthodes "habituelles" et analogue à celle des complexes préparés par les solutions hypertoniques[3, 95, 96, 97].

L'hydrolyse acétique directe des germes desséchés par l'acétone[24] fournit le même polyoside[25, 26] qu'il est alors plus facile de purifier [26, 27, 23]. Un tel polyoside ne contient plus qu'une toute petite partie de la glucosamine présente dans le lipopolyoside et qui en général fait uniquement partie de la fraction lipidique A; avec certains germes, elle peut même presque totalement disparaître*. Il est toujours complètement atoxique pour la souris.

L'hydrolyse alcaline des germes desséchés par l'acétone[28, 29, 6, 23] fournit par contre un autre polyoside plus complexe qui contient encore des hexosamines et quelques acides gras. Il se fixe sur les globules rouges et élimine une grande partie des agglutinines, par précipitation du sérum et surtout après absorption sur les globules rouges sensibilisés[6]. Il reste toutefois presque toujours dans les sérums des lapins faiblement immunisés une petite quantité d'agglutinines qu'on ne peut pas éliminer. Bien que possédant la glucosamine et quelques acides gras, ce polyoside est complètement atoxique pour la souris (5 mg ne tuent pas une souris de 20 g).

Aucun de ces polyosides injecté aux lapins ne provoque la formation d'agglutinines si on a soin de les purifier suffisamment pour enlever toute trace d'antigène complet. Néanmoins, il est possible d'obtenir des anticorps antipolyoside agglutinants en injectant le polyoside "acétique" couplé à une protéine par diazotation[21, 23], ou le polyoside "alcalin" fixé sur des stromas[23].

Nous avons dit que les polyosides extraits par acétolyse des germes ou de l'antigène complet portaient tout ou partie de la spécificité de tous les facteurs O du tableau de KAUFFMANN-WHITE. Des études récentes[30, 31, 32, 33] confirmant des résultats antérieurs[34, 35, 36, 37] ont montré que les différents facteurs O décelés par le sérologiste sur une même Salmonella n'étaient pas portés par des polyosides distincts, mais constituaient les "sites antigéniques" d'une même molécule de polyoside. Il semble donc raisonnable de substituer le terme de facteur O à celui d'antigène utilisé pendant longtemps et qui évoque faussement l'idée d'une grosse molécule autonome.

L'utilisation des techniques nouvelles de chimie (chromatographie sur colonne et sur papier, électrophorèse) et d'immunologie (étude quantitative des précipitations et inhibitions spécifiques) a permis aux immunochimistes de progresser considérablement depuis 10 ans dans la connaissance de ces facteurs**.

b. Etude des sucres constitutifs des facteurs O

L'analyse systématique des sucres constitutifs menée de front

---

* Elle demeure toutefois en quantité importante lorsqu'elle joue un rôle dans la spécificité du polyoside lui-même.
** Une revue détaillée de cette question doit paraître prochainement dans les Bacteriological Reviews, aussi ne donnerons nous ici que les lignes directrices et les conclusions de ces études[23].

dans plusieurs laboratoires[38, 39, 40], a permis de reconnaître un certain nombre de sucres communs (faibles quantités de heptose, glucose, galactose, souvent mannose, glucosamine, chez les lipopolyosides) et d'autres propres seulement à certains sérotypes parmi lesquels on trouve du ribose, des 6-désoxyhexoses (fucose, rhamnose) la galactosamine, l'acide colominique et les 3—6-didésoxyhexoses (abéquose, tyvélose, paratose, colitose) dont l'analyse et la synthèse ont été poursuivies en commun par les équipes des Professeurs E. LEDERER et O. WESTPHAL et qui ont fait l'objet d'une revue antérieure[41, **]. Cette étude a permis de proposer une classification des Salmonella en "types chimiques" (chemotype) où sont groupés tous les sérotypes possédant les mêmes sucres[39, 40]. On voit (Tabl. II) que des sérotypes très variés sont présents dans le même "type chimique". Il fallait donc adjoindre une analyse immunologique à l'analyse chimique pour acquérir une meilleure connaissance des facteurs O. C'est ce que permit de faire la technique des inhibitions spécifiques par les sucres simples et leurs glycosides qui établit tout d'abord le rôle prépondérant du tyvélose dans la spécificité du facteur 9, puis celle d'autres sucres dans les autres facteurs figurant sur les schémas de la fig. 2[43, 32, 31, 45]. Néanmoins, là encore, l'analyse n'était pas suffisamment poussée puisque deux facteurs distincts dans le tableau de KAUFFMANN-WHITE peuvent très bien posséder le même sucre terminal.

Il est vrai que ce tableau a été établi au moyen de sérums de lapin et que les facteurs 8 et 4 par exemple donnent une réaction croisée faible mais nette avec des sérums de cheval.

TABLEAU I.

Sucres présents dans les antigènes O des Salmonella

Sérotypes	Groupe	Facteurs O	Galactosamine Glucosamine	Heptose (s)	Galactose	Glucose	Mannose	Fucose Rhamnose	Abéquose Colitose Paratose Tyvélose
S. paratyphi A	A	1, 2, 12	+	+	+	+	+	+	+
S. paratyphi A var. durazzo		2, 12	+	+	+	+	+	+	+
S. kiel		1, 2, 12	+	+	+	+	+	+	+
S. abortus equi	B	4, 12	+	+	+	+	+	+	+
S. paratyphi B		4, 5, 12	+	+	+	+	+	+	+
S. java		4, 5, 12	+	+	+	±	+	+	+
S. schleissheim		4, 12, 27	+	+	+	±	+	+	+

** Il est possible que cette liste soit encore incomplète comme le suggère la toute récente découverte du 3-désoxy-2-céto-octonate des E. coli O: 111: B4, ignoré jusqu'à présent dans les microorganismes[52] et qui serait un constituant présent dans les lipopolysaccharides de toutes les entérobactéries.

TABLEAU I (suite)

Sérotypes	Groupe	Facteurs O	Galactosamine	Glucosamine	Heptose (s)	Galactose	Glucose	Mannose	Fucose	Rhamnose	Abéquose	Colitose	Paratose	Tyvélose
S. typhi murium		4, 5, 12		+	+	+	+	+		+	+			
S. paratyphi C	C₁	6, 7		+	+	+	+	+						
S. paratyphi C (Vi)		6, 7		+	+	+	+	+						
S. cholerae suis		6, 7		+	+	+	+	+						
S. decatur		6, 7		+	+	+	+	+						
S. isangi		6, 7		+	+	+	+	+						
S. montevideo		6, 7		+	+	+	+	+						
S. thompson		6, 7		+	+	+	+	+						
S. bareilly		6, 7		+	+	+	+	+						
S. muenchen	C₂	6, 8		+	+	+	±	+		+	+			
S. newport		6, 8		+	+	+	+	+		+	+			
S. virginia	C₃	(8)		+	+	+	±	+		+	+			
S. kentucky		(8), 20		+	+	+	+	+		+	+			
S. sendai	D₁	1, 9, 12		+	+	+	+	+		+				+
S. miami		1, 9, 12		+	+	+	+	+		+				+
S. onarimon		1, 9, 12		+	+	+	+	+		+				+
S. typhi		9, 12		+	+	+	+	+		+				+
S. typhi (Vi)		9, 12		+	+	+	+	+		+				+
S. ndolo		9, 12		+	+	+	+	+		+				+
S. enteritidis		1, 9, 12		+	+	+	+	+		+				+
S. dublin		1, 9, 12		+	+	+	+	+		+				+
S. gallinarum		1, 9, 12		+	+	+	+	+		+				+
S. strasbourg	D₂	(9), 46		+	+	+	±	+		+				+
S. haarlem		(9), 46		+	+	+	+	+		+				+
S. fresno		(9), 46		+	+	+	+	+		+				+
S. anatum	E₁	3, 10		+	+	+	+	+		+				
S. uganda		3, 10		+	+	+	±	+		+				
S. newington	E₂	3, 15		+	+	+	+	+		+				
S. binza		3, 15		+	+	+	±	+		+				
S. illinois	E₃	(3), (15), 34		+	+	+	+	+		+				
S. senftenberg	E₄	1, 3, 19		+	+	+	+	+		+				
S. chittagong		(1), 3, 10, (19)		+	+	+	+	+		+				
S. aberdeen	F	11		+	+	+	+	+		+				
S. rubislaw		11		+	+	+	+	+		+				
S. friedenau	G	13, 22	+	+	+	+	+		+					
S. poona		13, 22	+	+	+	+	+		+					
S. mississippi		1, 13, 23	+	+	+	+	+		+					
S. worthington		1, 13, 23	+	+	+	+	+		+					
S. carrau	H	6, 14, 24		+	+	+	+	+						
S. onderstepoort		(1), 6, 14, 25		+	+	+	+	+						
S. boecker		6, 14		+	+	+	+	+						
S. brazil	I	16	+	+	+	+	+	+	+					
S. hvittingfoss		16	+	+	+	+	+	+	+					
S. gaminara		16	+	+	+	+	+	+	+					
S. kirkee	J	17		+	+	+	±							
S. berlin		17		+	+	+	+							
S. michigan		17		+	+	+	+							

TABLEAU I (suite)

Sérotypes	Groupe	Facteurs O	Galactosamine	Glucosamine	Heptose (s)	Galactose	Glucose	Mannose	Fucose / Rhamnose	Abéquose / Colitose / Paratose / Tyvélose
S. usumbura	K	18	+	+	+	+	+	+		
S. cerro		18	+	+	+	+	+	+		
S. siegburg		6, 14, 18	+	+	+	+	+	+		
S. ghana	L	21	+	+	+	+	+			
S. minnesota		21	+	+	+	+	+			
S. magwa		21	+	+	+	+	+			
S. halle	M	$28_1, 28_2$	+	+	+	+	+		(x)	
S. tel-aviv		$28_1, 28_2$	+	+	+	+	+		(x)	
S. ezra		$28_1, 28_2$	+	+	+	+	+		(x)	
S. dakar		$28_1, 28_2$	+	+	+	+	+		+	
S. urbana	N	30	+	+	+	+	+		+	
S. godesberg		30	+	+	+	+	+		+	
S. donna		30	+	+	+	+	+		+	
S. adelaide	O	35		+	+	+	+			+
S. monschaui		35		+	+	+	+			+
S. inverness	P	38	+	+	+	+	+			
S. perth		38	+	+	+	+	+			
S. wandsworth	Q	39	+	+	+	+	+	+	+	
S. champaign		39	+	+	+	+	+	+	+	
S. riogrande	R	40	+	+	+	+	+	+		
S. bukavu		1, 40	+	+	+	+	+	+		
S. bulawayo		1, 40	+	+	+	+	+	+		
S. waycross	S	41		+	+	+	+	+		
S. offa		41		+	+	+	+	+		
S. kampala	T	1, 42		+	+	+	+		+	
S. weslaco		42		+	+	+	+		+	
S. milwaukee	U	43	+	+	+	+	+		+	
S. kingabwa		43	+	+	+	+	+		+	
S. niarembe	V	44		+	+	+	+			
S. guinea		44		+	+	+	+			
S. deversoir	W	45		+	+	+	+		+	
S. dugbe		45		+	+	+	+		+	
S. bergen	X	47		+	+	+	+			
S. kaolack		47		+	+	+	+			
S. dahlem	Y	48		+	+	+	+		ac. colominique	
S. djakarta		48		+	+	+	+		ac. colominique	
S. greenside	Z	50	+	+	+	+	+			+
S. treforest		51	+	+	+	+	+			
S. utrecht		52		+	+	+	+			
S. humber	53	53	+	+	+	+	+		+	
S. uccle	54	54		+	+	+	+	+	+	
S. tranoroa	55	55	+	+	+	+	+			
S. artis	56	56	+	+	+	+	+		Ribose	
S. locarno	57	57	+	+	+	+	+	+		
S. basel	58	58		+	+	+	+			
S. betioky	59	59		+	+	+	+		+	

TABLEAU II.

Classiffication des Salmonella en types chimiques

Type chimique	Nombre des sucres constitutifs	Hexosamine		Heptose (s)	Hexoses			6-Desoxy-hexoses		(X) ou Ribose	3, 6-Didesoxy-hexoses				Groupes O
		Galactosamine	Glucosamine		Galactose	Glucose	Mannose	Fucose	Rhamnose		Colitose	Abéquose	Paratose	Tyvélose	
I	4		o	o	o	o									J(17); V(44); X(47); Y(48); 52; 58
II	5	⊙	o	o	o	o									L(21); P(38); 51; 55
III	5		o	o	o	o	⊙								C$_1$ (6, 7); H (6, 14; 6, 14, 24; 1, 6, 14, 25); S (41)
IV	6	⊙	o	o	o	o	⊙								K (18; 6, 14, 18); R (40; 1, 40)
V	5		o	o	o	o		⊙							W (45)
VI	6	⊙	o	o	o	o		⊙							G (1, 13, 23; 13, 22); N (30); U (43)
VII	5		o	o	o	o			⊙						T (42; 1, 42); 59
VIII	6	⊙	o	o	o	o			⊙						M (28$_1$, 28$_3$); 53; 57
IX	6	⊙	o	o	o	o				⊙					M (28$_1$, 28$_2$); 56
X	5		o	o	o	o					⊙				O (35)
XI	6	⊙	o	o	o	o					⊙				Z (50)
XII	7	⊙	o	o	o	o	⊙	⊙							I (16), Q (39)
XIII	6		o	o	o	o	⊙		⊙						E (3, 10; 3, 15; 1, 3, 19; 1, 3, 10, 19; 3, 15, 34); 54; F (11)
XIV	7		o	o	o	o	⊙		⊙			⊙			B (4, 12; 4, 5, 12; 1, 4, 5, 12; 4, 12, 27); C$_2$ (6, 8); C$_3$ (8; 8, 20)
XV	7		o	o	o	o	⊙		⊙				⊙		A (2, 12; 1, 2, 12)
XVI	7		o	o	o	o	⊙		⊙					⊙	D$_1$ (9, 12; 1, 9, 12); D$_2$ (9, 46)

c. Etude des oligosides responsables de la spécificité des facteurs O

Conformément aux résultats obtenus par d'autres auteurs avec les antigènes artificiels[46, 47], les dextranes [48, 49, 50] et les substances spécifiques des groupes sanguins[50, 51], la spécificité des facteurs O des Salmonella dépend donc de la constitution de l'oligoside terminal et pas seulement du sucre terminal. Quelques uns de ceux présents dans les polyosides des groupes B, D, E, O et G ont été analysés. Les résultats de cette étude ont été résumés sur le tableau III et les notes qui l'accompagnent indiquent leur caractère hypothétique ou définitif. Un certain nombre de faits en ressortent:

1° — 3 ou 4 sucres au moins déterminent la spécificité des facteurs O. Ceci semble en défaut dans le cas du facteur 1 commun aux groupes A, B, D et E$_4$. En réalité, il ne s'agit pas vraiment d'un facteur distinct comme le pensaient les sérologistes, mais d'anticorps anti-19

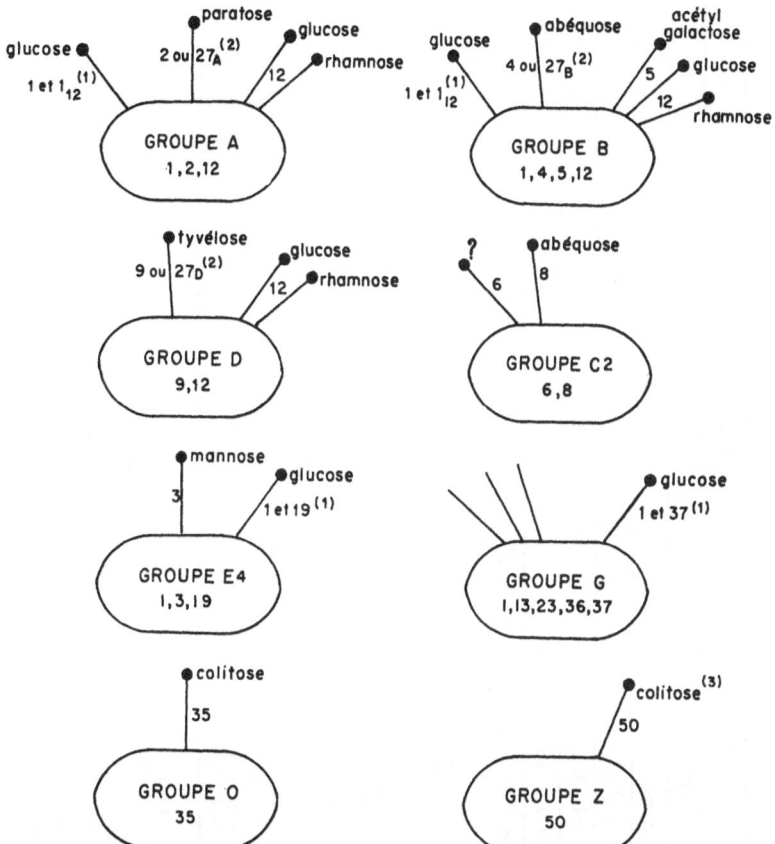

Fig. 2. Représentation schématique des sucres terminaux des facteurs O, déterminés au moyen de l'inhibition de la précipitation spécifique par les sucres simples.

(1) Pour la présence simultanée de 2 facteurs sur une même chaîne voir le texte p. 154.

(2) Pour l'alternative de 2 facteurs portant le même sucre terminal voir le texte p. 153.

(3) La position terminale de ce colitose n'a pas été déterminée par les études d'inhibition mais par la préparation d'antigènes artificiels.

et anti-1_{12} capables de s'unir au facteur hétérologue (1_{12} pour les anticorps anti-19, 19 pour les anticorps anti-1_{12}). On sait en effet qu'un même site antigénique peut donner naissance à toute une série d'anticorps adaptés à une partie plus ou moins longue de la chaîne polysaccharidique. Ces anticorps ont même été séparés récemment par KABAT dans le cas des dextranes[53]. Les anticorps anti-1 des groupes A sont donc des anticorps anti-1_{12} dont la spécificité est limitée à l'extrémité du facteur 1_{12}. Ceux du groupe E_4 sont des

TABLEAU III.

Nature chimique des oligosides responsables de la spécificité de quelques facteurs O

Groupe	Facteur	Constitution chimique	Références
B	4	abéquose α-(mannose-rhamnose)[1]-	43
	27B	abéquose α-(mannose + rhamnose)[2]-	44
	5	acétyl[4]-galactose α-(mannose-rhamnose-galactose-mannose)[3]-rhamnose-	32
	1_{12}	glucose α1—6 galactose α1—2 mannose 1-(3 ou 4) rhamnose-[7]	31, 61, 63
D	9	tyvélose α-(mannose-rhamnose)[1]-	42, 43, 66_b
	27D	tyvélose α-(mannose + rhamnose)[2]-	44
	12_2	glucose α 1—4 galactose α-mannose-rhamnose-[7]	66_a
E	E_1 3	mannose 1—(4 ou 5) rhamnose—	55
	10	acétyl[5]-galactose α 1—6 mannose 1-(4 ou 5) rhamnose-	55
	E_2 15	galactose β 1—6 mannose 1-(4 ou 5) rhamnose-	55
	E_3 34	glucose α 1—4 galactose β 1—6 mannose 1-(4 ou 5) rhamnose-	55
	E_4 19	glucose α 1—6 galactose-[6] mannose-rhamnose-	45
B et E_4	1	glucose α 1—6 galactose	45
G	37	glucose α 1—6 galactose (?) -X-	45

1. La nature des sucres entre parenthèses n'est encore que dubitative.
2. La nature des sucres entre parenthèses n'est encore que dubitative, mais ils sont vraisemblablement les mêmes que dans (1) et liés différemment.
3. La nature des sucres entre parenthèses est certaine mais leur ordre n'est que dubitatif.
4. Probablement sur la position 2 du galactose.
5. Probablement sur la position 6 du galactose.
6. La liaison galactose-mannose en cours de détermination n'est pas encore connue mais elle est sûrement différente de la liaison du facteur 1_{12}.
7. La présence du rhamnose sur la chaîne polyosidique est certaine mais nous ne savons pas actuellement s'il joue ou non un rôle dans la spécificité du facteur.

anticorps anti-19 dont la spécificité est limitée à l'extrémité de la chaîne 19.

Le cas des facteurs 4 et 27_B est un autre exemple de ces réactions croisées qui ont amené les sérologistes à attribuer à certaines Salmonella un facteur qu'elles ne possèdent pas. En effet, l'étude quantitative de la précipitation d'un sérum anti: 4, 12 par un polyoside: 4, 27, 12 a montré qu'en réalité il n'existait pas de facteur 4 sur ce polyoside, mais que seulement une grande partie des anticorps anti: 4 réagissaient avec le facteur 27[44].

2° — Avec les anticorps de lapin la présence d'un dioside terminal commun est suffisante, comme nous venons de le voir, pour créer une réaction croisée entre 2 oligosides. Elle est aussi nécessaire puisque nous avons vu que 2 facteurs terminés seulement par le même sucre terminal ne donnent pas de réaction croisée, contrairement aux anticorps de cheval, de chèvre, et de poule. Corrélativement, tous facteurs qui donnent une réaction croisée plus ou moins intense avec des anticorps de lapin doivent nécessairement porter au moins le même dioside terminal ou des diosides de configuration très voisine. C'est ce qui a motivé sur le tableau III la position du même mannose derrière l'abéquose dans les facteurs 4 et 27_B. De plus, il est très probable que tous les facteurs 1 présents dans le tableau de Kauffmann-White sont liés à la présence du dioside α-glucose-1-6-galactose à l'extrémité de divers facteurs. La position du galactose (α et β) ainsi que son mode de liaison au sucre suivant et la nature de celui-ci seraient alors responsables des parentés plus ou moins étroites des différents facteurs 1 qu'une analyse plus approfondie a nettement différenciés[54].

3° — Puisque la spécificité des facteurs O repose sur la nature et le mode de liaison d'au moins 3 ou 4 sucres terminaux, il s'en suit que toute modification portant sur ces sucres, ou leurs liaisons, fera apparaître une spécificité nouvelle, d'où un facteur nouveau. Le rapport entre ce facteur et le précédent dépendra de la place où aura lieu cette modification. S'il s'agit du sucre terminal ou de son mode de liaison au sucre subterminal, la spécificité du nouveau facteur sera complètement distincte de celle du facteur original. C'est le cas du facteur 5 apparu par simple addition d'un acétyl sur le galactose terminal. La suppression ou même la répression du système enzymatique responsable de cette acétylation expliquerait donc la mutation 4, 5, 12 → 4, 12.

C'est aussi le cas des facteurs 10 et 15 qui ne diffèrent que par la nature de leur sucre terminal: α-acétyl-galactose pour le facteur 10, β-galactose pour le facteur 15. Cette fois, le changement du système enzymatique porte sur la présence d'un galactose β au lieu du galactose α, accompagné ou non de la suppression (ou répression) de l'acétyl-transférase[55]. Ces changements . accompagnent d'ailleurs la présence d'un prophage dans la bactérie puisque toutes les Salmonella: 3, 15 sont lysogènes pour le phage ε 15[56], [57].

C'est encore à la présence d'un prophage (PLT 22)[58], [59], [60] qu'il faut rattacher l'apparition du facteur 1_{12}

$$\alpha\text{-gluc } 1\text{---}6\alpha \text{ gal} - \text{man} - \text{rham} -$$

par suite de la présence d'un α-glucose sur le carbone 6 du galactose terminal du trioside[61], [62], [63]:

$$\alpha \text{ gal } 1\text{---}2 \text{ man} - \text{rham}$$

Dans les Salmonella des groupes A, B et D, ce trioside peut exister tel quel dans les souches non lysogènes ou être lié au glucose dans le tétraoside du facteur 12

$$\alpha \text{ gluc } 1 - 4\alpha \text{ gal} - \text{man} - \text{rham} -$$

qui semble disparaître après l'apparition du facteur 1_{12}.

Enfin, c'est toujours à l'addition d'un α glucose sur un galactose terminal mais cette fois sur le carbone 4 qu'il faut rattacher l'apparition du facteur 34 sur les Salmonella 3, 15 converties par le phage ε 34[55] (Tab. IV).

Si le changement de liaison ou de sucre porte après le dioside terminal, la modification de spécificité est moins profonde et le nouveau facteur "croise" fortement avec le facteur original. C'est le cas du facteur 27 B présent dans les souches du groupe B après lysogénisation par le phage 27[64], [65] et qui remplace le facteur 4 comme nous l'avons vu. De nombreux arguments permettent de penser que là encore le rôle du phage se limiterait à un changement dans la liaison interne mannose-rhamnose (Tabl. IV). Ce qui expliquerait l'apparition de facteurs différents 27_A, 27_B, 27_D sur les Salmonella des groupes A, B, D lysogénisés par le même phage 27. Ces 3 facteurs sont terminés par des sucres différents (cf. Tabl. III) mais pourraient posséder la même liaison subterminale man-rham.

La conclusion de toutes ces recherches en pleine évolution encore, est que l'apparition ou la disparition des facteurs O peut être due à de faibles modifications des systèmes enzymatiques bactériens. Certaines de ces modifications ont pu déjà être rattachées à la présence de prophages. La liste n'en est certes pas exhaustive et bien d'autres s'y ajouteront certainement au fur et à mesure que se poursuivront les recherches dans cette direction.

d. Place des facteurs O sur le polyoside somatique

Nous avons dit que les facteurs O constituaient les différents "sites antigéniques" d'un même polyoside. Puisqu'il s'agit sûrement de courts oligosides, il nous reste à savoir s'il s'agit de courtes chaînes latérales reliées à une longue chaîne immunologiquement inactive (I fig. 3) ou d'extrémités de longues chaînes reliées à une courte chaîne immunologiquement inactive (II), ou de longues chaînes reliées à une courte chaîne immunologiquement inactive comme II, mais où les facteurs seraient portés par les divers segments

TABLEAU IV.

Rôle des phages dans la conversion des facteurs O des Salmonella

Facteurs O de la souche sauvage	Rôle du phage	Facteurs O de la souche convertie
4,12	gal α 1—2 man 1—(3 ou 4) rham $\xrightarrow{\varphi}$ gluc α 1—6 gal α -man-rham-	1, 4, 12
3, 15	gal β 1—6 man 1—(4 ou 5) rham $\xrightarrow{\varphi}$ gluc α 1—4 gal β 1—6 man 1—(4 ou 5) rham-	3, 15, 34
10	acétyl gal α 1—6 man 1—(4 ou 5) rham $\xrightarrow{\varphi}$ gal β 1—6 man 1—(4 ou 5) rham-	15
4(*)	abéquose α -man-rham $\xrightarrow{\varphi}$ abéquose α -man-●-rham-	27$_B$

(*) L'action du phage sur le changement de la liaison mannose-rhamnose n'a encore été avancée qu'à l'état d'hypothèse et attend la confirmation d'une étude chimique actuellement en cours.

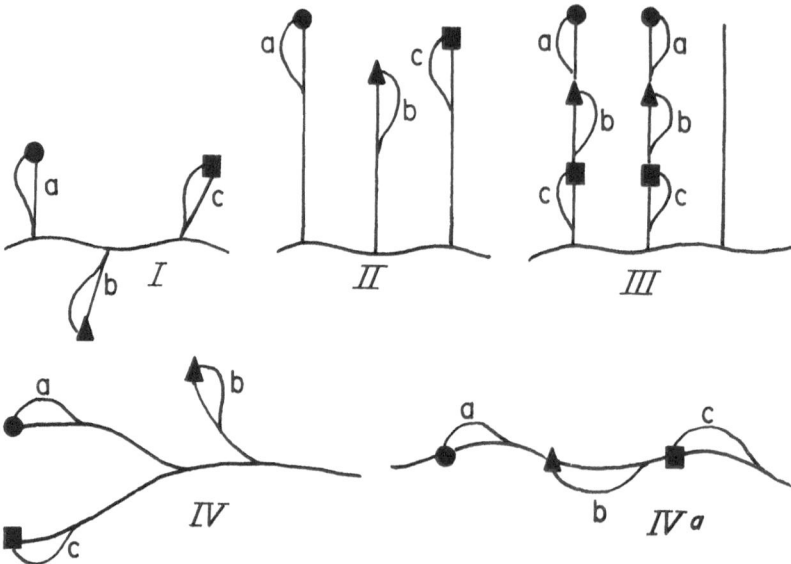

Fig. 3. Schémas possibles pour la constitution des polyosides porteurs des facteurs O. a, b, c, oligosides responsables de la spécificité des facteurs O ▲ ● ■ sucres terminaux différents.

de la longue chaîne (III), ou enfin d'une longue chaîne, branchée ou non, portant les différents facteurs sans qu'il y ait de chaîne centrale inactive (IV. IV$_a$). Disons tout de suite que les quelques renseignements que nous possédons sur les polyosides entiers présents dans les groupes D[66] et E[55] sont en contradiction formelle avec la première hypothèse (I) et avec la dernière (IV$_a$), il s'agit sûrement de polyosides très ramifiés et à longues chaînes. Un certain nombre de faits, qui seront analysés dans l'étude des antigènes R, sont en faveur d'un polyoside central avec de longues chaînes selon les schémas II et III, le polyoside central pouvant être le polyoside commun à toutes les souches R. Soulignons que si cette conception est vraie, il faut que les chaînes portant les facteurs soient très longues. On sait en effet aujourd'hui que les oligosides porteurs des spécificités O sont en réalité des monomères dont on a pu isoler des di- et même des tri-mères, ce qui permit à ROBBINS & UCHIDA[55] de proposer pour les polyosides des groupes E$_1$, E$_2$ et E$_3$ les formules du tableau V. Des chaînes très similaires semblent exister dans les polyosides des groupes B et D où l'on retrouve les oligosides responsables des spécificités 1, 4, 9 et 12. Pour ROBBINS & UCHIDA, chacun des monomères jouerait un rôle dans la spécificité. Il est toutefois certain qu'une petite partie seulement des sucres qui jouent un rôle prépondérant dans la spécificité des oligosides peuvent s'unir avec les anticorps correspondants[11], soit parce qu'ils sont vraiment situés à l'extrémité

TABLEAU V.

Constitution probable des chaînes de quelques polyosides

Groupe			
E$_1$ (3, 10)	10 Acétyl | gal α -man-rham-	3 | gal β -man-rham-	gal β -man-rham-
E$_2$ (3,15)	15 gal β -man-rham-	3 gal β -man-rham-	gal β —man-rham-
E$_3$ (3, 15, 34)	15 gal β -man-rham-	34 gluc α 1 | 4 gal β -man-rham-	gluc α 1 | 4 gal β -man-rham- 3
B (1, 4, 12)	1$_{12}$ gluc α 1 | 6 gal α -man-rham- abéquose |	gluc α 1 | 6 gal α -man-rham- 4 abéquose |	gluc α 1 | 6 gal α -man-rham- abéquose |
D (9, 12)	12 gluc α 1 | 4 gal α -man-rham- tyvélose |	gluc α 1 | 4 gal α -man-rham- 9 tyvélose |	gluc α 1 | 4 gal α -man-rham- tyvélose |

des chaînes latérales comme dans le schéma II, soit parce que seuls, ils sont accessibles aux anticorps, les autres oligosides étant par exemple bloqués par un anticorps voisin si l'on accepte le schéma III.

Quelle que soit la réalité, l'examen du tableau V nous permet de souligner, comme l'étude des facteurs, les faibles différences qui séparent les polyosides extraits de Salmonella appartenant non seulement aux sous-groupes d'un même groupe (E$_1$, E$_2$, E$_3$,), mais encore aux différents groupes (E, B et D). Il est vrai que si les sé-

158

quences sont les mêmes, les liaisons entre les sucres diffèrent certainement entre les groupes E et B. Mais il ne s'agit encore là que de modifications restreintes des systèmes enzymatiques qui, nous l'avons vu, peuvent même apparaître au cours de l'infection par un phage.

Antigène Rough et variantes diverses dépourvues de l'antigène somatique O

La variation S → R peut être définie au point de vue morphologique (différences dans l'aspect des colonies, dans la stabilité des suspensions bactériennes, etc...) ou au point de vue sérologique. Dans ce dernier cas, on appelle "Rough" les variantes qui ont perdu un certain antigène caractéristique des formes Smooth. Comme l'a fait remarquer WILKINSON[67], si on considère les divers antigènes de surface élaborés par les microorganismes, on est amené à distinguer les antigènes capsulaires ou microcapsulaires (antigènes K) peu liés à la paroi et correspondant aux variantes "muquoïdes"* des antigènes somatiques (O ou R) liés à la paroi cellulaire.

Les variantes Rough proprement dites des entérobactéries en général et des Salmonella en particulier, peuvent alors être définies comme les variantes chez lesquelles l'antigène polysaccharidique O des formes S a été remplacé par un autre antigène polysaccharidique, plus simple, l'antigène R. KRÖGER[68, 69, 70] après PRIGGE[71, 72] a proposé de symboliser cette variation antigénique par la notation O/o, la variation morphologique par S/R, mais cette notation n'a pas été universellement acceptée.

Les variantes Rough proprement dites se trouvent ainsi définies non seulement par l'absence de l'antigène O, mais par la présence d'un nouvel antigène polysaccharidique: l'antigène R.

Il existe en effet une série de variantes différentes qui ont perdu l'antigène O, et dont la dénomination reste confuse, car les antigènes correspondants ont été pendant longtemps mal caractérisés. Si on prend en considération les données relativement récentes obtenues sur le mécanisme de la biosynthèse des polysaccharides des entérobactéries et sur les déficiences observées chez certains mutants bien caractérisés, on peut essayer de dresser un schéma général des variantes "par défaut" (minus variantes) qui ont été décrites (fig. 4 d'après WESTPHAL ET LÜDERITZ[72a]).

Les variantes (rho) observées il y a de nombreuses années par WHITE[73] auraient d'après DAVIES[38], une paroi réduite au mucopeptide, qui représente la couche rigide de WEIDEL[74]. L'existence de mu-

* Les variantes dites "Smooth" du pneumocoque sont alors des variantes muquoïdes, les formes Rough du pneumocoque correspondant aux variantes Smooth et Rough des entérobactéries[67].

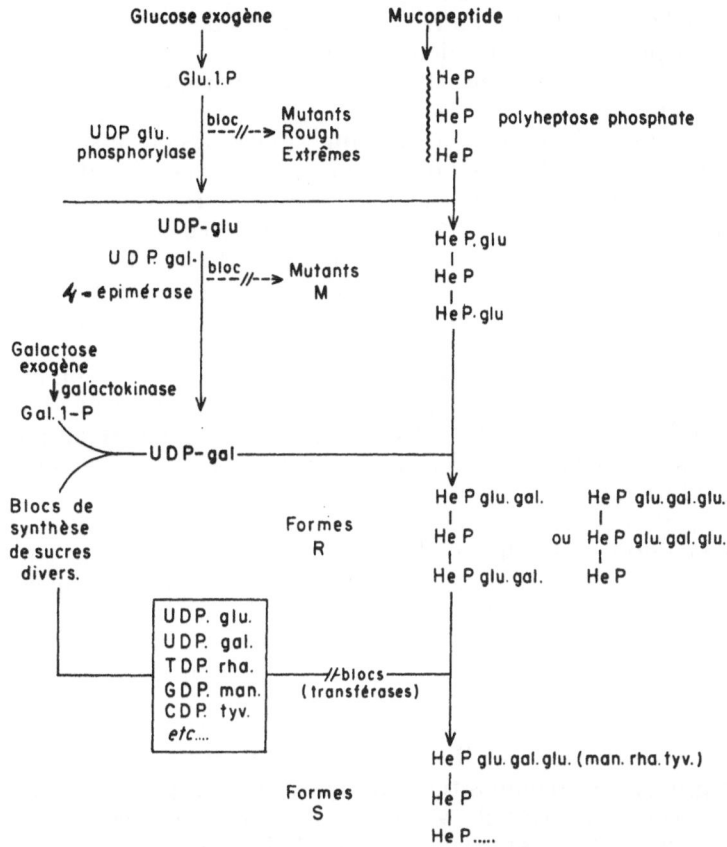

Fig. 4. Déficiences enzymatiques responsables de la production des différents
S → R → M → extra R, d'après WESTPHAL & LÜDERITZ[72a].

tants Rough Extrêmes dont le polysaccharide ne contient qu'un
heptose, a été établi dans le cas de quelques germes, en particulier
chez certains mutants de E. coli K[12], déficients en U D P G —
pyrophosphatase[75, 76]. ROTHFIELD et al. ont obtenu un polysaccharide analogue chez un mutant de S. typhi-murium déficient en phos-
phoglucose isomérase et cultivé sur fructose[77].

Les mutants "M*" décrits il y a longtemps par MURASE[78] ont été
analysés en ce qui concerne leurs déficiences enzymatiques par
NIKAIDO, FUKASAWA et al. [79, 80, 81, 82, 83, 84, 85, 86] et divers autres

---

* Ces mutants "M" (mutabile) n'ont rien de commun avec les mutants muquoïdes
que certains auteurs désignent par la même abréviation M. [89]bis.

auteurs[87, 88, 89]. Ces mutants sont incapables de fermenter le galactose: de plus, le galactose ajouté à une culture en milieu ne contenant
pas ce sucre, provoque leur lyse, après une courte période de latence
(20 à 30 minutes).

L'analyse enzymatique montre que ces germes ont une déficience
très particulière: ils contiennent de la galactokinase (K) de la galac-
tose-1-phospho-uridyl-transférase (T), et de l'uridine-diphosphoglu-
cose pyrophosphorylase, mais sont déficients en uridyl-diphospho-
galactose-4-épimerase (E). Lorsqu'on les fait pousser sur du glucose,
ils peuvent synthétiser certains des intermédiaires du métabolisme
du galactose, mais sont incapables d'effectuer la synthèse d'uridyl-
diphospho-galactose, à partir de l'uridyl-diphosphoglucose. La
synthèse du polysaccharide de la paroi s'arrête au stade où devrait
s'incorporer le galactose, et les autres sucres ne peuvent plus être
"ajoutés", par défaut du "substrat" nécessaire. Si on les fait pousser
en présence de galactose, (et qu'on recueille les germes peu de temps
après cette addition, avant la lyse), on constate la présence, dans la
paroi, d'un polysaccharide complexe, analogue au polysaccharide
O de la forme Smooth dont ces mutants dérivent. L'apport exogène
de galactose a permis de compenser la déficience en épimérase,
l'uridyl-diphosphogalactose ayant été synthétisé par l'intermédiaire
de la transférase. La synthèse du polysaccharide doit donc se faire
par additions successives, dans un ordre génétiquement déterminé,
des divers sucres, par l'intermédiaire de transférases spécifiques. On
conçoit toutes les variations possibles par mutation dans un système
de ce type. Les formes Rough proprement dites constituent un de
ces groupes. La composition en sucres de leur polysaccharide est
relativement simple: ils appartiennent tous au chemotype I de
KAUFFMANN & WESTPHAL[39, 90, 91] (Tabl. II), caractérisé par la présence de glucosamine, d'un heptose, de glucose et de galactose. Ces
sucres ont été seuls retrouvés dans les produits d'hydrolyse des formes Rough des divers "antigènes" et des diverses "toxines" solubles
que l'on a préparées à partir des entérobactéries Rough[3, 38, 92, 93, 94].

L'heptose présent a été identifié dans quelques cas de façon certaine (cf. [38]). Chez Escherichia coli B, WEIDEL[92] a trouvé le L-
glycero-D manno-heptose. WESTPHAL[94] a été ainsi amené à considérer le polysaccharide Rough, comme un polysaccharide de base,
toujours présent dans toutes les formes d'une espèce donnée de
Salmonella (ou d'entérobactéries) Smooth ou Rough. Ce polysaccharide aurait toujours la même structure pour les germes appartenant
à une espèce donnée. En raison de leur similitude générale, ces polysaccharides R correspondant à des espèces diverses présentent entre
eux des réactivités croisées plus ou moins importantes. Sur cette
structure de base, viendraient se greffer les chaînes latérales dont la
structure conditionne la spécificité des divers facteurs O.

Les formes Rough présenteraient une déficience enzymatique qui rendrait impossible la fixation sur le polysaccharide central de ces chaînes latérales. La nature exacte de cette déficience reste à préciser: elle se manifeste par l'accumulation chez certaines formes Rough, des produits du métabolisme de ces synthèses, en particulier des diphosphonucléosides des sucres divers présents dans les polysaccharides O et absents dans le polysaccharide R (STROMINGER, cité d'après[94]).

LÜDERITZ, BECKMANN & WESTPHAL (cités d'après [94]), ont observé que les lipopolysaccharides R donnent entre eux des réactions croisées intenses, lorsqu'on les étudie par hémagglutination passive après qu'ils aient subi une légère hydrolyse alcaline. Les polysaccharides O de chemotype I conservent une réactivité R résiduelle dans ce test, car leurs déterminants R n'y seraient pas complètement masqués, les chaînes supplémentaires caractéristiques des déterminants O y étant peu nombreuses. Sur les polysaccharides O de chemotypes plus complexes, les chaînes ou les déterminants supplémentaires seraient disposées de façon à "masquer" la réactivité R. Cette dernière peut être démasquée dans certains cas particuliers et favorables par une hydrolyse ménagée du lipopolysaccharide. C'est ce qu'ont réalisé LÜDERITZ, BECKMANN & WESTPHAL (cités d'après [94]), dans le cas de l'antigène O:35 de *S. adelaïde*.

Si nos connaissances relatives aux diverses mutations s'accompagnant de la perte de la réactivité O, dans le cas des entérobactéries, ont fait de grands progrès, il faut cependant faire un certain nombre de remarques importantes qui permettent de mieux préciser les nombreux points encore mal, ou non élucidés.

1°. La structure chimique des déterminants des divers polysaccharides de base n'est pas connue. On sait depuis KAUFFMANN *et al.*[39], que l'heptose est présent sous forme d'un polyphosphate d'heptose. La liaison ester phosphorique dans cette combinaison présente une résistance relative à l'hydrolyse acide (4 heures à $100°$ en SO_4H_2 N/1). La nature des autres liaisons reste à déterminer.

2°. Contrairement aux conceptions anciennement admises, les formes Rough contiennent, comme les formes Smooth, une endotoxine. Cette dernière peut être obtenue sous forme soluble par diverses méthodes: solutions hypertoniques[95, 96, 97], mélange phénol eau[9, 98], hydroxylamine[99]. Les germes Rough ont une teneur en endotoxine plus faible que celle des formes Smooth, mais l'endotoxine une fois obtenue sous forme soluble, est par unité de poids, aussi toxique que l'endotoxine que l'on peut obtenir à partir des formes Smooth [3, 96, 97].

3°. Le polysaccharide R présent sur les bactéries Rough, ne constitue qu'un des antigènes de surface du germe[3, 100].

Nous avons pu[3] par emploi de sérums de chevaux hyperimmunisés par injection de corps microbiens tués d'une variante Rough

($S.\ typhi$ souche R_2), distinguer par précipitation spécifique en gel, 7 antigènes distincts, dont 4 plus abondants, dans les extraits bactériens bruts de $S.\ typhi$ Rough. L'un d'entre eux (que nous avons désigné par antigène r4), correspond vraisemblablement à l'antigène polysaccharidique R que l'on peut étudier par les réactions d'hémagglutination passive. Les autres antigènes n'ont pas encore été caractérisés chimiquement, mais 3 d'entre eux ne sont pas spécifiques des variantes Rough, et se retrouvent dans les formes Smooth. Il existe donc des antigènes communs aux variantes Smooth et Rough dont la nature chimique demande à être précisée.

Antigène Vi

L'antigène Vi intéresse les microbiologistes à un double titre: par ses propriétés immunologiques (outre l'agglutination Vi, il est en partie responsable de la virulence des souches de $S.\ typhi$ chez l'homme), et par sa propriété de fixer les phages Vi [101, 102, 103]. Ceux-ci semblent toutefois se fixer indifféremment sur toutes les $S.\ typhi$ Vi+, quel que soit leur type phagique, leur différence ne s'accusant que dans les stades suivants (pénétration ou développement?).[104]

Si les antigènes O et H des Salmonella possèdent des spécificités multiples, l'antigène Vi présent sur quelques Salmonella ($S.\ typhi$, $S.\ paratyphi\ C$), Escherichia $(E.\ coli)$ et sur les Citrobacter $(Ballerup)$ semble toujours porter les mêmes sites antigéniques à condition qu'il n'ait pas été dégradé par les traitements chimiques subis au cours de sa préparation[105]. Néanmoins, cette simplicité apparente cache en réalité certainement une complexité de sites antigéniques plus grande qu'on ne la soupçonnait naguère. En effet, on sait aujourd'hui que des anticorps de différentes spécificités sont précipités par l'antigène Vi aussi bien des immunsérums de cheval que de ceux du lapin. Le traitement par la soude de l'antigène Vi détruit certains sites antigéniques[106, 107] avant de dégrader complètement la molécule. De même l'hydrolyse acétique préconisée par LANDY et coll.[108] pour la purification, détruit certainement quelques sites antigéniques qui jouent un rôle essentiel pour la fixation des phages Vi [109]. Il ne semble pourtant pas que ces sites jouent un rôle dans l'activité biologique de l'antigène Vi, puisque l'antigène hydrolysé préparé par LANDY vaccine la souris contre l'infection massive par $S.\ typhi$ et que les anticorps des lapins immunisés par cette préparation protègent les souris contre la même infection[110]. Peut-être pourrait-on rattacher par contre à la destruction d'un des sites de l'antigène Vi, l'absence d'anticorps protecteurs dans les sérums d'animaux immunisés avec des bactéries formolées présentant par ailleurs un taux élevé d'agglutinines Vi[111, 112]. Toutefois aucune action du formol n'a pu être observée jusqu'à présent $in\ vivo$ sur la spécificité de l'antigène Vi.

C'est justement en vue de cette activité protectrice des anticorps Vi, admise ou contestée suivant les auteurs, qu'un grand nombre de travaux ont été entrepris pour extraire et purifier cet antigène. La difficulté de le séparer des antigènes O ou R a pendant longtemps arrêté les chercheurs. C'est à WEBSTER, LANDY & CLARK[108] que nous devons nos premières connaissances sur la nature chimique de l'antigène Vi; leurs préparations, très purifiées, étaient malheureusement légèrement dégradées, nous le savons maintenant, par l'utilisation de l'hydrolyse acétique.

Pour éviter cette dégradation, BAKER, WHITESIDE et coll.[113, 114] ont précipité l'antigène Vi des extraits par le "cétavlon" qui ne précipite pas l'antigène O. Par contre, on précipite aussi l'acide nucléique qu'il faut éliminer soit par précipitation alcoolique fractionnée[113], soit par élution fractionnée avec le NaCl[109].

Un moyen élégant a été aussi proposé par KOJINSKI et coll.[115] et repris par TAYLOR & TAYLOR[116]. Ces auteurs utilisent la propriété de l'antigène Vi de se fixer sur les globules rouges de diverses espèces (en particulier du mouton) pour adsorber sélectivement l'antigène Vi qu'ils éluent ensuite.

Ces deux dernières méthodes fournissent des préparations d'antigène Vi qui ont des propriétés immunologiques pratiquement identiques: elles précipitent toutes deux plus d'anticorps des immunsérums anti Vi que les préparations de LANDY. Elles inhibent aussi toutes deux le phage Vi 2, beaucoup plus activement que celles-ci. Enfin, l'hydrolyse par l'acide acétique de la substance préparée par le cétavlon lui fait perdre 95% de son activité antiphagique et concurremment détruit un (ou plusieurs) site antigénique[109].

Depuis les travaux de CLARK, MCLAUGHLIN & WEBSTER, on sait que l'antigène Vi est un polyoside acide essentiellement (sinon totalement) constitué par des unités d'un acide hexuronique aminé[117] polymérisé à des degrés divers selon les préparations et même le plus souvent dans une même préparation[110, 109]. HEYNS et coll. ont établi qu'il s'agissait d'acide galactaminuronique[119]. BAKER a montré qu'une partie des groupements acyls étaient labiles en milieu faiblement alcalin et devaient par conséquent être des groupements-O-acyls. Ceci a été confirmé et la nature acétyl de ces groupements établie par GERFAUX et coll.[109]. De plus, une analyse plus poussée a montré que les groupements O-acétyls étaient plus nombreux (environ 2 fois) que les groupements N-acyls dont la nature exacte n'a d'ailleurs pas encore été déterminée. Rien ne nous permet de savoir actuellement si les autres groupements NH_2 sont libres ou engagés dans quelqu'autre liaison. De même nous ne savons pas si tous les groupes-COOH sont libres et nous ignorons si les unités sont liées en 1—3 avec acétylation du OH porté sur le carbone 4 ou s'il s'agit d'une liaison 1—4 avec estérification du OH porté par le carbone 3 (schéma de la fig. 5).

164

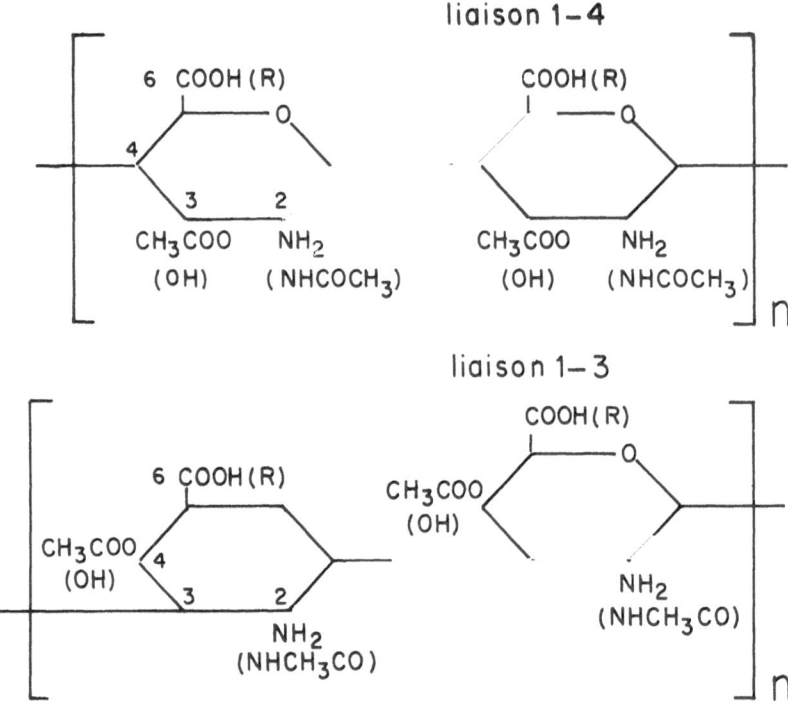

liaison 1–4

6 COOH (R)

4

3   2

CH₃COO   NH₂
(OH)   (NHCOCH₃)

COOH(R)

CH₃COO   NH₂
(OH)   (NHCOCH₃)

liaison 1–3

6 COOH(R)

CH₃COO
(OH)

4

3   2

NH₂
(NHCH₃CO)

COOH(R)

CH₃COO
(OH)

NH₂
(NHCH₃CO)

Fig. 5. Formules possibles pour l'antigène Vi (en supposant l'acide galactaminuronique sous forme pyranose)

Sur le carbone 6 nous ne savons pas si tous les -COOH sont libres ou si une partie est estérifiée (R).

Sur le carbone 2 peu des groupements aminés sont acétylés, les autres sont-ils libres?

Sur le carbone 3 ou 4 certains des OH sont acétylés, quelques uns ne le sont pas.

La complexité de l'unité du polymère qui constitue l'antigène Vi peut expliquer la complexité des sites antigéniques portés par cet antigène; malheureusement il est impossible actuellement de rattacher avec certitude une constitution chimique définie à une spécificité immunologique et encore moins une activité biologique aux anticorps correspondants. Il semble en effet bien acquis que les groupes O-acétyls jouent un rôle dans la spécificité mais la destruction de la molécule entière par la soude avant l'élimination de tous les O-acétyls empêche pour le moment d'obtenir des anticorps purement anti-O-acétyl et d'étudier leur pouvoir protecteur. La labilité de certains sites vis-à-vis de l'acide acétique est acquise mais nous ignorons encore tout de leur nature chimique si importante pourtant, sinon pour l'activité vaccinante (sur la souris) du moins pour l'activité antiphagique de l'antigène Vi.

Enfin, nous ne savons rien du rôle des autres facteurs (COOH, NH₂ libres s'ils existent) dans la spécificité de l'antigène, sans mentionner même l'éventualité d'un rôle des acides aminés toujours décelés jusqu'à présent dans les hydrolysats d'antigène Vi.

Antigène H**

Nos connaissances sur la structure des antigènes H sont beaucoup plus limitées que celles sur la structure des antigènes O ou R et Vi: il s'agit en effet non plus de polyosides mais de protéines et l'on sait que les chimistes commencent seulement à relier la spécificité des sites antigéniques de ces molécules à des structures chimiques bien déterminées, voire même dans certains cas à la présence de certains acides aminés[120, 121, 122].

Les récentes revues de WEIBULL[123] et de KOFFLER[124] ne donnent que la composition en acides aminés de certaines "flagellines". Tel est le nom, en effet, attribué à la substance purifiée extraite des flagelles et qui est constituée pour 99% au moins de protéine*. Ce sont les résultats de travaux encore plus récents en partie inédits qui nous permettent aujourd'hui d'entrevoir les différences chimiques qui accompagnent les différences sérologiques, en attendant de les analyser en détail.

Tout d'abord il ressort des travaux de STOCKER et de ses collaborateurs que les antigènes H dénombrés dans le tableau de KAUFFMANN-WHITE sont certainement beaucoup plus complexes qu'ils ne le paraissent d'après leur désignation: c'est ainsi que JOYS & STOCKER[125] ont pu obtenir par mutation 7 formes sérologiquement distinctes de l'antigène i de *S. typhi-murium* en phase 1.

L'étude de la composition en acides aminés des flagellines, extraites de différents antigènes, actuellement en cours[126], indique d'ores et déjà des différences considérables entre les antigènes g et i par exemple; bien plus, il a été possible de constater une différence sérologique entre les flagelles de *S. typhi-murium* ne différant que par la présence ou l'absence de N-méthyl-lysine[127].

Toutefois des techniques d'analyse plus fines que la simple analyse des acides-aminés permettent de différencier chimiquement les antigènes flagellaires 1, 2 et i par exemple, portés tous deux par la même Salmonella: *S. typhi-murium*. Les biochimistes ont actuellement la possibilité de séparer ou tout au moins de dénombrer les

---

* Les traces d'antigène O présentes dans ces préparations ne perturbent en rien les conclusions de l'analyse chimique bien qu'elles soient suffisantes pour provoquer toujours la formation d'anticorps O avec les anticorps H lorsqu'on immunise les lapins avec ces substances purifiées.

** Nous remercions le Dr. STOCKER de nous avoir fourni toute la documentation de ce chapitre ainsi que des résultats personnels inédits et le Dr. AMBER de nous avoir permis de faire état de ses résultats personnels inédits.

166

polypeptides d'une protéine après digestion enzymatique de cette dernière. Une électrophorèse sous haut voltage, suivie d'une chromatographie, permet d'établir la "carte"* de ces polypeptides révélés sous forme de "taches" par la ninhydrine. De telles "cartes" établies avec les antigènes i et 1, 2 sont nettement différentes puisque 16 seulement des 30 taches de i se retrouvent dans 1, 2[128].

Les différences deviennent encore moins marquées lorsque l'on compare les cartes des 7 mutants de l'antigène i isolés par JOYS & STOCKER, sérologiquement distincts, bien que présentant des réactions croisées toujours très fortes. Les cartes des polypeptides obtenus avec 4 de ces mutants, ont montré la disparition ou la modification d'un polypeptide. Pour les 3 autres aucune différence n'a pu encore être décelée[128].

Tous ces travaux évoquent donc très fidèlement le dénombrement des oligosides entre deux mutants ou 2 sérotypes porteurs de facteurs O propres et communs, et l'on peut espérer que sous peu l'identification de ces polypeptides permettra de rattacher telle mutation à telle modification chimique comme dans l'étude des facteurs O.

Conclusion

Dans ce très court résumé de nos connaissances actuelles des antigènes O, R, H et Vi, nous nous sommes efforcés de dégager les notions actuelles de spécificité "chimique" puisque dans tous ces antigènes nous pouvons relier la spécificité immunologique de quelques antigènes particulièrement étudiés, à des structures chimiques bien définies pour les antigènes somatiques, moins nettement déterminées encore dans le cas des antigènes H et Vi, mais qui le seront certainement sous peu.

Ce chapitre a souligné aussi l'unité de composition des divers antigènes dans le groupe des Salmonella: unité des antigènes Vi et R, connue déjà depuis longtemps, différences de composition et de mode de combinaison seulement pour les antigènes O dont les sucres constitutifs se limitent à un très petit nombre et dont la séquence peut ne varier aussi que très peu d'un sérotype à l'autre, même lorsqu'ils sont sérologiquement aussi éloignés que les groupes B et E; différences de la séquence des acides aminés souvent identiques pour les antigènes H.

Ces différences mineures dans la constitution chimique des antigènes propres à chaque sérotype suggèrent donc, comme les études génétiques, l'unité de l'espèce des Salmonella.

---

* On dit aussi "finger printing".

BIBLIOGRAPHIE

1. BRODHAGE H., Z. Hyg. InfektKr., 1962, 148, 208.
2. WHITESIDE R. E. & BAKER E. E., J. Immunol., 1962, 88, 650.
3. RAYNAUD M., DIGEON M. & NAUCIEL C., "Symposium on endotoxins" Rutgers University — New Brunswick, 1963. DIGEON M., Thèse Faculté des Sciences, Paris, 1955.
4. BOIVIN A. & MESROBEANU L., C.R. Soc. Biol., 1933, 114, 307. BOIVIN A. & MESROBEANU L., Rev. Immunol., 1935, 1, 553.
5. RAISTRICK H. & TOPLEY W. W. C., Brit. J. exp. Path., 1934, 15, 113.
6. STAUB A. M., Ann. Inst. Pasteur, 1954, 86, 618. DAVARPANAH C., Thèse Faculté Paris, 1956.
7. BOIVIN A. & MESROBEANU L., Rev. Immunol., 1938, 4, 197.
8. MORGAN W. T. J. & PARTRIDGE S. M., Biochem. J., 1940, 34, 169. MORGAN W. T. J. & PARTRIDGE S. M., Biochem. J., 1941, 35, 1140.
9. WESTPHAL O., LÜDERITZ O., EICHENBERGER E. & KEIDERLING W., Z. Naturforsch., 1952, 7b, 536.
10. WESTPHAL O. & LÜDERITZ O., Angew. Chem., 1954, 66, 407. WESTPHAL O., Ann. Inst. Pasteur, 1960, 98, 789.
11. STAUB A. M., Résultats inédits.
12. LANDY M., JOHNSON A. G., WEBSTER M. E. & SAGIN J. F., J. Immunol., 1955, 74, 466.
13. MESROBEANU I., GEORGESCA M., IEREMIA T., PAPAZIAN E., DRAGHICI D & MESROBEANU L., Arch. roum. Path. exp., 1960, 19, 345. MESRO-BEANU L., MESROBEANU I. & MITRICA N., Arch. roum. Path. exp., 1961, 20, 399. MESROBEANU I., MESROBEANU L., GEORGESCO M., DRAGHICI D., ALAMITA E. & IEREMIA T., Arch. roum. Path. exp., 1962, 21, 19. MESROBEANU L., MESROBEANU I. & MITRICA N., Arch. roum. Path. exp. 1962, 21, 31.
14. BHATNAGAR S. S., Brit. J. exp. Path., 1935, 16, 375.
15. BRAUN H. & NODAKE R., Zbl. Bakt., I. Abt. Orig., 1924, 92, 429.
16. BIOZZI G., STIFFEL C., LE MINOR L., MOUTON D. & BOUTHILLIER Y. Ann. Inst. Pasteur, 1963, 105, 635.
17. ARKWRIGHT J. A., J. Path. Bact., 1927, 30, 345.
18. STAUB A. M. & LIEBERMAN. R., Résultats inédits donnés dans ([23]).
19. BIOZZI G., STIFFEL C., HALPERN B. N. & MOUTON D., Rev. franç. Etud. Clin. biol., 1960, 5, 876.
20. BOIVIN A. & DELAUNAY A., Ann. Inst. Pasteur, 1946, 72, 139.
21. GRABAR P. & OUDIN J., Ann. Inst. Pasteur, 1947, 73, 627.
22. Yugoslav Typhoid Commission, Bull. O.M.S. 1957, 16, 897.
23. WESTPHAL O. & STAUB A. M., Bact. Rev., 1964 (en préparation).
24. WHITE P. B., J. Path. Bact., 1931, 34, 325.
25. HORNUS J. P. & GRABAR P., Ann. Inst. Pasteur, 1941, 66, 136.
26. STAUB A. M. & COMBES R., Ann. Inst. Pasteur, 1951, 80, 21.
27. FREEMAN G. G., Biochem. J., 1942, 36, 340.
28. WHITE P. B., J. Path. Bact., 1929, 32, 85.
29. THOMAS J. C. & MENNIE A. T., Lancet 1950, 259, 745.
30. STAUB A. M. & PON G., Ann. Inst. Pasteur, 1956, 90, 441.
31. STOCKER B. A. D., STAUB A. M., TINELLI R. & KOPACKA B., Ann. Inst. Pasteur, 1960, 98, 505.
32. KOTELKO K., STAUB A. M. & TINELLI R., Ann. Inst. Pasteur, 1961, 100, 618.
33. LÜDERITZ O., O'NEILL G. & WESTPHAL O., Biochem. Z., 1960, 333, 136.
34. FURTH J. & LANDSTEINER K., J. exp. Med., 1929, 49, 727.
35. MEYER K., Ann. Inst. Pasteur, 1939, 62, 282.
36. NAKAYA R. & FUKUMI H., Jap. J. med. Sci. Biol., 1953, 6, 17.
37. COHEN H. H., J. Immunol., 1958, 81, 445.

168

38. DAVIES D. A. L., *Advanc. Carbohyd. Chem.*, 1960, **15**, *271*.
39. KAUFFMANN F., LÜDERITZ O., STIERLIN H. & WESTPHAL O., *Zbl. Bakt.*, I. Abt. Orig., Parasit., 1960, **178**, *442*.
40. KAUFFMANN F., JANN B., KRUGER L., LÜDERITZ O. & WESTPHAL O., *Zbl. Bakt.* I. Abt. Orig., 1962, **186**, *509*.
41. WESTPHAL O. & LÜDERITZ O., *Angew. Chem.*, 1960, **72**, *881*.
42. STAUB A. M. & TINELLI R., *Bull. Soc. Chim. biol.*, 1957, **39**, Suppl. 1, *65*.
43. STAUB A. M., TINELLI R., LÜDERITZ O. & WESTPHAL O., *Ann. Inst. Pasteur*, 1959, **96**, *303*.
44. STAUB A. M. & FOREST N., *Ann. Inst. Pasteur*, 1963, **104**, *371*.
44a. STAUB A. M. & BAGDIAN G., (Résultats inédits).
45. STAUB A. M. & GIRARD R., *Ann. Inst. Pasteur* (en préparation).
46. LANDSTEINER K., "The specificity of serological reactions" Harvard University Press, 1947.
47. KARUSH F., *Trans. N.Y. Acad. Sci. Ser.* II, 1958, **20**, *581*.
48. KABAT E. A., *J. Cell. comp. Phys.*, 1957, **50**, Suppl. 1. *79*.
49. KABAT E. A. & MAYER M. M., "Experimental Immunochemistry" Ch. C. Thomas, Springfield 1961 (2° édition) Chap. 5, 241.
50. KABAT E. A., *Fed. Proc.*, 1962, **21**, *694*.
51. WATKINS W. M. & MORGAN W. T. J., *Vox Sang.*, 1962, **7**, *129*.
52. HEATH E. C. & GHALAMBOR M. A., *Biochem. Biophys. Res. Comm.*, 1963, **10**, *340*.
53. SCHLOSSMAN S. F. & KABAT E. A., *J. exp. Med.*, 1962, **116**, *535*.
54. LE MINOR L., ACKERMANN H. W. & NICOLLE P., *Ann. Inst. Pasteur*, 1963, **104**, *469*. *Ann. Inst. Pasteur*, 1963, **105**, *879*.
55. ROBBINS P. W. & UCHIDA T., *Biochemistry*, 1962, **1**, *323*. ROBBINS P. W. & UCHIDA T., *Fed. Proc.*, 1962, **21**, *702*. UCHIDA T, ROBBINS P. W. & LURIA S. E., *Biochemistry* 1963, **2**, *663* et communication personnelle.
56. ISEKI S. & SAKAI T., *Proc. Jap. Acad.*, 1953, **29**, *121*.
57. UETAKE H., NAKAGAWA T. & AKIBA T., *J. Bact.*, 1955, **69**, *571*.
58. ISEKI S. & KASHIWAGI K., *Proc. Jap. Acad.*, 1955, **31**, *558*. ISEKI S. & KASHIWAGI K. *Proc. Jap. Acad.*, 1957, **33**, *481*.
59. STOCKER B. A. D., *J. gen. Microbiol.*, 1958, **18**, *IX*.
60. ZINDER N. D., *Science*, 1957, **126**, *1237*.
61. STAUB A. M. & TINELLI R., *Bull. Soc. Chim. biol.*, 1960, **42**, *1637*.
62. STAUB A. M., *Path. et Microbiol.*, (Basel) 1961, **24**, *890*.
63. TINELLI R., *Proc. 5th Int. Cong. Biochem.* (Moscou 1961), **9** (n° 20—48) *483*.
64. LE MINOR L., LE MINOR S. & NICOLLE P., *Ann. Inst. Pasteur*, 1961, **101**, *571*.
65. LE MINOR L., *Ann. Inst. Pasteur*, 1962, **103**, *684*.
66. TINELLI R., *Bull. Soc. Chim. biol.*, 1961, **43**, *357*.
66a. TINELLI R. & STAUB A. M., *Bull. Soc. Chim. biol.*, 1960, **42**, *583*.
66b. TINELLI R. & STAUB A. M., *Bull. Soc. Chim. biol.*, 1960, **42**, *601*.
67. WILKINSON J. F., *Bact. Rev.*, 1958, **22**, *46*.
68. KROGER E., *Z. Naturforsch.*, 1953, **8b**, *133*.
69. KROGER E., *Zbl. Bakt.* I. Abt. Orig., 1953—54, **160**, *242*.
70. KROGER E., *Naturwissenschaften*, 1955, **42**, *629*.
71. PRIGGE R., *Zbl. Bakt.* I. Abt. Orig., 1939, **144**, *4*.
72. PRIGGE R. & KICKSCH L., *Z. Hyg. InfektKr.*, 1941, **123**, *417*.
72a. WESTPHAL O. & LÜDERITZ O., *Naturwissenschaften*, 1963, **50**, *413*.
73. WHITE P. B., *J. Path. Bact.*, 1933, **36**, *65*.
74. WEIDEL W., FRANK H. & MARTIN H. H., *J. gen. Microbiol.*, 1960, **22**, *158*.
75. FUKASAWA T., JOKURA K. & KURAHASHI K., *Biochem. Biophys. Res. Comm.*, 1962, **7**, *121*.

76. SUNDARARAJAN T. A., RAPIN A. M. C. & KALCKAR H. M., *Proc. Nat. Acad. Sci.*, 1962, **48**, *2187*.
77. ROTHFIELD L., FRAENKEL D. & OSBORN M. J., *Ann. Soc. exp. Biol.* Meeting Atlantic City 1963. Abstract. N° 1823.
78. MURASE W., *Jap. J. Bact.*, 1932, N° 440 *975* (en japonais) cité d'après: FUKASAWA T. & NIKAIDO H. in: *Nature, Lond.* 1959, **183**, *1131*.
79. FUKASAWA T. & NIKAIDO H. *Nature, Lond.* 1959, **184**, *1169*.
80. FUKASAWA T. & NIKAIDO H., *Virology*, 1960, **11**, *508*.
81. FUKASAWA T. & NIKAIDO H., *Genetics* 1961, **46**, *1295*.
82. FUKASAWA T. & NIKAIDO H., *Biochem. biophys. Acta*, 1961, **48**, *470*.
83. NIKAIDO H. & FUKASAWA T., *Biochem. Biophys. Res. Comm.*, 1961, **4**, *338*.
84. NIKAIDO H., *Biochim. biophys. Acta*, 1961, **48**, *460*.
85. NIKAIDO H., *Proc. Nat. Acad. Sci.*, 1962, **48**, *1337*.
86. NIKAIDO H., *Proc. Nat. Acad. Sci.*, 1962, **48**, *1542*.
87. KALCKAR H. M. & SUNDARARAJAN T. A. *Cold Spring Harbor Symp. quant. Biol.*, 1961, **26**, *227*.
88. OSBORN M. J., ROSEN S. M., ROTHFIELD L. & HORECKER B. L., *Science*, 1962, **136**, *328*.
89. ROSEN S. M. & HORECKER B. L., *Ann. Soc. exp. Biol. Med.*, Meeting Atlantic City, 1963, Abstract. N° 1822.
89 bis. BRAUN (W), *Bact. Rev.*, 1947, **11**, *75*.
90. LÜDERITZ O., KAUFFMANN F., STIERLIN H. & WESTPHAL O., *Zbl. Bakt.*, I. Abt. Orig., 1960, **179**, *180*.
91. KAUFFMANN F., KRUGER L., LÜDERITZ O & WESTPHAL O., *Zbl. Bakt.* I. Abt. Orig., 1961, **182**, *57*.
92. WEIDEL W., *Z. physiol. Chem.*, 1955, **299**, *253*.
93. DAVIES D. A. L., *Biochim. biophys. Acta*, 1957, **26**, *151*.
94. WESTPHAL O. & LÜDERITZ O., *Path. et Microbiol.*, Basel 1961, **24**, *870*.
95. RAYNAUD M. & DIGEON M., *C.R. Acad. Sci.*, 1949, **229**, *564*.
96. DIGEON M., RAYNAUD M. & TURPIN A., *Ann. Inst. Pasteur*, 1952, **82**, *206*.
97. DIGEON M. & RAYNAUD M., *Ann. Inst. Pasteur*, 1957, **92**, *642* et *790*. DIGEON M. & RAYNAUD M., *Ann. Inst. Pasteur*, 1957, **93**, *91* et *390*.
98. WESTPHAL O. & LÜDERITZ O., *Atti VI Congr. Int. Microbiol.*, Rome, 1953, **2**, *22*.
99. RAYNAUD M. & DIGEON M., *C. R. Acad. Sci.*, 1960, **251**, *985*.
100. RAYNAUD M., Les Endotoxines (à paraître aux "Frontières de la Biologie Médicale" (Fasquelle Ed.) Flammarion).
101. BARON L. S., FORMAL S. B. & SPILMAN W., *J. Bact.*, 1955, **69**, *177*.
102. BERNARD R. & STAUB A. M., in (62).
103. TAYLOR K. & TAYLOR A., *Acta microbiol. polon.*, 1963, **12**, *97*.
104. CRAIGIE J. & BRANDON K. F., *J. Path. Bact.*, 1936, **43**, *233*. SCHOLTENS R. TH., *J. Hyg.*, 1936, **36**, *452*. SCHOLTENS R. TH., *J. Hyg.*, 1937, 37, *315*. SERTIC V. & BOULGAKOV N., *C. R. Soc. Biol.*, 1936, **122**, *35*.
105. WHITESIDE R. E. & BAKER E. E., *J. Immunol.*, 1961, **86**, *538*.
106. WHITESIDE R. E. & BAKER E. E., *J. Immunol.*, 1960, **84**, *221*.
107. LANDY M., JOHNSON A. G. & WEBSTER M. E., *Amer. J. Hyg.*, 1961, **73**, *55*.
108. WEBSTER M. E., LANDY M. & FREEMAN M. E., *J. Immunol.*, 1952, **69**, *135*.
109. GERFAUX G., BERNARD R. & STAUB A. M., *Comm. Soc. Chim. biol.* 1er Mars 1963.
110. LANDY M. & WEBSTER M. E., *J. Immunol.*, 1952, **69**, *143*.
111. LE MINOR L., Communication personnelle.
112. FELIX A. & BHATNAGAR S. S., *Brit. J. exp. Path.*, 1935, **16**, *422*.
113. BAKER E. E., WHITESIDE R. E., BASCH R. & DEROW M. A., *J. Immunol.*, 1959, **83**, *680*.

114. WHIDESIDE R. E. & BAKER E. E., *J. Immunol.*, 1959, **83**, *687.*
115. KOZINSKI A. W., MACIEREWICZ M., MIKULASZEK E. & OPARA Z., *Bull. Acad. polon. Sci., Cl.* II 1954, **2**, *33.*
116. TAYLOR A., *Bull. Acad. polon. Sci.*, (en préparation).
117. CLARK W. R., MC LAUGHLIN J. & WEBSTER M. E., *J. biol. Chem.*, 1958, **230**, *81.*
118. WEBSTER M. E., SAGIN J. F., ANDERSON P. R., BREESE S. S., FREEMAN M. E. & LANDY M., *J. Immunol.*, 1954, **73**, *16.*
119. HEYNS K., KIESSLING G., LINDENBERG W., PAULSEN H. & WEBSTER M. E., *Chem. Ber.*, 1959, **92**, *2435.*
120. HARRIS J. I. & KNIGHT C. A., *J. biol. Chem.*, 1955, **214**, *215.*
121. ANDERER F. A., *Biochim. Biophys. Acta*, 1963, **71**, *246.* ANDERER F. A. & HANDSCHUCH D., *Z. Naturforschung*, 1962, **17b**, *536.*
122. LEVINE L., *Fed. Proc.*, 1962, **21**, *711.*
123. WEIBULL C., "The bacteria" 1960, vol. **I**, 153. (Academic Press, N.Y., Editeur).
124. KOFFLER H., *Bact. Rev.*, 1957, **21**, *227.*
125. JOYS T. M. & STOCKER B. A. D., *Nature, Lond.* 1963, **197**, *413.*
126. AMBLER R. P. & REES M. W., *Nature, Lond.* 1959, **184**, *56.* AMBLER R. P., (comm. pers.).
127. STOCKER B. A. D., Communication personnelle de résultats inédits.
128. MC. DONOUGH M. W., *Biochem. J.* 1962, **84**, *114P.*

DIE SALMONELLOSEN DER HAUSTIERE UND IHRE EPIDEMIOLOGISCHE BEDEUTUNG

VON

HANS FEY

Veterinär-Bakteriologisches Institut der Universität Bern

Von zahlreichen Autoren wird immer wieder betont, dass die Salmonellose von Mensch und Tier in den letzten 10 Jahren in stetem Zunehmen begriffen ist[26, 40, 47, 63, 101, 106, 107, 117, 123]. Einen Begriff von der Häufigkeit menschlicher Infektionen geben die Zahlen von COCKBURN in England, wo dank dem Publich Health Service am ehesten ein einigermassen reales Bild zu erwarten ist, obwohl auch dort nur ein Bruchteil der tatsächlichen Erkrankungen dem Health officer bekannt wird (MC CALL). COCKBURN rapportierte aus den Jahren 1949–58 über 4580 allgemeine Ausbrüche, 4800 Familieninfektionen und 46000 sporadische Fälle. Es kann also damit gerechnet werden, dass in Ländern wie England, Deutschland und neuerdings auch Holland und Belgien (KAMPELMACHER[62], VAN OYE) jährlich einige 10000 Personen an Salmonellose erkranken.) Diese Zunahme erstreckt sich aber nicht auf sämtliche Salmonellaspecies gleichmässig. Vielmehr beobachtete SEELIGER[106] in Deutschland eine dauernde Abnahme von *S. typhi* und *S. paratyphi B* (1959), dagegen eine Zunahme der andern Salmonellen bei Mensch und Tier. Dasselbe berichtete EDWARDS[26] aus den USA. Nach übereinstimmender Meinung machte man für diese Verschiebung der Specieshäufigkeit den nach dem Krieg gewaltig angestiegenen Import von Eiprodukten für die menschliche und von Futtermehlen für die Tierernährung verantwortlich[9, 26, 31, 40, 47, 61, 63, 83, 84, 96, 97, 104, 106, 107, 117, 129]. EDWARDS[26] spricht von einer oft dramatischen Zunahme von Serotypen, die früher bei Mensch und Tier kaum oder gar nicht angetroffen wurden und sieht eine direkte Korrelation zwischen den Salmonellaspecies von Mensch und Tier in einer gegebenen Gegend. SCHOFIELD (zit. n. BUXTON[14]) betonte schon 1945, dass zwischen menschlicher und tierischer Salmonellose keine Grenze mehr anzuerkennen sei, dass ferner den sog. tierischen Stämmen eine ständig steigende Bedeutung für die Infektion des Menschen zukomme. Auch MÜLLER[84] sah in Dänemark eine gleichzeitige Zunahme von ungewöhnlichen Species bei Geflügel und Mensch. NEWELL[86] zeigte an 2 Beispielen die Frequenzabhängigkeit einer Salmonellaspecies beim Menschen von der tierischen Infektion: *S. thompson* trat beim Menschen im Anschluss an Importe kontaminierten chinesischen Eipulvers auf und ging zurück, nachdem die Einfuhr gestoppt wurde. Dasselbe wurde für *S. heidelberg* in Nordirland im Zusammenhang mit importiertem Schweinefutter beobach-

tet. In Deutschland[107] scheinen die getroffenen Massnahmen
(Ausschaltung von Salmonellen in Lebens- und Futtermitteln,
systematische Erfassung von Ausscheidern in Lebensmittelbetrie-
ben) Früchte zu tragen, trat doch 1957 und noch deutlicher 1958
ein Rückgang der menschlichen Salmonellainfektionen ein (1956:
9928, 1958: 6785), und auch bei Tieren war ein 12%iger Rückgang
zu verzeichnen.

Die skizzierte Zunahme tierischer Infektionen mit immer neuen
Salmonellaspecies, die offenbar auf der ganzen Welt zu beobachten
ist, führte nun aber nicht zu breiten Epidemien beim Menschen,
sondern in der grossen Mehrheit der Fälle zu sporadischen Ausbrüchen (EDWARDS[25], COCKBURN). Der Kieldoktrin, nach der es beim
Menschen neben Typhus und Paratyphus noch wohldefinierte
Massenausbrüche von Lebensmittelvergiftung gibt, wurde deshalb
von HORMAECHE (zit.n. EDWARDS[25]) die Montevideo-Doktrin gegen-
übergestellt, in der die sporadische Natur der Fälle und die Häufigkeit der Infektionen bei Kindern hervorgehoben wurden. Jedenfalls
hat es sich auch unter den neuen Verhältnissen gezeigt, dass das
Tier nach wie vor eine verhängnisvoll beherrschende Rolle als Reservoir menschlicher Salmonellainfektionen spielt, ja, dass die epidemiologische Erfassung der tierischen Infektionsquelle gegenüber
früher wohl noch schwieriger wurde, seitdem die symptomlosen
Ausscheider ständig an Häufigkeit zunehmen und demzufolge tierische Lebensmittel vermehrt Salmonellen beherbergen.

NEWELL[86] hat überzeugend dargetan, dass die Salmonellaquelle
(immer abgesehen vom Mensch-adaptierten Typhus-Paratyphus)
letztendlich beim Tier liegt und dass die Verhinderung der menschlichen Salmonellose auf lange Sicht durch Unterbruch des Tier-Tier-
Kontaktes und nicht des Tier-Menschkontaktes anzustreben sei.
Das Tier steht somit im Mittelpunkt der Salmonellaepidemiologie,
aber trotzdem glaube ich, dass die tierischen Salmonellosen heute
nicht mehr in erster Linie ein veterinär-medizinisches Problem
darstellen, sondern vor allem ein solches der öffentlichen Gesundheit!
Natürlich entstehen durch gewisse Salmonellainfektionen z.T.
schwere wirtschaftliche Schäden an Tierpopulationen (Pullorumruhr
der Kücken, S. cholerae-suis-Infektionen der Schweine, S. dublin-
Infektionen der Rinder und Kälber), aber diese sind mit genügend
Energie und Finanzen verhältnismässig gut unter Kontrolle zu
bringen, bzw. auszumerzen. Die ständig zunehmende Bedrohung des
Menschen durch die tierischen Salmonellosen ist dagegen ein viel
schwierigeres hygienisches Problem. Der Tierarzt darf sich deshalb
keinesfalls auf die Erkennung, Bekämpfung und Verhütung der
tierischen Salmonellosen beschränken. Sein Ziel muss es sein, die
Infektionen des Menschen zu verhindern, und es ist zu hoffen,
dass ihm bei seinen Bestrebungen die volle Unterstützung der
Humanmedizin zuteil werde, damit wirklich eine gemeinsame epi-

demiologische Forschung zustandekomme, die über die getrennt betriebene Diagnostik hinausführt.

Es ist aus diesem Grunde beabsichtigt, im folgenden bei der Beschreibung der tierischen Salmonellosen zwar die veterinär-medizinische und landwirtschaftliche Bedeutung dieser Krankheiten gebührend darzustellen, darüber hinaus aber das Hauptgewicht auf die Darstellung der Infektkette Tier-Mensch zu legen. Vieles, das unter den Abschnitten Rinder- und Kälbersalmonellose aufgeführt wird, hat allgemeine Gültigkeit und wird deshalb bei den andern Tierarten nicht mehr erwähnt.

Die Rindersalmonellose

Rind und Kalb waren lange Zeit die wichtigsten Salmonellaträger unter den Schlachttieren, stehen jetzt aber in vielen Ländern der Welt in bezug auf Häufigkeit hinter dem Geflügel und Schwein [24, 26, 63, 103, 119]. Wie bei nahezu allen Tieren ist S. typhi-murium die am häufigsten isolierte Species[22, 37, 60, 102], jedoch nicht in allen Ländern. Im Norden Deutschlands, in Holland, Dänemark, Schweden, Italien, Südafrika, Venezuela und Brasilien dominiert die an das Rind weitgehend adaptierte S. dublin, besonders beim Kalb [38, 46, 53, 63, 71, 97]. EDWARDS, BRUNER & MORAN[24] weisen darauf hin, dass S. dublin in den USA nur im Westen vorkommt. In der Schweiz (eigene Untersuchungen) wurde bislang S. dublin nicht nachgewiesen, hingegen S. enteritidis. S. enteritidis wurde früher zu häufig diagnostiziert, viele Stämme entpuppten sich bei der Faktorenanalyse als S. dublin. Neben diesen Hauptspecies wurden beim Rind viele weitere isoliert, vor allem bei Untersuchungen in Schlachthausmaterial. BRUNER & MORAN[12] fanden 15 Species, darunter S. cholerae-suis, nach S. dublin und S. typhi-murium an dritter Stelle. BUXTON[13] präsentiert eine Liste von 28 Species.

Klinik und Pathogenese

Rindvieh jeden Alters infiziert sich mit Salmonellen[38, 46], wobei es zur schweren klinischen Salmonellose mit Septikämie, fieberhaftem profusem Durchfall, Annorrhexie, allgemeiner Abmagerung, Milchrückgang, event. Abortus und u.U. tödlichem Ausgang kommen kann oder zu einem allerdings eher labilen Gleichgewicht zwischen Infekt und Körperabwehr[13, 67, 127], das zu Keimträger- bzw. symptomlosem Ausscheidertum führt. In schweren Fällen werden auch Euter, Uterus, Lungen und Gelenke infiziert, wobei natürlich die infizierte Milch eine schwere Gefahr für den Konsumenten in sich birgt. Die Lebensmittelinfektionen mit Salmonella-haltiger Milch werden denn auch als besonders explosiv beschrieben (McCALL, HENNING[53]).

An sich wird anerkannt, dass die schwer erkrankte Milchkuh Sal-

monellen mit der Milch ausscheiden kann (AVERBECK), aber die
wirklichen Euterinfektionen sind die Ausnahme. Meist wurde die
Milch beim Melkakt mit faekalen Salmonellen kontaminiert[14, 26,
52, 81, 127], und viele Milchepidemien wurden beschrieben, bevor
die symptomlosen Darmausscheider in ihrer hygienischen Bedeutung
voll erkannt waren (MEYER).

An sich gehört vor allem die durch S. dublin, aber auch durch
S. enteritidis verursachte schwere klinische Salmonellose des erwachsenen Rindes zu den primären Salmonellosen (vergleichbar dem
Typhus und dem durch S. parathyphi A, B und C verursachten Paratyphus des Menschen (CLARENBURG[18], LERCHE). Die Konzeption
der primären und sekundären Salmonellose geriet aber etwas ins
Wanken, seit die sog. streng wirtsspezifischen tierischen Salmonellen
beim Menschen gefunden wurden und umgekehrt (RASCH[93]). Davon
sei im Abschnitt über die Epidemiologie einiger wichtiger Species
noch die Rede. Nach TERPSTRA (zit. n. KOLLER) findet man auch in
primären Fällen, dass disponierende Momente, wie Ernährungsfehler,
Parasitenbefall, Transportschäden dem klinischen Ausbruch vor-
angingen, andererseits kann auch schon ein geringfügiges Grund-
leiden eine sekundäre Salmonellose in Gang bringen.

Es wird allgemein, und nicht nur für das Rind, angenommen,
dass die Salmonellen über den Verdauungstrakt in den Organismus
gelangen[7, 13]. Nachdem aber FEY et al.[36] für die Colisepsis des
Kalbes mit Hilfe von Oesophagotomie und nasaler Infektion bewiesen haben, dass die parenterale Infektion im Nasenrachenraum
wohl die grössere Rolle spielt, sei hiemit angeregt, den Infektions-
weg, z.B. der S. dublin-Infektion bei Rind und Kalb einer experimen-
tellen Neuüberprüfung zu unterziehen. Jedenfalls findet die erfolgreiche Infektion, die bei jüngeren Tieren leichter angeht als bei
älteren, in ihrer klinischen Manifestation immer eine septikämische
Ausbreitung[7, 18, 46, 53, 127].

Diese Form der Krankheit ist epidemiologisch, d.h. für den Fleisch-
konsumenten nicht so heimtückisch wie man annehmen möchte,
obwohl die bakteriologische Untersuchung in allen Organen und dem
Fleisch massenhaft Salmonellen nachweist. Es kommt zu einer
Notschlachtung, Tierarzt und Fleischbeschauer sind gewarnt, die
Vornahme einer bakteriologischen Fleischuntersuchung, die in
vielen Ländern zwingend vorgeschrieben ist, drängt sich auf und
ergibt die septikämische Verteilung des Erregers, was die Konfiskation nach sich zieht. Bei Missachtung der gebotenen Sorgfalts-
pflicht des Fleischbeschauers, der bei Allgemeinstörungen und allgemein konsumierenden Krankheiten, wie Enteritis, Peritonitis,
Puerperalkomplikationen, Panaritien u.ä. unter allen Umständen
eine bakteriologische Fleischuntersuchung veranlassen sollte, kann
es allerdings zur Katastrophe kommen, wenn solches Fleisch in den
Konsum gerät (ALLENSPACH).

Viel gefährlicher aber, weil mit klinischen und fleischbeschauli-
chen Mitteln nicht nachzuweisen, ist das Keimträgertum. Nach der
septikämischen Phase kommt es zur Absiedelung der Erreger in gewissen Organen, wobei der Galle-Leberkomplex bevorzugt wird
[7], [18], [37], [52], [60], [67]. RASCH[94] fand bei 325 wegen *S.
dublin* aus-
gemerzten Rindern 315 mit Salmonellen im Lebersystem, bei 201
Tieren (64%) war die Leber der einzige Fundort. Nun gelangen zwar
die grossen Parenchyme nicht zum Rohgenuss, aber die Möglichkeit
der Kontamination von Schlachtgeräten besteht. Bedenklicher ist
die Residualinfektion in den Fleisch- und Mesenteriallymphknoten.

Die von der Salmonellose klinisch geheilten Tiere bleiben, zum
mindesten bei *S. dublin*, für unbeschränkte Zeit, möglicherweise
lebenslänglich, Keimträger und konstante oder intermittierende
Ausscheider[7], [16], [18], [37], [46], [52], [67], [94], [102], [127]. Man hat versucht,
den Begriff des Dauerausscheiders zu definieren. Nach FIELD[38] ist
ein Rind als aktiver Ausscheider zu betrachten, wenn in 7–14
Tagen 3mal nacheinander eine positive Faeceskultur erhoben wurde.
Auch besteht ein deutscher Ministerialerlass vom 2.12.42 [7],
wonach ein Tier, welches länger als 30 Tage Enteritiserreger mit den
Faeces ausscheidet, als Dauerausscheider zu deklarieren und auszumerzen ist.

Die Treffsicherheit der Faeceskultur ist aber für solche Reglemen-
te doch wohl etwas überfordert. Die Kotausscheidung ist zwar meist
massiv und gestattet oft den Massennachweis auf der Direktplatte
[7]. RIEVEL (zit. n. AVERBECK) fand aber bei 4 Ausscheidern unregel-
mässige Befunde. Diese klinisch gesunden Keimträger und -aus-
scheider bilden eine Hauptsorge für den Fleischbeschauer, weil diese
Tiere mangels spezifischer Läsionen unerkannt geschlachtet und frei
verkauft werden[71], [22]. Auch ein sorgfältiger Fleischbeschauer er-
kennt nur Läsionen bei Tieren mit generalisierter Infektion[22], [67],
aber selbst bei klinisch gesunden Dauerausscheidern lassen sich nach
RASCH (zit. n. FROMME) bei 90% gar keine oder nur spärliche pathologische Veränderungen finden [18].

Obwohl nach KARSTEN (zit. n. AVERBECK) die bakteriologische
Untersuchung der Galle das sicherste Mittel zur Auffindung von
Dauerausscheidern darstellt, sind damit die Keimträger mit ander-
weitigen Lokalisationen, z.B. Lymphknoten, nicht erfasst, und wir
stehen vor der immer mehr beunruhigenden Tatsache, dass es praktisch unmöglich ist, den zur Schlachtung kommenden Keimträger
zu erkennen und dessen Fleisch vom Konsum fernzuhalten (BUX-
TON[14], RASCH[94]), denn eine generelle bakteriologische Untersuchung
aller Schlachttiere kommt kaum in Frage.

Diagnose

Nach BEJERS (zit. n. KOLLER) ist bei jeder fieberhaften Enteritis
des Rindes mit einer Salmonellose zu rechnen, auch sollte man so

weit kommen, dass bei jeder Notschlachtung nach schwerer All-
gemeinerkrankung vorsichtigerweise die bakteriologische Fleischuntersuchung angeordnet wird. Bei der klinisch manifesten Salmonellose mit septikämischer Verbreitung des Erregers und Enteritis
ist der bakteriologische Nachweis kein Problem. Die Keime sind aus
Organen und Faeces meist auf der Primärplatte in Massen nachzuweisen. Immer sollte aber der Leber-Gallekomplex in die bakteriologische Untersuchung einbezogen werden[7, 37, 60, 127]. Eventuell
können Milch und Harn mituntersucht werden [79]. Die Blutkultur
führt nicht sehr weit und ist unter Praxisverhältnissen nicht populär.
Sie wird 36-48 Stunden nach der Infektion positiv, die Erreger verschwinden aber bald wieder aus der Blutbahn, wenn die Septikämie
nicht tödlich verläuft (AVERBECK) und entziehen sich damit dem
Nachweis.

Nach der Etablierung des Keimträger- und Dauerausscheider-
tums beschränkt sich der bakteriologische Nachweis auf die Faeceskultur, wobei wegen der schwankenden Keimmengen, die eine einmalige Kotuntersuchung unzuverlässig machen, auf Anreicherungs-
medien nicht verzichtet werden darf[16, 67]. Dabei ist es wohl eine
allgemeine Erfahrung, dass die Salmonellaausbeute mit der Anzahl
der verwendeten Anreicherungs- und Umzüchtungsmedien steigt.
Jedes Laboratorium muss aber entsprechend seinen Mitteln und
auch der Wichtigkeit des einzelnen Falles für sich selbst entscheiden,
wo es die Kompromissgrenze zwischen Ausbeute und Arbeitsauf-
wand ziehen will.

Bei Schlachtvieh-Reihenuntersuchungen müssen neben Galle und
Leber auch Darmlymphknoten mitverarbeitet werden, bei Abortus
der Foetus, bzw. Fruchthüllen oder Lochialflüssigkeit (GIBSON).

Die serologische Untersuchung ist unzuverlässig bis un-
brauchbar zur Entdeckung von Keimträgern und Ausscheidern [60].
Die aktiv infizierte Kuh hat hingegen gewöhnlich einen hohen
Bluttiter [38]. Eine negative Agglutination schliesst eine aktive Infektion aber nicht aus[38], der Grenztiter ist nicht abgeklärt[67].
CLARENBURG, VINK & SCHUURMANS stellen fest, dass S. dublin O-
Titer von weniger als 1/40, bzw. H-Titer von weniger als 1/320
kaum auf eine bestehende Infektion schliessen lassen. In Preussen
wurde 1937 der H-Grenztiter bei 1/200-300 festgelegt [37].

Über die Immunologie des erwachsenen salmonellainfizierten
Rindes konnte ich ausser den erwähnten serologischen Reaktionen
keine konkreten Angaben finden. An sich darf aus dem Umstand,
dass ein rekonvaleszentes Rind u.U. lebenslänglich massenhaft
Salmonellen im Kot ausscheidet und doch nicht mehr klinisch an
Salmonellose erkrankt, doch auf eine wohl vollständige Grundim-
munität geschlossen werden. HENNING[53] verwendete zum Schutze
des neugeborenen Kalbes über das Colostrum die Vakzination der
Muttertiere.

Bei der Therapie muss wohl unterschieden werden zwischen nur klinischer Genesung und bakterieller Sanierung. HOWARTH (zit. n. PARKE & DAVIS) heilte mit Chloramphenicol von 25 Rindern mit S. *dublin*-Infektion 20 prompt, 3 Tiere mussten nachbehandelt und 2 notgeschlachtet werden. Von SKAV (zit. n. AVERBECK) klinisch erfolgreich mit Chloramphenicol behandelte Tiere blieben aber alle Ausscheider. KÜNG hatte bei Kühen mit S. *newport* mit Chloramphenicol guten klinischen Erfolg, machte aber nur eine einmalige bakteriologische Kontrolle zum Beweis der Salmonellafreiheit der Faeces.

Epidemiologie

Die Infektion des erwachsenen Rindes kann im Stall, auf der Weide oder auf Märkten erfolgen. Als Infektionsquelle ist hauptsächlich das klinisch gesunde Ausscheidertier zu betrachten, welches entweder durch Zukauf neu in den Bestand kam[18, 46, 127] oder durch Futtermittel infiziert wurde. Von diesen Ausscheidern ausgehend kann sich eine eigentliche Stallenzootie entwickeln (KÜNG-S. *newport*, AVERBECK-S. *dublin*) mit gewaltiger Salmonellakontamination der Umgebung und Gerätschaften. Es darf dabei in Analogie zu den Verhältnissen bei der Coli-Säuglingsenteritis und Coli-Kälbersepsis sowie Salmonella-Lebensmittelvergiftung mit Keimzahlen von 10^{6-10}/g Faeces gerechnet werden (SCOTT THOMSON, FEY et al.[36]).

Das übliche ist aber doch das sporadische Auftreten bei einem oder wenigen Tieren des Bestandes (FIELD[38]). Eine weitere Infektionsmöglichkeit, vor allem in Niederungsgebieten mit häufigen Überschwemmungen, ist der Weidegang und das von diesen Weiden abgeerntete Futter[1, 7, 22, 67, 71, 93, 103]. Die auf den Weiden befindlichen Ausscheider erhöhen mit ihren Abgängen die Kontagiosität des Bodens (RASCH[93]), und es ergibt sich ein Circulus Wasser Boden - Rind - Wasser. Eine weitere Belastung des Bodens ergibt sich durch Dünger menschlicher und tierischer Herkunft[93, 126, 131], während dem salmonellahaltigen Industriedünger, dessen organisches Material meist mit Chemikalien vermischt wird, kaum eine epidemiologische Bedeutung zukommt. Die grösste Gefahr der organischen Düngemittel ergibt sich für die Arbeiter[131].

Das Wasser wird kontaminiert durch Mensch und Tier und unterhalb der Städte ist die Salmonellabelastung der Vorfluter sehr stark[26, 40, 63, 71, 91, 93, 102]. In dieser Hinsicht hat sich das Abwasser der Schlachthöfe als besonders keimreich erwiesen[40, 73, 102]. Die während der Schlachtperiode einströmenden blutigen Abwasser liessen in Flensburg den Salmonellatiter auf 100/ml ansteigen, in den Schlachtpausen fiel er auf ein Minimum ab[73]. Wegen dieser starken Belastung genügen u.U. weder die städtische Abwasserklärung noch die biologische Selbstreinigung sicher für die Eliminierung der Keime[102].

Die erwähnten epidemiologischen Verhältnisse im Freiland werden ermöglicht durch die relativ hohe Tenazität der Salmonellen (LERCHE). Sie bleiben in kontaminierter Erde oder auf Gras 150–307 Tage am Leben, in Faecesspritzern an Mauern bis 10 Monate, in infizierten Faeces sind sie wahrscheinlich in 6 Monaten tot (GIBSON). FIELD (zit. n. KOLLER) fand aber in getrockneten Rinderfaeces noch nach 1000 Tagen lebende Salmonellen.

Die Übertragungsmöglichkeiten von Salmonellen vom Rind auf den Menschen sollen zusammen mit der Epidemiologie der Kälbersalmonellose besprochen werden. Auch gewisse allgemein gültige Fragen (Prophylaxe, Bekämpfung u.a.) sollen am Schluss des Aufsatzes erörtert werden.

Die Kälbersalmonellose

Auch beim Kalb ist in vielen Ländern (nicht in allen, z.B. Schweiz) S. dublin die wirtsadaptierte dominante Species. In andern stehen S. typhi-murium und S. enteritidis im Vordergrund[46, 63, 71], und es werden ebenfalls mit zunehmender Häufigkeit ungewöhnliche Species isoliert[84]. In Deutschland[40] steht unter den Tierarten das Kalb mit 84% positiven Befunden an der Spitze.

Klinik und Pathogenese

Der Verlauf ist meist akut, mit Septikämie bei einer Letalität von 50% und mehr, aber auch chronisch (FIELD[38], HENNING[53]). Die Kälber können kurze Zeit nach der Geburt erkranken, meistens aber doch im Alter von mehr als 2 Wochen (1.—3. Monat, [46]). Bei rasch tödlichem Verlauf können pathologische Veränderungen fehlen, bei prolongiertem Verlauf kommt es zu Pneumonie und Arthritis [46] und disseminierten nekrotischen Herden in Leber und Milz. Äusserlich manifestiert sich die Krankheit in schwerer Hinfälligkeit, Exsiccation, profusem Durchfall, verminderter Sauglust und ist somit keineswegs charakteristisch.

Beim Kalb besteht im Gegensatz zum erwachsenen Rind die Tendenz zur raschen Ausbreitung der Infektion unter den Kälbern eines Bestandes, wobei allerdings auch Infektionen ohne klinische Zeichen vorkommen (FIELD[38]). Hingegen ist die Ausscheidung begrenzt und klinisch geheilte Kälber verlieren meist die Infektion [38, 46, 52]. Die Ausscheidung kann sogar bei der enteritischen Form intermittieren [46].

Die Infektion erfolgt auch hier meist peroral und nach der Besiedelung der Lymphknoten schliesst sich die Septikämie an, die in 3—5 Tagen zum Tode führen kann. Bei den Kälbern, die schon 2—3 Tage nach der Geburt sterben, wurde auch eine intrauterine Infektion vermutet (GIBSON). Dafür existieren aber ebenso wenig stichhaltige Beweise wie auf dem Gebiete der immer wieder behaupteten haematogenen Aussaat von Colibakterien bei Mastitis und im Puerperium (FEY[30]). Im übrigen abortieren trächtige Kühe mit klinischer Salmonellose üblicherweise, wobei die Salmonellen aus dem toten Foet zu isolieren sind (GIBSON), aber beim bakteriämischen Muttertier herrschen wahrscheinlich andere Permeabilitätsverhältnisse als beim gesunden Trägertier.

Keimträger gibt es unter den klinisch geheilten Kälbern nur ausnahmsweise (GIBSON). HENNING[52] fand, dass behandelte Kälber Träger bleiben können.

Diagnose

Es hat sich immer wieder gezeigt, dass die Bagatellisierung von Kälberaufzuchtkrankheiten grosse wirtschaftliche Schäden oder gar eine Gefährdung der öffentlichen Gesundheit nach sich zieht. Der Tierarzt sollte deshalb seinen ganzen Einfluss geltend machen, dass der Tierbesitzer schon den ersten Todesfall der Sektion und bakteriologischen Untersuchung zuführt. Es gilt als Faustregel, dass Todesfälle in der ersten Lebenswoche vorwiegend durch Colibakterien, ab der zweiten Lebenswoche durch Pneumokokken und Salmonellen bedingt sein können. Die bakteriologische Untersuchung ergibt beim toten Kalb normalerweise einen septikämischen Befund mit Massenkultur aus allen Organen, die Rectum-Faeces brauchen aber trotz muco-hämorrhagischer Enteritis nicht positiv zu sein (GIBSON). Wir haben hier erneut eine Parallele zur Colisepsis des Kalbes, bei der FEY & MARGADANT[35] zeigten, dass zwar im Dünndarm häufig eine Colienteritis mit Nachweis des spezifischen Sepsistyps vorliegen kann, dass aber trotzdem der Sepsistyp im Rectum fehlt. Dies hat natürlich eine grosse diagnostische und z.T. auch epidemiologische Bedeutung, indem der Faecesbefund nicht notwendigerweise repräsentativ ist für das Enteritisgeschehen. Auch NEWELL et al.[85] fanden beim Schwein im Caecum einen andern Befund als im Rectum.

Seit der breiten Anwendung von Antibioticis in der Veterinärmedizin sind die Schwierigkeiten bei der bakteriologischen Erfassung von Sepsisbefunden stark gestiegen. Es kommen immer wieder Tiere mit ausgesprochenem Sepsisbefund zur Untersuchung, bei denen wegen massiver Antibioticamedikation ante mortem der Nachweis des spezifischen Erregers misslingt. Darauf haben SOMPOLINSKI für den Schweinerotlauf, ferner CLARENBURG & KAMPELMACHER[17] für B. anthracis und S. dublin hingewiesen. 1959 bearbeitete FOLTIN das Problem des Salmonellanachweises bei bakteriologischer Untersuchung von antibioticabehandelten Mäusen und Kaninchen und kam zu ähnlichen Resultaten. Er konnte wenigstens zeigen, dass die Antibioticabeifütterung die postmortale Diagnose nicht beeinträchtigt. Zur Ausschaltung der Penicillinwirkung gibt es zwar Penicillinase-Nährböden, die aber meist nicht greifbar sind,

180

wenn man sie braucht. Unsere Erfahrungen mit der Colisepsis haben aber ergeben, dass man den Sepsiserreger (*E. coli* oder Pneumokokken) bei behandelten Tieren mit negativem Organbefund u.U. noch im Gehirn oder in der Gelenksflüssigkeit findet, wo das Antibioticum in nur ungenügender Menge hingelangte.

Die Serodiagnostik spielt erst bei älteren Kälbern und es werden hauptsächlich H-Agglutinine gebildet. Dies verwundert an sich nicht, da doch die Kälber agammaglobulinämisch zur Welt kommen und erst im Verlauf von mehreren Wochen langsam mit der Eigenproduktion von γ-Globulin beginnen (STECK).

Erstaunlich ist es eher, dass es HENNING[52] gelang, Kälber in der ersten Lebenswoche mit einer Aluminiumhydroxyd-adsorbierten Vollvakzine erfolgreich gegen *S. dublin*-Infektion zu schützen und H-Titer von 1/25—3200 erzielte. Mit Pneumokokken kann man das ganz junge Kalb jedenfalls nicht aktiv immunisieren (FEY & RICHLE, HAMMER[51]). Immerhin zweifelt FIELD[38] an der Solidität der durch Kälbervakzination zu erzielenden Immunität. HENNING[53] erwähnt selber, dass das neugeborene Kalb nicht immer gut anspricht und versuchte deshalb die Immunisierung der trächtigen Kuh. Solche Mütter scheiden grosse Mengen von Antikörpern mit dem Colostrum aus und ermöglichen damit eine gute passive Immunisierung des Kalbes. Zur genügenden Resorption von Gamma-Globulin aus dem Colostrum muss das Kalb innerhalb der ersten 24 Lebensstunden getränkt werden. Inwieweit die von FEY & MARGADANT[33] bei der Colisepsis entdeckte persistierende pathologische Agammaglobulinämie trotz Colostrumaufnahme auch für Salmonellainfektionen disponierend wirkt, müsste erst nachgeprüft werden.

Die Therapie mit Chloromycetin verspricht bei experimenteller wie spontaner Salmonellose gute Erfolge [46, 50, 55]. Dem kommt allerdings die Tendenz der überlebenden Kälber zur spontanen Elimination der Salmonellen entgegen. Die Antibioticatherapie beherrscht aber nur die Symptome, die Bakterienausscheidung wird beim behandelten Tier wie beim Menschen gegenüber dem spontan geheilten nicht verkürzt [46]. Die Faeceskulturen der von HENNING[50] behandelten Kälber waren sogar länger positiv als diejenigen der unbehandelten überlebenden Kontrollen. HOWARTH et al. empfehlen Chloromycetin sogar für die Routineprophylaxe. Dabei muss aber mit Resistenzentwicklung gerechnet werden (GIBSON). Neben Chloromycetin empfiehlt HENNING[50] auch Phthalylsulphathiazol. Die Verwendung von Mitteln ist aber kein Freipass für fehlende Hygiene [46, 55].

Epidemiologie

Die beim erwachsenen Rind aufgeführten epidemiologischen Fakten sind auch für das Kalb gültig. Das Kalb aquiriert die Salmonellen meist vom erwachsenen Dauerausscheider[46, 52, 53].

Danach verselbständigt sich die Salmonellose vor allem in Milch-
mästereien unter den Kälbern und diese infizieren sich in direktem
Kontakt durch Ausscheidung mit Faeces, Nasenschleim, Tränen,
Speichel und Urin (ACHMEDOW). Dabei kommt es wie bei Coliinfek-
tionen zu einer massiven Kontamination der Umgebung, der Boxen-
wand und sämtlicher Gerätschaften [22, 46]. Auch die Milch kann
Infektionsquelle sein und zwar, wie oben erwähnt, eher die faeces-
kontaminierte als die aus echt infiziertem Euter stammende [52].
Futtermittel sind für ältere Kälber ebenfalls Vehikel[47, 132]. Diese
enthalten zwar selten S. dublin[46], dagegen zahlreiche andere Species[47].

Für andere Species als S. dublin (z.B. S. typhi-murium) kommen
auch Hausfliegen in kontaminierten Futtermittellagern (STEINIGER
zit. n. FROMME, HENNING[53]), ferner andere Tiere als Vehikel in
Frage (Geflügel, Hund, Katze, Vögel [46].

Der Handel mit Schlacht- oder Mastkälbern spielt eine vorherrschende Rolle, wobei vor allem Verteilungszentren zum Verhängnis
werden können. Sehr oft kaufen Händler einzelne Kälber beim
Bauern, stellen viele Tiere kurz zusammen und verteilen sie grup-
penweise an kleinere oder grössere Mastbetriebe. Wahrscheinlich
kam auf solche Weise im Kanton Wallis, Schweiz eine grosse
S. typhi-murium-Epidemie mit ca. 1000 Kranken zustande. Gerade
für S. typhi-murium ist unter solchen Umständen die Lysotypie eine
unentbehrliche Hilfe, wie ANDERSON[5, 6] am Beispiel einer von
einem Oxforder Verteilzentrum ausgegangenen Kälberepidemie mit
Fleischvergiftung beim Menschen aufzeigte.

Selbst der Mensch kann das Tier direkt infizieren (LERCHE), wie
aus einem Bericht von MESSERLI hervorgeht: In einem Bauernhof
erkrankte die Frau an einer "Darmgrippe", die als S. typhi-murium-
Infektion verifiziert wurde. 2 Monate später waren 4 von 5 Familien-
mitgliedern infiziert, desgleichen 3 neugeborene Kälber, von denen
2 starben.

Damit soll versucht werden, darzustellen, auf welche Weise der
Mensch durch die Rinder- und Kälbersalmonellose gefährdet ist.
Eine Möglichkeit besteht im direkten Kontakt, den das Farm- und
Metzgereipersonal, aber auch der Tierarzt mit dem massiv Salmo-
nellen ausscheidenden kranken oder gesunden Tier pflegt (GIBSON).
In 3–6% der diagnostizierten Ausbrüche, eventuell mehr, geht
S. typhi-murium vom Kalb auf den Menschen über (GIBSON). Es
empfiehlt sich deshalb für den Tierhalter wie für den Tierarzt, im
Umgang mit diarrhoischen Tieren desinfizierbare Kleider zu tragen
(Gummischürze), weil die Faeces doch beträchtliche Keimmengen
enthalten (ca. 10^6/g). Man beachte auch immer wieder, dass Kinder
und ältere Leute besonders infektionsanfällig sind (MEYER, NEWELL
[86]). Die Beobachtung solcher Kontaktinfektionen im Bauernhof, die
z.T. unzweifelhaft akzeptiert werden können, bedeutet eine gewisse

Einschränkung der Doktrin, wonach Salmonellen (exkl. Typhus-Paratyphus) nicht von Mensch zu Mensch übertragbar seien (BADER). Wenn Direktübertragungen Tier-Mensch vorkommen, so sind bei ähnlichen Ausscheidungsverhältnissen auch Mensch-Menschübertragungen denkbar (NEWELL[86]).

Die menschliche Salmonellose ist deshalb zumeist nahrungsmittelbedingt, weil 1. grosse Keimzahlen für die Infektion notwendig sind und 2. das Lebenmittel häufig die einzige Verbindung zwischen dem Menschen und dem Salmonellareservoir Tier darstellt (NEWELL[86]). So ist der städtische, zentral versorgte Haushalt u.U. in grösserer Gefahr als der Bauernhaushalt, obwohl die Bauernfamilie direkten Tierkontakt hat. Bei der Pflege oder Behandlung von klinisch schwer an Salmonellose erkrankten Tieren ist aber mit einer erheblichen Keimaufnahme zu rechnen. Konkrete Vorstellungen über die für die Infektion des Menschen benötigten Keimmengen verdanken wir McCULLOUGH & EISELE: Sie fütterten *S. meleagridis*, *S. anatum*, *S. newport*, *S. derby* und *S. bareilly* an menschliche Freiwillige. Zur Auslösung klinischer Erkrankung brauchten sie $24 — 50 \times 10^6$ *S. meleagridis*, $1.2 — 67 \times 10^6$ *S. anatum* je nach Stamm, 1.25×10^5 *S. bareilly* und $1.5. \times 10^5$ *S. newport*. Mit viel kleineren Mengen (z.B. 12×10^3 *S. meleagridis*, bzw. *S. anatum*) erhielten sie Infektion mit Ausscheidung ohne Krankheit. Solche Keimmengen nimmt man im Umgang mit infizierten Tieren ohne weiteres auf. Dass die vom Tier inapparent infizierten Menschen für ihre Umgebung wegen Lebensmittelkontamination eine dauernde Gefahr darstellen, liegt auf der Hand.

Sehr leicht verständlich ist die indirekte Übertragung im infizierten Bestand, die über Tierausscheidung - Geräte - Hand - Küchengeschirr - Lebensmittel mit Anreicherung auf den Menschen führt. Auch Inhalationsinfektion durch kontaminierten Staub ist in einem Stall nicht unwahrscheinlich (NEWELL[86]). WALKER[126] erwähnt ferner den organischen Dünger als Infektionsquelle.

Abgesehen von Lebensmittelvergiftungen lässt sich ein Phagentyp von *S. typhi-murium*, der bei Kälbern in einem bestimmten Gebiet erscheint, in kurzer Zeit beim Menschen in der Umgebung nachweisen (ANDERSON[6]). Als häufigster Kalbtyp gilt 20 a, auch U 9 ist in erster Linie ein Kälbertyp. Wenn diese Typen somit beim Menschen auftreten, lässt sich daraus mit Wahrscheinlichkeit auf eine bovine Quelle schliessen (ANDERSON[6]). Da bei *S. dublin* meist nur 1 Phagentyp vorliegt, lässt sich die Lysotypie für diese Species epidemiologisch nicht auswerten (GIBSON).

Eine wichtige Infektionsquelle für den Menschen ist natürlich die Milch. Milchepidemien haben meist explosiven Charakter, weil die Milch bei ungeeigneter Lagerung eine massive und vor allem gleichmässige Vermehrung der Erreger gestattet[14, 18, 19, 67, 81, 94]. Die Salmonellamastitis ist zwar selten, meist wird die Milch beim

Melken durch tierische Faeces kontaminiert[18, 25, 46, 52, 127].
McCALL beschrieb einen eindrücklichen Ausbruch bei Schulkindern
mit *S. dublin*. Eine einzige Kuh mit Mastitis oder Faecesausschei-
dung kann gerade über die Schulmilch eine unverhältnismässig
grosse Zahl von Personen gefährden. Es ist deshalb zu fordern, dass
Betriebe, die Vorzugsmilch für Kinder produzieren, nicht nur auf
Bang- und Tuberkulosefreiheit untersucht, sondern mit Hilfe der
Faeceskultur periodisch auf Salmonellen geprüft werden (FROMME).

Die salmonellakontaminierte Milch ergibt auch gefährliche Milch-
produkte, sofern diese nicht mit pasteurisierter Milch hergestellt
werden[14, 18, 67, 113]. Einer der grössten *S. typhi-murium*-Aus-
brüche in den USA wurde durch Käse verursacht (TUCKER zit. n.
EDWARDS[25]). Es wäre abzuklären, ob Salmonellen im Hartkäse
(Emmentaler u.ä.) während der Reifung absterben, wie das bei
M. tbc und Brucellen der Fall ist, wohingegen diese in Weichkäse
(Camembert u.a.) praktisch unbegrenzt *(M. tbc)*, bzw. mehr als 57
Tage überleben (Brucellen) (KÄSTLI, RAUSCH). Es darf in diesem
Zusammenhang nicht unerwähnt bleiben, dass umgekehrt auch der
menschliche Ausscheider die Milch im Stall oder auf deren Weg zum
Konsumenten kontaminieren kann und damit wiederum den Men-
schen gefährdet. Aber dies gilt für alle Lebensmittel und wird des-
halb an anderer Stelle dieses Buches behandelt[18, 25, 26, 70, 81].

Die wichtigste Quelle menschlicher Salmonellainfektion ist nach
wie vor das Fleisch, welches beträchtliche Keimmengen enthält,
sofern es von klinisch kranken Tieren stammt. Deshalb werden
Notschlachtungen mit Recht als besonders gefährlich betrachtet
und müssen mit der gebotenen Vorsicht fleischschaulich beurteilt
werden[1, 4, 18, 22]. Die Lebendviehbeschau, bzw. eine genaue
Anamnese des vorbehandelnden Tierarztes zusammen mit genauer
pathologisch-anatomischer Beurteilung der Organe und breite Anwendung der bakteriologischen Fleischuntersuchung vermögen
aber diese Gefahr weitgehend zu bannen.

Viel heimtückischer sind die symptomlosen Ausscheider, die bei
völliger Gesundheit und völlig negativem Organbefund u.U. fast
Reinkulturen von Salmonellen mit ihren Faeces ausscheiden (RASCH
[94]). Von diesen Tieren werden intravital negatives Fleisch und
Gerätschaften, Behälter, Tische u.a. oberflächlich kontaminiert
[14, 18, 22, 26, 71, 115]. Auf diese Weise entstand wahrscheinlich
die grosse Lebensmittelvergiftung 1953 in Schweden mit 8845
Kranken und 90 Todesfällen (LUNDBECK et al.). Die Epidemie nahm
ihren Ausgang in der zentralen Schlächterei in Alvesta, von der aus
grosse Teile des Landes mit Fleisch versorgt wurden. Die Kontrolle
ergab *S. typhi-murium*, Lysotyp 8 in Kälberlymphknoten und in
Kuhfleisch. Die Umgebungsuntersuchung förderte denselben Typ
bei 15% der im Schlachthof tätigen Personen zutage, sowie bei
1,01% der Schlachttiere (Mesenteriallymphknoten, Organe). Hin-

gegen waren 15% der Fleischproben oberflächlich kontaminiert. Es wurde nicht befriedigend geklärt, ob die verantwortliche Kontamination vom Tier oder vom Menschen stammte, doch wurde als sicher angenommen, dass irgendwo im Verarbeitungsprozess eine Keimvermehrung stattfand. Eine ungewöhnlich hohe Temperatur und ein Streik im Schlachthof mit anschliessender Überlastung und dementsprechend ungenügender Kühlmöglichkeit wurden als begünstigende Faktoren für die Keimvermehrung betrachtet.

Das Beispiel von Schweden hat mit erschreckender Deutlichkeit gezeigt, dass in einem Land mit zentraler Lebensmittelversorgung ein einzelner *Salmonella*-Herd ein katastrophales Ausmass nehmen kann und dass die ständig zunehmende Zahl latent infizierter Schlachttiere ein sehr ernstes Problem darstellt. Die konsequente Kühlung der Lebensmittel von der Produktion bis zum Verbrauch ist deshalb auf breiter Basis zu propagieren (NEWELL[86]). DOLMAN meint, der Anfall von Lebensmittelvergiftungen wäre in den USA mit dem grossen *Salmonella*-Reservoir (Geflügel) bedeutend grösser, wenn nicht eine sozusagen ununterbrochene Kühlkette vom Produzenten bis zum Konsumenten existierte. STRUCK weist allerdings darauf hin, dass sich Salmonellen selbst bei Kühltemperaturen im Fleisch vermehren können.

Die Frage der Fleischkontamination im Schlachthof soll bei der Besprechung der Salmonellose der Schweine nochmals aufgegriffen werden.

Die bakteriologische Fleischuntersuchung bei Rinder-, Kälber- und anderen Schlachtungen stösst vorläufig noch auf eine Reihe organisatorischer und gesetzlicher Schwierigkeiten. Wohl in den meisten Ländern hat der tierärztliche Fleischbeschauer die Kompetenz, die bakteriologische Untersuchung bei einem bestimmten Fall nach seinem fachlichen Gutdünken anzuordnen. Damit ist es aber nicht getan. Im Anschluss an die Eruierung eines positiven Tieres sollte unbedingt die Möglichkeit einer Umgebungsuntersuchung im Herkunftsbestand und der bakteriologischen Reihenuntersuchung der aus diesem Bestand stammenden Schlachttiere bestehen. In der Schweiz, wo wie in anderen Ländern die tierische Salmonellose nicht meldepflichtig und damit auch nicht entschädigungsberechtigt ist, besteht für solche Massnahmen keine eindeutige Grundlage. (In Deutschland ist seit 1942 die Rindersalmonellose meldepflichtig (LERCHE)). Es steht dem uneinsichtigen Tierbesitzer frei, Kühe und Kälber aus einem verseuchten Bestand frei in den Handel und damit irgendwo im Lande auf die Schlachtbank zu bringen, wo in Unkenntnis der Gefahr und mangels Indikation eine bakteriologische Fleischuntersuchung mit aller Wahrscheinlichkeit unterbleibt. Die Gefahr für die Konsumentenschaft ist evident.

Der Tierbesitzer interessiert sich verständlicherweise nur für das kranke Einzeltier. Der symptomlose Ausscheider hat für ihn, solange weiter nichts Ernsthaftes passiert, etwas Unreales. Jedenfalls will er nicht die Kosten für zusätzliche Kotuntersuchungen und bakteriologische Fleischuntersuchungen tragen, die hauptsächlich im Interesse der öffentlichen Gesundheit zu veranlassen sind. Es ist deshalb auch billig, dass die öffentliche Hand die Kosten solcher epidemiologischer und sanitärer Erhebungen übernimmt, wofür die gesetzlichen Grundlagen zu schaffen sind.

Die Schaf- und Ziegensalmonellose

Beide Tierarten spielen eine geringe Rolle in der Salmonellaepidemiologie trotz der ausgesprochenen Massenhaltung der Schafe [12, 14, 25, 46, 47, 119, 127]. Fälle von Fleischvergiftungen nach Genuss von Schaffleisch sind praktisch unbekannt (KOLLER), immerhin fanden WILSON et al. bei 3% der Schaffleischproben auf Detailmärkten von Cincinnati Salmonellen. Unter den Salmonellastämmen aus Schafen von EDWARDS[26] dominierte S. typhi-murium wie bei anderen Tieren auch. Über einen akuten Ausbruch mit einem Dutzend Todesfällen unter 1000 australischen Schafen pro Tag berichtet JEBSON. Als Infektionsquelle wird die kontaminierte Weide oder das Trinkwasser vermutet.

Die Schweinesalmonellose

Die steigende Tendenz der Salmonellaisolierungen ist beim Schwein und Geflügel besonders auffällig (HAMMER[47], EDWARDS et al.[24]). Schweine sind auf der ganzen Welt besonders wichtige Salmonellaträger[14, 22, 63, 104, 106, 119], wobei im Material von BRUNER & MORAN 37 verschiedene Salmonellaspecies vorkamen, eine Mannigfaltigkeit der Typenverteilung, wie sie nur noch bei Mensch und Geflügel vorkommt. 935 von 1056 Stämmen gehörten zu S. cholerae-suis. KAMPELMACHER et al.[63] fanden 34 verschiedene Species, S. typhi-murium, S. dublin und S. cholerae-suis standen im Vordergrund.

Auch beim Schwein muss die klinisch manifeste Salmonellose vom blossen Trägertum grundsätzlich unterschieden werden. Die erste Gruppe, beherrscht durch S. cholerae-suis, kann erhebliche wirtschaftliche Schäden, vor allem in Mastbetrieben, verursachen, die zweite Gruppe ist aber epidemiologisch heimtückischer und für die menschliche Gesundheit eine ständige Bedrohung.

Klinik

S. cholerae-suis verursacht vornehmlich beim Schwein im Alter von 2—4 Monaten spontane Ausbrüche mit akutem, subakutem bis chronischem Verlauf und ausgesprochener Tendenz zur Septikämie (primäre Salmonellose)[38, 127]. Verminderte Resistenz, z.B.

Virusschweinepest, verstärkt die Symptome. Die akute Form äussert sich in Fieber, Inappetenz, Husten, Mattigkeit, Cyanose, event. Diarrhoe. Meist endigt die Infektion letal besonders bei Pneumonie, oder es entwickelt sich eine chronische, herdförmige Infektion mit Kümmern der Tiere[38], [67], [127]. Die chronische Infektion eines Bestandes kann sich über Monate, ja Jahre hinziehen und stellt dessen Wirtschaftlichkeit in Frage. JÖRG untersuchte in Zürich 1951–55 8000 Tiere aus Schweinepestbeständen auf *S. cholerae-suis*. 320 = 4% waren positiv.

Diagnose

Bei hoch fieberhaften Gruppenerkrankungen von Schweinen im kritischen Alter muss baldmöglichst mit den Mitteln der Autopsie und Bakteriologie auf Virus-Schweinepest mit oder ohne Komplikation durch *S. cholerae-suis* untersucht werden. Zu Unrecht wurde lange Zeit *S. cholerae-suis* als unselbständiger Begleiter der Viruspest angesehen. Der Erreger ist aber durchaus imstande, eine primäre Bestandesinfektion *sui generis* zu setzen. Die bakteriologische Untersuchung ergibt im allgemeinen eine septikämische Verbreitung (selten im Darm, EDWARDS et al.[24]), bei latent infizierten Tieren aber meist eine Lokalisierung im Leber-Darmsystem (Mesenteriallymphknoten), der Kot ist auch dann selten positiv (RUTQVIST et al.[98]).

Bei der Züchtung muss berücksichtigt werden, dass *S. cholerae-suis* in selenit-oder tetrathionathaltigen Selektivmedien nicht angereichert werden kann und dass die Direktkultur, z.B. auf Brillantgrünagar oder die Anreicherung in Brillantgrün-MacConkey-Bouillon vorzunehmen ist. Dieser Umstand gibt möglicherweise eine Erklärung dafür, dass in Lymphknoten und Darminhalt von latent infizierten Schweinen *S. cholerae-suis* nur ausnahmsweise isoliert wird, denn solche Untersuchungen werden meist mit einem Selektivmedium gestartet (SMITH zit. n. FIELD[38]). Die bakteriologische Fleischuntersuchung ist bei Viruspestschlachtungen wegen der möglichen Komplikation durch *S. cholereae-suis* eine Notwendigkeit, beträgt doch die Letalität von *S. cholerae-suis* für den Menschen, wenn es überhaupt zur Infektion kommt, 21%, während sie für *S. oranienburg* 4,7%, *S. typhi-murium* 4% und *S. anatum* 2% beträgt (SAPHRA[100]).

Therapie

Die aus Mesenteriallymphknoten der Schweine isolierten Salmonellen sind chloramphenicolempfindlich (JENSEN), auch gelingt es mit Nitrafurazon klinische *S. cholerae-suis*-Infektionen beim Schwein günstig zu beeinflussen, die bakteriologische Heilung ist aber schwierig zu erzielen (RUTQVIST et al.[98]). Wir selbst haben schon in Schweinebeständen, in denen wegen *S. cholerae-suis*-Verseuchung die Abschlachtung unumgänglich wurde, die Anwendung des Sul-

fonamids Diazil 2 Tage vor der Schlachtung empfohlen, um wenigstens das Fleisch salmonellafrei zu bekommen. Es frägt sich aber, ob
diese Massnahme nicht einfach wegen Bakteriostase trotz Salmo-
nellaanwesenheit im Fleisch zu einem negativen bakteriologischen
Befund führt, was einer Vertuschung der tatsächlichen Infektion
gleichkommt. Jedenfalls ist die Keulung von Viruspestbeständen
mit *S. cholerae-suis*-Infektion wegen der auszusprechenden Konfiskation ein schweres wirtschaftliches Problem.

Epidemiologie

Es ist erstaunlich, dass es in Schweinebeständen mit *S. cholerae-
suis*-Befall nicht häufiger zur Infektion des Menschen kommt. Die
Schweine infizieren sich mit der an sie adaptierten Salmonella-
species sehr leicht durch direkten und indirekten Kontakt. Die Ausscheidung erfolgt mit Bronchialschleim und Harn, weniger häufig
mit dem Kot. Für die experimentelle Infektion sind allerdings
grosse Keimmengen nötig. Am besten gelang dies SLAVIN (zit. n.
FIELD[38]), wenn er das Gras des Auslaufes mit *S. cholerae-suis* be-
sprengte. Der Erreger wird meist durch Zukauf in einen Bestand
eingeschleppt [127]. Im Gegensatz zu den übrigen Salmonellaspecies
spielen Futtermittel für *S. cholerae-suis* (wie auch für *S. dublin* beim
Kalb) als Infektionsquelle kaum eine Rolle[131, 132]. Beim Menschen sind Lebensmittelvergiftungen mit *S. cholerae-suis* wegen
Genuss von infiziertem Schweinefleisch bekannt (EDWARDS[26]), das
übliche sind aber die Einzelerkrankungen mit unbekannter Infektionsquelle, die, wie erwähnt, wegen der ausgesprochenen invasiven
Tendenz von *S. cholerae-suis* häufig tödlich, zum mindesten aber
sehr schwer verlaufen (SAPHRA[99, 100], MEYER). Diese erstaunliche
Virulenz für den Menschen kann bei Primaten durch Fütterungs-
infektionen nicht gemessen werden (MEYER).

In Pökelfleisch scheinen Salmonellen nicht abzusterben (MÜLLER
[82]). GALTON et al.[43] fanden in 23% frischen und 12,5% geräucherten
Schweinewurstwaren Salmonellen (nicht *S. cholerae-suis*). Eine
Übertragung auf den Menschen ist auf diese Weise möglich.

Immerhin muss in Erinnerung gerufen werden, dass die *S.
cholerae-suis*-Infektion im allgemeinen eine schwere klinisch manifeste Krankheit auslöst, die fleischbeschauliche Massnahmen nach
sich zieht. Dieser Umstand mag für die Tatsache mitverantwortlich
sein, dass *S. cholerae-suis*-Fleischvergiftungen beim Menschen zu den
Seltenheiten gehören.

Das Farmpersonal ist zwar erheblich exponiert, aber nimmt offenbar doch nur ausnahmsweise eine kritische Menge dieser Keime auf,
die auf das Schwein spezialisiert sind. Experimentelle Infektion von
Freiwilligen mit *S. cholerae-suis* zur Feststellung der infektiösen
Dosis im Sinne von MCCULLOUGH und EISELE können nicht ver-
antwortet werden.

Die Unklarheit über den Übertragungsmodus von *S. cholerae-suis* auf den Menschen führte u.a. zur Annahme, dass Hund und Katze, vielleicht nach Fütterung von rohem Schweinefleisch, für gewisse Fälle verantwortlich seien (KOLLER, BRUNER & MORAN). Tatsächlich wurde *S. cholerae-suis* recht oft aus dem Hund isoliert (SAPHRA [100], BRUNER & MORAN). Der Kuriosität halber sei der Nachweis von *S. cholerae-suis* aus einer Zuchtforelle, die mit Schweineabfällen gefüttert worden war, erwähnt (HAMMER[47]). Die grosse Mehrheit der Stämme kommt aber überall aus dem Schwein [12].

Andere Salmonellaspecies beim Schwein

Andere Species erzeugen im Gegensatz zu *S. cholerae-suis* nur ausnahmsweise Krankheit beim Schwein (NEWELL[85]). Das aus solchen Infektionen resultierende Träger- und Ausscheidertum ist ein grosses, auf lange Zeit hinaus unlösbares Problem.

Die Lage wurde alarmierend, als nach dem Kriege Futtermittelimporte in den europäischen Ländern massiv gesteigert wurden. Nach WINKLE & ROHDE basiert in Deutschland die Tierfütterung nur zur Hälfte auf inländischem Futter. Durch diese Massnahme wurden vor allem die Schweine- und Geflügelbestände breit latent infiziert[9, 47, 62, 63, 71, 83, 84, 85, 96, 97, 106, 107, 109, 117, 119]. Auch gewisse vegetabilische Futtermittel erwiesen sich als salmonellahaltig (HAUGE et al., JENSEN).

Da diese Futtermittel (auch Lebensmittel wie Eipulver) sehr oft eine ganze Reihe verschiedener Salmonellaspecies beherbergen, sind Mehrfachinfektionen beim Tier und in der Folge beim Menschen sehr verbreitet, was bei Zweituntersuchungen von Faeces mitunter diagnostische Verwirrung stiftet[78, 92, 107, 109]. Jedenfalls widerspiegelt die Speciesverteilung im Schwein (und auch Geflügel) diejenige in den Importfuttermitteln[9, 47, 61, 83, 84, 86, 88, 96, 104, 109]. Dem stimmen allerdings v.D. SCHAAF et al.[124] nicht ohne weiteres zu, weil sie in einigen Fällen von Salmonellosis bei Kälbern, Kanarienvögeln und Schweinen im verdächtigten Futter keine Salmonellen nachweisen konnten. Das Futter braucht aber im Zeitpunkt des Salmonellanachweises beim Tier nicht mehr positiv zu sein, ein Umstand, der die Rückverfolgung der Infektionsquelle vom Tier zum Futter meist verunmöglicht. Nach NEWELL et al.[85] verstreichen zwischen der Einführung einer bestimmten Futtercharge und der Passage der Salmonellen durch das Tier auf den Menschen einige Monate. Es können somit eher Aussagen gemacht werden, wann eine frisch mit Futter- oder Lebensmitteln importierte Salmonellaspecies voraussichtlich bei Tier und Mensch nachzuweisen sein wird. NIELSEN[88] vergleicht diesen Zustand mit "einer Zeitbombe, die ins Land eingeführt wurde".

Das Haften der Futtersalmonellen im Schwein hängt offenbar von der Species und der Dosis ab:

SMITH[109] fand in Angola-Fischmehl 50 Salmonellen/100 g, im Pakistan-Knochenmehl 700/100 g. Mit diesem Futter wurden Schweine erfolgreich infiziert, ohne dass sich je klinische Störungen ergeben hätten. Nach Absetzen der kontaminierten Nahrung wurden die Schweine bald salmonellafrei und blieben es. Durch Entzug der salmonellahaltigen Nahrung kurze Zeit vor der Schlachtung könnte somit die Gefahr der Übertragung auf den Menschen reduziert werden (SMITH[109], NEWELL[86]).

Der Zusammenhang zwischen Dosis und Ausscheidungsdauer (kleine Dosis, kurze Ausscheidung) ist aber noch nicht genügend sicher etabliert (NEWELL et al.[85]). NEWELL[86] zeigte, dass in einer Farm aus Futtermehl S. orion durch die Schweine selektioniert wurde, während das Geflügel der gleichen Farm S. infantis "bevorzugte". So scheint es neben den bekannten wirtadaptierten Salmonellaspecies weitere zu geben, die eine gewisse Affinität zu bestimmten Tierarten aufweisen. Solche Species erhalten sich im Tierkörper, andere sterben ab, sobald der Nachschub unterbleibt.

Auf diese Weise kam es dazu, dass das Schwein in einem hohen Prozentsatz latenter Salmonellaträger und eventuell Ausscheider wurde, worauf HORMAECHE & SALSAMENDI (zit. n. EDWARDS[26]) 1936 als erste aufmerksam machten[18, 21, 22, 32, 61, 62, 63, 64, 67, 77].

Bei der Schlachtung bildet nun nicht das Tier die Hauptgefahr, welches schon in vivo Salmonellen im Fleisch beherbergt, vielmehr ist es das Ausscheidertier, welches noch im Leben, nach der Schlachtung und bei der Verarbeitung Geräte, Einrichtungen, die Hände der Arbeiter und damit weitere primär negative Tierteile mit Salmonellen kontaminiert. In der Tiefe ist solches Fleisch salmonellafrei, aber der Abstrich der Oberfläche ist positiv[14, 18, 22, 26, 32, 42, 61, 72, 85, 127]. Ähnliches gilt für das Rind und Kalb[46, 97].

Es ist das Verdienst der Gruppe von GALTON, SMITH, MCELRATH & HARDY[42], schon 1954 den Weg der Salmonellakontamination in einem Schweineschlachthof aufgezeigt zu haben. In den Farmen waren 7,2% der Analabstriche positiv, nach dem Salmonellatransport zum Schlachthof waren es in den grossen Boxen 78%. Das Trinkwasser in den Trögen, in denen sich die Schweine wälzen konnten, war zu 75% positiv. Nachher wurde mit immer hoch positivem Befund Kot der getöteten Tiere, das Brühwasser, die Enthaarungsmaschine sowie die Haut der gebrühten Schweine untersucht. Die Folge dieser umfassenden Kontamination war daraus ersichtlich, dass in Florida eine starke Ähnlichkeit der Salmonellaspecies bei Mensch und Schwein bestand, was in Anbetracht der Tatsache, dass 23% der frischen Schweinewürste auf dem offenen Markt von Jacksonville Salmonellen enthielten, nicht erstaunt (GALTON[43]). MCDONAGH & SMITH machten in Bradford (England) die gleiche Erfahrung.

Nachdem in Holland 1959 eine "eigentliche nationale Epidemie"

190

ausbrach und 6% aller Hackfleischproben Salmonellen beherbergten, unternahmen KAMPELMACHER, GUINÉE, HOFSTRA & v. KEULEN[62, 64] ähnliche Untersuchungen wie GALTON et al.[42] wobei deren Resultate bestätigt wurden. Im Brühwasser von 62° C sterben die durch Kotpartikel geschützten Salmonellen nicht ab. Die Enthaarungsmaschine kontaminiert viel stärker als der Handschaber, das Flambieren der Haut bei 1200–1400° desinfiziert diese nicht zuverlässig! Die beim Transport der Kadaver in die tieferen Hautschichten eingedrungenen Salmonellen werden durch eine hitzekoagulierte Proteinschicht abgeschirmt.

Faeces, Mesenteriallympknoten und Organe gesunder Schlachtschweine waren zu 25,3–30,1% positiv. Der Transport von der Farm zum Schlachthof verursachte vermehrte Salmonellakontamination, allerdings findet diese auch schon auf der Farm statt. Die Salmonellaverteilung in den Lympknoten war unregelmässig, somit ist ein einzelner Lympknoten nicht repräsentativ, auch nicht eine einzelne Faecesprobe, da die Ausscheidung und die Besiedelung der verschiedenen Darmabschnitte unregelmässig sind. Auch bei guter Desinfektion und Hygiene wird somit ein Schlachthof immer wieder massiv von aussen mit Salmonellen belastet.

Die Ursache dieser erschreckend breiten Verseuchung der Schweinebestände mit nachfolgender Vermehrung der positiven Befunde durch Kontamination ist im Importfutter zu suchen, in dem sich Salmonellen vermehren können, wenn Reste davon über Nacht in den Trögen bleiben.

Jede Bekämpfung muss daher in der Farm beginnen und die Kontrolle der Futtermittel oder besser deren systematische Resterilisation, wie in Dänemark und Schweden (MÜLLER[83], RUTQVIST & THAL[97]), muss vorangetrieben werden, wo es noch nicht der Fall ist. Jedenfalls ist der Zustand bedenklich, dass die Fleischschaubehörden praktisch machtlos einen hohen Prozentsatz unerkannt salmonellapositiver Schlachtschweine dem freien Verkauf übergeben müssen, trotzdem sie wissen, dass bei Zusammentreffen ähnlich ungünstiger Bedingungen wie in Schweden 1953 jederzeit ein grosser Ausbruch von Lebensmittelvergiftung zu befürchten ist.

Die Geflügelsalmonellose

Die Verhältnisse liegen hier ähnlich wie beim Schwein. Wirtschaftliche Schäden entstehen fast ausschliesslich durch die primäre Salmonellose der Pullorumruhr, bzw. des Geflügeltyphus, die indessen durch systematische, energische Kontroll- und Ausmerzmassnahmen gut zu kontrollieren sind[10, 26, 38, 59, 74, 108, 114, 122, 127]. Da ausserdem S. *pullorum* nur ausnahmsweise beim Menschen als Erreger auftritt[13, 45, 93, 107] — gelegentlich zwar die Ursache von Lebensmittelvergiftung war[40] — sei auf die Besprechung dieser ausgesprochenen Tiersalmonellose verzichtet.

Viel wichtiger und für den Menschen eine Gefahr sind auch beim Huhn die latent infizierten Individuen, die z.T. ebenfalls mit den Faeces ausscheiden [67]. Infektionen mit anderen Salmonellen als *S. pullorum* führen allerdings bei jungen Hühnern und Truten recht oft zu klinischen, z.T. tödlichen Ausbrüchen, die von Pullorumruhr nicht zu unterscheiden sind (BUXTON[15]). Der deutsche Geflügelbestand ist wie der Schweinebestand massiv infiziert (SEELIGER[106], HAMMER[47]) und zwar wie anderswo in erster Linie durch früher seltene Typen[71, 84, 106]. Ein Beispiel dafür ist *S. blockley*, die 1958 in Baden erstmals in grossem Umfang von HAMMER[47] aus Geflügel isoliert wurde. Ein ausgesprochener Spezialist für Geflügel ist *S. thompson*[89, 131]. Die Untersuchungen von EDWARDS[26] veranlassten diesen, das Geflügel als das wahrscheinlich grösste amerikanische Salmonellareservoir zu bezeichnen. Deswegen soll auch eine Durchmischung von Vieh mit Schweinen, Hühnern, Enten und Truthühnern vermieden werden (DOLMAN).

Zum selben Schluss wie EDWARDS[26] kommen THAL, RUTQVIST & KARLSSON[97, 120]. 2–3% der schwedischen Hühnerbestände müssen als verseucht betrachtet werden. GRETE ELLEMANN[29] fand dagegen in Dänemark nur 0,9% Faecesausscheider unter 2835 Schlachthühnern.

Seit jeher waren Enteneier als gefährlich bekannt [25]. LINSERT fand bei 3,4% von 353 Schlachtenten *S. typhi-murium* im Lebersystem. PERELLI-MINETTI (zit. n. EDWARDS[26]) fand bei 41,4% aller Truthühner in Kalifornien Salmonellen. Als Infektionsquelle wird auch hier hauptsächlich das Importfutter betrachtet. Die Ausbreitung von Bestand zu Bestand erfolgt durch den Handel mit infizierten Jungtieren, die wegen der Belastung durch den Transport beim Käufer erkranken oder sogar sterben. Dieser Zusammenhang wurde durch Lysotypie von *S. montevideo* geklärt (RUTQVIST & THAL[97]). Die Salmonellaausscheidung dieser Kücken ist unregelmässig und sistiert meistens nach einigen Monaten (z.T. bis 18 Monate, BUXTON[15]). Demgemäss ist die bakteriologische Kotuntersuchung unzuverlässig. Sie ist höchstens dann beweisend, wenn sie während mehreren Monaten 4–6mal negativ ausfiel.

Auch die Klinik, die pathologische Anatomie und Serologie solcher infizierter Jungtiere ist völlig nichtssagend. Die parenchymatösen Organe, Galle, Knochenmark, Muskulatur sind negativ, die Blinddärme dagegen positiv [97, 120].

Der wirtschaftliche Schaden solcher Salmonellainfektionen ist zu vernachlässigen, da nur ausnahmsweise Krankheit oder Tod die Haltung beeinträchtigen. Hingegen sind solche Geflügelbestände eine Gefahr für die menschliche Gesundheit [84]. Die Übertragung von *S. montevideo* vom Huhn auf den Menschen wurde wiederum durch Lysotypie bewiesen. Die Lysotypie hat ferner bewiesen, dass 74% der Geflügelstämme von *S. typhi-murium* zum einzigen Typ 14 gehören (ANDERSON et al.[6]). (97% der tierischen *S. typhi-murium*-Stämme Englands stammen vom Rind oder Geflügel.) Typ 14 wird auch am häufigsten aus englischen Eiern gezüchtet und ist ferner der häufigste Typ beim Menschen. Die Ausdehnung der Lysotypie auf tierische Stämme erlaubt somit Rückschlüsse auf die Herkunft menschlicher *S. typhi-murium*-Stämme (Typ 14 vom Huhn, Typ 20a vom Kalb). Weitere bekannte Geflügelspecies sind *S. anatum*, *S. newport, S. thompson*.

Die hygienische Bedeutung dieser Keimträger liegt, wie beim Schwein und Rind, weniger in der Häufigkeit ihres individuellen Vorkommens und der Menge der intermittierend ausgeschiedenen Keime, sondern wiederum in der Möglichkeit der Kontamination der Umgebung und des Fleisches bei der Schlachtung. Immerhin haben ABELSETZ & ROBERTSON sowie HEDSTRÖM und HINSHAW et al. (alle zit. n. BUXTON[13]) Beweise dafür, dass der Umgang und direkte Kontakt mit ausscheidendem Geflügel für den Menschen ein Risiko darstellt. HAMMER[47] berichtet von 9 Personen, die nach unmittelbarem Kontakt mit infiziertem Schlachtgeflügel z.T. septisch an *S. blockley* erkrankten. In Geflügelschlachtstellen sind Salmonellen weitverbreitet auf Tischen, Gestellen und den mechanischen Einrichtungen, und es ist selbstverständlich, dass dadurch auch die Hände des Personals mit Salmonellen beschmutzt sind (EDWARDS[26]). Diese Kontaminationsgefahr durch Faeces veranlasste 1958 RUTQVIST & THAL[97], der schwedischen Regierung vorzuschlagen, dass Geflügel nur ohne Kopf, Füsse und Eingeweide verkauft werden dürfe. Eine weitere Infektionsgefahr besteht beim Geflügel ganz besonders durch kontaminierten Staub und Flaum (BUXTON[13], HOFMANN et al.[54]). Der Übergang des Infektes mit dem Staub ohne Zwischenmedium oder Vermehrungsstadium auf den Menschen ist allerdings noch nicht bewiesen, aber Anhaltspunkte dafür bestehen (NEWELL[86]). Trockene salmonellahaltige Exkremente kontaminieren auch die Brutapparate und Aufzuchtbatterien und damit die Eischalen, ferner Futter und Trinkwasser, womit ein gefährlicher Herd für die Kücken geschaffen wird (FIELD[37], BUXTON[15]), der durch periodische Verdampfung von Formaldehyd in den Brutapparaten unter Kontrolle gebracht werden kann (BUXTON[15]).

Wohl die wichtigste Infektionsquelle sowohl für den Embryo wie für den Menschen ist das Ei, welches entweder vom Follikel her mit Salmonellen besiedelt wird (vorwiegend bei Pullorumruhr) oder von der Schale her bei der Salpinx- und Kloakenpassage oder bei der Bebrütung eine Kontamination erfährt[15, 27, 28, 74, 81, 89].

Bei hoher Feuchtigkeit penetrieren die Salmonellen durch die intakte Eischale und zwar umso rascher, je höher die Temperatur ist (bis 30° C)[27, 28]. Die enorm wichtige Rolle, die salmonellahaltige Eier und Eiprodukte als Lebensmittel spielen, wird an anderer Stelle dieses Buches besprochen. Es sei noch erwähnt, dass auch der Mensch über Futter, Trinkwasser und Küchenabfälle das Geflügel infizieren kann[13, 89], wonach der Erreger möglicherweise über das Geflügel wieder auf andere Menschen zurückwechselt.

Bei Enten, Gänsen und Truthühnern bestehen im allgemeinen ähnliche epidemiologische Verhältnisse wie bei Huhn. S. *typhimurium* ist der wichtigste Erreger bei Enten und Gänsen[10, 13, 59] und deren Eier waren die Ursache vieler Lebensmittelvergiftungsausbrüche[18, 19]. Die Kontamination der Eier ist unter den Umständen, unter denen diese Vögel gehalten sind, noch verstärkt [15], diese können ihrerseits Fische infizieren und umgekehrt (GAUGUSCH). Auf dänischen Farmen, wo oft Hühner und Wassergeflügel frei herumspazieren, bestehen so Quellen von Kreuzinfektion. Damit ist auch der Kontakt mit Wildvögeln gegeben, die auf diese Weise Salmonellen aquirieren oder zutragen (NIELSEN[87]).

Bei den Tauben gehört die Salmonellose neben Ornithose und Schnupfenkrankheiten zu den verlustreichsten Seuchen. Die mit Arthritis oder Durchfall mit Allgemeinstörungen, eventuell nervösen Schäden einhergehende primäre Salmonellose wird fast ausschliesslich durch S. *typhi-murium* erzeugt[49, 59]. Jungtauben sterben nach 8—14 Tagen[40]. Die Infektion erfolgt durch direkte Übertragung, offenbar unabhängig vom Grad der Bestandeshygiene, trotzdem ist laufende Desinfektion in der Bekämpfung nicht zu vernachlässigen [11]. Die Behandlung ist problematisch, Keimträger und Ausscheider werden nicht zuverlässig beseitigt (HAUSER). BRÜCKER & SUTER erreichten mit einer massiven Aureomycinkur einen beachtlichen Erfolg, ohne aber alle Ausscheider zu erfassen. Solche sind auszumerzen, damit Reinfektionen des Bestandes vermieden werden. Auch eine Nager- und Spatzenbekämpfung ist empfehlenswert [11].

S. *typhi-murium* aus Tauben nimmt eine Sonderstellung ein. Die var. *copenhagen* (4,12: i: 1,2) wurde erstmals von KAUFFMANN beschrieben (zit. n. SØRUM). Es fehlen ihr die O-Antigene 1 und 5.

Taubenepizootien können durch Staub und Faeces eine direkte Gefahr für den Menschen bilden [13]. Die Variante 4, 12: i: 1, 2 wurde denn auch beim Menschen in Norwegen nachgewiesen (SØRUM), auch sind Fälle menschlicher Infektion nach Genuss von Taubenfleisch und -eiern bekannt (FROMME).

Die Pferdesalmonellose

FIELD[37] unterteilt in 1.) Stutenabortus und Fohlenarthritis mit S. *abortus-equi* als Erreger und 2.) akute Gastroenteritis, meist durch S. *typhi-murium* bedingt. Die aetiologische Bedeutung von S. *abortus-equi* für die primäre Salmonellose des Stutenabortus ist nicht unbestritten. Nach WAGENER & MITSCHERLICH kann diese Species beim Pferd unter folgenden Umständen auftreten: 1.) als

Saprophyt im gesunden Keimträger, 2.) als Erreger von Eiterungen im Bewegungsapparat, 3.) als Begleitbakterium von Virusinfektionen, wie dem ansteckenden Katarrh der oberen Luftwege und dem Virus-Stutenabort, 4.) als alleiniger Erreger von ansteckendem Verfohlen.

Der Erreger kommt nur in seltenen Fällen beim Menschen vor, ferner bei Truthuhn, Taube und Huhn [125].

Für den Menschen ist es deshalb wichtiger, dass das Pferd auch andere Salmonella-Species im Zusammenhang mit Gastroenteritis oder meist als Keimträger beherbergen kann (S. *typhi-murium,* [12], [60]. In Deutschland steht das Pferd bezüglich Salmonellahäufigkeit hinter Kalb, Rind und Schwein an 4. Stelle [103]. In Baden handelt es sich nur um Einzelfälle [47]. In Süddeutschland dominiert *S. typhi-murium*, in Norddeutschland kommt *S. dublin* noch dazu [103]. Die Ansteckung mit diesen Species erfolgt über kontaminiertes Futter und Trinkwasser. Pferde, bzw. Fohlen gaben wegen dieser manifesten oder latenten Infektion relativ häufig Anlass zu Lebensmittelvergiftungen (KOLLER, eigene Beobachtung).

Die Salmonellose von Hund und Katze

Auch hier liegt das Schwergewicht auf dem gesunden Keimträger[12, 18, 22, 26, 41, 47, 67, 81, 123]. Die Infektionshäufigkeit ist z.T. sehr erheblich: Im Material von BRUNER & MORAN ist der Hund nach dem Schwein der wichtigste asymptomatische Keimträger mit 26 verschiedenen Species, wovon 40% *S. typhi-murium*. Die gleichen Autoren vermuten Hund und Katze als Überträger von *S. cholerae-suis* auf Kinder. ADLER fand in Honolulu 13,2% von 294 gesunden Hunden aus allen Teilen der USA salmonellainfiziert. In England sind es ca. 1% und 0–1,4% der Katzen (BUXTON[14]). Ausgedehnte Untersuchungen von MILDRED GALTON et al.[41, 78, 116], in Florida ergaben eine enorme Verseuchung mit 52 verschiedenen Species und Mehrfachinfektionen bei normalen Hunden (15%), in Greyhound-Zwingern, abhängig vom Hygienezustand (32–55%), in Tierspitälern (15,1%) und sogar bei gesunden Familienhunden (15%). Der gleichen Gruppe gelang auch die häufige Isolierung von Salmonellen aus Schweine- und Rinderlebern für Hundefutter des lokalen Schlachthauses [43].

Trotz dieser erstaunlichen Verseuchung fand sich kein sicherer Hinweis einer epidemiologischen Bedeutung dieser Salmonellaträger unter den Familienhunden für die menschliche Gesundheit [78b].

Die meisten positiven Hunde waren gesund, einige erkrankten fieberhaft an Diarrhoe, gelegentlich mit Respirationsinfektion, ganz ähnlich wie Kinder [41].

Welpen erkranken allerdings häufig an Gastroenteritis oder

sterben an Sepsis[15], [67], [130] und disponierende Momente (Staupe, Parasitenbefall) komplizieren den Verlauf[67], [130]. Hunde infizieren sich vornehmlich durch direkten Kontakt, durch Fleischabfälle [43], durch Milch, Ausmerzeier, die dem Hundefutter zugefügt werden (WOLFF et al.), Katzen durch Milch und Mäuse.

Tetracycline und Sulfadiazin beeinflussen die Infektionsrate nicht (STUCKER et al.). Der besonders enge Kontakt dieser Haustiere mit dem Menschen macht sie zu einer potentiellen Gefahr (NEWELL[86]). K. F. MEYER zitiert einen grossen *S. typhi-murium*-Ausbruch beim Menschen im Anschluss an eine Katzeninfektion, direkte Übertragung vom Hunde auf den Menschen ist bekannt (EDWARDS[26]). KAUFFMANN (zit. n. WOLFF et al.) berichtete über Gastroenteritis bei 6 Personen einer Familie und dem Hund *(S. glostrup)*.

Die Nagersalmonellose

Wir wollen Mäuse und Ratten nicht zu den Haustieren rechnen, aber der enge Kontakt, der zwischen beiden besteht, macht es notwendig, die Nager als Salmonellareservoir zu erwähnen.

Ratten und Mäuse werden traditionellerweise als wichtige Salmonellaträger betrachtet und mit vielen Ausbrüchen in Zusammenhang gebracht[22], [37], [41], [52], [63], [81]. *S. typhi-murium* und *S. enteritidis* sind die Hauptspecies[15], [66]. 1956 zeigte EDWARDS[25], dass Nager nicht nur *S. typhi-murium* und *S. enteritidis* beherbergen (was für Labortiere nach wie vor stimmt), sondern eine ganze Menge von Serotypen. Es bestehe deshalb kein Grund mehr, Nager vor allem für die *S. enteritidis*-Infektion des Menschen verantwortlich zu machen, da Ratten und Mäuse ebensogut andere Salmonellaspecies übertragen können.

Die Rolle von Nagern als Quelle menschlicher Salmonellaausbrüche ist umstritten und nur für Infektionen mit dem Ratten- "virus" sicher festgestellt (MEYER). Der Nachweis, ob die Nager die primäre Quelle oder vielmehr Empfänger einer menschlichen Infektion sind, dürfte kaum beweiskräftig zu erbringen sein, in beiden Fällen wäre ja der Lysotyp derselbe. KETZ fand in Leipzig 0,15% von 2005 Ratten mit Salmonellen infiziert, er betrachtet sie deshalb nicht als ernste Gefahrenquelle für Lebensmittelvergiftungen. In Anbetracht der Prozentzahlen, die bei den gesunden schlachtbaren Haustieren gefunden werden, kann man ihm beipflichten.

Sicher können Nager tierische Futtermittel mit ihren Ausscheidungen kontaminieren und damit die Haustiere infizieren[2], [9], [46], [64], [71], [89], [127], umgekehrt können auch sie aber durch die Ausscheidungen der Haustiere infiziert werden [15].

Lange Zeit wurden für die Ratten- und Mäusebekämpfung auf Ködern ausgelegte Salmonellakulturen verwendet. Ursprünglich handelte es sich um einen Stamm von *S. enteritidis danysz*, Ratin

genannt. Später kamen auch gewöhnliche Stämme von *S. enteritidis*
und *S. typhi-murium* dazu, die man in irreführender Weise oft auch
als "Rattenvirus" bezeichnet (BUXTON[15], FEY eig. Unters.). Unter
günstigen Umständen kann aus einer solchen künstlichen Infektion
eine Enzootie unter der Nagerpopulation entstehen, meist sind aber
die Resultate unbefriedigend (MEYER), und es ergibt sich eine
Häufung von äusserlich gesunden Tieren mit latenter Infektion
[15, 20, 133].

Viel bedenklicher ist aber die durch die absichtliche Verbreitung
von lebenden Salmonellen vom Fleischvergiftertyp entstehende
Gefahr für Mensch und Tier, die allerdings von den Herstellern be-
stritten wird.

LÜTJE[133], BABO (zit. n. LÜTJE), HENNING[52], DOLMAN, DATHAN
et al., K. F. MEYER berichten und zitieren über z.T. schwere Infektionen mit Salmonellen aus Nagervertilgungsmitteln beim Menschen
und beim Tier.

Deutschland (1936, KETZ) und andere Länder haben deshalb den
Vertrieb dieser Mittel untersagt, in andern sind sie nach wie vor in
Gebrauch. In einer Zeit, in der überall in der Welt der Kampf gegen
die Salmonellose von Mensch und Tier intensiviert wird, drängt sich
das allgemeine Verbot dieser anachronistischen Nagerbekämpfungs-
methode gebieterisch auf.

Bekämpfungsmassnahmen

Es wäre wohl unrealistisch anzunehmen, eine Ausrottung der
Salmonellose läge im Bereiche der Möglichkeiten. Aber es gibt doch
eine Reihe von Massnahmen, deren konsequente Befolgung zu einem
Rückgang der Häufigkeit führen kann. Nach SEELIGER[107] sind z.B.
in Westdeutschland die Gruppen- und Massenerkrankungen deutlich
zurückgegangen, seit menschliche Salmonellaausscheider in Lebens-
mittelbetrieben systematisch erfasst und neutralisiert werden.

Die Massnahmen zur Kontrolle der primären Tiersalmonellosen
wurden bereits angedeutet, Die Diagnose muss mit bakteriologischen und serologischen Mitteln möglichst schnell und umfassend
angestrebt werden, damit die Möglichkeit zur Isolierung und wenn
möglich Ausmerzung der befallenen Tiere gegeben ist[40, 46, 98].
Die Ausmerzung ist vor allem in Geflügel- und Schweinebeständen
und bei Rindern, die Dauerausscheider sind, ein dringendes, wenn
auch wirtschaftlich hartes Gebot und erlaubt die schnellste und
wohl auch billigste Sanierung.

Bei Kälbern, die ja meist spontan ausheilen, kann unter Wahrung
aller hygienischen Vorsichtsmassnahmen die Selbstreinigung abge-
wartet werden (Isolierung), aber gerade die wichtige Absonderung
ist aus baulichen Gründen häufig nicht zu realisieren. Das gleiche
gilt für die, besonders bei Schweinezukäufen in hohem Masse wünsch-

bare Quarantäne, wie überhaupt bei Zukäufen alle Vorsicht geboten ist. Therapieversuche mögen diese Bestrebungen unterstützen, bei erwachsenen Dauerausscheidern sind sie zu unzuverlässig und verleiten höchstens zu einer falschen Sicherheit.

Endlich ist die Sanierung, bzw. das Meiden gefährdeter Weiden sowie eine gründliche Stalldesinfektion erforderlich. Der grossen Gefahr der Salmonellaübertragung durch die Milch kann durch die Pasteurisation gesteuert werden. Die Vakzination wurde von HENNING[52] für Kälber und trächtige Kühe empfohlen, bei Schweinen mit S. cholerae-suis ist sie wertlos [60, 127].

Einen bedeutenden Fortschritt würde bestimmt die Meldepflicht der tierischen Salmonellosen bringen, die für die menschlichen Salmonellosen in vielen Ländern bereits besteht. Sie wird von zahlreichen Autoren[18, 26, 46, 79] gefordert und existiert bereits in Dänemark, Italien, Libanon, Tschechoslovakei u.a. [79] und in Deutschland für das Rind [71].

Eine solche gesetzliche Grundlage würde nicht nur dem in der Fleischbeschau tätigen und beamteten Tierarzt den Rücken stärken, sondern den Tierärzten allgemein erst die Möglichkeit geben, die Salmonellose energischer als bisher anzugehen. Es ist dringend notwendig, die menschliche und tierische Salmonellose nicht nur vom klinischen, individuellen Standpunkt aus zu betrachten, sondern sie grundsätzlich als epidemiologische Entität zu begreifen. Die Aufdeckung eines Ausbruchs oder eines sporadischen Falles sollte automatisch eine Umgebungsuntersuchung bei Mensch und Tier auslösen, wozu die gesetzlichen und damit die finanziellen Grundlagen zu schaffen sind[7, 19, 38, 40, 71, 85, 86, 105]. Der sporadische Einzelfall darf nicht länger bagatellisiert werden, denn er ist nur das sichtbare Zeichen einer umfassenden Infektion der menschlichen und tierischen Population [19].

Solche gezielten Umgebungsuntersuchungen führen unweigerlich zur Entdeckung von Ausscheidertieren (und -Menschen), die potentiell jederzeit einen grösseren Ausbruch auslösen können. Da die Salmonellaquelle letzten Endes beim Tier liegt und die menschliche Salmonellose nur eine mehr oder weniger zufällige Konsequenz der tierischen Infektion ist (NEWELL[86]), muss die Ausmerzung möglichst vieler tierischer Ausscheider sich günstig auf die Zahl menschlicher Infektionen auswirken, auch wenn man nie ganz zu Rande kommt. Es geht im wesentlichen darum, zu verhindern, dass Salmonellen, auch nicht in kleiner Zahl, in die Küche gelangen. Die Veterinärmedizin hat hier eine beträchtliche Verantwortung, in der sie allerdings von Seiten der Humanmedizin vermehrt unterstützt werden sollte.

Die Anzeigepflicht würde auch die Überwachung von Schweine- und Geflügelbeständen ermöglichen, für die HAMMER[47] und THAL[120] eintreten.

Es müssen auch alle Anstrengungen zur Verhinderung der Infektion der Tierbestände unternommen werden. Dazu gehört in erster Linie die Kontrolle der Importfuttermittel[9, 18, 47, 61, 83, 84, 86, 97, 104, 120, 127].

Diese besteht entweder in der stichprobeweisen bakteriologischen Kontrolle der Futtermehle, die naturgemäss viele Lücken aufweist. Positive Befunde, verbunden mit entsprechenden Sanktionen gegen den Exporteur, führen aber erfahrungsgemäss zu Verbesserungen in der Produktion und damit der hygienischen Qualität der Futtermehle. Die konsequenteste Lösung ist natürlich die prinzipielle Resterilisation der Futterstoffe bei der Einfuhr, wie das in Dänemark und Schweden der Fall ist [83, 84, 97, 120].

Ist der Ausscheider erkannt, so ist bei der Schlachtung Vorsicht geboten. Aus vielen Arbeiten geht hervor, dass die Kontamination von Geräten und Lebensmitteln durch den Darmträger das beherrschende Moment der Salmonellaepidemiologie ist (MEYER). Die Verbesserung der Schlachthofhygiene vom Antransport der Tiere aus der Farm bis zum Ausstoss des Fleisches und der Fleischwaren ist deshalb ein brennendes, aber zugleich sehr schwer zu lösendes Problem[61, 67, 127]. Die Aufklärung des Personals über Infektionshygiene gehört unbedingt dazu[25, 26, 127]. Von der bakteriologischen Fleischuntersuchung muss in der tierärztlichen Praxis wie im Schlachthof vermehrt Gebrauch gemacht werden. Ihr Indikationsbereich ist in manchen Ländern gesetzlich geregelt [79].

Die Massnahmen, die bei einem salmonellapositiven Schlachttier zu treffen sind, sind nicht unbestritten [67]. STRUCK plädiert sicher zu Recht auf Konfiskation und Beseitigung aller Tierkörper, bei denen Salmonellen nachgewiesen werden und lehnt den GLÄSSERschen[115] Vorschlag ab, wonach Fleisch zu sterilisieren sei, auch wenn Salmonellen in mehreren Organen und Lymphknoten festgestellt werden. Eine ähnliche Lösung wie GLÄSSER[115] empfiehlt FIELD[38] für S. dublin beim Rind.

Da Kontaminationen nie zu vermeiden sind, muss eine möglichst ununterbrochene Kühlkette zwischen Fleischproduktion und Konsument erstrebt werden[61, 127]. In Schlachthöfen und Futterlagern empfiehlt sich ferner die Nagerbekämpfung[98].

Die Salmonellose bei Mensch und Tier ist zu einem weltweiten, ausserordentlich komplexen Problem geworden. Eine erfolgreiche Bekämpfung kann deshalb nur von einer intensiven Zusammenarbeitet zwischen Tierarzt, Arzt und den zuständigen Behörden erwartet werden.

LITERATUR

1. ACHMEDOW, A. M.: Die sanitäre epidemiologische Rolle des Paratyphus der Kälber bei der Entstehung von Nahrungsmitteltoxiko-Infektionen vom Typus der Salmonella-Infektion. *Proc. 2. Symp. Int. Assoc. Vet. Food Hyg. 255—257* (1960).
2. ADAM, A.: Über die Herstellung und Pasteurisierung von Fischmehl 1955—56 in Hamburg. *Berl. Münchn. tierärztl. Wschr.* **70**, *49—51* (1957).
3. ADLER, H. E.: Incidence of Salmonella in apparently healthy dogs. *J. Amer. Vet. med. Ass.* **118**, *300—304* (1951).
4. ALLENSPACH, V.: Die Enteritis-Gärtner Epidemie in Gontenschwil. *Schweiz. Arch. Tierheilk.* **94**, *80—90* (1952).
5. ANDERSON, E. S.: Special methods used in the laboratory for the investigation of outbreaks of Salmonella food poisoning. *Roy. Soc. Health J.* **80**, *260—267* (1960).
6. ANDERSON, E. S., MAUREEN, E. & WILSON, J.: Die Bedeutung der S. typhi-murium Phagentypisierung in der Human- und Veterinärmedizin. *Zbl. Bakt.* I. Orig. **181**, *368—373* (1961).
7. AVERBECK, W.: Zur Epidemiologie und Bekämpfung der S. enteritidis-Infektion des Rindes. Diss. Vet. Unters. amt Hannover 1958.
8. BADER, R. E.: Die Salmonellosen. In GRUMBACH, A. & KIKUTH, W. Die Infektionskrankheiten des Menschen und ihre Erreger. G. Thieme Stuttgart 1958.
9. BISCHOFF, H.: Ein Beitrag zur Klärung der Frage nach der Herkunft seltener Salmonellatypen. *Berl. Münchn. tierärztl. Wschr.* **68**, *306—307* (1955).
10. BLAXLAND, J. D., SOJKA, W. J. & SMITHER, A. M.: Avian Salmonellosis in England and Wales 1948—56 with comment on its prevention and control. *Vet. Rec.* **70**, *374—382* (1958).
11. BRÜCKER & SUTER, P.: Zur Salmonellose der Tauben. *Schweiz. Taubenztg.* Nr. **28** (1958).
12. BRUNER, D. W. & MORAN, A. B.: Salmonella infections of domestic animals. *Cornell Vet.* **39**, *53—63* (1949).
13. BUXTON, Salmonellosis in animals. A review. Commonwealth Agricultural Bureaux, Farnham Royal, Bucks. 1957.
14. BUXTON, A.: Public Health aspects of Salmonellosis in animals. *Vet. Rec. London.* **69**, *105—109* (1957).
15. BUXTON, A.: Salmonellosis in "Infectious diseases of animals". Vol. 2, 481—528. Butterworths Scientific Publ. London 1959.
16. CLARENBURG, A., VINK, H. H. & SCHUURMANS, R.: Salmonella-Dauerausscheider bei Rindern. *Tijdschr. Diergeneesk.* **75**, *435* (1950).
17. CLARENBURG, A. & KAMPELMACHER, E. H.: Die Bedeutung von Antibiotika für die bakteriologische Fleischschau mit besonderer Berücksichtigung der Milzbrand- und Salmonelladiagnostik. *Berl. Münchn. tierärztl. Wschr.* **70**, *203* (1957).
18. CLARENBURG, A.: Die Epidemiologie der Salmonellose bei Mensch und Tier. *Wien. tierärztl. Mschr.* **48**, *339—348* (1961).
19. COCKBURN, W. CH.: Reporting and incidence of food poisoning. *Roy. Soc. Health. J.* **80**, *249—253* (1960).
20. DATHAN, J. G., MCCALL, A. J., ORR-EWING, J. & TAYLOR, JOAN: Salmonella enteritis infection associated with use of anti-rodent "virus". *Lancet* **1**, *711—713* (1947).
21. DE LA CRUZ, E.: Salmonellosis epidemiology in Costa Rica. I. Salmonellosis in pigs. II. Salmonellae in processed meat. *Rev. Biol. Trop.* (S. José) **6**, *37—41* (1958).
22. DOLMAN, C. E.: The epidemiology of meat-borne diseases. In "Meat Hygiene". World Hlth. Org. Genf 1957.

23. EDWARDS, P. R., BRUNER, D. W. & MORAN, ALICE B.: The genus Salmonella: Its occurrence and distribution in the United States. *Kentucky Agric. exp. Stat., Lexington. Bull.* **525**, *1—60* (1948).
24. EDWARDS, P. R., BRUNER, D. W. & MORAN, ALICE B.: Further studies on the occurrence and distribution of Salmonellatypes in the United States. *J. inf. Dis.* **83**, *220—231* (1948).
25. EDWARDS, P. R.: Salmonella and Salmonellosis. *Ann. N.Y. Acad. Sci.* **66**, *44—53* (1956).
26. EDWARDS, P. R.: Salmonellosis: Observations on incidence and control. *Ann. N.Y. Acad. Sci.* **70**, *598—613* (1958).
27. ELLEMANN, GRETE & TERP, P. L.: Egg-borne Salmonellosis (hen's eggs). 8. *Nord. Vet. mötet Helsinki*, *810—818* (1958).
28. ELLEMANN, GRETE: Salmonella infection of eggs. 16. Int. Vet. Congr. Madrid. *767—769* (1959).
29. ELLEMANN, GRETE: Untersuchungen von Hühnern und Kücken in einer Geflügelschlächterei auf den Befund von Salmonellabakterien in der Kloake. *Nord. vet. Med.* **12**, *47—53* (1960).
30. FEY, H.: Zum klinischen Begriff der "septischen" coliformen Mastitis. *Schweiz. Z. allg. Path. Bakt.* **21**, *926—934* (1958).
31. FEY, H. & WIESMANN, E.: Die Gefahr des Salmonellaimportes mit Eiprodukten und tierischen Futtermitteln. *Schweiz. med. Wschr.* **90**, *791—800* (1960).
32. FEY, H. & VALLETTE, H.: Nachweis von Salmonellen in Fluss- und Abwässern sowie bei gesunden Schlachtschweinen in Genf. *Schweiz. Arch. Tierheilk.* **103**, *519—529* (1961).
33. FEY, H. & MARGADANT, ANITA: Hypogammaglobulinämie bei der Colisepsis des Kalbes. *Path. Microbiol.* **24**, *970—976* (1961).
34. FEY, H. & RICHLE, R.: Serologische Untersuchungen an Kälbern und Kühen nach Vakzination mit Pneumokokken. *Schweiz. Arch. Tierheilk.* **103**, *349—358* (1961).
35. FEY, H. & MARGADANT, ANITA: Zur Pathogenese der Kälber-Colisepsis. I. Verteilung des Sepsistyps in den Organen. *Zbl. Bakt.* I. Orig. **182**, *71—79* (1961).
36. FEY, H., LANZ, E. & MARGADANT, ANITA: VI. Experimentelle Infektion zum Beweis der parenteralen Genese. *Dtsch. tierärzt. Wschr.* **69**, *581—586* (1962).
37. FIELD, H. I.: Salmonelloses des equidés, des bovidés et des oiseaux. *Bull. Off. Int. Epiz.* **34**, *338—358* (1950).
38. FIELD, H. I.: Salmonella infections in cattle and in pigs. In "Infectious Diseases of Animals". Vol. 2, *528—556*, Butterworths Scientific Publ. London. 1959.
39. FOLTIN, H. W.: Zur Frage des Salmonellanachweises bei der bakteriologischen Fleischbeschau nach der Antibiotikabeifütterung bzw. Therapie. Diss. München 1959.
40. FROMME, W.: Zur Epidemiologie der Salmonelleninfektion. *Erg. Mikrobiol.* **32**, *161—195* (1959).
41. GALTON, MILDRED M., SCATTERDAY, J. E. & HARDY, A. V.: Salmonellosis in dogs. *J. inf. Dis.* **91**, *1—5* (1952).
42. GALTON, MILDRED M., SMITH, W. V., McELRATH, H. B. & HARDY, A. B.: Salmonella in swine, cattle and the environment of abattoirs. *J. inf. Dis.* **95**, *236—245* (1954).
43. GALTON, MILDRED M., LOWERY, W. D. & HARDY, A. V.: Salmonella in fresh and smoked pork sausage. *J. inf. Dis.* **95**, *232—235* (1954).
44. GAUGUSCH, Z.: Les recherches sur les animaux à sang froid comme porteurs de Salmonelles. 2. Congr. Int. Ass. Vet. Food Hyg. Nizza 1961.
45. GAUMONT, R.: Les Salmonelloses aviaires dans le Nord de la France. *Ann. Inst. Past. Lille.* **3**, *140—149* (1950).

46. GIBSON, E. A.: Salmonellosis in calves. *Vet. Rec.* **73**, *1284—1296* (1961).
47. HAMMER, D.: Die Entwicklung der Salmonellose bei Haustieren in Baden unter Berücksichtigung gesicherter Infektketten. *Berl. Münch. tierärztl. Wschr.* **74**, *64—70* (1961).
48. HAUGE, ST. & BØVRE, K.: Forekomst av salmonellabakterier i importert vegetabilsk proteinkraftfor og kraftforblandinger. *Nord. Vet. Med.* **10**, *255—262* (1958).
49. HAUSER, K. W.: Die Salmonellosis der Tauben. *Berl. Münchn. tierärztl. Wschr.* **72**, *126—129* (1959).
50. HENNING, M. W.: A preliminary report on the therapeutic value of Chloromycetin for the treatment of calf paratyphoid. *J. South Afr. vet. med. Ass.* **23**, *86—87* (1952).
51. HAMMER, D. Die Immunisierung trächtiger Rinder gegen Pneumokokken-Polysaccharide und biologische Bedeutung der im Kolostrum ausgeschiedenen spezifischen Antikörper. *Zbl. vet. Med.* **8**, *369—402, 405—450* (1961).
52. HENNING, M. W.: Calf Paratyphoid. *Onderstepoort J. vet. Res.* **26**, *3—23, 25—44, 45—59* (1953).
53. HENNING, M. W.: Paratyphoid in calves. In "Animal diseases in South Africa". 3rd ed. Central News Agency Ltd. South Africa. 1956.
54. HOFMANN, P., HÖRCHNER, F. & WOLL-JOHN, R.: Einschleppung seltener Salmonellen durch importierte Geflügelfedern. *Zbl. Bakt. I. Orig.* **178**, *484—491* (1960).
55. HOWARTH, J. A., CORDY, D. R. & BITTLE, J.: S. bredeney infection of calves and prophylaxis with Chlormycetin and Streptomycin. *J. Amer. vet. med. Ass.* **124**, *43—46* (1954).
56. JEBSON, J. L.: Salmonellosis in sheep. *Austr. vet. J.* **26**, *256—258* (1950).
57. JENSEN, P. TH.: Salmonella infection in mesenteric glands of swine and in imported vegetable feeds with sensitivity testing of strains. *8. Nord. Vet. mötet,* Rapport **8**, *819—826* (1958).
58. JÖRG, A.: Über Vorkommen, Diagnose und fleischbeschauliche Beurteilung von Infektionen mit B. suipestifer bei Viruspest. *Schweiz. Arch. Tierheilk.* **99**, *99—104* (1957).
59. JØRGENSEN, A. & MARTHEDAL, H. E.: Über Geflügelsalmonellosen. Serologische, kulturelle und epidemiologische Untersuchungen. *Nord. vet. Med.* **3**, *271—296* (1951).
60. JOSLAND, S. W.: Salmonella infections of animals in New Zealand. *Austr. vet. J.* **26**, *249—253* (1950).
61. KAMPELMACHER, E. H. & GUINÉE, P. A. M.: Salmonellaträger bei Schweinen und ihre Bedeutung für die Fleischhygiene. 2. Symp. Int. Ass. Vet. Food. Hyg. 258—261 (1960).
62. KAMPELMACHER, E. H., GUINÉE, P. A. M., HOFSTRA, K. & VAN KEULEN, A.: Studies on Salmonella in slaughter-houses. *Zbl. vet. Med.* **8**, *1025—1042* (1961).
63. KAMPELMACHER, E. H., GUINÉE, P. A. M. & CLARENBURG, A.: Salmonella organisms isolated in the Netherlands during the period from 1951 to 1960. *Zbl. Bakt. I Orig.* **185**, *490—502* (1962).
64. KAMPELMACHER, E. H., GUINÉE, P. A. M., HOFSTRA, K. & VAN KEULEN, A.: Further studies on Salmonella in slaughter-houses and in normal slaughter pigs. *Zbl. vet. Med.* im Druck.
65. KÄSTLI, P.: Die Übertragung des Typus bovinus auf den Menschen. *Schweiz. Z. Tbc.* **6**, *353—363* (1949).
66. KETZ, A.: Zur Frage der postmortalen Infektion von Lebensmitteln durch Ratten mit Fleischvergiftungserregern. *Berl. Münchn. tierärztl. Wschr.* **65**, *195—196* (1952).

67. KOLLER, R.: Die durch Salmonellen verursachten bakteriellen Lebens-mittelvergiftungen. *Wien. tierärztl. Mschr.* **37**, *248—261, 329—344* (1950).
68. KÜNG, W.: Weiterer Beitrag zur Salmonellose bei Mensch und Tier. *Schweiz. Arch. Tierheilk.* **105**, *81—86* (1963).
69. LINSERT, H.: Salmonellabefunde bei Schlachtenten. *Berl. Münchn. tierärztl. Wschr.* **72**, *192* (1959).
70. LENK, V., RASCH, K. & BULLING, E.: Über das Vorkommen von S. paratyphi B bei Tieren. *Zbl. Bakt.* I Orig. **180**, *304—309* (1960).
71. LERCHE, M.: Prophylaxie des Salmonelloses communes à l'homme et aux animaux. *Bull. Off. Int. Epiz.* **48**, *177—186* (1957).
72. LUNDBECK, H., PLAZIKOWSKI, U. & SILVERSTOLPE, L.: The Swedish Salmonella outbreak of 1953. *J. appl. Bact.* **18**, *535—548* (1955).
73. LÜTJE, F.: Zusammenstellung des jüngeren Schrifttums über die Frei-landbiologie der Salmonellen, der Salmonellosen der Möwenvögel und ihre Beziehung zum Abwasser und zum Menschen. *Berl. Münchn. tierärztl. Wschr.* **68**, *249—252* (1955).
74. MARTHEDAL, H. E.: Studier over Pullorum-Disease. *Nord. vet. Med.* **4**, *201—224* (1952).
75. MCCALL, A. M.: An explosive outbreak of food-poisoning caused by Salmonella dublin. *Lancet* **264**, *1302—1304* (1953).
76. MCCULLOUGH, N. B. & EISELE C. W.: Experimental human Salmonello-sis. *J. inf. Dis.* **88**, *278—289* (1951), **89**, *209—213, 259—265* (1951).
77. MCDONAGH, V. P. & SMITH, H. G.: The significance of the abattoir in Salmonella infection. *J. Hyg. (Lond.)* **56**, *271—279* (1958).
78a. MCELRATH, H. B., GALTON, MILDRED M. & HARDY, A. V.: Salmonello-sis in dogs. *J. inf. Dis.* **91**, *12—14* (1952).
78b. MACKEL, D. C., GALTON, M. M. & GRAY, H.: *J. inf. Dis.* **91**, *15—18* (1952).
79. MERLE A.: Les Salmonelloses dans les abattoirs et dans les élevages d'après leurs rapports avec la santé humaine. *Bull. Off. Int. Epiz.* **49**, *441—482* (1958).
80. MESSERLI, W.: Salmonellose bei Mensch und Tier im gleichen Gehöft. *Schweiz. Arch. Tierheilk.* **104**, *294—297* (1962).
81. MEYER, K. F.: Food poisoning. *New England J. Med.* **249**, *765—773, 804—812, 843—852* (1953).
82. MÜLLER, M.: Übertragung latenter Paratyphusinfektionen bei Schwei-nen auf den Menschen durch Genuss rohen Schinkens. *Dtsch. Schlacht-hofz.* **11**, *114* (1931).
83. MÜLLER, J.: Le problème des salmonelloses au Danemark. *Bull. Off. Int. Epiz.* **48**, *323—336* (1957).
84. MÜLLER, J.: On the epidemiology of Salmonella infections with special reference to cattle. *Nord. Vet. mötet* **8** (1958).
85. NEWELL, K. W., MCCLARIN, R., MURDOCK, C. R. & MACDONALD, W.N.: Salmonellosis in Northern Ireland with special reference to pigs and salmonella-contaminated pig meal. *J. Hyg. (Camb.)* **57**, *92—105* (1959).
86. NEWELL, K. W.: The investigation and control of Salmonellosis. *Bull. Wld. Hlth. Org.* **21**, *279—297* (1959).
87. NIELSEN, B. B.: Salmonella typhi murium carriers in seagulls and mal-lards as a possible source of infection to domestic animals. *Nord. vet. Med.* **12**, *417—424* (1960).
88. NIELSEN, F. W.: Les infections à Salmonella dans les troupeaux de bovins du Danemark. *Bull. Off. Int. Epiz.* **42**, *746—774* (1954).
89. NORDBERG, B. K. & EKSTAM, M.: Salmonellen bei Hühnern. *Nord. vet. Med.* **2**, *23—30* (1950).
90. PARKE DAVIES & Co.: Über Chloromycetin. Vet. Kurzinformationen München 1962.

91. PIENING, C.: Beitrag zur Klärung der Aetiologie des sogenannten Kälberparatyphus. *Berl. Münchn. tierärztl. Wschr.* **67,** *277—281* (1954) *Städtehygiene* **9,** *180—181* (1955).

92. POMEROY, B. S., FENSTERMACHER, R. & ROEPKE, M. H.: Sulfonamides in the control of Salmonelloses of chicks and poults. *J. Amer. vet. med. Ass.* **112,** *296—303* (1948).

93. RASCH, K.: Über das Verhalten der Salmonellen in der Aussenwelt, ein auch für die Verhütung der Fleischvergiftung wichtiger Faktor. *Arch. Lebensmittelhyg.* **6,** *1—3* (1955).

94. RASCH, K.: Die Bekämpfung der Rindersalmonellose im Blickpunkt der Lebensmittelhygiene. *Berl. Münchn. tierärztl. Wschr.* **70,** *161—162* (1957).

95. RAUSCH, R.: Die Lebensfähigkeit von Bangbakterien in verschiedenen Käsesorten. *Schweiz. Milchz.* Nr. **95,** *1—28* (1957).

96. ROHDE, R.: Über den Nachweis eines neuen Salmonellatypes aus Fischmehl in Hamburg. *Zbl. Bakt.* I Orig. **163,** *570—571* (1955).

97. RUTQVIST, L. & THAL, E.: Salmonella isolated from animals and animal products in Sweden during 1956—1957. *Nord. vet Med.* **10,** *234—244* (1958).

98. RUTQVIST, L., THAL, E., PETRELIUS, T. & SWAHN, O.: Salmonella cholerae suis Infektion und Nitrofurazonbehandlung. *Nord. vet. Med.* **13,** *3—19* (1961).

99. SAPHRA, I.: Fatalities in Salmonella infections. *Amer. J. med. Sci.* **220,** *74—77* (1950).

100. SAPHRA, I.: S. cholerae suis. A clinical and epidemiological evaluation of 329 infections identified between 1940 and 1954 in the New York Salmonella Center. *Amer. J. med. Sci.* **228,** *525—533* (1954.)

101. SAVAGE, W.: Problems of Salmonella food-poisoning. *Brit. med. J.* **2,** *317—323* (1956).

102. SCHAAL, E.: Schlachthofabwässer und ihre hygienische Bedeutung. *Berl. Münchn. tierärztl. Wschr.* **72,** *66—70* (1959).

103. SCHELS, H.: Ein Beitrag zur Brucellose und Salmonellose des Pferdes. *Tierärztl. Umschau* **13,** *341—343* (1958).

104. SCHÜTZ, G.: Über das Vorkommen seltener Salmonellatypen in der Galle und im Kot gesund geschlachteter Rinder und Schweine. *Berl. Münchn. tierärztl. Wschr.* **72,** *192* (1959).

105. SEDLMEIER, H., KOTTER, L. & TERPLAN, G.: Zum Vorkommen von nicht hühnerspezifischen Salmonellen bei Hühnern. *Berl. Münchn. tierärztl. Wschr.* **70,** *433—435* (1957).

106. SEELIGER, H. P. R.: Salmonellosis in Western Germany and Berlin (1945—1957). *Proc. 6th. Congr. Trop. Med. and Malaria* **4,** *43—48* (1959).

107. SEELIGER, H. P. R., HOFMANN, S. & ROHDE, R.: Jahresbericht über die Salmonellosen in der Bundesrepublik Deutschland und West-Berlin 1958. *Zbl. Bakt.* I Orig. **182,** *357—403* (1961).

108. SMITH, H. W.: The immunity to Salmonella gallinarum infection in chickens produced by live cultures of members of the Salmonella genus. *J. Hyg. (Camb.)* **54,** *433—439* (1956).

109. SMITH, H. W.: The effect of feeding pigs on food naturally contaminated with Salmonella. *J. Hyg. (Camb.)* **58,** *381—389* (1960).

110. SØRUM, L.: Infektion mit S. typhi murium var. Copenhagen bei Tauben in Norwegen. *Nord. vet. Med.* **5,** *385—400* (1953).

111. SOMPOLINSKI, D.: Influence of Penicillin-therapy on bacteriological diagnosis. *Nord. vet. Med.* **6,** *442—448* (1949).

112. STECK, F.: Die Übertragung von Gammaglobulinen auf das neugeborene Kalb mit dem Colostrum. *Schweiz. Arch. Tierheilk.* **104,** *525—536, 593—607* (1961).

204

113. STEINIGER, F.: Zur Freilandbiologie der Salmonellen im Bereich des westlichen Mittelmeeres. *Zbl. Bakt.* I Orig. **166**, *245—265* (1956).
114. STOCKMAYER, W.: Die Pulloruminfektion der Hühner und ihre Bekämpfung. *Tierärztl. Umschau.* **19/20**, *354* (1951).
115. STRUCK, M.: Kann es bei Tierkörpern, bei denen nur in den Organen Salmonellen festgestellt wurden, nachträglich zur Anreicherung von Salmonellen kommen? *Berl. Münchn. tierärztl. Wschr.* **70**, *163—164* (1957).
116. STUCKER, C. L., GALTON, MILDRED M., COWDERY, J. & HARDY, A. V.: Salmonellosis in dogs. *J. inf. Dis.* **91**, *6—11* (1952).
117. TAYLOR, JOAN: The diarrhoeal diseases in England and Wales. *Bull. Wld. Hlth. Org.* **23**, *763—779* (1960).
118. TAYLOR, JOAN: Salmonella and Salmonellosis. *Roy. Soc. Health J.* **80**, *253—259* (1960).
119. THAL, E., RUTQVIST, L. & HOLMQVIST, H.: Salmonella isolated from animals in Sweden during the years 1949 to 1956. *Nord. vet. Med.* **9**, *822—830* (1957).
120. THAL, E., RUTQVIST, L. & KARLSSON, K. A.: Salmonella bei Hühnern in Schweden. *2. Symp, Int. Ass. Vet. Food Hyg.* *250—254* (1960).
121. SCOTT, THOMSON: The numbers of pathogenic bacilli in faeces in intestinal diseases. *J. Hyg.* **53**, *217—244* (1955).
122. VAN OYE, E. & LAFONTAINE, A.: Etat actuel du problème des salmonelloses humaines en Belgique. *Arch. Belg. Méd. soc. Hyg., Méd. du Travail, Méd. leg.* **20**, *503—514* (1962).
123. VAN DER SCHAAF, A.: Salmonellosis in carnivorous domestic animals and fur animals. *Tijdschr. Diergeneesk.* **86**, *99—110* (1961).
124. VAN DER SCHAAF, A., VAN ZIJL, H. J. M. & HAGENS, F. M.: Meal of animal origin and Salmonellosis. *Tijdschr. Diergeneesk.* **87**, *211—221* (1962).
125. WAGENER, K. & MITSCHERLICH, E.: Die Erregerwirkung von S. abortus equi beim Pferd. *Mh. vet. Med.* **7**, *141—144* (1951).
126. WALKER, H. H. C.: Organic fertilisers as a source of samonella infection. *Lancet* **10**, *283* (1957).
127. World Health Organization: Comité mixte d'experts des Zoonoses. *Rapp.* Nr. **169**, *9—21* (1959).
128. WILSON, ELIZABETH, PAFFENBARGER, R. S., FOTER, M. J. & LEWIS, K. H.: Prevalence of Salmonellae in meat and poultry products. *J. inf. Dis.* **109**, *166—171* (1961).
129. WINKLE, S. & ROHDE, R.: Über die Gefahr der bakteriologisch unkontrollierten Importes ausländischer Tierfuttermittel mit bes. Berücksichtigung der Schweinezucht. *Berl. Münchn. tierärztl. Wschr.* **70**, *442—448* (1957).
130. WOLFF, A. H., HENDERSON, N. D. & McCALLUM, GRACE, L.: Salmonella from dogs and the possible relationship to salmonellosis in man. *Amer. J. Pub. Hlth.* **38**, *403—408* (1948).
131. Working party of the Public Health Laboratory Service: Salmonella organisms in animals feeding stuffs and fertilizers. *Monthly Bull. Min. Hlth.* **18**, *26—34* (1959).
132. Working party of the Public Health Laboratory Service: Salmonella organisms in animal feeding stuffs. *Monthly Bull. Min. Hlth.* **20**, *73—85* (1961).
133. LÜTJE, F.: Über die Salmonellose unserer Schlachttiere. *Tierärztl. Umschau* **7**, *385—390* (1952).

TRANSMISSION OF SALMONELLAE AND PATHOGENESIS OF SALMONELLOSIS IN MAN

BY

F. N. SICKENGA

*The Hague**

Introduction

This article will deal with the transmission of salmonellae and the pathogenesis of salmonellosis in man, with the exclusion of S. *typhi* and S. *paratyphi B* and the diseases caused by them.

The "other salmonelloses" can be considered principally as "zoönoses". Zoönoses are defined by the WHO/FAO expert committee as "diseases and infections which are naturally transmitted between vertebrate animals and man".

As in the case of salmonellosis the transmission from animals to man (as a rule by animal food) is much more frequent than that from man to man or from man to animals, there is reason to expect that eradication of the salmonella-reservoir in animals will lead to a very important reduction of salmonellosis in man.

This does not mean that transmission from man to man is impossible, and it happens now and then indeed, but is epidemiologically less important than that from animals to man. This transmission is seldom direct; as a rule man contracts it by eating contaminated meat or animal products. These products are especially dangerous when they have been kept under circumstances that favour the multiplying of salmonellae. These circumstances occur more frequently during hot weather; therefore in countries with moderate climate on the northern hemisphere the frequency of registered cases tends to increase greatly in summer-time.

Exceptionally other media serve as the source of contagion, e.g. drinking-water or imported vegetable products, such as cocos and peanuts.

Sometimes other substances than food or drinking-water have to be considered as sources of contagion, e.g. manure or organical fertilizers and dust.

The following terms will be used:

By contamination is meant the presence of salmonellae in or on dead material (meat, butchers' tools, dust, etc.).

By contagion is meant the presence of salmonellae in or on man

---

* The author is indebted to Dr. E. H. KAMPELMACHER, Utrecht, and Dr. J. E. MINKENHOF, Amsterdam, for revising the manuscript and giving some valuable suggestions.

or animals, without considering the question whether they have caused disease or not.

By infection is meant salmonellosis as a disease.

We will deal first with the different ways of transmission of salmonellae to man and subsequently with the pathogenesis of human salmonellosis.

The transmission of salmonellae to man

Transmission by food

This way of transmission is undoubtedly the most frequent and the most important. The great majority of cases of "food-poisoning" are caused by the ingestion of food contaminated with salmonellae.

In England in 1960 salmonellae were found in 95% of all cases of "food-poisoning", the cause of which could be traced. The number of reported "incidents",* caused by salmonellae amounted to 3576 in 1954, 5132 in 1959, 4105 in 1960, and 3771 in 1961 [1, 1a, 1b]. In 1961 the most frequently found types were:

	group			group
1. *S. typhi-murium*	B	4. *S. saint paul*		B
2. *S. heidelberg*	B	5. *S. newport*		C_2
3. *S. enteritidis*	D_1	6. *S. meleagridis*		E_1

Most outbreaks could be attributed to the ingestion of contaminated meat- and egg-products.

Among 200 outbreaks of salmonellosis associated with food, 5 were associated with fresh meat, 87 with processed or made-up meat, 41 with shell eggs (mainly duck's eggs), and egg products, 23 with cream confectionery, 10 with milk, and 34 with a wide variety of foods (Joint WHO/FAO expert committee on zoönoses [49]).

In several countries the presence of salmonellae was established in different foodstuffs.

In Holland in three consecutive years minced meat, sold in butcher's shops was investigated on the presence of salmonellae. This had the following results:

year	butcher's shops investigated	number of shops where salmonellae were found in minced meat	
			%
1959	159	15	9.4
1960	139	11	7.9
1961	136	16	11.7

These investigations are continued.

Most of the minced meat contains pork.

Pork is contaminated much more frequently than beef. This was confirmed frequently in slaughter-houses.

---

* These "incidents" comprise general outbreaks, family outbreaks and sporadic cases, the latter including a number of symptomless excreters.

In Holland in 1960 in $\pm$ 25% of 2100 slaughtered pigs in different slaughter-houses salmonellae were found in either the portal or mesenteric lymphnodes and/or the faeces, or in both (KAMPELMACHER et al.[3]). In some regions the percentage was even higher (ranging from 14 to 38). In animals from one specialized modern pig-farm it was 98 notwithstanding very good hygienic conditions (KAMPELMACHER et al.[2]).

On the contrary in 1956 in 1600 slaughtered cattle, of which samples of the liver, gall-bladder and portal lymph-nodes were cultured, salmonellae were only found in 9 cases (0.56%). (KAMPELMACHER[4]).

In England a similar experience was made. TAYLOR[5] never found salmonellae in 1518 samples of fresh beef, but repeatedly in processed meat-products. Pork-saucages furnished the greatest variety of types.

Imported and cooled beef and veal are more liable to contamination with salmo-nellae than fresh inland ditto. In 1959 at Smithfield market (London) salmonellae were found in 17.5% of the samples of imported boneless veal (HOBBS & GREEN-WOOD WILSON[6]).

Second in frequency are the egg-products, mostly applied in cakes or custards, etc.

HARVEY et al.[7] reported a rather serious epidemic which could be attributed to the consumption of trifle by school-children at a Christmas-party. From the cake, used for the trifle *S. typhi-murium*, phage-type 2c and *S. thompson*, phage-type 4, were isolated.

The former type could be cultured from the stools of 117 patients, the latter from those of 23 patients, and both from those of 69 patients.

In Holland from 1952 to 1959 2956 samples of products from hen's eggs were investigated; 132 (4.5%) contained salmonellae, belonging to 18 types (POLAK[8]).

Before 1953 in Holland epidemics not seldom occurred which could be attributed to duck's eggs or duck's egg products, contaminated by *S. typhi-murium* (HEMMES[9]). This changed since in 1953 legal prescriptions concerning the pasteurization of duck's eggs products and the cooking of duck's eggs were given.

In Western Germany on April 1st, 1957 a regulation came into operation "zum Schutze gegen Infektion durch Erreger der Salmonella-Gruppe in Eiprodukten". KRESSMANN & ALBERT [10] reported about 144,732 samples of imported egg-products tested at the Hamburg Veterinary Institute in the period from April 1, 1957 to April 30, 1961. Salmonellae belonging to 48 types were found in 7.0% of these samples, viz. in 9.5% of dried, in 3.9% of frozen and in 1.0% of fluid egg products. *S. thompson* was most frequently found, followed by *S. typhi-murium*.

Also in other slaughtered animals and animal products salmonel-lae were found, e.g. in game and fowls (ROHDE & PACHL[11]), and in a pancreatic hormone product (HUISMAN & DANIËLS-BOSMAN[12]). In developed countries milk as a rule is not an important source of contagion, because the great majority of the population consumes pasteurized milk. Where raw milk is still used, it may cause an epi-demic of salmonellosis (PARRY[16]).

The chance of transmission by fish and other water-animals is small. SEMPLE[15] found salmonellae in prawns, imported from Japan.

Transmission by direct contact

Besides transmission by food, transmission by direct contact is also possible, as well from man to man as from animal to man.

208

Epidemiologically, this way of transmission is less important than that by food, but not altogether negligible.

The concept "direct contact" is used here in a large sense. By this term is meant either direct corporeal contact, or contact through the intermedium of a single object, (food excluded), e.g. the teat of a nursing-bottle.

The spread of salmonellae by direct contact is undoubtedly favoured by unhygienic habits, especially in families of low social standard with a quiver full of children, and in institutions for mentally defective children. Clinically ill persons are more dangerous as a source of contagion than healthy excreters of bacteria.

Transmission by direct contact from man to man

It is nearly impossible to furnish a scientific proof that in a certain case the contagion was effectuated by direct contact and not through the intermedium of food or perhaps by inhalation of salmonella-containing dust (see page 210). The danger of contagion by direct contact with adults must be small, as was proved by Mc CULLOUGH & EISELE[17, 17a], in their experiments with volunteers (prisoners). Only in certain circumstances, e.g. when mothers and their babies are concerned, the danger of contagion by direct contact is greater, as well between themselves as regarding other persons in their neighbourhood.

In the experiments of Mc CULLOUGH & EISELE not one person who was in contact with the experimentally infected contracted a salmonellosis, or became an excreter of salmonellae, even if they shared the same cell, and notwithstanding only 12 of 32 persons who became clinically ill were transported to hospital. The only precaution taken was, that the infected individuals were excluded from foodhandling.

NOORDAM & POSTMA[18] found in 1959 and 1960 among 805 butchers at Amsterdam 26 salmonella-excreters (3.2%). They examined also the families of these butchers. In 21 families none of the housemates excreted salmonellae; one family-member excreted another type; in two families one excreter of the same type was found and only in the two remaining families mild illnesses, due to the same type occurred. In 1961 another 387 butchers were examined, among whom 10 excreters were found. 29 Family-contacts of these excreters were all free from salmonellae (NOORDAM[19]).

We noticed that the picture is different where infants and their mothers are concerned: in this case the chance of contagion by direct contact is greater.

LEEDER[21] described an epidemic of *S. panama* in a maternity unit, comprising 138 individuals (asymptomatic excreters included) among whom 59 newborn infants and babies, 2 hospitalized mothers of newborn infants, one woman who underwent an operation in the same hospital, 5 members of the personnel, and not less than 72 individuals among other members of the families into which a *S. panama*-positive infant had been introduced. 17 Infants and one adult died from *S. panama*-meningitis.

Another epidemic in a maternity unit was reported by MURRAY & WALKER[38]; here *S. heidelberg* could be incriminated. 5 mothers and 21 babies suffered from enteritis and 24 mothers and 13 babies were symptomless excreters. Five immature babies died from enteritis; *S. heidelberg* was isolated from four of them. Whilst the contagion rate was similar in babies and in adults, the morbidity rate was nearly five times higher in babies. The babies also produced only about half the proportion of symptomless excreters when compared with mothers. Deaths only occurred in premature babies; the symptoms in normal babies were of a minor nature.

During an outbreak in a children's ward, caused by *S. bovis morbificans* also mentioned at the foot of this page it was certified that diapers of babies suffering from salmonellosis were not disinfected by immersing them in a 40% solution of benzalkonium chloride during 12 hours, (JELLARD et al.[55]). It seems that this disinfectant does not penetrate sufficiently into the stools.

Some nurses who handled these diapers became salmonella-excreters and the authors suppose that contagion took place during the mangling of the diapers that followed the immersion.

MARSEILLE[20] mentioned two cases of presumed contagion of infants by salmonella-excreting mothers during delivery. One baby, delivered by forceps, died from *S. heidelberg*-meningitis; the other, delivered without surgical intervention, remained healthy, but became an excreter of *S. paratyphi B* of the same phage-type as excreted by his mother.

The question may be raised, whether transplacental transmission of salmonellae from the mother to the fetus is possible. This cannot be excluded with certainty but is not very probable.

NETER[31] mentions a woman in childbed who harboured *S. cholerae-suis* in the blood for at least 6 days prior to delivery, and whose blood culture was still positive on the day of the birth of the child. The placenta contained numerous salmonellae, but the blood and mouth cultures of the baby were sterile.

Thus the placenta seems to be an effective barrier, and the contagion of newborn children by their mothers who suffer from salmonellosis is most probably established by direct contact, either during delivery or in the next days. Even if the child is taken from the mother directly after birth, it may be infected already.

NETER[31] mentions another case where the mother of a premature baby had had very mild diarrhoea one week prior to delivery and was excreting *S. oranienburg*. The baby was taken immediately after birth to the premature nursery and had no more contact with the mother. It developed diarrhoea on the third day and high fever on the fifth day; on that day a blood culture and a feces culture were positive for *S. oranienburg*. Attempts to isolate salmonellae from the attendants at the premature nursery failed. But the infection spread to two other infants, notwithstanding careful precautions. As this child harboured *S. oranienburg* in the nasopharynx, (where it was even the predominant organism), it is supposed that the two others were infected by air-borne transmission (see page 211).

Transmission by direct contact from animal to man

This way of contagion is apparently rare; nevertheless a few cases have been mentioned in literature.

HEMMES[23] reported cases of *S. bareilly*-salmonellosis in children who had played with infected chickens on a chicken-farm.

Contagion of children with *S. typhi-murium* by playing with infected ducklings was also observed (RUYS[24]).

JELLARD et al.[55] described an outbreak in a children's ward, caused by *S. bovis morbificans* the origin of which could be traced to a girl of 6 months old, suffering from diarrhoea with blood and mucus who had been admitted with the provisional diagnosis of intussusception, and whose father drove a lorry for transporting cattle. The same type was found on the floor of this lorry (even after it had been cleaned with cresol and lysol) and in the feces of a nine years old brother of the patient who sometimes travelled in this lorry and who probably infected her. In the feces of the parents no salmonellae were found.

It may be possible that in farmers' children and in farmpersonnel, transmission from animal to man occurs more often than is supposed.

Transmission by other media

Salmonellae have been found in animal feeds, manure and other fertilizers, dust, sewage and surface-water.

Imported animal feeds are not seldom contaminated with salmonellae (KAMPELMACHER & GUINEÉ[27]: RACKOW & WIESE[28]). It is probable that contagion of the people handling them happens now and then, but presumably not to such an extent that illness follows. The same applies to manure and other fertilizers.

Fresh excrements of farm-animals, and especially of pigs often contain salmonellae, but in manure that is allowed to heat in a dung-hill they die as a rule within a few days, except during rather severe frost (BAETGEN[60]).

Other organic fertilizers may also contain salmonellae, e.g. bone-meal (WALKER[26]) and sludge from sewage-disposal plants (NOORDAM[19]).

These facts are of more importance when the ways of transmission to animals are studied than when direct contagion of man is concerned.

In the latter context the possibility of contagion by salmonella-containing dust deserves special interest. This dust may of course contaminate human food in the second instance and there the salmonellae may multiply under certain conditions.

But we are not sure about the importance of direct contagion of man by salmonella-contaminated dust. It might be that this way of transmission is not altogether harmless for infants, especially for premature babies and for those who are weakened by other causes.

The importance of the presence of contaminated dust as a source of contagion in a children's ward was firmly established by BATE & JAMES[40]. In this case dissemination by food or by human carriers could be excluded. The authors described an epidemic of salmonellosis, caused by S. typhi-murium phage type 2 in a ward, used for the isolation of infants under one year of age not suffering from gastroenteritis. This epidemic lasted eleven months, although the ward was closed several times, and washed down and disinfected thoroughly before being reopened. All feeds were prepared in a central milk-kitchen which operated under constant bacteriological control and which supplied feeds to other wards unaffected by the infection. During the epidemic, all children were isolated in single cubicles. Moreover the whole medical, nursing and domestic staff, and the porters of the hospital were examined. In the beginning some symptomless excreters were found among the personnel, but during the last four outbreaks no human carrier could be detected. Nevertheless ever and again new cases of gastroenteritis occurred among newly admitted children, and each time S. typhi-murium, phage type 2, was cultured from the stools. At last this type was found in abundance in the contents of the dust-bag of a vacuum-cleaner that had been used constantly in the ward; the bag had been emptied infrequently. The epidemic ceased when the dust was collected in disposable paper-bags that were destroyed after each cleaning. The salmonellae still remained viable in the original dust-bag 10 months after the removal of the last case from the ward.

In this context we have to consider the question whether respiratory infection plays a role in human salmonellosis-pathology. This question is still open; the possibility cannot be absolutely discarded.

It was proved that animals are liable to direct infection of the respiratory tract, when they are exposed to an aërosol containing salmonellae even in a lower dose than that required for enteral infection.

CLEMMER et al.[29] succeeded in inducing salmonellosis in chickens by exposing them to a spray containing extremely small quantities of salmonellae; in some experiments not more than 20 germs sufficed. For infection per os quantities of bacteria were required that were a 100 to a 1000 times greater. Especially after inhalation of small quantities of S. pullorum extensive foci of pulmonary infection were seen.

DARLOW et al.[30] could induce extensive pneumonitis in mice exposed to an aërosol containing S. typhi-murium. The quantities of bacteria used in these experiments were much larger than those administered to the chickens, but also in these animals the lethal dose by inhalation was much smaller than that required in case of ingestion.

In man, salmonellae have been found in the upper respiratory tract, and also cases of salmonella-pneumonitis as a sole or preponderant site of infection have been observed, although rather seldom.

Among 7779 cases of salmonella-infection in man, identified at the New York salmonella Center, there were 572, where focal infections were the predominant and frequently exclusive clinical signs, and 85 of them were localized in the respiratory tract, mostly as a lobar pneumonia or a bronchopneumonia. The type of salmonella most frequently found in these cases was S. cholerae-suis, followed by S. typhimurium. A great deal of the patients were aged people, or persons who suffered from other diseases (SAPHRA & WINTER[35]). In a few cases there was a salmonella-tonsillitis or otitis media.

When salmonella-pneumonitis occurs in man, the route of infection of the lungs is not clear; it may be as well aerial as hematogenic. SAPHRA & WINTER[35] are of opinion that a transient bacteremia is the rule rather than the exception in salmonella infections with chills and fever. They mention 100 cases of acute gastroenteritis or acute focal salmonellosis with positive blood-cultures.

A third route of infection may be considered as well especially in weakened infants, namely contagion of the upper respiratory tract by salmonella-containing dust or droplets and subsequently infection of the intestine by swallowed secretions.

Even a fourth route of infection is possible, namely the conjunctival route.

MOORE[39] produced evidence that in guinea-pigs the conjunctival route is far more effective than the oral route in producing systemic salmonella infection. He found also that conjunctival swabbing was more effective than rectal swabbing for detecting the spread of contagion.

It is not known if this applies also to human pathology. It would be interesting to examine the conjunctivae of persons who are exposed to salmonellae-containing dust or droplets, and in the same time the secretions of their nose and throat and their stools. It is conceivable that salmonellae may be conveyed from the conjunctivae through the naso-lacrimal ducts to the nose and throat; and perhaps further to the intestinal tract if the secretions are swallowed.

NETER[31] cultured *S. oranienburg* and *S. cholerae-suis* from the upper respiratory tract of two children and holds that one of them infected two others by aerial route.

DATTA & PRIDIE[32] isolated salmonellae from the nose, pharynx and sputum of patients in hospital. They reported a persistant hospital-epidemic which they ascribe, at least partially, to the presence of salmonella-contaminated dust.

RUBINSTEIN & FOWLER[22] reported two epidemics in infants in an obstetrical unit, caused by the routine use of delivery-room resuscitators that were contaminated by salmonellae. These resuscitators were used for the administration of oxygen and carbon dioxide and for mechanical suction. In both cases salmonellae were recovered in pure culture from the water trap fluid of the resuscitator; in one hospital *S. montevideo* was the culprit, in the other *S. bareilly*. It was apparent that contamination of the fluid in the trap had resulted in a spray of organisms into the atmosphere of the delivery room when the resuscitator was being used for suction. As in the case of MURRAY & WALKER (page 208), the contagion spread to family contacts outside the hospital.

Winding up the argument, we conclude that salmonella-contaminated dust or sprayed droplets may cause contagion of human individuals, and that this contagion sometimes leads to infection. But we don't know exactly what is the link between the two. It is possible that different routes are followed in different circumstances.

Moreover we must remember that salmonella-epidemics within a hospital may originate from quite other sources than from dust. Epidemics as described by BATE & JAMES take a chronic course, and befall especially infants. Their food is pasteurized and can be excluded as a source of infection, and perhaps they are more disposed to air-borne infection than older children and adults.

The picture of food-borne infections in hospital is quite different. Here the outbreak is acute and as a rule a great number of adults fall ill at the same time.

In 1959 two such epidemics occurred in Amsterdam hospitals. In one of them ± 100 persons were concerned, in the other 30. These outbreaks were caused by salmonella-contaminated meat.

In cases as those mentioned here the course of events is obvious, but this is not always so. WILLIAMS et al.[36] in their book on hospital-infections state that "with salmonella infections, as with so much of hospital infections the relative importance of possible routes of transfer cannot yet be judged on any truly scientific basis".

Direct contagion of man by surface-water is seldom, at least in countries with a controlled supply of drinking water. Where this is lacking, and where the hygienic conditions are not first rate, for instance in camps, epidemics of salmonellosis caused by drinking water may occur.

An example of such an epidemic was observed by WILSON & BAADE[63] in a small town in Alaska that got its drinking water from a nearby lake, populated in autumn by thousands of sea-gulls. *S. manhattan* was found as well in the patients as in the excretions of the gulls.

The risk of contracting salmonellosis by bathing in salmonella-containing surface-water seems to be practically non-existent; evidently the dose ingested in this way is so small that it does not give rise to morbid symptoms.

The pathogenesis of salmonellosis in man

The factors that determine the course of events after transmission

After transmission of salmonellae to man irrespective of the route they followed we have to consider three alternatives (RUYS[37]).

Either the germs cannot be traced back at all, or the person subject to contagion becomes during a certain time a carrier or excreter of bacteria, thereby remaining in good health, or morbid symptoms develop.

Which of these alternatives is realised, depends upon:
1) The dose of the transmitted bacteria
2) The resistance of the host
3) The type of Salmonella, and the virulence of an individual strain of a certain type.

MC CULLOUGH & EISELE[17, 17a, 17b] were able to analyse these factors for a great deal in their experiments with volunteers (prisoners).

In their first experiments they used three strains of *S. meleagridis* and three of *S. anatum*, which had been isolated from egg-powder. The germs were administered by mouth.

The minimum dose required for causing morbid symptoms in some of the volunteers were for the three strains of *S. meleagridis* respectively 24 million, 10 million and 7.7 million germs; for those of *S. anatum* respectively 44.5, 1.26 and 0.59 million. Temporary excretion of bacteria occurred in five of ten subjects of experiment already after an ingestion-dose of 12000, and when the dose was augmented this happened more frequently, though with exceptions. In two persons who ingested 10 million *S. meleagridis* and one who ingested 23.9 million *S. anatum* the germs could not be recovered from the stools.

The length of the period during which salmonellae were excreted was extremely variable, in ill persons between 2 and 72 days, in healthy excreters between 2 and 128 days.

Also there was a great difference in the seriousness of the symptoms of the patients, varying between slight diarrhoea during one or two days to serious gastro-enteritis with chills and high fever, requiring two to three weeks for complete recovery. This happened sometimes even after ingestion of a relatively low dose, for instance of 0.86 million *S. anatum*, strain I.

The incubation-time varied between 8 and 72 hours; in three quarters of the cases between 20 and 48 hours.

There was also a very great difference in the agglutination titres: in the ill persons between 0 and 1:640; in the healthy excreters between 0 and 1:1280; in those who did not excrete bacteria from 0 to 1:80.

In a second series of experiments *S. bareilly*, *S. newport* and *S. derby* were used. The pathogenicity of the first two of these proved to be even greater than that of *S. meleagridis* and *S. anatum*. In one person 125,000 germs of *S. bareilly* sufficed to cause a serious illness, and in another 152,000 of *S. anatum* to produce mild symptoms. The other findings were of the same order as in the first experiments.

With *S. pullorum* on the contrary a much larger dose was required to cause morbid symptoms. This dose amounted in three strains, isolated from market-samples of spray-dried whole egg, to respectively 10,000,000,000, 7,640,000,000 and 6,750,000,000 bacteria, and in one strain of human origin to 1,280,000,000 bacteria.

It may be that these results were influenced by the fact that the persons in this experiment had been vaccinated against typhoïd fever.

It was difficult to recover *S. pullorum* from the stools. When the dosage was below the level producing illness, they could never be traced back, and in patients only during the first one or two days or not at all. The agglutination-titers in the patients varied from 0 to 1 : 320.

In none of Mc Cullough & Eisele's experiments blood cultures were made.

When such different effects result from the administering of salmonellae to healthy adults who live in the same circumstances, we need not wonder that in society as a whole the differences are still greater.

The influence of the transmitted dose

It is obvious that transmission of a large dose of a certain type of bacteria has, ceteris paribus, a greater effect than that of a small one. This is simple logic, and has been proved by countless animal experiments.

Therefore one wonders somewhat about the outcome of the cited experiments in man, according to which the individual resistance of the tested persons to a special strain of salmonella seemed to be of at least equal importance as the dose.

The influence of the resistance of the host.

We have seen that the resistance of the host shows a great variety, even in healthy adults who live in the same circumstances. This variety will be still greater, if other factors, occurring in free society, are taken into account.

Let us try to analyse the different factors that may influence the resistance of the host. These are:

(a) acquired immunity
(b) natural resistance (of genetic origin)
(c) age
(d) debilitating factors.

Acquired immunity

Our knowledge about acquired immunity in salmonellosis is scanty. We know that it exists to a certain degree against salmonellae which have a specific affinity for man, such as *S. typhi* and *S. paratyphi B*, either after clinical disease or as a result of vaccination. Also, only in these forms of salmonellosis, the benefit of vaccination has been firmly established. There is no convincing evidence that acquired immunity is of practical importance in other salmonelloses. This may be due to three circumstances.

The "other salmonelloses" are caused by a range of various types that differ largely in the composition of their antigens. When infection with a certain type has given rise to antibody-production against its specific antigens, these will not protect the individual against infection with another type with different ones. Only when nearly related types are concerned, some degree of cross-immunization may be expected.

There is some evidence that this may occur. LEVINE et al.[52] published an interesting report on the incidence of "other salmonelloses" on the island of Oahu, Hawaii, where the entire population had been vaccinated against typhoïd and paratyphoïd A and B in 1942, and where from that year onwards this so-called TAB-vaccination was made compulsory at 3 years of age. In 1951, salmonellosis was made reportable. (The antigenic formula of *S. typhi* is 9, 12, Vi:d:-; that of *S. paratyphi A*: 1,2, 12:a:-, and that of *S. paratyphi B*: 1, 4, 5, 12:b:1,2. *S. typhi* belongs to group D, *S. paratyphi A* to group A and *S. paratyphi B* to group B. Group A contains only *S. paratyphi A* and the rare type *S. kiel*. Therefore it may be expected that TAB-vaccination provokes the formation of antibodies against the O-antigens of the groups (A), B and D. All members of group B have the O-antigens 4 and 12 in common, and all members of group D_1* the O-antigens 9 and 12. In salmonellae of the groups C and E these antigens are absent. In group C the O-antigens 6, 7 and 8 preponderate and in group E the O-antigens 3, 10, 15 and 19).

Now the authors found that among the Oahu population cases due to other salmonellae of groups B and D were 25.9 times as frequent among unvaccinated infants than among the vaccinated population of 4 years and older, whereas those due to types in groups C and E were only 5.7 times as frequent.

This phenomenon may be explained most likely by assuming a certain degree of cross-immunity against infection with salmonellae belonging to the same groups as those that were used for vaccination.

However that may be, it does not seem that such cross-immunity plays an important part in natural conditions.

This may have still another reason. When salmonellae provoke a typhoïd or a septicemic syndrome, there is as a rule a clearly perceptible antibody-production, but in gastroenteritic forms of salmonellosis this production varies considerably or is absent.

We may refer here to the great difference in agglutination-titers observed in volunteers who swallowed salmonellae (page 213). It was also shown that in healthy carriers there was never a significant rise in these titers when compared before and after the experimental feeding, and only in some of the patients a slight rise was observed.

Perhaps this may be explained by assuming that salmonellae do not give rise to the production of antibodies as long as they remain confined to the digestive tract and its appendices and that they do so only when they reach the blood-circulation.

A third reason for the reduced significance of acquired immunity in salmonelloses may be that this immunity, when it exists, probably does not last long. To maintain a sufficient immunity against infec-

---

* Group D_2 contains the O-antigens (9), 46, but comprises only rare types.

tion with *S. typhi* the vaccination has to be repeated every year. Even chronic carriers of *S. typhi* often lack the specific H and O-agglutinins in their serum.

When immunity were common in pregnant women, we would expect that antibodies passed to the foetus, and that new-born children were more or less protected against infection at least during the first months of life, as is supposed to be the case in other infectious diseases, e.g. infectious hepatitis. In salmonelloses (at least in other than typhoïd fever) no such phenomenon has ever been observed. On the contrary the first year of life is one of the age-groups with the highest morbidity, and also a rather high letality*.

In the Netherlands in 1959 the rate of notified cases of other salmonelloses than typhoïd and paratyphoïd B (SCHOTTMÜLLER) in the first year of life was 2.6 × the average for all ages. For infectious hepatitis this figure was 0.1 (BEKKER[53]).

We mentioned already the deleterious effects of epidemics of salmonellosis in maternity units (page 208).

We may conclude that evidently acquired immunity seldom contributes to resistance of the host against salmonella-infection in natural circumstances.

Natural resistance

It is probable that the degree of natural resistance in an individual determines for a great deal his susceptibility to salmonella infection, and that this resistance depends chiefly on hereditary factors. This provides the most plausible explanation for the differences in resistance that were observed in the experiments with volunteers (prisoners). In this case the environmental factors were practically equal. The persons concerned were all healthy male adults who got the same food and lived in the same housing.

Also it is a well-known fact that different breeds of the same animal may differ largely in their susceptibility to infectious agents. Even animals of the same breed differ if the breed is not absolutely pure.

For salmonellae the influence of genetic factors in the host was demonstrated a.o. by SCHNEIDER[41] in mice.

Age factors

There is ample evidence that, in man as well as in animals, young individuals are more susceptible to salmonella infection than adults, and especially so in the first period of life.

---

* It is not allowed to suppose that the transmission of immunizing antibodies from mother to child during pregnancy (or through the milk after delivery) is the only possible explanation for a low morbidity of an infectious disease during the first year of life, but it is probable that it plays an important part.

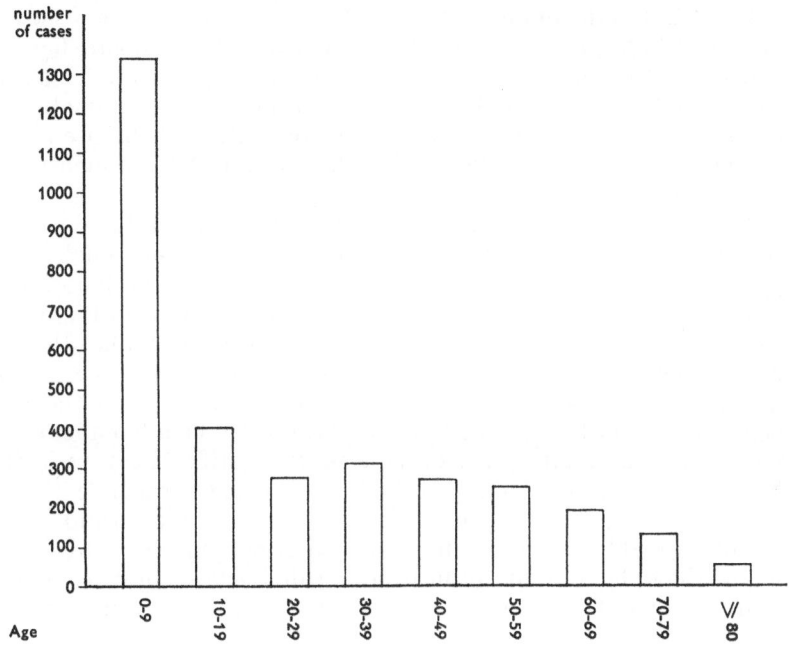

Fig. 1. Notifications of ,,other salmonelloses'' in the Netherlands in 1958,
per age-group (absolute figures).

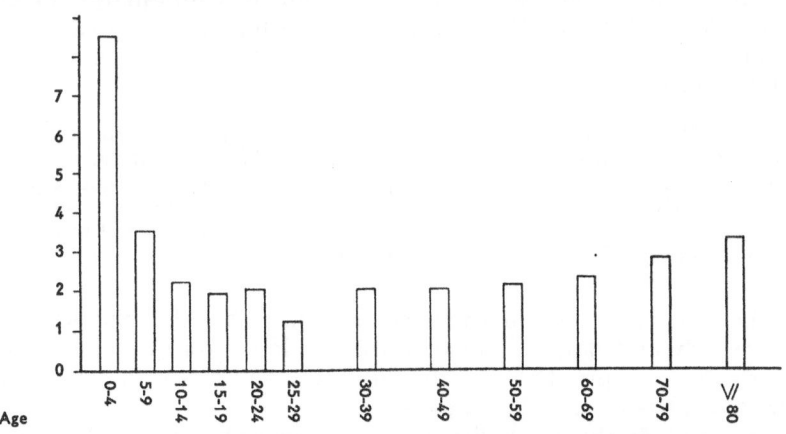

Fig. 2. Notification-rate per 10,000 inhabitants of ,,other salmonelloses''
per age-group in the Netherlands in 1958.

Fig. 1 shows the number of notified cases of "other salmonelloses" in different age-groups in the Netherlands in 1958 in absolute figures, and Fig. 2 the rate per 10,000 persons of each age-group. This rate was 8.5 in the age-group 0—4; 3.4 in the group 5—9, and fluctuated about 2 from the age of 10 years onwards, with a slight rise above the age of 70. In the first year of life the rate was 9.0 (not mentioned in the figure).

The objection may be raised that perhaps other factors than susceptibility influence this age-distribution. First, one might suppose that exposition to contagion is greater in early childhood than at a later age. This is certainly not so in the first year of life, when the risk of contagion by contaminated food is even smaller than in later periods. Infants are fed either at the breast or with pasteurized babyfeeds; their diet contains no meat, at least in the first eight months of life. So the most important source of contagion by food has been eliminated. Nevertheless the notification-rate in the first year of life is among the highest. We may conclude that a greater exposition in early childhood than in older children and adults is not probable. Secondly it might be supposed that physicians follow the obligation of notification of salmonellosis more conscientiously when they meet cases of gastro-intestinal disturbances in early childhood than when their patients are older, or that they are more frequently consulted by anxious mothers.

We have no statistical means to control the value of this hypothesis. If this supposition were true, it might be explained partially by the more serious symptoms that usually develop in infants, and so would not furnish an argument against a greater susceptibility in early childhood.

But we must conclude that we lack sufficient information about the relation between the real and notified number of cases of salmonellosis in general, and we don't know either if this relation differs in different age-groups.

STOCKS[54] tried to examine the completeness of notification of certain infectious diseases in England and Wales and came to the conclusion that it varies greatly in different diseases. He estimated that probably four fifths of the cases of typhoïd and paratyphoïd fever are notified, but gave no evaluation for the other salmonelloses. We may be pretty sure that they belong to the same category as dysentery, of which he says: "Notification only fractional".

Leaving this question aside, we may state that the experience in maternity units and children's wards provides strong arguments for a greater susceptibility in early childhood, and this is in agreement with what is observed in animals.

In old age the morbidity-rate rises somewhat again, but here special debilitating factors may intervene. These will be reviewed in the next paragraph.

Debilitating factors

Under this heading we may range the influence of malnutrition, concurrent diseases, prematurity in infants, very old age.

The influence of nutrition was clearly demonstrated by SCHNEIDER[41] in mice.

As regards the connection with other diseases, besides the weakening influence of these diseases in general, some specific relations have been established.

SAPHRA[43] analysed 174 fatal cases of salmonellosis. A good deal of these fatalities occurred in patients who suffered already a long time from chronic ailments, such as other infections, malignant tumors, arteriosclerosis, nephritis, diabetes, anemia, alcoholism.

Apart from these debilitating diseases in a general sense the following examples of specific relations with certain pathological conditions may be mentioned:

REITHER *et al.*[42] noted the enhancing effect of staphylococcal enterotoxin in salmonella infections.

Patients with sickle-cell disease show a predisposition for acquiring a salmonella-caused osteomyelitis (VAN OYE[44]).

Corticosteroïd-therapy seems to decrease the resistance to salmonellae (WOMACK[45]).

VAN FURTH & KLEIN[46] pointed out the enhancing influence of achylia gastrica and of endocrine disorders.

WADDELL & KUNZ[47] observed the association of salmonellosis with operations on the stomach.

The pathogenicity of different types of salmonellae, and of different strains of the same type.

We will have to consider next the pathogenicity of the salmonella-type in question and of different strains of the same type. Not all salmonella-types are equally pathogenic for man, although none of them can be considered as essentially harmless. Our knowledge of this subject shows still many gaps, but our insight is growing, especially since we know something more about the biochemical constitution of the salmonella-antigens.

KAUFFMANN et al.[48] suppose that the pathogenicity of the different types of salmonellae for man and/or for animals depends for a great deal on the chemical constitution of their O-antigens. These are localised in the cell-wall of the bacteria, and contain polysaccharids, which are composed of different sugars. The O-antigens of all types of salmonellae have in common 4 sugar-materials, viz. glucosamin, heptose(s), galactose and glucose, whereas the most complicated contain 7 of them. It seems that the presence of 6-desoxyhexoses (fucose and rhamnose) enhances the pathogenicity of a salmonella-type, and that the presence of 3-6-didesoxyhexoses (colitose, abequose, paratose and tyvelose) does so to an even higher degree. The groups of salmonellae with desoxyhexoses and didesoxyhexoses contain together $\pm$ 71% of all known types and $\pm$80% of all types isolated from man and animals. They nearly all belong to the groups designated by the first letters of the alphabet, (A to G, except the groups C_1 and C_4 which do not contain them).

In this field much work remains to be done and we may expect from it further fundamental elucidation of the problem of virulence in general. The biochemical interaction between the constituents of the cells of the host and those of the bacteria can be seen as the core of this problem.

Apart from these general aspects of pathogenicity we find that some types of salmonellae have a special affinity for a certain animal-species.

Such an affinity exists in S. *typhi* and S. *paratyphi B* for man, in S. *gallinarum/pullorum* for chickens* in S. *dublin* for cattle (and especially for calves) in S. *abortus ovis* for sheep, in S. *abortus-equi* for horses (especially for the genital tract of these animals).

S. *cholerae-suis* is nearly equally pathogenic for man and for pigs, but less for other species.

It is not allowed to consider these more or less specific types as absolutely harmless for other species; if the dose is large enough, they may infect them as well. S. *pullorum* for instance is nearly harmless for man, except when ingested in very large quantities (Mc Cullough & Eisele[17b]).

Many other types do not show a marked preference for a certain species; they are pathogenic for a large range of warm-blooded animals. The prototype of this group is S. *typhi-murium*.

This holds good for the type S. *typhi-murium* in general. But when the different phage-types of this salmonella are considered some affinity of special phage-types to certain animal species comes to light. For instance in England 74% of the S. *typhi-murium*-strains, isolated from poultry belonged to phage-type 14, which was never found in cattle. On the contrary in man there was a much more equal distribution of the different phage-types. Phage-type 14 counted for 14% of the S. *typhi-murium*-strains isolated from man. In this field more research is needed.

Not seldom the simultaneous presence of two or more salmonella-types in man (Juenker[50]), and in animals (Kampel-macher[2]) is observed, as a rule only in healthy carriers. When these cases are followed up, it seems as if within the human or animal organism some selection may take place. This could partly account for the preponderance of certain types above others. But it has to be kept in mind this is not more than a hypothesis which has yet to be confirmed.

From the experiments of Mc Cullough & Eisele in volunteers may be deduced that there are also differences in pathogenicity of individual strains of the same type. The authors did not make use of phage-typing, but whether the strains used in the experiments belonged to the same phage-type or not, we evidently

---

* S. *gallinarum* and S. *pullorum* have roughly the same antigenic formula, characterized by the total absence of H-antigens (1,9, 12:— : —) only their 0-12-antigens are somewhat different. They are the only known salmonella-types that lack flagellae and as a consequence are non-mobile. The two types differ in their biochemical reactions; for instance S. *gallinarum* ferments dulcitol in pepton-broth and S. *pullorum* does not. Also there is a difference in the clinical picture in chickens.

meet here some subtle genetic varieties that are as yet insufficiently elucidated. Perhaps studies of the lysogenicity of the strains would help us further.

All these differences in virulence are very interesting from a scientific point of view, but for practical purposes in the combat against salmonellosis they are not of great avail. The best policy is to start from the principle that food for man and animals must be free from living salmonellae of any type whatever.

The clinical patterns of salmonellosis in man.

Introduction

It is probable that contagion of man with salmonellae follows the oral route in the great majority of cases; at least in adults and older children. The significance of the respiratory and conjunctival routes is still problematic, although possibly they play a more important role, especially in infants, than we supposed until recently.

Anyhow we must recognise the fact that acute gastroenteritis is by far the prevailing symptom in salmonellosis in man.

SAPHRA & WINTER[35] analysed 7779 cases of salmonellosis (excluding typhoïd fever, but including paratyphoïd A, B and C); 1209 of these cases concerned healthy carriers.

The other cases could be ranged as follows according to the only present or prevailing clinical symptoms.

1° gastroenteritis: 80.9%
2° typhoïdal or septic syndrome: 10.4%
3° foci of local infection: 8.7%.

The infecting strains were all diagnosed at the New York salmonella center.

CLYDE[56] found among 100 patients with salmonellosis in a children's hospital (also including paratyphoid A, B and C), a somewhat other distribution, especially among children above 6 months:

Age groups	symptoms			total
	gastro-intestinal	septic	local	
0— 6 months	41	3	2	46
7—12 months	17	4	2	23
1— 6 year	14	6	3	23
7—16 year	1	5	2	8
Totals	73	18	9	100

60% of the children with gastro-intestinal symptoms were treated on an out-patient basis because they were mild or of moderate degree.

It is probable that the preponderance of the gastroenteritic syndrome is still very much greater than these figures suggest, because in many of these cases either the physician is not called for or a bacteriologic investigation is not thought necessary. As a rule typhoid or septicemic syndromes and local foci of infection are reported to a doctor and it is also much more likely that these cases will be subjected to bacteriologic control.

The question may be raised whether clinical salmonellosis can occur without penetrating of the salmonellae into the blood-circulation. When a typhoïd or septicemic syndrome exists we may be sure that they did so. Also some local manifestations cannot be plausibly explained otherwise than having originated from hematogenic infection, e.g. endo- and pericarditis, osteomyelitis and osteoarthritis, meningitis. In other cases direct infection via the digestive or respiratory tract is also conceivable, e.g. in appendicitis, cholecystitis, periproctal and perineal abscesses, tonsillitis and otitis media; perhaps also in pneumonia. Urinary infection may be as well hematogenic as ascending. The latter route is probably sometimes followed in females who excrete salmonellae in their stools, especially if catheterization is applied (CLYDE[56].)

The same problem is met in colipyelitis of children. In the first year of life this is nearly equally frequent in boys and girls; among older children it is more frequent in girls. It seems that infants are more susceptible to hematogenic spread than older children, and girls are certainly more predisposed to ascending infection of the urinary tract than boys, because of the anatomy of their urethra.

Cases of salmonella meningitis occur very seldom in children above one year of age; it seems that the younger the infants, the more susceptible they are to this localisation (SAPHRA & WINTER[35]). In early infancy the blood-c.s.f.-barrier is evidently more easily crossed than at a later age. Before the advent of the antibiotics all these cases were fatal, but even now the prognosis remains dubious.

A curious case of separate infection of two organ-systems in the same person by different types of salmonellae was mentioned by NAUMANN & RÖHRS[57]. Here the infection of the urinary tract was most likely of hematogenic origin. In a male patient suffering from Hodgkin's disease the urine contained continuously only *S. typhi-murium* and the stools only *S. blockley*. At autopsy *S. typhi-murium* was cultured from both kidneys and from the spleen, and *S. blockley* from all parts of the intestinal canal from the stomach to the rectum, and from the gall-bladder and the liver. It was thought possible that *S. typhi-murium* had been introduced into the circulation during one of the many direct blood-transfusions from donor to patient that had been given.

The salmonellae are always more easily isolated from the stools than from the blood. The chance to get a positive blood-culture is greatest during the febrile phase, but succeeds sometimes even in afebrile patients (MINKENHOF[58]).

The most probable hypothesis is that in all febrile cases, irrespective of the prevailing of septicemic, typhoïd, or gastroenteritic symptoms, salmonellae penetrate into the blood-circulation. Perhaps in afebrile cases with slight diarrhoea the infection remains localized to the intestinal tract and its appendices (including the mesenteric and periportal lymph-nodes), and it is doubtful whether the salmonellae reach the blood-circulation in symptomless excreters, but the question is not absolutely settled. Some light on this question is thrown by investigations in salmonella-carriers among healthy pigs.

KAMPELMACHER et al.[2] made interesting observations on salmonella-carrying healthy pigs that were slaughtered in Dutch abattoirs. They investigated among others two groups of animals, one consisting of 566 individuals, comprising 214 salmonella-carriers (series A), and another of 115 individuals, comprising 98 salmo-nella-carriers (series B). The animals of series A were reared by different farmers and slaughtered in a public abattoir after being collected in a lorry. The duration of the transport was from 1 to 6 hours and they were put up in the abattoir during 2 to 4 hours before being slaughtered. The animals of series B were reared in a specialized model-pig-farm and slaughtered directly after a short transport ($\pm$ 25 km) in a meat-processing factory. After slaughtering specimens of different organs and of the faeces were investigated on the presence of salmonellae. In group B an extra differentiation was made between six sub-divisions of the mesenteric lymph-nodes, the first ones draining the lymph from the proximal parts of the small in-testine, the last ones from the distal parts. Also a differentiation was made between the faeces from a proximal and a distal part of the colon and from the rectum.

The most strict precautions were taken to prevent secondary contamination of the samples.

The results of these observations can be resumed as follows:

1. The positive findings of series B remained confined to the digestive tract and its appendices (liver, gall, regional lymphnodes), with the exception of 2 cases (2%) where salmonellae were found in the spleen.

2. In series A there were 59 positive findings in the spleen (27.6% of all positive cases), and moreover in 21 cases (9.9%) salmonellae were found in the diaphragm and in 9 cases (4.2%) in a piece of muscle of the fore-leg.

3. Thus in series A the lymph-blood-barrier must have been passed rather fre-quently, whereas in series B this happened evidently seldom.

4. It is probable that the duration of the transport has some influence on these findings (greater and longer stress?).

5. The mesenteric lymphnodes yielded in both series the highest percentage of positive findings (in series B in 83.7% of the positive cases), and these became gra-dually more frequent in proportion to the more distal position of the section of the small intestine drained by them. This would suggest that the salmonellae multiply during their passage through the small intestine even in symptomless carriers.

6. The percentage of positive findings in the faeces in different parts of the colon and the rectum showed no striking differences; they were even somewhat more frequent in those from the colon ascendens than in'those from the rectum. Thus the milieu of the colon seems to be less favourable to the multiplication of salmonellae than that of the small intestine, at least in healthy pigs.

7. When the frequency of the different types in the mesenteric lymphnodes and the faeces was compared it became apparent that some types penetrate more easily into the lymphnodes than other ones.

In series B the most frequently observed types were S. cubana, S. bovis morbificans and S. typhi-murium.

The first type was found much more frequently in the faeces than in the mesen-teric lymphnodes but for the other two the proportion was inverse (KAMPEL-MACHER)[51]. It seems that the latter types have more affinity to the animal organism than the first one, perhaps in connection with their antigenic structure (S. bovis morbificans and S. typhi-murium contain di-desoxyhexoses in their O-antigens and S. cubana only a desoxyhexose).

If the intestinal conditions in man are comparable to those in pigs we may conclude that in healthy salmonella-carriers the bacteria may remain confined to the intestinal tract and its direct appendices, including the mesenteric lymph-nodes, but that the lymph-blood-barrier may be passed in more or less harmful circumstances.

We will now consider more closely the different clinical manifes-tations.

The gastro-enteritic syndrome

In the great majority of cases the only or preponderant symptoms are those of a gastro-enteritis, but there is a wide range in the severeness of these symptoms. At one end or the scale are the mild ambulant forms and at the other end a most severe cholera-like syndrome with rapid dehydration convulsions and death or a dysentery-like syndrome with bloody, slimy discharges and tenesmus (SAPHRA & WINTER[35]). Nausea and vomiting are common in the beginning, and the diarrhoea is often accompanied by abdominal cramps. As a rule there is a rather great loss of weight.

In most cases diarrhoea precedes the fever, in contrast to shigelloses where as a rule the sequence is inverted. In salmonellosis typhimurium the stools often resemble thin vegetable soup. Fever usually lasts only a few days and is commonly of moderate degree. In adults there is in most cases some degree of leukopenia and relative lymphocytosis (MINKENHOF[58]). On the contrary in children usually normal leucocyte counts or even a certain degree of leucocytosis are found (CLYDE[56]).

Excretion of salmonellae during convalescence is common. RUBINSTEIN et al.[59]) found it in 43% in the fourth week of illness; 18% in the eighth week and 11% in the tenth week. In infants as a rule it persists longer than in older children and adults; the more so the younger they are.

The typhoidal or septic syndrome

This is rather common in salmonelloses caused by S. paratyphi A, B and C, and by S. cholerae-suis, but rare in other salmonelloses ($\pm$ 6% in the material of SAPHRA & WINTER[35]).

Sometimes this syndrome is preceded by a gastroenteritis some days or weeks before, which was apparently healed; in other cases it develops without any prodromes.

It is characterized by high fever, usually with spiking tops, more rarely continuous. It lasts from a few days to several weeks. It may be accompanied by severe headache and sometimes loss of conscience. Enlargement of the spleen is common, and in a few cases roseolae are seen. If diarrhoea is present, it is mild and of short duration; constipation is more frequent. Blood-cultures are positive in the great majority of cases.

Focal manifestations

These occurred with an overall frequency of 8.7% in SAPHRA & WINTER's clinical cases. In salmonelloses caused by S. cholerae-suis the frequency was highest, followed by those where S. bredeney, resp. S. panama were the infecting agents.

In their material there was a great variety of localizations. (see table I).

Table I.

572 Focal Manifestations in the Most Frequent Salmonellas
(from SAPHRA & WINTER[35])

Type	No. of cases	Bacterial & Endocarditis Pericarditis (no. of cases)	Appendicitis Cholecystitis Salpingitis & Peritonitis (no. of cases)	Pneumonia & Pleurisy (no. of casse)	Urinary-Tract Infection (no. of cases)	Osteomyelitis & Osteo-arthritis (no. of cases)	Meningitis (no. of cases)	Abscess (no. of cases)
S. paratyphi B & S. st. paul	19	—	10	2	1	—	2	4
S. san diego	8	—	1	1	—	—	3	3
S. derby	12	—	4	2	2	—	2	2
S. typhi-murium	117	4	35	13	16	14	9	26
S. paratyphi C & S. cholerae-suis	129	13	8	41	6	22	9	29
S. montevideo	36	1	13	5	3	3	1	10
S. oranienburg	56	—	25	7	7	4	3	10
S. bareilly	15	—	4	—	1	1	1	8
S. tennessee	8	—	2	5	1	—	3	2
S. newport	32	—	11	2	2	4	1	9
S. muenchen & S. manhattan	14	1	2	2	1	1	4	4
S. enteritidis	23	—	4	2	1	3	7	5
S. panama	28	—	6	1	1	1	12	7
S. anatum & S. newington	14	—	4	2	2	—	1	5
Other types	62	1	17	2	5	6	19	12
Totals	572	20	146	85	49	59	77	136

It is striking that *S. paratyphi C* and *S. cholerae-suis* are for a great deal responsible for focal manifestations that are necessarily of bacteriemic origin, such as endocarditis and osteomyelitis, whereas *S. typhi-murium* and some other types prevail in inflammations in organs that may have been infected directly from the digestive tract, e.g. appendicitis and cholecystitis.

Symptomless salmonella-carriers

There are three varieties of symptomless salmonella-carriers:

a. those who excrete salmonellae during a certain time after being exposed to contagion, but never show any morbid symptoms,

b. those who temporarily excrete salmonellae after being healed from a clinical salmonellosis,

c. permanent salmonella-excreters.

The first category has little epidemiological significance because as a rule the excretion is sporadic and of rather short duration. The second is of somewhat more importance, it contains for instance children who may spread the contagion in and outside a hospital. The third category is rare but dangerous from an epidemiologic point of view.

In permanent salmonella-excreters there is most likely a reservoir somewhere in the body where the bacteria reproduce themselves continuously; in most cases the gall bladder, but possibly in some cases also an intestinal diverticulum or an anatomical anomaly of the urinary tract may function as such.

There are persons who excrete salmonellae only in the urine, and others, who excrete them as well in the faeces as in the urine.

Antibiotic therapy is of little avail in salmonella-excreters. In permanent faecal excreters cholecystectomy may be indicated, but only after verifying by roentgenologic control and sounding of the duodenum that the gall-bladder is in a pathological condition (VAN FURTH & KLEIN[46]); MAIN[33]). Also it must be verified that there is no excretion in the urine.

Conclusions

1 The "other salmonelloses", i.e. the salmonelloses caused by other salmonellae than *S. typhi* and *S. paratyphi B* can be considered as zoönoses, in the definition of "diseases and infections which are naturally transmitted between animals and man".

2 In the "other salmonelloses", the transmission from animals to man is more important from an epidemiological viewpoint than that from man to man or from man to animals.

3 The transmission from animals to man is seldom direct; the most common route is the ingestion by man of contaminated meat or animal products.

4 These products are especially dangerous when they have been

kept under circumstances that favour the multiplication of salmonellae.

5 The great majority of cases of "food-poisoning", (regarding individual cases as well as outbreaks), are caused by salmonellae, at least in England.

6 Among animal products responsible for salmonellosis in man processed or made-up meat ranks in the first place, followed by eggs and egg-products.

7 Pork is more frequently contaminated by salmonellae than beef.

8 Transmission of salmonellae by direct contact from man to man is possible but not frequent among adults and older children. The danger is greater in unhygienic conditions and where mothers and their babies are concerned.

9 It is not probable that salmonellae pass the placental barrier from the mother to the foetus.

10 Some cases have been reported of direct transmission of salmonellae from animal to man. Perhaps this happens more often than is supposed in farmer's children and farm-personnel.

11 Salmonellae have repeatedly been found in the upper respiratory tract of hospital patients.

12 It is possible that infection of infants and perhaps also weakened aged persons may result from the inhalation of salmonella-contaminated dust.

13 The possibility of infection by the conjunctival route has also to be considered.

14 There are two types of salmonellosis epidemics in hospitals. The first one is caused by the ingestion of a rather heavily contaminated meal and comprises many persons at once, mostly patients as well as personnel. It ends after the elimination of the incriminated food. In an epidemic of the second type, which is encountered especially in children's wards, other routes of transmission have to be considered. It has a chronic course (not seldom during several months), and affects one patient after the other. It ends only after a thorough cleaning and sterilization of the whole surroundings.

15 After transmission of salmonellae to man three alternatives have to be considered:

(a) The germs cannot be traced back at all.

(b) The person subject to contagion becomes during a certain time a carrier or excreter of bacteria, thereby remaining in good health.

(c) Morbid symptoms develop.

16 Which of these alternatives is realised, depends upon:

(a) The dose of the transmitted bacteria.

(b) The resistance of the host.

(c) The type of Salmonella and the virulence of an individual strain of a certain type.

It has been possible to analyse these factors in experiments with volunteers.

17 The resistance of the host is influenced by different factors, namely

(a) acquired immunity
(b) natural resistance (of genetic origin)
(c) age
(d) debilitating factors.

18 The acquiring of immunity against "other salmonelloses" is hampered by the fact that the infection may be due to a great variety of salmonella-types with different antigenic composition. But there is some evidence that a certain degree of cross-immunity may exist versus salmonellae with a related antigen-frame. It is not probable that such cross-immunity plays an important part in natural conditions.

19 The hypothesis is expressed that salmonellae do not give rise to the production of antibodies as long as they remain confined to the digestive tract and its appendices, or at least only to a very small and varying degree.

20 Transmission of immunizing antibodies from mother to foetus is probably very rare in salmonellosis.

21 It is probable that the degree of natural resistance in an individual determines for a great deal his susceptibility to salmonella-infection, and that this resistance depends chiefly on hereditary factors.

22 There is ample evidence that, in man as well as in animals, young individuals are more susceptible to salmonella-infection than adults, especially so in the first period of life.

23 Debilitating factors that enhance the appearance of clinical salmonellosis in case of contagion are:

(a) malnutrition
(b) concomitant diseases
(c) prematurity in infants
(d) very old age.

24 Besides the weakening influence of concomitant diseases in general, the special influence of certain pathological conditions was established. Examples of these conditions are mentioned in the text.

25 Not all salmonella-types are equally pathogenic for man, although none of them can be considered as essentially harmless.

26 It seems that the presence of 6-desoxyhexoses in the polysaccharids of the cell-wall of salmonellae enhances their pathogenicity for man and/or animals, and that the presence of 3—6-didesoxyhexoses has a still greater enhancing effect.

27 There is some evidence that in case of contagion of man or animals with more than one salmonella-strain a certain selection within the organism of the host may take place.

28 Various strains of the same type may differ in pathogenicity. This depends probably on subtle differences in their genetic material. Perhaps these problems may be elucidated by phage-typing and studies in lysogenicity.

29 These differences in virulence of types and of strains of the same type are very interesting from a scientific point of view but not of great avail in the combat against salmonellosis. The best policy is to start from the principle that food for man and animals must be free from any salmonellae of any type whatever.

30 Acute gastroenteritis is by far the prevailing symptom in salmonellosis in man. Next in frequency is the typhoidal or septic syndrome, and then follow the foci of local infection.

31 The pathogenesis of some focal infections cannot be explained otherwise than by assuming a hematogenic spread of the bacteria. In others direct infection from the respiratory of digestive tract is also conceivable. In urinary infection both the hematogenic and the ascending route have to be considered.

32 Salmonella-meningitis is very rare in children above one year of age. In early infancy the blood-c.s.f. barrier is evidently more easily crossed than at a later age.

33 There is some evidence that in healthy salmonella-carriers the bacteria may remain confined to the intestinal tract and its direct appendices, including the mesenteric lymph-nodes but that the lymph-blood-barrier may be passed in more or less harmful circumstances.

34 The gastro-enteritic syndrome varies from a very benign to a most severe form. Excretion of salmonellae during convalescence is common; on an average it lasts longer in infants than in older children and adults.

35 The typhoïdal or septic syndrome is rather common in salmonelloses, caused by S. paratyphi A, B and C and by S. cholerae-suis but rare in other salmonelloses.

36 Focal manifestations show a great variety of sites (named in the text). S. paratyphi C and S. cholerae-suis are for a great deal responsible for focal infections that are necessarily of bacteriemic origin, such as endocarditis and osteomyelitis whereas S. typhi-murium and some other types prevail in inflammations in organs that may have been infected directly from the digestive tract, e.g. appendicitis and cholecystitis.

37 There are three varieties of symptomless salmonella-carriers:

(a) those who excrete salmonellae during a certain time after being exposed to contagion but never show any morbid symptoms;

(b) those who temporarily excrete salmonellae after being healed from a clinical salmonellosis;

(c) permanent salmonella-excreters.

From an epidemiologic viewpoint the second group is more impor-

230

tant than the first, and the third more important than the second. In persons belonging to the third group there is most likely a reservoir somewhere in the body where the bacteria reproduce themselves continuously, usually the gall bladder. In these cases surgical therapy may be considered.

LITERATURE

1. Food poisoning in England and Wales 1959 (1960). *Mth. Bull. Minist. Hlth. Lab. Serv.* **19**, *224—237.*
1a. Idem 1960 (1961). *Mth. Bull. Minist. Hlth. Lab. Serv.* **20**, *160—171.*
1b. Idem 1961 (1962). *Mth. Bull. Minist. Lab. Serv.* **21**, *180—195.*
2. KAMPELMACHER, E. H., P. A. M. GUINÉE, K. HOFSTRA & A. VAN KEULEN (1962). Verdere onderzoekingen over Salmonella in slachthuizen en bij normale slachtvarkens. *T. Diergeneesk.* **87**, *1486—1520.*
3. KAMPELMACHER, E. H., P. A. M. GUINÉE, K. HOFSTRA & A. VAN KEULEN. (1962). Salmonella-onderzoek in slachthuizen. *T. Diergeneesk.* **87**, *77—94.*
4. KAMPELMACHER, E. H. (1957). Onderzoekingen omtrent salmonella-dragers bij normale slachtrunderen. *Versl. Volksgezondh.* **1957**, *1293—1299.*
5. TAYLOR, J. (1960). Salmonella and salmonellosis. *Roy. Soc. Hlth. J.* **80**, *253—259.*
6. HOBBS, B. C. & J. GREENWOOD WILSON. (1959). Contamination of wholesale meat supplies with salmonellae and heat-resistant Cl. Welch. *Mth. Bull. Minist. Hlth. Lab. Serv.* **18**, *198—206.*
7. HARVEY, R. W. S., T. H. PRICE, A. R. DAVIS & R. B. MORLEY-DAVIES. (1961). An outbreak of salmonella food poisoning attributed to bakers' confinery. *J. Hyg., Camb.* **59**, *105—108.*
8. POLAK, M. F. (1960). Kippe-eiprodukten in bakkerijen als mogelijke bron van salmonellose bij de mens. *Ned. T. Geneesk.* **104**, *2660—2663.*
9. HEMMES, G. D. (1952). Salmonellose, veroorzaakt door S. typhi-murium. *Ned. T. Geneesk.* **96**, *1450—1454.*
10. KRESSMANN, H. & O. H. ALBERT. (1961). Vier Jahre Salmonella-verordnung und ihre Auswirkung in der Praxis. *Berl. Münch. tierärztl. Wschr.* **74**, *359—363.*
11. ROHDE, R. & K. PACHL. (1961). Die Salmonellose des Schlachtgeflügels, ein ernstes Problem in der tierärztlichen Lebensmittelüberwachung. *Mh. vet. Med.* **16**, *351—354.*
12. HUISMAN, J. & M. S. M. DANIËLS-BOSMAN. (1961). Salmonella-verontreiniging van een orgaan-preparaat. *Ned. T. Geneesk.* **106**, *371.*
13. GALBRAITH, N. S., B. C. HOBBS, M. E. SMITH & A. J. TOMLINSON. (1960). Salmonellae in desiccated coconut. *Mth. Bull. Minist. Hlth. Lab. Serv.* **19**, *99—106.*
14. HUISMAN, J. & M. S. M. DANIELS-BOSMAN. (1961). Infected coconuts and peanuts. *Lancet* **1961** I, *1113.*
15. SEMPLE, A. B. (1960). Some recent problems of imported food. *Med. Offr.* **104**, *101—105.*
16. PARRY, W. H. (1962). A milk-borne outbreak due to Salmonella typhimurium. *Lancet* **1962** I, *475—477.*
17. Mc CULLOUGH, N. B. & C. W. EISELE. (1951). Human experimental salmonellosis. *J. infect. Dis.* **88**, *278—289.*
17a. Idem. *J. infect. Dis.* **89**, *209—213.*
17b. Idem. *J. infect. Dis.* **89**, *259—265.*

231

18. Noordam, A. L. & C. Postma. (1961). Varkensvlees als bron van para-typhus ("andere salmonellosen"). *Ned. T. Geneesk.* **105**, *1421—1425*.
19. Noordam, A. L. (1962). Personal communication.
20. Marseille, A. (1961). Enige ongewone lokalisaties van salmonella-infec-ties. *Ned. T. Geneesk.* **105**, *1579—1580*.
21. Leeder, F. S. (1956). Epidemic of salmonella panama infections in infants. *Ann. N.Y. Acad. Sci.* **66**, *54—60*.
22. Rubinstein, A. D. & R. N. Fowler. (1955). Salmonellosis of the new-born by delivery room resuscitators. *Amer. J. publ. Hlth.* **45**, *1109—1114*.
23. Hemmes, G. D. (1952). Salmonellose, veroorzaakt door S. bareilly. *Geneesk. Gids* **30**, *93—102*.
24. Ruys, A. Ch. (1936). Het gevaar van eenden-paratyphose voor den mensch. *Ned. T. Geneesk.* **80**, *3272—3274*.
25. Steiniger, F. (1956). Zur Freiland-biologie der Salmonellosen im Bereich des westlichen Mittelmeeres. *Zbl. Bakt.* I. Abt. Orig. **166**, *245—265*.
26. Walker, J. H. C. (1957). Organic fertilizers as a source of salmonella infection. *Lancet* 1957 II, *283—284*.
27. Kampelmacher, E. H. & P. A. M. Guineé. (1960). Onderzoek betref-fende salmonellosen. *Versl. Volksgezondh.* 1960, *1681—1722*.
28. Rackow, H. G. & H. Wiese. (1961). Ergebnisse und Erfahrungen 3-jähriger bakteriologischer Einfuhr-untersuchungen von Futter-mitteln tierischer Herkunft auf gesetzliche Grundlage. *Berl. Münch. tierärztl. Wschr.* **74**, *340—342*.
29. Clemmer, D. I., J. L. S. Hickey, J. F. Bridges, D. J. Schlussmann & M. F. Shaffer. (1960). Bacteriologic studies of experimental airborne salmonellosis in chicks. *J. infect. Dis.* **106**, *197—210*.
30. Darlow, H. M., W. R. Bale & G. B. Carter. (1961). Infection of mice by the respiratory route with Salmonella-typhi-murium. *J. Hyg. Camb.* **59**, *303—308*.
31. Neter, E. (1950). Observations on the transmission of salmonellosis in man. *Amer. J. publ. Hlth.* **40**, *929—933*.
32. Datta, N. & R. B. Pridie. (1960). An outbreak of infection with Sal-monella typhi-murium in a general hospital. *J. Hyg. Camb.* **58**, *229—240*.
33. Main, G. (1961). Treatment of the chronic alimentary enteric carrier. *Brit. med. J.* 1961 I, *328—332*.
34. Darlow, H. M. & W. R. Bale. (1959). Infective hazards of waterclosets. *Lancet* 1959 I, *1196—1200*.
35. Saphra, I. & J. W. Winter. (1957), Clinical manifestations of salmo-nellosis in man. *New Engl. J. Med.* **256**, *1128—1134*.
36. Williams, R. E. O., R. Blowers, L. P. Garrod & R. A. Shooter. (1960). Hospital infection, causes and prevention, p. 106. Lloyd-Luke Ltd., London, publishers.
37. Ruys, A. Ch. (1960). De gevaren van ziekteverwekkende micro-organis-men in levensmiddelenbedrijven. *Voeding* **21**, *540—549*.
38. Murray, J. O. & J. H. C. Walker. (1958). An outbreak of enteritis (Salmonella heidelberg) in a maternity unit. *Med. Offr.* **100**, *221—222*.
39. Moore, B. (1957). Observations pointing to the coniunctiva as the portal of entry in salmonella-infection of guinea-pigs. *J. Hyg. Camb.* **55**, *414—433*.
40. Bate, J. G. & U. James (1958). Salmonella typhi-murium infection dust-borne in a children's ward. *Lancet* 1958 II, *713—715*.
41. Schneider, H. A. (1956). Nutritional and genetic factors in the natural resistance of mice to Salmonella infections. *Ann. N.Y. Acad. Sci.* **66**, *337—347*.
42. Reitler, R. D., Yarom & R. Seligmann (1960). The enhancing effect

232

of staphylococcal enterotoxin on salmonella infection. *Med. Offr.* **104,** *181.*
43. SAPHRA, I. (1950). Fatalities in Salmonella infections. *Amer. J. med. Sci.* **220,** *74—78.*
44. VAN OYE, E. (1960). Sur l'association entre ostéomyélite à Salmonella et hémoglobinopathie chez l'enfant africain. *Bull. Soc. Path. exot.,* **53,** *89—100.*
45. WOMACK, A. M. (1959). Acute urinary infection due to Salmonella typhimurium in a patient receiving steroid therapy. *Mth. Bull. Minist. Health. Lab. Serv.* **18,** *184—187.*
46. VAN FURTH, R. & F. KLEIN. (1961). Salmonellosen, enige klinische aspecten. *Ned. T. Geneesk.* **105,** *1426—1431.*
47. WADDELL, W. R. & J. KUNZ. (1956). Association of salmonella enteritis with operations on the stomach. *New Engl. J. Med.* **255,** *555—559.*
48. KAUFFMANN, F. (1961). Die Bakteriologie der Salmonella-species, p. 209—226. Einar Munksgaard, Copenhagen, publisher.
49. Joint WHO/FAO expert committee on zoönoses, second report. (1959). *Wld. Hlth. Org. techn. Rep.* Ser. **169,** *8—19.*
50. JUENKER, A. P. (1957). Infections with multiple types of salmonellae. *Amer. J. clin. Path.* **27,** *646—651.*
51. KAMPELMACHER, E. H. (1962). Personal communication.
52. LEVINE, M., J. R. ENRIGHT & G. CHING (1962). Salmonellosis in TAB-vaccinated population, island of Oahu, Hawaii. *Publ. Hlth. Rep. (Wash.)* **77,** *293—300.*
53. BEKKER, B. V. (1962). Personal communication.
54. STOCKS, P. (1949). Sickness and the population of England and Wales 1944—1947. Studies on med. and pop. subj. No. 2, H. M. Stationary Office, London.
55. JELLARD, C. H., H. JOLLY & R. N. BROWN. (1959). An outbreak of S. bovis morbificans infection in a children's ward. *Lancet* **1959** I, *390—392.*
56. CLYDE, W. A. JR, (1957). Salmonellosis in infants and children. *Pediatrics* **19,** *175—183.*
57. NAUMANN, P. & H. RÖHRS. (1958). Ein Beitrag zur Frage der Salmonella-Doppelinfektion. *Z. Hyg. Infekt. Kr.* **145,** *103—110.*
58. MINKENHOF, J. E. (1961). Kliniek en behandeling van de "andere salmonellosen". *Ned. T. Geneesk.* **105,** *1435—1437.*
59. RUBINSTEIN, A. D., R. F. FEEMSTER & H. M. SMITH. (1944). Salmonellosis as a public health problem in wartime. *Amer. J. Publ. Hlth.* **34,** *841—853.* (cited after CLYDE (56)).
60. BAETEN, D. (1962). Das Absterben von Bakterien der TPE und Ruhrgruppe in kompostiertem Mull und in kompostiertem Schlachthofmist. *Arch. Hyg. Bakt.* **146,** *292—320.*
61. WATANABE, T. (1963). Infective heredity of multiple drug resistance in bacteria. *Bact. Rev.* **27,** *87–115.*
62. LEBEK, G. (1963). Über die Entstehung mehrfachresistenter Salmonellen. *Zbl. Bakt. I Abt. Orig.,* **188,** *494-505.*

Addendum

A new concept in bacteriology is the infective heredity of bacterial drug resistance, exchanged between different genera of enterobacteriaceae by means of so-called episomes (WATANABE).

LEBEK published an observation on the occurrence of multiple drug-resistant *S. typhi murium* in a child and demonstrated the transfer of this resistance from *E. coli* to *S. typhi murium*.

OSTÉOMYÉLITE À SALMONELLA ET HÉMOGLOBINOPATHIE

PAR

E. van OYE* & P. VASSILIADIS**

* *Institut d'Hygiène et d'Epidémiologie, Bruxelles.*
** *Hôpital Général "Reine Frederica" du Pirée, Grèce.*

La pathologie des salmonelloses a été étudiée d'une manière si exhaustive au cours des dernières décades qu'il semblait que tout devait être connu, du moins tout ce qui pourrait présenter une certaine importance, Or, récemment différents auteurs ont décrit une entité clinique entièrement nouvelle caractérisée par l'association entre deux états pathologiques bien distincts: une ostéomyélite à *Salmonella* d'une part et une hémoglobinopathie d'autre part. Il est maintenant établi que cette association n'est pas l'effet du seul hasard et le mécanisme de l'interaction entre les deux affections a été décelé.

Dès 1937, DIGGS et coll. affirment que les malades souffrant de sicklanémie sont particulièrement susceptibles de faire des infections osseuses. En 1951, LAMBOTTE-LEGRAND au Congo-Leopoldville et HODGES & HOLT aux Etats-Unis, attirent l'attention sur la très curieuse association entre une ostéomyélite à *Salmonella* et une anémie à cellules falciformes, et ils se demandent s'il ne faut pas accepter une relation causale entre ces deux affections. SMITH (1953) se pose la même question, tandis que VANDEPITTE et coll. (1953) affirment d'une façon nette, et à notre connaissance pour la première fois, que "l'association entre sicklanémie et infection osseuse à *Salmonella* est hautement significative et ne peut pas être le résultat d'une simple coïncidence".

La majorité des cas signalés dans la littérature concerne des noirs souffrant de sicklanémie (= homozygotie pour le gène déterminant la production de l'hémoglobine S) et l'aire de distribution géographique de cette association couvre donc principalement l'Afrique équatoriale et les Amériques. Mais des cas d'ostéomyélite à *Salmonella* ont également été signalés chez des malades qui souffraient d'une hémoglobinopathie autre que la sicklanémie vraie. La plupart de ceux-ci étaient néanmoins marqués de la tare sicklémique (= hétérozygotie pour le gène déterminant la production de l'hémoglobine S, celle-ci étant alors associée à une autre hémoglobine anomale, C, D, E, etc. ou bien à l'hémoglobine normale A).

Il faut malheureusement convenir que chez certains de ces derniers cas la nature de l'hémoglobinopathie n'est pas bien précisée. Ainsi GIACCAI & IDRISS (1952) publient l'histoire d'une fille arabe atteinte de nombreux foyers d'ostéomyélite à *S. enteritidis* et souffrante d'une anémie hémolytique dont la nature n'est pas spécifiée.

Les deux cas de CHOREMIS et coll. (1955) doivent être mentionnés ici. Il s'agit de deux garçons grecs âgés de 6 et de 7 ans, atteints très probablement de thalasso-drépanocytose, c'est à dire d'une dyshémoglobinose déterminée par la présence à la fois du gène de la thalassémie et de celui de la sicklémie. Il est vraisemblable que le cas de GOLDENBERG (1955) fait aussi partie de cette catégorie. Il s'agit d'un garçon sicilien de 11 ans qui présentait de nombreux foyers d'ostéïte à S. enteritidis et chez qui le diagnostic de thalassémie avec sicklanémie fut posé, mais il n'y eut ni enquête familiale ni étude électrophorétique des hémoglobines.

Nous sommes mieux renseignés et plus certains en ce qui concerne un autre garçon d'origine sicilienne dont parlent SILVER et coll. (1957) et chez qui la nature de l'anémie, une thalasso-drépanocytose, a été établie. J. et C. LAMBOTTE-LEGRAND (1958) décrivent l'histoire d'un enfant noir chez qui la première phalange du médius droit ainsi que les tarses et métatarses droits étaient atteints d'ostéïte à S. kisangani, tandis que HUET et coll. (1958) signalent le cas d'un enfant arabe de 6 ans qui présentait une atteinte ostéomyélitique à S. tunis du fémur droit. Ces deux derniers malades étaient atteints de la même anémie hémolytique que celui de SILVER et coll. (1957) qui mentionnent par ailleurs la première observation faite chez un malade atteint d'hémoglobinose sans signes de falciformation: une fillette noire de 5 ans dont les érythrocytes contiennent les hémoglobines C et D et qui fut atteinte d'une ostéomyélite d'un humérus et des deux tibias. Le germe en cause était S. typhi-murium.

Enfin, quelques cas ont été décrits chez qui une hémoglobine S était associée à une autre hémoglobine anormale. Chez le malade n° 4 de ROBERTS & HILBURG (1958) il s'agit d'un garçon noir de 11 ans, souffrant d'une hémoglobinose S/C et d'une ostéïte à S. infantis de l'épiphyse humérale gauche. Un second cas avec hémoglobinose S/C est décrit par WIDEN & CARDON (1961): une jeune fille noire de 16 ans atteinte d'ostéomyélite à S. typhi-murium.

Par contre, des doutes planent sur deux cas qui concernent des Jamaïquains adultes chez qui chaque fois S. paratyphi C a été isolée. Le premier vivait à la Jamaïque même et souffrait d'un syndrome de sicklanémie (GAY & GRANT, 1956). Nous ne pouvons nous empêcher d'avoir un certain scepticisme au sujet de ce diagnostic parce qu'un sicklanémique vrai atteint rarement l'âge adulte. Le deuxième habitait l'Angleterre et la nature de son hémoglobinopathie n'est pas clairement spécifiée (WHITE & MEYNELL, 1956). VANDEPITTE & VAN OYE (1956) ont émis l'hypothèse que chez ce dernier malade il pouvait s'agir d'une hémoglobinose S/C. ROBERTS & HILBURG (1958) émettent la même hypothèse en ce qui concerne un troisième Jamaïquain chez qui LEVER & BARKER (1945) avaient trouvée *S. typhi* dans une infection ostéomyélitique du crâne.

Nous ne nous étendrons pas plus longuement, notre but n'est pas de faire une revue générale avec énumération de tous les cas qui ont été décrits mais d'analyser les divers aspects du problème à la lumière de nos propres observations. Pour résumer: s'il est certain que la grande majorité des cas d'ostéomyélite à *Salmonella* avec hémoglobinopathie concerne des malades atteints de sicklané-mie vraie, il y a néanmoins un certain nombre d'observations se rapportant à des cas chez lesquels l'hémoglobine S est associée à une autre hémoglobine anormale. Enfin, un seul cas a été décrit avec certitude chez un malade atteint d'hémoglobinose sans signes de falciformation.

Ajoutons pour terminer ce chapitre que parmi tous les cas men-tionnés dans la littérature, un seul concerne un malade de race blanche atteint de sicklanémie vraie: il s'agit d'un garçon d'origine turque, âgé de 10 ans, chez lequel *S. muenchen* a été isolée de plusieurs foyers d'ostéite (Aksoycan et coll., 1958). Peu avant, Russo (1957) a décrit le cas d'un garçon de Palermo (Sicile), âgé de 3 ans, chez lequel plusieurs métacarpiens de la main droite furent atteints d'une infection à *S. typhi*. Les deux parents étaient sicklémiques et les globules rouges du malade, également sicklé-mique, montraient à l'électrophorèse 81,2% d'hémoglobine S.

* * *

Vandepitte et coll. (1953) expliquent l'association entre ostéomyélite à *Salmonella* et sicklanémie de la façon suivante: D'une part, la moelle osseuse constitue un milieu de prédilection pour les *Salmonella**; il est en effet connu depuis longtemps que ces germes y restent parfois présents pendant une période très longue après la guérison clinique d'une salmonellose. D'où l'habitude prise dans certains pays de recourir à des médullocultures pour la recherche des *Salmonella* aussi bien chez les malades que chez les porteurs de germes. D'autre part, le processus de falciformation *in vivo* crée des petits thrombi dans divers organes et tissus et notamment dans la moelle osseuse où le courant sanguin est ralenti, ce qui provoque un état d'anoxie relative. Les oblitérations vasculaires qui en résul-tent sont à l'origine de zones plus ou moins étendues d'infarctus et de nécrose qui sont autant d'endroits où d'éventuelles *Salmonella* trouveront un excellent milieu de culture; en s'y multipliant elles formeront des microabcès. Ceux-ci provoqueront finalement l'appa-rition d'une ostéomyélite avec ou sans périostite et fistulisation concomitantes.

Certains auteurs croient que la diminution des défenses de l'or-ganisme, due entre autres à l'état d'anémie et à une "auto-splénec-

---

* La preuve expérimentale en a été donnée par Chantemesse & Widal dès 1893.

tomie'' jouerait le rôle de facteur prédisposant. Si cela est vrai, ce facteur ne peut intervenir que chez les enfants d'un certain âge déjà. Il nous semble par conséquent qu'il ne faille pas attacher beaucoup d'importance à l'intervention éventuelle d'un tel facteur.

Il est évidemment impossible de prouver expérimentalement que l'explication proposée par VANDEPITTE et coll. (1953) reflète la réalité. Il est néanmoins généralement admis que c'est de cette façon que les choses se passent. C'est ainsi que COLE (1955, cité par GAY & GRANT, 1956) affirme:''...Thrombosis occurring in sickle cell crisis must also influence the localization of infection''. Et ROBERTS & HILBURG (1948) arrivent à la conclusion: ''We postulate that the bone changes caused by sicklemia may result in *Salmonella* osteomyelitis from previously dormant medullary organisms or from transient bacteremia.''

Il faut noter que VANDEPITTE et coll. (1953) passent sous silence la question de la porte d'entrée des *Salmonella*. Ce problème avait déjà été abordé par HODGES & HOLT (1951): ''Perhaps the characteristic capillary thrombotic phenomena occurring in the intestinal circulation of patients with sicklanemia in some way predisposes to remote infection in bone by common intestinal organisms such as Paratyphoid-B''. Il n'est pas possible de réfuter d'emblée cette façon de voir, mais elle ne nous satisfait pas entièrement parce qu'elle n'explique pas pourquoi ce sont précisément des germes du groupe *Salmonella* qui provoquent l'apparition d'une ostéomyélite chez les sicklanémiques. A ce propos HUET et coll. (1960) ont sans doute raison de se demander ''si la présence, très transitoire évidemment, de *Salmonella* dans le courant sanguin, ne serait pas un phénomène beaucoup plus fréquent qu'on ne le pense.'' Ces auteurs rejoignent par ailleurs, dans leur supputation, ROBERTS & HILBURG (1958) que nous avons déjà cités.

Quoi qu'il en soit, et quelque puisse être le mécanisme qui favorise et déclenche une infection osseuse à *Salmonella* chez un sicklanémique, il reste acquis que cette association entre deux affections bien distinctes et qui n'ont à premier vue aucun point commun, n'est pas le fait d'un simple hasard. VAN OYE (1960a) a cité récemment des chiffres très éloquents à l'appui de cette affirmation. Nous exposerons maintenant nos propres données statistiques:

Au cours d'une période qui va de 1947 à 1962 nous avons eu connaissance de 98 cas d'ostéomyélite à *Salmonella* observés en Afrique Centrale, notamment 94 cas au Congo-Léopoldville et 4 au Congo-Brazzaville (CHARMOT et coll., 1958).

1. Un seul cas concerne un enfant européen de Léopoldville. Nous ne possédons à son sujet aucun renseignement d'ordre clinique ou hématologique et nous savons uniquement que le germe en cause était *S. kisangani*, ce qui laisse supposer que cet enfant s'est infecté au Congo même.

2. Un seul cas concerne un enfant congolais, de Léopoldville également, avec hémoglobines absolument normales A/A. Le germe en cause fut ici *S. dublin*.

3. Nonante-six cas concernent des congolais atteints d'une hémoglobinopathie, prèsque tous de sicklanémie. Nous reviendrons plus loin sur l'aspect hématologique de la question pour l'examiner ici sous son aspect statistique.

Dans la population centre-africaine, l'incidence de la sicklémie est d'environ 25%. Un mariage sur 16 se fait donc entre partenaires marqués de cette tare et 1 enfant sur quatre, né d'une telle union, sera donc un sicklanémique homozygote S/S. Ceci revient à dire que sur 64 nouveau-nés, 1 sera un sicklanémique vrai. Ces chiffres théoriques ont été confirmés par les faits, notamment par VANDE-PITTE (1954) qui a trouvé que sur 10.000 naissances à Léopoldville il y avait 158 nouveau-nés homozygotes S/S. Donc, si l'association entre ostéomyélite à *Salmonella* et sicklanémie était fortuite nous devrions trouver 64 fois plus de cas chez les enfants normaux que chez ceux qui sont atteints de drépanocytose. Pour la nonantaine de cas d'ostéomyélite à *Salmonella* observés chez des sicklanémiques, on devrait donc en avoir observé plus de 6.000 chez des enfants sans hémoglobinopathie. Or, on n'en observé qu'un seul!

Nous avons la conviction que chez les noirs d'Afrique, et peut-être dans une moindre mesure également chez les noirs des Amériques, il y a chaque année des centaines de cas présentant l'association d'une ostéomyélite à *Salmonella* avec une hémoglobinopathie, plus particulièrement avec une sicklanémie. Mais pour deux raisons majeures ils ne sont pas reconnus. D'abord, et ceci est surtout vrai pour l'Afrique, la plupart des malades avec infections osseuses se présentent chez le médecin dans des conditions qui ne lui permettent pas de procéder à un examen bactériologique, *a fortiori* à une détermination du germe. Ensuite, jusqu'à une date récente aucun médecin ne se doutait de ce que la recherche de la falciformation *in vitro* pouvait présenter un certain intérêt chez un malade souffrant d'une infection osseuse.

Notre expérience personnelle dans ce domaine a été pour nous concluante: au total, nous avons eu connaissance de 92 cas observés chez des autochtones du Congo-Léopoldville. Ce chiffre est remarquable non seulement quand on le compare au nombre total des cas connus parmi les populations noires d'Afrique et d'Amérique, mais plus encore quand on pense que ce n'est qu'au cours des toutes dernières années que l'attention des médecins a été attirée sur l'intérêt qu'il y a à demander un examen hématologique complet dans un cas d'infection osseuse. Signalons à ce propos que durant la décade de 1947 à 1956 nous avons réçu pour détermination 25 souches seulement de *Salmonella* isolées dans des cas d'ostéomyélite, tandis que nous en avons reçu 57 au cours des seules années 1957,

1958 et 1959. Les évènements survenus au Congo en 1960 ont mis fin aux recherches plus ou moins systématiques, ce qui n'a pas empêché qu'une dizaine de cas aient encore été observés depuis lors dans la seule ville de Léopoldville.

Il n'est pas sans intérêt de signaler ici que dans son livre, "A l'orée de la forêt vierge" (Lausanne, 1925), le Dr ALBERT SCHWEITZER mentionne les ostéomyélites parmi les sept maladies les plus fréquemment rencontrées en Afrique noire.

Les ostéomyélites à *Salmonella* figurent pour plus de 6% dans nos statistiques sur les salmonelloses au Congo-Léopoldville (VAN OYE, 1959 — VASSILIADIS, 1960) et pour 5,5% dans une statistique publiée au Nigeria (COLLARD & SEN, 1958). Il n'y a pour nous pas de doute que ces pourcentages déjà si élevés sont inférieurs à la réalité car, nous venons de le dire, des recherches tant soit peu systématiques n'ont été entreprises qu'au cours des dernières années, et encore dans quelques rares centres uniquement.

Deux très intéressantes observations méritent d'être soulignées. La première a été faite à Stanleyville: deux soeurs, une âgée de 5 ans et l'autre de 18 mois, toutes les deux sicklanémiques ont fait en même temps une ostéomyélite à *S. enteritidis* var. *chaco*. Cette observation d'une infection familiale est, pour autant que nous sachions, unique. La deuxième a été faite à Léopoldville chez un garçonnet de 18 mois et chez une fillette de 3 ans qui ont fait chacun une infection ostéomyélitique causée par deux espèces différentes de *Salmonella*. Chez le premier de ces malades on a isolé une souche de *S. typhi-murium* et une souche de *S. wangata* du pus de ses foyers d'ostéomyélite, — tandis qu'une hémoculture révéla en plus une bactérémie à *S. stanleyville*. Chez le second malade ont été isolées successivement deux cultures de *S. oranienburg* dont l'une avait des propriétés biochimiques normales tandis que l'autre était anaérogène et ne produisit de l'H_2S que tardivement. Il y a des raisons de croire que, malgré le traitement, la malade s'est surinfectée au cours de son séjour à la clinique. Ces observations d'infections mixtes sont, à notre connaissance, également uniques.

Les quatre cas que nous venons de signaler soulignent la très grande sensibilité des sicklanémiques aux infections à *Salmonella*. Ils posent le problème de savoir si, en dehors du processus pathogénique tel que le voient VANDEPITTE et coll. (1953) il ne faut pas admettre l'hypothèse de l'existence d'un ou éventuellement de plusieurs facteurs favorisants. Rappelons que, dès 1951, HODGES & HOLT ont pensé à une plus grande perméabilité de la paroi intestinale chez les sicklanémiques.

Un examen plus détaillé des données que nous possédons a révélé certains faits que nous voudrions exposer brièvement:

1. Le sexe nous est connu de 91 de nos malades: il y a parmi eux 39 filles (55%) et 32 garçons (45%). La différence est statistiquement

non-significative. D'après les données de la littérature il semble d'ailleurs qu'il n'y ait pas de différence marquée dans la distribution des cas entre les deux sexes et qu'il n'y ait donc aucune prédisposition particulière soit chez les filles, soit chez les garçons.

2. La fréquence par groupe d'âge fait ressortir que c'est surtout au cours des toutes premières années de la vie que l'association que nous décrivons se rencontre. Ceci n'a rien d'étonnant vu la grande mortalité qui frappe les enfants sicklanémiques (VANDEPITTE, 1954 — LAMBOTTE-LEGRAND, 1955).

Nous avons pu obtenir des renseignements précis sur l'âge de 66 malades: parmi eux 28% n'avaient pas un an, 61% étaient âgés de moins de deux ans, 73% de moins de trois ans et 81% de moins de quatre ans. Le malade le plus jeune n'avait qu'un mois, tandis que le plus âgé était une adolescente de 12 ans. Parmi les 17 cas dont parle VASSILIADIS (1960), 16 étaient âgés de moins de trois ans et un entre 5 et 6 ans. Notre documentation contient bien le nom d'une femme adulte, mais il s'agit d'un cas de thalasso-drépanocytose. Nous avons déjà fait remarquer que les sicklanémiques vrais n'ont pour ainsi dire aucune chance d'atteindre l'âge adulte.

3. Nous avons reçu pour identification 92 cultures de *Salmonella* isolées chez des africains souffrant d'une ostéomyélite. Elles proviennent de 90 malades vu les deux cas d'infection double dont nous avons déjà parlé. De deux autres cultures nous savons seulement qu'il s'agit de *Salmonella* dont l'une appartient au groupe sérologique B et l'autre au groupe D. La liste ci-après, qui comprend également les 4 souches isolées à Brazzaville (CHARMOT et coll., 1958), mentionne les 20 espèces et les 2 variétés différentes de *Salmonella* qui ont été trouvées dans des cas d'ostéomyélite diagnostiqués chez des autochtones de l'Afrique centrale: *S. typhi-murium* (43, dont 6 var. *copenhagen*), *S. enteritidis* (14, dont 3 var. *chaco*), *S. kisangani* (6), *S. oranienburg* (6), *S. bovis morbificans* (5), *S. newport* (4), *S. heidelberg* (3), *S. paratyphi-C* (2), *S. dublin* (2), *S. léopoldville* (1), *S. infantis* (1), *S. uganda* (1), *S. stanleyville* (1), *S. braenderup* (1), *S. amersfoort* (1), *S. paratyphi B* (1), *S. saint-paul* (1), *S. wangata* (1), *S. akanji* (1), et *S. brazzaville* (1).

Cette énumération illustre encore une fois l'énorme richesse en sérotypes différents qui caractérise la flora microbienne en Afrique équatoriale dans le domaine des *Salmonella* (VAN OYE, 1960 b), et elle confirme l'affirmation de ROBERTS & HILBURG (1958): " . . . osteomyelitis due to *Salmonella* in association with sickle cell disease is not caused by a predominant species." Une constatation identique a été faite au Nigéria par HENDRICKSE & COLLARD (1960). Il faut cependant souligner que *S. typhi* n'est pour ainsi dire jamais isolée d'un cas d'ostéite: en Afrique Centrale où cette espèce est responsable de la moitié environ des cas de salmonellose humaine, *S. typhi* n'a pas été trouvée une seule fois dans un cas d'ostéomyélite.

4. Une atteinte multiple des os a été observée chez la plupart des malades et non pas d'un seul os comme c'est le plus souvent le cas chez les non-sicklanémiques. Cette particularité qui frappe les enfants atteints d'anémie drépanocytaire a été notée par plusieurs auteurs: GIACCAI & IDRISS (1952), HUGHES & CARROLL (1957), ROBERTS & HILBURG (1958), LAMBOTTE-LEGRAND (1958).

5. Il existe une autre particularité sur laquelle LAMBOTTE-LEGRAND ont attiré l'attention dès leurs premières publications (1951): "... les os des extrémités (phalanges, métacarpiens, métatarsiens) sont surtout entrepris chez les nourrissons, tandis que chez l'enfant plus grand il s'agit avant tout d'atteinte des os des deux segments proximaux des membres, et les lésions plus ou moins symétriques ne sont pas exceptionnelles." HENDRICKSE & COLLARD (1960) soulignent également une "remarkable bilateral symmetry in patients with sicklanemia; this was not evident in the non-sickler."

6. Nous avons déjà fait allusion à la possibilité qu'un ou éventuellement plusieurs facteurs favorisants pourraient intervenir dans la pathogénèse des ostéites à *Salmonella* chez les sicklanémiques. Certains auront peut-être tendance à penser aux traumatismes. Or, il semble bien qu'aucun auteur n'ait eu connaissance qu'un trauma quelconque ait joué un rôle dans l'apparition de l'ostéite chez un de ses malades.

7. La maladie est généralement grave et longue. Beaucoup de cas observés en Amérique ont néanmoins été guéris sans sequelles, bien que parfois après de nombreuses rechutes. Le pronostic semble plus sombre en Afrique et cela malgré une thérapeutique rationnelle et énergique. Il nous paraît logique d'admettre que les enfants noirs en Afrique sont dans un état physique général moins favorable que ceux d'Amérique car nous ne croyons pas qu'on puisse incriminer une différence dans la virulence des germes en cause.

L'exposé ci-dessus n'a pas la prétention d'être complet. Il ne saurait pas l'être d'ailleurs car, bien qu'on en soit arrivé à saisir pourquoi deux maladies tellement différentes puissent être liées par une interdépendance telle que leur association semble presque normale, certains faits restent encore sans interprétation satisfaisante.

Les réponses à ces questions ne pourront probablement être trouvées que dans l'analyse d'observations cliniques détaillées et complètes et qui devront nécessairement comprendre un examen hématologique avec notamment une électrophorèse de l'hémoglobine, non seulement du malade lui-même mais aussi, dans la mesure du possible, de ses parents car la distinction entre homo- et hétérozygotie requiert une enquête familiale. L'identification du germe est de toute évidence indispensable et contribuera à fournir aux épidémiologistes des renseignements d'une utilité certaine. Par ailleurs, l'attention devrait se porter davantage sur les hémoglobinopathies autres que la sicklanémie qui a fait l'objet de recherches très poussées.

BIBLIOGRAPHIE

AKSOYCAN, N., ÖZSOYLU, S. & GÜLMEZOGLU, E., Salmonella muenchen osteomyelitis in a white boy with sickle cell disease. *Turkish J. Pediatr.* 1958, **1**: *162—167.*

CHANTEMESSE, M. & WIDAL, F., Des suppurations froides consécutives à la fièvre typhoïde. Spécificité clinique et bactériologique de l'ostéomyélite typhique. *Bull. Mém. Soc. méd. Hôp. Paris*, 1893, **10**: *779—792.*

CHARMOT, G., REYNAUD, R., AUBERT, & RAVISSE, P., Anémie drépanocytaire et ostéomyélite à salmonelles: 4 observations. *Bull. méd. A.O.F.*, 1958, **3**: *340—343.*

CHOREMIS, K., YANNAKOS, D. & BONTA, CH., 2 Fälle von ausgedehnter Osteomyelitis bei Sichelzellanämie. *Helv. paediat. Acta*, 1955, **10**: *478—483.*

COLLARD, P. & SEN, R., Salmonella types isolated in Ibadan, Nigeria. *Trans. Roy. Soc. trop. Med. Hyg.*, 1958, **52**: *283—287.*

DIGGS, L. W., PULLIAN, H. N. & KING, J. C., Bone changes in sickle-cell anemia. *Southern med. J.*, 1937, **30**: *249—259.*

GAY, K. & GRANT, L. S., Paratyphoid C osteomyelitis. *West Ind. med. J.* 1956, **5**: *284—288.*

GIACCAI, L. & IDRISS, H., Osteomyelitis due to Salmonella infection. *J. Pediat.*, 1952, **41**: *73—78.*

GOLDENBERG, I. S., Sickle cell anemia, Salmonella enteritidis osteomyelitis, and remote postoperative wound abscess: report of case. *Surgery*, 1955, **38**: *758—763.*

HENDRICKSE, R. G. & COLLARD, P., Salmonella osteitis in Nigerian children. *Lancet*, 1960, i: *80—82.*

HODGES, F. J. & HOLT, J. F., Editorial comment. In "The 1951 Year Book of Radiology". Edited by F. J. HODGES et al., 394 pp., Chicago, Year Book Publ., Inc., 1951, p. 89.

HUET, M., SEBAG, A., BEN BRAHEM & FARHAT, M., A propos d'un cas d'ostéomyélite à Salmonella tunis chez un enfant drépanocytaire. *Arch. Inst. Pasteur de Tunis*, 1960, **37**: *227—230.*

HUGHES, J. G. & CARROLL, D. S., Salmonella osteomyelitis complicating sickle cell disease. *Pediatrics*, 1957, **19**: *184—191.*

LAMBOTTE-LEGRAND, J. & C., L'anémie à hématies falciformes chez l'enfant indigène du Bas-Congo. *Mém. Inst. Roy. Col. Belge*, 1951, XIX, fasc. 75, 93 pp.

LAMBOTTE-LEGRAND, J. & C. L'anémie à hématies falciformes chez l'enfant indigène du Bas-Congo. *Ann. Soc. belge Méd. trop.*, 1951, **31**: *207—234.*

LAMBOTTE-LEGRAND, J. & C., Le pronostic de l'anémie drépanocytaire au Congo Belge (à propos de 300 cas et de 150 décès). *Ann. Soc. belge Méd. trop.*, 1955, **35**: *53—57.*

LAMBOTTE-LEGRAND, J. & C., Notes complémentaires sur la drépanocytose (Sicklémie). III. Les salmonelloses dans l'anémie drépanocytaire et microdrépanocytaire. *Ann. Soc. belge Méd. trop.*, 1958, **38**: *535—545.*

LEVER, J. M. & BARKER, G. B., Osteomyelitis of the skull due to Salmonella typhi. *Brit. med. J.*, 1945, **2**: *459—460.*

VAN OYE, E., Les salmonelloses humaines au Congo belge et au Ruanda-Urundi: statistiques sur 1000 cas. *Acta trop.*, 1959, **16**: *158—165.*

VAN OYE, E., Sur l'association entre ostéomyélite à Salmonella et hémoglobinopathie chez l'enfant africain. *Bull. Soc. Path. exot.*, 1960 (a), **53**: *89—100.*

VAN OYE, E., Répertoire général et revisé des Salmonellae du Congo et du Ruanda-Urundi. *Acad. Roy. Sci. d'Outre-Mer*, Classe Sci. nat. et méd., Mém. in—8°, 1960(b), Nouv. Série, Tome XI, fasc. 6, 49 pp.

242

ROBERTS, A. R. & HILBURG, L. E., Sickle cell disease with Salmonella osteomyelitis. *J. Pediat.*, 1958, **52**: *170—175*.

RUSSO, G., Osteomielite da "Salmonella tiphi" in bambino affetto da anemia drepanocitica. *Minerva Pediat.*, 1957, **9**: *16—20*.

SILVER, H. K., SIMON, J. L. & CLEMENT, D. H., Salmonella osteomyelitis and abnormal hemoglobin disease. *Pediatrics*, 1957, **20**: *439—447*.

SMITH, W. S., Sickle cell anemia and Salmonella osteomyelitis. *Ohio State med. J.*, 1953, **49**: *692—695*.

VANDEPITTE, J., COLAERT, J., LAMBOTTE-LEGRAND, J. & C. & PERIN, F., Les ostéïtes à Salmonella chez les sicklanémiques: à propos de 5 observations. *Ann. Soc. belge Méd. trop.*, 1953, **33**: *511—522*.

VANDEPITTE, J., Aspects quantitatifs et génétiques de la sicklanémie à Léopoldville. *Ann. Soc. belge Méd. trop.*, 1954, **34**: *501—516*.

VANDEPITTE, J. & VAN OYE, E., Salmonella osteomyelitis. *Lancet*, 1956, i: *966* (Correspondence).

VASSILIADIS, P., Les Salmonellae du Congo belge (Huitième Rapport). *Ann. Soc. belge Méd. trop.*, 1960, **40**: *423—428*.

WHITE, G. & MEYNELL, M. J., Paratyphoid-C osteomyelitis of the tibia. *Lancet*, 1956, i: *362—363*.

WIDEN, A. L. & CARDON, L., Salmonella typhi-murium osteomyelitis with Sickle-cell — Hemoglobin C disease: a review and case report. *Ann. intern. Med.*, 1961, **54**: *510—521*.

PART II
EPIDEMIOLOGY OF THE SALMONELLOSES

EPIDEMIOLOGIE DER SALMONELLOSEN
IN EUROPA
1950–1960
(ausgenommen Osteuropa)

VON

H. P. R. SEELIGER
unter Mitarbeit von A. E. MAYA

Hygiene-Institut der Universität Bonn/Rhein
(Direktor: Prof. Dr. H. Habs)

(mit 8 Fig.)

A. EINLEITUNG

Beim Versuch einer epidemiologischen Analyse der Salmonellosen in Europa ist zu berücksichtigen, dass die vorhandenen Unterlagen aus verschiedenen Ländern einen wechselnden Grad von Genauigkeit aufweisen. So werden erst seit einigen Jahren allgemein Typhus- und Paratyphuserkrankungen getrennt gemeldet. Die übrigen Salmonellosen erscheinen in den offiziellen Statistiken nicht getrennt oder oft unter den Rubriken "Paratyphus" bzw. "Lebensmittelvergiftungen". Meist werden in den amtlichen Berichten auch die Verdachtsmeldungen erfasst; aber nur einige Länder bereinigen nachträglich ihre Statistik.

Hinsichtlich des Wertes und der Vollständigkeit der statistischen Unterlagen seien Jusatz & El Rubaie zitiert. Danach kommen in Ländern mit primitiven Verhältnissen wesentlich mehr Typhus- und Paratyphusfälle vor, als amtlich erfasst werden, wogegen in Ländern mit fortschrittlichem Gesundheitswesen die meist geringen Erkrankungszahlen vollständiger ermittelt werden. Somit dürften die Unterschiede in den einzelnen Ländern noch grösser sein, als aus den Statistiken erkennbar ist.

Für die anschliessende Darstellung erschien es zweckmässig, den Abdominaltyphus von den verschiedenen Arten des Paratyphus sowie den übrigen *Salmonella*-Infektionen abzugrenzen. Dies rechtfertigt sich nach dem klinischen und epidemiologischen Verhalten der Erreger. Eine gewisse epidemiologische Sonderstellung nehmen die Infektionen des Säuglingsalters ein; sie werden daher zusammenfassend abgehandelt.

Bezüglich der geschichtlichen Entwicklung sei auf die umfassenden Arbeiten der medizinischen Geographie einschliesslich der vorhandenen Seuchenkarten (Bader, Donle, Habs, Jusatz & El Rubaie, Rodenwaldt & Jusatz u.a.) verwiesen.

Aus Gründen der Übersichtlichkeit war eine Beschränkung auf allgemeine Angaben unter weitgehendem Verzicht auf Detailberichte unerlässlich.

B. TYPHUS ABDOMINALIS

Die folgenden Ausführungen stützen sich auf die Darstellungen der Typhus-Epidemiologie von Habs sowie Jusatz & El Rubaie im Weltseuchenatlas von Rodenwaldt & Jusatz und ergänzen diese durch weitere Unterlagen der nationalen Gesundheitsdienste sowie der Weltgesundheitsorganisation ab 1950.

Der Abdominaltyphus ist in Europa heimisch und zeigt eine gewisse räumliche Stetigkeit. In den befallenen Ländern ist die jährliche Zahl der Neuerkrankungen jeweils annähernd gleich; doch sind die Befallsquoten unterschiedlich. Insgesamt ist über längere Perioden eine sinkende Tendenz erkennbar.

Infektionsmodus

Das epidemiologische Bild ist durch den bekannten Infektionsmodus (primäre Infektionsquelle ist stets der infizierte Mensch = Erkrankter, Keimträger bzw. seine Ausscheidungen — Kot, selten Urin —) bestimmt. Die Übertragung erfolgt teils durch Kontakt, überwiegend aber durch verunreinigtes Wasser und Lebensmittel. Primitive Lebensführung und ungenügende Abfallbeseitigung sind wesentliche Faktoren für die Endemizität. Mangelnde Lebensmittelhygiene und die Mitwirkung von Keimträgern bei der Nahrungsgewinnung wie -zubereitung spielen bei der Entstehung von begrenzten Ausbrüchen eine wichtige Rolle. In Kriegszeiten, bei grossen Bevölkerungsbewegungen durch Flucht oder Aussiedlung, beim Zusammenpferchen von Menschenmassen in Lagern mit unzureichenden sanitären Verhältnissen, als Folge von Überschwemmungen usw. entwickeln sich leicht ausgedehnte Epidemien, wie es die Jahre 1940 bis 1947 gezeigt haben. Da die Typhusbakterien in der Umwelt nicht lange lebensfähig bleiben, sind solche epidemischen Ausbrüche aber meist nur von kurzer Dauer. Doch hinterlässt jede Epidemie eine grössere Zahl von Ausscheidern, die das Seuchengeschehen meist noch für mehrere Jahre ungünstig beeinflussen und oft noch lange danach plötzlich wieder explosionsartige Ausbrüche bewirken können.

Geographische Verbreitung

Da die Übertragungsbedingungen im wesentlichen von den wirtschaftlichen und zivilisatorischen Gegebenheiten abhängen, unter denen eine Bevölkerung lebt, ist es gerechtfertigt, die Epidemiologie dieser Krankheit unter Berücksichtigung der politischen Grenzen zu erörtern. Dort, wo die besten wirtschaftlichen, technischen und gesundheitspolizeilichen Voraussetzungen zu einer systematischen Bekämpfung der Seuche bestehen, sind Typhus-

Morbidität und -Mortalität am niedrigsten. Dies erklärt, warum diejenigen Länder, die von den beiden Weltkriegen und ihren Folgen verschont blieben, heute zu den Gebieten zählen, in denen der Typhus bereits zu den Seltenheiten gehört.

Dementsprechend findet man bei einer Gesamtbetrachtung Europas einen typhusarmen bis typhusfreien Norden und eine relativ starke Verseuchung im Süden. Die Endemizität nimmt von Westen nach Osten bzw. Südosten zu. In Fig. 1 ist nach Angaben von JUSATZ & EL RUBAIE die durchschnittliche Typhusmorbidität, bezogen auf 100.000 Einwohner, für den Zeitraum von 1948 bis 1957 wiedergegeben. Diese Zahlen vermitteln für manche Länder, wie Frankreich und die Bundesrepublik Deutschland, ein etwas zu ungünstiges Bild, da die Durchschnittswerte noch durch die hohen Krankheitszahlen der unmittelbaren Nachkriegszeit beeinflusst werden. Auch ist zu beachten, dass für den

Fig. 1. Durchschnittliche Typhusmorbidität (für Frankreich, Spanien, Griechenland und Zypern: Typhus und Paratyphus) in Europa (ausgenommen östliche Staaten) von 1948 bis 1957 nach Zahlen von JUSATZ & EL-RUBAIE.

Gemeldete Neuerkrankungen an Typhus abdominalis in Europa 1951—1960, zu-

	1951	1952	1953	1954
Dänemark	18	15	31	31
Finnland	129	59	90	123
Norwegen	15	10	17	6
Schweden	20	18	10	12
England & Wales	206	135	101	122
Irland	46	47	44	52
Nordirland	22	38	14	13
Schottland	12	13	22	14
Deutschland (BRD und Saar)	4154	3240	3651	3785
Deutschland ("DDR")	4588(+P)	3272(+P)	4001(+P)	3808(+P)
Österreich	1138	680	608	618
Schweiz	84	74	90	173
Belgien	318	356	235	300
Frankreich	5393(+P)	5944(+P)	3954(+P)	3953(+P)
Luxemburg	11	6	14	10
Niederlande	137	123	162	79
Polen	6354	5864	6174	5541
Rumänien	5694	4828	3404	2966
Tschechoslowakei	2978	2266	2098	1882
Ungarn	1555	1332	1611	957
Azoren & Madeira	255	382	579	
Griechenland	3891(+P)	3949(+P)	3470(+P)	2229(+P)
Italien	26125	22072	21259	20499
Jugoslawien	5465	4703	4075	3418
Malta u. Gozzo	180(+P)	118(+P)	132(+P)	107(+P)
Portugal	3005	2500	2691	2529
Spanien	15950(+P)	11938(+P)	13384(+P)	11329(+P)
Türkei	6583	6850	4694	6689
Zypern	371	231	138	112

(+P) Typhus und Paratyphus gemeinsam gemeldet.

gewählten Zeitraum manche Staaten keine getrennten Typhus-
und Paratyphus-Statistiken führten, z.B. Spanien. Nach MERINO
(briefl. Mitt. 1963) sollen etwa 90% der dort amtlich gemeldeten
Erkrankungen Typhusfälle sein.

Die absolute Typhushäufigkeit, gemessen an der Zahl der Neuerkrankungen in den Jahren 1951 bis 1960, ist aus Tabelle I ersichtlich.

Die allgemein sinkende Tendenz ist am Beispiel Italiens,

I.

sammengestellt nach den Statistical Reports der Weltgesundheitsorganisation.

1955	1956	1957	1958	1959	1960
9	12	15	12	11	4
114	71	80	94	37	31
15	10	9	1	2	1
18	21	9	17	21	14
193	136	164	147		
27	48	23	25	93	10
70	17	11	7	46	4
30	19	13	19	18	6
2913	2235	2074	1727	1859	1510
2925(+P)	1707	1581	1200	1136	819
503	393	406	366	354	331
102	75	82	62	56	50
165	155	100	92	185	85
4537(+P)	3678(+P)	3167(+P)	2391(+P)	2067(+P)	2263(+P)
5	6	1	4	3	4
100	87	87	77	63	52
6201	4797	5171	4633	4254	3464
2642	2305				
1363	1290	1177	1463	1043	788
909	748	815	608	509	513
437	367	209	159	261	189
4129(+P)	2630(+P)	2711(+P)	2123(+P)	2430(+P)	1947(+P)
19551	18015	18141	19344	19871	15068
3022	3122	3388	4258	3701	3123
109(+P)	131(+P)	124(+P)	60(+P)	80(+P)	54(+P)
3257	2322	1852	2798	3011	2092
14394(+P)	12259(+P)	10043(+P)	12464(+P)	11345(+P)	10001(+P)
7629	5492	6121	8382	6275	6884
120	50	52	42	62	67

der Bundesrepublik Deutschland (Fig. 2) und Frankreichs (Fig. 3) dargestellt. Ab 1952 tritt fast überall ein deutlich verlangsamter Rückgang der Neuerkrankungen ein. Plötzliche Epidemien bewirken gelegentlich einen, meist allerdings nur vorübergehenden Umschwung oder eine Verzögerung dieser Entwicklung. Trotz der abnehmenden Morbidität steigt die Zahl der erfassten Dauerausscheider weiter an, offenbar als Folge besserer Umgebungs- und Kontrolluntersuchungen. So waren in der Bundesrepublik Deutschland (ohne Saarland) und Westberlin 1953 4410, 1957 hingegen 5131 und

1960 (mit Saarland) 5322 Dauerausscheider amtlich bekannt.
Die Verteilung der Typhusbakterienausscheider nach Geschlecht
und Lebensalter ist am Beispiel der Bundesrepublik Deutschland

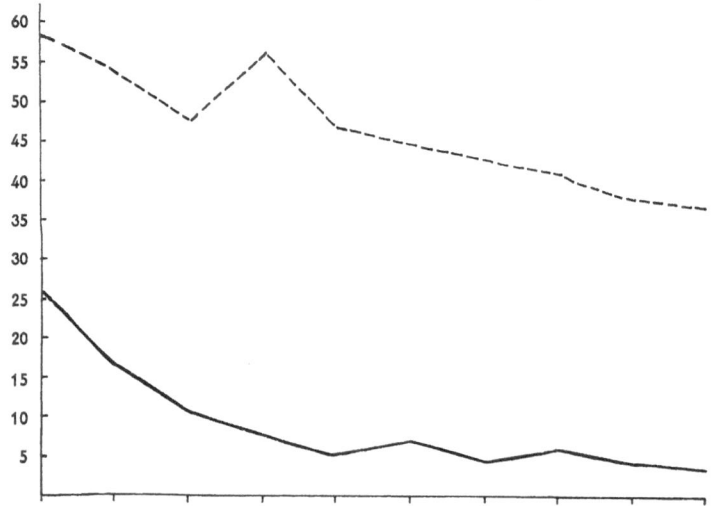

Fig. 2. Rückgang der Typhusmorbidität in Italien (obere Kurve) und in der
Bundesrepublik Deutschland (untere Kurve) (bezogen auf 100.000 Einwohner)
1948—1957 (nach Jusatz & El-Rubaie).

dargestellt (Fig. 4). Danach lassen sich drei Altersgruppen erkennen:
Bis zum Alter von 20 Jahren ist die Zahl der Ausscheider verschwin-
dend gering (maximal 1,7 pro 100.000); sie nimmt dann bis zum
Erreichen des 45. Lebensjahres deutlich zu (4,3—11,9 pro 100.000)
und erreicht ihr Maximum jenseits des 45. Jahres mit Quoten zwischen 14,1 und 22,5 pro 100.000 der Wohnbevölkerung. Der Anteil
der weiblichen Ausscheider ist dabei 3 bis 17 mal so hoch wie der
der männlichen.

Nach den Zusammenstellungen von Meyer-Oschatz sowie
Jusatz & El Rubaie sind in Europa Milch- und Molkereipro-
dukte für die Mehrzahl der Typhusepidemien verantwortlich; ihnen
folgen Trinkwasser bzw. Wasser und in grösseren Abständen
Fische und andere Meerestiere, die roh oder ungenügend
erhitzt genossen werden.

Bei einer Gegenüberstellung der Erkrankungs- und Sterbe-
ziffern an Abdominaltyphus in 26 europäischen Staaten hat Habs
für die Jahre 1931-1935 vier Gruppen unterschieden. Vergleicht
man die Typhusmortalität jener Periode mit der des Zeitraums
zwischen 1955 und 1960, ergibt sich das in Tab. II niedergelegte
Bild.

In der ersten Gruppe mit der niedrigsten Sterblichkeit, welche

251

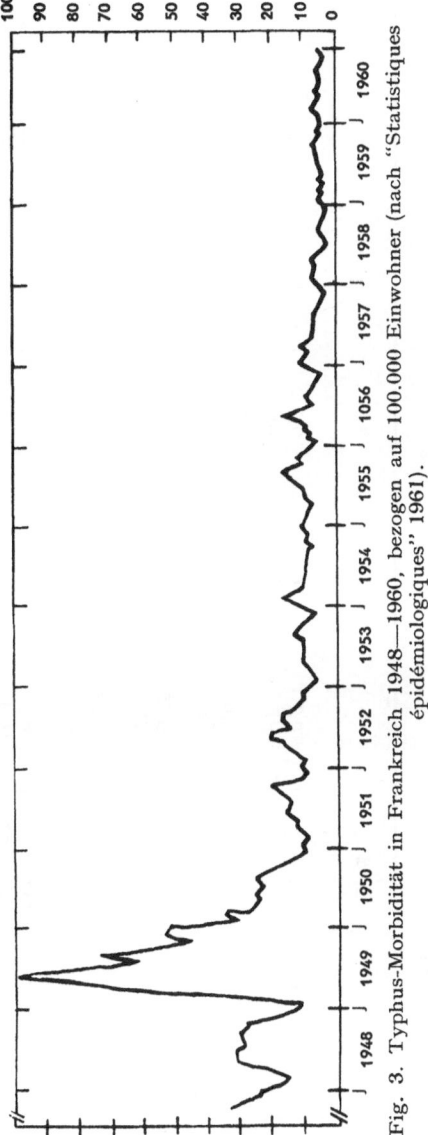

Fig. 3. Typhus-Morbidität in Frankreich 1948—1960, bezogen auf 100.000 Einwohner (nach "Statistiques épidémiologiques" 1961).

Tabelle II.

Typhus-Mortalität in Europa 1955—1960 verglichen mit den Durchschnittswerten 1931—1935 (Habs) (bezogen auf 100.000 Einwohner).

	1913 (nach Habs)	1931—1935 nach Habs	1955	1956	1957	1958	1959	1960
Dänemark		0,52	—	—	0,0.. (1)*	—	—	—
Schweiz	0,9	0,74	0,1 (4)	0,0.. (2)	0,1 (6)	0,1 (3)	0,1 (5)	
Norwegen		0,73	0,0.. (1)	—	—	—	0,0.. (1)	—
England u. Wales	0,6	0,44	0,05 (7)	0,0.. (2)	0,05 (8)	0,0.. (2)	0,0.. (3)	0,0.. (2)
Niederlande		0,72	0,05 (4)	0,1 (9)	0,05 (4)	0,0.. (2)	0,05 (5)	0,0 (2)
Belgien		keine Angaben	0,1 (11)	0,1 (9)	0,1 (7)	0,0 (4)	0,05 (5)	0,0.. (4)
Deutschland (BRD)	0,8	0,64**	0,2*** (92)	0,2 (81)	0,1 (62)	0,1 (55)	0,1 (40)	0,1 (46)
Schweden	4,1	0,58(+P)	—	0,0.. (2)	—	—	—	—
Schottland		0,53	0,0.. (1)	0,0.. (2)	—	0,0.. (1)	—	—

Finnland	2,56	0,0.. (1)	0,1 (6)	0,0.. (2)	0,1 (4)	0,1 (3)	0,0 (1)
Österreich	1,32	0,4 (30)	0,2 (16)	0,4 (26)	0,2 (13)	0,2 (16)	0,2 (14)
Frankreich	keine Angaben	0,2 (91)	0,2 (82)	0,15 (62)	0,1 (33)	0,1 (48)	0,1 (49)
Irland	2,45	0,2 (6)	0,1 (3)	—	0,1 (2)	0,0.. (1)	—
Tschechoslowakei	3,84	0,3 (45)	0,3 (43)	0,4 (50)	0,3 (39)	0,3 (34)	0,1 (17)
Polen	3,39	1,1 (290)	0,7 (196)	0,7 (195)	0,5 (196)	0,5 (132)	
Portugal	7,21(+P)	1,0 (92)	0,9 (77)	0,7 (61)	0,8 (73)	0,8 (76)	0,6 (53)
Griechenland	5,55	0,2 (97)	0,9 (74)	0,6 (51)	0,5 (41)	0,4 (37)	
Spanien	9,84(+P)	1,3 (37,5)	0,9 (267)	0,8 (244)	0,7 (218)	0,6 (187)	
Italien	10,79	1,0 (476)	0,8 (386)	0,6 (322)	0,6 (325)	0,5 (240)	
Ungarn	27,2 (+P)	0,4 (35)	0,2 (23)	0,2 (16)	0,1 (13)	0,2 (16)	0,1 (11)

(+P) = einschliesslich Paratyphus.
* Absolute Zahl der Typhustodesfälle.
** Deutsches Reich, Grenzen 1937.
*** Bundesrepublik Deutschland ohne West-Berlin, aber einschliesslich Saarland.

einst nur die skandinavischen Länder, die britischen Inseln, die
Niederlande, Be~~~~~~~~~~~~~~~~~~~t die Mortalität
von durchschnit~~~~~~~~~~~~~~~~~~~~~~~~~~~~~~~~~~niger gesunken.

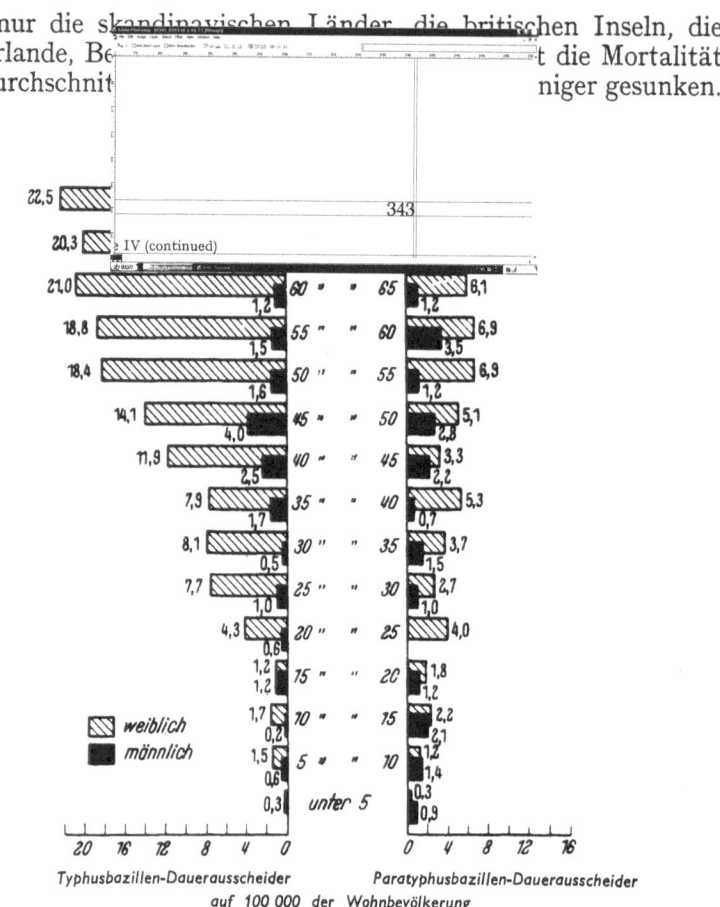

Fig. 4. Verteilung der Typhus- und Paratyphus B-Bakterien-Ausscheider nach
Alter und Geschlecht unter der Bevölkerung von Bayern (nach K. BINGOLD im
Handbuch d. inn. Med., IV. Aufl., S. *1472.* 1952, Springer-Verlag).

In Österreich (Gruppe 2 nach HABS) — früher 1,3 pro 100.000 – liegt
sie immer noch zwischen 0,4 und 0,2. Finnland, Frankreich und
Irland zeigen 1955–1960 — im Gegensatz zu früher — eine ähnliche
Typhussterblichkeit wie die Länder der ersten Gruppe. Im Vergleich
dazu weisen Polen und die Tschechoslowakei trotz eines Rückganges
um das Fünf- bis Zehnfache weiterhin eine höhere Typhusmortalität auf. Das eindrucksvollste Absinken verzeichnet Ungarn, dessen
Typhusmortalität ab 1956 dem westeuropäischen Bild entspricht.
Die höchsten Mortalitätsquoten werden nach wie vor aus den
südeuropäischen Ländern berichtet. Diese stellen bei dem
steigenden Reise- und Güterverkehr derzeit das wichtigste Typhus-

reservoir für West-, Mittel- und Nordeuropa dar, was u.a. auch durch die Ergebnisse der Lysotypie bestätigt wird.

Gemessen an der Morbidität und Mortalität hat sich somit die von HABS für 1947 und JUSATZ für 1957 geschilderte Lage bis 1960 weiter verbessert. Die typhusarme Zone hat sich vom Norden her wieder über Mittel- und Westeuropa ausgedehnt und greift bereits auf den Balkan über, während der Süden dieser Entwicklung erheblich langsamer folgte.

Tabelle III.

Gipfelmonate für Typhus und Paratyphus B in verschiedenen europäischen Ländern.

Land	Häufigster Gipfelmonat für Abdominaltyphus			Häufigster Gipfelmonat für Paratyphus B
	1951-60	1926-1931 (bzw. 1931-35) [nach HABS]	Verschiebung um Monate	1951-1960
Griechenland (+P)	7	8	1	siehe Typhus
Österreich	(6)* 7/8	10	2—3	8
Schweiz	7/8	9	1—2	8
Belgien	8	8/9	1/2	8
Dänemark	8	10	2	8
Niederlande	8	9	1	9
Schottland	[8]	9	1	[7]
Schweden	8	8	—	7
Tschechoslowakei	8	9	1	9
Ungarn	8	10	2	6/7 (1956-60)
Deutschland (BRD)	[9]	10	1	[9]
England und Wales	[9]	9	1	[9]
Frankreich (+P)	9	11	2	siehe Typhus
Irland	[9]	12	2—3	[9/10]
Italien	9	9	—	8/9
Jugoslawien	9	10	1	8
Nordirland	[9]	9	—	[10]
Spanien (+P)	9	10	1	siehe Typhus
Türkei	9	10	1	9
Portugal	10	10	—	9

(6)* = bedingt durch lokale Epidemie 1951
[ ] = Meldungen in 4-Wochen-Perioden
(+P) = Typhus- und Paratyphus gemeinsam gemeldet.

Jahreszeitliche Einflüsse

In Europa ist eine gewisse Bindung der Typhusmorbidität an die Jahreszeit erkennbar. Die meisten Neuerkrankungen treten im Spätsommer oder Herbst auf. Doch bleiben auch die Wintermonate nicht von Ausbrüchen verschont.

Ermittelt man nach HABS den jeweiligen Monat mit der höchsten Krankheitszahl, den sog. Gipfelmonat, ergeben sich für die Periode von 1951 bis 1960 in einer Reihe von Ländern Verschiebungen, wenn man den häufigsten Gipfelmonat mit dem der Zeitspannen von 1926 bis 1931 bzw. 1931 bis 1935 vergleicht (Tab. III). Während sich einst die Typhuserkrankungen im Norden Europas etwa zwei Monate früher häuften als auf dem Balkan, ist eine solche Ordnung im vergangenen Jahrzehnt nicht mehr erkennbar. Nicht selten wird jetzt der Höhepunkt allgemein ein bis zwei Monate früher erreicht. In manchen Ländern, z.B. Frankreich, ist der Gipfelmonat nur wenig ausgeprägt. An seine Stelle tritt ein breiter Saisongipfel, der schon in der Jahresmitte beginnt und sich über fünf Monate erstreckt. Andererseits äussern sich lokale Epidemien meist in einer steilen Gipfelbildung (England, 1948; Schottland und Nordirland, 1955). Die Ursachen für die erkennbare Tendenz zu einer Vorverlegung des Gipfelmonats sind noch nicht geklärt.

Untersucht man weiter die Intensität der jahreszeitlichen Schwankungen, kommt man durch Berechnung, wieviele von jährlich 1.200 bzw. 1.300 Erkrankungen auf den Gipfelmonat entfallen, zum sog. Gipfelwert (Tab. IV).

Für die Zeit von 1931 bis 1935 hatte HABS noch folgern können, daß der Saisonrhythmus der Typhusmorbidität in Westeuropa nur mässig, in der Mitte wie im Süden etwas mehr und im Osten relativ stark ausgeprägt sei. Doch zeigen die Werte der vergangenen Dekade 1951–1960 bei Ländern mit hohen Durchschnittszahlen kein eindeutiges Verhalten mehr, teilweise sogar Verschiebungen, die fast einer Umkehr gleichkommen. Galt früher, daß Länder mit einem starken Gegensatz der Jahreszeiten häufig auch die höchsten jahreszeitlichen Typhusschwankungen aufweisen, so trifft dies in der Berichtsperiode nur noch für Ungarn zu.

Epidemiologische Folgerungen und Ausblick

Die Entwicklung der Typhusvorkommen Europas zeigt, daß die räumliche Verteilung des Abdominaltyphus mit dem sozialen Niveau und den allgemein-hygienischen Verhältnissen der Staaten im Zusammenhang steht. Jede Besserung der allgemeinen Lage führt meist zu einem Rückgang der Typhusverbreitung, jede Verschlechterung zum Gegenteil. Keimträger und Neueinschleppung verzögern

Tabelle IV.

Gipfelwerte nach HABS für Typhus und Paratyphus B in europäischen Ländern [mit jährlichen Erkrankungszahlen von > 100 (Typhus) und 200 oder mehr. (Paratyphus B)] 1951—1960.

Land	Typhus abdominalis			Paratyphus B	
	Gipfelwert			Gipfelwert	
	1951—1960		1931—1935	1951—1960	
Polen	155	(5.245)*	207	192	(645)*
Italien	158	(19.994)	224	172,6	(4.477)
Portugal	163	(2.605)	187	187	(116)
Tschechoslowakei	166	(1.633)	175	245	(361)
Deutschland ("DDR")	163,3	(1.288)		181	(266)
Deutschland (BRD			222		
mit Saar)	168	(2.616)		178	(3.424)
Türkei	176		—	172	(514)
Jugoslawien	179	(3.827)	229	299	(1.342)
Ungarn	228	(945)	266	255	(191)***
Schweden	zu kleine			108	(1.379)**
Griechenland	Durchschnitts-			152	(472)
Finnland	zahlen			175	(839)
Österreich	179,9 (	539)		206	(904)
Belgien	177,9 (	209)		217	(209)
Niederlande	184,9 (	96,7)		266	(307)

Gipfelwerte für Typhus und Paratyphus gemeinsam

Frankreich	124	(3.734)	142	
Spanien	162	(1.239)	157	} nach HABS
Griechenland	164	(2.950)	175	

```
*   = Jahresdurchschnittszahl 1951—1960.
**  = Wert durch fast 9000 Fälle seit 1953 belastet.
*** = (1956—1960).
```

aber auch in den typhusarmen Ländern die völlige Ausrottung der Seuche, die immer wieder aufflammt, sobald unzureichende hygienische und sanitäre Bedingungen die Voraussetzung hierfür bieten. Im Gegensatz zu früher spielen Kontakt und Verbreitung durch verunreinigtes Wasser eine abnehmende Rolle, während Lebensmitteln, insbesondere Milch und Milchprodukten, eine steigende epidemiologische Bedeutung zukommt. Solange das Problem der Sanierung von Keimträgern nicht endgültig gelöst ist, wird die Seuchenprognose ebenso von der politischen und wirtschaftlichen Entwicklung wie von der Verbesserung der Wasser-, Abwasser- und Lebensmittelhygiene bestimmt werden.

C. PARATYPHUS A

Bei der Schilderung des Paratyphus A wird die Darstellung BADER's für die Zeit der beiden Weltkriege sowie die Periode von 1900 bis 1950 lediglich um einige neuere Daten erweitert.

Infektionsmodus

Prinzipiell gilt für den Paratyphus A der gleiche Infektionsmodus wie für den Abdominaltyphus; denn auch hier ist nur der infizierte Mensch als eigentliches Erregerreservoir anzusehen. Die Infektion erfolgt vorzugsweise durch direkten Kontakt oder durch verunreinigte Lebensmittel.

Geographische Verbreitung

In Europa zeigt der Paratyphus A eine besondere Vorliebe für die Balkan-sowie die südöstlich gelegenen Mittelmeerländer und damit eine weitgehende regionale Gebundenheit. In den übrigen Ländern ist er selten bzw. wieder selten geworden. Während Skandinavien und die britischen Inseln praktisch Paratyphus A-frei sind, wird er in Mittel-und Westeuropa noch sporadisch gefunden. So weisen die Niederlande 1957 und 1960 je einen, 1956, 1958 und 1959 je zwei Fälle auf. Dieser Rückgang gegenüber früher ist wohl mit dem Nachlassen der Verbindungen nach Indonesien, wo Paratyphus A häufig ist, zu erklären. Auch in Belgien wurde der Erreger nach VAN OYE (briefl. Mitteilung 1963) nur selten isoliert, und zwar meistens im Raume Lüttich, wo sich viele Fremdarbeiter aus dem Mittelmeergebiet aufhalten. In der Bundesrepublik Deutschland wurden von SEELIGER zwischen 1949 und 1952 375 Erkrankungsfälle ermittelt; 1953 bis 1957 waren es insgesamt nur noch 17. Auch hier zeigt zich, dass mit dem Aufhören der Einschleppung aus Ost- und Südosteuropa durch heimkehrende Soldaten, Kriegsgefangene und Flüchtlinge die Zahl der Neuerkrankungen auf ein Minimum zurückging, obwohl es zu Zeiten sogar zu einer beachtlichen Verseuchung des Freilandes (vgl. STEINIGER) gekommen war. Keimträger und Ausscheider waren in den Nachkriegsjahren nicht selten. In Berlin wurden zwischen 1947 und 1953 insgesamt 27 Ausscheider ermittelt, und in Hamburg fand ROHDE Paratyphus A-Träger u.a. beim Personal von Milchhandlungen. — Ähnliches gilt auch für Österreich, wo 1951 fünf, 1955 sechs, aber 1957, 1959 und 1960 keine Fälle mehr gesichert wurden (ROSCHKA, briefl. Mitteilung 1963).

Während in Spanien Paratyphus A relativ selten zu sein scheint (MERINO, briefl. Mitteilung 1963), wurde er früher in Frankreich häufiger gefunden. Nach DONLE soll dort im Raume von Montpellier ein kleines Endemiegebiet existiert haben. 1942/43 wurden 218 Erkrankungsfälle gezählt, 1949 etwa 100 (vgl. BADER); 1958 waren es 18 und 1959 noch 13. Der Abbau der Beziehungen zu Indochina und die Verminderung der Kontakte mit Nordafrika dürfen mit dieser Entwicklung in Zusammenhang gebracht werden. In der französischen Orientarmee und der Fremdenlegion sollen regelmässig, wenn auch nicht gehäuft, Paratyphus A-Fälle vorgekommen sein.

Aus Italien, wo die Krankheit während der 30er Jahre in kleinen Ausbrüchen auftrat, liegen keine Meldungen vor, die auf eine stärkere Endemizität hindeuten. Obwohl anzunehmen ist, dass die Endemizität des Paratyphus A im Balkan auch für Griechenland gilt, wurden dort nur wenige Fälle gesichert (ALIVISATOS, briefl. Mitteilung 1963). Die Schliessung der Balkangrenzen seit Beendigung des Bürgerkrieges (1949) hat möglicherweise auch zu einer Verminderung der Ausbreitung in südlicher Richtung beigetragen. Doch beweist das Auftreten von Paratyphus A-Fällen bei Gastarbeitern aus den Mittelmeerländern, dass dort noch grössere Herde existieren müssen. — Aus der Türkei wird 1956 berichtet (AKYAY), dass 2,4% von insgesamt 1000 isolierten *Salmonella*-Stämmen zu *S. paratyphi* A gehörten; aber es gibt auch dort nach BERKIN (briefl. Mitteilung 1963) keine zusammenfassenden Unterlagen.

Epidemiologische Folgerungen und Ausblick

In Nord-, Mittel- und Westeuropa ist der Paratyphus A eine Fremdseuche. Die überwiegende Mehrzahl der meist sporadisch auftretenden Fälle wird aus anderen Gebieten eingeschleppt. Die Zahl der aus früheren Erkrankungsfällen und kleinen Ausbrüchen zurückgebliebenen Ausscheider ist zu gering, um bei den gegebenen Verhältnissen ein Aufleben der Krankheit zu bewirken. Der zunehmende Reiseverkehr in die südosteuropäischen und am Mittelmeer liegenden Endemiegebiete sowie das Millionenheer der aus dem Süden kommenden Fremdarbeiter sorgen jedoch dafür, daß der Paratyphus A in den übrigen Ländern Europas gelegentlich noch auftritt.

D. PARATYPHUS B

Über die kontinentale Verbreitung des Paratyphus B liegen noch keine Seuchenkarten oder umfassende monographische Darstellungen vor. Der Bericht stützt sich auf die zahlreichen Einzelarbeiten vieler Autoren, die Berichte der WHO und die Zusammenstellungen der nationalen Salmonella-Zentralen. Im übrigen gelten sinngemäss die einleitenden Ausführungen zum Abschnitt B (s. S. 246). — Trotz mancher Rückschläge ist die Gesamt-tendenz der Paratyphus B-Morbidität seit 1950 zwar rückläufig, aber nicht im gleichen Masse wie beim Typhus. Allgemein zeigt sich die Verbesserung der hygienischen Situation in europäischen Ländern zunächst in einem steilen Abfall der Typhus-, aber nur in einem verzögerten Rückgang der Paratyphusmorbidität.

Infektionsmodus

Der Mensch ist in der überwiegenden Mehrzahl der Fälle als primäre Infektionsquelle anzusehen; doch wird *Salmonella paratyphi B* gelegentlich auch vom Tier ausgeschieden und auf Nahrungsmittel übertragen. Letzteres ist in Europa selten; hingegen tritt beim Tier häufig die antigengleiche *Salmonella java* auf, die beim Menschen jedoch enteritische Infektionen bewirkt. Da *S. paratyphi B*-Keime gegen äussere Einflüsse widerstandsfähiger sind als beispielsweise Typhusbakterien, halten sie sich in der Aussenwelt länger und werden deshalb häufig im Abwasser und in den Vorflutern gefunden. In den Küstengebieten der Nordsee stellen die Krabben und Muscheln ein wichtiges Keimreservoir dar, das seit jeher in diesem Raum für zahlreiche Einzelfälle und Epidemien verantwortlich ist. Die Infektion dieser Meerestiere geht auf die stark verseuchten Vorfluter zurück, die in das Meer münden (STEINIGER, MEYER). Im letzten Jahrzehnt wurde die Epidemiologie des Paratyphus B vorwiegend durch Lebensmittelinfektionen, dabei gelegentlich Importwaren, bestimmt. Eingeführte Eikonserven, wie Flüssigei und Eialbumin, spielen hierbei eine gewisse Rolle. Mittels der Lysotypie konnte nachgewiesen werden, dass chinesische Eiprodukte, die Paratyphus B-Erreger enthielten, in Deutschland und England sporadische Infektionen und kleinere Ausbrüche verursachten (WINKLE & ROHDE). In England und Wales wurden mehrfach Ausbrüche durch infizierte "synthetische" Crèmes beobachtet. Der vermehrte Touristenverkehr und der Zuzug von Arbeitskräften bewirken kontinuierlich Neueinschleppungen aus dem Süden nach verschiedenen Teilen Europas. Trinkwasserinfektionen durch *S. paratyphi B* werden immer seltener. Die Zahl der menschlichen Dauerausscheider ist hoch; allein in der Bundesrepublik Deutschland (mit Saarland) wurden 1953 amtlich 3986 solcher Ausscheider, 1957 sogar 4671 erfasst; man muss jedoch in Wirklichkeit mit einer noch höheren Zahl rechnen.

Die Verteilung der Paratyphus B-Bakterienausscheider nach Alter und Geschlecht ist am Beispiel Bayerns (Deutschland) ersichtlich (Fig. 4). Beim Vergleich mit den auf Seite 250 besprochenen Werten für die Typhusbakterienausscheider fällt auf, dass die altersmässige Gebundenheit des Ausscheidertums viel weniger ausgeprägt ist. Obwohl auch hier die Quote der weiblichen Dauerausscheider deutlich überwiegt, ist ihr Anteil doch geringer als bei den Typhusbakterienausscheidern.

Geographische Verbreitung

In den Kriegs- und unmittelbaren Nachkriegsjahren kam es in weiten Gebieten Europas zur epidemischen Ausbreitung des Paratyphus B, der in Deutschland und Österreich etwa ein Drittel der Typhusmorbidität erreichte. In den Niederlanden stieg die Erkrankungszahl von 343 (1943) über 1322 im Jahre 1945 auf 1153 im Jahre 1946 an. In Italien verdoppelte sich 1946 die Zahl von 4000 bis 5500 jährlichen Fällen (Vorkriegszeit) auf annähernd 8000. Demgegenüber lag die Typhus-Paratyphus-Morbidität in Spanien 1945/46 niedriger als in der Periode 1937 bis 1942, d.h. nach dem Bürgerkrieg (vgl. S. 253). Auch in Griechenland erreichte die Seuche 1945 ihren Gipfel mit 1548 erfassten Fällen. Der Befall war auch während des anschliessenden Bürgerkrieges hoch (1946: 999; 1947: 613 Fälle). Selbst die Schweiz blieb 1945 nicht verschont. Finnland erfuhr 1948/49 eine ungewöhnliche Winterepidemie mit 3600 Fällen, nachdem 1946 etwa 4000 und 1945 sogar über 8500 Neuerkrankungen aufgetreten waren. Während die britischen Inseln 1943 bis 1945 durchschnittlich 376 Neuerkrankungen verzeichneten, stieg die Zahl 1946 auf über 1000 an. Zusammenfassend ergibt sich, dass in den Kriegs- und Folgejahren eigentlich nur die neutral gebliebenen Länder von der allgemeinen Paratyphus B-Verseuchung ausgespart waren. Doch blieben auch sie nicht auf die Dauer verschont.

Tab. V gibt die Zahl der gemeldeten Neuerkrankungen zwischen 1951 und 1960 wieder. Skandinavien erlebte in dieser Zeit einen schweren Rückschlag; denn 1953 brach in Schweden eine Epidemie mit fast 9000 Fällen aus, deren Folgen von da ab das Seuchengeschehen beeinflussten, so dass sich seither die durchschnittliche Zahl der gemeldeten Fälle gegenüber 1950 etwa verdoppelt und gegenüber 1951/52 vervierfacht hat. Im benachbarten Finnland trat 1954/55 ebenfalls eine erhebliche Zunahme auf, die nur von einem zögernden Rückgang gefolgt war. Während in Norwegen und Dänemark kein wesentlicher Paratyphus B-Anstieg erfolgte, brachte das Jahr 1956 Grönland einen Höhepunkt mit 180 Erkrankten. In Irland erfolgte 1958 eine sprunghafte Zunahme. Von April bis Juni 1955 trat sogar im sonst typhusfreien Island in einem Vorort von Rejkjavik eine lokalisierte Epidemie auf, deren Ursache unklar blieb (DUNGAL, briefl. Mitteilung 1963). Auf den britischen Inseln ist dagegen ab 1950 ein stetiger, wenn auch langsamer Rückgang erkennbar; in England und Wales ist Paratyphus etwa viermal häufiger als Typhus. Ähnliches gilt für Deutschland und Österreich, wo sich die Typhus- und Paratyphus B-Fälle — abweichend von früher —

Tabelle V.

Gemeldete Neuerkrankungen an Paratyphus-B in Europa (ausgenommen Teile von Osteuropa) 1951 bis 1960, zusammengestellt nach den Statistical Reports der Weltgesundheitsorganisation.

	1951	1952	1953	1954	1955	1956	1957	1958	1959	1960
Dänemark	51	40	35	64	57	59	16	15	35	9
Finnland	935	557	649	1971	1277	1549	625	717	811	303
Island	—	—	2	42	11	—	—	—	—	—
Norwegen	3	12	24	9	8	15	4	6	25	5
Schweden	140	129	8760	800	685	456	1026	545	629	628
Grönland	16	23	4	37	45	180	33	39	25	9
England & Wales	1095	1039	353	548	876	440	310	200	483	—
Irland	15	11	10	47	10	9	13	276	83	37
Nordirland	47	8	6	8	56	9	4	13	5	3
Schottland	124	80	56	90	136	95	47	105	97	59
Deutschland (BRD und Saar)	5489	3980	4080	3174	3837	3121	2877	2370	2914	2399
Deutschland ("DDR")	—	—	—	—	—	868	564	341	649	244
Österreich	1346	1186	1066	1186	814	731	1075	572	487	581
Schweiz	90	108	—	155	182	207	972	204	136	95
Belgien	267	317	157	143	138	121	179	142	193	437
Luxemburg	124	66	61	48	25	41	88	30	13	143
Niederlande	290	262	308	619	313	219	317	277	306	163
Polen	373	558	1151	790	921	619	713	750	448	422
Rumänien	151	147	86	120	95	—	—	—	—	—
Tschechoslowakei	356	390	647	382	723	241	186	217	235	239
Ungarn	155	418	174	81	310	156	158	147	117	199
Griechenland	—	—	—	497	732	396	456	506	368	350
Italien	6352	6152	5874	5727	4491	4444	3596	3488	2802	2306
Jugoslawien	1619	1197	1090	1297	980	1349	1072	1147	1699	1978
Portugal mit Azoren & Madeira	213	189	127	127	152	78	54	84	71	84
Türkei	465	627	442	621	702	469	508	546	380	382

zahlenmässig jetzt die Waage halten. In der Schweiz ist der deutliche Anstieg im Jahre 1957 bemerkenswert. Die Benelux-Staaten zeigen insgesamt eine Verminderung der Paratyphus-Morbidität, doch ist diese von einzelnen Anstiegen, wie 1954 in Holland, unterbrochen. Aus Frankreich werden für 1959 2019 Fälle gemeldet. In Spanien und Portugal hat sich das Bild gegenüber früher nicht wesentlich geändert. Typhus soll dort aber erheblich häufiger sein als Paratyphus B. Während in Italien eine sinkende Tendenz erkennbar ist, kam es in Jugoslawien ab 1959 wieder zu einer Zunahme (1960 fast 2000 Fälle). In Griechenland und der Türkei ist in der gesamten Berichtszeit keine wesentliche Veränderung in den Zahlen gemeldeter Erkrankungen zu erkennen.

Paratyphus B und Jahreszeit

Ein Vergleich der jahreszeitlichen Häufung des Typhus und Paratyphus B (vgl. Tab. III) zeigt nur unwesentliche Differenzen. Generell tritt das Maximum der Neuerkrankungen zwischen Juli und Oktober auf. In Belgien, Dänemark, Deutschland, England und Wales, Irland und der Türkei fallen die Gipfelmonate für beide Krankheiten zusammen, während der Paratyphus B-Gipfel in Schottland, Schweden, Ungarn, Italien, Jugoslawien und Portugal meist früher, dafür aber in Österreich, Schweiz, Holland, der Tschechoslowakei und Nordirland etwas später erreicht wird. Es scheint nicht möglich, diese Befunde mit irgendwelchen besonderen Gegebenheiten in Zusammenhang zu bringen; doch mögen klimatische Besonderheiten für die fallweise Verschiebung der Gipfelmonate innerhalb der Berichtsperiode von Bedeutung sein.

In ähnlicher Weise lässt sich auch keine Übereinstimmung oder Correlation der Gipfelwerte (s.S. 257) bei Typhus und Paratyphus B erkennen, wenn man sie (vgl. Tab. IV) einander gegenüberstellt. Nur Ungarn bietet insofern vielleicht noch den Sonderfall, daß beide Krankheiten einen der höchsten Gipfelwerte aufweisen und damit ein relativ ausgeprägter Saisonrhythmus auch beim Paratyphus B erkennbar ist. Dies gilt in gewissen Grenzen auch für die Tschechoslowakei und für Österreich. Man kann den Saisonrhythmus des Paratyphus B aber nicht allein mit den jahreszeitlichen Klimadifferenzen erklären; denn die Niederlande mit ihrem Seeklima weisen bei einem ähnlichen Jahresdurchschnitt der Fallzahl einen noch höheren Gipfelwert auf als Ungarn mit seinem Kontinentalklima. Eine direkte Beeinflussung der Intensität des Paratyphus-Saisonrhythmus durch das Makroklima wird hierdurch unwahrscheinlich.

Epidemiologische Folgerungen und Ausblick

Im Gegensatz zum Abdominaltyphus und dem Paratyphus A ist

somit der Paratyphus B in ganz Europa endemisch. Dies äussert sich im Spätsommer und Herbst in zahlreichen Einzelerkrankungen und Ausbrüchen. Obwohl diese meist örtlich gebunden sind, kommt es gelegentlich, vor allem bei Lebensmittelinfektionen, zu ausgedehnten Epidemien, deren grösste 1953 in Schweden auftrat. Die relativ grosse Zahl von Dauerausscheidern bedingt — analog dem Abdominaltyphus (s.S. 256) — eine ähnlich unsichere Seuchenprognose. Es ist wahrscheinlich, dass sich der Paratyphus B bei dem derzeitigen Grad der Verseuchung in Europa noch länger halten wird.

E. PARATYPHUS C

Die nachfolgende Besprechung schliesst sich an die Darstellung der globalen Verbreitung des Paratyphus C (1915–1945) von BADER an. Zu den Erregern des Paratyphus C zählen zwei eng verwandte Salmonella-Keime (S. paratyphi C = S. hirschfeldii und S. choleraesuis var. kunzendorf). Hiervon abzugrenzen sind der Enteritiserreger S. cholerae-suis und die zahlreichen übrigen Serotypen der C-Gruppe. Fälschlicherweise werden menschliche Infektionen durch letztgenannte Keime manchmal noch als Paratyphus C bezeichnet, was bei epidemiologischen Analysen leicht zu irrigen Folgerungen verleitet.

Infektionsmodus

Der Infektionsmodus des Paratyphus C wird dadurch bestimmt, dass seine Erreger, insbesondere der Kunzendorf-Typ, zu jenen Salmonellen gehören, deren Standort primär nicht beim Menschen, sondern beim Schlachttier (besonders Schweinen) zu suchen ist. Menschliche Infektionen verlaufen deshalb sowohl unter den Erscheinungen einer akuten Lebensmittelvergiftung als auch typhusähnlich. Das typhöse Krankheitsbild manifestiert sich aber meist erst im Zusammentreffen mit anderen Krankheiten, wie Malaria, Hepatitis usw. Nach BADER scheint bei der Verbreitung die Ausscheidung durch den Urin des Infizierten — zumindest in bestimmten Gebieten — von grösserer Bedeutung zu sein.

Geographische Verbreitung

Während des ersten Weltkriegs war der Paratyphus C im gesamten östlichen Mittelmeerraum heimisch; doch nahm später die Zahl der beobachteten Fälle stark ab. Erst im zweiten Weltkrieg ist die Krankheit dann wieder häufiger unter den deutschen Truppen auf dem Balkan aufgetreten (HABS, BADER). Aus neuerer Zeit liegen nur wenige einschlägige Berichte vor, z.B. aus der Türkei, wo 9 von insgesamt 1000 Salmonella-Stämmen als S. para-

typhi C diagnostiziert wurden. Alle 9 Stämme wurden aus menschlichem Untersuchungsmaterial gezüchtet (GIRIS, AKYAY). AKSOYCAN berichtet für die Zeit von 1955 bis 1960 von drei *S. paratyphi C*-Kulturen unter insgesamt 693 Stämmen. — Nach MERINO (briefl. Mitteilung 1963) wurden an der Salmonella-Zentrale in Madrid innerhalb der letzten 20 Jahre nur drei Fälle verifiziert. Im nördlichen, mittleren und westlichen Europa ist der Paratyphus C praktisch unbekannt. Die wenigen sporadischen Fälle, z.B. 1959 einer in Schottland (STEVENSON, briefl. Mitteilung 1963), gehen wohl immer auf Einschleppung zurück. In neuerer Zeit konnten in den Salmonella-Zentralen der mittel- und westeuropäischen Länder relativ häufig Stämme von *S. cholerae-suis*, darunter gelegentlich auch die Variatio *kunzendorf*, identifiziert werden, ohne dass hierdurch menschliche Paratyphus C-Erkrankungen ausgelöst wurden. In Deutschland hat es sich fast stets um Fälle von Nahrungsmittelvergiftungen gehandelt, die durch Genuss von infiziertem Fleisch bzw. Fleischprodukten ausgelöst waren.

Epidemiologische Folgerungen und Ausblick

Zusammenfassend ist zu folgern, dass der Paratyphus C in den hier besprochenen Gebieten Europas nicht heimisch ist und nur ausnahmsweise zur Beobachtung gelangt; stets handelt es sich um eingeschleppte Erkrankungen.

F. ANDERE SALMONELLOSEN

Allgemeines

Dank der Einführung des KAUFFMANN-WHITE-Schemas ist es möglich geworden, die Epidemiologie und Epizootologie der Salmonellosen Europas weitgehend aufzuklären. Der Wert einer konsequent durchgeführten, biochemischen und serologischen Feinanalyse der Salmonellen (Typendifferenzierung) wurde von HABS (1935) erkannt. Ungeachtet neuerer Bestrebungen um die Vereinfachung des diagnostischen K-W-Schemas muss der Epidemiologe trotz der fast unübersehbaren Vielfalt von *Salmonella*-Species, -Typen und -Varianten zunächst auch weiterhin an einer möglichst exakten Feindiagnostik festhalten, da ihre Preisgabe das epidemiologische Bild verwischen oder verzerren würde. Gerade das vergangene Jahrzehnt hat den grossen Wert des geübten Vorgehens erwiesen und die Erkennung vieler, bislang verborgener Zusammenhänge ermöglicht. In jüngster Zeit haben LILLENGEEN, ANDERSON (Lit. 21), BRANDIS, RISCHE & KRETZSCHMAR u.a. durch Einführung der Lysotypie die Feindiagnostik für die in Europa vorherrschenden Typen *S. typhi-murium, S. enteritidis, S. dublin, S. gallinarum-pullorum* usw. verbessert und dadurch neue Möglichkeiten der epidemiologisch-epizootologischen Analyse erschlossen. Da die oft komplizierte, aufwendige und zeitraubende Typendiagnostik der *Salmonella*-Stämme die Möglichkeiten der meisten diagnostischen Laboratorien überschreitet, sind auf Anregung KAUFF-MANN'S und unter Förderung der nationalen Gesundheitsdienste sowie der Weltgesundheitsorganisation praktisch in jedem Staate Europas n a t i o n a l e Salmonella-Zentralen entstanden, deren Arbeiten die Grundlagen für epidemiologische und epizootologische Studien liefern. Dem gleichen Zweck dient die Tätigkeit der Zentralen für *Salmonella*-Lysotypie.

Das zur *Salmonella*-Epidemiologie in Europa entstandene Schrifttum ist kaum mehr zu übersehen und kann in diesem Rahmen nicht im Einzelnen berücksichtigt werden. Der nachfolgende Versuch einer zusammenfassenden Darstellung verzichtet bewusst auf viele Einzelheiten. Der Interessierte sei u.a. auf die umfassenden Darstellungen von DRÄGER (1951, 1958), BUXTON (1957), FROMME (1959), COCKBURN, TAYLOR, ANDERSON & HOBBS (1962) sowie DOBINSKY (1962) verwiesen.

Herkunft der Salmonellen in Europa

Im Gegensatz zu den Erregern des Abdominaltyphus und der verschiedenen, weiter oben besprochenen Formen des menschlichen Paratyphus handelt es sich bei den übrigen Salmonellen um eine Vielzahl von Bio-, Sero- und Phagentypen. Hiervon sind bisher allein an die tausend Serotypen näher beschrieben. Ein beachtlicher Teil dieser Typen wurde erstmalig in Europa festgestellt; doch ist Europa selbst nicht als eigentliches Herkunftsgebiet der meisten dieser Keime anzusehen. Vielmehr handelt es sich häufig um Stämme, die aus den tropischen Ländern eingeschleppt wurden und noch werden. Die nach dem Ort der Erstisolierung benannten Keime haben demnach mit diesem Ort selbst meist nichts zu tun. Trotz europäischer und amerikanischer Namen lässt das Studium der Fundorte dieser Typen vielmehr die Deutung zu, dass diese oft bereits vor Jahrzehnten in Europa, Nord-, Mittel- oder Südamerika entdeckten Stämme in Wirklichkeit auch dort Fremdkeime waren, die wahrscheinlich schon damals mit Futtermitteln aus anderen Kontinenten in die verschiedenen Länder eingeschleppt wurden und sich dann in den Tierbeständen ihrer neuen Heimat eingenistet haben. Vor 1925 war die Zahl der in Europa heimischen *Salmonella*-Typen relativ gering. Dies konnte u.a. erst kürzlich vom Seniorautor (unveröffentlicht) bei der Nachuntersuchung der umfangreichen *Salmonella*-Stammsammlung des Instituts für Mikrobiologie der Tierärztlichen Hochschule in Hannover, die zwischen 1925 und 1935 angelegt worden war, bestätigt werden.

Die vorherrschenden und für Europa als e n d e m i s c h anzusehenden Typen sind *S. typhi-murium (B. enteritidis* Breslau*), S. chole-rae-suis (suipestifer)* (s. vorstehenden Abschnitt), *S. thompson, S. newport, S. enteritidis* (Gärtner), *S. dublin* (Kiel), *S. gallinarum-pullorum, S. anatum* und einige weitere. Erst ab 1930 nimmt in Europa, und zwar nicht nur als Folge besserer Diagnostik, die Zahl der isolierten *Salmonella*-Typen stetig (LERCHE & BARTELS), ab 1950 sogar rasch zu, was auf die ständige Einschleppung anderenorts endemischer Keime zurückgeht (COCKBURN et al., SEELIGER, WINKLE & ROHDE u.a. m.). Dieser Import blieb aber zunächst unerkannt.

Standort, Infektiosität und Symptomatik

Im Unterschied zu den oben besprochenen Seuchenerregern findet sich der primäre Standort der übrigen *Salmonella*-Typen im Tierreich. In Abhängigkeit von jeweils wechselnden Gegebenheiten verursachen diese Keime bei ihrer oder ihren Tierarten enterale bzw. septische Infektionen, auch Ausscheidertum, und gelangen erst sekundär durch Genuss infizierter Lebensmittel auf den Menschen. Die Aufnahme nur kleiner Keimmengen kann klinisch ohne Folgen bleiben, führt aber zu einer mehr oder minder lange dauernden Ausscheidung im menschlichen Kot mit allen sich daraus ergebenden weiteren epidemiologischen Folgen. Vielfach entwickelt sich jedoch nach oraler Aufnahme der Keime sowie ihrer Stoffwechsel- und Zersetzungsprodukte das Krankheitsbild der akuten, fieberhaften Salmonella-Enteritis bzw. Gastro-Enteritis, die klinisch als Brechdurchfall verläuft. Dieses Krankheitsbild wird meist als bakterielle Lebensmittelvergiftung (bacterial food poisoning, toxi-infection alimentaire) bezeichnet und dort, wo eine Meldepflicht für solche Erkrankungen besteht, als solche amtlich erfasst. In dieser Gruppe von Erkrankungen ist der Anteil von Salmonellosen (60% und mehr) hoch. Da diese Salmonellosen aber in nur wenigen europäischen Ländern getrennt gemeldet werden, ist ihre Häufigkeit bestenfalls schätzbar.

Manchmal verläuft die Erkrankung des Menschen unter dem Bild einer Paratyphus-ähnlichen Infektion. Solche Fälle werden dort, wo die amtlichen Meldungen nicht durch den Erregernachweis verifiziert werden, als Paratyphus registriert (vgl. Kapitel "Paratyphus C", Seite 264). Gelegentlich werden im europäischen Raum auch eitrige *Salmonella*-Infektionen des Menschen mit wechselnder klinischer Symptomatologie beschrieben, z.B. unter dem Bild einer Strumitis, Osteomyelitis, Otitis, Meningitis, eines Lungenabszesses und anderer akuter oder chronischer Infektionen (Lit. s. DOBINSKY).

Eine Anzahl von epizootologisch wichtigen *Salmonella*-Typen ist an bestimmte Wirte so adaptiert, dass ihr Übergang auf einen anderen Wirt, wie den Menschen, nicht oder nur ausnahmsweise, zu erkennbaren klinischen Folgen führt. So steht die Häufigkeit von *S. galli-narum-pullorum*-Befunden in den europäischen Geflügelbeständen in gar keinem Verhältnis zu den ausgesprochen seltenen Infektionen des Menschen mit diesem Typ (Popp, 1956; Seeliger, 1960). Ähnliches gilt, wenn auch mit Einschränkungen, für *S. abortus-ovis* und einige andere Aborterreger der *Salmonella*-Gruppe. Doch muss offenbleiben, ob nicht gewisse Schwierigkeiten des bakteriologischen Nachweises gerade solcher Erreger manchmal dafür verantwortlich sind, dass weniger menschliche Infektionen diagnostiziert werden, als tatsächlich vorkommen.

Das epidemiologisch wichtige Problem des unterschiedlichen Haftens von Salmonellen, d.h. der Ansiedlung, Vermehrung und Ausscheidung mit dem Kot, ist sowohl für den Menschen als auch für die in Europa befallenen Tierarten noch ungenügend erforscht. Die Beobachtungen in der vergangenen Dekade deuten vorerst darauf hin, dass es einem Grossteil der aus den Tropen eingeschleppten Stämme, Typen oder Arten offensichtlich bisher nicht gelungen ist, sich in Europa einzunisten. Viele *Salmonella*-Befunde blieben deshalb vereinzelt oder gar einmalig und damit epidemiologisch-epizootologisch ohne Bedeutung. Bei anderen Stämmen, Typen oder Arten zeigte sich jedoch eine ganz andere Entwicklung, da sich an ihr erstmaliges Auftreten im europäischen Raum eine Serie von Befunden — sowohl sporadischer als auch epidemischer Natur — oft über Jahre hinweg in einem oder mehreren Ländern anschloss. Beispiele: *S panama* (ab 1950 — vgl. Seeliger 1956, 1957), *S. bareilly* (ab 1952/53, vgl. Bonitz, 1953, 1957), *S. blockley* (ab 1955/56, vgl. Handloser, 1956; Klein, 1959) u.a.m. Es ist heute noch nicht zu übersehen, ob diese Entwicklung nur dem Zufall oder etwaigen Besonderheiten im Wirt-Gast-Verhältnis zuzuschreiben ist.

Neben den Stämmen, die nur einmalig gefunden werden, steht eine zahlenmässig bedeutsame Gruppe von Typen, die beim Tier — insbesondere Schlacht- und Haustieren — meist nur subklinische Infektionen mit Befall der Mesenteriallymphknoten verursacht (Bøvre, Buxton, Jensen, Kampelmacher et al., Linke, Smith). Solche Tiere sind klinisch unauffällig, und ihr Fleisch wird bei der Schlachtung in der Regel nicht beanstandet. Mehr als von krankgeschlachteten Tieren droht jedoch gerade hier Gefahr für den Menschen, da sich die Keime postmortal vermehren und dann durch Schmierinfektion in die Verarbeitungsbetriebe und dort auf andere Lebensmittel übertragen werden können, von wo aus dann der den Menschen betreffende Teil der Infektkette (s.u.) mit wiederum wechselnder Symptomatik beginnt.

Infektketten

Es würde zu weit führen, in diesem Abschnitt alle äusseren und inneren Infektketten im Sinne von GOTSCHLICH, HABS u.a. detailliert zu behandeln, umsomehr, als wesentliches aus anderen Kapiteln entnommen werden kann. In Fig. 5 ist der Versuch gemacht, das derzeit vorherrschende Infektionsschema für die Salmonellosen in Europa auf einen allgemeinen, vereinfachten Nenner zu bringen.

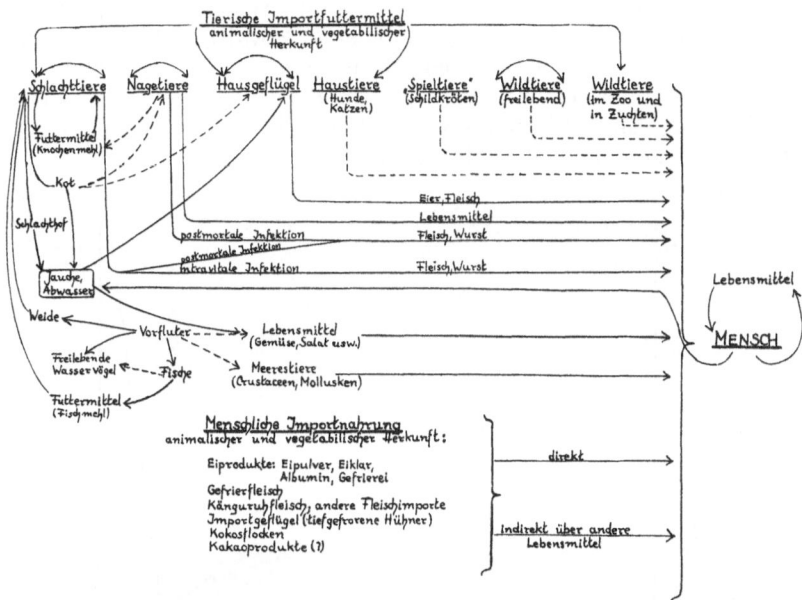

Fig. 5. Infektketten der Salmonellosen in Europa (vereinfacht).

Wesentliche Teile dieses Schemas dürften auch für andere Kontinente Gültigkeit haben; doch scheinen einige Infektketten für Europa — beispielsweise im Vergleich zu Nordamerika — charakteristisch zu sein. Der in Mittel-, West- und Nordeuropa ausgesprochen häufige Genuss von rohem Hackfleisch (Beefsteak-Tartar) bedingt z.B. eine besondere Gefährdung des Europäers durch intravital oder postmortal in das Rohfleisch gelangte Salmonellen. — Ähnliches gilt auch für den vor allem in Westeuropa häufigen Verbrauch von Enteneiern, zumal die Ente eines der wichtigsten *Salmonella*-Reservoire des Kontinents darstellt. Schon in den Dreissiger Jahren wurde deshalb in Deutschland versucht, die Verbreitung der Salmonellen durch solche Nahrungsmittel durch gesetzliche Maßnamen (Verordnung vom 24. 7. 1936) zu unterbinden. — Wegen der geringen Zahl der für solche Infektionen in Frage kommenden endemischen Typen waren früher die Infektketten meist einfach, d.h. auf jeweils einen Typ beschränkt und blieben übersehbar.

Das Infektionsrisiko war meist auf Einzelpersonen oder kleine Gruppen begrenzt. Dies gilt heute nur noch ausnahms-weise in abgelegenen Gegenden und kleineren Gemeinden, ausge-dehnt wohl nur noch in Südeuropa.

Nachdem schon Anfang der Dreissiger Jahre von KNORR fest-gestellt worden war, dass Eipulver chinesischer Herkunft reich-lich Salmonella-Keime enthielt, brachten die nordamerikanischen Eipulver-Importe ab 1942 in England und Wales (vgl. Fig. 6), ab 1944/45 auch auf dem Kontinent eine gewaltige Zunahme der Salmonella-Infektionen auf der Basis einer vorher seltenen und kaum beachteten Infektkette. Die Infektionen betrafen die Zivil-bevölkerung Englands, dann die anglo-amerikanischen Truppen und ihre Kriegsgefangenen, später weite Teile der Bevölkerung West-europas.

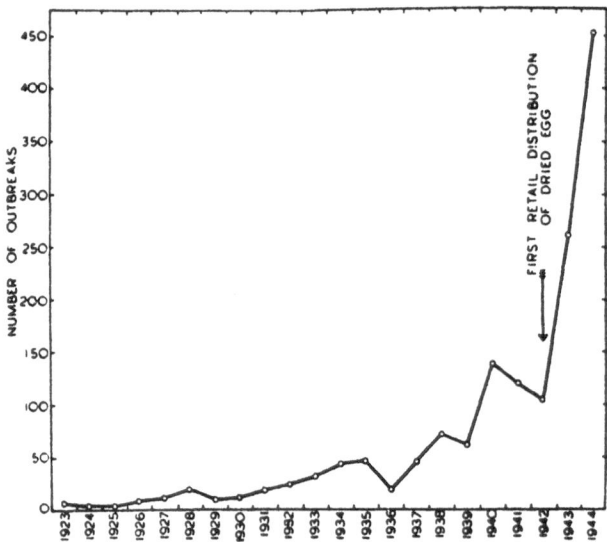

Fig. 6. Ausbrüche von Salmonella-Nahrungsmittelvergiftung in England und Wales 1923—1944. (Aus: Great Britain, Medical Research Council, H.M. Stationary Office, 1947).

Diese Entwicklung wiederholte sich ab 1950 mit der Zunahme der Importe von Eipulver, Flüssigeigelb, Eialbumin, Gefriervollei usw., z.T. aus China, z.T. aus anderen Ländern einschliesslich der USA (vgl. COCKBURN et al., DRÄGER, FEY & WIESMANN, FROMME, KELCH, BULLING u.a.). Unseres Wissens ist eine Verschlechterung der epidemiologischen Lage in einem solchen Ausmass wohl nur in Europa erfolgt, wo man die von verunreinigten Hühnereiprodukten ausgehenden Gefahren nicht rechtzeitig erkannte und sich die bisher übliche Küchentechnik der Zubereitung und Aufbewahrung als oft unzureichend erwies. Da es durch solche Eiprodukte auch zur Einschleppung von Salmonellen in Bäcker- und Konditoreibetriebe mit dem Befall von Crèmes, Speiseeis, Kuchenfüllungen usw. kam, ergab sich eine neue, ebenfalls für Europa — mit seinem reichlichen Genuss von leicht verderblichem Backwerk — charakteristische Infektkette. Diese wurde ab 1959 durch die Importe von Salmonellen-haltigen Kokosflocken bzw. Kokosraspeln aus Ceylon und Australien noch erweitert und kompliziert (vgl. GALBRAITH et al., TAYLOR: in COCKBURN et al., ANDERSON, GÄRTNER & BÖHLCK, WINKLE et al.). Auf Süsswaren bleiben *Salmonella*-Keime ungewöhnlich lange lebensfähig (KUDICKE).

Salmonellen-infizierte Schweinefleischimporte aus Südost-Europa (GÄSSLEIN, HEPP u.a.), Gefrierfleisch aus Südamerika, Känguruhfleisch aus Australien und tiefgefrorene Brathähnchen und Truthühner aus Übersee geben in jüngster Zeit den *Salmonella*-Infektketten Europas ein zusätzliches Gepräge. Dabei ist zu berücksichtigen, daß sich diese Infektketten keineswegs auf die aus dem Tierreich stammenden Salmonellen beschränken, sondern oft durch Typhus- und Paratyphusbakterien verursachte Krankheitsfälle eingestreut sind, woran deutlich wird, daß diese multipel verseuchten Lebensmittel nicht nur mit tierischen Salmonellen, sondern auch mit denen menschlicher Dauerausscheider verunreinigt werden. Dies lässt auf entscheidende Mängel bei der Herstellung schliessen.

Alle diese Infektketten werden jedoch überschattet durch eine seit 1950 ständig um sich greifende latente Verseuchung der Nutzvieh- und Geflügelbestände (Hühnerfarmen, Truthühner) durch *Salmonella*-haltige Importfuttermittel tierischer Herkunft. Diese geht auf importierte aussereuropäische Produkte zurück, vor allem Fischmehl, Blutmehl, Knochenschrot, Tierkörpermehl und durch die auf dieser Basis hergestellten Mineralstoffmischungen nebst Kraftfutter (BISCHOFF & ROHDE, BØVRE, BUXTON, CHRISTIE et al., COCKBURN et al., DOBINSKY, DIXON & WILSON, HARVEY & PRICE, NEWELL et al., RASCH, ROHDE & BISCHOFF, SEELIGER, TAYLOR, WINKLE & ROHDE u.a.m.). Diese anfangs verborgen gebliebene Salmonellose-Verbreitung führte in Europa zur Verseuchung vieler Schlachthöfe (McDONAGH & SMITH), der Fleischereien und Grossbetriebe für Geflügel-, Fleisch- und Wurstverarbeitung (DIXON & POOLEY, KAMPELMACHER et al.) und in der Folge zu zahlreichen sporadischen Fällen und epidemischen Ausbrüchen unter der Bevölkerung. In den Betrieben kam es zu einer hohen Ausscheiderquote beim Personal. Durch Rohverfütterung intravital oder postmortal infizierten Pferdefleisches traten *Salmonella*-Infektionen beim Hund (vgl. GALBRAITH et al.) sowie bei Wildtieren zoologischer Gärten auf (vgl. DOBINSKY).

Ausgehend von der latenten Verseuchung der Tierbestände kam

es gelegentlich auf recht verschlungenen Wegen zu Massenausbrüchen beim Menschen. Als Beispiel für mehrere sei die ausgedehnte *S. bareilly*-Epidemie im norddeutschen Raum mit über 10 000 klinischen Fällen erwähnt. Sie wurde durch Camembert-Käse verursacht, der durch *Salmonella*-haltigen Leim der aufgeklebten Etiketten infiziert worden war. Dieser Leim war von einer *S. bareilly*-Ausscheiderin verunreinigt worden, die auf einem Gut lebte und arbeitete, auf dem Schweine mit *S. bareilly*-haltigem Proteinfutter aufgezogen worden waren (BONITZ 1953, 1957).

Diese neuen, exogen bedingten Infektketten beim Schlachttier und Geflügel überlagern und durchmischen sich in Europa mit jenen der endemischen Salmonella-Typen und gestalten das Bild oft völlig unübersichtlich.

Neben *Salmonella*-verseuchten Futtermitteln tierischer Herkunft werden aber auch solche vegetabilischen Ursprungs zunehmend nach Westeuropa importiert, nicht nur die bereits erwähnten Kokosflocken, sondern auch Tapioka- und Maniokmehl (SONTAG), extrahierte Sonnenblumen- und Baumwollsamen aus der UdSSR (JENSEN, 1958) sowie gemahlene Nuss-, Sonnenblumen-, Kokosnuss- und Sojabohnenkuchen, ferner Sojabohnen und Rapsmehl. Bei Kontrolluntersuchungen erwies sich das aus Mexiko stammende Baumwollsamenkuchenmehl als besonders häufig Salmonellen-haltig (HAUGE & BØVRE).

Diese vegetabilischen Proteinkraftfutter stellen ebenfalls einen nicht zu gering einzuschätzenden Faktor bei der Verseuchung von Haustieren dar. — Als Sonderfall, der wohl für andere Kontinente bedeutsamer ist als für Europa, sei die Salmonellose-Verseuchung von Schildkröten erwähnt, ein hierzulande von Kindern begehrtes "Spieltier". Kontrollen bei mehreren hundert Schildkröten aus Reptilienhandlungen und Tierparks ergaben eine 36%ige Verseuchung mit Salmonellen, insbesondere bei griechischen Landschildkröten (KIESEWALTER et al.).

Alle diese Infektketten münden auf den verschiedensten Wegen in das Abwasser und die Vorfluter, deren Salmonellen-Gehalt in Europa ein Spiegelbild der ansteigenden Verseuchung von Mensch und Tier liefert (Salmonella-Kataster) (vgl. JØRGENSEN, KAMPELMACHER, POHL, POPP, SCHAAL, SCHMIDT & LENK, SEELIGER, SEELIGER et al., u.a.m.). Über die Vorfluter haben sich weitere Infektketten gebildet, die sowohl Infektionen des Schlachtviehs über Weidegebiete, der Vogelwelt — vor allem Möven usw. — (vgl. HENZE; KLOSE et al., RASCH, PETZELT & STEINIGER, STEINIGER), der Crustazeen im Küstengebiet (vgl. DOBINSKY, STAACK) und selbst der Süsswasserfische, wie Karpfen (MATZKE), in Fischteichen einschliessen.

Endlich sei erwähnt, dass Salmonellen in wechselnder Häufigkeit auch in Kunstdünger gefunden wurden, dem zwecks Anreicherung tierische Stoffe zugesetzt worden waren (GALBRAITH et al., WALKER). Doch besteht begründeter Anlass zur Annahme, dass eine Gefährdung von Mensch und Tier nur dann gegeben ist, wenn solche Düngemittel im Laden zum Verkauf feilgeboten werden, in dem sie andere Lebensmittel verunreinigen können. Die relativ lange Lebensdauer von Salmonellen im Freiland (vgl. MAIR & ROSS) zeigt gleichwohl, dass die Gefahren nicht unterschätzt werden dürfen.

Seit jeher haben in Europa Nagetiere mit ihren eigenen *Salmonella*-Epizootien, aber auch als Überträger, eine erhebliche Rolle bei dem Salmonellose-Infektionszyklus des Stallviehs und auf dem Umweg über Lebensmittel bei menschlichen Infektionen gespielt. Die bisherigen Befunde geben noch keinen Anlass zur Annahme, dass sich die im vergangenen Jahrzehnt neu eingeschleppten *Salmonella*-Typen unter den Nagetieren eingenistet haben.

Insgesamt erweisen sich somit die Infektketten der Salmonellosen in Europa als ungewöhnlich komplex und verschlungen mit mancherlei Eigenarten, die sich aus der wirtschaftlichen und sozialen Struktur des Kontinents ergeben. Von besonderem Interesse ist jedoch die Tatsache, dass es gerade die wohlhabenden Länder mit einer dichten Bevölkerung sind, die sich neuerdings durch eine besondere Stärke der Verseuchung und zunehmende Vielfalt der Infektionsmöglichkeiten auszeichnen.

Vorkommen und Verbreitung der Salmonellosen

Obwohl für viele Länder der hier besprochenen Teile von Europa die gleichen Gesetzmässigkeiten der Salmonellose-Verbreitung zutreffen, zeigen sich vor allem bei einer Gegenüberstellung von Nord-, Mittel- und Westeuropa mit den südlichen Ländern beträchtliche Unterschiede.

Nordeuropa: Obwohl MÜLLER schon 1949–1952 in Dänemark auf die Gefahren aufmerksam gemacht hatte, die sich aus dem von ihm erstmalig festgestellten Salmonellengehalt in importierten Fleisch- und Knochenmehlen usw. ergeben, ist es trotz der sofort einsetzenden Abwehrmassnahmen nicht gelungen, die Ausbreitung der Salmonellen beim Schlachtvieh, besonders Schweinen, und unter dem Geflügel zu verhindern (vgl. JENSEN, JØRGENSEN). Dies ist, wie sich aus der zunehmenden Zahl von Salmonellose-Erkrankungen und der dabei isolierten Serotypen zwischen 1950 und 1958 ergibt (vgl. Tabelle VI) nicht ohne Folgen für den Menschen geblieben. *S. typhi-murium* beherrscht das epidemiologische Geschehen, wobei regelmässig eine Anzahl von Fällen in den Monaten Januar und Februar aufzutreten pflegt, wahrscheinlich ausgehend von länger gelagerten Eiern (TULINIUS, pers. Mitteilung).

Ein ganz ähnliches Bild zeichnet sich in Schweden ab, wo in den

Tabelle VI.

Salmonellosen des Menschen in Dänemark 1950—1958
(nach Jørgensen, 1962).

Jahr	Zahl der Fälle			Akute infektiöse Gastroenteritis	
	Gesamt	S. typhi-murium	Andere Serotypen	Gesamt	pro 100.000
1950	105	67	11 *(newport)*, 27 andere Typen	44,740	1040
1951	73	50	23	50,885	1180
1952	143	94	24 *(newport)*, 25 (8 Typen)	47,212	1010
1953	494	458	23 *(jena)*, 13 (9 Typen)	48,574	1110
1954	584	543	11 *(jena)*, 30 (12 Typen)	53,866	1223
1955	1332	1224	20 *(jena)*, 88 (10 Typen)	59,949	1350
1956	456	392	64 (22 Typen)	62,167	1392
1957	529	487	42 (16 Typen)	49,961	1113
1958	334	291	43 (17 Typen)	65,761	1456

Jahren 1956/57 312 *Salmonella*-Stämme (15 Serotypen) unter Tier-beständen festgestellt und weitere 161 (32 Serotypen) aus Import-futtermitteln gezüchtet wurden. Bei der bakteriologischen Import-kontrolle von Rind-, Kalb- und Pferdefleisch, Tierorganen usw. wur-den 230 Stämme (18 Serotypen) isoliert (Rutqvist & Thal). Vor allem beim Geflügel (Thal et al.) tritt zu dieser Zeit eine deutliche Vermehrung der Infektionen mit anderen Typen als *S. gallinarum-pullorum* in Erscheinung (z.B. *S. montevideo*).

Dem entspricht eine vergleichbare Situation in Norwegen (Bøvre).

Auch in Finnland ist eine erhebliche Zunahme des *Salmonella*-Befalls der Schlacht- und Nutztiere zu verzeichnen. Von 1950–1960 hat sich die Zahl der beim Tier nachgewiesenen Infektionen etwa verzwanzigfacht. Während bei Kälbern *S. typhi-murium*-Infektionen vorherrschen, wurden bei Schweinen und beim Geflügel zunehmend neue Typen (entsprechend denen im Importfutter) gefunden. Bei Pelztieren ist die Salmonellose als Sekundärinfektion weit verbreitet (Stenberg & Vasenius). Insgesamt scheint die Situation (relative Typenarmut) in Finnland zur Zeit aber noch günstiger zu sein als in den anderen Teilen Skandinaviens.

Britische Inseln und Nordirland: Die Salmonellose-Epi-zootologie in der Tierwelt der britischen Inseln wird ausführlich von Buxton sowie Cockburn et al. behandelt. Pferde und Esel sind nur von relativ wenigen, für diese Tierarten spezifischen Salmonellen befallen, vorzugsweise von *S. abortus-equi;* doch sind in jüngster Zeit weitere Typen hinzugekommen. Beim Kalb und Rind überwiegen, distriktsweise verschieden, *S. typhi-murium*, *S. enteritidis* und *S. dublin*. Auch hier hat sich die Typenhäufigkeit erheblich vermehrt (1957–1960 beim Rindvieh 24 weitere Typen). Beim Schaf und der Ziege überwiegen Infektionen durch *S. abortus-ovis* und *S. typhi-murium*. Beim Schwein wurden in dieser Periode 52 verschiedene Typen nachgewiesen. Eine deutliche Zunahme ist auch bei Hunden und Katzen festzustellen. Neben *S. gallinarum-pullorum* und *S. typhi-murium* wurden bei Hühnern zwischen 1957 und 1960 27 weitere, bei Truthühnern 24 und bei Enten 9 verschiedene Serotypen festgestellt. Die latente Verseuchung der Schweine ist hoch, nach SMITH 12% von 500 untersuchten Tieren, ebenfalls die der Hunde und Katzen (4,5% und 2,5% bei jeweils 200 Tieren). In den Mesenteriallymphknoten und Innereien von Schweinen wurden 17 verschiedene Serotypen nachgewiesen, am häufigsten *S. typhi-murium*, *S. anatum* und *S. cholerae-suis*. *S. typhi-murium* scheint besonders bei den Truthühnern stark verbreitet zu sein. In zwei Verarbeitungsbetrieben fanden DIXON & POOLEY vorwiegend den Phagtyp 1a, var. 1.

Das von CALLOW eingeführte neue Lysotypieschema für *S. typhi-murium* wurde von ANDERSON (vgl. COCKBURN et al.) in epidemiologischen Untersuchungen angewandt. Dabei zeigte sich, dass der Phagtyp 14 für 74% der Infektionen unter Hühnchen und Geflügel verantwortlich ist, während die Typen 20a (28%), 15a (24,2%) und 18 (15,0%) bei Kälbern und Rindern überwiegen, bei denen der Phagtyp 14 nicht gefunden wurde. Dies deutet darauf hin, dass Infektketten zwischen Geflügel und Rindvieh nicht in grösserem Ausmass bestehen.

Nach TAYLOR (vgl. auch COCKBURN et al.) wurden von 1958–1960 allein in tierischen Futtermitteln über 1000 mal Salmonellen (97 verschiedene Serotypen) nachgewiesen und in Kokosflocken (1959 bis Juni 1961) 57 Serotypen. In letzteren fanden sich am häufigsten *S. waycross* (157 mal), gefolgt von *S. paratyphi B* (154 mal) und *S. bareilly* (122 mal). Nicht weniger eindrucksvoll sind die 1956–1960 erhobenen Befunde an inländischen und importierten Eiprodukten:

Herkunftsland:		
Australien	5784 positive Befunde	
China	2021 positive Befunde	
England	405 positive Befunde	insgesamt
Amerika	315 positive Befunde	56 verschiedene
Dänemark	153 positive Befunde	Serotypen,
Holland	107 positive Befunde	am häufigsten
Neuseeland	75 positive Befunde	*S. typhi-murium*
Canada	61 positive Befunde	
Südafrika	31 positive Befunde	
Schweden	29 positive Befunde	

Diese Befunde finden ihre Parallele bei den menschlichen Infektionen. 1953–1960 wurden beim Menschen nicht weniger als 140 verschiedene Serotypen nachgewiesen, davon 45 allerdings nur einmal. Dabei nahmen bis 1955 die durch S. typhi-murium bedingten Erkrankungen (stets als Vorkommen und nicht als Einzelfälle registriert) zu, danach ab. Doch wird dieser Rückgang durch das Hervortreten anderer Serotypen kompensiert (1954: 15%, 1960: 28%). Die Zahlen der Vorkommen aus dieser Periode sind in Tab. VII wiedergegeben.

Tabelle VII.

Salmonella-Infektionen in England und Wales 1953—1960* (zusammengestellt nach Angaben des englischen Gesundheitsdienstes und ergänzt durch Cockburn et. al. 1962).

Salmonella-Vorkommen**	***1953	1954	1955	1956	1957	1958	1959	1960
Total	3171	3576	5383	4412	4278	4952	4864*	3925*
S. typhi-murium	2438	3038	4276	3245	2973	3406	3241	2943

*   = Zahlen durch Weglassung einiger Einzelbefunde nicht vollständig.
**   = Nicht gleich Zahl der Einzelfälle.
*** = Die Zahlen von 1953—1958 enthalten einige Befunde von symptomlosen Ausscheidern.

Die Zahl der symptomlosen menschlichen Ausscheider ist beachtlich (2,5 pro 1000). Eine Aufklärung der Infektketten wird dadurch erschwert, dass die gleichen Serotypen sowohl bei Mensch und Tier als auch in Lebensmitteln, Eiern und Eiprodukten sowie in Tierfutter und Düngemitteln nachgewiesen wurden, z.B. S.bovis-morbificans, S. muenchen, S. senftenberg, S. anatum, S. derby, S. heidelberg, S. newport und S. typhi-murium (vgl. auch Niederlande und Deutschland). In einigen Fällen half bei der Aufklärung der Zusammenhänge die Lysotypie, z.B. bei einem durch Phagtyp 20a verursachten Ausbruch (Anderson). Während hier die herkömmliche Typenbestimmung durch Antigenanalyse am Ende ihrer Möglichkeiten ist (vgl. S. 266), bewährte sie sich in anderen Fällen, z.B. bei S. saint-paul-Infektionen des Menschen durch infiziertes Geflügel (Dixon & Pooley 1961, 1962, Galbraith et al., 1962).

Aus einem Bericht von Newell et al. ist zu folgern, dass in Nordirland im Prinzip ähnliche epidemiologische Verhältnisse herrschen wie in England und Wales. Die Zahl der gemeldeten Fälle bei bakteriellen Lebensmittelvergiftungen erreichte 1956 mit 126 ihren Höhepunkt (1958/59 77 bzw. 70), während in Schottland der Gipfel im Jahre 1957 (1247) lag (1959: 777).

Mittel- und Westeuropa: Die Salmonellose-Befunde in der

Bundesrepublik Deutschland liegen bis 1960 fast lückenlos vor (SEELIGER, SEELIGER et al., HOFMANN et al., ROHDE et al.). Dazu kommen die Befunde aus Westberlin 1948–1957 (MARCUSE et al.) und für 1952–1954 (MEYER-OSCHATZ) aus den ost- und mitteldeutschen Gebieten. Die Epidemiologie der bakteriellen Lebensmittelvergiftung (vorzugsweise Salmonellosen) wird von ANDERS behandelt.

Die Gesamtsituation ähnelt weitgehend der, die weiter oben für die britischen Inseln beschrieben wurde. Deutschland ist als Importland für Lebensmittel weitgehend auf die Einfuhr ausländischer und überseeischer Produkte angewiesen und hat dementsprechend in derselben Weise wie die westlichen Nachbarländer unter einem zunehmenden Salmonellenbefall zu leiden. Dazu kommt eine um sich greifende Verseuchung der Tierbestände, zumal die Bundesrepublik ihren Bedarf an Tierfuttermitteln aus mindestens 24 verschiedenen Ländern deckt (vgl. DOBINSKY, LENTZE, WINKLE & ROHDE u.a.m.) so 72.766 t Futtermittel im Jahre 1956 und 91.115 t 1957 gegenüber nur 3005 t im Jahre 1948. Dementsprechend vermehrte sich die Zahl der *Salmonella*-Infektionen bei Mensch und Tier gewaltig. Während von 1937–1941 bei Tieren insgesamt 17 verschiedene *Salmonella*-Typen (einschl. *S. paratyphi B*) nachgewiesen wurden (fast ausschliesslich die damals in Mitteleuropa heimischen Erreger – LERCHE & BARTEL), nahm in den Jahren 1953 bis 1959 die Typenhäufigkeit zu:

1953	16 verschiedene Serotypen
1954	19 verschiedene Serotypen
1955	37 verschiedene Serotypen
1956	37 verschiedene Serotypen, davon 7 je einmal
1957	41 verschiedene Serotypen, davon 14 je einmal
1958	55 verschiedene Serotypen, davon 17 je einmal
1959	49 verschiedene Serotypen, davon 14 je einmal

Dabei machen jedoch die 6 vorherrschenden Serotypen *S. typhimurium, S. dublin, S. gallinarum-pullorum, S. abortus-ovis, S. choleraesuis* und *S. enteritidis* jährlich rund 90% der bei 3000 bis 4000 Erstisolierungen liegenden Gesamtzahlen aus. Die überwiegende Mehrzahl der Infektionen durch andere Typen bezieht sich auf Routinebefunde bei der bakteriologischen Fleischkontrolle bzw. latent infizierten Tieren. An der Spitze liegen stets die Serotypen *S. typhimurium* und *S. dublin* mit jeweils über 30% aller Erstisolierungen.

Dem Typenreichtum bei den Tieren entspricht die Vielzahl der aus Importfuttermitteln isolierten Stämme (vgl. BISCHOFF & ROHDE, DOBINSKY, FROMME, SEELIGER et al., WINKLE & ROHDE u.a.). Um einer weiteren Verseuchung der Tierbestände entgegenzuwirken, wurde von der Deutschen Bundesregierung am 14.12.1956 eine *Verordnung über die Ein- und Durchfuhr von Futtermitteln tierischer*

Herkunft aus dem Ausland erlassen, wonach nur solche Produkte eingeführt werden dürfen, die nach der Trocknung einem Salmonellen abtötendem Erhitzungsverfahren unterzogen wurden. Werden bei bakteriologischer Kontrolle am Platz der Einfuhr Salmonellen nachgewiesen, müssen die Importwaren unter amtlicher Aufsicht zuerst ein Erhitzungsverfahren durchlaufen, das Salmonellen vernichtet. Obwohl damit eine Infektionsquelle verstopft werden konnte, haben sich die endemischen und hinzugekommenen Salmonellen so ausgebreitet, dass auch die im Inland fabrizierten Tierfuttermittel tierischer Herkunft (Knochenfutter, Tierkörpermehl, Mineralstoffgemische, Hundekuchen usw.) zunehmend Salmonellen enthalten (Lit. s. DOBINSKY, SONTAG). — Als Sonderfall sei noch der 1960 gelungene Salmonellen-Nachweis in ostasiatischen Entenfedern erwähnt (HOFMANN et al.). Die Autoren folgern, dass unerhitzte Federnabfälle weder als Düngemittel noch zur Herstellung von Pressplatten abgegeben werden sollten.

Die Häufigkeit menschlicher Salmonellosefälle hat in der Bundesrepublik, West-Berlin, aber auch in Mitteldeutschland seit 1950 ein vorher ungeahntes Ausmass erreicht. Einzelnen Epidemien mit mehreren tausend Fällen (z.B. 1950 *S. panama*-Epidemie in Nordrhein-Westfalen und im Rheinland, 1950 *S. london*-Epidemie in Berlin, 1953 *S. bareilly*-Epidemie in Norddeutschland, 1953 *S. kirkee*-Epidemie in Hamburg u.a.m.) (Lit. s. FROMME) stehen zahllose kleinere Gruppenerkrankungen und Einzelfälle gegenüber, die amtlich nur teilweise erfasst wurden, da die infektiöse Enteritis erst seit 1960 meldepflichtig ist und sporadische Fälle von Nahrungsmittelinfektionen wohl nur selten registriert werden.

Die für 1953–1959 vorliegenden Sammelstatistiken aus 60 humanmedizinischen Untersuchungsstellen repräsentieren etwa 95% der bakteriologisch gesicherten Fälle (vgl. Tab. VIII).

Tabelle VIII.

Menschliche *Salmonella*-Infektionen (Erkrankungsfälle und Ausscheider - ohne *S. typhi*, *S. paratyphi* A und B) in der Bundesrepublik Deutschland 1953—1959 und Westberlin.

Jahr	Fallzahl	*S. typhi-murium*	*S. enteritidis*	Zahl der anderen Typen
1953	n.b.	875	307	38
1954	n.b.	975	456	53
1955	n.b.	1674	633	51
1956	6808	2610 (38%)	402 (5,9%)	80
1957	5514	1919 (34%)	296 (5,4%)	84
1958	4843	1988 (41%)	329 (6,8%)	79
1959	5629	1941 (34%)	248 (4,4%)	90

n.b. = nicht bekannt.

Auffallend ist in Deutschland die starke Zunahme der *S. typhimurium*-Infektionen ab 1955, die von 1956–1959 mehr als ein Drittel aller Einzelfälle ausmachen. Der Anteil der *S. enteritidis*-Infektionen *(S. dublin)* mit 4,4–6,8% ist ebenfalls relativ konstant. Da die einzelnen Vorkommen nicht gesondert registriert wurden, muss offenbleiben, ob die Anteile der Vorkommen den Befundzahlen entsprechen. Wahrscheinlich hat sich aber auch hier eine Änderung vollzogen, indem die absolute Zahl der Vorkommen für die beiden vorherrschenden Typen zugunsten der übrigen Typen, die sich gegenüber 1953 mehr als verdoppelt haben, etwas zurückgegangen sein dürfte. Die epidemiologische Auswertung einzelner Vorkommen wird dadurch erschwert, dass menschliche Infektionen z.T. symptomlos verlaufen sind, z.T. auch durch Mehrfachinfektionen mit verschiedenen *Salmonella*-Typen bedingt waren (SEELIGER, WINKLE & ROHDE). Die 1956 erlassene *Verordnung zum Schutze gegen Infektionen durch Erreger der Salmonella-Gruppe in Eiprodukten* hat keinen wesentlichen Einfluss auf die Salmonellose-Morbidität der Bevölkerung gehabt, wenn man vielleicht von den Paratyphus-B Infektionen absieht, die seitdem deutlich zurückgegangen sind.

Wenn genaue Einzelheiten über die Situation in Mitteldeutschland (DDR) auch nicht bekannt wurden, so deutet doch der 1952/54 betreffende Bericht von MEYER-OSCHATZ für 23 Institute darauf

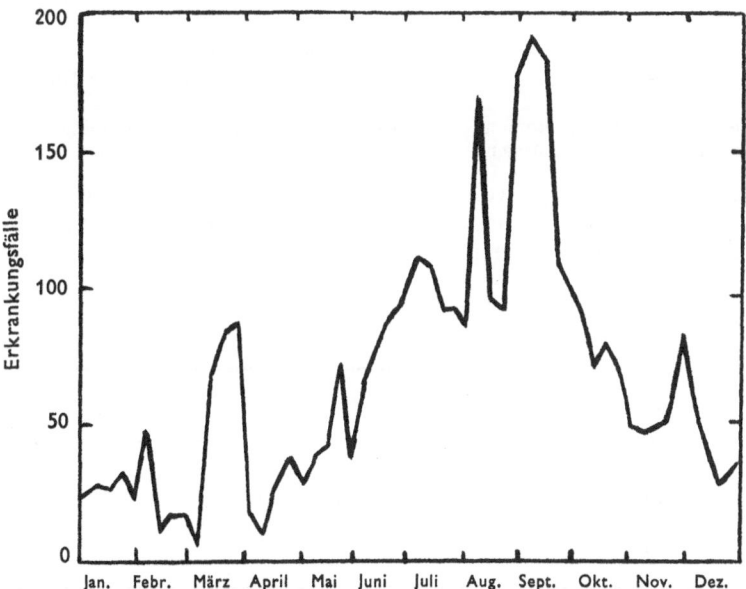

Fig. 7. Erkrankungshäufigkeit an "bakterieller Lebensmittelvergiftung" in der Bundesrepublik Deutschland, aufgeschlüsselt nach den einzelnen Wochen des Jahres 1959 (nach W. ANDERS, *Bundesgesundheitsblatt* **3**, *243* (1960).

hin, dass auch in diesem Teil Deutschlands eine Zunahme des Sal-
monellose-Befalls stattgefunden hat; in diesem Sinne sprechen auch
die Typisierungsbefunde aus Ost-Berlin (KIESEWALTER).

Wenngleich sich die *Salmonella*-Infektionen auf alle Jahreszeiten
verteilen, so ist eine Häufung der Einzelfälle und Vorkommen im
Spätsommer und Herbst unverkennbar. Diese jahreszeitliche
Gebundenheit geht der Erkrankungshäufigkeit an bakterieller
Lebensmittelvergiftung parallel (Fig. 7).

Die von Mensch und Tier stammenden Salmonellen finden sich in
wechselnder Häufigkeit und Verteilung im Abwasser und in den
Vorflutern, in Deutschland meist Flüssen und Flussmündungen
sowie Küstengebieten, wieder, deren starke Verseuchung durch
Stichprobenuntersuchungen in verschiedenen Landesteilen erforscht
wurde (POPP, SCHMIDT & LENK, STEINIGER, STAHN, weitere Lit. s.
DOBINSKY). Der wiederholte Nachweis von Typen, die in den jeweiligen Berichtsperioden weder vom Menschen noch vom Tier isoliert
werden konnten, im Abwasser beweist, dass die tatsächliche Verseuchung noch grösser ist, als die bakteriologische Befunde besagen.
Die in Berlin von SCHMIDT & LENK durchgeführten Abwasseranaly-
sen erbrachten, dass aus dem Ergebnis die Durchseuchung einer Bevölkerung zwar qualitativ, aber nicht quantitativ ermittelt werden
kann. DOBINSKY teilt Deutschland aufgrund der hydrogeologischen
Verhältnisse und bezüglich des Auftretens von *S. dublin* und *S.
typhi-murium* in drei Zonen ein:

Zone 1: starkes Vorkommen: Norddeutschland, nördlich Hannover und Osna-
                              brück, besonders Küstengebiet der Nord- und
                              Ostsee.
Zone 2: mässiges Vorkommen: südlich dieses Raumes bis zur Mainlinie.
Zone 3: geringes Vorkommen: südlich des Mains bis zu den Alpen.

Eine im Prinzip wohl ähnliche Entwicklung spielte sich in
Österreich ab, wo nach KNIEWALLNER folgende Verteilung bei
Lebensmittelvergiftungen festgestellt wurde:

Tabelle IX.

Salmonellose-Häufigkeit beim Menschen in Österreich.

Jahr	*S. paratyphi B*	Infektionen durch andere Typen
1954	755	64
1955	563	135
1956	558	336
1957	924	278

Vorerst zeigen aber die beim Tier erhobenen Befunde, 1951–57,
eine erheblich geringere Typenhäufigkeit als in Deutschland.

Demgegenüber hat sich in der Schweiz die Zahl der menschlichen Infektionen von 1952 bis 1959 verdoppelt bis verdreifacht (1952: 171, 1956: 558, 1959: 396). Doch meinen FEY & WIESMANN, dass die wirkliche Morbidität um ein Vielfaches höher gewesen sein dürfte. Ursächlich dürften die bereits geschilderten Gegebenheiten in gleicher Weise wirksam geworden sein, zumal auch die Schweiz ein Importland für Lebensmittel und Tierfutter ist.

Nicht weniger Beachtung wurde dem Problem der Salmonellose-Epidemiologie in den westeuropäischen Küstenländern geschenkt, vor allem in den Niederlanden. Aus der wachsenden Zahl von Berichten seien hier lediglich die zusammenfassenden Darstellungen von KAMPELMACHER et al. herausgegriffen.

Fig. 8 gibt einen Eindruck von der zunehmenden Ausbreitung der

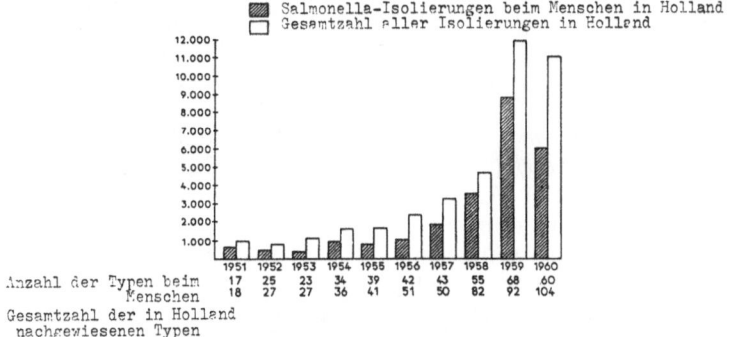

Fig. 8. Salmonella-Isolierungen in den Niederlanden 1951—1960 (nach KAMPEL-MACHER, GUINÉE & CLARENBURG, Zbl. Bakt. I Orig. 185, 490—502 (1962)).

menschlichen Salmonellosen in diesem Land. Wie Tabelle X ausweist, macht S. typhi-murium ab 1956 50 und mehr Prozent aller Fälle aus, während ihr Anteil am Beginn dieser Dekade — ebenso wie in Deutschland — nur etwa ein Viertel aller Fälle einschloss. Auf diesem relativ kleinen Raum wurde innerhalb von 10 Jahren aus menschlichen Ausscheidungen rund ein Achtel aller bisher bekannten Salmonella-Typen isoliert.

Bei Rindern und Kälbern wurden in der Zehnjahresperiode 1951–1960 32 verschiedene Serotypen gefunden. Davon machte S. dublin 1951 99%, 1959 und 1960 aber nur noch 96 bzw. 92% aus. Demgegenüber erhöhte sich der S. typhi-murium-Anteil von 0,8 auf 6,0%. Die latente Verseuchung der Schweinebestände stellt für Holland ein ernstes Problem dar. In der Berichtszeit wurden von 1951 bis 1960 bei diesen Tieren 34 verschiedene Typen gefunden. 1951 waren es jährlich 4, 1960 nach der Intensivierung der Untersuchungstätigkeit aber 573 positive Befunde. 60% wurden durch

Tabelle X.

Jährlicher prozentualer Anteil der beim Menschen vorherrschenden *Salmonella*-Typen in Holland (1951—1960) (nach KAMPELMACHER et al. 1962).

	1951	1952	1953	1954	1955	1956	1957	1958	1959	1960
S. typhi-murium	23	12	20	33	36	48	53	62	56	55
S. paratyphi B	20	20	14	9	14	11	14	6	5	5
S. bareilly	14	37	18	21	5	15	4	4	2	2
S. heidelberg	—	0,2	—	4	3	1	2	4	5	6
S. bredeney	0,3	0,4	0,2	2	4	1	1	0,9	6	6
S. newport	0,4	0,2	0,2	0,9	0,4	2	4	6	5	2
S. dublin	1	3	8	3	4	2	1	0,9	1	0,9
S. typhi	3	8	8	2	8	3	2	1	0,9	1
S. muenchen	4	0,9	0,4	4	1	1	1	1	1	0,8
S. bovis morbific.	20	2	—	4	1	0,3	4	1	2	2
S. enteritidis	0,9	2	9	1	4	3	1	0,7	0,6	0,2
S. senftenberg	0,1	4	0,8	0,8	1	0,3	0,1	0,2	0,03	0,03
S. stanley	9	0,9	—	0,1	1	0,2	0,1	3	0,2	0,08

die Typen *S. typhi-murium* (29,5 %), *S. dublin* (20,9%) und *S. cholerae-suis* var. *kunzendorf* (10,6%) gestellt, die übrigen 40% durch 31 weitere Typen. Der Aufklärung dieser Verseuchung dienten ausgedehnte Kontrollen in Schlachthäusern (KAMPELMACHER et al. 1961, 1962). Nicht weniger als 80% eines ausgewählten Bestandes von 120 Tieren erwiesen sich bereits vor dem Tode als verseucht, 85% von 115 Tieren nach der Schlachtung. Ursächlich geht diese Entwicklung auf die Verfütterung von Tierfuttermitteln verschiedener Provenienz zurück, vor allem auf importierte Tier- und Fischmehle, aus denen am häufigsten *S. binza, S. newington, S. oranienburg, S. senftenberg* und *S. anatum* gezüchtet wurden, ferner noch 74 weitere Serotypen.

Demgegenüber bietet sich beim Geflügel — im Gegensatz zu anderen Ländern — ein wesentlich günstigeres Bild, zumal es gelungen ist, die *S. gallinarum-pullorum*-Verseuchung unter Kontrolle zu bringen. Die vorübergehend häufigen Infektionen der Hühner durch *S. bareilly* sind inzwischen erheblich zurückgegangen. Ein Rhamnose-negativer Biotyp von *S. typhi-murium*, ferner *S. enteritidis* var. *essen, S. bareilly, S. anatum* und *S. meleagridis* wurden aber häufig von Enten isoliert. Nach CLARENBURG & VINK (zit. nach KAMPELMACHER et al.) betrug 1950–1952 die Befallsquote bei Enten rund 10% von 600 untersuchten Tieren, bei Hühnern (2000 Tieren) aber nur 0,4%. Holländische Hühnereiprodukte waren andererseits zu einem hohen Prozentsatz von Salmonellen befallen (*S. bareilly* 32,5%, *S. typhi-murium* 27% der untersuchten Proben). Doch haben hier die letzten Jahre, bedingt durch bessere Verarbeitungsmethoden der Exportindustrie, einen wesentlichen Rückgang gebracht. Dies

gilt jedoch nicht für Enteneiprodukte, die nach wie vor häufig
S. typhi-murium enthalten.

Eine ähnliche Situation findet sich auch in Belgien, wo nach
VAN OYE & LAFONTAINE innerhalb der letzten Jahre 90 verschiedene Salmonella-Typen nachgewiesen wurden, davon 65 beim Menschen, 18 bei Tieren und 47 in Lebens- und Futtermitteln. Doch ist,
wie die Autoren betonen, die Herkunft der menschlichen Salmo-
nellose-Fälle in diesem Lande noch nicht sicher aufgeklärt, wenngleich aufgrund der Typendiagnostik der gezüchteten Stämme die
Vermutung naheliegt, dass Schlachttiere und Geflügel sowie die
bereits erwähnten Lebens- und Futtermittel ursächlich anzuschul-
digen sind. Eine grosse Zahl von Typen wurde aber bisher ausschliesslich vom Menschen isoliert, zwei davon (S. mikawasima und S.
mampeza) bei Ausscheidern, die aus dem Congo bzw. Brasilien
zurückgekehrt waren. Wie in den übrigen europäischen Ländern ist
S. typhi-murium der vorherrschende Erreger menschlicher Infektionen.

Kaum anders ist es um die Salmonellose-Verseuchung in
Frankreich bestellt; denn auch hier weisen die Berichte der
Pariser Salmonella-Zentrale eine ständige Vermehrung der gefunde-
nen Typen auf (vgl. 1955–1957 BONNEFOI & LE MINOR., LE MINOR,
pers. Mitteilung 1960–1963, PANTALEON). Neuere Befunde bei Säuglingen (vgl. folgendes Kapitel) sprechen im gleichen Sinne.

Als Besonderheit sei hinzugefügt, dass bei der Untersuchung von
310 in Frankreich eingefangenen Schlangen 5 mal (= 1,8%) S. java
und 134 mal (= 43,2%) 16 verschiedene Arizona-Typen (Subgenus
III der Gattung Salmonella) nachgewiesen wurden (LE MINOR et al.).

Mittelmeerländer: Im Gegensatz zu den bisher dargestellten Er-
gebnissen sind wir über die Salmonellose-Verbreitung in den Mittelmeerländern nur ungenügend unterricntet. Soweit die wenigen
Mitteilungen aus Spanien und Portugal einen Rückschluss zulassen, sind dort die für Europa als endemisch anzusehenden Typen
ebenfalls weit verbreitet. So konnte STEINIGER im Küstengebiet von
Barcelona und in der Umgebung des Hafens von Palma in Wasser-
proben, Muscheln und sonstigen Seetieren reichlich S. typhi, S. java
und S. bareilly finden, in der vogelreichen Camargue hingegen vorzugsweise S. typhi-murium in Reihern, S. paratyphi B in Nachtrei-
hern und Staren, sowie S. bredeney und S. give in Staren. Die das
Hafenwasser nicht aufsuchenden Krickenten waren salmonellafrei.
Während im Wasser des Hauptsiels von Barcelona keine Salmonellen
gefunden wurden, traten diese erst nach Verdünnung des Sielwassers
mit Hafenwasser auf (5 bis 10 Salmonellen pro ml Hafenwasser).

Dass Nahrungsmittelinfektionen in Italien keine Seltenheit sind,
gehört zu den wenig angenehmen Erfahrungen vieler Betroffener,
die jährlich mit dem Millionenheer der Touristen nach dem Süden
reisen. Da in diesem Lande die für West-, Mittel- und Nordeuropa

dargestellten Gegenbenheiten aufgrund seiner andersartigen wirtschaftlichen und sozialen Struktur noch nicht in gleichem Ausmasse zum Tragen gekommen sind, entspricht — soweit die bisherigen Berichte einen Schluss zulassen — die *Salmonella*-Typenhäufigkeit vorerst noch dem Stand, der Europa bis Ende des 2. Weltkriegs kennzeichnete. Immerhin zeigen die zunehmenden Befunde "seltener" Typen, wie *S. thompson, S. bareilly* und *S. binza* in Tauben, Kanarienvögeln und Fischmehl, dass sich in Italien eine ähnliche Entwicklung anbahnt. In einem Sammelbericht über 124 Ausbrüche von Nahrungsmittelvergiftungen in Süditalien und Sizilien zwischen 1955 und 1961 wurde festgestellt (GULLOTTI & SPANO), dass *Staphylococcus aureus* als Ursache überwog (58 Ausbrüche verursacht durch Crèmekuchen, Speiseeis, Käse, Milch usw.). 32 Ausbrüche waren durch Salmonellen bedingt *(S. enteritidis* 16 mal, *S. typhimurium* 12 mal, *S. paratyphi B* 2 mal und *S. dublin* 2 mal). Diese waren auf Kalb-, Rind-, Schweine- und Pferdefleisch sowie Taubenfleisch, Quark und Crèmekuchen zurückzuführen. Die Zunahme der *S. typhi-murium*-Infektionen wird auf die Einfuhr infizierter Tiere aus anderen Ländern zurückgeführt. Die Salmonellose-Ausbrüche zeigen hier — im Gegensatz zu den im August und September häufigen Staphylokokken-Lebensmittelvergiftungen — eine gewisse Häufung in den Monaten zwischen Oktober und Februar (15 von 32 Ausbrüchen, davon 10 im Oktober und November). — Die Lysotypie von 130 *S. typhi-murium*-Stämmen aus 65 epidemischen Ausbrüchen in einigen Teilen Italiens ergab ein Überwiegen des Lysotyps 1a var. 1 (= Typ 3 der neuen Nomenklatur) in Nord- und Mittelitalien, während im Südteil und in Sizilien vier andere Phagtypen vorherrschten (SPANO & GULLOTTI).

Im Prinzip ähnliche Verhältnisse wird man für die zahlreichen Mittelmeerinseln sowie für Griechenland (PAPAVASSILIOU, pers. Mitteilung) und die Türkei (vgl. AKYAY, AKSOYCAN, BERKMEN) unterstellen dürfen. Ebenso wie in anderen europäischen Ländern sind die Hafenbecken mit den Abwässern der Menschen und Schlachthöfe die Sammelstelle für die ausgeschiedenen Salmonellen. Sie gefährden, vor allem mit ihrem Reichtum an Typhus- und Paratyphusbakterien, die oft in der Nähe gelegenen Badestrände, deren Wasser meist einen hohen Colititer aufweist. Nahrungsmittelinfektionen, durch die im Lande heimischen *Salmonella*-Typen der Schlachttiere, des Geflügels und durch menschliche Ausscheider verursacht, sind häufig; doch sind zuverlässige Zahlen nicht erhältlich, da sich die bakteriologische Diagnostik meist nur auf grössere Städte beschränkt.

Prognose

In gleicher Weise wie sich der Umbruch Europas von der Individualgesellschaft zur Wohlstands- und Massengesellschaft vollzieht, verändern sich auch die Infektketten der Salmonellosen. Am Beginn der Entwicklung steht der individuelle Erzeuger *eines* Lebensmittels (z.B. Fleisch oder Speiseeis) mit einem eng begrenzten Verbreitungsgebiet (Familie, örtliche Kundschaft). Ist ein Lebensmittel mit Salmonellen infiziert, dann meist nur mit einem *Salmonella*-Typ. Sein Genuss trifft eine kleine, begrenzte Gruppe von Menschen. Das Infektionsrisiko ist daher relativ klein. Am Ende der Entwicklung steht die organisierte Massenproduktion unter schärfster hygienischer Kontrolle. Das Infektionsrisiko ist dann für die Masse weitgehend ausgeschaltet oder minimal. Zur Zeit befinden sich aber viele Berufszweige der Lebensmittelproduktion Europas im Stadium der Strukturänderung, wobei zahlreiche *Halbfertigprodukte*, oft mit einer Vielzahl von *Salmonella*-Typen verseucht, zur Verarbeitung gelangen, z.B. in Fleischereien, Konditoreien, Speiserestaurants und Grosskantinen. Für den Verbraucher solcher Produkte ist zur Zeit das Salmonellose-Infektionsrisiko am grössten; denn viele Hersteller sind sich der Gefahren, die von der Verwendung *Salmonella*-haltiger Lebensmittel ausgehen, nicht bewusst, und die Produktion wie küchentechnischen Methoden sind hygienisch oft unzureichend. In der gleichen Situation ist die Hausfrau, die zwar das Risiko der Verwendung von Hackfleisch und Enteneiern kennt, im Eipulver, in Kokosraspeln, in Backhalbfertigprodukten, Importgeflügel oder im Futter für ihre Haustiere keine Gefahrenquelle vermutet. Am Beispiele Englands, Deutschlands und Italiens lässt sich erkennen, dass der früher riskante Genuss von Speiseeis mit Übergang zur kontrollierten Grossproduktion immer gefahrloser wird, zumal der einstige Kleinhersteller eines oft dubiösen Produkts nunmehr weitgehend zum Verteiler einer hochwertigen, salmonellafreien Ware geworden ist. Das gilt auch für die Importe von Grundnahrungsmitteln für Mensch und Tier aus tropischen Ländern, deren bereits diskutierte epidemiologischen Folgen zu scharfen Kontrollmassnahmen auf nationaler und übernationaler Basis zwingen. Die Salmonellose-Verbreitung hat sich somit im letzten Jahrzehnt von einem lokalen zu einem regionalen und kontinentalen Problem entwickelt, das eine globale Betrachtungsweise erfordert. Es fragt sich unter den gegenwärtigen Umständen, ob und wie lange noch die bisherige epidemiologisch-epizootologische Analyse des Salmonellose-Geschehens durch exakte Typisierung und Feststellung von Infektketten aussichtsreich bzw. sinnvoll bleiben wird. Nach unserem Dafürhalten ist diese Form des Studiums der Salmonellose-Epidemiologie nur noch dort erfolgreich, wo sichere Infektketten durch jeweils einen Typ mit hinreichender Sicherheit rekonstruiert werden können. Wo hingegen — wie in vielen europäischen Ländern — ein nicht mehr übersehbares Durcheinander von multiplen Infektketten mit oft zahlreichen *Salmonella*-Typen in der gleichen Infektionsbahn auftritt, wird die epidemiologische Feinanalyse unzweckmässig, ja sogar zwecklos, da sie auf die Bekämpfung der Salmonellosen ohne Einfluss bleibt. In diesem Stadium kann eine Verbesserung der Situation bzw. Sanierung nicht mehr durch eine genaue Aufklärung der Infektketten — so interessant dies im Einzelfall sein mag — bewirkt werden, sondern nur durch eine konsequente Anhebung der hygienischen Forderungen und Verstärkung der allgemeinen Bekämpfungsmassnahmen. Diese müssen einen Verbrauch Salmonellenfreier Lebens- und Futtermittel ebenso sichern wie eine Verhinderung ihrer Neueinschleppung und Weiterverbreitung. Da dieser Prozess der produktionstechnischen Umstellung und der Verhinderung der Neueinschleppung langwierig ist und von zahlreichen wirtschaftlichen und politischen Momenten beeinflusst wird, ist zu erwarten, dass die Salmonellose-Verbreitung, selbst wenn sie den Kulminationspunkt schon erreicht haben sollte, noch für Jahre ein wichtiges Seuchenproblem Europas bleiben wird. Eine Zunahme ist vor allem in den Mittelmeerländern zu befürchten, wenn sie in gleicher Weise von der wirtschaftlichen Entwicklung erfasst werden und die Aufzucht ihrer eigenen Tierbestände durch Importfuttermittel verbessern wollen. Die in diesen Ländern noch vielfach anzutreffenden, hygienisch rückständigen Verhältnisse könnten zu schlimmeren Folgen führen, als sie in anderen Ländern bereits eingetreten sind.

Das Salmonellose-Problem lässt sich bei der immer stärker werdenden wirtschaftlichen und politischen Verflechtung weiter Teile Europas nicht mehr auf örtlicher oder nationaler, sondern nur auf übernationaler Basis lösen.

G. SALMONELLOSEN BEIM SÄUGLING

Die Salmonellosen des Säuglings nehmen insofern eine Sonderstellung ein, als die Unterschiede der klinischen Erscheinungen zwischen typhösen, paratyphösen und enteritischen Verlaufsformen oft nicht so ausgeprägt sind wie beim Erwachsenen. Nicht selten verursachen die Enteritis-Erreger nämlich paratyphöse Erkrankungen. Auch ist der klinische Verlauf oft viel schwerer (SCHMIDT, TAYLOR). Septische Infektionen, wie Meningitis, verschleiern gelegentlich die wahre Ätiologie.

Infektionsmodus

Nur ausnahmsweise kommt es zu intrauterinen Infektionen, die u.a. von OSTENFELD (zit. nach SCHMIDT) diskutiert werden. Meist erfolgt die Infektion des Säuglings und Kleinkindes durch Lebensmittel oder verunreinigte Gebrauchsgegenstände. Erkrankungen durch unzureichend erhitzte Milchnahrung sind relativ selten. Im Vordergrund stehen sekundäre Infektionen der fertigen Nahrung durch den Menschen, Spülwasser oder verunreinigte Küchengeräte (SCHMIDT); auch Obst wird ursächlich angeschuldigt.

Besonders gefürchtet sind Heim- bzw. Hausinfektionen auf Säuglingsstationen, deren Bekämpfung sich als ausserordentlich schwierig erweist, da die kleinen Patienten auch nach ihrer klinischen Genesung oft für Wochen und Monate *Salmonella*-Keime ausscheiden und so weiter ihre Umgebung gefährden (STAHN). Bei Kontrolluntersuchungen in mehreren englischen Städten wurde z.B. eine Ausscheiderquote von 2 pro 1000 bei gesunden Kindern unter 5 Jahren ermittelt (SPICER). Charakteristisch für das Säuglings- und Kleinkindesalter ist oft eine relative Schutzlosigkeit gegen *Salmonella*-Infektionen. SCHMIDT spricht von einer besonderen Altersdisposition.

Geographische Verbreitung

Im Ganzen gesehen lehnt sich die Verbreitung der Salmonellosen beim Säugling eng an die übrige Salmonellose-Häufigkeit in den einzelnen Ländern an, so dass hier verallgemeinernde Hinweise genügen mögen. Generell überwiegen in den europäischen Ländern Infektionen durch *S. typhi-murium*; auch ist *S. paratyphi B* relativ häufig. Während in den skandinavischen Ländern — analog zur übrigen Bevölkerung — Salmonellose-Erkrankungen der Säuglinge relativ selten sind, werden sie in England und Wales recht zahlreich beobachtet. So stammten 153 (= 12%) der 1949–1951 dort isolierten *S. typhi-murium*-Stämme von Kindern unter einem Jahr (TAYLOR). Die Krankheitsdauer (bei 37% über 14 Tage) war in dieser Altersgruppe deutlich verlängert, und nur 9% wurden schon nach drei Tagen wieder gesund. Eine deutliche Zunahme wird auch in Belgien festgestellt, wo man beim Säugling neben *S. typhi-murium* auch *S. brandenburg* und andere Typen nachwies (GRAFFAR, FEILLEU). Während PINTELON in Gent von 1954–1956 keine einschlägigen Fälle beobachtete, wurden von Juni 1956 bis Ende 1958 neben *S. typhi*, *S. paratyphi B*, *S. typhi-murium* und *S. enteritidis* aus 108 Säuglingsstühlen nicht weniger als 22 mal *S. bredeney*, 18 mal *S. london*, 8 mal *S. bareilly* und noch weitere 6 Typen isoliert. Eine ähnliche Situation zeichnet sich in Frankreich ab (MARIE, VILLEMIN), wo mehrere Epidemien durch *S. typhi-murium* und eine weitere bei Säuglingen einer Wochenstation durch *S. brandenburg* verursacht wurden (VILLEMIN et al., zit. nach SCHMIDT). GERMAIN et al. fanden in den Jahren 1956–1961 bei der Untersuchung von 39 Säuglings- und Kleinkinderstühlen nicht weniger als 40 *Salmonella*-Stämme, die zu 18 Typen gehörten, dabei je 7 mal *S. bredeney* und *S. coeln*, 6 mal *S. montevideo* usw. 29 der betroffenen Kinder waren im ersten Lebensjahr, und eines starb an der Salmonellose; acht weitere Fälle verliefen schwer. Erwähnenswert sind Doppel- oder Folgeinfektionen zwischen Salmonellen und pathogenen *E. coli-*

Typen. Eine analoge Entwicklung bietet sich in der Bundesrepublik Deutschland, wo z.B. in der Heidelberger Kinderklinik von 1950–1959 neben 254 Typhusfällen auch 187 andere Salmonellose-Erkrankungen behandelt wurden (ACKER & SCHREIER). BRANDIS & FUNK fanden in Frankfurt/Main vorwiegend *S. typhi-murium*; von 28 positiven Stuhlbefunden im dortigen Hygiene-Institut stammten 18 von Säuglingen und Kleinkindern. SCHMIDT führt weiter Enteritiden durch *S. paratyphi B* und eine Reihe von Typhus-, *S. enteritidis*- und *S. panama*-Infektionen im norddeutschen Gebiet (Rostock, Magdeburg usw.) auf. Besonders zu erwähnen sind zwei Fälle von angeborenem Paratyphus B (SCHMITZ, zit. bei SCHMIDT) und eine *S. newington*-Epidemie bei Frühgeborenen (LORENZ). Auf gleicher Ebene liegen die Befunde in Süd- und Südosteuropa.

Epidemiologische Folgerungen und Ausblick

Für die Bekämpfung der Salmonellosen beim Säugling und Kleinkind ergibt sich die Notwendigkeit peinlicher Sauberkeit bei der Reinigung aller Pflegegegenstände. Doch lassen sich diese Forderungen, wie vielfältige Erfahrungen gerade in den hochentwickelten Gebieten Europas zeigen, oft nur ungenügend realisieren. Deshalb ist zu folgern, dass eine gezielte Prophylaxe der Salmonellosen in dieser Altersgruppe erst dann einen durchschlagenden Erfolg haben dürfte, wenn es gelungen sein wird, der allgemeinen Salmonellose-Verseuchung Herr zu werden. Davon ist man aber in Europa wohl noch weit entfernt, da die Gefahren, die von diesen Keimen ausgehen, vielfach noch nicht genügend erkannt und gewürdigt werden.

LITERATURVERZEICHNIS*

1. ACKER, R. & SCHREIER, K. Erfahrungen mit der Chloramphenicol-Behandlung von 3468 Patienten einer Kinderklinik. *Med. Klin.* **57**, *474—480* (1962).
1a. AKSOYCAN, N.: Sur les bacilles paratyphiques isolées à Ankara. *Rev. Turque Hig. Biol. expér.* **1955**, No. 2.
2. AKYAY, N.: Salmonella-strains isolated in Turkey. *Türk. Ijiyen* **16**, *34—45* (1956).
2a. AKYAY, N. & FISEK, N. H.: Typhoid and paratyphoid fevers in Turkey. Türk. Ijiyen. **16**, 14—33 (1956).
4. ANDERS, W.: Epidemiologie der bakteriellen Lebensmittelvergiftung. *Bundesgesundheitsblatt* 3, *241—245*, (1960).
4a. ANDERSON, E. S.: Special methods used in the laboratory for the investigation of outbreaks of Salmonella food poisoning. *Roy. Soc. Health J.*, **80**, *260—266* (1960).
4b. ANDERSON, E. S.: cf. COCKBURN et al. *loc. cit.*
5. BADER, R. E.: Paratyphus C 1900–1945. Weltseuchenatlas **I**, 5, (1952 Falk-Verlag, Hamburg.

---

* Auszugsweise.

6. BADER, R. E.: Paratyphus A in Europa 1900–1950. Weltseuchenatlas II, 24, (1956) Falk-Verlag, Hamburg.
7. BADER, R. E.: Die Epidemiologie des Paratyphus C. *Verh. Naturhist. Med. Verein, Heidelberg* 19, 29 (1953).
8. BERKMEN, L.: Vorkommen von Salmonellen in Fleisch und Fleischwaren in Mittelanatolien. *Arch. Lebensmittelhyg.* 8, *278*, (1957).
9. BERKMEN, L.: Bevölkerung und Kultur. Statistisches Bundesamt, Wiesbaden, Reihe 7 "Gesundheitswesen" (1957—1960). Kohlhammer-Verlag, Stuttgart.
10. BISCHOFF, J. & ROHDE, R.: Salmonellen in Fisch- und Fleischmehlen ausländischer Herkunft. *Berl. Münch. tierärztl. Wschr.* 69, *50—53*, (1956).
11. BONITZ, K.: Salmonella bareilly-Epidemie durch Lebensmittelvergiftung im norddeutschen Raum. *Dtsch. med. Wschr.* 78, *1412—1413*, (1953).
12. BONITZ, K.: Über die Epidemiologie von Salmonella bareilly unter besonderer Berücksichtigung infizierten Camembert-Käses. *Zbl. Bakt.* I Orig. 168, *244—256* (1957).
13. BONNEFOI, A. & LE MINOR, S.: Activités du Centre Français des Salmonella de l'Institut Pasteur. *Rev. Hyg. Med. Soc.* 6, *721—730* (1958).
14. BØVRE, K.: Latent salmonella infeksjon hos slaktedyr i Norje. *Nordisk. vet. Med.* 9, *855—867* (1957).
15. BRANDIS, H.: Die Anwendung von Phagen in der bakteriologischen Diagnostik mit besonderer Berücksichtigung der Typisierung von Typhus- und Paratyphus B-Bakterien sowie Staphylokokken. *Erg. Mikrobiol.* 30, *96—159* (1957).
15a. BRANDIS, H. & FUNCK, L.: Über Breslau-Infektionen bei Säuglingen und Kleinkindern. *Z. Kinderhlk.* 65, *368—371* (1948).
16. BULLING, E.: Lebensmittelvergiftungen durch Eiprodukte, Geflügel und Hühnereier. *Arch. Lebensmittelhyg.* 10, *179—181*, (1959).
17. BULLING, E.: Verbreitung und Bedeutung der Tiersalmonellen in der Bundesrepublik. *Zbl. Vet. med.* 10, *216—225* (1963).
18. BUXTON, A.: Salmonellosis in animals. A review. CAB, (1957).
19. CALLOW, B. R.: A new phage-typing scheme for Salmonella typhimurium. *J. Hyg.* 57, *346—359* (1959).
20. CHRISTIE, D. R. et al.: Salmonella organisms in animal feeding stuffs and fertilizers. *Monthly Bull. Min. Health and Publ. Health. Lab. Serv.* 18, *26—35* (1959).
21. COCKBURN, W., TAYLOR, J., ANDERSON, E. S. & HOBBS, B. C.: Food poisoning. The Royal Society of Health ,London (1962).
22. DIXON, J. M. S. & POOLEY, F. E.: Salmonellae in a poultry-processing plant. *Monthly Bull. Min. Health.* 20, *30—36* (1961).
22a. DIXON, J. M. S. & POOLEY F. E.: Salmonellae in two turkey processing factories. *Monthly Bull. Min. Health* 21, *138—141* (1962).
22b. DIXON, J. M. S. & WILSON, F. N.: Salmonellae in fertilizers containing superphosphate. *Monthly Bull. Min. Health and Publ. Health. Lab. Serv.* 19, *79—82* (1960).
23. DOBINSKY, H.: Salmonellosen durch tierpathogene Typen und ihre Epidemiologie. Inaug. Diss. Frankfurt/M. 1962.
23a. DONLE, W.: Paratyphus A in: Seuchenatlas, herausgeg. v. H. ZEISS, Gotha 1943.
23b. DONLE, W.: Über eine Paratyphus A-Epidemie in Wien. *Z. Hyg.* 124, *683* (1943).
24. DRÄGER, H.: Diagnostik der Bakterien der Salmonella-Gruppe. Akademie-Verlag Berlin (1951).
25. DRÄGER, H.: Entstehung und Verhütung von Lebensmittelvergiftungen durch Salmonellabakterien. Gustav Fischer-Verlag, Jena (1958).
26. Epidemiologica and Vital Statistical Reports 1950—1961, WHO Genf.

290

27. FEY, H. & WIESMANN, E.: Die Gefahr des Salmonellenimports mit Eiprodukten und tierischen Futtermitteln. *Schweiz. med. Wschr.* **90**, *791* (1960).
28. FROMME, W.: Zur Epidemiologie der Salmonelleninfektion. *Erg. Mikrobiol.* **32**, *161—195* (1959).
29. GALBRAITH, N. S., TAYLOR, C. E. D., CAVANAGH, P., HAGAN, J. G. & PATTON, J. L.: Pet foods and garden fertilizers as sources of human salmonellosis. *Lancet* **I**, *372—374* (1962).
30. GALBRAITH, N. S., MAWSON, K. N., MATON, G. E. & STONE, D. M.: An outbreak of human salmonellosis due to Salmonella saint-paul associated with infection in poultry. *Monthly Bull. Min. Health and Publ. Health Lab. Serv.* **21**, *209—215* (1962).
30a. GALBRAITH, N. S., HOBBS, B. C., SMITH, M. E. & TOMLINSON, A. J. H.: Salmonellae in desiccated coconut. *Monthly Bull. Min. Health* **19**, *99—105* (1960).
30b. GÄRTNER, H. & BÖHLCK, I.: Kokosraspeln als Infektionsquelle für Paratyphus-Erkrankungen. *Dtsch. med. Wschr.* **34**, *1609—1610* (1961).
30c. GRAFFAR, M.: Contribution à l'étude de la pathologie digestive du nourrisson. *Acta med. belg.* 1950 zit. nach O. H. BRAUN: *Erg. Inn. Med.* **4**, *51—194* (1963).
31. GÄSSLEIN, F.: Salmonella-Lebensmittelvergiftungen durch ungarische Import-Schweine. *Öff. Ges. Dienst* **19**, *327* (1957).
32. GERMAIN, D., COURTIEU, A. L., MOULIN & LE TELLIER, H.: Les salmonella de serotype rare en pathologie pédiatrique. *Pédiatrie* **17**, *389—407* (1962).
33. GULLOTTI, A. & SPANO, C.: Ulteriori osservazione condotte dal centro enterobatteri patogeni par l'Italia meridionale in tema die tossinfezioni alimentari (1955—1961). *Igiene Mod.* **55**, *133—145* (1962).
34. HABS, H.: Zur Typendifferenzierung in der Salmonellagruppe. *Z. Hyg. Inf.* **116**, *537—549* (1934).
35. HABS, H.: Abdominaltyphus. Weltseuchenatlas **I**, *1—3* (1952) Falk-Verlag, Hamburg.
36. HANDLOSER, M.: Epidemiologische Beobachtungen bei einer Masseninfektion durch Salmonella blockley. *Arch. Hyg.* **140**, *569—580* (1956).
37. HARVEY, R. W. S. & PRICE, T. H.: Salmonella serotypes and Arizona paracolons isolated from Indian crushed bone. *Monthly Bull. Min. Health and Publ. Health Lab. Serv.* **21**, *54—58* (1962).
38. HAUGE, S. & BØVRE, K.: Forekomst av salmonella-bakterier i importert vegetablisk proteinkraftfor og kraftforblandninger. *Nord. vet. Med.* **10**, *255—262* (1958).
39. HENZE, B.: Das Auftreten von Salmonellen bei Lachmöven auf Berliner Rieselfeldern. *Ges. Wes. u. Desinfekt.* **1961**, H. 11.
40. HEPP, L.: Eine Salmonellose beim Menschen durch den Genuss von Schweinefleischkonserven jugoslavischer Herkunft. *Berl. Münch. tierärztl. Wschr.* **68**, *221—228* (1955).
41. HOFMANN, P., HÖRCHER, F. & WOLLE-JOHN, R. Einschleppung seltener Salmonellen durch importierte Geflügelfedern. *Zbl. Bakt.* I Orig. **178**, *484—491* (1960).
42. JENSEN, P. T.: Forekomst of salmonella bakterier, svinets mesenteriallymfekirtler og i importerede vegetabilske fodderstoffer, samt resistensbestimmelse af fundne stammer. VIII. Nordiska veterinärmötet, Sektion E, Rapport 8 (1958).
43. JØRGENSEN, B. V.: The occurrence of Salmonella in domestic sewage and abattoir wastes with special reference to the epidemiology. The Royal Vet. and Agricult. College, Yearbook, Copenhagen **1962**, 1—38.
44. JUSATZ, H. J. & EL RUBAIE: Weltseuchenatlas, **III**, 146 (1956 ff) Falk-Verlag, Hamburg.

45. KAMPELMACHER, E. H. & GUINÉE, P. A. M.: Salmonellosen. Rijks Instituut voor de Volksgezondheid, Utrecht (1960).
46. KAMPELMACHER, E. H., GUINÉE, P. A. M., HOFSTRA, K. & VAN KEU-LEN, A.: Studies on Salmonella in slaughterhouses. Zbl. Vet. Med. 8, 1025—1042 (1961).
47. KAMPELMACHER, E. H., GUINÉE, P. A. M., & CLARENBURG, A.: Salmonella-organisms isolated in the Netherlands during the period from 1951 to 1960. Zbl. Bakt. I Orig. 185, 490—502 (1962).
48. KELCH, F.: Über das Vorkommen von Salmonellen in chinesischem Gefriervollei. Berl. Münch. tierärztl. Wschr. 69, 307 (1956).
49. KIESEWALTER, J., RUDAT, K. D. & SEIDEL, G.: Salmonellen aus Reptilien. 1. Mitteilung: Untersuchungen an Schildkröten. Zbl. Bakt. I Orig. 180, 503—509 (1960).
50. KLEIN, H.: Salmonella blockley als Ursache einer Lebensmittelvergiftung. Arch. Lebensmittelhyg. 10, 6—7 (1959).
51. KLOSE, F., KNOTHE, H. & STEINIGER, F.: Die epidemiologische und nahrungsmittelhygienische Bedeutung des Keimträgertums von Paratyphus B und anderen Salmonellen bei wildlebenden Vögeln. Ärztl. Wschr. 7, 824—829 (1952).
52. KNIEWALLNER, K.: Über das Vorkommen von Salmonellen in Österreich. Wien. tierärztl. Wschr. 45, 710—716 (1958).
53. KNORR, M.: Über die Herkunft des Paratyphus C in Deutschland. Arch. Hyg. 105, 237—244 (1931).
54. KUDICKE, H.: Über die Lebensdauer von Salmonella-Bakterien auf Süsswaren. Öffentl. Ges. Dienst 23, 109—111 (1961).
55. LE MINOR, L., FIFE, M. A., EDWARDS, P. R. & CHARIÉMARSAINES, C.: Recherches sur les Salmonella et Arizona hébergés par les vipères de France. Ann. Inst. Pasteur 95, 326—333 (1958).
56. LENK, V., RASCH, K. & BULLING, E.: Über das Vorkommen von S. paratyphi B (d-Tartrat negativ) bei Tieren. Zbl. Bakt. I Orig. 180, 304—309 (1960).
57. LENTZE, F.: Zur "modernen" Problematik der Salmonellosen. In: Aktuelle Themen der Inneren Medizin und ihrer Grenzgebiete. F. Enke-Verlag, Stuttgart (1959).
58. LERCHE, M. & BARTEL, H.: Fünf Jahre Typendifferenzierung. Dtsch. Tierärztl. Wsch./Tierärztl. Rundschau 51/49, 41—48 (1943).
59. LINKE, H.: Über das Vorkommen von Salmonellen in den Mesenteriallymphknoten gesund geschlachteter Pferde. Arch. Lebensmittelhyg. 8, 244 (1957).
59a. LILLEENGEN, K.: Typing of Salmonella dublin and Salmonella enteritidis by means of bacteriophage. Acta path. microbiol. scand. 27 (1950).
59b. LILLEENGEN, K.: Typing of Salmonella gallinarum and Salmonella pullorum by means of bacteriophage. Acta path. microbiol. scand. 30, 194—202 (1952).
59c. LILLEENGEN, K.: Typing of Salmonella typhi-murium by means of bacteriophage. An experimental bacteriologic study for the purpose of devising a phage-typing method to be used as an aid in epidemiologic and epizootologic investigations in outbreaks of typhi-murium infections. Herausgegeb. Ivar Hoeggströms, Stockholm 1948.
60. LORENZ, C.: Mitteilung über eine Salmonella newington-Epidemie bei Frühgeborenen. Kinderärztl. Praxis 26, 145—150 (1958).
61. MAIR, N. S. & ROSS, A.: Survival of Salmonella typhi-murium in the soil. Monthly Bull. Min. Health and Publ. Health Lab. Serv. 19, 39—41 (1960).
62. MARCUSE, K., HENZE, B. & POHLE, H. D.: Das Vorkommen von Salmonellen in West-Berlin. Zbl. Bakt. I Orig. 169, 478—492, 493—514 (1957).

62a. MARIE, J., SERINGE, Ph. et al.: *Sem. hôp. Paris*, **27**, *15* (1951) (zit. nach SCHMIDT, loc. cit. 83).

63. MATZKE, W.: Ist eine Salmonellen-Übertragung von Wassergeflügel auf Speisekarpfen möglich? *Mh. Vet. Med.* **15**, *849—853* (1960).

64. McDONAGH, V. P. & SMITH, H. G.: The significance of the abattoir in Salmonella infection in Bradford. *J. Hyg.* **56**, *271—279* (1958).

64a. MEYER, R.: Über die Verwendbarkeit des Salmonellenquotienten in der Ortshygiene und Epidemiologie. *Arch. Hyg.* **144**, *95—118* (1960).

64b. MEYER, R.: Auswertung zweigipfliger Überlebenskurven der Salmonella paratyphi B im Grundwasser. *Arch. Hyg.* **144**, *564—568* (1960).

65. MEYER-OSCHATZ, W.: Über die Häufigkeit und Verbreitung der Salmonellen in Deutschland. *J. Hyg. Epid. Mikrobiol. Imm.* **1**, *190—206* (1957).

66. MEYER-OSCHATZ, W.: Erfahrungen über Typhus- und Paratyphus-epidemien im Blickwinkel der Milchwirtschaft. *Zbl. Bakt.* I Orig. **170**, *146—150* (1957).

67. MÜLLER, J.: Bacteriological examination of imported meat and bone meal and the like. *Nord. vet. Med.* **4**, *290—295* (1952).

68. NEWELL, K. W., McCLARIN, R., MURDOCK, C. R., MacDONALD, W. N. & HUTCHINSON, H.L.: Salmonellosis in Northern Ireland with special reference to pigs and Salmonella contaminated pig meal. *J. Hyg.* **57**, *92—105* (1959).

69. PETZELT, K. & STEINIGER, F.: Salmonellen bei der Vogelwelt von Kläranlagen. *Arch. Hyg.* **8**, *605* (1961).

69a. PINTELON, S.: Résultats d'une recherche systématique de Salmonella à la clinique universitaire de Pédiatrie de Gand. *Acta paediat. belg.* **14/4**, *161—177* (1960).

70. POHL, G.: Vorkommen von Salmonellen in geklärten Abwässern, ihren Vorflutern, Rieselfeldabflüssen und -drainagen und Klär- und Faulschlamm. *Berl. Münch. tierärztl. Wschr.* **68**, *163—167* (1955).

71. POPP, L.: Über eine menschenpathogene Varietät der S. gallinarum Klein. *Zbl. Bakt.* I Orig. **152**, *358—366* (1947/48).

72. POPP, L.: Der Salmonella-Kataster eines Flussgebiets. *Ges. Ing.* **78**, *333* (1957).

73. PUBLIC HEALTH LAB. SERVICE: The contamination of egg products with salmonellae, with particular reference to S. paratyphi B. *Monthly Bull. Min. Health and Publ. Health Lab. Serv.* **17**, *36—51* (1958).

74. RASCH, K.: Über das Verhalten der Salmonellen in der Aussenwelt, ein auch für die Verhütung von Fleischvergiftungen wichtiger Faktor. *Arch. Lebensmittelhyg.* **6**, *1—3* (1955).

75. RASCH, K.: Salmonellen in Knochenschrot. *Berl. Münch. tierärztl. Wschr.* **68**, *213—214* (1955).

76. RISCHE, H.: Lysotypie von Salmonella paratyphi A. *Zbl. Bakt.* I Orig. **171**, *568—572* (1958).

77. ROHDE, R. & BISCHOFF, J.: Die epidemiologische Bedeutung salmonellainfizierter Tierfuttermittel (insbesondere Knochenschrot und Fischmehl) als Quelle verschiedener Lebensmittelvergiftungen. *Zbl. Bakt.* I Ref. **159**, *162* (1956).

78. RODENWALDT, E. & JUSATZ, H. J.: Weltseuchenatlas, **I, III**, (1956 ff) Falk-Verlag, Hamburg.

79. RÖHR, W.: Zur Infektion der Haustiere mit S. paratyphi B. *Zbl. Bakt.* I Orig. **182**, *276—278* (1961).

80. RUTQVIST, L. & SWAHN, O.: Epizootologiska och bakteriologiska undersökningar vich mjältsbrandsepizootien i Sverige 1957. *Nord. vet. Med.* **9**, *641—663* (1957).

80a. RUTQVIST, L. & THAL, E.: Salmonella isolated from animals and animal products in Sweden during 1956-1957. *Nord. vet. med.* **10**, *234-244* (1958).

81. SCHAAL, E.: Schlachthofabwässer und ihre hygienische Bedeutung. *Berl. Münch. tierärztl. Wschr.* **72**, *66—70* (1959).
82. SCHMIDT, B. & LENK, V.: Der Nachweis von Salmonellen im Abwasser als möglicher Massstab für die Seuchenlage einer Bevölkerung. *Zbl. Bakt.* I Orig. **178**, *459—483* (1960).
83. SCHMIDT, E. F.: Salmonella- und Shigella-Enteritis. In: Säuglingsenteritis. Von A. ADAM, Thieme-Verlag, Stuttgart (1956).
85. SEELIGER, H. P. R.: Salmonellosis in Western Germany and Berlin, 1945—1957. Proc. 6th Int. Congr. Trop. Med. Mal. Vol. IV, *43—48* (1958).
86. SEELIGER, H. P. R.: Jahresbericht über die Salmonellosen in Deutschland 1957. *Zbl. Bakt.* I Orig. **174**, *327—347* (1959).
87. SEELIGER, H. P. R., HOFMANN, S. & ROHDE, R.: Jahresbericht über die Salmonellosen in der Bundesrepublik Deutschland und Westberlin 1958. *Zbl. Bakt.* I Orig. **182**, *357—403* (1961).
88. SEELIGER, H. P. R.: Food-borne infections and intoxications in Europe. *Bull. Wld. Health Org.* **22**, *469—484*, (1960).
89. SMITH, H. W.: The isolation of Salmonellae from the mesenteric lymphnodes and faeces of pigs, cattle, sheep, dogs and cats and from other organs of poultry. *J. Hyg.* **57**, *266—273* (1959).
90. SONTAG, M.: Über den Salmonellabefall pflanzlicher und tierischer Futtermittel unter besonderer Berücksichtigung von inländischem Knochenfuttermehl und Futterknochenschrot. Inaug. Diss. Giessen (1962).
91. SPANO, C. & GULLOTTI, A.: Phage typing of S. typhi-murium in food borne infections occurred in various areas of Italy. *Zbl. Bakt.* I Orig. **181**, *391—394* (1960).
92. SPICER, C. C.: Frequency of carriers of Salmonellae, Shigellae and pathogenic coliform organisms in normal children under 5 years. *Monthly Bull. Min. Health. Publ. Health Lab. Serv.* **18**, *86—91* (1959).
93. STAACK, H. H.: Die Bedeutung der Salmonellen-Erkrankungen unter besonderer Berücksichtigung der Verhältnisse in Schleswig-Holstein. *Arch. Hyg.* **142**, *105—127* (1958).
93a. STAHN, I.: Zur Epidemiologie der Salmonellosen im Lande Bremen von 1956—1961. *Arch. Hyg.* **147**, *481—494* (1963).
93b. STAHN, I.: Über die epidemiologische Beziehung zwischen den in den Vorflutern und Abwässern Bremens isolierten Salmonellen vom Typ manchester und dem Befall der Bevölkerung. *Arch. Hyg.* **147**, *598—608* (1963).
94. STEINIGER, F.: Paratyphus-B-Bakterien im Nordseewasser. *Zbl. Bakt.* I Orig. **157**, *52—56* (1951).
95. STEINIGER, F.: Der Zyklus Vogel-verunreinigtes Wasser in der Freilandbiologie der Salmonellen. *Zbl. Bakt.* I Orig. **160**, *80—84* (1953).
96. STEINIGER, F.: Zur Freilandbiologie der Salmonellen im Bereich des westlichen Mittelmeeres. *Zbl. Bakt.* I Orig., **166**, *245—265* (1956).
97. STENBERG, H.: Salmonella infections encountered during the last years from the veterinary point of view. *Eripainos Suomen El.* **64**, *567—573*, (1958).
97a. STENBERG, H. & VASENIUS, H.: The increase in Salmonella infections in domesticated animals. *Maatalousja Koetoiminta* **15**, *284—290* (1961).
98. TAYLOR, J.: The diarrheal diseases in England and Wales. *Bull. WHO* **1960**, *763—779*.
99. THAL, E., RUTQVIST, L. & KARLSSON, K. A.: Salmonella bei Hühnern in Schweden. Kongressbericht des 2. Symp. der I.A.V.F.H. Basel (1960).
100. VAN OYE, E. & LAFONTAINE, A.: Etat actuel du problème des salmonelloses humaines en Belgique. *Arch. belges Méd. soc.* **20**, *503–514* (1962).

101. WALKER, J. H.: Organic fertilizers as a source of Salmonella infection. *Lancet* **2**, *283* (1957).
102. WINKLE, S. & ROHDE, R.: Über das häufige Auftreten von Paratyphus A in Deutschland und die Frage, ob es auch bei dieser Salmonellose eine "Infektbahnung" gibt. *Arch. Hyg.* **139**, *165—173* (1955).
103. WINKLE, S. & ROHDE, R.: Über die Gefahr des bakteriologisch unkontrollierten Importes ausländischer Futtermittel mit besonderer Berücksichtigung der Schweinezucht. *Berl. Münch. tierärztl. Wschr.* **70**, *243—249* (1957).
104. WINKLE, S. & ROHDE, R.: Die Bedeutung von Mischinfektionen durch mehrere Salmonella-Typen für Diagnostik und Epidemiologie. *Zbl. Bakt.* I Orig. **173**, *153—158* (1958).
105. WINKLE, S., ROHDE, R. & ADAM, W.: Salmonellabefallene importierte Kokosraspeln als eine weitere gefährliche Infektionsquelle. *Zbl. Bakt.* I Orig. **179**, *583* (1960).
106. WUNDT, W.: Zur Ausbreitung des Typhus abdominalis, des Paratyphus und der Salmonellenenteritis. *Hippokrates* **12**, *481—486* (1962).

NACHTRAG

107. GADEHOLT, H. & MADSEN, S. T.: Clinical course, complications and mortality in typhoid fever as compared with paratyphoid B. A survey of 2647 cases. *Act. Med. Scand.* **174**, *753—760*.
108. HOFMANN, S., ROHDE, R. & SEELIGER, H. P. R.: Jahresbericht über die Salmonellosen in der Bundesrepublik Deutschland einschl. Berlin (West). *Bundesgesundheitsblatt* **1963**, *361—369*.
109. PANTALEON, J.: Présence de salmonelles dans les viandes. *Ann. Inst. Pasteur* **104**, *598—620* (1963).
110. RISCHE, J. & KRETZSCHMAR, W.: Biochemotypen und Lysotypen von S. typhimurium. *Arch. Hyg.* **146**, *530—539* (1962).

SALMONELLA INFECTIONS IN EAST EUROPE

BY

KAZIMIERZ LACHOWICZ

State Institute of Hygiene, Warsaw

(with 2 figs.)

Introductory remarks

East Europe, as referred to in this article, comprises Czechoslovakia, Poland, the USSR, Hungary, Roumania, Yugoslavia and Bulgaria. This, however, is an arbitrary definition as East Europe is not a distinctly delineated geographical concept. Furthermore, when discussing problems relating to the Soviet Union, as it is the case, one has necessarily to refer to matters beyond Europe's boundaries. On the other hand, such a division of the European continent has some justification. Along with East Germany and Albania, the above mentioned countries have a similar economic and social life, different from the remaining European countries. The similarity extends also over public health institutions and many more fields important from the epidemiological point of view. They differ, however, in some ways and it would be a mistake to consider them all to be an epidemiological entity. Their starting level had been different, different are their living, housing and social-hygiene standards, different also their traditions and customs.

A student of epidemiological events in Eastern Europe faces many difficulties.

One of them is the language barrier. Though the territory is mostly inhabited by Slavic peoples, even a Slav speaking like the author three Slavic languages, has to overcome great difficulties when reading papers in other Slavic languages, not to mention the Hungarian and Roumanian languages. And it is important to read them. Most of the countries are publishing scientific papers in one or more world languages, but it is rather rare that they publish materials on domestic matters. English, French, German or Russian summaries, if any, are not informative enough.

East Europe has undergone many political changes since the beginning of the century. Bulgaria had been declared an independent state in the first decade of the century. Poland, Czechoslovakia and Hungary regained their independence only after World War I. Before that, Poland was partitioned between Russia, Prussia and Austria, while Czechoslovakia and Hungary formed a part of Austria (the Austro-Hungarian monarchy). Yugoslavia was created after World War I by federation of two independent Balkan states

(Serbia and Montenegro) and other nations chiefly deriving from the Austro-Hungarian monarchy.

Both, the states already in existence at the beginning of the century and those created subsequently, had since then had their frontiers changed, some of them more than once and most markedly. As a result, it is rather difficult to trace epidemiological events back to the past in regard to the actual state of the territories. The more so, as it is not only frontiers that had changed. The populations of large territories, too, had been moved, along with the political changes. These occurrences are to be kept in mind when considering *Salmonella* infections in their historical course in these areas.

Scientific and professional medical publications being almost exclusively the only source of the author's information it should be mentioned that some of them stopped being published or altered their subjects or titles what did not make easier the task of looking for epidemiological data. Furthermore, as a result of political and social events, the collections of journals available in the libraries are often far from being complete, many bibliographic items being altogether unobtainable.

Last but not least, there was a period in this part of Europe when data of epidemiological importance were secret, and if published at all, were void of figures, names of places and other definite particulars. It is not so now, but some of the countries do not yet publish epidemiological data, or do so only irregularly and incompletely. The gap resulting has not been filled up, yet. This adds greatly to the difficulties.

All that was bound to impair the presentation of the subject, no matter how great an effort had been made to collect scattered figures and facts, to complete them and to try to compose as clear and as instructive a picture as possible. Apart from the author's efforts, much more, and more complete data are needed, compiled according to a uniform scheme and from every country concerned. This, consequently, requires also greater collaboration between human and veterinary epidemiologists.

It is proposed to consider here typhoid fever, paratyphoids, salmonella food-poisoning (salmonella gastroenteritis) as well as participation of *Salmonella* in infantile diarrhoeas. Though not much could be said about the last one it seems right to treat it as a separate epidemiological problem.

Official or unofficial notification figures were sometimes supplemented by figures taken from different publications or even by figures conjectured. It was not possible to cite the source of every particular figure or of the source of every conjecture made. It was not possible to check the reliability of some scattered data collected, either. Some of them may be subjected to criticism as they were taken from articles other than scientific reports. It was felt, however, that they might be, nevertheless, of some use.

General view

Typhoid fever

Though typhoid and paratyphoids will be treated separately in the following, it seems reasonable to have first a look at them taken together. The more so, as they were being notified together in East Europe for rather a long time and there is hardly any possibility to separate them in the past.

Fig. 1 presents the typhoid and paratyphoid morbidity rate in East Europe in its trend throughout the century. It is based on various sources and far from being complete nor very accurate either. Even these rough estimates, however, appear to give a quite clear idea of what is going on in regard to the typhoid fever in the area, paratyphoids often being only rather a small fraction of the number of cases notified.

It may be said in the first place that the first two decades of the century witnessed a very high level of the typhoid incidence and that there has been a steady improvement since. During the following two decades (1921—1940) the curves are swinging at the de-

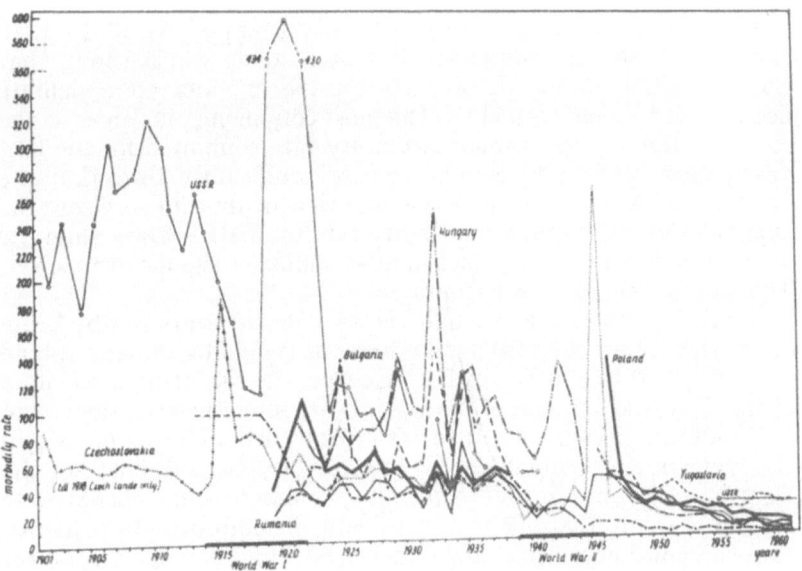

Fig. 1. Typhoid and paratyphoid morbidity rate per 100.000 population in East Europe.

finitely lower but still high level and then, in the last two decades
an impressive drop occurred to a rather low morbidity rate. Particular countries, though, participated in the improvement in various
degrees.

Another view of the trend followed by typhoid and paratyphoid
incidence in East Europe is demonstrated in table I. By means of

Table I.

Average annual typhoid and paratyphoid morbidity rates in the countries of East
Europe (number of cases per 100 000 population).

	1921—1940	1941—1960	1951—1960	1961
Czechoslovakia	45.2	32.0	15.5	6.2
Poland	57.7	32.7	21.4	11.0
U.S.S.R.	132.8	about 40—50	about 20—25	about 20—25
Hungary	101.3	34.9	12.5	6.9
Roumania	38.0	25.5	16.2	8.6
Yugoslavia	36.9	35.6	29.6	28.0
Bulgaria	71.4	7.9	3.1	1.1

Note. In figures for Poland and Yugoslavia years of World War II are omitted
for lack of data. Figures for U.S.S.R. are based on fragmentary data, or conjectured.

average annual morbidity rates for twenty-year periods, 1921—1940
and 1941—1960, a comparison has been made which shows that
excepting Yugoslavia all of East Europe experienced a definite
decrease in typhoid incidence. The most convincing, however, is the
fact that the average annual morbidity rate computed for the ten-
year period 1951—1960 is in every case significantly lower than the
one for the last two-decade period and that in almost every case the
typhoid and paratyphoid morbidity rate for 1961 is lower than the
latest decade's average, in fact in most countries significantly lower.
This appears to be very encouraging.

Another point is that though the two World Wars resulted in a
great, sometimes even tremendous rise in typhoid and paratyphoid
incidence in the countries of East Europe, the rise being most often
at its highest at the end of the war or just after the war, they seem
to divide the period concerned into three parts differing greatly in
the incidence of typhoid fever. Any direct beneficial effect of the
wars is, of course, unthinkable, but so many other events were
related to them, that one may consider some indirect effects (political, economic and social changes inclusive). However, one may detect
the outset of the decrease in typhoid incidence just before the World
War II. Thus it could be presumed that, given sanitary and economic conditions for a steady decrease in typhoid incidence, even the

calamity of war can bring only a temporary deterioration which ceases as soon as the previous sanitary and economic conditions are restored.

The most impressive drop in typhoid incidence was noted in Bulgaria, the least in Yugoslavia, Poland showing a more definite improvement only in the last five years. The Soviet Union's achievement, however resulting in not a too low level, is nevertheless considerable, taking into account the very high and unfavourable starting point. The authorities as well as scientists and practitioners concerned are prone to ascribe the favourable and promising state arrived at, to both the control measures undertaken (vaccination campaings, compulsory hospitalization of cases and suspects, carrier detection and supervision, improvement of sanitary and hygiene standard etc.) and to the general improvement ot the living conditions and cultural level of the population[1, 2]. However, Czechoslovakia is not applying mass vaccination as a control measure against typhoid[3], and Hungary, in fact, stopped doing so since 1951[4]. On the other hand, the children being mostly involved now, there is a tendency, e.g. in Bulgaria[5], not only to vaccinate children before their admission to school but also to begin antityphoid vaccination in the second year of life.

Some factors seem to hamper a further drop in typhoid incidence. It is being said that typhoid occurring more and more rarely, physicians stopped considering it when meeting with a typhoid case, causing thus the right diagnosis to come late and therefore also the control measures. With growing use of antibiotics a possibility exists that typhoid cases are being treated undiagnosed without any consideration for preventive measures; furthermore, new typhoid carriers may evade detection.

Much attention is being paid in the countries of East Europe to the problem of typhoid carriers as a key to the control of typhoid fever. Generally two processes are applied simultaneously. Food handlers, water work technical employees etc. are checked bacteriologically before assuming their work and later are reexamined periodically. People with a typhoid history are examined as to their possible carrier state whatever their profession; in some countries the typhoid history is traced as far back as 30 years. However, the process of detecting typhoid carriers is far from being complete and by far not all carriers existing are known. It is pointed out in the U.S.S.R., Czecholovakia and Bulgaria that each year a number of new typhoid carriers are detected, previously unknown. Does Hungary's typhoid carrier rate figure amounting to about 20 per 100 000 population[6] or Poland's amounting to about 14[7], if near the truth, appear to be a fairly promising starting point for further progress? (compare[36]).

A very serious problem in the control of typhoid is sanitation, especially that of the water supply and the sewage disposal in the geographical regions where water sources are scarce and people are forced to use surface waters (e.g. some Middle Asia Republics of the U.S.S.R.). Further developments depend greatly upon a successful solution of this problem.

However difficult it would be to tell for certain what had caused the almost dramatic decrease in typhoid incidence in most countries of East Europe at about the midcentury, there is a reason to believe that the falling trend of the typhoid incidence will continue or at least that it will remain at its present level. In two instances of a halt noted recently (U.S.S.R. and Yugoslavia) there are, it is felt, possibilities of breaking it. It had been explained, in the Soviet Union e.g., that over a vast territory the typhoid incidence rate is below 10.0 per 100 000 population and a steady further improvement is being achieved; and it is only because of some distinct areas, chiefly in the Middle Asia Republics, with a high typhoid endemicity level and poor water supply, that the country average incidence is above that and no further progress has been observed recently [8, 9]. An improvement in those areas may affect the country average significantly. The same to some extent may be said of Yugoslavia.

On the other hand, any temporary deterioration of the sanitary condition and of the hygiene and living standard, like those related to wars, may result in great typhoid epidemics comparable to those which occurred at the end of World War II.

According to various reports [10-16] four phage types of S. typhi are predominant all over East Europe: El, A, F1 and D1, with the C1 type following them rather closely. There are two exceptions only: U.S.S.R. where D1 type is substituted by F2 type, rather exceptionally found in the other countries, and Bulgaria from where no mentioning of C1 type finding has been encountered. However, most of the information relating to U.S.S.R. were reported in the early fifties and only scanty data were published recently, being rather incongruous with the earlier ones [17].

Paratyphoids

As far as may be judged from Eastern European notification figures available, paratyphoids are a more serious and growing problem only in Yugoslavia. In the other countries paratyphoid incidence is going down or appears to be rather stabilized. Rough estimates show that in most of the countries the paratyphoids' share is about 15—20% of the "typhoid" cases. One exception is Roumania where paratyphoids constitute a smaller fraction (about 5%), the other being Yugoslavia with her 25% paratyphoid share, rising recently even higher, to more than 30%. So far, the last estimate known for all the U.S.S.R. territory comes from the interwar period and is as low as 5—10% [18]. More recent figures refer to some distinct areas of the U.S.S.R. and some of them amount to a much higher percentage, but they, maybe, do not influence the country's average share of paratyphoids.

Paratyphoid A has been only rarely met with in Czechoslovakia, Hungary, Yugoslavia and Bulgaria after World War II. In Czechoslovakia an epidemic of paratyphoid A was notified in a mental hospital in 1948 with 37 cases involved [19]. There were three other paratyphoid A cases reported since, one in 1954 and two in 1961. No paratyphoid A epidemic has occurred in Hungary, Yugoslavia or Bulgaria, and only few sporadic cases. More affected have been Poland, Roumania and U.S.S.R. In Poland about 40 paratyphoid A cases a year were confirmed bacteriologically in 1945—1956 [20]; a canteen and a mental hospital epidemic as well as some small family epidemics have been reported too. It is in more densely inhabited regions of Poland that most of paratyphoid A cases occurred. In Roumania paratyphoid A strains constituted about 9% of all *Salmonella* material of 1948—1951, and in a hospital, paratyphoid A cases amounted to 11% of all paratyphoid cases in a 12 year period [21]. Galati town was cited as the place of a paratyphoid A epidemic as well as an endemic focus where paratyphoid A amounted to 10% of all typhoid-paratyphoid cases [22]. As to the U.S.S.R. there was a rather great increase in paratyphoid A incidence during the World War II and just afterwards. It constituted about 25—30% of enteric-like cases then [23], but has decreased since. In a Moscow hospital e.g. the percentage of paratyphoid A cases dropped from 8.7 in 1948 to 3.3 in 1957 [24]. But there are regions or towns where the paratyphoid A incidence has remained high. Generally, it is the southern and south-eastern regions of the U.S.S.R. where paratyphoid A is more frequent.

Paratyphoid B is the main paratyphoid problem in all of East Europe. In Czechoslovakia, Hungary, Yugoslavia and Bulgaria it is practically the only paratyphoid problem of the country. Therefore, what has been said of the paratyphoid problem as a whole, relates to paratyphoid B. Thus, paratyphoid B seems to be in retreat almost everywhere except for Yugoslavia. However, in most instances the typhoid fever incidence decreases more quickly than that of paratyphoid B, and perhaps the time is ahead when the paratyphoid B problem in East Europe will be as important as the typhoid one, whatever their future weight may be.

Paratyphoid B epidemics occur in all the countries, chiefly water-born or milk-born, but the bulk of the paratyphoid incidence is formed by sporadic cases. Usually they are not evenly distributed all over the country. In the Soviet Union it was the northern, north-western and central parts of the country where paratyphoid B was most common [26], and they probably continue to be so. In Yugoslavia it is the Slovene Republic where the paratyphoid

B incidence grew higher than the one of typhoid fever [27], [28].

Not many figures are allowed to be known as to the incidence of paratyphoid B carriers in the countries. When given, the number of paratyphoid B carriers reported for an area is usually much lower than that of typhoid carriers, but instances were quoted showing an opposite proportion.

In most of the countries phage type Taunton predominates, others more often met with, being to varying degree, B.A.O.R., 1 most common, Dundee, 3aI var. 1 and var. 2^{14}, [29], [30]. Data on paratyphoid B phage type distribution in U.S.S.R. are very scanty and so far quite different from those in neighbouring Eastern European states [31], [32], [33].

Paratyphoid C is from the epidemiological point of view certainly the least important paratyphoid problem in East Europe. It is, at the time, not very easy to deal with. There was a habit throughout East Europe to call any case due to a *Salmonella* organism of the group C a paratyphoid C case. E.g. a table was published in 1956 showing the geographical distribution of paratyphoid C cases in the U.S.S.R. [34]. But, at least for the postwar time, most of the papers referred to, if not all, are dealing with infections due to group C *Salmonella* organisms other than *S. paratyphi C*.

As far as is known to the author no paratyphoid C case has been reported from Czechoslovakia, Hungary or Bulgaria since World War II, and probably no one in the interwar period either. Only one case of a healthy paratyphoid C carrier was registered in Czechoslovakia (Slovakia) in 1955 [35]. There were two bacteriologically confirmed paratyphoid C cases in Poland in 1951 and 1952 [36]. The list of *Salmonella* types isolated in Roumania in the period of 1948—1951 does not include *S. paratyphi C* [101]. However, 5 strains of *S. paratyphi C* were reported from a Roumanian pediatric clinic to have been isolated from diarrhoeal cases in children in the period 1951—1954 [37], and the infectious clinic at Cluj mentioned one paratyphoid C case among its 100 paratyphoid cases observed in a 12-year period [21]. In Yugoslavia 5 paratyphoid C cases were reported from a Sarayevo clinic among 400 typhoid cases between 1950 and 1958 [38], and one paratyphoid C among 23 typhoid patients in the region of Knin in 1952 [39]; *S. paratyphi C* isolations in a children's hospital have been reported from Ljubljana, too.

A few more paratyphoid C cases or *S. paratyphi C* isolations have been reported from the U.S.S.R. In a hospital laboratory in Leningrad [40] the number of the *S. paratyphi C* strains isolated (together with the untypable ones) amounted to about 30% of the *Salmonella* group C isolations, *S. cholerae-suis* var. *kunzendorf* being the first of about 67%. However, no absolute figures were allowed to be known. In other Leningrad hospitals 3 *S. paratyphi C* strains were isolated from 14437 children admitted, due to gastrointestinal disorders in

1951— 1960 [41], and out of 100 enteric fever patients one was diagnosed as a paratyphoid C case [42]. Some figures from Moscow were reported too. Among 1449 *Salmonella* strains isolated in 1948—1957 as many as 124 (!) were diagnosed as *S. paratyphi C* strains [24], and so were 7 strains among 2539 *Salmonella* strains isolated in 1957—1959 [43] and 4 strains among further 210 *Salmonella* strains [44]. One *S. paratyphi C* strain was found in Crimea in 1958—1959 among 101 *Salmonella* strains [45]. Cases of *S. paratyphi C* carrier state among food handlers were encountered as well [46].

TOPLEY & WILSON's remark suggests that it is in Eastern Europe that paratyphoid C has been met with most often [47]. This may be true, though it does not seem to occur there often. It is most certainly an accident, but TOPLEY & WILSON's remark has been translated into Russian as "Western Europe" [48].

It may be added here that *S. sendai* was isolated in U.S.S.R. in 1942—1944 from two human cases [49]. Other five *S. sendai* strains were isolated in Crimea in 1958—1959 [45], but no indication has been found as to the condition of the people they were isolated from.

Salmonella food poisoning

Salmonella organisms remain one of the most important bacterial agents, if not the most important one, in both outbreaks and sporadical cases of food poisoning in East Europe. It has been often stressed that the most characteristic actual feature of *Salmonella* food poisoning in Eastern European countries is the growing number of cases per outbreak, the growing number of *Salmonella* types involved, and a growing proportion of sporadic cases. The figures available, however scanty and incomplete, seem to congrue with the opinion. In Czechoslovakia e.g. in 1939—1946 the average number of cases per food poisoning outbreak was 41.5, while in 1946—1951 it amounted to 260 [3]; in Hungary the proportion between the interwar and postwar period average was approximately about 1 : 3 [50].

The growing number of cases per *Salmonella* food poisoning outbreak may be explained by the expansion of communal feeding. The other two characteristics must be regarded as closely related to the ecological aspects of salmonellosis, various "new" *Salmonella* types being more and more often encountered outside man too.

Material collected so far does not allow to define exactly and completely the frequency distribution of various *Salmonella* types in East Europe. Thus, the list of other than typhoid and paratyphoid *Salmonella* types isolated therein (fig. 2) should not be considered as completely up-to-date. It is only exceptionally based on full accounts of National Salmonella Centres concerned, most of the information being taken from separate or collective communications published at various times. Besides, not much information from

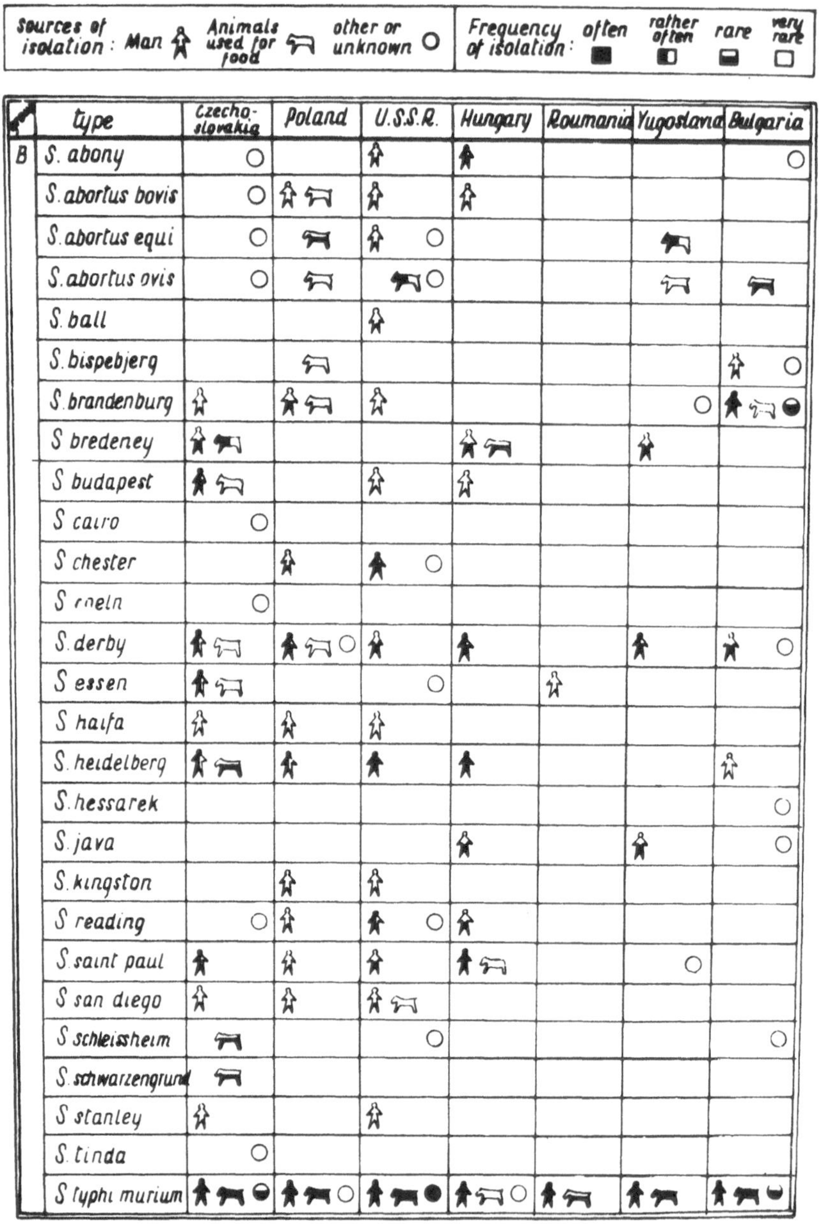

Sources of isolation:	Man 🚹	Animals used for food 🐾	other or unknown O	Frequency of isolation:	often ■	rather often ▥	rare ▤	very rare □

	type	Czecho-slovakia	Poland	U.S.S.R.	Hungary	Roumania	Yugoslavia	Bulgaria
B	S. abony	O		🚹	🚹			O
	S. abortus bovis	O	🚹🐾	🚹	🚹			
	S. abortus equi	O	🐾	🚹 O			🐾	
	S. abortus ovis	O	🐾	🐾 O			🐾	🐾
	S. ball			🚹				
	S. bispebjerg		🐾					🚹 O
	S. brandenburg	🚹	🚹🐾	🚹			O	🚹🐾●
	S. bredeney	🚹🐾			🚹🐾		🚹	
	S. budapest	🚹🐾		🚹	🚹			
	S. cairo	O						
	S. chester		🚹	🚹 O				
	S. coeln	O						
	S. derby	🚹🐾	🚹🐾 O	🚹	🚹		🚹	🚹 O
	S. essen	🚹🐾		O		🚹		
	S. haifa	🚹	🚹	🚹				
	S. heidelberg	🚹🐾	🚹	🚹	🚹			🚹
	S. hessarek							O
	S. java				🚹		🚹	O
	S. kingston		🚹	🚹				
	S. reading	O	🚹	🚹 O	🚹			
	S. saint paul	🚹	🚹	🚹	🚹🐾		O	
	S. san diego	🚹	🚹	🚹🐾				
	S. schleissheim	🐾		O				O
	S. schwarzengrund	🐾						
	S. stanley	🚹		🚹				
	S. tinda	O						
	S. typhi murium	🚹🐾●	🚹🐾 O	🚹🐾●	🚹🐾 O	🚹🐾	🚹🐾	🚹🐾 O

Fig. 2. *Salmonella* types isolated in Eastern European countries (*S. typhi* and *S. paratyphi* A, B, C exclusive).

1st part.

	type	Czecho-slovakia	Poland	U.S.S.R.	Hungary	Roumania	Yugoslavia	Bulgaria
C	S. amersfoort			👤				
	S. bareilly	👤🐄 ○		👤	👤			
	S. blockley	👤					👤🐄	
	S. bovis morbificans	👤	👤🐄	👤 ○	○	👤		👤 ○
	S. braenderup	○						
	S. cholerae suis	🐄	👤	👤🐄 ○		○	👤	
	S. cholerae suis var. kunzendorf	👤🐄 ○	👤🐄 ○	👤🐄 ●	👤	👤🐄	👤🐄	👤🐄 ○
	S. colorado	👤						
	S. concord	○		👤				👤
	S. curacao			👤				
	S. duesseldorf	○						
	S. gatuni			👤				
	S. gdansk		👤					
	S. glostrup	👤	👤					
	S. hadar		👤					
	S. infantis	👤🐄		👤🐄				👤
	S. irumu			👤				👤
	S. kentucky	○		👤				
	S. kottbus	○ 👤		👤	👤			👤🐄
	S. litchfield			👤				
	S. livingstone	○						
	S. lomita	👤						
	S. manhattan		👤		👤		○	👤
	S. mikawasima							👤
	S. mission	👤	👤	👤🐄				👤🐄
	S. montevideo	👤🐄	👤			👤		
	S. muenchen	👤🐄	👤	👤			👤	👤🐄

Fig. 2 – 2nd part.

veterinary sources could be found in relation to some countries, especially Roumania and Hungary. Thus, the picture conveyed by the list may lead to some conclusions not quite correct presently. One example only: the list of *Salmonella* types encountered in Czechoslovakia is in fact SEDLAK's list [3] summarizing discoveries

Group	type	Czecho-slovakia	Poland	U.S.S.R.	Hungary	Roumania	Yugoslavia	Bulgaria
C	S. newport	🧍	🧍🐄	🧍 ●	○🧍🐄	🧍	🧍	
	S. norwich	○						
	S. oranienburg	○	🧍	🧍	🧍	🧍	🧍	○
	S. oslo			🧍				
	S. potsdam	○	🧍	🧍				
	S. praha	○						
	S. richmond	○					🧍	
	S. takoradi	○						
	S. tananarive	○						
	S. tennessee	○	○	🧍	🧍			
	S. thompson	🧍🐄		🧍 ○		🧍		🧍 ○
	S. tosamanga	○						
	S. typhi suis		🐄	🧍 ○				
	S. virchow	🐄	🧍	🧍 ○				
	S. virginia			🧍				
D	S. berta	○						
	S. dublin	🧍🐄	🧍🐄 ○	🧍🐄 ●	🧍		🐄	🧍🐄
	S. eastbourne	○		🧍	•			🧍
	S. enteriditis	🧍🐄 ○	🧍🐄 ○	🧍🐄 ●	🧍 ○	🧍🐄	🧍🐄 ○	🧍🐄 ○
	S. goettingen	🧍						
	S. javiana	🧍						
	S. moscow	○		🧍🐄 ○				
	S. onarimon			🧍				
	S. panama	🧍	🧍	🧍				
	S. pullorum gallinarum	🐄 ○	🧍🐄	🧍🐄 ○			🐄	🐄
	S. rostock	○	🧍	🧍				
	S. saarbrücken		🧍					

Fig. 2 – 3th part.

up to 1957, which contains 77 types; however, it was later reported that 95 *Salmonella* types had been isolated in Czechoslovakia up to 1958, but no list of the types was attached [51]. Lacks of this or another kind have not been avoided in regard to some other coun-

	type	Czecho-slovakia	Poland	U.S.S.R.	Hungary	Roumania	Yugoslavia	Bulgaria
D	S. sendai			🧍				
E	S. anatum	🧍🐕○	🧍	🧍🐕	🧍🐕	🧍	🧍	🧍
	S. binza		🧍					
	S. butantan		🧍					
	S. cambridge		🧍					
	S. give		🧍					🧍
	S. hamilton	○						
	S. lexington		🧍	🧍				
	S. london	🧍🐕	🧍	🧍🐕○			○	🧍
	S. meleagridis	🧍🐕○	🧍	🧍	🧍🐕			
	S. muenster			🧍				🧍🐕○
	S. nchanga	○	🧍					
	S. new brunswick		🧍		🧍			
	S. new-haw		🧍					
	S. newington	○	🧍🐕					
	S. orion	○	🧍	🧍				
	S. senftenberg	🧍🐕	🧍	🧍			○	○
	S. taksony	○	🧍					
	S. uganda						🧍	
	S. vejle			🧍				🧍
	S. weltevreden		🧍					
	S. westhampton	○						
F	S. aberdeen	🧍		○				
	S. solt				🧍			
G	S. cubana		🧍					
	S. poona	🧍						
H	S. neves		🧍		🧍			

Fig. 2 – 4th part.

tries too. Nonetheless, it is felt, the list may serve the purpose to a certain extent.

In all, according to reports found and information received, 119 *Salmonella* types have been isolated in the countries of East Europe. The great majority amounting to 103 types belong to the group

№	type	Czecho-slovakia	Poland	U.S.S.R	Hungary	Roumania	Yugoslavia	Bulgaria
I	S.huttingfoss	O						
	S.szentes				O			O
J	S.berlin			⚹				
	S.kirkee	O						
K	S.minnesota	O						
M	S.nashua							O
	S.vinohrady	O						
40	S.johannesburg		⚹					
	S.riogrande	O						
41	S.waycross	O						
42	S.detroit							O

Fig. 2 – 5th part.

B, C, D or E, the groups F, G, H, I, J, K, M, 40, 41 and 42 being represented by one or two types each. The greatest number of types were reported from Czechoslovakia (78 types), then from U.S.S.R. (63 types), Poland (55 types) and Bulgaria (37 types). The lowest figures came from Hungary (25 types), Yugoslavia (20 types) and Roumania (11 types). It cannot be excluded that, apart from real ecological relations, the activity of the National Salmonella Centres, too, plays a certain role in the number of types found, the most important factors being the quality of diagnostic sera and their availability as well as more or less close collaboration between the National Salmonella Centre and the country's diagnostic laboratories, the veterinary laboratory service inclusive.

Though quantitative relations are expressed rather indistinctly, it will be seen from the list that the old, well-known food-poisoning causes, S. typhi-murium, S. cholerae-suis var. kunzendorf and S. enteritidis, are far from going to quit the scene, S. typhi-murium usually keeping the first place in the countries concerned. There are other types whose distribution is growing steadily wider, like S. derby, S. heidelberg, S. bovis-morbificans, S. newport, S. oranienburg, S. anatum, S. dublin, and other ones, which are limited to a region or two, but with a quite significant frequency, like S. brandenburg in Bulgaria, S. saint-paul and S. bareilly in Hungary and Czechoslovakia, S. newington in Poland etc. A number of other Salmonella types seem to play an important role in human salmonellosis in East Europe, but are limited to a period or an area like e.g. S. chester, S. reading, S. london in U.S.S.R., S. blockley in Czechoslovakia and

Yugoslavia, *S. kottbus, S. abony, S. meleagridis* in Hungary, *S. thompson* in Roumania.

Most of the rest of the *Salmonella* types isolated in East Europe look so far like ephemerids and no indication can be given as to their probable origin. In two instances evidence has been given that imported food, animal feeding stuffs or fertilizers may be responsible for the presence of new salmonellas in the country, as it was in U.S.S.R. with egg products [52] and in Czechoslovakia with fish meal from Angola [53], but in both instances no wider spread of the types imported was observed. A possible role of migratory birds has been considered and examined without conclusive results [54, 55].

A growing importance of human carriers in the dissemination of salmonella food-poisoning may be suggested by some findings.

Salmonella in infantile diarrhoea

It has been estimated in Hungary that out of the cases of intestinal disorders in the Hungarian capital, among which about 40% were infants, only 5% were examined bacteriologically in 1958, and only 9% in 1959 [56]. Percentages may change from one country to another, the fact remaining unchanged that in most of East Europe as a rule only a small proportion of infantile diarrhoea cases are being examined bacteriologically. Thus, the role of salmonellas in etiology of infantile diarrhoeas in the countries of East Europe may be only conjectured on the basis of more or less fragmentary figures.

Studies carried out in children's hospitals, mostly medical school clinics, on the bacterial etiology of infantile diarrhoeas, however limited to a period of time and area, give some insight into the place of salmonellas in the etiology of infantile diarrhoeas. Reports of more or less extensive research of this kind are available from most of East Europe. Perhaps the most extensive research thereon was made in Poland in 1952—1953 [57, 58]. Fifteen research teams composed of pediatricians and microbiologists in eleven towns of Poland were involved, representing nearly all parts of the country. They worked according to a unified plan and used the same examination methods to make the results as comparable as possible. Their results are in accord with those reported from the other countries and may be summarized like this. *Salmonella* organisms play a varying role in bacterial etiology of infantile diarrhoeas in different areas and at different time. While at the time of the studies *Salmonella* findings in infantile diarrhoeas in Poland amounted on the average to about 10%, there were some regions where salmonellas were not encountered at all as the cause of infantile diarrhoeas and there were others where the positive findings amounted to as much as about 30%. Usually such a high incidence was an evidence of an epidemic wave which fell afterwards to a lower incidence rate. *S. typhi-murium* predominated to as much as nearly 9/10 of the *Salmonella* share, the other types being, in the frequency order, *S. derby*, *S. enteritidis*, *S. paratyphi B*, *S. typhi* and *S. heidelberg*.

Lower figures for salmonellas in infantile diarrhoeas were reported from Bulgaria: 0—0.5%[59, 60], Hungary: 1.3%[61], Roumania: 0—2.31%[62, 63, 64], and Yugoslavia: 0—2.2%[65, 66]. Figures more similar to those from Poland came from U.S.S.R. where, too, besides quite negative results figures as high as 8% were reported[67, 98, 99]. Many various *Salmonella* types were encountered as a cause of infantile diarrhoeas, but as a rule *S. typhi-murium* was the type most often met with, the other more common causative agents being *S. saint-paul*, *S. enteritidis* and *S. heidelberg* in Czechoslovakia, *S. heidelberg*, *S. newport*, *S. cholerae-suis* var. *kunzendorf* in U.S.S.R., *S. enteritidis* in Bulgaria and Roumania, and *S. paratyphi B* and *S. cholerae-suis* var. *kunzendorf* in Yugoslavia.

The importance of *Salmonella* organisms in infantile diarrhoeas cannot be estimated only on the basis of their share in sporadic cases. Many institutional epidemics of infantile diarrhoeas were reported, too, where salmonellas had been found as the causative agent, and it is why attention is often called to salmonellas while dealing with the epidemiology of the illness. *S. cholerae-suis* var. *kunzendorf*, *S. bovis-morbificans*, *S. enteritidis*, *S. derby*, *S. typhi-murium*, *S. heidelberg*, *S. newport* were the types reported as the causes of epidemics, but not much light has been thrown on the way they enter a child community. Stress has been only laid on the great ease with which they spread in a child community as well as on the rather high incidence of healthy carriers at the time of an epidemic.

Some facts concerning the various countries

The tables below illustrate the incidence of notifiable *Salmonella* infections in the 15 year post-war period. Possible omissions therein are due to lack of data. Countries are cited by geographical order.

Czechoslovakia

As a whole, Czechoslovakia shows an almost continuously decreasing typhoid incidence and to a certain extent a decreasing paratyphoid incidence as well (see table II). As to other salmonelloses the position is just reverse and the number of cases is almost continuously increasing. The picture varies significantly when Czech lands and Slovakia are considered separately[3, 71]. However typhoid morbidity rate has been decreasing in both, and more definitely in Slovakia, the level of the typhoid incidence is on the average twice as high in Slovakia, the average annual morbidity rate in 1959—1961 being 10.6 per 100 000 population as against 4.6 in the

Table II.

Salmonellosis notifications in Czechoslovakia. Number of cases (c) and morbidity rate per 100 000 population (r).

Years		1947	1948	1949	1950	1951	1952	1953	1954	1955	1956	1957	1958	1959	1960	1961
Typhoid	c	...	3405	2543	3337	2978	2266	2098	1882	1363	1290	1177	1434	1033	747	748
	r	38.4	27.6	20.6	26.9	23.8	17.9	16.4	14.5	10.4	9.8	8.8	10.4	7.7	5.4	5.5
Paratyphoids	c			486	1372	356	390	647	382	723	241	186	226	223	217	101
	r			3.9	10.5	3.0	3.1	5.0	3.1	7.5	1.8	1.4	1.6	1.6	1.6	0.7
Other salmo-nelloses	c					296	634	1044	1200	2134	2511	2287	3275	2936	4462	...
	r					2.4	5.0	8.1	9.8	16.3	19.0	17.1	24.3	21.6	32.7	27.4

Note: Since 1949 paratyphoid cases have been reported separately and since 1951 a distinction has been made in notification between paratyphoids and other salmonelloses.

Czech lands. No distinct difference in paratyphoid incidence could be demonstrated between the two parts, the average annual morbidity rate in 1959—1961 amounting to 1.43 per 100 000 population in both of them. On the other hand, the incidence of other salmonelloses is definitely higher in Czech lands, the average annual morbidity rate per 100 000 population for the same period amounting to 33.6 as against 12.1.

Typhoid fever

The unprecedented World War II typoid epidemic of 1945 (see fig. 1) has been followed by an almost continuous decrease in typhoid incidence approaching actually a 5.0 per 100 000 population morbidity rate. Mostly sporadic cases are being recorded but typhoid epidemics, too, have been reported, most of them limited to a small number of persons, and mainly water-born. It has been computed in Slovakia that the average number of cases per typhoid epidemic in 1959 amounted to 4.5 and the average number of cases per typhoid focus to 1.5[73]. Typhoid morbidity rate in small Slovak country villages was five times as high as in bigger ones[74]. Children are affected more often; in 1957 about 35% of typhoid cases were children up to 14 years of age[72].

No mass typhoid vaccination has been carried out as a control measure in Czechoslovakia[3]. No figures of typhoid carrier incidence in Czechoslovakia have been met with. However, new typhoid carriers are being repeatedly detected[75].

Paratyphoids

Paratyphoid is steadily decreasing in Czechoslovakia and has reached a morbidity rate below 1.0 per 100 000 population. It constitutes one eighth to one third of the typhoid incidence. Apart from sporadic cases, paratyphoid epidemics were recorded[75, 74, 76] and usually they were water-born or milk-born. Paratyphoid in Czechoslovakia is practically limited to paratyphoid B, since no paratyphoid C case had been notified in the post-war period and only 3 paratyphoid A cases, apart from a paratyphoid A epidemic in a mental hospital in 1948[19].

Salmonella food poisoning

Salmonellosis notification figures have grown about ten times in Czechoslovakia since 1951 (see table II), the increase chiefly reflecting the epidemiological position in Czech lands. In 1939—1951 salmonella food poisoning in Czechoslovakia was mostly recorded as outbreaks, the increasing number of sporadic cases being observed only recently[3]. Up to 1946 the majority of the outbreaks were connected with meat or meat produce consumption, but in 1946—1951 only about one third were so, eggs and pastries being predominant. The size of salmonella food poisoning outbreaks rose greatly, the number of cases per outbreak in the second period amounted to 260 as against 41.5 in the first one. *S. typhi-murium* was most often met with, then *S. enteritidis*; *S. senftenberg* and *S. goettingen* being single findings.

In 1960 salmonellas were the infective agent in 15 out of 36 food poisoning outbreaks of bacterial origin[75]. The individual types responsible were (in frequency order): *S. typhi-murium*, *S. enteritidis*, *S. newport*, *S. blockley* and *S. bredeney*. An interesting example of a "new" *Salmonella* type established on a territory has been reported from Olomouc district. In 1953 two *S. bareilly* strains were isolated there and in 1954 six. Further yearly isolations amounted to 78, 51, 102, 133, 169 and in the first half of 1960 the number of *S. bareilly* isolations was as high as 187[77]. The great number of *Salmonella* types isolated in Czechoslovakia recently has been explained by imported food, animal feeding stuffs and fertilizers[3]. Many various *Salmonella* types have been isolated from domestic animals and meat products, too[78, 79].

Poland

However the post-war figures for typhoid were lower than those of the inter-war period, the decrease trend was rather meagre until the midfifties when a more definite fall in the incidence occurred and the typhoid morbidity rate turned out below the level of 10 cases per 100 000 population (table III). As far as may be judged from paratyphoid notification figures the post-war paratyphoid incidence was higher than before and it is only in the last three years that a decrease trend is observed. Salmonella food poisoning morbidity rate is lower in Poland than in Czechoslovakia or Hungary and looks rather stabilized. Many sporadic cases of salmonella food poisoning, however, probably escape notification yet.

Typhoid fever
From the annual average of about 50 cases per 100 000 population in the inter-war period through that of a little below 20 in the first half of the fifties, the typhoid incidence in Poland entered a period of continuous decrease which resulted in overcoming the 10 per 100 000 population morbidity rate. And the trend so far continues. Though sporadic cases prevail, epidemics occur every year, some of them of fairly large dimensions. The number and size of typhoid epidemics, however, seems to decrease recently. While reports from the fifties include descriptions of such events like the Łódź and Zgierz milk-born epidemic of 570 typhoid cases in 1950[80], or the Cracow University canteen epidemic of 176 cases in 1957[81] and others of a similar extent, four of the largest typhoid epidemics of 1960 amounted to 86, 35, 34 resp. 15 cases and two of those in 1961

Table III.

Salmonellosis notification in Poland. Number of cases (c) and morbidity rate per 100.000 population (r).

Years		1947	1948	1949	1950	1951	1952	1953	1954	1955	1956	1957	1958	1959	1960	1961
Typhoid	c	11748	7975	6191	7215	5984	5317	6174	5541	6201	4797	5717	4633	4254	3464	2904
	r	49.3	33.5	25.6	29.1	23.9	20.8	23.7	20.5	22.6	17.0	18.3	16.0	14.5	11.7	9.4
Paratyphoids	c					373	558	1151	790	921	619	713	750	448	422	400
	r					1.48	2.18	4.41	2.99	3.3	2.2	2.5	2.6	1.5	1.4	1.3
Salmonella food poisoning	c					58	550	1675	1590	597	1083	875	821	764	1213	950
	r					0.23	2.1	6.4	6.0	2.1	3.8	3.0	2.8	2.5	4.0	3.1

Note: In 1947—1950 typhoid and paratyphoids were notified together.

numbered 16 resp. 15 cases. Contaminated milk and water, or food prepared by a typhoid carrier were most often incriminated. Some differences in typhoid distribution exist between various parts of the country. It may be stated that it is the centre of the country, from the Baltic sea southwards to the arch of Vistula, that comprises provinces whose annual typhoid morbidity rate regularly exceeded the country's average in the last five years, the other provinces being almost regularly below the average. In 1958 about 42.5% of typhoid cases were children up to 14 years of age. The proportion has not changed much since.

More than 4200 chronic typhoid carriers were on record by the end of 1961 which would mean about 13.7 carriers per 100 000 population. There are most probably more typhoid carriers yet undetected.

Typhoid vaccine is largely used as a control measure. About 10 million people a year are usually vaccinated. The vaccination is statutory and universal except for the youngest age groups, and it is the province administration which decides every year if the typhoid vaccination is to be carried out[7].

Paratyphoids

In the inter-war period paratyphoid notification figures were very low and the morbidity rate in 1931—1938 did not exceed 0.5 per 100 000 population. After the war a separate notification of paratyphoids began in 1951. On the average, incidence figures used to be at least 3 or 4 times as high and gave a picture of stabilization, though, since 1959 a sort of decrease trend may be observed. As a rule paratyphoid incidence distribution is rather irregular and variable. There were paratyphoid epidemics of 40—70 cases each recorded in the fifties, but recently mostly sporadic cases occur. About 2/3 of all the paratyphoid cases were urban[82].

Most of paratyphoid cases in Poland in the post-war period were due to *S. paratyphi B*. In 1951—1956 the paratyphoid B incidence was ten times as high as that of paratyphoid A, paratyphoid C with its two bacteriologically confirmed cases being negligible[36]. Epidemics of both paratyphoid A and paratyphoid B were recorded but as a rule they occurred sporadically. It was in the provinces of Warsaw, Cracow, Katowice, Łódź and Wrocław that paratyphoid A occurred more often, the cases being usually encountered in towns and other places with a dense population[20].

Salmonella food poisoning

Salmonellas constitute the most important bacterial agent in food poisoning in Poland, the next one being staphylococci with less than a half of the number of outbreaks resp. cases due to salmonellas[83]. In 1951—1956 the number of food poisoning outbreaks registered amounted to 512 with 20944 cases involved[84]. Of these 146 were known as due to salmonellas (28.5%) with 5553 cases involved (26.5%). From 173 salmonella food poisoning outbreaks recorded in Poland in 1946—1956 as many as 129 were due to S. typhi-murium (74.5%), the next being S. enteritidis (12.0%), S. dublin (5.8%), S. cholerae-suis var. kunzendorf, S. bovis-morbificans, S. newport and S. heidelberg. Meat or meat products were incriminated in 65.3% of the outbreaks, in 31.2% of the outbreaks the infection sources remaining unknown. About 80% of salmonella food poisoning outbreaks occurred during the warm months of the year (V-IX).

U.S.S.R.

In relation to a state of the dimensions of the U.S.S.R. (more than a 200 million population) any figures for the country as a whole may give a quite inaccurate picture whatever the precision of the figures would be. Thus, the fact that no tabular presentation of the typhoid, paratyphoid or other salmonellosis incidence in the U.S.S.R. is given is of less importance than the fact that only scattered numerical data were available as to the salmonella infections in particular republics. It is, however, believed that some picture has been composed which is near the truth, though some conjectures or even guesses were unavoidable.

Typhoid fever
Before World War II typhoid and paratyphoid incidence in the U.S.S.R. had decreased to a level of about 50 cases per 100 000 population, as judged from figures available in respect to the Ukrainian S.S.R. in 1940[85]. There was a great increase in typhoid and paratyphoid incidence during the war but it ceased soon, and not only the inter-war level but also a further decrease in the incidence has been achieved. It may be illustrated again by the typhoid morbidity rate per 100 000 population in the Ukrainian S.S.R., which amounted to 57 in 1947, to 18 in 1950, to 13 in 1953, and to 12.3 in 1956. A decrease trend like this, however, has not been observed in all the republics. As a result the average typhoid morbidity rate for all the country in the mid-fifties constituted about a half of that in the last inter-war years[86], probably about 25 per 100 000 population. The decrease trend ceased in the fifties. Three groups of the republics can be revealed according to the typhoid incidence level and trend in the fifties[9]; in the Russian Federal Republic as well as Belorussian, Ukrainian, Estonian, Moldavian and Latvian S.S.R. the typhoid incidence was relatively low and decreasing, in the Lithuanian, Georgian, Kazakh, Azerbaidzhan and Armenian S.S.R. it was higher, and in the Kirghiz, Tadzhik, Uzbek and Turkmen S.S.R. it was high and increasing. But even in the republics with

a relatively low typhoid incidence there were regions with a high incidence.

About 60% of the typhoid cases were recorded among town population. Some towns are known as usually more affected: Tashkent, Volgograd (Stalingrad), Ashkhabad, Saratov, Kuibyshev among others. In the Baltic republics seasonal increase in typhoid incidence has stopped being a rule, in other European republics the seasonal increase of 3—4 months' duration and in Middle-Asian republics that of 6—7 months' duration is observed. Young age groups participate in the typhoid incidence to a greater degree, but instances were reported of places with adult incidence prevailing [25, 87]. Water is often incriminated as the source of infection, especially in the Middle Asian republics. Milk contaminated by typhoid carriers has, too, been rather often reported as the source of infection.

Typhoid vaccines are widely used as a control measure along with other prophylactic methods. Typhoid carriers are looked for with varying intensity[9]. According to fragmentary reports the typhoid carrier incidence may be rather high[88, 89].

The stabilization of the country's typhoid morbidity rate in the fifties calls for more effort in sanitation investment, especially in the Middle Asian republics, where the water supply is scanty. As the state pays much attention to the public health problems, there is no doubt that this will be achieved and thus conditions created for a further decrease in typhoid incidence.

Paratyphoids

Fragmentary figures demonstrate a steady decrease in the paratyphoid incidence at least in some parts of the U.S.S.R. In the Belorussian S.S.R. the paratyphoid morbidity rate per 100 000 population dropped from 5.0—6.0 in the early post-war years to 3.7 in 1957[90]. A similar trend show the annual morbidity rate figures in Tatar A.S.S.R. in 1956—1960, viz. 2.7, 3.0, 2.8, 1.3 and 0.6[81], or in the town of Lvov from 24.6 in 1946 to 2.0 in 1956 and 0.2 in 1957[92]. The same sources demonstrate that at least in the western parts of U.S.S.R. paratyphoids' share in the enteric fever incidence is greater than that shown by the average inter-war index[18]. Both paratyphoid A and B take part in the paratyphoid incidence to a varying degree in various parts of the country; paratyphoid B being usually most common in the north-west, paratyphoid A in the south-east. The two types occur both in epidemic and sporadic form, water, milk and food being usually incriminated. No quantative data of comparative value may be cited on paratyphoid C. It is understood, however, that its share in the general paratyphoid incidence in the U.S.S.R. is rather small if not negligible. Some more information on paratyphoids in the U.S.S.R. has been already given above.

318

Salmonella food poisoning

It has been stressed in relation to the Russian Federal Republic which comprises more than one half of the country's population, that about 2/3 of all food poisonings are of bacterial origin, and that approximately 9—12% of those are due to salmonellas[93, 94]. In the period of time referred to, however, the percentage was higher, dropping from 22.5 to 15. Meat and meat products were incriminated in about 40%, milk in about 25% and fish in about 10%. The average number of cases per outbreak was growing.

S. typhi-murium was the Salmonella type most often met with in food poisoning in the U.S.S.R., in fact in more than 50% of cases. The next most common, S. enteritidis and S. cholerae-suis (probably var. kunzendorf) were a fourth of that number each.

Leningrad is one of the few Soviet towns where sporadic salmonella gastro-enteritis notification is statutory beside salmonella food poisoning outbreaks[95]. A growth of the salmonellosis incidence has been observed there, the figures in 1958 being twice as high as in 1956. Though S. typhi-murium was responsible for about one half of the cases, too, the next type most often met with was S. heidelberg (13.4%), with S. anatum, S. london, S. bovis-morbificans and others following; in all 40 Salmonella types.

While in the Georgian S.S.R. in 1948—1949 most food poisoning outbreaks were due to S. cholerae-suis var. kunzendorf, and only few to S. newport, S. enteritidis, S. dublin, S. moscow or S. paratyphi C, in 1961 most of them were due to S. typhi-murium, the rest being divided between S. cholerae-suis, S. oranienburg, S. dublin and S. newport[97, 96].

There are many various reports on salmonella food poisoning in the U.S.S.R. with many fragmentary figures or other data. In all the opinion is rather generally shared that enteric disorders of salmonella origin grow in number and so do sporadic cases of the condition. A significant incidence of healthy carriers detected among food handlers has been reported too[45, 46].

Hungary

Hungary enjoys a rather good position as to the typhoid incidence, showing a steady decrease trend; its paratyphoid incidence is not high, but no definite decrease trend has been observed so far. Salmonella food poisoning seems to be a growing problem, but its more precise evaluation will be possible when more progress in notification of the condition is achieved.

Typhoid fever

Up to 1937 the typhoid morbidity rate fluctuated at about 100 cases per 100 000 population jumping up significantly in the epide-

319

mic years and going down in the favourable ones. In the most unfavourable epidemic year of 1932 it amounted to as high as 253 per
100 000 population. During World War II a significant rise in typhoid incidence occurred, but a steady improvement has been
noted since and almost every year's notification figures have been
a low level record (see table IV). In the fifties the morbidity level of
10 per 100 000 population was passed, in the sixties a further drop
under 5.0 per 100 000 population was observed and the decrease
trend seems to continue.

As a rule sporadic typhoid cases are being notified in recent years,
and only small local epidemics occur from time to time. No region
of definitely higher typhoid morbidity rate could be shown, but in
the fifties the average annual typhoid morbidity rate calculated for
the large and intermediate towns exceeded that of the country as a
whole[4]. The capital, Budapest, was an exception and, apart from the
war time, steadily enjoyed a typhoid morbidity rate below the country's. Increased participation of children in the typhoid incidence
has been observed for several years now.

It is believed there, that it was chiefly the vaccination campaign
carried out for several years since 1935, first with Kolle vaccine then
with a bacterial preparation after BOIVIN, adsorbed on aluminium
hydroxide, that resulted in the spectacular fall in typhoid incidence[1].
Since 1951 it stopped, however, to be a mass campaign; no area
could be found with a typhoid morbidity rate justifying a mass
vaccination. About 2000 typhoid carriers have been detected and
supervised and more effort is being made to detect the yet unknown.

Paratyphoids

Paratyphoid notification figures suggest an incidence stabilized
within a range of 1.0—5.0 cases per 100 000 population. Not too
much reliance, however, may be placed upon the figures yet. Even
after 1959, which was the first year of statutory notification of
salmonellosis gastroenterica in Hungary, it was suspected that some
of these had been reported as paratyphoids[68]. As no cases of paratyphoid C and only few of paratyphoid A have been diagnosed in the
post-war time[50], paratyphoid in Hungary is practically a synonym
of paratyphoid B. Sporadic cases predominate, but paratyphoid
B epidemics have been recorded too, as e.g. one at Szombathely
and vicinity in 1952, comprising 386 cases[69] of two smaller ones
in 1960, one of them waterborn. The number of paratyphoid B
carriers detected amounts to about 5% of the total of the known
typhoid-paratyphoid carriers, and is four times as high as that of
paratyphoid A carriers.

Salmonella food poisoning

Since "Salmonellosis gastroenterica" is being reported separately

Table IV.

Salmonellosis notifications in Hungary. Number of cases (c) and morbidity rate per 100 000 population (r).

Years		1947	1948	1949	1950	1951	1952	1953	1954	1955	1956	1957	1958	1959	1960	1961
Typhoid	c	2854	1974	1629	1354	1555	1232	1611	957	909	748	815	608	509	513	480
	r	31.4	21.6	17.6	14.5	16.5	13.0	16.8	9.9	9.3	7.6	8.3	6.2	5.1	5.1	4.8
Paratyphoids	c	279	367	110	499	153	418	174	81	310	156	158	147	117	199	92
	r	3.1	4.0	1.2	5.4	1.6	4.4	1.8	0.8	3.2	1.6	1.6	1.5	1.2	2.0	0.9
Salmonellosis gastroenterica	c													564	677	972
	r													5.7	6.8	9.7

Note: Statutory notification of Salmonellosis gastroenterica introduced in 1959.

only recently, the growing number of cases notified in the three year period may be a result of notification improvement as well as a real incidence growth. Both sporadic cases and outbreaks are included. Three major outbreaks of the period numbered from 72 to 106 cases and were caused by *S. typhi-murium*. This organism was the infective agent in more than a half of 42 salmonella food poisoning outbreaks recorded in Hungary in 1936—1944 and 1947—1957, too, the infective agent in the rest of them being *S. enteritidis, S. java, S. anatum, S. saint-paul, S. cholerae-suis, S. heidelberg,* and *S. kottbus*[50].

The growing incidence of *Salmonella* types other than typhoid and paratyphoid organisms among the healthy population (positive findings amounting to 0.32% in 1956—1958 as against 0.006% in 1936 —1944 and 0.085% in 1953) is considered as an evidence of the growing spread of various *Salmonella* types in Hungary[70].

Roumania

As may be seen from table V the typhoid incidence is gradually declining in Roumania and since 1958 the morbidity rate per 100 000 population has always been under 10. Paratyphoid incidence is quite low and constitutes about 1/15-th of that of the typhoid fever. It may be calculated for salmonella food poisoning in Roumania that the average annual morbidity rate in 1951—1956 amounted to about 8 per 100 000 population. This is quite certainly much more than formerly, but the calculation was based on figures related to outbreaks only, sporadic food poisoning being probably unnotifiable at that time.

Typhoid fever
In the inter-war time the average annual typhoid-paratyphoid morbidity rate amounted to about 38.0 per 100 000 population. It dropped significantly in war-time but rose to the previous height in the early post-war years. Since 1953 a gradual decrease has continued and a comparatively low incidence level was reached. Typhoid epidemics occur as well as places with a higher endemicity level, but no detailed information as to the typhoid incidence distribution has been obtained. Vaccination as a control measure has been widely used, particularly in the fifties. Civilian population was vaccinated from the 12-th year of age. Kolle TAB vaccine was applied, but apart from the standard Ty_2 strain five other local typhoid strains of various phage types were used for vaccine production[100].

Paratyphoids
Paratyphoid notification figures, though low, show a fluctuation which may be a result of epidemics. Paratyphoid B predominates, but paratyphoid A has been recorded too. No evaluation of its

Table V.

Salmonellosis notifications in Roumania. Number of cases (c) and morbidity rate per 100 000 population (r).

Years		1947	1948	1949	1950	1951	1952	1953	1954	1955	1956	1957	1958	1959	1960	1961
Typhoid	c	7579	7052	5981	5596	5694	4828	3402	2966	2642	2305	1976	1662	1559	1638	1555
	r	46.0	42.8	36.2	32.0	32.6	27.4	19.2	15.8	14.1	12.4	11.3	9.2	8.5	8.8	8.4
Paratyphoids	c											111	58	54	295	46
	r											0.6	0.3	0.3	1.6	0.2

Note: Some typhoid figures comprise paratyphoids as well.

Table VI.

Salmonellosis notifications in Yugoslavia. Number of cases (c) and morbidity rate per 100 000 population (r).

Years		1947	1948	1949	1950	1951	1952	1953	1954	1955	1956	1957	1958	1959	1960	1961
Typhoid	c	6828	5811	3954	4583	5465	4708	3679	3320	3028	3122	3310	4175	3791	3134	3208
	r	44.3	36.8	24.6	28.2	33.1	28.1	21.6	19.5	17.2	18.0	18.4	23.2	21.0	16.8	17.2
Paratyphoids	c	963	1349	1047	966	1619	1192	928	1240	974	1299	1166	1137	1662	1978	2033
	r	6.2	8.6	6.6	5.9	9.9	7.2	5.5	7.3	5.5	6.8	7.0	6.8	9.9	10.6	10.9

frequency distribution can, however, be made, since data are scanty and a marked variation occurs. So, e.g. among *Salmonella* strains other than *S. typhi*, isolated in Roumania in 1948—1951, *S. paratyphi A* strains amounted to 9.09% of the total as against 20.45% of *S. paratyphi B*[101]. Out of 100 paratyphoid cases treated at the Cluj clinic of infectious diseases in a 12 year period only 11 were paratyphoid A, the rest, except for three, being paratyphoid B cases[21]. Among 500 children examined *S. paratyphi A* and *S. paratyphi B* were encountered in 0.4% of the total each[102]. The case of the town of Galati has already been mentioned before. Paratyphoid C probably occurs in Roumania, but only very rarely and it seems to play no role in the paratyphoid problem of the country.

Salmonella food poisoning
No systematic notification figures concerning salmonella food poisoning in Roumania are known to the author. From a lecture delivered by Roumanian authors in Warsaw in 1957[103] it is known that from 144 food poisoning outbreaks involving 5361 cases in 1905—1950 as many as 76.4% were due to salmonellas, and from 84 outbreaks involving 11623 cases in 1951—1956 salmonellas' share amounted to 63.1%. The majority of outbreaks were caused by *S. typhi-murium* (66%), then by *S. enteritidis* (14%); *S. cholerae-suis*, *S. thompson*, *S. newport*, *S. oranienburg*, *S. bovis-morbificans*, and *S. montevideo* were encountered in 14% of the outbreaks all together. It was meat and meat products which were incriminated most often, but dairy products and eggs were recorded, too.

Yugoslavia

This is to a certain extent an exception among the Eastern European states as there has been only little change in its typhoid-paratyphoid morbidity rate during the last 40 years (see fig. 1). If, however, typhoid and paratyphoids are considered separately, a decrease of the typhoid incidence and an increase in the paratyphoids is apparent (table VI). No data for a tabular presentation of salmonella food poisoning incidence could be found.

Typhoid fever
After a period of a comparatively high typhoid incidence in the early post-war years the typhoid morbidity rate has been stabilized since 1953 at a level of about 20 cases per 100 000 population, in fact more often below than above that. A comparatively often occurrence of typhoid epidemics and existence of places and regions of a rather high typhoid endemicity level is a counterpart of the position[104, 105, 106, 107]. There are considerable differences in typhoid incidence between particular republics of the country as well as

324

between particular regions of the republics. E.g. Slovenia's typhoid morbidity rate in its decrease trend amounted in 1958 to 6.1 cases per 100 000 population as against 23.3 of the country as a whole[108], or, in Serbia the AKMO province has greatly exceeded the UZA Srbija province in typhoid morbidity rate and has demonstrated quite a different trend too[107]. Water has been most often incriminated as the source of infection in epidemics and endemics. Typhoid carriers and milk have been referred to as well, but rather rarely.

Paratyphoids

As mentioned above, paratyphoids have demonstrated recently an increase trend in Yugoslavia. They constitute about 1/3 of all the enteric cases and paratyphoid epidemics are no rarity. It is chiefly paratyphoid B that is responsible for the paratyphoid incidence and increase trend in Yugoslavia. Paratyphoid A is rather a rare event there. In a hospital at Zagreb 12 paratyphoid cases were treated in 1946—1952[109], and at Ljubljana Institute 2 *S. paratyphi A* strains were isolated from human cases in 1954 as against 63 *S. paratyphi B* and 18 *S. typhi* strains[27]. Paratyphoid C has occurred even less often in Yugoslavia. At a Sarayevo clinic 5 paratyphoid C cases were diagnosed among 400 enteric cases[38]. A single paratyphoid C case was recorded at Knin region in 1952[31] as well as rare isolations of *S. paratyphi C* strains at a children's hospital in Slovenia in 1956—1957.

Apart from being almost a synonym for paratyphoid in Yugoslavia, paratyphoid B is the most often met with enteric fever in Slovenia[27, 28], in fact more that twice as often as typhoid. This is not the case in other parts of the country. At Ossijek e.g. at the time of epidemic in 1958 the number of *S. paratyphi* B strains isolated amounted to 21 as against 99 of *S. typhi*[105], or, at a Zagreb hospital in 1946—1952 there were 223 paratyphoid B cases as against 1327 typhoid ones[109]. Along with that, paratyphoid B carriers are probably numerous. In Croatia e.g. from 47602 samples of faeces of food handlers 61 *S. paratyphi* B strains were isolated in 1959 as against 90 *S. typhi* strains[110].

Salmonella food poisoning

If figures from Serbia may be considered as referring to all the country a conclusion should be drawn that *Salmonella* organisms are not a frequent cause of food poisoning outbreaks of bacterial origin in Yugoslavia, since out of 72 outbreaks in Serbia in 1946—1953 only 6.9% were due to salmonellas[111].

This is, however, probably not the case, at least not in recent years, as reports on salmonella food poisoning outbreaks in Yugoslavia are not so rare recently. *S. typhi-murium* and *S. cholerae-suis* var. *kunzendorf* were found most often as the cause of food poisoning outbreaks, the others being *S. derby, S. blockley, S. bredeney, S. java,*

S. enteritidis, S. oranienburg and *S. newport*. It has been stressed that *S. enteritidis* was an often finding formerly[109]. As to *S. blockley*, which was a "new" type in Yugoslavia, it has been reported[111, 112] that the organism was first isolated from animals and only after a month or so from human cases.

Bulgaria

To some extent this state may be considered to be a special case in the epidemiology of typhoid fever in East Europe, as since 1949, with the only exception of 1952, Bulgaria's typhoid morbidity rate has always been below 5 cases per 100 000 population, and since 1954 even less than 2.0 (table VII). The paratyphoid incidence, however low, does not seem so unprecedented. No tabular presentation of salmonella food poisoning incidence could be drawn for lack of figures.

Typhoid fever
The drop in typhoid incidence that began in Bulgaria in 1938 has continued ever since although not uninterrupted by transitory fluctuations. There was, however, no war-time increase in typhoid incidence; on the contrary, a further marked drop occurred during World War II. In spite of the low typhoid morbidity rate typhoid epidemics occur, but they are less frequent now and usually rural and water-born[113, 114, 115]. As a rule most typhoid cases are sporadic [116]. A shift of typhoid incidence to the youngest age groups has been observed. E.g. in Burgas region in 1952—1960 as much as 51.6% of typhoid cases were children up to 7, and 24.7% were children between 8 and 14[5].

Typhoid vaccination has been widely used as a control measure in Bulgaria and it is believed, there, that it is the reason, along with sanitation improvement especially in rural districts, that success in typhoid control has been achieved[117]. Typhoid carrier incidence is considered to be unsufficiently known since new carriers have been repeatedly detected.

Paratyphoids
The paratyphoid morbidity rate per 100 000 population is low in Bulgaria and always below 1.0 in the last five years. On the average the paratyphoid incidence is four times as low as that of all enteric cases.

Paratyphoid A has been met with very rarely in the post-war years. Only one paratyphoid A case was treated along with 174 enteric cases at a Sofia clinic lately[117], and no paratyphoid A case has been recorded at the Plovdiv infectious clinic in 1947—1957[118]. Only one instance of the isolation of a *S. paratyphi A* strain in Bulgaria

Table VII.

Salmonellosis notifications in Bulgaria. Number of cases (c) and morbidity rate per 100 000 population (r).

Years		1947	1948	1949	1950	1951	1952	1953	1954	1955	1956	1957	1958	1959	1960	1961
Typhoid	c	882	736	340	349	177	520	207	320	204	153	215	168	115	141	79
	r	11.9	10.4	4.7	4.8	2.4	7.2	2.8	4.2	2.7	2.0	2.8	2.2	1.5	1.8	1.0
Paratyphoids	c											22	29	16	71	9
	r											0.3	0.4	0.2	0.9	0.1

Note: Typhoid and paratyphoids were notified together up to 1950; no paratyphoid figures, however, were available to the author up to 1956. It is possible that the typhoid figures from 1951 to 1956 contain paratyphoid notifications as well.

in the period of 1956—1960 has been reported from the Institute of Epidemiology and Microbiology in Sofia[119]. As no paratyphoid C case has been reported lately, paratyphoid B is practically responsible for all the paratyphoid cases in Bulgaria. As a rule only sporadic cases are reported. A small paratyphoid B epidemic occurred in 1958; 14 persons from 5 places were involved, and vegetables were incriminated as the source of infection.

Salmonella food poisoning

It has been announced[120] that food poisoning incidence in Bulgaria decreases year by year recently and that in 1961 the notification figure was as low as one eighth of that in 1954. About 85% of the food poisoning incidence used to be due to bacterial contamination of food; how much of that is to be ascribed to salmonellas would be difficult to tell. In a Sofia hospital 240 patients suffering from food poisoning of bacterial origin were treated in 1954—1955 and as few as 23 (9.6%) were due to salmonellas[121]. It would denote a rather lesser part played by salmonellas in food poisoning in Bulgaria. But, from 1957 to 1959 there were 138 enterocolitis cases treated in a Sofia hospital (the same?) and in 71 of them salmonella etiology was established[126]. This indicates rather a greater share of salmonellas in food poisoning etiology. To the same conclusion may lead the study of papers on individual food poisoning outbreaks as well as reports from the National Salmonella Centre, Institute of Epidemiology and Microbiology, Sofia[119, 122]. Thus, in 1956—1960 there were 316 *Salmonella* strains from acute gastroenteritis cases on record in the Institute. As many as 123 strains were diagnosed as *S. typhi-murium*, the next place being occupied by *S. brandenburg* with 82 strains, then by *S. enteritidis* with 46 strains, *S. muenster* with 34 strains, *S. cholerae-suis* var. *kunzendorf* with 10 strains, and *S. paratyphi B, S. muenchen, S. concord, S. manhattan, S. thompson, S. infantis, S. irumu, S. bovis-morbificans, S. dublin* and *S. uganda* with a smaller number of strains down to a single isolation. It may be added that, besides, as many as 50 *S. brandenburg* strains, 24 *S. muenster* strains, 11 *S. enteritidis* strains and 10 *S. typhi-murium* strains were isolated from healthy people.

Meat and meat products were most frequently incriminated, especially those provided from emergency slaughter. A presumption has been made that some *Salmonella* types may be rather widely distributed in animals and human environment in Bulgaria without causing so far food poisoning in man. Thus, *S. senftenberg* has been found in doves at Plovdiv and on restaurant utensils in Sofia without being isolated from human cases[123].

Concluding remarks

We were able to present here only a vaguely outlined picture of the distribution and the trend adopted in recent times, of salmonella infections in East Europe.

No attempt was made to give an answer to any substantial problems. The belief has been expressed only that, where typhoid fever is concerned, a further improvement in the epidemiological position can be reasonably expected without, though, specifying the reason for such an optimistic view. It is only too tempting to forecast a further decrease in the typhoid incidence in East Europe on the basis of the decrease experienced up to now. But as long as the reason for the noted decrease remains unknown for certain, every forecast remains a guess, if not wishful thinking.

It may seem strange that research in the countries concerned has paid little attention only to the reasons of the spectacular decrease in East Europe's typhoid incidence. True, some research has been carried out on the evaluation of typhoid vaccines in field trial [25, 124, 125]. However valuable these studies may be, and there is little doubt as to their contribution to the problem, one has to be careful to expect too much from their results. They refer to one factor only in the typhoid spread: the immunity level of the population. This, however, is insufficient.

A great reservoir of the pathogen exists in the army of carriers, whose number is so far difficult to evaluate. The low typhoid incidence in East European countries reduces the reservoir. This, however, is a very slow process. A continuous effort to cut the routes of transmission and thus eliminate the sources of infection ought to remain the chief task of the health authorities for years ahead; at least as long as no feasible means for typhoid carrier treatment are available.

As to the salmonella food poisoning which appears to be a growing health problem in all of East Europe but Bulgaria, much more information is required in order to assess its real position. It has appeared that the most valuable information thereon may be provided by the National Salmonella Centre accounts when compiled in accordance with the requirements of the epidemiologist as well, i.e. if the Centres were preoccupied not only with strains and their proper diagnosis but also with the cases, outbreaks, sources of infection and routes of transmission. In most of the countries of East Europe National Salmonella Centres exist and there are a few examples of a fairly broad activity. In that field, however, the requirements of East Europe are exceedingly greater.

REFERENCES

1. PETRILLA, A.: Aktuelle epidemiologische Fragen in Ungarn. *Acta microbiol. Acad. Sci. hung.*, 1954, 1, *297—305*.
2. VERBEV, P. E.: Epidemiological effects of the Great October Socialist Revolution. *Khig. Epidem. Microbiol.*, 1957, 1, 5, *24—33*. (Bulg).
3. SEDLÁK, J.: Zur Problematik der Salmonellosen in der ČSR. *Z. ges. Hyg.*, 1959, 5, *91—102*.
4. BAKÁCS, T.: Typhoid fever and communal hygiene in Hungary. *O.K.I. Működése*, 1960, *159—179*. (Hung.).
5. TYUFEKCHIEV, T.: Control and eradication of infectious diseases in Burgas region. *Khigijena*, 1961, 4, 2, *1—4*. (Bulg.).
6. BAKÁCS, T.: Personal communication.
7. KOSTRZEWSKI, J.: Infectious diseases in Poland in 1919—1962. Warsaw. (In preparation). (Polish).
8. ZHDANOV, V. M.: Speed up the process of control and eradication of infectious diseases. *Klin. Med. (Mosk.)*, 1960, 41, 8, *13—20*. (Russ.).
9. KADEN, M. M., M. K. KHAZANOV & Z. V. PANFILOVA.: Typhoid and paratyphoids in U.S.S.R. and ways to a further decrease of their incidence. *Sovetsk. Med.*, 1960, 24, 5, *17—21*. (Russ.).
10. MATĚJOVSKÁ, V. & J. JELÍNEK: On phage typing of *S. typhi*. *Čs. Epidem.*, 1959, 8, *168—172*. (Czech).
11. BUCZOWSKI, Z.: Report on the activity of the National Phage Typing Centre (unpublished).
12. KRYLOVA, M. D.: Phage typing of typhoid bacteria. *Zh. Mikrobiol. (Mosk.)*, 1962, 33, 10, *135—139*. (Russ.).
13. EÖRSI, M.: Phage-Types of *S. typhi* strains isolated in Hungary and relevant investigations made from 1950 to 1954. *Acta microbiol. Acad. Sci. hung.*, 1956, 3, *285—298*.
14. CIUCA, M., N. NESTORESCO, M. POPOVICI & A. TUPA: Observations sur la lysotypie de *S. typhi* et *S. paratyphi* B en Roumanie. *Arch. roum. Path. exp.*, 1962, 21, *351—356*.
15. LEVI, E. & P. TOMAŠIĆ: (in SEDLÁK & RISCHE's Enterobacteriaceae-Infektionen, Leipzig, 1961, p. 412).
16. AWDJIEW, G. & S. ATANASSOVA: Sur les lysotypes des bacilles typhiques en Bulgarie. Tagungsbericht des 4. Colloquium über Fragen der Lysotypie. Wernigerode/Harz, 8.–11. Juni 1960, p. *66—68*.
17. PANKOV, N. V.: Distribution of *S. typhi* phage types in Turkmenian S.S.R. *Zh. Mikriobiol. (Mosk.)*, 1962, 33, 6, *58—60*. (Russ.).
18. GROMASHEVSKII, L. N. & G. M. VAINDRAKH: Special Epidemiology, Moscow, 1947, p. *38*. (Russ.).
19. BÁRDOŠ, V.: Water-born paratyphoid A epidemic. *Čs. Epidem.*, 1953, 2, *55—63*. (Czech).
20. BUCZOWSKI, Z.: Paratyphoid A in Poland. *Biul. Inst. Med. morsk. Gdańsk.*, 1959, 10, *131—139*.
21. GAVRILĂ, I., L. COMES, R. JOSAN, C. PIRVU & M. GHIDALI: On the incidence of paratyphoid fever and its diagnosis in relation to 100 paratyphoid cases. *Rev. Igiena*, 1955, 2, *70*. (Roum.).
22. FLAX, A. & B. LEIBOVICI: A water-born mixed epidemic of typhoid, paratyphoid A and dysentery. *Microbiologia (Buc.)*, 1958, 3, *547—550*. (Roum.).
23. BRAUDE, I. R.: On pathology and clinic of paratyphoid A. *Vrach. Dyelo*, 1947, 27, *746—748*. (Russ.).
24. MELNIK, E. G., E. A. GINZBURG & E. G. BURKOVA: Bacteriological and serological diagnosis of salmonellosis on the basis of laboratory findings in one Moscow infectious clinic. *Zh. Mikrobiol. (Mosk.)*, 1960, 31, 1, *143—147*. (Russ.).

25. KHEIFETS, L. B., V. A. KILESSO, A. E. KAPLAN, G. S. GURALEVICH, YA. E. TIMAN, A. V. SKROZNIKOVA & YU. I. GUSEVA: Study of poly-vaccine in field trial. *Zh. Mikrobiol. (Mosk.)*, 1958, **29**, 10, *44—48*. (Russ.).
26. DOBREITSER, I. A.: Geographical distribution of paratyphoid fevers in U.S.S.R. *Zh. Mikrobiol. (Mosk.)*, 1934, **3**, 2, *1—15*. (Russ.).
27. ZAJC-SATLER, J.: Most common enteric pathogens. *Zdrav. Vestn.*, 1955, **24**, *400—403*. (Slov.).
28. PLANINŠEK, F.: Typhoid in Celje region. *Zdrav. Vestn.*, 1956, **25**, *358—360*. (Slov.).
29. LALKO, J. & K. PIETKIEWICZ: Experiments on bacteriophage and biochemical typing of *Salm. paratyphi B. Biul. Inst. Med. morsk. Gdańsk.*, 1962, **13**, *23—30*.
30. MILCH, H. & V. G. LÁSZLÓ: Phage typing of *Salmonella paratyphi B. Acta microbiol. Acad. Sci. hung.*, 1961, **8**, *311—319*.
31. KADZINOVA, E. A.: Phage typing of typhoid and paratyphoid B organisms. *Zh. Mikrobiol. (Mosk.)*, 1953, **24**, 7, *81*. (Russ.).
32. KNYAZHANSKII, O. M. & O. M. KOLODII: Vi-phage types of typhoid and paratyphoid B strains and their use in epidemiological practice. *Zh. Mikrobiol. (Mosk.)*, 1953, **24**, 7, *78—79*. (Russ.).
33. MIRISEVA, E. E.: On phage typing of typhoid and paratyphoid B organisms. *Zh. Mikrobiol. (Mosk.)*, 1956, **27**, 2, *108*. (Russ.).
34. GUREVICH, E. S.: Group C paratyphoids (salmonelloses). Moscow, 1956, p. *67—69*. (Russ.).
35. TARABČÁK, M.: Rare *Salmonella* types in Eastern Slovakia. *Lék. Obz.*, 1957, **6**, *231—235*. (Czech).
36. BUCZOWSKI, Z.: in SEDLÁK and RISCHE'S Enterobacteriaceae-Infektionen, Leipzig, 1961, p. *104* and *126*.
37. ELIAS, H., Š. BRUCKNER & B. BRICMAN: Salmonelloses in children. *Pediatria (Buc.)*, 1954, **3**, 2, *41—58*. (Roum.).
38. TEFTEDARIJA, M.: Our experience in chloromycetin treatment of sal-monelloses. *Med. Arkh. (Sarayevo)*, 1958, **12**, 4, *111—120*. (Serb.).
39. CVJETANOVIĆ, B. & B. TEODOROVIĆ: Sanitary deficiencies as the cause of a water-born typhoid epidemic. *Higijena*, 1957, **9**, *12—21*. (Serb.).
40. GUREVICH, E. S.: *l.c.* p. *252—253*. (Russ.).
41. LAPAKHA, A. A., E. M. PIK-LEVONTIN & N. I. SHEKHINA: Salmonella infection in children, especially in infants. *Pediatriya*, 1962, **40**, 2, *16—21*. (Russ.).
42. AVSARKISYAN, S. O.: Bacteriemia in typhoid-paratyphoid conditions in children of various age. *Pediatriya*, 1956, **6**, *54—55*. (Russ.).
43. MELNIK, E. G. & YU. M. MIKHAILOVA: Clinical and laboratory parallels in salmonella food poisoning. *Zh. Mikrobiol. (Mosk.)*, 1961, **32**, 10, *122—127*, (Russ.).
44. VILSHANSKAYA, F. L. & L. B. BOGUYAVLENSKAYA: Studies on *Salmonella* strains isolated in Moscow in 1957. *Zh. Mikrobiol. (Mosk.)*, 1960, **31**, 1, *137—140*. (Russ.).
45. TSEYUKOV, S. P.: On salmonellas and salmonella carriers. *Zh. Mikrobiol. (Mosk.)*, 1960, **31**, 12, *112—116*. (Russ.).
46. TSEYUKOV, S. P.: Salmonella carriers among food handlers and prophy-laxis of food poisoning. *Gig. i Sanit.*, 1961, **26**, 3, *65—68*. (Russ.).
47. WILSON G. S. & A. A. MILES: Topley and Wilson's Principles of Bac-teriology and Immunity, London, 1955, p. *841*.
48. GUREVICH, E. S.: *l.c.* p. *19*. (Russ.).
49. TARASOVA, A. P.: On the serological identification of Salmonella or-ganisms in U.S.S.R. according to Kauffmann-White. *Zh. Mikrobiol. (Mosk.)*, 1946, **4**, *11—16*. (Russ.).
50. RAUSS, K.: On salmonellosis. *Orv. Hetil.*, 1960, **101**, *181—188*. (Hung.).

51. Raška, K.: On some aspects of the epidemiology of enteric infections. *Čas. Lék. Čes.*, 1959, **98**, *737—745*. (Czech).
52. Shapiro, S. E., I. S. Zhdanov & L. R. Chapovskaya: Hen egg products as a source of paratyphoid B. *Gig. i Sanit.*, 1961, **26**, 1, *112—114*. (Russ.).
53. Dvořak, J.: Isolation of salmonellas from fish meal imported from Angola. *Veterinářstvi*, 1957, **7**, *310*. (Czech).
54. Bykova, Z. A. & V. M. Gusev: Bacteriological examination of birds in Daghestan A.S.S.R. *Zh. Mikrobiol. (Mosk.)*, 1959, **30**, 9, *126*. (Russ.).
55. Mikhailova, R. S. & V. S. Gusev: Salmonelloses of wild birds. *Zh. Mikrobiol. (Mosk.)*, 1960, **31**, 6, *110—111*. (Russ.).
56. Ferenczi, E., K. Stoll & G. Viragh: Epidemiological problems of infectious enterocolitis in the capital. *Népegészségügy*, 1960, **41**, *160—168*. (Hung.).
57. Brokman, H. & K. Lachowicz: Bacterial etiology of diarrhoeal conditions in children in the light of collective studies. *Postępy Pediat.*, 1955, **1**, *5—34*. (Polish).
58. Brokman, H. & K. Lachowicz: Les diarrhées infantiles en Pologne (Recherches cliniques et microbiologiques). *Pédiatrie (Lyon)*, 1956, **11**, *861—869*.
59. Atanasova, S. A., M. Lotova & S. Stefanov: Studies on infantile enterocolitis. *Trudy NIIEM (Sofiya)*, 1960, **7**, *71—74*, (Russian edition).
60. Bakalov, I., S. Spasov & Ts. Pavlova: Studies on enterocolitis in infants. *Vop. Pediat. Akusher. Ginekol. (Sofiya)*, 1959, **3**, 6, *27—34*. (Bulg.).
61. Rauss, K., L. Gyengési & Gy. Újváry: Studies on etiology of sporadic infantile diarrhoeas. *Népegészségügy*, 1952, **33**, *275—280*. (Hung.).
62. Nicolau, I., A. Gaiginschi, M. Camner, A. Bombea, G. Teodorovici, E. Duca, C. Feldi, N. Lescinschi, I. Pencea & A. Cheptea: Bacteriological investigations of nutritional disorders in infants. *Pediatria (Buc.)*, 1954, **3**, 2, *32—35*. (Roum.).
63. Hurmuzache, E., T. Mărculescu, R. Bărbută, M. Burdea, A. Stuleanu, M. Haimovici, S. Apostol, V. Mihul, T. Vexler, E. Gugles, R. Cozorok & D. Gurăndeanu: The incidence of dysenteric etiology in acute nutritional disorders in infants. *Pediatria (Buc.)*, 1954, **3**, 2, *36—40*. (Roum.).
64. Nestorescu, N., M. Popovici, S. Novac, S. Librescu & Elian: Studies on the incidence of infantile enterocolitis in our country. *Pediatria (Buc.)*, 1956, **5**, *214—221*. (Roum.).
65. Bratelj, Z.: Nutritional disorders in infants. *Arhiv Zašt. Majke i Djet.*, 1958, **2**, 6, *20—21*. (Serb.).
66. Čupić, V. A., A. Čvorić & M. Čemerikić: Etiology and treatment of summer diarrhoeas. *Glas. soc. Pediat.*, 1953, **4**, *62—71*. (Serb.).
67. Chernova, V. N. & N. S. Vasileva: On salmonellosis in infants. *Zh. Mikrobiol. (Mosk.)*, 1947, **8**, *31—32*. (Russ.).
68. Egészségügyi Ministérium statisztikai osztálya.: Hungary's health problems in 1961. *Népegészségügy*, 1962, **43**, *225—249*. (Hung.).
69. Kneffel, P.: Epidemiological studies on a paratyphoid B epidemic due to icecream. *Népegészségügy*, 1952, **33**, *355—359*. (Hung.).
70. Mihályfi, I., E. Kende, E. Jonas & G. Vámos: Salmonella Untersuchungen in Budapest in den Jahren 1956—1958. *Acta microbiol. Acad. Sci. hung.*, 1961, **8**, *35—42*.
71. Jaroš, J.: Demographic development and state of health of the Czechoslovak population in 1961. *Čs. Zdrav.*, 1962, **10**, *308—319*. (Czech).

72. ŠKOVRÁNEK, V.: Epidemiological position of the most important infectious diseases in Czechoslovakia in 1957. *Čs. Pediat.*, 1958, **13**, *673—683*. (Czech).
73. MASÁR, I. & A. NOVÁK: Effects of 15 year battle against infectious diseases in Slovakia (1945—1959). *Lék. Obz.*, 1960, **9**, *486—498*. (Slovak).
74. ČERVENKA, J.: Typhoid incidence in Slovakia in 1920—1958. *Čs. Epidem.*, 1959, **8**, *413—420*. (Slovak).
75. KAZMAR, A. & J. RUDNÝ: Epidemiological position of ČSSR in 1960. *Čs. Zdrav.*, 1961, **9**, *291—297*. (Czech).
76. BURIAN, V.: Studies on the incidence of paratyphoid B in Liberec region in 1950—1952. *Čs. Epidem.*, 1953, **2**, *447—455*. (Czech).
77. ROZTOMILY, A., Salmonellosis bareilly. *Prakt. Lék. (Praha)*, 1961, **41**, *828—829*. (Czech).
78. MARTINů, K.: Actual problems in epidemiology of salmonellosis. *Prakt. Lék. (Praha)*, 1959, **39**, *597—601*. (Czech).
79. KŘIVINKA, J.: Poultry salmonellosis. *Veterinářstvi*, 1953, **3**, *160—161*. (Czech).
80. DRYL, L., E. GORZELAK & W. PRAŻMOWSKI: Milk-born typhoid epidemic at Zgierz and Lódź in 1950. *Przegl. Epidem.*, 1961, **15**, *33—40*. (Polish).
81. KOSTRZEWSKI, J.: Typhoid epidemic at student hotels in Cracow in 1957. *Przegl. Epidem.*, 1959, **13**, *223—248*. (Polish).
82. WIÓR, H.: Paratyphoid A, B and C in Poland (in preparation as a chapter in KOSTRZEWSKI's: Infectious diseases in Poland in 1919—1962. Warsaw). (Polish).
83. LEWANDOWSKA, E.: Food poisoning in Poland in 1952—1956. *Przegl. Epidem.*, 1958, **12**, *249—252*. (Polish).
84. BUCZOWSKI, Z.: Salmonelloses of man diagnosed in the years 1946—1956 in Poland. *Biul. Inst. Med. morsk. Gdańsk*, 1961, **12**, *51—71*.
85. BIRKOVSKII, Yu, E.: Effects of the 40 year struggle against infectious diseases in Ukraine. *Zh. Mikrobiol. (Mosk.)*, 1958, **29**, 10, *139—142*. (Russ.).
86. ZHDANOV, V. M.: Health condition of the population of the U.S.S.R. and the task of the Public Health Service. *Zdravookhr. Beloruss.*, 1958, **4**, 6, *3—5*. (Russ.).
87. MARIEV, A. N. & G. K. GURBANOV: On the epidemiology of typhoid fever. *Zh. Mikrobiol. (Mosk.)*, 1957, **28**, 3, *32—34*. (Russ.).
88. TOVAROV, S. L.: Typhoid and paratyphoid carrier incidence according to findings in corpse. *Zh. Mikrobiol. (Mosk.)*, 1951, **5**, *60*. (Russ.).
89. LEZNIK, A. I.: Some ways in prophylaxis of enteric infections. *Zh. Mikrobiol. (Mosk.)*, 1958, **29**, 4, *94—97*. (Russ.).
90. BELATSKII, D. P.: Urgent problems in prophylaxis of acute enteric infections. *Zdravookhr. Beloruss.*, 1958, **4**, 4, *23—26*. (Russ.).
91. MUKHUTDINOV, I. Z.: For further decrease and eradication of infectious diseases. *Kazan. med. Zh.*, 1961, **4**, *15—19*. (Russ.).
92. UKHOV, A. YA.: Epidemiological characteristics of the typhoid-paratyphoid fever. *Vrach. Dyelo*, 1958, *719—722*. (Russ.).
93. KRASNITSKAYA, E. S.: Some data on food poisoning in the Russian Federal Republic in 1955—1957. *Gig. i Sanit.*, 1959, **24**, 4, *30—33*. (Russ.).
94. KRASNITSKAYA, E. S.: Food poisoning in the Russian Federal Republic in 1956. *Gig. i Sanit.*, 1958, **23**, 3, *49—53*. (Russ.).
95. ARBUZOVA, V. M.: Some data on epidemiological characteristics of salmonelloses in Leningrad. *Zh. Mikrobiol. (Mosk.)*, 1960, **31**, 6, *124—126*. (Russ.).

96. Tsereteli, E. V.: On defining the etiology of human salmonelloses observed in Georgian S.S.R. (The author's summary). Tbilisi, 1954. (Russ.).

97. Tsereteli, E. V. & Sakanderidzhe: Studies on the etiological structure of salmonelloses in Georgian S.S.R. (Summary of the lecture read at the Scientific Conference at Baku in 1962). (Russ.).

98. Klimenko, E. P., L. S. Lazareva & F. A. Zismanova: On epidemiology of enteric disorders in children in the light of the data from a district sanitary epidemiological station of the city of Moscow. *Zh. Mikrobiol. (Mosk.)*, 1962, 33, 11, *153—157*. (Russ).

99. Nisevich, N. I. & A. G. Avanasova: On coli infections in children. *Vop. Okhrany Materin. Dets.*, 1959, 4, 2, *12—16*. (Russ.).

100. Ciucă, M., N. Nestoresco, M. Popovici, A. G. Ciplea, C. Ciurea, E. Alexenco, L. Ianco, I. R. Vladocianu & S. Libresco: Contribution à l'étude de l'immunisation anti-typhoïde. *Arch. roum. Path. exp.*, 1958, 17, *321—358*.

101. Ciucă, M., C. Combiescu, N. Nestorescu, M. Popovici, Zilişteanu, M. Nicolescu, S. Toma & T. Meitert: Contribution to studies of salmonellas. Incidence, geographical distribution and characteristics of strains in Roumania. *Stud. Cercet. Infvamicrobiol.*, 1953, 4, *441—448*. (Roum.).

102. Flax, A., G. Tirnoveanu, G. Ionescu & G. Trifan: Examination method in prophylaxis of acute and chronic nutritional disorders. *Pediatria (Buc.)*, 1957, 6, 5, *447—450*. (Roum.).

103. Nestoresco, N. & M. Popovici: Contribution à l'étude de l'étiologie et de l'incidence des toxi-infections alimentaires dans la Republique Populaire Roumaine, Communication en Colloque de Santé Publique Vétérinaire, Varsovie, 1957.

104. Gaon, J. & R. Pavlović: Water-born typhoid epidemic in Drvar in 1949—1950. *Med. Arkh. (Sarayevo)*, 1952, 6, 1, *29—45*. (Serb.).

105. Nevidal, A., A. Gran, A. Merdžo, S. Jurišić & M. Miling: Typhoid and paratyphoid epidemic in Ossijek-Donje grad in 1958. *Vojnosanit. Pregl.*, 1961, 18, *273—280*. (Serb.).

106. Škokljev, A.: Water-born typhoid epidemic in Titov Veles in 1955—1956. *Vojnosanit. Pregl.*, 17, *895—900*. (Serb.).

107. Borjanović, S.: Infectious diseases in NR of Serbia — Problems and methods of control. *Narod. Zdrav.*, 1962, 18, *302—309*. (Serb.).

108. Kalčić, S.: Epidemiological problems in Slovenia. *Zdrav. Vestn.*, 1959, 28, *160—164*. (Slovene).

109. Fališevac, J.: Human salmonellosis from the clinical point of view. *Rad. med. Fak. Zagrebu.*, 1955, 1, *5—60*. (Serb.).

110. Emili, H.: Laboratory findings at detecting carriers among people having or applying for a health book. *Narod. Zdrav.*, 1961, 17, *296—299*. (Serb.).

111. Aleraj, D.: Salmonellosis due to *S. blockley* in this country. *Liječn. Vjesn.*, 1960, 82, *709—711*. (Serb.).

112. Vodopija, J. & B. Tompak: Epidemiological observations on cases due to *S. blockley. Liječn. Vjesn.*, 1960, 82, *713—717*. (Serb.).

113. Awdjiew, G. & S. Atanassova: Die Lysotypen der Typhusbakterien in Bulgarien. Tagungsbericht des 2. Colloquium über Fragen der Lysotypie, Wernigerode/Harz, 7.—10. Oktober 1956, *53—54*.

114. Tachev, B. & Kh. Odisseev: Incidence and clinic of typhoid in the region of Burgas in 1951—1958. *Vop. Pediat. Akusher. Ginekol. (Sofiya)*, 1960, 4, 9. *50—53*. (Bulg.).

115. Avdzhiev, G., R. Koen, G. Futskov, M. Todeva & A. Pancheva: Water-born typhoid epidemic. *Khigiena*, 1961, 4, 2, *21—24*. (Bulg.).

116. AVDZHIEV, G., A. TRIFONOVA, S. ATANASOVA, R. KOEN & P. PETEV: Problems of bacterial enteric infections in this country and prospects for the control. *Khigiena*, 1961, **4**, 2, *15—20*. (Bulg.).

117. RADEV, I.: Studies on typhoid and paratyphoid fever in this country. *Nauch. Tr. vissh. med. Inst. Sofiya*, 1961, **40** (8), 3, *161—182*. (Bulg.).

118. LEFTEROV, N. & Z. ZAPRYANOVA: Relapses and complications in typhoid and paratyphoid B treated with antibiotics. *Sûvr. Med.*, 1959, **10**, 2/3, *63—69*. (Bulg.).

119. KOEN, R., P. PETEV, P. GINCHEV, P. RAIKOV, G. APOSTOLOV, A. DISHKOV, I. RANDEV, Z. ZAKHARIEV, M. MITEVA, D. PETROV & S. ZHEKOVA: Etiological structure of salmonelloses in this country in 1956—1960. 3-rd communication. *Khigiena*, 1963, **6**, 5, *53—58*. (Bulg.).

120. CHUCHKOV, N.: The practice of struggle against food poisoning. *Vet. Sbir.*, 1962, **59**, 6, *23—26*. (Bulg.).

121. KARALAMBEV, N., N. KOVACHOVA, M. KAZAKOVA & V. NEDELCHEVA: On the etiology of food poisoning of bacterial origin in Sofia. *Sûvr. Med.*, 1957, **8**, 11, *89—96*. (Bulg.).

122. KOEN, R. S., P. P. PETEV, P. GINCHEV, P. RAIKOV, I. DIMOV, G. APOSTOLOV, A. DISHKOV & Z. ZAKHRIEV: Etiological structure of salmonelloses in Bulgaria in 1958—1959. *Tr. Inst. Epidem. Mikrobiol. Sofiya*, 1960, **7**, *65—70*. (Russian edition).

123. VESELINOV, V., B. KRUSHEV & P. PENKOV: The occurrence of *S. senftenberg* in this country. *Vet. Sbir.*, 1959, **56**, 4, *41—42*. (Bulg.).

124. Yugoslav Typhoid Commission: Field and Laboratory Studies with Typhoid Vaccines. *Bull. Wld. Hlth. Org.*, 1957, **16**, *897—910*.

125. KOSTRZEWSKI, J.: Evaluation of typhoid vaccines and effectiveness of the vaccinations. *Przegl. Epidem.*, 1963, **17**, *1—12*. (Polish).

126. VERBEV, P., S. ZHELYAZKOV, M. GEORGIEVA, V. MONEV, R. MANOLOV & A. EFREMOVA: Etiologic studies on 1776 enterocolitis cases. *Nauch. Tr. vissh. med. Inst. Sofiya*, 1961, **40** (8), 3, *129—145*. (Bulg.).

SALMONELLOSIS IN ISRAEL

BY

W. SILBERSTEIN & CH. B. GERICHTER

Government Central Laboratories (National Salmonella Centre of Israel), Ministry of Health, Jerusalem.

(with 1 fig.)

Introduction

Israel's epidemiological climate is characterized by the fact that in a comparatively small country (20,850 km²) with extremely varied geographical and meteorological conditions the population rose since the establishment of the state in 1948 and in the wake of mass-immigration, from about 800,000 to 2,320,000 at the end of 1962. Thus a situation has arisen in which elements old and new came into close contact in a limited area and people living in most primitive conditions and hailing partly from underdeveloped areas of the east, bringing with them diseases and various cultural habits characteristic of their countries of origin, have come to live side by side with people who, for a long period, had been enjoying all the comforts of civilization and technological attainments of the twentieth century. This peculiar composition entails on one hand a high incidence of such diseases which usually spread under conditions of backward personal hygiene, while it leads, on the other hand, to a low incidence of diseases which are apt to be controlled by general public health measures. Owing to the limited area of the country and in view of the high standard of medical care, the health authorities are generally aware and well-informed of the changing epidemiological situation. At any rate it is quite remarkable that in the years of mass-immigration following the establishment of the State, when scores of thousands of newcomers were living in extremely crowded conditions together with people suffering from various communicable diseases, there were no large outbreaks of epidemics.

This general statement applies, as well, to all infections constituting the group of enteric fevers and other Salmonella infections. In this connection it should be noted that particularly typhoid fever, at least since the days of the Ottoman Empire, had been endemic in this country. Nevertheless it is true that in the thirty years preceding the establishment of Israel a gradual improvement of the health situation among the population has taken place, both in the wake of Jewish settlement and as a result of the endeavours of the British Mandatory Government. As a matter of fact the prophylactic vaccination of all immigrants and of contacts of actual enterics came into force as early as 1920. However, despite all the progress made especially with regard to the struggle against malaria, trachoma and other diseases, Typhoid fever remained an endemic disease with an incidence rate of over 20 per 10,000 as late as 1948.

Salmonella research has received a considerable stimulus as a result of a series of basic discoveries made in this country. It may be said that in this respect research workers in Palestine have gained outstanding achievements. We should here mention STUART & KRIKORIAN (1928) who demonstrated the practical importance of the qualitative receptor analysis initiated by FELIX; the introduction by OLITZKY (1928) of the S. typhi 0901 strain as a standard for the Widal reaction and finally the proof of the practical importance of the serological Vi-antibody test in the blood serum for the detection of Typhoid carriers by FELIX, REITLER & KRIKORIAN (1935). As early as 1939 KLOPSTOCK established the first Salmonella Centre in Tel-Aviv. Later in 1949, after the foundation of the State it was transferred to Jerusalem as the National Salmonella Centre.

Typhoid and Paratyphoid Fever

Table I summarizes the incidence of Typhoid fever from 1949 to the end of 1962.

Table I.

Reported cases of typhoid fever and rates per 10,000; deaths in brackets, Jews 1949—62.

	1949	1950	1951	1952	1953	1954	1955
Cases	689 (21)	707 (6)	841 (11)	968(9)	474 (2)	440 (4)	427
Rates	3.0	6.0	6.4	6.8	3.2	2.9	2.7

	1956	1957	1958	1959	1960	1961	1962
Cases	342	306 (1)	292 (1)	271 (4)	218 (2)	175 (2)	304 (2)
Rates	2.3	1.7	1.6	1.4	1.1	0.7	1.5

Typhoid fever is a mild disease in Israel compared to the classical accounts of this disease, and case fatality is low. Almost all cases are hospitalized and presumably over 90% of the cases are notified to the health authorities. Table I shows that in the years following the establishment of Israel, the incidence of Typhoid fever ranged between 3.0 and 6.8 per 10,000 (in 1952). The highest incidence occurred in new-immigrant camps as well as in rural districts obviously because of the lack of sanitary facilities. In addition there is good evidence that in some cases it was caused by direct contact of persons with food and cooking utensils. In 1951 an attempt was made to vaccinate the entire population with T.A.B. vaccine. However, only about one third of the population has been actually inoculated, although the percentage among the new immigrants was higher. It is quite obvious that the effect of this measure can hardly be fully assessed. In 1953 many residents of the larger immigrant camps were transferred to permanent housing and a system of contact-tracing and follow-up was set up. The incidence dropped by 50% and in 1955 the rate among the immigrants was not higher than among the rest of the population. From then on the rate continued to decrease steadily, and in 1961 it reached 0.7 per 10,000. It is still premature to conclude whether the increase of Typhoid cases in 1962 indicates a real change for the worse, but in epidemiology one ought never to indulge in complacency.

Several factors were responsible for the gradual decrease in the incidence of Typhoid fever in this country, among them the introduction of improved sanitary facilities in the homes, purification of water and a proper sewage disposal system. In addition the threat by carriers must still be seriously considered and control can be achieved only by the decisive effect of maintaining rigid personal and public hygiene.

On the other hand, the case rate among the non-jewish population remained static, ranging between 3 and 4 per 10,000 even in recent years. Taking into account that among this sector of the population we are obliged to make allowance for a certain measure of under-reporting, this figure is not far removed from the figure of 11 per 10,000 for the Moslem population of Palestine in the years 1935—1945. There is no doubt, however, that the situation since the establishment of Israel has improved greatly and that the health state of the Israeli Arab is superior to that of any of his neighbours across the borders.

The degree of accuracy of the diagnosis is high and in no more than 5% of the cases in 1956 relied solely on clinical observation. In 46.5% the diagnosis was based on positive culture and in 42% — on serological evidence.

Phage-typing has proved a very important means for the tracing of a number of outbreaks. It is carried out in the Public Health Laboratories in Tel-Aviv.

Following we present a survey showing the findings of Typhoid phage-typing for 1961, kindly supplied by Prof. EYLAN, Director of the Public Health Laboratories in Tel Aviv*.

E_1 and A are the most frequent phage-types occurring in this country, following them, in order of frequency are C_1, 40, 28, F_1.

Of some interest is the age distribution of the Typhoid cases. Among the Jewish population the highest percentage is recorded in the age group of 1—9 (3—4.5 per 10,000 in 1959) while among the non-Jewish population the peak was in the age group 10—19 (1.7—1.5 per 10,000 in 1959).

---

* We are taking this opportunity to express our gratitude to him for his valuable contribution.

Findings of Typhoid phage typing in 1961

Focuses	No. of strains	Phage-type
18	40	A
17	42	C_1
2	2	D_1
1	2	D_4
1	1	D_9
22	48	E_1
10	20	F_1
1	1	F_4
4	7	G_1
1	3	O
6	12	T
1	1	25
4	7	27
8	25	28
3	8	38
10	32	40

Generally the annual incidence shows a decrease in spring and a peak in the late summer. Of some significance is the distribution by types of settlement. The incidence is somewhat lower in urban areas and in kibbutzim (collective settlements) (0.1—0.7 in 1959). It is comparatively high in Arab villages (4.0 per 10,000 in 1959).

Table II.

Reported cases of paratyphoid fever and rates per 10,000; deaths in brackets, Jews 1949—62.

	1949	1950	1951	1952	1953	1954	1955
Cases	357 (1)	617 (1)	490 (2)	606 (1)	619 (1)	217 (2)	154
Rates	4.0	5.5	3.7	4.2	4.2	1.5	1.0

	1956	1957	1958	1959	1960	1961	1962
Cases	147	107	72	59	39	21	40
Rates	0.9	0.6	0.4	0.3	0.2	0.1	0.2

In contrast to Typhoid, Paratyphoid fever, as late as the mandatory period, shows considerable variations in the annual rate of incidence, for instance 27 per 10,000 Jews in 1941 as against 1.8 in 1942 and again 22 in 1943. It may be safely assumed that these figures also include numerous cases of other Salmonella types and of food-borne Paratyphoid fever. As is shown in Table II, the rate in 1949 was 4 per 10,000 and so it remained until 1953. Since 1954 other Salmonelloses are listed quite separately from Paratyphoid. This is the reason that, as from that year, the incidence of Paratyphoid fever dropped considerably (1.5 to 0.1 in 1961). Again it should be stressed that about 1956 70% of the cases were diagnosed on the evidence of positive culture, while the rest — serologically. 38% of the cases were *S. paratyphi A*, 49% — *S. paratyphi B* (the majority of *S. paratyphi B* strains belong to the phage-types Taunton or Dundee)* and only 6% were *S. paratyphi C*. Clinically, Paratyphoid fever too is a mild disease in this country. The case fatality is very low. The seasonal peak occurs during the summer months, the age group most affected is under 1 year.

Other Salmonelloses

Israel provides a classical example of interaction: Typhoid versus other Salmonella, decrease of Typhoid owing to improvement of sanitary conditions, increase of other Salmonella infections on account of the rapid development of food industries etc. This fact clearly emerges from the data given both in Table III and in Fig. 1.

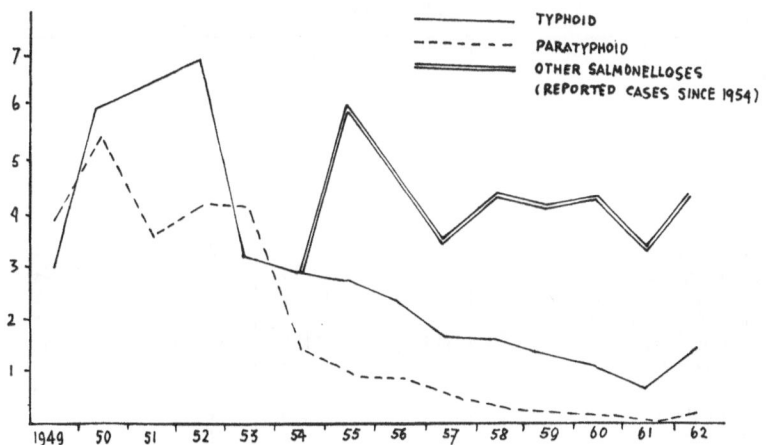

Fig. 1. Reported cases of typhoid fever, paratyphoid fever and other salmonelloses. Rates p. 10.000 pop. 1949—1962

It should be mentioned that notification of cases of other Salmonella infections is incomplete as part of them are not included in the notifications of the health authorities.

Table III.
Other Salmonelloses

Year	1954	1955	1956	1957	1958	1959	1960	1961	1962
Cases	475	956	774	627	785	778	827	686	914
Rate/10,000	3.1	6.1	4.7	3.6	4.4	4.2	4.3	3.4	4.5

* Personal communication by Prof. EYLAN.

Therefore it is not surprising that in the course of the last few years the number of Salmonella infections not only kept increasing as such but that since the introduction of the KAUFFMANN-WHITE scheme in Salmonella research, side by side with its growing extension, the number of *Salmonella* types prevalent in this country grew as well. While OLITZKY in his survey (1944) listed 26 different *Salmonella* types, both in humans and animals, that figure then rose as early as 1952 to 44 (SILBERSTEIN & GERICHTER 1952).

Table IV.

Salmonella findings in man in Israel during the years 1949—62
Figures indicating frequency of occurrence of species without
referring to number of incidents or outbreaks.

Species	1949-1951	1952-1954	1955-1957	1958-1960	1961-1962	Totals
Group A						
S. paratyphi A	44	68	40	43	12	207
Group B						
S. hessarek	—	3	—	—	—	3
S. sofia	—	2	—	4	3	9
S. paratyphi B	50	74	85	84	147	440
S. abony	—	1	5	10	8	24
S. wien	—	—	1	18	6	25
S. schwarzengrund	—	—	1	—	—	1
S. saint-paul	3	1	3	20	22	49
S. san-diego	—	—	3	2	6	11
S. reading	1	—	11	9	3	24
S. derby	—	—	1	—	1	2
S. essen	—	—	—	—	1	1
S. typhi-murium	549	769	1138	1568	1437	5461
S. bredeney	—	3	4	—	—	7
S. brandenburg	—	—	1	2	—	3
S. heidelberg	4	1	5	4	5	19
S. shubra	—	—	3	2	—	5
S. haifa	—	—	—	2	62	64
S. stanleyville	1	5	20	19	2	47
Group C$_1$						
S. oslo	10	14	53	63	39	179
S. edinburg	1	—	—	110	386	497
S. paratyphi C	17	26	3	4	1	51
S. lomita	—	1	24	92	53	170
S. braenderup	4	83	511	470	217	1285
S. montevideo	111	121	284	298	182	996
S. oranienburg	103	47	49	43	8	250
S. thompson	—	1	—	7	4	12
S. concord	34	9	10	2	3	58
S. irumu	—	5	26	15	12	58
S. bonn	—	—	1	—	—	1
S. potsdam	1	56	3	1	—	61
S. colorado	—	—	—	1	1	2

Table IV (continued)

Species	1949-1951	1952-1954	1955-1957	1958-1960	1961-1962	Totals
S. ness-ziona	3	—	—	10	8	21
S. virchow	—	1	7	4	1	13
S. infantis	—	—	74	161	84	319
S. bareilly	—	—	—	—	1	1
S. hartford	1	1	3	9	—	14
S. aequatoria	—	—	—	—	18	18
S. tennessee	—	92	289	123	31	535
Group C$_2$						
S. muenchen	11	45	35	26	10	127
S. manhattan	3	1	13	8	3	28
S. newport	639	319	299	138	102	1497
S. kottbus	5	4	40	36	50	135
S. lindenburg	—	—	—	26	18	44
S. takoradi	—	3	—	—	—	3
S. bonariensis	—	—	—	1	—	1
S. blockley	—	—	—	25	74	99
S. litchfield	—	—	7	—	—	7
S. manchester	5	—	—	—	—	5
S. bovis-morbificans	12	16	11	1	12	52
S. hadar	2	10	192	79	22	305
S. glostrup	11	4	—	—	—	15
S. uno	—	—	2	—	—	2
Group C$_3$						
S. emek	9	32	138	317	174	670
S. kentucky	8	38	79	111	35	271
S. corvallis	—	—	2	—	—	2
Group C$_4$						
S. jerusalem	17	2	37	53	17	126
Group D$_1$						
S. onarimon	—	—	—	7	—	7
S. frintrop	—	—	—	1	—	1
S. eastbourne	—	3	23	3	2	31
S. enteritidis	60	156	165	271	301	953
S. dublin	2	13	4	38	13	70
S. rostock	—	3	9	—	—	12
S. moscow	—	1	—	—	—	1
S. panama	1	—	4	—	—	5
S. goettingen	—	—	—	4	—	4
S. wangata	—	—	1	1	—	2
S. canastel	—	2	—	—	—	2
Group D$_2$						
S. baildon	—	—	—	1	—	1
Group E$_1$						
S. butantan	—	—	—	1	—	1
S. vejle	11	—	12	4	—	27
S. muenster	—	8	15	12	18	53
S. anatum	5	20	30	72	37	164

Table IV (continued)

Species	1949-1951	1952-1954	1955-1957	1958-1960	1961-1962	Totals
S. meleagridis	52	188	102	97	31	470
S. zanzibar	—	9	18	68	11	106
S. london	3	3	7	30	31	74
S. give	1	1	19	10	9	40
S. uganda	—	—	—	—	2	2
S. amager	—	—	—	—	1	1
S. orion	—	—	—	9	—	9
Group E$_2$						
S. newington	7	—	1	—	—	8
S. cambridge	1	2	—	—	—	3
S. new-brunswick	1	—	—	—	—	1
Group E$_4$						
S. senftenberg	6	90	278	70	30	474
S. taksony	15	39	42	5	6	107
S. liverpool	—	—	—	3	—	3
S. simsbury	—	—	1	8	2	11
Group F						
S. chingola	—	—	1	—	—	1
S. pretoria	1	—	—	—	—	1
S. abaetetuba	—	—	—	—	1	1
S. rubislaw	—	—	1	—	—	1
S. tel-hashomer	1	—	—	—	—	1
Group G$_2$						
S. mishmar-haemek	—	2	—	—	1	3
S. tel-el-kebir	—	—	—	5	—	5
S. wichita	—	4	—	—	—	4
S. havana	7	2	—	3	1	13
S. nachshonim	—	12	1	8	3	24
S. worthington	—	1	1	5	—	7
S. cubana	1	20	7	2	—	30
Group H						
S. lindern	—	—	—	—	1	1
S. caracas	—	—	—	1	—	1
S. madelia	—	—	—	1	—	1
S. sundsvall	4	1	1	—	—	6
Group I						
16: a-monophasic	—	—	—	2	—	2
S. hvittingfoss	1	—	2	1	—	4
S. gaminara	1	—	—	—	—	1
S. weston	—	—	—	1	1	2
Group J						
S. carmel	2	—	—	16	1	19
Group K						
S. cerro	—	—	3	19	9	31
S. siegburg	—	—	—	3	50	53

Table IV (continued)

Species	1949-1951	1952-1954	1955-1957	1958-1960	1961-1962	Totals
Group L						
S. minnesota	—	2	—	—	—	2
Group M						
S. halle	—	—	—	—	1	1
S. ilala	—	—	20	2	—	22
S. tel-aviv	1	—	3	10	14	28
Group N						
S. urbana	—	1	—	—	—	1
S. godesberg	—	—	—	1	1	2
Group R						
S. shikmonah	—	—	—	1	—	1
S. riogrande	—	—	1	—	—	1
S. johannesburg	—	3	1	38	3	45
S. millesi	—	—	—	2	—	2
S. degania	—	—	7	20	15	42
Group T						
S. uphill	—	—	—	1	—	1
Totals	1843	2444	4293	4878	3877	17335

Table IV presents a summary of the various *Salmonella* types in Israel during the years 1949—1962 (excluding *S. typhi*). It is based on the combined findings of 2 laboratories: The National Salmonella Centre, Jerusalem, and the Central Laboratory of the Kupat Cholim (Sick Fund of the General Federation of Jewish Workers in Israel, Haifa, Director: Prof. Dr. W. HIRSCH)*. For the sake of completeness, the table also includes a large part of the isolated types of *S. paratyphi A*, *S. paratyphi B* and *S. paratyphi C*.

Table IV shows that in the years 1949 till 1962 altogether 122 different species were identified in Israel. Altogether the following 18 new species (from man or animals) were found during this period:

S. jerusalem	(1950)	KAUFFMANN & SILBERSTEIN
S. emek	(1950)	HIRSCH, HENIG & SAPIRO
S. haifa	(1950)	SAPIRO & HIRSCH
S. tel-hashomer	(1950)	KAUFFMANN, SILBERSTEIN & GERICHTER
S. ness-ziona	(1950)	KAUFFMANN, SILBERSTEIN & GERICHTER
S. israel	(1953)	EDWARDS, ALTMANN & McWORTER
S. nachshonim	(1953)	KAUFFMANN, SILBERSTEIN & LUBLING
S. carmel	(1954)	HIRSCH, HIRSCH & SAPIRO-HIRSCH
S. hadar	(1954)	HIRSCH, GERICHTER, BREGMAN, LUBLING & ALTMANN

* Our special thanks are due to Prof. HIRSCH for having made this material available.

S. mishmar-haemek	(1954)	SILBERSTEIN, GERICHTER & REITLER
S. ahuza	(1955)	HIRSCH & SAPIRO-HIRSCH
S. ezra	(1955)	HIRSCH & SAPIRO-HIRSCH
S. degania	(1957)	HIRSCH & SAPIRO-HIRSCH
S. vitkin	(1957)	HIRSCH & SAPIRO-HIRSCH
S. negev	(1957)	SAPIRO-HIRSCH, ALTMANN & HIRSCH
S. hofit	(1958)	SAPIRO-HIRSCH & HIRSCH
S. ramat-gan	(1959)	SAPIRO-HIRSCH, ALTMANN & HIRSCH
S. shikmona	(1960)	SAPIRO-HIRSCH & HIRSCH

It also shows that a series of types occurred throughout that period with an equally high incidence, while other types remained definitely rare. The most frequent types are found almost equally distributed all over the country. 23 of the rarer species were found during these 14 years (1949—1962) only once, while 17 other types occurred two or three times each.

Table V.

The 20 most frequent Salmonella species in Israel during 1949—1962.

No.	Species	Number of strains	%
1.	S. typhi-murium	5461	32.8
2.	S. newport	1497	9
3.	S. braenderup	1285	7.7
4.	S. montevideo	996	6
5.	S. enteritidis	953	5.7
6.	S. emek	670	4
7.	S. tennessee	535	3.2
8.	S. edinburg	497	3
9.	S. senftenberg	474	2.8
10.	S. meleagridis	470	2.8
11.	S. infantis	319	1.9
12.	S. hadar	305	1.8
13.	S. kentucky	271	1.6
14.	S. oranienburg	250	1.5
15.	S. oslo	179	1.1
16.	S. lomita	170	1
17.	S. anatum	164	1
18.	S. kottbus	135	0.8
19.	S. jerusalem	126	0.7
20.	S. taksony	107	0.6

In table V 20 of the most frequent species are listed (*S. paratyphi A*, *S. paratyphi B* and *S. paratyphi C* have not been included). It should be stressed that 16 out of the 20 species are also the most frequently encountered in hospitalized cases (details follow).

Table VI.

Salmonella strains in Tortoises 1953—1962.

S. abony	73	S. richmond	27
S. rubislaw	18	S. ezra	5
S. bareilly	16	S. meleagridis	1
S. kottbus	54	S. typhi-murium	3
S. ahuza	16	S. nachshonim	1
S. sofia	24	S. gaminara	2
S. hvittingfoss	8	S. singapore	1
S. uphill	13	S. vitkin	3
S. cubana	1	S. hofit	5
S. potsdam	25	S. ilala	5
		S. kokomlemle	2
		Total	303

Table VI presents the occurrence of some *Salmonella* species in tortoises in the years 1953 till 1962*. These findings deserve consideration in view of the fact that 14 of these species have been found in Israel in humans too. Hence the possibility of infection, especially by children cannot be discarded.

Food-poisoning due to Salmonelloses

Along with the refinements of diagnostic methods and the growing consciousness of causal connections in all outbreaks affecting two or more persons after sharing the same contaminated meal, the part played by the various *Salmonella* species as causes of such outbreaks has been increasingly recognized in recent years. One important and surprising finding is the fact that the largest number of outbreaks originated from infections contracted at home.

Table VII.

Outbreaks of food poisoning by place of outbreak; 1953—1960.

Place	No. of outbreaks	No. of cases	Agent involved			
			Staphylo-coccus	Salmo-nella	Other	?
Family meal	143	859	63	24	21	35
School, Work-camp, Factory	25	2207	13	6	2	4
Kibbutz, Farm	53	2566	7	31	10	5
Hospital, Hotel, Institution, Restaurant	19	425	7	8	—	4
Festive Meal	17	505	7	6	—	4
Ship	8	730	4	4	—	—
Totals	265	7292	101	79	33	52

* Relevant data were kindly supplied by Prof. W. HIRSCH, Haifa.

Table VII shows that staphylococci preponderate in home-outbreaks, while Salmonellae are mainly responsible for outbreaks in kibbutzim as well as farms. It is noteworthy that the total share of Salmonellae as causal agents of food-poisoning has been recently increasing. While during the period 1953 to 1957, 41% of outbreaks were due to staphylococcus enterotoxin and only 14% to Salmonellae, the part played by the latter as the cause of outbreaks in the period of 1953 to 1960 rose to 29.8% (as against 38.1% staphylococcus enterotoxin).

It is well known that the sources of Salmonellae are human carriers, clinical cases or domestic animals, used as pets or sources of food. We may be permitted to quote in this connection, an interesting illustration of the role of human carriers.

For some time the local health authorities had been rather perturbed by the frequent occurrence of Salmonella infections aboard liners of a shipping company plying between Haifa and other mediterranean ports. A thorough three months survey conducted in 1957 of the kitchen personnel of one of the major ships revealed the surprising fact, that 10% of the personnel were carriers of different Salmonella strains. The following table VIII summarizes the findings of this survey.

Table VIII.
Salmonella-carriers on board of a Mediterranean Liner, 1957.

Month	Number of examined	Positive for Salmonella	%	Types of Salmonella (Number of strains)
June	107	8	7.4	1 S. paratyphi B 4 S. braenderup 1 S. meleagridis 2 S. senftenberg
July	48	8	16.6	4 S. paratyphi B 4 S. oslo
August	44	4	9.1	3 S. paratyphi B 1 S. senftenberg
Totals	199	20	10	

Clinical Observations

It is interesting to compare the experience gained in a major general hospital with our overall statistics. We refer here to a report, contained in a recently published paper by ALTMANN & GERICHTER (1963)*.

In the years 1954—1960, 33 cases of Paratyphoid fever were

---

* We are grateful to the authors for their permission to quote some data from this report.

observed in a hospital located in a densely populated region of the country.

Table IX.

Paratyphoid in a General Hospital 1954—60 (From ALTMANN & GERICHTER)

	Cases	Carriers	Cholecystitis
S. *paratyphi* A	22	4	3
S. *paratyphi* B	10	11	1
S. *paratyphi* C	1	2	—
Totals	33	17	4

Table IX shows that the number of these cases was relatively small. The course of the disease was extremely mild, there were no complications and no fatalities. All carriers were faecal carriers. On 4 carriers (all women aged 39 to 65 yaers,) with a chronic infection of the gall-bladder, cholecystectomy was performed. S. *paratyphi* A (3 strains) and S. *paratyphi* B (1 strain) were isolated from the bile during the operation.

Table X.

Summary of Isolations of 48 Salmonella species from 502 Patients
(From ALTMANN & GERICHTER).

	No. of Patients	Gastro-enteritis	Septi-cemia	Focal infec-tion	Hospi-tal infec-tion	Car-riers	No in-forma-tion
Group B							
S. wien	3					3	
S. saint-paul	2	1					1
S. reading	2	1	1				
S. san-diego	1	1					
S. typhi-murium	159	121	8	6	24	16	8
S. heidelberg	1	1					
S. stanleyville	2	1					1
S. haifa	1	1					
Group C₁							
S. oslo	4	2		1			1
S. lomita	6	6					
S. braenderup	60	36	2	1	6	18	3
S. montevideo	47	34	3		12	10	
S. oranienburg	12	10		1	6	1	
S. thompson	1	1					
S. irumu	1	1					
S. tennessee	15	7			2	6	2
Group C₂							
S. muenchen	2	1					1
S. manhattan	2	2					
S. newport	11	6			1	3	2

Table X (continued)

	No. of Patients	Gastro-enteritis	Septi-cemia	Focal infec-tion	Hospi-tal infec-tion	Car-riers	No in-forma-tion
S. kottbus	8	7			2	1	
S. lindenburg	1	1					
S. bonariensis	1	1					
S. bovis-morbificans	1		1				
S. hadar	6	4		1		1	
Group C$_3$							
S. emek	40	31	1		8	7	1
S. kentucky	17	14			7	2	1
Group C$_4$							
S. jerusalem	3					1	2
Group D							
S. loma-linda	1	1					
S. eastbourne	3	3					
S. enteritidis	51	31	8	3	7	2	7
S. dublin	1						1
Group E$_1$							
S. anatum	7	6			2		1
S. meleagridis	6	4			2	2	
S. zanzibar	1	1					
S. london	4	2				2	
S. orion	4	2			1	2	
Group E$_4$							
S. liverpool	2	1				1	
S. senftenberg	7	5			1	2	
S. simsbury*	2	2			1		
S. taksony	1	1					
Group G-T							
S. havana	1	1					
S. siegburg	4	2				1	1
S. carmel	2		2				
S. tel-aviv	1					1	
S. johannesburg	10	8			2	2	
S. degania	2	1					1
S. negev	1	1					
S. uphill	1	1					
Total	521	364	26	13	84	84	34

Table X shows clinical cases caused by 48 species in 502 patients, (patients with double infections are listed twice and one who had a triple infection was listed three times). In the sum total of patients, cases of hospital infections were not included.

---

* According to KAUFFMANN R-Phase of S. *senftenberg* var. *newcastle*.

Gastroenteritis is the most common manifestation of Salmonella infection (345 patients from whom 42 different species were isolated), 17 were infected with 2 different types while 1 patient had a triple infection. The most common species were: S. *typhi-murium* (121 patients), S. *braenderup* (36), S. *montevideo* (34), S. *emek* (31), and S. *enteritidis* (31), i.e. these 5 species were the cause of 70% of all cases of gastroenteritis.

These frequency figures correspond well with our above reported findings over a period of 14 years. The age distribution shows that 40% were infants below one year of age, while 55% were children below 10 years.

26 patients had Salmonella septicemia without any primary or subsequent localized infection. The clinical picture was that of enteric fever though much milder and shorter than Typhoid or Paratyphoid.

As may be seen from Table XI, 3 infants aged 1 to 2.5 months had meningitis, 2 patients aged 7 months and 12 years suffered from osteomyelitis, 3 patients (84, 60 and 1 year) — from empyema of the pleura, 3 patients from cholecystitis. From 1 patient with a peri-anal abscess S. *typhi-murium* was isolated in pure culture from pus drained from the abscess. One year later the same organism was again cultured from this abscess. Finally from one patient aged 30 years, S. *typhi-murium* was several times isolated from the urine. On nephrectomy of the right kidney a tuberculous process with secondary pyonephritis due to this Salmonella was found.

84 patients were found to excrete Salmonella in their faeces without any clinical symptoms. 3 of them, infants, 2 to 5 months of age, were shown to excrete S. *emek* for 54, 36 and 30 days respectively. One infant excreted S. *newport* for 75 days.

Sensitivity to antibacterial substances

According to the findings of ALTMANN & GERICHTER (1963), Salmonella strains have retained sensitivity to antibiotics. Over 95% of all strains tested were found to be sensitive to Chloramphenicol and Tetracycline. All strains were sensitive to Furazolidin and Penbritin, while Kanamycin-sulphate and Paramomycine-sulphate (Humatin) were apparently less effective.

In conclusion it can be stated that a steadily increasing yield of Salmonellae during the last few years has been observed in Israel. There are many reasons for this, the most important of them being the continuous flow of immigrants, particularly from underdeveloped countries. Only a combined effort of all people involved in planning and prevention may finally succeed by means of health education especially among minority villagers, by improvement of rural health and sanitary conditions in new villages, and by sanitary

Table XI.

Focal Infections. (From ALTMANN & GERICHTER)

	Age of patient	Species	Source of isolation	Underlying disease	Remarks
Meningitis	1 month	*S. typhi-murium*	CSF	Hydrocephalus	Exitus
Meningitis	2 months	*S. enteritidis*	CSF & Blood	Atresia of bile duct	Exitus
Meningitis	2½ months	*S. hadar*	CSF	—	Exitus
Osteomyelitis	7 months	*S. typhi-murium*	Bone	—	—
Osteomyelitis	12 years	*S. typhi-murium*	Bone	Hodgkin's disease	Exitus
Empyema of pleura	84 years	*S. enteritidis*	Pleural fluid	—	—
Empyema of pleura	1 year	*S. enteritidis*	Pleural fluid	—	—
Empyema of pleura	60 years	*S. oslo*	Pleural fluid	Adenocarcinome of Breast	Exitus
Cholecystitis	69 years	*S. typhi-murium*	Bile, stools & pus	—	—
Cholecystitis	50 years	*S. braenderup*	Bile & stools	—	—
Cholecystitis	14 years	*S. oranienburg*	Bile	—	—
Perianal abscess	30 years	*S. typhi-murium*	Pus	—	—
Pyonephritis	30 years	*S. typhi-murium*	Urine	Tb of right kidney	—

351

food-handling, in restraining the further spread of Salmonellosis in this country.

The statistical data presented in this paper, unless otherwise stated, have been compiled from "Health Services in Israel, A Ten Year Survey 1948—1958", edited by Prof. Th. GRUSHKA, published by the Ministry of Health, Jerusalem, 1959; The Statistical Monthly of Israel published by the Central Bureau of Statistics, December 1961; and by information kindly supplied by the Department of Epidemiology, Ministry of Health, Jerusalem.

REFERENCES

ALTMAN, G. & GERICHTER, CH. B. (1963) Salmonellosis in a general hospital. *Isr. med. J.* **21**, 243.
EDWARDS, P. R., ALTMAN, G. & McWORTHER, A. C. (1953) A new Salmonella Type (S. israel). *Publ. Health Labor,* **11,** *141.*
FELIX, A., KRIKORIAN, K. S. & REITLER, R. (1953) The occurrence of Typhoid Bacilli containing Vi Antigen in cases of Typhoid fever and of Vi Antibody in their sera. *J. Hyg.*, 35 No. 3, *421—427.*
HIRSCH, W., HENIG, E. & SAPIRO, R. (1950) A new Salmonella Type (S. emek). *J. Bact.,* **60,** *213.*
HIRSCH, W., HIRSCH, MARG. & SAPIRO-HIRSCH, R. (1954) A new Salmonella Type (S. carmel). *Acta med. Orient.,* **13,** *41.*
HIRSCH, W., GERICHTER, CH. B., BREGMAN, E., LUBLING, P. & ALTMAN, G. (1954) A new Salmonella Type (S. hadar). *Acta med. Orient.,* **13,** *42.*
HIRSCH, W. & SAPIRO-HIRSCH, R. (1955) Two new Salmonella Types, isolated from tortoises (S. ezra and S. ahuza). *Acta med. Orient.,* **14,** *298— 299.*
HIRSCH, W., & SAPIRO-HIRSCH, R. (1955) A new Salmonella Type (S. degania). *Acta. med. Orient.,* **14,** *297.*
HIRSCH, W. & SAPIRO-HIRSCH, R. (1957) A new Salmonella Type (S. vitkin). *Acta med. Orient.,* **15,** *29.*
HIRSCH, W. & SAPIRO-HIRSCH, R. (1958) Two new Salmonella Types (S. uno and S. hofit). *Acta med. Orient.,* **17,** *80.*
HIRSCH, W. & SAPIRO-HIRSCH, R. (1960) A new Salmonella Type (S. shikmona). *Isr. med. J.,* **19,** *36.*
KAUFFMANN, F. & SILBERSTEIN, W. (1950) A new Salmonella Type (S. jerusalem). *Acta path. microbiol., scand.,* **27,** *78.*
KAUFFMANN, F., SILBERSTEIN, W., & GERICHTER, CH. B. (1950) A new Salmonella Type (S. ness-ziona). *Acta pathol. microbiol. scand.,* **27,** *829.*
KAUFFMANN, F., SILBERSTEIN, W., & GERICHTER, CH. B. (1950) A new Salmonella Type (S. tel-hashomer). *Acta pathol. microbiol. scand.,* **27,** *888.*
KAUFFMANN, F., SILBERSTEIN, W., & LUBLING, P. (1953) A new Salmonella Type (S. nachshonim). *Acta pathol. microbiol. scand.,* **33,** *79.*
OLITZKI, L. (1928) Die Beziehungen zwischen dem Auftreten Stabilotropen-Agglutinins im Blute von Typhus Kranken und dem klinischen Verlaufe des Typhus. *Zbl. Bact.,* 1 Abt. Orig., **106,** *247—259.*
OLITZKI, L. (1944) Salmonella strains in Palestine from different sources. *Palestine Vet. Surgeons Ass. Mag.* **1,** *1—8.*
SAPIRO, R. & HIRSCH, W. (1950) A new Salmonella Type (S. haifa). *J. Bact.,* **60,** *101.*

352

Sapiro-Hirsch, R., Altman, G. & Hirsch, W. (1957) A new Salmonella Type
(S. negev). *Acta med. Orient.*, **16,** *274.*
Sapiro-Hirsch, R., Altman, G. & Hirsch, W. (1959) A new Salmonella Type
(S. ramat-gan). *Isr. med. J.*, **18,** *135.*
Silberstein, W. & Gerichter, Ch. B. (1952) Salmonella findings in Israel.
Harefuah. **43,** *1—4.*
Silberstein, W., Gerichter, Ch. B. & Reitler, R. (1954) A new Salmo-
nella Type (S. mishmar-haemek). *Acta med. Orient.*, **13,** No. *1—2.*
Stuart, G. & Krikorian, K. S. (1928) Serological diagnosis of the Enterica
by the method of qualitative receptor analysis. *J. Hyg.* **28,** *105—126.*

LES SALMONELLOSES EN AFRIQUE

AVANT-PROPOS

Ce n'est qu'après la dernière guerre mondiale que la plupart des pays d'Afrique ont entrepris l'étude des salmonelloses basée sur l'isolement et l'identification complète (biochimique, sérologique et, en ce qui concerne *S. typhi*, aussi bactériophagique) des *Salmonella*.

Ceci est aisément compréhensible: l'évolution de ce continent a été freinée, dans certains domaines même arrêtée, tant que duraient les hostilités pour prendre ensuite un essor immédiat et rapide. En de nombreuses localités de nouveaux laboratoires médicaux et vétérinaires sont alors construits et leur équipement en matériel et en personnel est souvent remarquable. Un apport bénéfique décisif a sans doute été l'expansion généralisée des moyens de communication par air grâce auxquels le chercheur le plus éloigné ne devait plus connaître cet isolement intellectuel dont ont souffert tant de ses aînés.

A ceci il faut ajouter d'autres avantages qui n'existaient pas auparavant: les milieux de culture préparés industriellement, la standardisation des techniques d'isolement et d'identification des germes, la possibilité de se procurer dans le commerce des sérums agglutinants convenables et, avant tout, l'aide et la collaboration qu'il fut possible de trouver auprès du Centre International des Salmonella et auprès de certains Centres Nationaux. Toutes ces circonstances favorables ont contribué à susciter de l'intérêt pour les *Salmonella* qui avaient bien dû être négligées jusque-là parce que les moyens pour les étudier manquaient et aussi, ne le perdons pas de vue, parce que les médecins qui oeuvraient en Afrique se trouvaient confrontés avec des maladies épidémiques beaucoup plus importantes que les salmonelloses et par conséquent prioritaires.

D'emblée le nombre aussi bien que la variété des découvertes furent remarquables et les publications se succédèrent à un rythme accéléré. Même celui dont c'est le métier ou l'agrément de les suivre éprouve des difficultés pour le faire. Aussi le moment est-il venu de dresser un premier bilan et d'en dégager les conclusions générales. C'est ce qui sera fait dans les pages qui suivent.

LES SALMONELLOSES EN AFRIQUE CENTRALE

PAR

EUGEEN van OYE

Institut d'Hygiène et d'Epidémiologie, 14, rue Juliette Wytsman, Bruxelles 5

Les territoires dont il sera question dans ce chapitre englobent la République du Congo-Léopoldville, la République du Rwanda et le Royaume du Burundi, c'est à dire donc l'ancienne colonie du Congo Belge et les anciens territoires sous mandat du Ruanda-Urundi. Le Congo comprend la majeure partie du bassin du fleuve Congo, vaste étendue de 2.343.930 km², couverte de fôrets équatoriales ou de savanes boisées et à climat essentiellement tropical. Le Rwanda et le Burundi, avec une superficie totale de 52.112 km², sont situés à l'Est du Congo et se trouvent presque tout entier en région de montagnes. Géographiquement ces pays sont situés entre les parallèles 5°20′ de latitude Nord et 13°27′ de latitude Sud et entre les méridiens 12°15′ de longitude Est à la côte atlantique et 31°15′ de longitude Est au Lac Albert. La population du Congo est estimée à 13 millions d'habitants, celle du Rwanda et du Burundi à environ 4 millions et demi.

Avant que ne débute, en 1946, l'étude systématique des salmonelloses de ces territoires seule *S. paratyphi C* avait fait l'objet de la curiosité de certains. Dès 1949 fonctionnait à Léopoldville un "Centre d'Etude et de Diagnostic des Entérobactéries pathogènes" auquel furent adressées les cultures de *Salmonella* isolées dans les laboratoires médicaux et auquel firent appel également divers laboratoires vétérinaires. Les évènements survenus en juillet 1960 ont mis fin brusquement à toute activité suivie. De ce fait, nos statistiques devront se borner aux données recueillies avant le 30 juin 1960. Mais d'autre part, des contacts ont pû être maintenus malgré tout et certaines observations intéressantes, bien qu'isolées, nous sont encore parvenues; elles sont reprises dans ce travail.

Au cours d'une période qui s'étend de 1946 à fin juin 1960, 3216 souches de *Salmonella* isolées chez des humains ont été étudiées, notamment 1659 cultures de bacilles typhiques et 1557 cultures de bacilles paratyphiques, ainsi que plusieurs centaines de souches de *Salmonella* isolées chez des animaux les plus divers, tant domestiques que sauvages. Enfin, 6 souches ont été isolées dans la nature.

Au total, 173 espèces et variétés différentes de *Salmonella* ont été identifiées à ce jour (31 décembre 1962), dont près du tiers — 57 exactement — sont nouvelles. Une seule espèce n'a pas été isolée au Congo même: il s'agit de *S. loma-linda* dont une souche, avec des propriétés biochimiques particulières, a été isolée en Suisse chez un malade venant du Congo.

Parmi ces 173 espèces et variétés il y en a 129 qui ont été trouvées chez des humains, 106 chez des animaux et 64 ont été trouvées chez les deux. Une espèce, *S. usumbura*, n'a encore été isolée jusqu'ici ni chez un humain ni chez un animal, mais uniquement de l'eau du Lac Tanganyika. Par ailleurs, cinq espèces ont été trouvées dans la nature: la première culture de *S. tinda* a été isolée de l'eau d'une source qui ravitaillait un camp minier; *S. mgulani* a été isolée de l'eau du Lac Tanganyika; *S. dublin* a été trouvée dans l'eau du Lac Mohasi (au Rwanda) à proximité d'un abreuvoir pour bétail, son hôte naturel; deux souches de *S. typhi* ont été isolées de l'eau du Lac Kivu recueillie dans les environs immédiats de Bukavu; enfin, *S. usumbura* a été découverte dans de l'eau du Lac Tanganyika prélevée au niveau d'une plage déserte fréquentée par des oiseaux migrateurs située près d'Usumbura.

Les Salmonelloses humaines

Aucune épidémie importante de salmonellose n'a été signalée en Afrique centrale au cours des 15 dernières années à l'exception d'une épidémie de fièvre typhoïde qui a sévi pendant plusieurs semaines à Léopoldville en 1958 et dont le caractère marquant est qu'elle fut causée par la variété centro-africaine du lysotype C_1 de *S. typhi* (PRUNET & NICOLLE, 1962). Pour ainsi dire tous les cas de salmonellose sont des cas sporadiques et nos statistiques donnent ainsi une image de l'endémie typho-paratyphique qui doit refleter de près la réalité.

Ce qui frappe en premier lieu est la grande diversité des espèces qui caractérise la flore microbienne de l'Afrique centrale. Pas moins de 129 sérotypes différentes ont été trouvés chez des humains et presque la moitié d'entre elles a été découverte en Afrique (cfr. Tableau I).

Mais par ailleurs, très peu de ces nombreux sérotypes se rencontrent fréquemment: sur un total de 3216 souches, 1659 sont des *S. typhi*, soit 51,5%. Si on envisage uniquement les cas de salmonellose à bacilles dits paratyphiques, on voit que la moitié d'entre eux sont dûs à quatre espèces seulement, notamment *S. typhimurium* (26,5%), *S. enteritidis* (10,2%), *S. kisangani* (8,6%) et *S. dublin* (4,8%). Seize espèces seulement atteignent ou dépassent une fréquence de 1% (cfr. Tableau II) et ensemble elles sont responsables de 78,3% de tous les cas de salmonellose.

Parmi les espèces rencontrées au moins dix fois on ne trouve plus que *S. mikawasima* (14), *S. java* (13) *S. wagenia* (11), *S. limete* (10) et *S. braenderup* (10). Les 16 espèces qui figurent au tableau II

356

Tableau I.
Espèces de Salmonella isolées chez des humains.

Groupe A	*S. garoli (1)	S. gallinarum-pullorum
S. paratyphi A (2)	*S. galiema (2)	(3)
Groupe B	*S. irumu (25)	Groupe E.1
*S. kisangani (135)	S. bonn (2)	*S. kalina (1)
S. arechavaleta (2)	S. potsdam (2)	S. butantan (1)
S. abortus-equi (3)	*S. makiso (8)	S. vejle (3)
*S. tinda (1)	S. virchow (21)	S. anatum (9)
S. java (13)	S. infantis (18)	S. zanzibar (1)
S. paratyphi B (7)	S. richmond (3)	*S. ruzizi (1)
*S. limete (10)	S. bareilly (2)	S. uganda (28)
S. abortus-bovis (1)	S. mikawasima (14)	*S. elisabethville (2)
*S. wagenia (11)	*S. businga (1)	*S. simi (2)
S. schwarzengrund (2)	*S. aequatoria (6)	S. amager (3)
S. saint-paul (8)	*S. mbandaka (1)	S. orion (2)
S. reading (1)	Groupe C.2	S. bolton (4)
S. derby (3)	S. nagoya (2)	*S. coquilhatville (3)
S. typhi-murium (388)	*S. banalia (2)	Groupe E.2
S. typhi-murium var.	S. muenchen (5)	S. newington (1)
copenhagen (27)	S. newport (55)	*S. kinshasa (5)
S. hato (7)	S. kottbus (3)	*S. binza (8)
S. bredeney (2)	S. bonariensis (1)	Groupe E.4
*S. kimuenza (2)	S. blockley (1)	S. senftenberg (7)
S. heidelberg (67)	S. bovis-morbificans (51)	Autres groupes
S. coeln (6)	S. akanji (1)	S. chandans (9)
*S. ruki (2)	S. chailey (1)	S. aberdeen (2)
S. shubra (1)	S. mapo (1)	S. rubislaw (1)
S. kiambu (1)	S. hadar (2)	S. mississippi (1)
*S. stanleyville (35)	Groupe C.3	S. ajiobo (1)
*S. kalamu (1)	*S. sanga (5)	S. cubana (1)
*S. ituri (1)	S. emek (4)	S. hull (2)
Groupe C.1	S. kentucky (1)	S. gaminara (1)
S. umhlali (2)	Groupe D	*S. matadi (1)
S. oslo (4)	S. loma-linda (1)	S. cerro (1)
S. brazzaville (1)	S. durban (4)	*S. blukwa (1)
S. edinburg (4)	*S. ipeko (2)	S. minnesota (2)
S. georgia (1)	S. typhi (1659)	*S. kibusi (22)
*S. leopoldville (5)	*S. ndolo (5)	S. pomona (1)
S. paratyphi C (50)	*S. zega (2)	*S. yolo (1)
S. kaduna (1)	S. enteritidis (83)	S. adelaide (1)
*S. mission var. isangi(7)	S. enteritidis var.	S. mgulani (5)
S. amersfoort (3)	chaco (77)	S. johannesburg (1)
*S. gombe (8)	S. dublin (75)	*S. bukavu (4)
S. livingstone (3)	S. panama (2)	S. waycross (5)
S. braenderup (10)	*S. kapemba (8)	S. nairobi (3)
S. montevideo (3)	*S. wangata (15)	*S. kingabwa (1)
S. oranienburg (48)		*S. niarembe (1)

* = espèce nouvelle découverte en Afrique centrale.

Les chiffres entre ( ) indiquent le nombre de souches isolées.

N.B. Pour des raisons que nous avons déjà exposées, cette statistique a été clôturée à la date du 30 juin 1960. Depuis lors, les espèces suivantes ont encore été trouvées chez des humains: *S. congo, S. kivu, S. luckenwalde, S. os* et *S. teshie.*

Tableau II.

Espèces dont la fréquence atteint ou dépasse 1%.

S. typhi-murium	415	S. oranienburg	48
S. enteritidis	160	S. stanleyville	35
S. kisangani	135	S. uganda	28
S. dublin	75	S. irumu	25
S. heidelberg	67	S. kibusi	22
S. newport	55	S. virchow	21
S. bovis-morbificans	51	S. infantis	18
S. paratyphi C	50	S. wangata	15

plus les 5 que nous venons de mentionner sont responsables de 82% de toutes les salmonelloses autres que les fièvres typhoïdes, les 18% des cas restants sont dûs à 107 espèces différentes. Il n'y a donc rien d'étonnant à ce que 38 d'entre elles n'aient été isolées qu'une seule fois.

La distribution par groupe sérologique nous montre que 95,5% des souches appartiennent à 4 groupes sérologiques, notamment au groupe B (47,3%), au groupe C (25,2%), au groupe D (17,8%) et au groupe E (5,2%). Ces derniers chiffres ne tiennent pas compte du nombre des S. *typhi* qui, nous l'avons déjà signalé, dépasse celui de toutes les autres *Salmonella* réunies.

La lysotypie des souches de S. *typhi* isolées au Congo, au Rwanda et au Burundi a été faite systématiquement au Centre Français de Lysotypie des Bacilles Entériques (Institut Pasteur de Paris) et elle nous a appris (VAN OYE & NICOLLE, 1953 et NICOLLE & HAMON, 1954) que:

1. La proportion des cultures qui peuvent être caractérisées par la méthode bactériophagique est remarquablement élevée: 98,5%.

2. Le nombre de lysotypes rencontrés est restreint: A, B_2, B_3, C_1, C_2, C_4, D_1, E_1, G, N, O et I + IV. Dans la région de l'Ouest, le type E_1 domine nettement. Dans la région centrale, les types E_1 et A tendent à s'équilibrer. Dans la région de l'Est, le type A est le plus fréquent.

3. Le type C_1 n'est par rare. Il est représenté exclusivement par la variété centro-africaine caractéristique pour l'Afrique Equatoriale et Madagascar (NICOLLE et coll., 1955 — PRUNET & NICOLLE, 1962).

4. Le lysotype O est relativement fréquent mais on ne le trouve que dans un seul foyer, celui de Niarembe situé tout au Nord-Est du Congo près de la frontière de l'Uganda et non loin du Soudan. Est-ce que le foyer de Niarembe s'étend dans ces deux pays? Nous ne possédons aucune indication à ce sujet.

5. Les autres types sont rares; il s'agit par ailleurs souvent de lysotypes ubiquitaires à fréquence généralement faible. Le type G, qui est un lysotype exotique, d'origine orientale, n'a été rencontré

que dans la région Est où il a peut-être été introduit par des Indiens ou par des Pakistanais qui sont nombreux dans cette région.

6. Enfin, coïncidence ou incidence curieuse?, le lysotype C₄, fréquent à Dakar, a été trouvé à trois reprises à Léopoldville après qu'un contingent sénégalais des forces de l'O.N.U. ait stationné dans cette ville en 1960 (NICOLLE, comm. pers.).

En ce qui concerne les espèces de *Salmonella* adaptées à l'homme et qui chez lui provoquent souvent des syndromes typhoïdes, la situation se présente ainsi:

S. paratyphi A, répandue dans tout l'Orient et l'Extrème-Orient, est pratiquement inconnue en Afrique centrale: nous n'avons reçu pour examen que 2 souches provenant une de la Province Orientale (Stanleyville) et l'autre de la Province de l'Equateur (Coquilhatville) et toutes deux isolées chez des congolais.

S. paratyphi B, espèce bien connue en Europe, est très rare en Afrique centrale: 7 cultures seulement y ont été isolées au cours des 15 dernières années.

S. paratyphi C est relativement fréquente: nous avons reçu 50 souches, soit 3,2% du total. Cette espèce se présente presque exclusivement sous sa forme "var. *east africa*", c'est à dire possédant l'antigène Vi et ayant des caractères biochimiques qui lui sont propres: rhamnose $+^2$, tréhalose $+^{2-3}$, arabinose et inosite —. Cette espèce semble douée d'une virulence nettement plus grande que les autres *Salmonella* car la plupart des cas au sujet desquels nous avons obtenu des renseignements cliniques étaient des cas graves.

Les très nombreuses autres espèces de *Salmonella* ont été isolées en majeure partie dans des cas sporadiques de gastroentérite. Il est plus que probable que ceux-ci étaient d'origine alimentaire, mais ce n'est qu'exceptionnellement que la preuve a pu en être fournie, ce qui n'a rien d'étonnant vu les conditions de travail en Afrique.

Il nous faut signaler ici que les enfants, plus particulièrement les enfants en bas âge et même les nourrissons, sont très sensibles aux infections à *Salmonella* mais que, par contre, ils ne sont pas souvent gravement atteints. Plusieurs nouvelles espèces de *Salmonella* ont été trouvées chez des petits enfants qui, en général, étaient déjà guéris de leur dérangement intestinal au moment où le résultat de la coproculture était connu.

Il est difficile de faire la comparaison entre la pathologie des salmonelloses en Afrique et en d'autres régions. Pourtant, nous avons la très nette impression que les infections localisées dues à une *Salmonella* se rencontrent plus fréquemment sur le continent noir. La toute grande majorité des souches proviennent bien entendu d'affections gastro-intestinales, mais nous en avons reçu de nombreuses autres qui avaient été isolées d'abcès ou de phlegmons superficiels ou profonds (et dans ces cas il s'agissait presque toujours d'une *S. typhi-murium* var. *copenhagen*), de cas de méningite, d'infections rénales ou pulmonaires, de cas d'arthrite, de pleurésie, de pyo-myosite, de péritonite, d'orchite, d'infections osseuses postopératoires ou bien se greffant sur une fracture, etc., etc., et, surtout, provenant de cas d'ostéomyélite survenant chez des enfants souffrant d'anémie falciforme. Près de 100 observations ont été faites au cours de quelques années et la découverte de l'association entre une ostéomyélite à *Salmonella* et une hémoglobinopathie représente pour nos connaissances de la pathologie des salmonelloses la contribution la plus importante apportée par les études poursuivies en Afrique centrale (VANDEPITTE et coll., 1953 — VAN OYE, 1960).

Les Salmonelloses animales

La présence de *Salmonella* a été recherchée chez un grand nombre d'animaux les plus divers: invertébrés, reptiles, oiseaux de basse-cour ou de volière, oiseaux sauvages, poissons, mammifères sauvages ou domestiques, animaux d'abattoir. La plupart de ces recherches ont été faites par curiosité scientifique, certaines l'ont été lors d'examens de routine, enfin, quelques enquêtes systématiques ont été poursuivies. Les unes comme les autres ont fourni une ample moisson de renseignements que nous passerons en revue maintenant.

I. *Animaux d'abattoir*:
Les espèces suivantes ont été trouvées chez:
A. Des bovidés: *S. aberdeen, S. ajiobo, S. amager, S. anatum, S. binza, S. bonn, S. bovis-morbificans, S. braenderup, S. chingola, S. cubana, S. dublin, S. enteritidis, S. gallinarum-pullorum, S. heidelberg, S. hull, S. infantis, S. kentucky, S. kisangani, S. manchester, S. miami*: les souches congolaises diffèrent de la culture originale (LE MINOR, 1955) et il a été question de les élever au rang d'espèce sous le nom de *S. bambesa* (VAN OYE, 1956), mais il a été décidé finalement de les rattacher à *S. miami* (KAUFFMANN, 1959), *S. mission* var. *isangi, S. poona, S. pretoria, S. reading, S. roan, S. schwarzengrund, S. senegal, S. stanleyville, S. tshiongwe*, *S. tuebingen, S. typhi-murium, S. vejle, S. wagenia, S. wangata.*
B. Des suidés: *S. abortus-equi, S. amager, S. anatum, S. bolombo*, *S. braenderup, S. chandans, S. chingola, S. cholerae-suis, S. cubana, S. dublin, S. enteritidis, S. gallinarum-pullorum, S. haifa, S. heidelberg, S. hull, S. infantis, S. inganda*, *S. ipeko*, *S. kapemba, S. kibusi, S. livingstone, S. mikawasima, S. ndolo, S. poona, S. reading, S. sanga, S. schwarzengrund, S. senftenberg, S. simi, S. tuebingen, S. typhi-murium, S. uganda, S. urbana, S. vejle, S. wagenia, S. wangata.*
C. Des asinés: *S. amager, S. anatum, S. banana* (est tombée en synonymie avec *S. california* mais ne fermente pas l'inosite), *S. bonn,*

---

* = Premier isolement de cette espèce.

S. *braenderup*, S. *bulawayo**, S. *chingola*, S. *gallinarum-pullorum*, S. *gwaai**, S. *hull*, S. *mobeni*, S. *onderstepoort*, S. *pretoria*, S. *roan*, S. *senegal*, S. *tshiongwe*, S. *typhi-murium*.
D. Des équidés: S. *abortus-equi*, S. *anatum*, S. *bredeney*, S. *hull*, S. *kibusi*, S. *newport*.
E. Des ovidés: S. *anatum*, S. *bovis-morbificans*, S. *infantis*, S. *kisara-we*, S. *moëro**, S. *newport*, S. *poona*, S. *schwarzengrund*, S. *typhi-murium*, S. *zanzibar*.
F. Des capridés: S. *abortus-equi*, S. *dublin*, S. *kibusi*, S. *newport*.

Deux enquêtes ont été faites chez des animaux d'abattoir, une première à Stanleyville (WIKTOR & VAN OYE, 1955) et une seconde à Elisabethville (VAN OYE et coll., 1957). Une troisième a débuté à Coquilhatville mais elle a dû être arrêtée suite aux évènements.

L'enquête de Stanleyville a porté sur 205 bovidés et 158 suidés en provenance de la région d'élevage de l'Ituri située à environ 1000 km au Nord-Est. Elle a permis d'isoler 12 souches de *Salmonella* chez les premiers (5,9%) et 28 chez les seconds (17,7%). En dehors de S. *typhi-murium* et de S. *dublin* qui sont des ubiquitaires, seuls ont été trouvés des sérotypes découverts au Congo et dont l'aire de répartition géographique est surtout centro-africaine. Il est donc évident que les infections chez les animaux de l'Ituri doivent avoir une origine locale. En outre, la concordance entre les espèces de *Salmonella* trouvées chez les animaux de boucherie et chez les humains à Stanleyville incite à penser qu'il est hautement probable qu'il existe un rapport causal entre les deux. Des améliorations apportées aux conditions de travail à l'abattoir de Stanleyville y ont par ailleurs réduit sensiblement le nombre de cas de salmonellose humaine.

L'enquête menée à Elisabethville a porté sur plusieurs groupes d'animaux qui étaient pour la plupart importés des Rhodésies ou de l'Afrique du Sud. Au total, 2137 animaux ont été examinés et leur degré d'infection était respectivement de 13,9% chez les bovidés, 12,2% chez les suidés, 5,5% chez les ovidés, 2,4% chez les capridés, 42,8% chez les équidés et 18,8% chez les asinés. La flore microbienne fut extrêmement variée et comprenait, outre les espèces cosmopolites, plusieurs *Salmonella* dont l'aire de distribution se situe principalement dans la partie Sud de l'Afrique. Les conditions de travail à l'abattoir d'Elisabethville étaient parfaites du point de vue hygiénique et nous croyons devoir attribuer à cette situation favorable le fait que les cas de salmonellose humaine ont toujours été rares dans cette ville: l'importation de tant d'animaux hébergeant des *Salmonella* n'y a jamais provoqué la moindre épidémie.

Nous avons l'impression que la situation était tout aussi favorable à Coquilhatville. Le service vétérinaire y a commencé à examiner les animaux abattus à l'abattoir local et il a constaté que ces animaux étaient souvent porteurs de *Salmonella*, mais d'espèces non-encore identifiées et qui n'avaient jamais été trouvées chez des humains comme *S. bolombo*, *S. inganda*, *S. ipeko*. Par ailleurs, les salmonelloses humaines n'ont jamais été nombreuses à Coquilhatville. Les circonstances n'ont malheureusement pas permis de poursuivre cette enquête.

II. *Volailles de basse-cour*: (cfr. VAN OYE & DEOM, 1958)

Les espèces suivantes ont été trouvées chez:

A. Des canards: *S. aberdeen*, *S. adelaide*, *S. cairina**, *S. chandans*, *S. elisabethville*, *S. emek*, *S. heidelberg*, *S. hull*, *S. infantis*, *S. irumu*, *S. ituri**, *S. kasenyi**, *S. korovi*, *S. makiso*, *S. mikawasima*, *S. new-port*, *S. simi*, *S. stanleyville*, *S. tinda*, *S. typhi-murium*, *S. zanzibar*, *S. zega**.

Les canards ont été étudiés surtout par FAIN (1953) qui, sur 347 de ces oiseaux en trouva 41 porteurs de *Salmonella*, soit une proportion de 11,8%. Ces *Salmonella* appartiennent à 19 espèces différentes parmi lesquelles 4 sont nouvelles pour la science (voir ci-desssus). FAIN (1953) et KAUFFMANN & FAIN (1953) ont souligné le danger que représente le canard domestique en tant que réservoir de *Salmonella*, ce qui d'ailleurs a été confirmé partout où le canard a été étudié.

Il est à remarquer que plus de la moitié des espèces mentionnées sont typiquement centro-africaines.

B. Des poules: *S. anatum*, *S. bolombo*, *S. bovis-morbificans*, *S. braenderup*, *S. chester*, *S. dublin*, *S. elisabethville*, *S. gallinarum-pullorum*, *S. infantis*, *S. kisangani*, *S. saint-paul*, *S. senegal*, *S. thompson*, *S. typhi*, *S. typhi-murium*, *S. vejle*.

DEOM (1956) et HUYGELEN, MORTELMANS & VERCRUYSSE (1958) ont attiré l'attention sur le danger que représentent les jeunes poussins pour la dissémination des *Salmonella*, surtout les poussins d'un jour qui sont expédiés par milliers d'Afrique du Sud vers d'autres régions africaines.

Contrairement à ce qui s'observe généralement, *S. gallinarum-pullorum* n'est pas dominant chez les gallinacés en Afrique centrale. Le nombre de souches de cette espèce qui y ont été isolées est plutôt restreint. Par contre, et ceci est très curieux, le nombre d'espèces différentes est relativement élevé et la plupart de ces espèces ne sont pas d'origine centro-africaine. La différence avec ce que nous avons vu chez les canards est frappante et nous ne voyons pas l'explication qu'il faut en donner.

C. Des dindons: *S. chester*, *S. infantis*, *S. gallinarum-pullorum*, *S. san-diego*, *S. thompson*.

D. Des pigeons: *S. dublin*, *S. gallinarum-pullorum*.

E. Oiseaux sauvages: ce n'est évidemment qu'exceptionnellement qu'un examen bactériologique a été pratiqué sur un oiseau non-domestique et il est d'autant plus remarquable que plusieurs espèces de *Salmonella* aient été isolées chez les rares oiseaux sauvages qui ont été examinés, notamment *S. california, S. djugu*, S. java, S. muenchen, S. newcastle, S. sundsvall, S. typhi-murium, S. bovis-morbificans, S. umhlali.*

III. *Reptiles*:

Dans toutes les régions du monde et à maintes reprises il a été observé que les animaux terrestres à sang froid sont fréquemment porteurs de *Salmonella* et cela sans qu'ils en ressentent le moindre inconvénient. Ces germes se conduisent donc chez eux en parfaits commensaux. L'Afrique étant un paradis pour les reptiles, grands et petits, plusieurs bactériologistes en ont tiré profit pour examiner les fèces d'animaux capturés soit occasionnellement soit lors d'une enquête (FULTON et coll., 1961). Ces coprocultures ont permis d'isoler les espèces suivantes: *S. banana**: a été découverte chez un serpent capturé à Banana (Cette espèce est maintenant rattachée à *S. california* dont la formule antigénique fut élargie à cette occasion), *S. braenderup, S. champaign, S. coeln, S. gatow, S. irumu, S. kibusi, S. kintambo**: a été découverte chez un lézard, *S. korovi**: a été découverte chez un serpent, *S. landau, S. leopoldville, S. limete, S. oslo, S. plymouth, S. ramat-gan, S. stanleyville, S. tel-aviv, S. vancouver, S. waycross.*

IV. *Animaux divers*:

Nous ne pouvons suivre aucun ordre ici et nous devons nous contenter de signaler simplement quelques données qui nous semblent intéressantes, par exemple la découverte de *S. ngozi** chez un chien du Burundi ou celle de *S. mampeza** chez un cobaye à Léopoldville, ainsi que les isolements de *S. johannesburg* de poissons du Lac Kivu, de *S. muenchen* d'un broyat de sangsues, de *S. dublin* de tiques prélevées sur du bétail au Rwanda, de *S. nairobi* chez un chat à Léopoldville.

De l'ensemble des recherches sur les *Salmonella* chez des animaux il ressort que les animaux d'abattoir et les volailles de basse-cour représentent les sources d'infection les plus importantes et les plus dangereuses. Ce fait étant établi, il appartient aux autorités sanitaires et principalement aux hygiénistes et aux vétérinaires d'en tirer les conséquences sur le plan de la pratique.

Contrairement à ce qui a été observé dans d'autres régions, les rongeurs ne semblent pas, en Afrique Centrale, constituer un réservoir de *Salmonella* dont il faut tenir compte. Mais nous devons ajouter immédiatement qu'aucune enquête n'a été menée ni dans les ports de mer Matadi, Boma ou Banana ni dans les ports des grands lacs du Centre-Afrique.

Les reptiles se sont avérés être fréquemment des porteurs de *Salmonella* mais leur rôle épidémiologique n'a pas pu être fixé. Des exemples fournis par d'autres territoires africains prouvent néanmoins qu'ils peuvent représenter un danger réel et qu'ils ne peuvent pas être ignorés; ils méritent au contraire d'être étudiés d'une manière plus approfondie.

Pour clore ce chapitre sur les salmonelloses chez les animaux il nous faut mentionner ici que MORTELMANS et coll. (1961) ont examiné les fèces de 872 chiens, notamment 682 en territoire d'Astrida (Rwanda) dont 11, soit 1,61%, étaient infectés de *Salmonella* et 190 en territoire de Ngozi (Burundi) dont 7, soit 3,67%, ont été trouvés infectés. Dans l'ensemble, il y avait donc 18 chiens sur les 872, c'est à dire 2,06%, qui étaient porteurs de germes. Les sérotypes suivants ont été isolés: *S. bovis-morbificans* (1), *S. dublin* (1), *S. java* (1), *S. ngozi** (1), *S. paratyphi* C (1), *S. senftenberg* (1), *S. stanleyville* (9), *S. typhi-murium* (1), *S. umhlali* (1) et *S.*/ 1,40:—:1,5,7 (1). Ces données incitent à considérer le chien comme une source non-négligeable de *Salmonella* et comme un danger permanent tant pour les humains que pour les autres animaux.

Nous avons donc pu étudier en moyenne par an une centaine de souches de *S. typhi* et autant d'autres aspèces de *Salmonella*. Ces chiffres donnent de la situation en Afrique centrale une image qui n'est certainement pas conforme à la réalité épidémiologique car ils font croire d'une part à une prédominance très marquée des fièvres typhoides et d'autre part à une importance prèsque négligeable des autres salmonelloses. Il est de toute évidence impossible qu'il n'y ait que quelques centaines de cas de salmonellose par an chez une population qui compte au total plus de 17 millions d'habitants particulièrement exposés aux infections intestinales et de par leur mode de vie (humains et animaux habitant la même hutte dans une promiscuité totale) et de par l'absence générale d'hygiène la plus élémentaire. L'explication de ce décalage entre la réalité et nos données numériques doit être cherchée en partie dans le fait que seuls les centres urbains importants disposent d'un laboratoire de bactériologie et que la masse des habitants de la brousse en est par trop éloignée. À ceci il faut ajouter que les africains ne consultent guère le médecin pour des troubles intestinaux peu graves; ils en ont tellement l'habitude. De son côté, le médecin isolé à l'intérieur du pays a acquis un sens clinique et une sûreté de diagnostic qui le dispensent de recourir à des examens de laboratoire, par ailleurs trop souvent impossibles à faire exécuter. Enfin, les sulphamidés, les antibiotiques et certains antiseptiques intestinaux sont là qui guérissent le malade souvent avant qu'une demande d'analyse n'ait pu atteindre un laboratoire.

Il n'en est pas tout à fait de même en ce qui concerne les fièvres typhoides qui sont le plus souvent suffisamment graves pour nécessi-

364

ter l'hospitalisation et pour justifier le recours à un laboratoire, quelque soit son éloignement. Le problème du diagnostic différentiel se pose ici avec plus d'acuité et ceci explique peut-être pourquoi les salmonelloses à *S. typhi* donnent l'impression de tellement dominer toutes les autres: étant plus graves il n'est que naturel que soient mis en oeuvre plus de moyens d'investigation clinique.

Il est hors de doute que le nombre des cas de salmonellose humaine doit être infiniment supérieur à celui des souches examinées. Nous ne possédons malheureusement aucune base sûre qui pourrait nous aider à fixer l'importance réelle ou même approximative des salmonelloses dans la hiérarchie des maladies infectieuses qui règnent en Afrique centrale. Il aurait fallu pour cela pouvoir faire des enquêtes coprologiques et sérologiques étendues sur des échantillonages représentatifs de la population dans diverses régions. Des impératifs d'ordre local joints à un manque de moyens techniques et de personnel ont empêché de procéder à de telles enquêtes qui doivent nécessairement être très vastes si on désire aboutir à des résultats auxquels on peut se fier.

BIBLIOGRAPHIE

1. DEOM, J., Occurrence of new Salmonella types from South Africa. *Nature, Lond.* 1956, **178**: *702.*
2. FAIN, A. – Importance du réservoir animal dans l'épidémiologie des salmonelloses au Congo Belge et au Ruanda-Urundi. *Ann. Soc. belge Méd. trop.*, 1953, **33**: *403—422.*
3. FULTON, M., SZAFRAN, PH. & LESKO, M. – Five less-common Salmonella serotypes from Congo reptiles. *Nature, Lond.* 1961, **189**: *240—241.*
4. HUYGELEN, C., MORTELMANS, J. & VERCRUYSSE, J. – Infekties door Salmonella saint-paul, Salmonella senegal, Salmonella vejle, Salmonella infantis, Salmonella braenderup en Salmonella thompson als oorzaak van kuikensterfte. *VI. Diergen. Tschr.*, 1958, **27**: *201—215.*
5. KAUFFMANN, F. & FAIN, A. – Three new Salmonella types (S. ituri, S. kasenyi and S. niarembe) from the Belgian Congo. (Occurrence of Salmonella types in ducks). *Acta path. microbiol. scand.*, 1953, **32**: *513—515.*
6. KAUFFMANN, F. – Supplement to the Kauffmann-White scheme (II). *Acta path. microbiol. scand.*, 1959, **45**: *411—416.*
7. LE MINOR, L. – Variantes biochimiques de S. miami et S. sendai. Etude de l'antigène a. *Ann. Inst. Pasteur*, 1955, **88**: *76—83.*
8. MORTELMANS, J., CIMPAYE, J., PINCKERS, F. & CLAEYS, R. – A propos des salmonelloses des chiens au Ruanda-Urundi. *Bull. epiz. Dis. Afr.*, 1961, **9**: *241—244.*
9. NICOLLE, P., PAVLATOU, M. & DIVERNEAU, G. – Subdivision de quelques types Vi fréquents de Salmonella typhi par des lysotypies auxiliaires. *C. R. Acad. Sci.*, 1953, **236**: *2453—2454.*
10. NICOLLE, P. & HAMON, Y. – Distribution des lysotypes du Bacille typhique et du Bacille paratyphique B en France, dans les territoires d'Outre-Mer et dans quelques autres pays. *Rev. Hyg. et Méd. soc.*, 1954, **2**: *424—463.*

11. NICOLLE, P., VAN OYE, E., CROCKER, CL. G. & BRAULT, J. – Sur une variété du lysotype C de Salmonella typhi rencontrée en Afrique Equatoriale et à Madagascar. *Bull. Soc. Path. exot.*, 1955, **48**: *492—510*.
12. VAN OYE, E. & NICOLLE, P. – La lysotypie des bacilles typhiques isolés au Congo belge. *Bull. Soc. Path. exot.*, 1953, **46**: *48—56*.
13. VAN OYE, E. – Les Salmonellae du Congo Belge. (Quatrième Rapport). *Ann. Soc. belge Méd. trop.*, 1956, **36**: *299—306*.
14. VAN OYE, E., DEOM, J., VERCRUYSSE, J. & FASSEAUX, P. – Recherches sur l'incidence des Salmonella chez les animaux de boucherie à Elisabethville. *Ann. Soc. belge Méd. trop.*, 1957, **37**: *551—558*.
15. VAN OYE, E. & DEOM, J. – Les salmonelloses chez les oiseaux de basse-cour au Congo Belge et au Ruanda-Urundi. *Office int. Epiz.*, 1958, **50**: *337—345*.
16. VAN OYE, E. – Sur l'association entre ostéomyélite à Salmonella et hémoglobinopathie chez l'enfant africain. *Bull. Soc. Path. exot.*, 1960, **53**: *89—100*.
17. PRUNET, J. & NICOLLE, P. – Recherches sur une variété du lysotype C_1 de Salmonella typhi particulière à l'Afrique Equatoriale et à Madagascar. *Ann. Inst. Pasteur*, 1962, **103**: *536—561*.
18. VANDEPITTE, J., COLAERT, J., LAMBOTTE-LEGRAND, J. & C. & PERIN, F. Les ostéïtes à Salmonella chez les sicklanémiques: à propos de 5 observations. *Ann. Soc. belge Méd. trop.*, 1953, **33**: *511—522*.
19. WIKTOR, T. & VAN OYE, E. – Importance des animaux de boucherie comme propagateurs de Salmonelloses humaines à Stanleyville. *Ann. Soc. belge Méd. trop.*, 1955, **35**: *825—832*.
N.B. Pour une bibliographie complète et détaillée, voir:
20. VAN OYE, E. – Répertoire général et revisé des Salmonellae du Congo et du Ruanda-Urundi. *Acad. Roy. Sci. d'Outre-Mer*, Classe Sci. nat. & méd., 1960, T. XI, fasc. 6, 49 pp.

LES SALMONELLA A MADAGASCAR, AU NIGERIA ET AU SENEGAL

PAR

L. LE MINOR

Centre des Salmonella de l'Institut Pasteur de Paris.

Il est impossible de présenter un bilan des *Salmonella* dans un continent aussi grand que l'Afrique. Dans certains pays, les recherches ne sont pas faites de manière systématique et suivie, dans d'autres, la fréquence des *Salmonella* dépend de la présence d'un bactériologiste intéressé par la recherche de ces Entérobactéries; aussi pensons-nous qu'il est plus judicieux, plutôt que d'essayer de faire une synthèse imparfaite, de ne citer comme exemples que les bilans effectués dans certains pays.

I. Les Salmonella à Madagascar

Pour l'ensemble de l'île, le plus grand nombre de *Salmonella* est isolé chez l'homme de décembre à mars, c'est-à-dire pendant la saison chaude. Le sérotype de beaucoup le plus fréquent chez l'homme est *S. typhi* dont 63% des souches sont du lysotype A, 32% du lysotype E_{1a} et 5% de lysotypes variés.

S. paratyphi A et B sont beaucoup moins fréquentes tandis que *S. paratyphi* C arrive en seconde position suivie par *S. newport* du groupe C_2. Il est à noter que les *Salmonella* les plus fréquemment rencontrées lors d'une enquête chez les porcs (*S. newport, S. anatum, S. cholerae-suis, S. london, S. typhi-murium*) sont aussi souvent retrouvées dans les infections humaines. Dans une enquête faite chez ces animaux en ensemençant leurs ganglions mésentériques, 10,1% des 168 prélèvements contenaient des *Salmonella*. Comme la viande de porc est volontiers consommée crue par les autochtones, elle peut constituer une source importante de dissémination de *Salmonella* chez les humains. [1]

Des recherches de *Salmonella* chez les caméléons malgaches [2] ont montré que ces bactéries y étaient fréquentes: 13 souches appartenant à 12 sérotypes différents furent trouvées dans le contenu intestinal de 33 de ces animaux.

BIBLIOGRAPHIE

1. NEEL, R., GRABAR, J. & LE MINOR, L. — Salmonelles et salmonelloses à Madagascar de décembre à mai 1949. *Ann. Inst. Pasteur*, 1950, **78**, *583*.
2. BRYGOO, E. & LE NOC, P. — Note préliminaire sur les Salmonelles de caméléons malgaches. — *Bull. Soc. Path. exot.*, 1961, **54**, *166*.
3. DODIN, A. — Les Salmonelloses à Madagascar. — *Arch. Inst. Pasteur Madagascar*, 1960, **28**, *19*.

Tableau I.

Bilan des Salmonella identifiées de 1949 à 1960 inclus.

Groupe	Sérotype	Nombre total	Origine humaine			Origine animale		
			Hémo culture	Copro culture	Divers	Porc	Divers sang chaud	Divers sang froid
A	S. paratyphi A	14	8	5	1			
B	S. paratyphi B	16	3	12	1			
	S. abony	1				1		
	S. typhi-murium	33	2	17	4	3	7	
	S. saint-paul	3		3				
	S. san diego	1						1
C₁	S. paratyphi C	26	6	20				
	S. cholerae-suis	8	1	1	2	4		
	S. thompson	2	2					
	S. oranienburg	2				1		1
	S. westerstede	1	1					
	S. montevideo	1						1
C₂	S. manhattan	1		1				
	S. muenchen	2		1		1		
	S. newport	17		8		8	1	
	S. tananarive	1				1		
	S. glostrup	1						1
D₁	S. typhi	186	178	8				
	S. durban	1		1				
	S. gallinarum-pullorum	8					8	
E₁	S. anatum	15		8		4	1	2
	S. give	1						1
	S. london	11	1	7		3		
E₃	S. new-brunswick	1						1
E₄	S. senftenberg	1		1				
I	S. hvittingfoss	1						1
L	S. minnesota	2						2
M	S. pomona							1
N	S. aqua	1						1

II. Les Salmonella au Nigéria

Le deuxième pays que nous citerons en exemple est le Nigéria, où COLLARD & SEN[1] ont fait une excellente étude sur la répartition des *Salmonella* et dont nous rapporterons ci-dessous les résultats de fin 1955 à 1959.

Tableau II.
Salmonella isolées chez l'homme.
(entre parenthèses le nombre de souches isolées).

Groupe A

S. paratyphi A	(1)		

Groupe B

S. africana	(2)	S. saint-paul	(12)
S. agama	(52)	S. typhi-murium	(14)
S. brancaster	(5)	S. stanleyville	(4)
S. bredeney	(3)	S. jericho	(1)
S. bury	(2)	S. kalamu	(1)
S. chester	(6)	S. teddington	(4)
S. derby	(6)	S. jos	(1)
S. duisberg	(1)	S. jaja	(1)
S. jacksonville	(1)	S. ayinde	(1)
S. hessarek	(1)	S. kamoru	(1)
S. kappstad	(7)	S. bradford	(1)
S. kingston	(11)	S. california	(1)
S. paratyphi B	(9)		

Groupe C₁

S. aequatoria	(2)	S. paratyphi C	(2)
S. edinburg	(1)	S. oritamerin	(6)
S. livingstone	(2)	S. infantis	(8)
S. nigeria	(5)	S. garoli	(1)
S. oranienburg	(10)	S. mission	(1)
S. virchow	(12)	S. colindale	(1)

Groupe C₂

S. aba	(1)	S. takoradi	(7)
S. alagbon	(1)	S. utah	(1)
S. chailey	(3)	S. mapo	(2)
S. edmonton	(1)	S. bukuru	(1)
S. hadar	(4)	S. akanji	(2)
S. kottbus	(2)	S. molade	(1)
S. nagoya	(1)		

Groupe D

S. berta	(1)	S. itutaba	(2)
S. dublin	(20)	S. ekotedo	(1)
S. enteritidis	(15)	S. neasden	(1)
S. penarth	(5)	S. lishabi	(1)
S. typhi	(70)	S. eastbourne	(2)
S. wangata	(13)	S. zega	(1)

Groupe E₁

S. adabraka	(2)	S. mokola	(1)
S. anatum	(2)	S. onireke	(2)
S. coquilhatville	(1)	S. shangani	(1)
S. elisabethville	(6)	S. okerara	(1)
S. give	(3)	S. suberu	(2)
S. okefoko	(1)	S. yaba	(1)
S. oxford	(5)	S. aminatu	(2)
S. weybridge	(1)	S. stockholm	(1)
S. butantan	(1)		

Groupe E$_4$

S. senftenberg	(1)	S. korlebu	(1)
S. gwoza	(1)		

Groupe F

S. chandans	(3)	S. rubislaw	(14)
S. marseille	(1)		

Groupe G

S. ajiobo	(4)	S. agbeni	(1)
S. cubana	(2)	S. ibadan	(3)
S. durham	(18)	S. okatie	(2)
S. poona	(19)	S. worcester	(1)
S. tel-el-kebir	(2)	S. clifton	(1)

Groupe H

S. albuquerque	(2)	S. garba	(2)
S. magumeri	(2)		

Groupe I

S. adeoyo	(1)	S. salford	(1)
S. brazil	(2)	S. amina	(1)
S. amunigun	(4)	S. mobeni	(1)

Groupe (16)

S. haddon	(1)

Groupe (28)

S. ona	(3)	S. dakar	(1)
S. patience	(5)	S. moero	(1)
S. taunton	(1)	S. vinohrady	(1)
S. chicago	(1)	S. tel-aviv	(1)
S. ank	(1)		

Groupe (30)

S. urbana	(3)	S. gege	(1)

Groupe (35)

S. agodi	(1)	S. adelaide	(3)
S. monschaui	(5)		

Groupe (40)

S. omifisan	(1)

Groupe (1, 40)

S. johannesburg	(3)

Groupe (41)

S. waycross	(2)	S. offa	(1)

Groupe (1, 42)

S. loenga	(1)

Groupe (43)

S. berkeley	(1)

Groupe (45)

S. dugbe	(2)	S. apapa	(1)

Groupe (47)

S. bere	(1)	S. bergen	(1)
S. luke	(1)		

Ici encore, on voit que le sérotype prédominant est *S. typhi*.
Mais, particularité du Nigéria, le sérotype suivant par ordre de
fréquence est *S. agama* qui avait été isolé pour la première fois d'un
lézard dans ce pays. Les sérotypes rencontrés dans les hémocultures
furent *S. typhi*, *S. dublin*, *S. enteritidis*, *S. typhi-murium*, *S. paratyphi A* et *B*, *S. stanleyville*, *S. stockholm* et *S. aminatu*. Un autre fait
remarquable est la relative rareté de *S. paratyphi C* et l'absence dans
cette statistique de *S. cholerae-suis*.

Il est intéressant de comparer ce tableau des salmonelloses humaines avec celui des 150 *Salmonella* isolées de prélèvements divers
pendant la même période.

Tableau III.

Groupe	Type	Source d'isolement	Total
B	S. africana	Volaille	1
	S. agama	Bétail, lézard, porc, cobaye	16
	S. bredeney	Volaille	2
	S. bury	Bétail	1
	S. chester	Mouche, rat	2
	S. jacksonville	Bétail	1
	S. hessarek	Souris (laboratoire)	1
	S. kaapstad	Lézard	1
	S. kingston	Lézard	5
	S. paratyphi B	Bétail	1
	S. saint-paul	Viande de marché	3
	S. typhi-murium	Bétail, lapin (laboratoire)	2
	S. stanleyville	Bétail	1
	S. jos	Chat	1
C_1	S. edinburg	Volaille	1
	S. livingstone	Porc	1
	S. nigeria	Bétail	2
	S. oranienburg	Bétail, rat	2
	S. virchow	Rat, austrolop, canard	3
	S. mission	Oeuf d'autruche	1
C_2	S. alagbon	Bétail	1
	S. hadar	Volaille	4
	S. takoradi	Lézard	1
	S. bukuru	cobaye	1
D	S. dublin	Bétail, viande de marché, chien	7
	S. wangata	Bétail	1
	S. lishabi	Maki	1
	S. eastbourne	Lézard	1
E_1	S. adabraka	Aliment cuit	1
	S. anatum	Eaux d'égout	1
	S. coquilhatville	Bétail (cas fatal)	1
	S. elisabethville	Lézard, bétail	5
	S. okefoko	Rat	1
E_4	S. gwoza	Lézard	1

Groupe	Type	Source d'isolement	Total
F	S. chandans	Bétail	1
	S. rubislaw	Bétail, volaille, viande de marché, canard mort	7
G	S. ajiobo	Rat	1
	S. durham	Lézard	1
	S. poona	Porc	16
H	S. magumeri	Lézard	1
	S. garba	Cobaye, lapin	2
I	S. brazil	Mouche	1
(28)	S. ona	Bétail	1
	S. taunton	Sol (plantation marécageuse)	1
	S. moero	Lézard	1
(30)	S. urbana	Viande de marché	2
(35)	S. monschaui	Bétail	1
(1, 40)	S. johannesburg	Bétail	2
(45)	S. dugbe	Viande de marché	2

On retrouve ici aussi une grande fréquence de *S. agama* qui constitue un exemple d'un sérotype très rare ou inexistant dans d'autres pays, même en Afrique, et qui, au Nigéria, est largement distribué tant chez l'homme que chez des animaux à sang chaud (chat, porc) ou à sang froid. *S. poona* est très fréquente chez le porc et chez l'homme. Ceci laisse supposer que cet animal peut être à l'origine des salmonelloses humaines dues à ce sérotype. Trois autres exemples de sérotypes fréquents chez les animaux et l'homme sont *S. dublin*, *S. elisabethville* et *S. kingston*.

BIBLIOGRAPHIE

1. COLLARD, P. & SEN, R. — Serotypes of Salmonella at Ibadan, Nigeria, with special note of the new serotypes isolated in Nigeria. *J. inf. Dis.*, 1960, **106**, 270—275.

III. Les Salmonella au Sénégal

Le relevé suivant des Salmonelles identifiées à l'Institut Pasteur de Dakar de 1950 à 1962 inclus nous a été communiqué par le Dr. CHAMBON, Directeur. (tableau IV)

Tableau IV.

Repertoire des Salmonelles identifiées à Dakar.

(de 1950 à 1962 inclus)

Types	Homme	Eau	Animaux									
			Bovidés	Reptiles	Porc	Equidés	Cobaye	Chien	Lapin	Souris	Panthère	Oiseaux
S. paratyphi A	9	—										
S. paratyphi B	6	—										
S. brandenburg	3	—	1									
S. bredeney		—		1	9	2						
S. chester	2	—										
S. derby	1	—										
S. kaapstad	1	—										
S. santiago	1	—										
S. stanleyville	9	—										
S. typhi-murium		—			1					1		
S. paratyphi C	13	—										
S. cholerae-suis		—			2		2					
S. cholerae-suis, var. kunzendorf	8	—		5	3							
S. corvallis	5	—										
S. infantis	3	—			10			1				
S. kentucky	1	—										
S. kottbus		—										
S. kralendijk	4	—										
S. montevideo	1	—			1				1			
S. ness-ziona		—										
S. oranienburg		—										
S. pikine		—										
S. virchow	8	—	1		6							

	C1	C2	C3	C4	C5	C6	C7
S. typhi	604						
S. berta	1						
S. dublin	4		3				
S. durban	5					3	
S. enteritidis	11	1					
S. goettingen	1	1					
S. ouakam	—			6			
S. portland	1			6	7		
S. saarbruecken	1						
S. chailey	1						
S. anatum	3	1			1		
S. butantan	—						
S. coquilhatville	1				1		
S. give	8		1	5			
S. llandoff	—						
S. london	3			1	1		
S. manila	3				10		
S. meleagridis	1				1		
S. muenster	1				1		
S. new brunswick	1						
S. shangani	—				12		
S. souza	3				3		
S. vejle	1				3		
S. 3,10:eh	1	2					2
S. abaetetuba	—						
S. chingola	—	1					
S. fann	—						
S. friedenau	3			5	3		
S. havana	2				2		
S. poona	36	1			1		
S. rubislaw	1	1					

374

Tableau IV (suite)

Types	Homme	Eau	Animaux									
			Bovidés	Reptiles	Porc	Equidés	Cobaye	Chien	Lapin	Souris	Panthère	Oiseaux
S. tel-el-kebir	2	—					1					
S. charity		—					1					
S. hull		—										
S. gaminara		—										
S. salford	2	1										
S. welikada		—	2	4	1+		1					
S. matadi	3	—										
S. niamey	2	—										
S. cerro	2	—										
S. minnesota	1	—										
S. chicago	1	—										
S. dakar		—										
S. nima		—		1	1		1					
S. pomona		—		2								
S. vinohrady	1	1		4	9							
S. urbana		—		4								
S. cambérène		—		4								
S. yoff		—			1							
S. mgulani		—		2	1							
S. thiaroye	1	—			5							
S. johannesburg		—			1							
S. karamoja	1	—										
S. tilène		—		46								
S. waycross	1	—			7							
S. m'bao	3	—										
S. n'gor	1	—										
S. santhiaba	1	—										
S. kaolack		—										

Tableau V.
Provenance des Salmonelles identifiées chez l'homme à Dakar.
(de 1950 à 1962 inclus).

Types	Sang	Fèces	Urines	LCR	Pus	Ganglions	Rhino-pharynx	Total
S. paratyphi A.	9							9
S. paratyphi B	6							6
S. brandenburg	1	2						3
S. chester		2						2
S. derby	1							1
S. kaapstad		1						1
S. santiago		1						1
S. stanleyville	5	4						9
S. typhi-murium	11	14			1			26
S. paratyphi C.	12	1						13
S. cholerae-suis, var. kunzendorf	8							8
S. corvallis		4			1			5
S. kentucky		3						3
S. kottbus		1						1
S. montevideo	4				1			5
S. oranienburg	1							1
S. virchow	2	5			1			8
S. typhi	583	18			2	1		604
S. berta	1							1
S. dublin	3	1						4
S. durban	1	4						5
S. enteritidis	9	2						11
S. goettingen			1					1
S. portland		1						1
S. saarbruecken		1						1
S. chailey		1						1
S. anatum		2		1				3
S. coquilhatville	1							1
S. give	1	7						8
S. london	2	1						3
S. meleagridis		1						1
S. muenster		1						1
S. new-brunswick		1						1
S. souza		3						3
S. vejle		1						1
S. 3, 10 : eh		1						1
S. friedenau	1	2						3
S. havana	1	1						2
S. poona	1	35						36
S. rubislaw		1						1

Tableau V (suite)

Types	Sang	Fecès	Urines	LCR	Pus	Ganglions	Rhino-pharynx	Total
S. tel-el-kebir		2						2
S. welikada						2		2
S. niamey		2			1			3
S. cerro		2						2
S. minnesota	1	1						2
S. chicago		1						1
S. dakar		1						1
S. urbana		1						1
S. johannesburg		1						1
S. tilène		1						1
S. m'bao							1	1
S. n'gor		3						3
S. santhiaba		1						1
S. kaolack	1			1	1			3
								820

Ici encore, comme à Madagascar, les *S. typhi* représentent le sérotype de beaucoup le plus fréquent. Le plus grand nombre des souches isolées n'est pas dû à de grandes épidémies. Tous les cas observés sont sporadiques, mais leur fréquence est telle tout au long de l'année que l'on peut parler d'une trame épidémique lâche mais continue. Ce sont le plus souvent les adultes, et en particulier les Africains, qui sont atteints. La distribution des cas de fièvre typhoïde montre nettement l'intérêt de la vaccination: la plupart des adultes africains et européens atteints sont des femmes et toujours des personnes non vaccinées. Si certains médecins européens, se basant sur la faible fréquence des typhoïdes dans leur pays, se croient autorisés à déconseiller la vaccination TAB, ceci constitue une lourde erreur pour ceux qui ont à voyager outre-mer car le chloramphénicol n'a pas résolu le problème des fièvres typho-paratyphoïdiques, comme on a trop facilement tendance à le croire.

Les *S. paratyphi* A et B sont relativement rares au Sénégal[1] et arrivent après *S. paratyphi* C, ce qui se passait déjà, nous l'avons vu ci-dessus, à Madagascar. Les *S. cholerae-suis* vraies n'ont pas été rencontrées chez l'homme, contrairement aux *S. cholerae-suis* var. *kunzendorf*. Les autres sérotypes ne sont que rarement trouvés. Il faut noter que les 36 isolements signalés de *S. poona* provenaient en grande majorité d'une même intoxication alimentaire. Une seule a été trouvée en hémoculture. Il nous paraît intéressant à ce propos de rapporter l'origine des *Salmonella* humaines identifiées à Dakar (tableau V).

Les deux lysotypes de *S. typhi* les plus fréquents sont E_1 (environ 50%) et A (plus de 20%).

Les souches du lysotype A présentent à Dakar la particularité d'être du biotype II de KRISTENSEN (xylose —).

Comme à Madagascar, le porc est très fréquemment porteur de *Salmonella* au Sénégal: ces germes ont été trouvés dans 20% des ganglions mésentériques prélevés à l'abattoir de Dakar.[2].

A de rares exceptions près, telles celles que nous avons signalées avec *S. poona*, les intoxications alimentaires *sensu strictu* sont rarement observées au Sénégal. Les salmonelloses sont généralement dues à des contaminations favorisées par les conditions d'hygiène et d'habitat, conditions qui permettent souvent la contamination des aliments ou du sol (où jouent les enfants) par les déjections des animaux porteurs de germes.

BIBLIOGRAPHIE

1. DARRASSE, H., LE MINOR, L., PIECHAUD, M. & NICOLLE, P. — Les entérobactéries pathogènes à Dakar. — *Bull. Soc. Path. exot.*, 1957, **50**, *257*.
2. KIRSCHE, P. & BAYLET, R. — Résultats d'une enquête sur les ganglions de porc à Dakar. — *Soc. méd. A.O.F.*, 9 juin 1958.
3. CHAMBON, L. — Communication personnelle des relevés de l'Institut Pasteur de Dakar 1962.

IV. Maroc

Nous ne mentionnerons ici qu'une étude pour la recherche des *Salmonella* chez les tortues[1] dont les résultats nous paraissent devoir être cités: 96,3% des animaux capturés loin des villes hébergeaient des *Salmonella* et 64% des animaux capturés en ville. En moyenne plus de deux sérotypes ont été trouvés par animal. Ces animaux qui font l'objet d'un commerce important sont donc un facteur non négligeable de dissémination des *Salmonella*. Ces pourcentages de résultats positifs sont, à notre connaissance, les plus élevés de ceux cités jusqu'à présent, suivis de celui de 76% de positivités chez les reptiles d'Ethiopie[2].

BIBLIOGRAPHIE

1 VINCENT, J., NEEL, R. & LE MINOR, L. - Les *Salmonella* des tortues. Contribution à l'étude des *Salmonella* du Maroc. *Arch. Inst. Pasteur Tunis*, 1960, **37**, *187*.
2 SERIE, C. & LE MINOR, L. - *Bull. Soc. Path. exot.*, 1959, **52**, *133*.

Conclusion

La répartition des sérotypes de *Salmonella* dans les pays d'Afrique cités en exemple, situés sur une diagonale S.E.-N.E. permet de faire certaines comparaisons et de mettre en évidence certaines analogies:

— *S. typhi* est, chez l'homme, le sérotype le plus fréquent;

— *S. paratyphi B* est relativement rare en Afrique;

— la répartition des *S. paratyphi C* et *S. cholerae-suis* subit de grandes fluctuations suivant les pays;

— certains sérotypes, inexistants dans certains pays sont très fréquents chez d'autres;

— il existe une certaine corrélation entre la distribution des sérotypes chez les animaux comestibles, les animaux domestiques, les animaux à sang froid et chez l'homme;

— les animaux à sang froid chez lesquels les *Salmonella* se comportent comme des commensaux constituent un réservoir de *Salmonella* chez lesquels l'abondance de ces entérobactéries peut être extraordinaire, comme par exemple chez les tortues du Maroc.

LES SALMONELLOSES AU SUD DE L'AFRIQUE

PAR

H. D. BREDE

Université de Stellenbosch, Bellville, C.P.

Dans ce chapitre nous parlerons des salmonelloses dans les territoires situés au Sud du Congo-Léopoldville et du Tanganyika-
Territory. Ils couvrent une superficie de 6.062.500 km² et ont une
population d'environ 35.500.000 habitants. Le nombre des médecins
par rapport à la population diffère énormément d'une région à une
autre. Il y a dans toute cette partie de l'Afrique à peu près 10.000
médecins, dont 9.000 exercent leur art dans la République de l'Afrique du Sud et plus ou moins 1.000 dans les autres territoires. La
superficie de ces derniers est estimée à environ 4.838.200 km². La
plupart des médecins-internistes et pour ainsi dire tous les spécialistes pratiquent dans les grands centres: Johannesburg, Durban, Le
Cap, Lourenço Marques, Luanda. Dans ces villes dont le niveau de
vie est comparable à celui des grands centres européens ou améri-
cains, il y a 1 médecin pour 700 habitants, tandis que le nombre des
praticiens est minime dans l'intérieur des territoires. Le chiffre
plutôt théorique d'un médecin pour 3.500 habitants n'est que rarement atteint. Ainsi, dans le Swaziland, le Basutoland et le Bechuanaland on ne trouve qu'un médecin pour 30.000 habitants En conséquence, il existe des différences marquées dans le fonctionnement
des services médicaux, et de ce fait nos connaissances des salmonelloses seront plus ou moins approfondies suivant les endroits ou les
régions.

Dans tous les territoires du Sud de l'Afrique les salmonelloses présentent une importance très grande. En tant que problème sanitaire,
elles suivent immédiatement la tuberculose.

Seuls les cas de fièvre typhoïde doivent être signalés aux autorités
médicales et sont consignés dans les statistiques officielles. Le diagnostic de la plupart de ces cas repose uniquement sur la symptomatologie clinique; une confirmation bactériologique n'étant possible
que dans quelques rares centres importants: Luanda (Angola),
Lourenço Marques (Mozambique), Pretoria (où se trouve le Centre
de Lysotypie des *S. typhi* pour la République sudafricaine), Onderstepoort (qui centralise les données sur les salmonelloses animales),
Johannesburg (avec son "South African Institute for Medical
Research" S.A.I.M.R.) dont les activités s'étendent sur la région
minière du Transvaal et sur celle de l'Etat Libre d'Orange (E.L.O.)
et où se trouve le Centre National des Salmonella), enfin, Durban

(Natal) et Le Cap (Province du Cap) qui possèdent des laboratoires universitaires et gouvernementaux.

Un diagnostic bactériologique précis n'est généralement fait que pour les cas qui se déclarent dans les environs immédiats de ces centres; ailleurs il est aléatoire. Les Services sanitaires des territoires portugais de l'Angola et du Mozambique sont centralisés resp. à Luanda et à Lourenço Marques et se trouvent sous la direction générale de l'Instituto de Medicina Tropical de Lisbonne. Par contre, les services sanitaires des autres territoires ont été organisés selon le système anglais: ils sont décentralisés, indépendants les uns des autres et ils fonctionnent sous la responsabilité des autorités municipales. Nous y trouvons des spécialistes du type "Clinical pathologist", c.à.d. des médecins théoriquement formés en histopathologie, microbiologie et biochimie mais qui sont en fait surtout des histopathologistes. Il en découle que dans ces vastes territoires qui comprennent notamment les Rhodésies, le Nyassaland, le Tanganyika-Territory, le Sud-Ouest-Africain, le Bechuanaland, le Basutoland et le Swaziland, on ne peut guère s'attendre à ce que la détermination des *Salmonella* soit faite avec la précision souhaitée. Les statistiques qui suivent sont par conséquent incomplètes: elles ne mentionnent que ce que quelques spécialistes ont trouvé.

La fièvre typhoïde règne à l'état endémique dans tout le Sud africain nonobstant toutes les mesures prophylactiques qui ont été prises par les divers services médicaux, et la morbidité réelle est sans doute supérieure à celle indiquée au Tableau I.

Tableau I.
Statistique officielle des cas de Fièvre typhoïde, modifié selon SIMMONS et coll., 1951.

Pays	Minimum et Maximum	(1945—1960)
Angola	20— 250	cas par an
Mozambique	30— 100	cas par an
Nyassaland	20— 50	cas par an
Rhodésie du Nord	30— 150	cas par an
Rhodésie du Sud	100— 200	cas par an
Basutoland	150— 350	cas par an
Bechuanaland	?	Aucune donnée publiée
Sud-Ouest-Africain	?	Aucune donnée publiée
Swaziland	?	Aucune donnée publiée
Rép. Afrique du Sud	3.000—7.000	cas par an, en majeure partie diagnostiqués cliniquement, sans examen de laboratoire.

Pour déterminer l'incidence de la fièvre typhoïde chez les différents groupes ethniques, nous avons établi des moyennes pour les années 1956, 1957 et 1958 en nous basant sur les statistiques officielles du Service de Santé de la République d'Afrique du Sud (voir Tableau II).

Tableau II.

Nombre de cas de Fièvre typhoïde par 100.000 habitants dans la République d'Afrique du Sud.

Province	Européens	Noirs	Colorés	Asiatiques
Du Cap	4,97	15,26	19,70	—
Natal	10,43	49,11	41,26	12,93
d'Orange	9,28	38,21	19,99	—
Transvaal	8,89	30,38	8,53	7,46
Totaux:	8,39	33,24	22,37	10,19

Dans tous les groupes ethniques la morbidité typhoïdique est la plus élevée au Natal, région à climat chaud et humide, et la plus basse dans la Province du Cap, région à climat également chaud mais sec.

L'incidence de la fièvre typhoïde chez les différentes races reflète leur situation sociale et n'est pas en rapport avec une immunité raciale. Les européens jouissent en général du niveau de vie le plus élevé et la morbidité typhoïdique est chez eux très basse. Suivent les Asiatiques qui se sont enrichis par le commerce. Le standard de vie des Colorés est meilleur que celui des Noirs qui, en grande partie, vivent encore dans leur milieu coutumier et qui connaissent le taux de morbidité le plus élevé.

La mortalité due à la fièvre typhoïde reflète encore mieux les différences du standing social (voir Tableau III).

Tableau III.

Mortalité due à la Fièvre typhoïde par 100.000 habitants dans la République d'Afrique du Sud.

Européens	0,25
Asiatiques	0,62
Colorés	1,60
Noirs	4,00

Les statistiques sur la morbidité et sur la mortalité dans les autres pays sud-africains sont trop inexactes et nous avons préféré ne pas les mentionner. Dans l'ensemble, il faut accepter que les statistiques sont basées sur des données recueillies principalement dans les centres et que, par conséquent, elles sont plutôt inférieures à la réalité.

La lysotypie des souches de *S. typhi* se fait au "Department of Microbiology" de l'Université de Pretoria: les cultures venant de toute l'Afrique du Sud y sont examinées dans le service du Dr. C.G. CROCKER qui travaille en collaboration avec le Dr. P. NICOLLE, Chef du Service des Bactériophages à l'Institut Pasteur de Paris. Plus de 6.000 souches ont déjà été étudiées à Pretoria, ce qui permet d'avoir une vue très bonne sur la distribution géographique des divers lysotypes (voir Tableau IV).

Tableau IV.

Lysotypes de *S. typhi* identifiés en Afrique du Sud. (Tableau composé d'après les rapports du DR. CROCKER).

Province:	du Cap	Natal	E.L. d'Orange	Transvaal
Type				
A — Montreal				+
Coquilhatville				+
Tananarive	+	+	+	+
Maracaibo	+			+
B1				+
C				+
D1	+	+		+
D4	+			
D7				+
E1	+	+	+	+
F1	+			+
G				+
L2				+
O				+
T				+
28	+	+		+
40	+			
45				+
46	+			

Les recherches sont faites au moyen de 78 phages différents et 19 lysotypes ont été identifiés jusqu'à présent: 16 ont été trouvés au Transvaal (où travaillent de nombreux mineurs venus du Mozambique, des Rhodésies et du Nyassaland), 9 dans la Province du Cap, 4 dans la Province du Natal et 2 seulement dans l'Etat Libre d'Orange.

Le lysotype A domine nettement (plus particulièrement les sous-types Tananarive et Maracaibo). Selon NICOLLE, la fréquence relative du lysotype A de *S. typhi* est de 1 à 23% en Afrique du Nord et de 35 à 62% en Afrique Noire. CROCKER signale dans ses rapports sur l'Afrique du Sud les chiffres suivants: 45,4% en 1960, 53% en 1961 et 60% en 1962. Les pourcentages les plus élevés ont été trouvés au Natal: 73% en 1960, 70% en 1961 et 81,7% en 1962. Dans cette Province, plus de 95% des souches de *S. typhi* appartenant au lysotype A sont du sous-type Tananarive. — Dans la Province du Cap prédominent le type 40 et la variété Maracaibo du type A. — Les lysotypes A, D1, E1, T et 28 sont cosmopolites, de même que le type B1, mais ce dernier est rare et il se confond facilement avec le groupe des cultures aliénosensibles (Vi-dégradées de FELIX). Le type B1 a été trouvé quelques fois au Transvaal. — Les lysotypes C, G et O n'ont été recontrés qu'à Johannesburg et leur origine reste inconnue. — Le type D4 n'est guère fréquent non plus mais son aire de distribution s'étend sur toute la région du Cap. — Les types 40 et 46 prédominent dans l'Ouest de la partie Sud de l'Afrique. — Les lysotypes B1, D4, 28, T, G et O ne représentent que 1,12% de la totalité. — Le lysotype 46 a été signalé pour la première fois dans la région de Windhoek et dans la Province du Cap en 1961–1962. — Les premières souches du type F1 ont été isolées à Potchefstroom à l'occasion d'une petite épidémie de fièvre typhoïde; elles fermentent toutes le maltose.

Les lysotypes les plus répandus sont les types A et E1. La prédominance de la varieté A-Tananarive dans les régions Sud-Est de l'Afrique nous fait croire à la possibilité qu'il s'agit ici d'un type ancien du massif paléolitique de Gondwana, car il domine surtout à Madagascar et aux Indes. Le sous-type Tananarive est fréquent jusqu'à Port-Elizabeth: au Nord et à l'Ouest de cette ville il est graduellement remplacé par d'autres sous-types A, surtout par la variété Maracaibo. Il est intéressant de signaler que cette dernière n'a été trouvée jusqu'ici que dans les Caraibes, le Venezuela, les Antilles, la Guyanne et dans la région Sud-Ouest de la Province du Cap.

La lysotypie des souches de *S. typhi* fournit des renseignements utiles pour les recherches épidémiologiques dans toutes les régions de l'Afrique du Sud à l'exception du Natal où presque toutes sont du type A var. Tananarive.

BOKKENHEUSER (1959a) signale que 90% des souches de *S. typhi* isolées à Johannesburg fermentent le xylose. LEWIN (1937) communique que 90,4% des souches fraîchement isolées possèdent à la fois les antigènes O et Vi, 7,9% ne possèdent que le seul antigène Vi et 1,9% uniquement l'antigène O.

Nous n'avons pu recueillir que peu d'informations au sujet des porteurs de germes. Dans son livre "Public Health in South Africa",

CLUVER écrit (p. 170) que la moitié environ des Noirs adultes a souffert de la fièvre typhoide au cours de son existence. Etant donné que 4% environ des convalescents deviennent des porteurs de germes, CLUVER pense que 2% de la population noire est composé de porteurs de S. typhi. Ces estimations nous semblent trop élevées. L'examen bactériologique des selles de plus de 5.000 personnes qui se sont présentées à l'Hôpital Karl Bremer à Bellville pour consultation médicale au cours des trois dernières années nous a donné les résultats suivants: des souches de S. typhi ont été isolées chez 0,1% des européens, chez 0,8% des colorés et chez 0,6% des Noirs. Il est évident que les cas de fièvre typhoïde ne sont pas inclus dans ces chiffres.

La recherche des agglutinines anti-Vi se pratique couramment en Afrique du Sud pour le dépistage des porteurs de germes. Selon BOKKENHEUSER (1959c), sur 20.000 réactions de Widal faites au "South African Institute for Medical Research" à Johannesburg, 3 à 5% étaient positives pour l'antigène Vi à un taux de 1:10. La coproculture n'a pas été faite systématiquement chez tous ces cas considérés comme positifs. Il est accepté que 5% des personnes chez lesquelles la réaction de Widal est ainsi positive sont des porteurs de germes. Cette opinion nous semble un peu spéculative et aux réactions sérologiques, dont l'exactitude n'est pas suffisamment prouvée, nous préférons les examens bactériologiques des selles répétés à intervalles hebdomadaires.

Le "Public Health Act of the Union of South Africa" de 1919 stipule les mesures qui doivent être prises pour combattre le danger que représentent les porteurs de germes. Ceux-ci sont exclus de toutes les professions qui les mettraient en contact avec des denrées alimentaires. Un traitement au Chloramphénicol est essayé et, le cas échéant, la cholécystectomie est conseillée.

Toutes les souches de S. typhi qui ont été isolées au département de microbiologie médicale de l'Université de Stellenbosch au cours des trois dernières années étaient sensibles au Chloramphénicol, au Kantrex, aux Tétracyclines et à la Néomycine; 12,5% étaient résistantes à la Streptomycine, 69% à l'Erythromycine et toutes à la Novobiocine.

Les fièvres typhoïdes se présentent en Afrique du Sud avec une symptomatologie classique. Toutefois, nous sommes frappés par la fréquence des cas d'endocardite subaiguë avec hémoculture positive à S. typhi sans que le diagnostic de fièvre typhoïde ait été posé auparavant. Nous rencontrons cette complication surtout chez des fermiers européens âgés hospitalisés pour endocardite subaiguë et chez lesquels l'étiologie est révélée seulement par l'hémoculture. Chez les Colorés on observe presque constamment des bronchites. Des hépatites se déclarent fréquemment au cours de la maladie mais indépendamment peut-être de l'infection microbienne: l'hépatite épidémique est très répandue en Afrique du Sud et il est fort possible que ce soit elle qui déclenche l'apparition clinique d'une salmonellose latente. Il est rare d'observer un ictère franc au cours d'une fièvre typhoïde. Des infections localisées peuvent se rencontrer, mais elles sont plutôt rares si on pense au grand nombre d'infections à *S. typhi*.

Les fièvres typhoïdes s'observent tout au long de l'année mais leur fréquence varie sensiblement selon les saisons: le nombre des cas diagnostiqués se double le plus souvent au cours de la saison des pluies, surtout dans la région Sud-Ouest, sur le plateau boer couvert de savanes et sur la bordure méridionale, haute muraille de massifs montagneux avec des plateaux en gradins. Comment expliquer cette augmentation? Elle est très probablement due au fait qu'après des pluies torrentielles les eaux de surface ne sont pas absorbées par la terre, elles ruissellent le long des pentes et elles inondent et polluent les citernes et réservoirs d'eau potable. Des épidémies parfois sérieuses de fièvre typhoïde peuvent s'en suivre.

Les fluctuations saisonnières se manifestent par ailleurs également pour les autres salmonelloses et il s'agit donc d'un phénomène général. Le Tableau V montre bien les différences observées à Johannesburg (où la saison des pluies s'étend d'octobre à février) et au Cap (où il pleut irrégulièrement durant toute l'année avec néanmoins une recrudescence au mois de mai).

Tableau V.

Tendances saisonnières des salmonelloses: pourcentages des Salmonella isolées suivant les mois

Mois	à Johannesburg*	au Cap**
Janvier	11,4%	6,2%
Février	6,8	10,2
Mars	7,1	11,2
Avril	6,3	10,9
Mai	6,2	21,0
Juin	6,1	7,4
Juillet	6,4	4,4
Août	6,8	7,2
Septembre	6,7	4,2
Octobre	12,3	4,7
Novembre	12,3	4,7
Décembre	11,7	7,5
Nombre total de cultures	962	339

* Selon BOKKENHEUSER (1959b).
** Observations de l'auteur.

Les transmissions par contact sont fréquentes en saison sèche; par contre, la dissémination par l'eau ou par le lait est assez rare durant cette période, mêmes dans les régions rurales où l'hygiène rurale laisse à désirer. Dans les agglomérations urbaines l'épuration des eaux de distribution et le contrôle des produits laitiers sont efficaces. La fabrication des crêmes glacées ou autres reçoit une attention toute spéciale de la part des services sanitaires.

Du vaccin endotoxoïd du type T.A.B., préparé en grandes quantités par le "South African Institute for Medical Research", est utilisé pour l'immunisation active de certains groupes de la population. Il semble donner des résultats satisfaisants, si l'on en croit les statistiques faites chez les mineurs du Transvaal et de l'Etat Libre d'Orange. Mais on peut toujours se poser la question: "Est-ce que l'amélioration générale des conditions de vie et de l'hygiène n'ont pas contribué dans une plus large mesure à faire diminuer les cas de fièvre typhoïde que le vaccin?" Quoi qu'il en soit, il est intéressant de noter que de 1934 à 1962 le nombre des mineurs est passé de 21.000 à plus d'un demi-million, tandis que le nombre des cas de typhoïde est tombé au cours de cette même période à un vingtième du chiffre original.

*
* *

Les cas de salmonellose à S. *paratyphi A* sont très rares en Afrique du Sud. Nous n'avons pu examiner qu'une seule souche au cours des quatre dernières années: elle fut isolée à Cape Town chez un marin étranger. Plus à l'intérieur du continent il n'y a, pour autant que nous sachions, pas de problème dû à S. *paratyphi A* ni dans les territoires portugais, ni dans les territoires anglais. Toutefois, à Johannesburg des cas de paratyphoïde A ont été diagnostiqués avec certitude: BOKKENHEUSER (1959b) signale qu'en 1957, 3% des souches de *Salmonella* isolées dans cette ville étaient des S. *paratyphi A*. Ceci est assez étonnant, car jusque là cette espèce était pour ainsi dire inconnue au Sud de l'Afrique. Originaire de l'Asie, elle a peut-être été introduite à Johannesburg par des immigrants chinois qui forment une communauté importante dans cette ville. Il nous paraît difficile d'incriminer la communauté indienne: celle-ci est établie principalement au Natal et les infections à S. *paratyphi A* ne jouent aucun rôle dans cette province.

Les cas de salmonellose à S. *paratyphi B* sont eux aussi pour ainsi dire totalement inconnus au Sud de l'Afrique. Dans la Province du Cap nous n'en avons pas encore observé un seul. Quelques-uns ont été signalés à Johannesburg, mais ce sont là apparemment les seuls qui sont connus.

Les isolements de S. *paratyphi C* sont d'une extrême rareté dans la partie Sud de l'Afrique. BOKKENHEUSER (1959b) signale avoir trouvé cette espèce une fois lors d'une coproculture et une fois lors d'une urinoculture mais nous ne savons pas s'il s'agit ici d'examens qui ont été faits pour le même malade ou pour deux malades différents. Garrow (1920) a décrit un cas au Mozambique: à Porto Amelio il a isolé par hémoculture une souche de *S. paratyphi C* var. *east africa* chez un soldat qui venait d'Afrique du Sud. A notre connaissance, il n'y a pas eu d'autres isolements signalés. Selon van Oye (1960) *S. paratyphi C* var. *east africa* serait surtout répandue en Afrique centrale et orientale.

* *

Les paratyphoïdes vraies à *S. paratyphi A*, *B* ou *C* sont donc exceptionnelles dans la partie Sud de l'Afrique. Par contre, les "autres" salmonelloses y sont très répandues et elles règnent à l'état endémique dans tous les territoires. Paradoxalement, il nous est impossible d'en estimer l'importance parce que, comme nous l'avons déjà dit, seuls les cas de fièvre typhoïde doivent être déclarés aux autorités sanitaires. Nous ne pouvons donc que nous baser sur les données fournies par quelques grands laboratoires pour avoir une idée approximative des fréquences relatives des diverses espèces de *Salmonella* rencontrées en Afrique du Sud. Ces données font ressortir en premier lieu des différences prononcées d'après les régions. Ceci ressort clairement des tableaux VI et VII qui concernent respectivement la région de Johannesburg (Transvaal) et celle du Sud-Ouest de la Province du Cap. (Pour les espèces qui ont été découvertes en Afrique du Sud nous avons mentionné entre parenthèses l'année de leur découverte.)

Le tableau VI mentionne 138 espèces différentes de *Salmonella* dont 23 ont été découvertes en Afrique du Sud; cette liste a été établie en ajoutant à celle publiée par Bokkenheuser (1959c) les espèces qui ont été identifiées pour la première fois au cours des dernières années. En fait, elle intéresse la région du Transvaal, l'Etat Libre d'Orange, une partie du Natal, le Nord de la Province du Cap et une partie du Sud-Ouest africain, c'est à dire toutes les régions dans lesquelles la saison des pluies se situe en été. Certaines des espèces de *Salmonella* qui y sont mentionnées ont probablement été importées par des ouvriers étrangers en provenance des pays situés au Nord et au Nord-Est du Transvaal: le Mozambique, le Nyassaland, les Rhodésies. On peut donc affirmer que le tableau VI intéresse en quelque sorte tout le Sud de l'Afrique à l'exception des régions Sud-Ouest qui connaissent une saison des pluies en hiver.

Le tableau VII mentionne au total 113 espèces différentes de *Salmonella*; parmi elles 31 ont été découvertes en Afrique du Sud, dont 21 à l'Institut de Microbiologie Médicale de l'Université de Stellenbosch à Bellville.

Les tableaux VI et VII ont été établis sur la base de données recueillis jusqu'au début de 1964. De la comparaison de ces deux

Tableau VI.

Espèces de Salmonella identifiées à Johannesburg.

Groupe	Espèce
A	S. paratyphi A.
B	S. abortus equi - S. paratyphi B - S. wagenia - S. stanley - S. duisburg - S. saint-paul - S. reading - S. kaapstad (1941) - S. chester - S. san-diego - S. derby - S. budapest - S. typhi-murium - S. bredeney - S. heidelberg.
C.1	S. stanleyville - S. san-juan - S. umhlali - S. edinburg - S. georgia - S. bloemfontein (1960) - S. paratyphi C - S. cholerae-suis - S. mission - S. amersfoort (1937) - S. livingstone - S. braenderup - S. montevideo - S. oranienburg - S. thompson - S. concord - S. irumu - S. colorado - S. infantis - S. bareilly - S. aequatoria - S. eschweiler - S. tennessee.
C.2	S. narashino - S. nagoya - S. muenchen - S. manhattan - S. labadi - S. newport - S. kottbus - S. baragwanath (1955) - S. germiston (1955) - S. lindenburg - S. takoradi - S. bonariensis - S. litchfield - S. fayed - S. bovis-morbificans - S. hidalgo - S. gold-coast - S. tananarive - S. praha - S. glostrup.
C.3	S. shipley - S. virginia - S. kentucky - S. amherstiana.
D.1	S. durban (1941) - S. typhi - S. ndolo - S. eastbourne - S. israël - S. enteritidis - S. dublin - S. pensacola - S. seremban - S. panama - S. göttingen - S. victoria - S. gallinarum-pullorum.
D.2	S. strasbourg.
E.1	S. vejle - S. münster - S. anatum - S. newlands - S. meleagridis - S. london - S. alexander (1957).
E.4	S. senftenberg - S. krefeld.
F	S. chandans - S. chingola (1953) - S. aberdeen - S. pretoria (1941) - S. tel-hashomer.
G.1	S. ibadan - S. borbeck - S. poona - S. roodepoort (1956).
G.2	S. mishmar-haemek - S. havana - S. natal (1961) - S. worcester (1952) - S. nachshonim - S. cubana.
H	S. florida - S. albuquerque - S. onderstepoort (1936) - S. carrau - S. homosassa.
I	S. hvittingfoss - S. gaminara - S. weston - S. mobeni - S. rowbarton - S. haddon (1958) - S. lisboa.
J	S. hillbrow (1956).
K	S. cerro
L	S. minnesota.
M	S. kibusi - S. pomona - S. umbilo.
N	S. urbana - S. landau - S. donna.
O	S. umhlatazana (1954?) - S. adelaide - S. alachua.
P	S. roan.
R	S. springs (1956) - S. riogrande - S. johannesburg (1952) - S. duval - S. boksburg (1957).
S	S. waycross.
T	S. uphill - S. rand (1957) - S. weslaco.
W	S. windhoek (1955).
Z	S. krugersdorp (1961) - S. greenside (1957).

tableaux il ressort que les espèces qui ont été trouvées à Johannesburg diffèrent sensiblement de celles qui ont été trouvées à Bellville. Les données numériques dont nous disposons, et que nous commu-

Tableau VII.

Espèces de Salmonella identifiées à Bellville.

Groupe	Espèce
B	S. sofia - S. saint-paul - S. reading - S. kaapstad (1941) - S. chester - S. san-diego - S. makumira - S. derby - S. caledon (1961) - S. typhi murium - S. bredeney - S. essen - S. heidelberg - S. kiambu - S. durban-ville (1962).
C.1	S. calvinia (1962) - S. bloemfontein (1960) - S. cholerae suis var. kunzendorf - S. mission - S. isangi - S. amersfoort - S. larochelle - S. lomita - S. braenderup - S. oranienburg - S. irumu - S. bonn - S. virchow - S. infantis - 6,7: z:z₄₂ (1963) - S. djugu - S. mbandaka - S. tennessee.
C.2	S. muenchen - S. manhattan - S. newport - S. kottbus - S. lindenburg - S. baragwanath (1955) - S. bovis-morbificans - S. uno.
C.3	S. kentucky.
D.1	S. loma-linda - S. durban (1941) - S. mjimwema - S. typhi - S. eastbourne - S. lindrick var. 1,7 (1962) - S. enteritidis - S. kuilsrivier (1962) - S. neasden - S. hamburg - S. pensacola - S. napoli - S. stellenbosch (1960) - S. wynberg (1963).
D.2	S. haarlem.
E.1	S. vejle - S. münster - S. anatum - S. nyborg - S. ruzizi - S. westpark.
E.2	S. cambridge - S. parow (1963).
E.4	S. accra - S. senftenberg.
F	S. grabouw (1962) - S. pretoria (1941) - S. nyanza.
G.1	S. poona - S. clifton - S. goodwood (1962).
G.2	S. acres (1962) - S. natal (1962) - S. okatie - S. worcester (1952) - S. kintambo - S. cubana.
H	S. onderstepoort (1936) - S. fischerkietz - S. homosassa - S. sundsvall.
I	S. hvittingfoss - S. bellville (1961) - S. mobeni - S. merseyside - S. rowbarton - S. salford - S. lisboa - S. elsiesrivier (1962) - S. woodstock (1961).
K	S. cerro.
L	S. minnesota - S. wandsbek.
M	S. chicago - S. ceres (1962).
N	S. urbana - S. landau.
O	S. adelaide.
Q	S. anfo.
R	S. johannesburg (1952) - S. alsterdorf - S. bukavu.
T	S. uphill.
U	S. mosselbay (43:g, s, t:z₄₂) (1963).
W	S. apapa - S. klapmuts (1962) (jusqu'à présent isolée chez des tortues et des vipères seulement).
X	S. chersina (1962) (idem) - 47₁, 47₃:z₆:1,6 (1963).
Y	48:d:- (1963).
Z	S. wassenaar.
51	S. roggeveld (1963) (idem).

niquons dans le tableau VIII, nous apprennent que des différences se manifestent également sur le plan quantitatif.

Le nombre total des espèces de *Salmonella* identifiées jusqu'à présent dans la République d'Afrique du Sud s'élève à 200. Pour

Tableau VIII.

Pourcentages des Salmonella les plus fréquentes isolées.

à Johannesburg*		à Bellville**	
1. S. adelaide	14%	1. S. reading	6%
2. S. typhi-murium	12%	2. S. typhi-murium	6%
3. S. labadi	11%	3. S. typhi	5%
4. S. london	8%	4. S. saint-paul	4%
5. S. montevideo	8%	5. S. manhattan	4%
6. S. anatum	6%	6. S. anatum	4%
7. S. typhi	5%	7. S. vejle	4%
8. S. thompson	4%	8. S. merseyside	4%
9. S. derby	3%	9. S. minnesota	4%
	71%		41%

* Selon BOKKENHEUSER (1959b).
** Données de l'auteur

l'entièreté de la partie Sud du continent africain ce chiffre est beaucoup plus élevé et dépasse largement les 200, car il convient d'y ajouter les espèces qui ont été découvertes en Angola, au Mozambique, dans les Rhodésies, au Tanganyika-Territory, ainsi que les *Salmonella* qui ont été trouvées dans ces pays mais pas dans la République sud-africaine. Nombre d'espèces nouvelles, bien que figurant dans le Schéma de KAUFFMANN-WHITE, n'ont pas encore été "officiellement" décrites. Certaines ont été découvertes en dehors des pays que nous venons de mentionner, comme par exemple *S. angola* et *S.benguella* (isolées en Hollande de lots de farines de poissons importées de l'Angola), *S. bulawayo* (isolée à Elisabethville, Katanga, chez un âne en provenance de Bulawayo, Rhodésie du Sud), *S. locarno* (isolée en Suisse chez un tatou, *Cordylus giganteus*, importé d'Afrique du Sud), *S. makumira* (isolée à Hambourg, Allemagne, d'une petite crotte de pigeon complètement desséchée provenant du Tanganyika), etc.

Il n'est pas sans intérêt de souligner qu'un tiers environ des *Salmonella* connues en Afrique du Sud est constitué par des espèces nouvelles dont la majorité appartient au sous-groupe II de KAUFFMANN.

Nous avons déjà dit que les salmonelloses du type gastroentérique sont très fréquentes dans toutes les régions du Sud de l'Afrique. Ceci est le cas non seulement pour les formes cliniques mais également pour ce que Charles NICOLLE a dénommé "les infections inapparentes". BOKKENHEUSER & RICHARDSON (1959) ont fait des coprocultures chez 1.565 noirs et colorés qui manipulent des denrées alimentaires et ils ont trouvé des *Salmonella* chez 67 (= 4,3%) d'entre eux. Nous avons trouvé un même pourcentage chez des ouvriers à Bellville. Il ne nous semble pas qu'on puisse parler ici de porteurs de germes chroniques, car chez la plupart d'entre eux les *Salmonella* avaient disparu en moins de trois mois . . . pour être remplacées peu après par d'autres espèces et celà d'une manière tout aussi passagère. Des observations analogues peuvent se faire chez des européens: 5% des infirmières du service de pédiatrie de l'Hôpital Karl Bremer de Bellville se sont révélées être des porteuses temporaires de *Salmonella*.

* *

Les premières recherches sur les salmonelloses animales ont été faites par HENNING (1939) à Onderstepoort. De 507 cultures de *Salmonella* isolées chez des veaux malades, 491 étaient des *S. dublin*, 11 des *S. typhi-murium*, 4 des *S. enteritidis* et 1 une *S. bovis-morbificans*. HENNING démontre aussi l'importance de *S. cholerae-suis* chez les suidés, de *S. typhi-murium* chez les volailles, les suidés et les ovidés, de *S. abortus-equi* chez les équidés et de *S. pullorum* chez les gallinacés.

Le même auteur découvra (1941) les nouvelles espèces *S. durban*, *S. kaapstad* et *S. pretoria* chez des animaux de boucherie. Ces trois *Salmonella* ont été isolées depuis lors à plusieurs reprises dans des cas de gastro-entérite.

En 1960, nous avons commencé à rechercher les *Salmonella* chez diverses espèces d'animaux sauvages et domestiques et nous avons e.a. examiné plus de 100 tortues capturées dans la région qui s'étend au Sud de l'embouchure du fleuve Orange jusqu'au Cap. Ces tortues appartenaient aux espèces *Chersina angulata*, *Homopus areolatus*, *Psammobates geometricus* et *Psammobates tentorius trimesis* et l'examen bactériologique de leurs fèces nous a entre autres permis de découvrir 21 espèces de Salmonella, c'est-à-dire: *S. durbanville*, *S. calvinia*, *S. bloemfontein*, 6,7:z:z$_{42}$, *S. mjimwema*, *S. lindrick* var. 1,7 (inositol positive), *S. neasden*, *S. westpark*, *S. ruzizi*, *S. grabouw*, *S. pretoria*, *S. mobeni*, *S. rowbarton*, *S. wandbek*, *S. chicago*, *S. bukavu*, *S. uphill*, *S. mosselbay*, *S. klapmuts*, *S. chersina* et *S. roggeveld*; à l'exception de *S. ruzizi*, *S. pretoria*, *S. chicago* et *S. bukavu*, toutes ces espèces appartiennent au sous groupe II de KAUFFMANN. Nous avons également procédé à des réactions de Widal sur les sérums de ces tortues mais toutes ont été négatives. Il n'empêche que VERGE & PLACIDI (1960) ont probablement raison d'affirmer: "Il est évident que les recherches sur les sérums des hétérothermes sont susceptibles d'apporter une intéressante contribution à divers problèmes d'immunologie générale et méritent d'être poursuivies." 30% des animaux examinés hébergaient des *Salmonella* dans leurs intestins mais aucun dans le sang ou dans un organe quelconque. Aucune tortue ne donna l'impression d'être malade. Les *Salmonella* semblent être des commensaux normaux des chéloniens et on est porté à croire qu'ils peuvent jouer un rôle certain dans la propagation de ces germes.

En Afrique du Sud, les tortues sont souvent couvertes d'un très grand nombre de tiques. Nous en avons recueilli de très nombreuses (des espèces *Haemophysalis leachi, Hyalomma hebraeum, Rhipicephalus* sp.,), nous les avons désinfectées à l'alcool-éther et ensuite broyées. Ces broyats ont été ensemencés dans du tétrathionate de sodium et, bien que nous ayons ainsi isolé certaines *E. coli* entéropathogènes, seulement *S. klapmuts* et *S. roggeveld* ont été être détectées de cette façon.

L'examen bactériologique de 138 vipères *(Naja nivea, Bitis arietans arietans, Psammophylax rhombeatus, Elaps lacteus* et *Dispholidus typus)* nous a permis de découvrir les 23 espèces de Salmonella suivantes:

S. reading, S. makumira, S. calvinia, S. braenderup, S. irumu, S. infantis, S. muenchen, S. manhattan, S. bovis-morbificans. S. neasden, S. pensacola, S. westpark, S. pretoria, S. nyanza, S. kintambo, S. salford, S. anfo, S. johannesburg, S. boksburg, S. alsterdorf, S. klapmuts, S. chersina, S. wassenaar.

Au contraire des découvertes obtenues chez des tortues, la majorité des Salmonella trouvées chez les vipères sont membres du sous groupe I, seulement 9 appartiennent au sous groupe II de KAUFFMANN.

A notre grande surprise, nous avons isolé à plusieurs reprises *S. natal* de cochenilles (= genre d'insectes hémiptères: *Dactylopius coccus)* que nous avons trouvées en grandes quantités sur des cactées dans les environs de Malmesbury (au Sud-Ouest du Cap).

Occasionnellement nous avons trouvé *S. natal* également chez des animaux domestiques et notamment à Bellville où pendant une certaine période des chats furent infectés par cette espèce et par *S. merseyside.*

Des souches de *S. typhi-murium* ont été isolées de ramiers et de pigeons domestiques; elles sont toutes du lysotype 1a tandis que les *S. typhi-murium* isolées chez des humains appartiennent aux lysotypes 1, 1 var. 2, 2b, 18 et 32.

NESER, KLEIN & SACKS (1957) ont démontré que le "Biltong" (viande de boeuf ou d'autruche desséchée) contient souvent des *Salmonella*, surtout de l'espèce *S. newport.*

De temps en temps nous avons isolé *S. senftenberg* de lots de farines de poissons d'origine sudafricaine.

* * *

En conclusion il est permis d'affirmer, maintenant que le paludisme a été vaincu sur de grandes étendues du Sud africain, que les

salmonelloses, après la tuberculose, représentent le problème sanitaire le plus important dans cette vaste région. De nombreuses
questions les concernant restent encore sans réponses et il est hautement souhaitable d'y multiplier les laboratoires de bactériologie
susceptibles de contribuer à leur solution par une collaboration
suivie à des recherches coordonneés.

BIBLIOGRAPHIE

Annual Report of the Department of Health. The Government Printer,
Pretoria, 1956, 1957, 1958.
BADER, R. E.: Paratyphus C 1915—1945. In: Weltgesundheitsatlas. Edité
par E. RODENWALDT & H. J. JUSATZ. Falk Verlag, Hamburg.
BOKKENHEUSER, V. (1959a) Salmonella and Shigella infections in Africa.
South Afr. med. J., 33: 36—37.
Idem. (1959 b) A review of Salmonellosis in South Africa. Ibidem. 33: 702—
706.
Idem. (1959 c) Epidemiology of Salmonelloses and Shigelloses in South Africa.
The Leech, 29: 167—172.
BOKKENHEUSER, V. & RICHARDSON, N. J. (1959) The bacteriology of the
Bantu food-handler: Enterobacteriaceae. South Afr. med. J., 33:
784—786.
CLUVER, E. H. (1948) Public Health in South Africa. Nasionale Pers, Cape
Town. 5th Edition.
CROCKER, C. G. (1957) Distribution of types of typhoid bacteria over Africa.
South Afr. med. J., 31: 169—172.
CROCKER, C. G. & VAN WYK, T. (1960), (1961), (1962) Report on phage
typing of Salmonella typhi during the calendar year 1960, (1961), (1962)
at the Institute for Pathology, University of Pretoria.
GARROW, R. P. (1920) The myth of atypical enteric fever. Lancet, 199:
886—891.
HENNING, M. W. (1939) The antigenic structure of Salmonellas obtained
from domestic animals and birds in South Africa. Onderstepoort J. vet.
Sci. animal Ind., 13: 79—189.
Idem. (1953) Calf Paratyphoid. I. A general discussion of the disease in re-
lation to animals and man.
NESER, A. T., KLEIN, S. & SACKS, I. (1957) Fatal Salmonella foodpoisoning
from infected Biltong. South Afr. med. J., 31: 172—174.
VAN OYE, E. (1960) Répertoire général et revisé des Salmonellae du Congo et
du Ruanda-Urundi. Acad. Roy. Sci. d'Outre-Mer. Mém. in-8°, Nouv.
Série, XI, fasc. 6, 49 pp.
SIMMONS, J. S., WHAYNE, T. F., ANDERSON, G. W. & HORACK, H. M. (1951)
Global Epidemiology, Vol. II. J.B. Lippincott Cy., London.
VERGE, J. & PLACIDI, L. (1960) Notes sur la physiologie et l'immunologie
des vertébrés inférieures. Leur rôle potentiel dans l'épidémiologie de
l'homme et des homéothermes. Maroc Méd., 39: 1287—1297.

CONCLUSIONS

Les études qui ont été faites sur les salmonelloses en Afrique ont contribué dans une mesure appréciable à augmenter nos connaissances de la pathologie de ces affections, des germes qui sont à leur origine et des modes de transmission et de dissémination de ceux-ci. La découverte de plusieurs dizaines de nouvelles espèces de *Salmonella* en est le témoignage le mieux connu. Il nous faut à ce sujet insister sur le fait que l'intérêt de ces nouvelles espèces dépasse le cadre limité en soi de la bactériologie pure: elles permettent de connaître l'image plus ou moins caractéristique de la flore microbienne des diverses régions africaines, ce qui donne aux hygiénistes notamment des renseignements des plus précieux.

En ce qui concerne les *Salmonella* dites humaines, il a été constaté sur l'ensemble du continent une grande fréquence et une répartition géographique presque uniforme de *S. typhi* tandis que *S. paratyphi A* et *S. paratyphi B* sont très rares partout, contrairement à *S. paratyphi C* qui sous sa forme "var. *east africa*" est assez fréquente, du moins dans les territoires du Centre et de l'Est africain. Les autres *Salmonella* sont représentées par des centaines d'espèces et de variétés différentes, pour la plupart caractéristiques pour le continent noir.

Nombreuses sont les espèces de *Salmonella* découvertes en Afrique qui possèdent des propriétés biochimiques qui les distinguent des espèces connues en Europe ou en Amérique, et ceci a incité KAUFFMANN (1960) à subdiviser le genre *Salmonella* en deux sous-genres dont le second comprend nombre de biotypes d'origine africaine qui représentent une transition entre les *Salmonella* classiques et les *S. arizonae* du sous-genre III dont la création a été proposée récemment par KAUFFMANN & ROHDE (1962).

La lysotypie des cultures de *Salmonella* en provenance des territoires d'Afrique permet de dresser une carte de la distribution géographique des divers lysotypes assez précise non seulement dans ces grandes lignes mais parfois également dans certains de ses détails. Ces études ont en outre révélé l'existence en Afrique Equatoriale et à Madagascar d'une variété du lysotype C_1 caractéristique pour ces régions centro-africaines (NICOLLE et coll., 1955) ainsi que de certaines variétés du lysotype A, notamment les variétés Coquilhatville, Douala, Leopoldville et Tananarive (NICOLLE, PAVLATOU & DIVERNEAU, 1953).

Il existe peu de régions au monde où tant d'espèces d'animaux des plus divers jouent un rôle effectif dans la dissémination des *Salmonella*. Citons à ce propos, en plus de ceux déjà mentionnés, les travaux de ZWART (1962) au Ghana et de COLLARD & SEN (1960) au Nigéria. Il a été démontré partout que les animaux de boucherie sont largement infectés ainsi que les volailles de basse-cour. Il en est de même très vraisemblablement en ce qui concerne les animaux domestiques.

Les reptiles méritent d'être mentionnés tout particulièrement. Il n'est plus permis de les considérer comme de simples réservoirs inoffensifs dans lesquels les bactériologistes trouvent une source quasi inépuisable d'espèces nouvelles. Des observations précises ont prouvé qu'ils constituent un danger potentiel non-négligeable. En voici deux exemples: MACKEY (1955) découvre que dans 130 maisons sur 276 de Dar es Salaam les crottes desséchées des gecko's, petits lézards domestiques qui sont présents partout, contiennent des *Salmonella* et il présume qu'elles peuvent se désintégrer pour former des poussières infectées capables de contaminer les aliments. DARRASSE, LE MINOR & LECOMTE (1959) isolent à Dakar 8 souches de *Salmonella* appartenant à 6 espèces différentes dans une eau de distribution; l'origine de cette contamination fut une citerne-réservoir souillée des déjections de lézards du genre *Agama*. Il est curieux, et non sans intérêt, de constater que les espèces de lézards pouvant jouer un rôle de réservoir de *Salmonella* sont très différentes d'une région à une autre: COLLARD & MONTEFIORE (1956) ont trouvé à Ibadan 11,2% des Agames infectées et par contre aucun gecko, contrairement à MACKEY que nous venons de citer.

Deux groupes d'animaux méritent que leur soit porté un intérêt plus attentif, les oiseaux sauvages et les poissons. Il est connu que les farines de poissons en provenance de l'Afrique sont souvent massivement contaminées par des *Salmonella*. D'autre part, FLOYD & JONES (1954) en Egypte et JADIN et coll. (1957) au Congo ont isolé des *Salmonella* chez des poissons de rivière et de lac. Etant donné la grande consommation qu'en font les populations africaines, il serait intéressant d'entreprendre des prospections systématiques en divers endroits afin de se fixer sur l'importance exacte des poissons en tant que réservoir de germes. Il en est de même pour les oiseaux sauvages, et notamment pour les oiseaux migrateurs capables de transporter des *Salmonella* sur de grandes distances; certains indices nous font en effet soupçonner que ce rôle pourrait être moins négligeable qu'on ne le croit.

Deux problèmes d'ordre plus général doivent encore être soulevés. Tous ceux qui se sont penchés sur l'épidémiologie des salmonelloses en Afrique ont constaté, non sans un certain étonnement parfois, l'absence pour ainsi dire totale d'épidémies. Tout au plus observe-t-on des petits foyers, souvent à caractère familial. Or, dans la plupart

396

des territoires l'organisation des services médicaux est telle qu'une épidémie de salmonellose n'échappe pas à la vigilance des autorités. Peut-on accepter que la plupart des africains présente un certain degré d'immunité suite aux infections contractées dans leur jeunesse et qui les protègerait contre des infections nouvelles? Ceci n'est pas impossible mais ne semble pourtant pas devoir intervenir d'une manière suffisamment généralisée pour pouvoir expliquer cette absence d'épidémies.

Ceci nous conduit au problème de l'efficacité des vaccinations anti-typho-paratyphiques. Celles-ci ne sont appliquées d'une façon systématique que chez les travailleurs et chez les membres de leurs familles habitants les régions minières. Le nombre de cas de salmonellose est incontestablement moins élevé dans les concessions minières que dans les centres urbains ou dans les régions agricoles. Seulement, l'hygiène générale y est également nettement meilleure et il est impossible de délimiter la part qui revient aux vaccinations spécifiques préventives et aux mesures d'hygiène générale dans la dimunition, la disparition même en certains endroits, de la salmonellose humaine chez les habitants des régions minières.

BIBLIOGRAPHIE

1. Collard, P. & Montefiore, D. – Agama agama as a reservoir of Salmonella infection in Ibadan. *West Afr. med. J.*, 1956, 5: *154—156.*
2. Collard, P. & Sen, R. – Serotypes of Salmonella at Ibadan, Nigeria, with special note of the new serotypes isolated in Nigeria. *J. inf. Dis.*, 1960, **106**: *270—275.*
3. Darrasse, H., Le Minor, L. & Lecomte, M. – Isolement de plusieurs Salmonella dans une eau de distribution: originalité de la contamination. *Bull. Soc. Path. exot.*, 1958, 52: *53—60.*
4. Floyd, T. M. & Jones, G. B. – Isolation of Shigella and Salmonella organisms from Nile fish. *Amer. J. trop. Med. Hyg.*, 1954, 3: *475—480.*
5. Jadin, J., Resseler, J. & Van Looy, G. – Présence de Shigella et de Salmonella chez les poissons et dans les eaux des Grands Lacs du Congo Belge et du Ruanda-Urundi. *Bull. Acad. Roy. Méd. de Belg.*, 1957, VI° Série, 22: *85—96.*
6. Kauffmann, F. – Two biochemical subdivisions of the genus Salmonella. *Acta path. microbiol. scand.*, 1960, **49**: *393—396.*
7. Kauffmann, F. & Rohde, R. – Eine Vereinfachung der serologischen Arizona-Diagnose. *Acta path. microbiol. scand.*, 1962, **54**: *473—478.*
8. Mackey, J. P. – Salmonellosis in Dar es Salaam. *East Afr. med. J.*, 1955, **32**: *1—6.*
9. Zwart, D. – Notes on Salmonella infections in animals in Ghana. *Research in Vet. Sci.*, 1962, 3: *460—469.*

SALMONELLOSIS IN CANADA

BY

E. T. BYNOE & J. A. YURACK

Laboratory of Hygiene, Ottawa, Ontario.

At the beginning of this century, "enteric fever" was a very common complaint in Canada, and as Dr. A. H. GORDON, a leading physician in Montreal at that time, said in an address in that city in 1922[1] was "for years the commonest medical disease in Montreal hospitals". In this respect, Canada was no different from most other countries. With the improvement of sanitation and hygiene, the wider distribution of pure (treated) community water supplies and of pasteurized milk, the limited use of T.A.B. vaccines, improved methods of bacteriological diagnosis and a concerted effort on the part of medical officers of health to search out 'carriers', the enteric fevers — typhoid and paratyphoid — have been slowly but steadily decreasing, while the less pathogenic strains of salmonellae, dependent possibly on other means of transmission, have been actually increasing in the country.

When in 1929, the federal Bureau of Statistics started collecting data on the incidence of "typhoid and paratyphoid" fevers on a national scale, the report for that year[2] showed 1882 cases in Canada (exclusive of Newfoundland, which was not part of Canada at that time), or a rate of 19.0 per 100,000 population. During the last year, 1962, the Bureau of Statistics reported only 268 cases of "typhoid and paratyphoid" for all Canada (including Newfoundland) or a rate of only 1.5 per 100,000 population.

It was towards the end of the 1930's that the decrease in the incidence of "typhoid and paratyphoid" became evident, and since then this decrease has been steady. Table I shows the number of cases and incidence rates per 100,000 population in Canada and the provinces in 5 year periods from 1929 to 1962. The rates progressively decrease from 23.0 to 18.7, 11.7, 7.0, 4.3, 2.4 and finally to 1.9. There has been a corresponding decrease in deaths from these diseases from 421 in 1931 (a rate of 4.1 per 100,000) to only 2 in 1961.

These statistics are not to be taken as an absolutely accurate statement of the incidence of these diseases in Canada. In presenting its report the Bureau mentions the limitations of the data; all cases do not come to the attention of either physicians or other reporting personnel, the reporting of diagnosed cases is not complete and the degree of completeness in reporting varies for different areas of the country (case-finding techniques, and the degree of cooperation between local and provincial health bodies affecting the reporting).

Table I.

Typhoid and Paratyphoid Fever in Canada.

(prepared from Bureau of Statistics Annual Reports 1929—1962).

(Rates per 100,000 population).

Year		Canada	Nfld.	P.E.I.	N.S.	N.B.	Que.	Ont.	Man.	Sask.	Alta.	B.C.	Yukon
1929—1933 (incl.)	Cases	11,891	.	30	156	622	5,323	3,969	613	402	345	431	.
	Rate	23.0		8.5	6.0	30.4	36.8	23.2	17.6	8.8	9.7	12.5	
1934—1938	Cases	10,216	.	73	170	584	5,721	1,940	422	550	307	449	.
	Rate	18.7		15.8	6.2	27.0	36.9	10.8	11.8	11.9	7.9	12.0	
1939—1943	Cases	6,733	.	16	95	351	4,216	851	397	367	188	252	.
	Rate	11.7		3.4	3.3	15.4	25.3	4.5	10.9	8.2	4.8	6.0	
1944—1948	Cases	4,310	.	2	67	122	2,722	453	145	120	214	465	.
	Rate	7.0		0.4	2.2	5.1	15.1	2.2	3.9	2.9	5.2	9.2	
1949—1953	Cases	2,999	50	4	15	124	1,839	289	34	80	119	445	.
	Rate	4.3	3.4	0.8	0.5	4.8	9.1	1.3	0.9	1.9	2.5	7.7	
1954—1958	Cases	1,888	40	4	11	64	1,113	280	22	74	82	197	1
	Rate	2.4	1.9	0.8	0.3	2.3	4.8	1.0	0.5	1.7	1.5	2.8	2.3
1959—1962 (4 years)	Cases	1,413	30	2	27	19	974	160	20	36	64	79	2
	Rate	1.9	1.6	0.5	0.9	0.8	4.8	0.6	0.5	1.0	1.2	1.2	3.8

. Not available.

While it is probable that most cases of typhoid fever are reported, it is also probable that many cases of less severe paratyphoid fever are not.

Typhoid and paratyphoid fever are, however, only a part of the total illness caused by the salmonellae.

In 1953, the laboratory directors of the provincial public health laboratories, in addition to referring strains of salmonellae to the National Salmonella Reference Centre at the Laboratory of Hygiene, Ottawa for identification, also reported to this Centre a list of the *Salmonella* serotypes isolated from new cases in their laboratories. From these reports and the identifications carried out at the National Reference Centre are compiled, in table II, the data for the incidence of the more important *Salmonella* serotypes isolated in Canada annually for the last ten years, 1953 to 1962 inclusive.

While these data have the same limitations of accuracy as those reported by the Bureau of Statistics they are relatively accurate and at least indicate the bacteriologically confirmed infections. Even if we assume a greater awareness of the problem on the part of the medical profession, better diagnoses and better reporting year by year, these factors are not sufficient to explain the increase to 2,462 cases in 1962 over the 675 cases in 1953. While there has been no increase in the strictly human parasites of the genus — S. typhi, S. paratyphi A and S. paratyphi B — there is a very marked increase in those members of the genus which are parasites of both man and animals.

Salmonella Serotypes which have been found in Canada

There has been a total of 101 serotypes reported in Canada, up to the end of 1962. RANTA & DOLMAN[3] in 1947, reviewed the literature up to that time and reported on their own observations at the Canadian Salmonella Typing Centre, Vancouver, British Columbia for the years 1945—1947. They reported that the literature recorded the isolation of 28 *Salmonella* types in Canada and they added another 9 types which they had identified in the material submitted to their typing centre. Since then (March 1947), BYNOE, BAILEY & LAIDLEY[4] in a further review of *Salmonella* types found in Canada up to the end of 1952, added another 19 types, while in the last decade 1953—1962, 45 or almost half of all types reported to-date in Canada, have been reported for the first time. A complete list of these serotypes is given at the end of the chapter. Most of these have been isolated from both man and animals, but some have so far been isolated only from man and others only from "non-human" sources. Twenty-six of the 101 serotypes have been isolated only once. S. london was probably imported from Asia and S. wien from Europe but the origin of most of the others has remained a mystery. Many types produce only a very small number of infections year after year, but never entirely disappear. There are only comparatively few that have become well established in either man or animals or both and continue every year to be the principal contributors to salmonella disease in the Country (see Table II).

Table II.

Salmonella serotypes isolated from man (new cases) and other sources in Canada, 1953—1962.

	Source	1953	1954	1955	1956	1957	1958	1959	1960	1961	1962
S. anatum	*H	1	2	1	24	6		2	4	4	13
	*A	1	2	6	4	3	1	1	6		
S. bareilly	H	15	35	22	8	9	13	31	13	7	7
	A	5	74	10	6	29	6	3	6	15	2
S. blockley	H				1		1	30	23	38	23
	A							5	2	23	3
S. brandenburg	H	2	2		3	4	6	11	2	9	6
	A										1
S. bredeney	H	5	7		1	5	3		5	4	18
	A	12	8	12	12	8	5	3	12	14	11
S. canada	H								59	3	1
	A										
S. cholerae-suis	H	3	1		1		1	5		2	17
	A	4	8	8	20	20	9	19	32	43	49
S. derby	H	3	5	7	1	4	39	27	7	13	2
	A	6	7	5		6	2	2	2	5	6
S. enteritidis	H	5	3	3	5	5	7	26	23	53	34
	A	5	2	2	6	2	19	4	1	11	1
S. heidelberg	H	24	47	58	174	183	143	258	280	384	550
	A	1	10	3	3	15	29	101	60	115	86
S. infantis	H		2		2	23	15	7	51	32	35
	A					1	1	4		6	45
S. kentucky	H	6	2	1	1	2	6	8	1	1	2
	A		9	1	2		4	2	5	1	6
S. manhattan	H	2		1	2	1	6	19	52	9	2
	A		1		4			2		1	
S. montevideo	H	6	12	15	149	64	125	64	72	21	28
	A		3	3	6	1	3	2	19	21	28
S. muenchen	H		5	3	11	3	1	10	9	5	16
	A				1			1			1
S. newington	H		2	2	1	1	2		3	6	6
	A	1	3	2	10	3	6	6	32	34	14
S. newport	H	21	31	56	50	100	160	115	100	77	206
	A	5	1		1	1	5	10	3	6	29
S. oranienburg	H	19	32	9	24	15	21	13	13	29	35
	A		1	2	2	2	1	5	17	11	3
S. panama	H				4	2	3	9	3	5	9
	A									2	
S. paratyphi A	H	3	2	3	1	5	16	1		6	1
	A										
S. saint paul	H				4	2	3	1	3	12	27
	A					1	1		1	3	3
S. san diego	H		1	2		5	14	16	11	11	9
	A	1						6	22	13	3

Table II (continued)

	Source	1953	1954	1955	1956	1957	1958	1959	1960	1961	1962
S. paratyphi B	H	107	119	88	116	78	82	86	92	135	106
	A						1			3	7
S. senftenberg	H		2		3	3	1		1	2	3
	A			1	3		3		9	5	19
S. tennessee	H	22	38	7	28	26	58	37	36	21	52
	A	2	1	2	3	2		5	5		11
S. thompson	H	15	13	23	56	65	32	18	256	554	690
	A	25	69	16	58	34	13	21	9	7	23
S. typhi	H	213	312	126	241	196	141	151	145	126	128
	A										
S. typhi-murium	H	194	384	304	290	401	497	483	423	435	395
	A	54	88	56	81	111	64	90	83	105	108
S. worthington	H		1	7					1	2	
	A					1			12	18	33
Other types	H	7	6	7	23	12	28	20	34	48	38
	A	1	8	1	3	2	1	8	13	10	42
Unidentified	H	2		3	2	2	2	10	8	8	3
	A						4	1	4	6	
Totals	H	675	1084	748	1226	1222	1426	1458	1730	2062	2462
	A	123	295	130	225	242	177	302	350	478	508

* H = Human A = Animal.

The distribution of salmonellae in man is inextricably linked with that in the animal population of man's environment. The epidemiologist who fails to consider the 'animal' factor in the epidemiology of salmonelloses in man is doomed to failure. In Canada, one province, Alberta, has been particularly interested in the problem of salmonellosis in animals, with the result that most of the animal isolates identified and reported in Table II are from that province. Of the total of 2,835 cultures of salmonellae isolated in Canada during the last ten years and reported to the National Salmonella Centre, 1,823 or 64.3% were from Alberta, while the two large provinces, Quebec and Ontario with considerably greater animal populations, have contributed only 56 cultures (1.9%) and 429 cultures (15.1%) respectively of the total animal isolates identified. The sampling, therefore, given in Table II, which are the best data available, is poorly representative of the *Salmonella* distribution in the animal population of the Country and there is at least a strong possibility that some if not many of the serotypes which so far have been found only in man would have been found with a larger sampling of the animal population.

While the case and the symptomless carrier are the primary sources of those salmonellae which are specific for man viz. *S. typhi*, *S. paratyphi A* and *S. paratyphi B*, animals particularly poultry appear to be the principal reservoir of most of the other important salmonella infections in man, viz. *S. typhi-murium*, *S. thompson*, *S. heidelberg*. Strict isolation and quarantine of 'typhoid — paratyphoid' cases and carriers and prevention of these shedders of typhoid-paratyphoid bacilli from contaminating water, food and their environment are the only effective measures of control. Active programs by medical officers of health to search out these carriers, particularly amongst food handlers, have been largely successful in slowly bringing these diseases in Canada under control.

This type of control program, however, has proven ineffective in controlling those other types of salmonella infections, in which the natural reservoir of the salmonellae is the animal population. The rapid and marked increase in *S. thompson* and *S. heidelberg* infections in man in recent years is clearly related to their presence in very large numbers in foods which have been prepared for general widespread use in the Canadian home. The sources of these food contaminants is the infected animal from which the food is produced — poultry and eggs being the chief offenders. An intensive search during recent years of a number of human foods on the market, particularly egg-containing processed foods has revealed a large number of *Salmonella* types. The majority of these are shown in Table III. The finding of large numbers of salmonellae in many prepared animal foods (bone meal and the like) is probably likewise responsible for spreading much infection among the animal population. Successful control is therefore in large measure dependent on stopping the spread of infection among the animals which are to be used for food, and on better processing and handling of the food prepared for human use. Successful control therefore requires the active participation and combined efforts of the veterinarian, the epidemiologist, the bacteriologist, the food processor and handler, and the sanitarian.

Table III.

The 12 most commonly found Salmonella serotypes in some processed (human) foods (esp. cake-mixes and various egg products) on the Canadian market.

	1960	1961	1962
S. bredeney			25
S. gallinarum-pullorum	5		11
S. heidelberg	7	7	88
S. infantis	7	6	9
S. manhattan	5		7
S. montevideo	19	3	18
S. oranienburg	23	5	11
S. senftenberg	2		13
S. tennessee	11	1	113
S. thompson	79	39	460
S. typhi-murium	2	1	9
S. worthington	4		26

Complete list of Salmonella Serotypes found in Canada to 1962

S. adelaide. A single culture of this type isolated in 1950 from the stool of an infant in British Columbia with vomiting and diarrhea was the first report of this type in Canada. In 1962, 2 isolations from human sources were reported in Ontario.

S. anatum, (Table II) reported by Gibbons[5] from Canadian egg powders, is found occasionally in both man and animals. In 1950 it was the agent of a food poisoning outbreak affecting 15 persons made acutely ill following a wedding reception at Lachine, Quebec. In 1956, it was responsible for another outbeak in a hospital in Montreal, involving both staff and patients, but the source was not determined. It has been isolated from chickens, turkeys, a mink, a dog and caused severe enteritis and several deaths in an experimental monkey colony.

S. bareilly, (Table II) is a not uncommon type in both man and animals. The most frequent source of its isolations has been poultry (chickens and turkeys) but it has also been isolated from a beaver, a pigeon and egg powders. In 1959 it was responsible for an outbreak of typical "food poisoning" among a group of student nurses following a picnic lunch. Turkey salad was the incriminated food.

S. berta. There have been only 5 isolations of this type reported — all in humans. The first isolation was in 1956 in Ontario.

S. binza, was first isolated in 1958 in Ontario and to-date there have been 6 isolations from man reported. In 1962, it was isolated from animals (including a dog), animal feeds and egg products.

S. blockley (Table II) was found only twice before 1959. It was first isolated in 1956 from a man in Ontario with acute gastroenteritis, who had recently returned from a visit to Atlantic City, U.S.A. There was another case in Ontario in 1958, and then in 1959 there was a marked increase in the incidence of this type, human cases appearing not only in Ontario but also in Saskatchewan, and as far west as British Columbia. Every year since it has caused over 20 cases of human infection. While the principal focus has remained Ontario, it has been reported from 5 of the provinces. It made its first appearance in animals in 1959 in poultry and has been isolated from "cake mixes".

S. bonariensis was isolated from hen's faeces by Chase[6] but during the last ten years has been reported only once — from a human case in Ontario in 1961.

S. bovis-morbificans was first reported in Canada in 1953. The culture was isolated from a boy in British Columbia presenting with pain in his right side, nausea, vomiting, loose stools and low fever. The disease was diagnosed as acute appendicitis but cleared without surgery. The following year it was isolated from poultry in

Alberta. During the past 10 years, only 7 isolates of this type have been reported, 5 from man and 2 from poultry.

S. braenderup. Like *S. bovis-morbificans,* this is an uncommon type in Canada. First isolated in 1959 in New Brunswick from a 2-year old girl with a malabsorption syndrome, it has been isolated from 2 cases each year since and in 1962 was isolated from a "cake mix".

S. brandenburg (Table II), was isolated in Ontario in 1944; this was the first report of this type in North America[7]. During the past ten years, a few cases due to this type have occurred every year, with the exception of 1955; 44 of the 45 cases reported in the past 10 years occurred in British Columbia. In 1962 one case was reported from the other side of Canada, Newfoundland, and during this year the first non-human isolate, from swine mesenteric lymph glands, was reported from British Columbia.

S. bredeney (Table II) is more commonly found in animals than in man. In Canada it has been isolated chiefly from poultry but also from dogs, swine, sheep, mink, a cow and from egg-powders and animal feeds. While most of the human illnesses are of the "food-poisoning" type, the organism was isolated from a subphrenic abscess in a woman with cholelithiasis.

S. budapest was isolated from the stool of a man with diarrhea in 1953 in British Columbia and is the only isolate of this type reported from humans. There was a single isolate of this type in poultry in 1960 and again in 1961.

S. california. A single isolate of this type was first made in 1952 from the faeces of a woman in a nurses' home in British Columbia. Since then, there has been a single isolation from man in 1956 in Saskatchewan and another in 1962 in Alberta. To date it has not been found in animals.

S. cambridge was first reported in Canada (and in North America) by RANTA & DOLMAN[3] in 1947. They isolated it in 1945 from a woman with severe gastroenteritis, living in Quesnel, B.C. They mention that in 1944, TAYLOR had first described this serotype from an R.A.F. cadet in Britain. The Canadian patient's son, with the R.C.A.F., had returned from England some weeks before his mother's illness but the association was not investigated. One other case — a baby with diarrhea — was reported in 1951, also from British Columbia; in 1957 and 1959 it was isolated from poultry and in 1962 from human food.

S. canada. (Table II) As the name implies this was a new Canadian serotype, described by YURACK et al.[8] in 1961. It was isolated simultaneously in two widely separate areas of the Country. From February 1960 to August 1960 this organism spread through 8 of the 10 provinces with a total of 63 isolations from cases and contacts. The elderly and especially the very young were particularly susceptible; 36 of the 63 infections were in children under 5 years of age. The children often remained carriers for a considerable time. The widespread occurrence and rapid dissemination of this organism are highly suggestive of some packaged food, with country-wide distribution, being the vehicle but attempts to discover such a food failed. Since 1960 there have been only 4 cases of this infection reported; apparently the agent of transmission has been eliminated.

S. *canoga*, has been reported only once in Canada — from a human source in 1949 in Ontario.

S. *cerro*. The first report of this type in Canada was in 1960. It was isolated from a woman with gastroenteritis in Ontario. Three additional cases were reported during 1962 in two other provinces. It has not so far been found in animals in Canada.

S. *chailey* has been found only once in Canada. This was from a War Veteran in a Montreal hospital in 1959.

S. *cholerae-suis* (Table II) is a common 'animal' type. While not entirely host-specific for swine, this animal species is definitely the host of choice. Of over 200 animal isolates of this type during the past decade, more than 80% of them were from swine, in which the organism caused much fatality. The organism was isolated from the liver, lung and gallbladder of a cougar in a Toronto zoo, and the remaining isolates were from poultry. BIGLAND[10] reported that S. *cholerae-suis* accounted for 40% of the animal isolations (excluding poultry) in Alberta during the years 1949—1960 but that all of these, with one exception, were from pigs. The organism is responsible for the occasional human case, many of which show unusual symptoms. It has been isolated at least twice from abscessed neck glands, from pockets of pus in the lungs of a man dying of cancer of the lungs, from the blood of a patient, before death, with acute endocarditis, and from the spleen of one and the gallbladder of another patient at surgery.

S. *coeln* has been reported only once in Canada. The single culture was isolated, in 1959, from the viscera of a turkey.

S. *cubana* is an uncommon type in Canada. The first isolations were made in 1957, when it caused colitis and diarrhea in a man in British Columbia and gastroenteritis in an airman at the Air Force Base in Newfoundland, on the other coast. There have been 11 other cases due to this type, 9 in Ontario and 1 in Alberta (1962). The single non-human isolate was from animal feeds in 1960.

S. *daytona* is another uncommon type in Canada, reported only 6 times. The first isolation was made in 1953 from a baby in British Columbia with severe diarrhea; the other 5 isolates, (one in 1957; three in 1960; one in 1961) were all from humans and all from British Columbia. It has not been found outside this province or in other than humans.

S. *denver* has been isolated only once in Canada — from a 17-

months old boy in British Columbia, with vomiting, diarrhea and fever.

S. derby (Table II). A few isolates of this type have been reported every year for the past 10 years from both man and animals. The large numbers from man for the years 1958—1959 shown in the Table are due to an outbreak of severe diarrhea in the babies of a hospital nursery in Ontario which lasted for almost two years. This type has been responsible for heavy losses in young chickens in Prince Edward Island and turkey poults in Saskatchewan.

S. dublin has been found only twice in humans in Canada — once in 1954 from the faeces, urine and blood of a man in a Quebec hospital and again in 1956 from another case, also in Quebec. In 1951 it was isolated once from an animal source in Ontario.

S. eastbourne was isolated from three human cases in 1950 in Ontario and not again until 1961, when it was isolated from another case in Ontario and one in Quebec. To-date it has not been found in animals in Canada.

S. enteritidis (Table II) is a not uncommon type in Canada. Rats and mice are the chief reservoirs of this serotype but it has also been isolated in Canada from foxes and mink[9], from a cow, a cat, a pig, guinea pigs and occasionally poultry. In 1957 it was responsible for 10% mortality in a flock of 1—4 day old ducklings. A number of human cases are reported every year — over 20 each year for the past four years; most of these present with typical "food poisoning". The type has also been found in egg-products prepared for human use.

S. escanaba has been found only once in Canada — from a human case in 1962 in Quebec.

S. gallinarum-pullorum. This is a very common poultry pathogen. BIGLAND[10] reported that between 1949 and 1960 in Alberta more than 1,000 isolations of this serotype were made. Control measures, as developed by the federal Department of Agriculture are concerned primarily with the detection and control or eradication of pullorum disease. This involves the compulsory testing of all breeder flocks of poultry by an agglutination test. Infected flocks are then rejected as sources of eggs for hatch purposes. In Canada during 1960 a total of 5,199 breeder flocks were tested serologically and 51 were rejected. Some index of the incidence of infection in the poultry population is given by the cases in Alberta as diagnosed bacteriologically. During the years 1959, 1960 and 1961 respectively there were 32, 23 and 37 cases or outbreaks due to *S. pullorum* and 18, 14 and 5 cases due to *S. gallinarum.* Each case represents one incident. This serotype has also been incriminated occasionally in human illnes.

S. give was first isolated in 1954 from two different turkey hatcheries in Alberta, in one of which it caused a loss of 61 of 350 three to four-day old birds. In 1956 it was isolated from a man with diarrhea in a Quebec hospital. One of the other 5 cases reported, since 1956, probably acquired his infection in Mexico. In 1962 the organism was found in egg-products manufactured for human consumption.

S. halmstad. The single isolate of this type reported in Canada was in 1960 from a specimen of animal feed.

S. hartford is another rare type in Canada, first isolated in 1957 in Quebec from a war veteran with colitis and diarrhea. This man had suffered from intermittent attacks of diarrhea since 1940 when he was stationed at Aldershot Camp, England. There have been only 3 other isolates of this type reported in Canada, 1 from a human case in Saskatchewan and 2 of human origin in Alberta.

S. havana. The single isolate of this serotype reported in Canada was in 1958 from a woman living in Lachine, Quebec.

S. heidelberg (Table II) is one of the most important serotypes in Canada. This type made its first appearance in Canada in Alberta in 1952, where it caused considerable loss in three flocks of turkey poults and one of baby chicks, but all these baby birds had originated in one hatchery in Southern Alberta. The turkey eggs had been imported from Oregon, U.S.A. and gave a very poor hatch[10]. The poultry isolations in May and June were followed by the first recorded human case in July in a baby boy with diarrhea. Of the 4 human isolations of this type during that year, three of them were from rural areas in the vicinity of the infected hatchery[10]. In September of that year a culture was isolated from a butcher with acute gastro-enteritis in the neighbouring province of British Columbia, but whether this butcher had handled infected birds from Alberta was not investigated. From these small beginnings, this serotype spread fairly rapidly across Canada in both the human and animal population. From being non-existent in 1951, it rose to being the third most frequently found type in human infections in 1954 and since then has continued each year to be one of the three most commonly occurring serotypes. In 1953 it spread from Alberta and British Columbia to the neighbouring province of Saskatchewan, and to Saskatchewan's easterly neighbour Manitoba. In 1954 it continued to spread easterly to Ontario and then to Quebec. In 1955 it spread to Nova Scotia and Newfoundland. In 1957 it appeared simultaneously in man and poultry in New Brunswick and finally in 1961 it appeared in Prince Edward Island. Coincident with this increase in human cases, a similar increase in animal isolates (especially poultry) of this type occurred. By 1958, it was the second most commonly found type in animals and in 1959 and 1961 it was the most commonly found type. BIGLAND[10] found *S. heidelberg* the most common serotype (exclusive of *S. gallinarum-pullorum*) in poultry in Alberta in the years 1958, 1959 and 1960, and while the evidence is lacking to demonstrate the exact means whereby this organism was spread across the country, there is little doubt that poultry have played a significant role in this spread. Recently the isolation of this type from animal feeds, cake mixes and egg-products suggests other means of distribution in both the human and animal population. During the past 10 years this type has been involved in many 'food poisoning' outbreaks — in which the incriminated foods have been pastry, cooked turkey, sausages, vegetable salad and turkey dressing. There have also been several outbreaks in institutions, such as hospitals, where carriers have been incriminated.

S. hvittingfoss. The single isolate of this type reported in Canada was in 1958 from a male patient hospitalized with gastroenteritis in British Columbia.

S. illinois was first reported in Canada in 1955. The culture was isolated in Quebec from an adult female whose history of illness dated back to 1948 when she underwent cholecystectomy. This is the only human case due to this serotype which has been reported in Canada. In 1960, it was isolated from animal feeds in Ontario.

S. indiana. There were five isolations of this organism in 1960. All were from poultry washings and the poultry had all originated from the same Ontario farm. There have been no other reports of this serotype in Canada.

S. infantis (Table II). This type made its appearance in Canada first in December 1954 in the faeces of a woman in Quebec and in a 2-months old baby girl with diarrhea and fever in Ontario. Since 1956 there have been a number of cases of illness due to this organism every year — over 30 each of the last three years. It has been reported from 7 of the 10 provinces and appears to be now a well established serotype in Canada. The organism was first isolated from non-human sources in 1957 from "feather meal" and while the principal animal reservoir would appear to be poultry, it has been isolated on several occasions from animal feeds and since 1960 not infrequently from cake mixes.

S. israel. The single isolate of this serotype reported in Canada was in 1962 from a human case in Ontario.

S. javiana is an uncommon type in Canada. The first report of this serotype in Canada was from a human source on 1949 in Ontario[11]. Since then, there have been 19 isolations of this type from human sources in Canada; most of these have been isolated from the very young and have been widely separated. Only one isolate from non-human sources has been reported. This was from a mink in Alberta in 1951. Interestingly enough, there have been no human cases reported from this province.

S. kaapstad has been isolated only once in Canada. This was in 1961 in Ontario from a human source.

S. kentucky (Table II) was first reported in Canada in 1949, when it was isolated from a man, being treated with chloramphenicol for typhoid fever[4]. A few cases of illness due to this organism have appeared every year since, and while it has never become one of the more common types it is quite widely distributed, having been reported from 7 of the 10 provinces. Soon after it was observed in human cases it began to be isolated from poultry and its distribution in the animal population is very similar to that in man. In more recent years it has been isolated from animal feeds, and from cake mixes and egg-products used for human food.

S. kotte has been isolated only once in Canada, from an adult male in Ontario, in 1960, who suffered a sudden severe attack of gastro-enteritis which lasted for about 8 hours.

S. lexington was first isolated in Canada in 1956, in Ontario, from the stool of a known 'typhoid' carrier in a routine examination. There was no history of recent illness. Two additional cases were reported in the same province in 1960. These are the only reported isolations of this type in Canada to-date.

S. lindenburg. The single isolate of this type, reported in Canada, was in 1960, in Ontario from a human source.

S. litchfield was first reported from the faeces of a 78-year old man in Nova Scotia. Since then, only single isolations have been reported for the years 1957, 1958, 1959, 1960 and 1962. It is curious that of the 6 isolations reported to date, only two were from the same province British Columbia, while the other four isolates were each from a different province, Nova Scotia, New Brunswick, Ontario and Alberta, so that not only have these cases been widely separated in time but also in location. One of the illnesses was associated with the eating of canned meat, which had been left open for 3 days before being consumed.

S. livingstone. Only 3 isolations of this serotype have been reported in Canada — all from Ontario and all from human sources; the first two were in 1961 and the third in 1962.

S. llandoff. The single isolate of this type reported in Canada was in 1962. The culture was submitted from a packing plant along with others which were isolated from "feed" ingredients and egg products.

S. loma-linda has been isolated only once in Canada, from the faeces of a woman in British Columbia in 1954.

S. london has also been isolated only once in Canada. This was in 1960, in Quebec, from the faeces of a 10-year old boy who had only very recently returned from Asia, and it is probable that the infection was acquired outside the Country.

S. madelia was first isolated in 1945 in British Columbia. Four strains of this type were isolated during the year in British Columbia from epidemiologically unrelated sources. The first was isolated from the faeces of a symptomless adult carrier in the routine check of contacts of his daughter suffering from an infection due to *S. typhi-murium*. Two others were from infants with severe diarrhea and the fourth from a woman with diarrhea. Consumption of a veal roast was suspected as the source of the latter infection but this was not proven. During the last ten years, there have been only 6 reported isolations of this type, all of human origin, one in 1958 and five in 1961. Five of these 6 cases were in Ontario and one in Quebec.

S. manhattan (Table II) was reported from human sources in 1945[12] and from egg-powders by GIBBONS & MOORE[13] in 1944. It has been found infrequently in Canada in both man and animals, with the exception of the years 1959 and 1960, when it accounted for 19 cases of human illness in 1959, and 52 in 1960. In 1950 and 1951 there were 58 and 27 isolations respectively from animal sources in Ontario. In 1960, of 27 isolations of this organism in Alberta, more than 20 were from persons involved in a food poisoning outbreak in Medicine Hat which was traced to a carrier among the food-handlers. While this is not one of the most commonly found serotypes it is quite widely distributed, having been reported in all of the provinces except New Brunswick and Nova Scotia. Poultry would appear to be the principal animal reservoir and in recent years it has been isolated from animal feeds and egg-containing foods for human use.

S. meleagridis. Only four isolations of this serotype have been reported in Canada. It first appeared in 1950 in two patients in a hospital in Montreal, both suffering from colitis. There had been no contact between these patients before their admission to hospital, which suggests a possible cross infection. The third isolate was from an infant in 1952 in British Columbia with diarrhea and the fourth in 1962 from a human case in Ontario.

S. mendoza. There have been two isolations of this serotype in Canada. The first was in 1956 from the faeces of a young man hospitalized, in Ontario, with acute vomiting, diarrhea and fever, and the second, two years later, also in Ontario.

S. miami. A single isolation of this type was first reported by RANTA & DOLMAN[3] from Saskatchewan of human origin. In 1951 there were two isolations in Quebec and in each of the years 1957, 1958 and 1961 there was a single isolation in Ontario — all from human sources.

S. mikawasima. This organism has been isolated only once in Canada, in 1961, in Ontario from a human source.

S. minneapolis was first reported in Canada in 1958, in poultry, in Alberta. The first and only human isolate was four years later, 1962, also in Alberta. During this year there were four isolations of this type in Ontario, from egg-products and cake mixes.

S. minnesota was isolated from Canadian egg-powders by GIB-BONS & MOORE[13] and from a human source in Quebec in 1949. Al-though reported from both man and animals it is a rarely-found type in Canada. From 1953—1962, there have been only 8 isolations of this organism from man and 16 from non-human sources. Of the 8 isolations from man, 3 were in patients attending the same clinic in Quebec, 3 were from Ontario, all unrelated, and 2 from Alberta, also unrelated. Of the non-human isolates, 8 were from poultry in Prince Edward Island, New Brunswick and Alberta, 1 from a pig in Alberta, 1 from a dog in Saskatchewan and 6 from animal feeds in Manitoba and Alberta.

S. mississippi has been isolated once in Canada — in 1956 from the stool of a human case in Quebec.

S. montevideo (Table II) is one of the more common serotypes in Canada. It has been reported from human sources in every province. During the five years 1956—1960 it was one of the 6 serotypes which were responsible for 70 to 80% of all human cases of Salmonellosis, being itself responsible for 4 to 12% of these infections. While it has been found in every province, Ontario has been the principal focus. In 1956 more than 100 of the isolations were from a long standing outbreak in an institution for mentally retarded children where the standards of hygiene were exceptionally poor. In 1957, there were two fairly large outbreaks of 'food poisoning' in Red Deer, Alberta but the responsible foods were not definitely identified. One isolate was from a baby with bronchopneumonia, another from a pelvic abscess and one was from a lesion on the vertebra. While most of the animal isolates were from poultry, it was also isolated from a mink, and during recent years it has been on several occasions isolated from animal feeds and human foods (cake mixes and egg products).

S. moscow. The first report of this type in Canada and in North America was in 1944 in Ontario[7], probably of human origin. The only other reported isolation of this serotype was from a flock of day-old ducklings in which there was a high mortality.

S. muenchen (Table II). Four isolates of this serotype, from human sources, were first reported in Canada by RANTA & DOLMAN[3] as *S. oregon*, three of these from British Columbia and the fourth from Alberta. This is an uncommon type in Canada, but has been isolated from human sources from five of the ten provinces. Most of the few non-human isolates have been from human food products (cake mixes and egg products), a couple from poultry and one from a mouse.

S. new brunswick. RANTA & DOLMAN[3] reported 5 isolations of this serotype from human sources in British Columbia during 1945—47. In 1949, 3 more isolations of this organism from human sources were reported from Quebec. The next isolation was in 1960 from animals in Alberta. In 1961 there was a single isolate in British

Columbia and another single isolate in 1962 in Saskatchewan, from human sources.

S. newington was reported from human sources in 1945[12] and by GIBBONS[5], 1947, from Canadian egg powders. It has been found in 7 of the 10 provinces in Canada, but is a rare cause of human illness (Table II). The principal reservoir appears to be poultry and since 1960 it has been isolated frequently from animal feeds and less so from human foods (cake mixes, egg products and canned meat). It has also been isolated from mink.

S. newport (Table II) is one of the six most commonly found serotypes in both man and animals in Canada. It is widely distributed, having been reported as the cause of illness in man in all provinces. In 1950, it was the cause of a food poisoning outbreak in British Columbia, involving 22 persons attending a picnic in which the incriminated food was turkey meat. In 1959 another outbreak of food poisoning involving 9 persons was attributed to uncooked ham. In 1962 an outbreak involving 130 cases occurred in a 'mental hospital' — food was suspected but the source was never identified. Also one of the more common 'animal' pathogens, it has been isolated from poultry, mink, foxes, swine, cows, a racoon and a monkey as well as, more recently, from animal feeds and from human food.

S. nyborg was first reported in North America by RANTA & DOLMAN[3] from a human case in Alberta. This represents the only reported isolation of this serotype in Canada.

S. oranienburg (Table II) is a not uncommon serotype in Canada. It is widely distributed, having been found in all 10 provinces, from human sources. Every year there are a few cases reported from several provinces — most of the cases being in the very young, presenting with acute diarrhea. In one baby with meningitis, this organism was isolated from the brain at autopsy. It is also a commonly found poultry strain. BIGLAND[10] lists it as the most common type (excluding *S. gallinarum-pullorum*) in poultry in Alberta during the years 1949, 1950 and 1951. GIBBONS & MOORE[13] isolated it from Canadian egg powders and in recent years it has been frequently found in cake mixes and egg products prepared for human use. It has also been isolated on several occasions from animal feeds.

S. panama (Table II) was first isolated in 1952 from the blood of two cases which appeared in the same month in two separate provinces, Ontario and Quebec. It is an uncommon type — only 34 isolations of this type from human sources being made during the years 1953—1962 and it has been found in only 5 of the 10 provinces. It has been isolated only twice from non-human sources and this was in 1961 and the source of both isolates was frozen pumpkin pie.

S. paratyphi A (Table II) is a rare type in Canada. Like *S. typhi* this serotype is host specific for man. A few cases have been reported every year for the past ten years, with the exception of 1960. Several of the cases have shown a recent sojourn outside the Country. At least 2 were sailors travelling between North America and the Far East. Two of the cases were recent immigrants from China, and one from Europe — another had returned from Mexico shortly before becoming ill. While most of the cases had typical paratyphoid fever, one isolate was from a scraping of the bone of a case of osteomyelitis of the chest bone and two were from patients presenting with 'acute gallbladder'.

S. paratyphi B (Table II). RANTA & DOLMAN[3] in reviewing the salmonella types reported in Canada up to that time stated, "But there seems little doubt that *S. typhi* and *S. paratyphi B* have been the types most frequently isolated to date from humans". Like *S. typhi* and *S. paratyphi A* this organism is host-specific for man even though on very rare occasions it has been found in animals. It has been isolated from mink and in 1962 there were 7 isolations from poultry, but man is the principal reservoir and spread of the infection is largely by means of water or food contaminated by cases or carriers or by direct contact with a case or carrier. For the past ten years 1953—1962, *S. paratyphi B* varied between the 3rd. and 6th. most frequently found salmonella in human cases. During the years 1950, 1951 and 1952 the disease was more frequent in females than males — the sex ratio being almost 2 to 1. This sex difference was not maintained, however, in subsequent years. There were 3 small outbreaks in the North West Territories in which phage typing helped to trace the source to a specific carrier. In one small family outbreak, in which the man, his wife and 2 children all became ill, the infection was traced to cream. Of 92 strains reported by DESRANLEAU to the International Typing Committee[14] 38.0% were type 3A, 16.3% Taunton, 11.9% 3A1, 10.8% type 1, 10.8% untypable, 9.7% 3b and 2.17% Dundee.

S. paratyphi C is very rarely found in Canada. During the ten years 1953—1962 only 5 cases of infection due to this salmonella have been reported and the symptomatology in 4 of these has been unusual. The organism seems to be particularly pyogenic. One culture was isolated from pus on the leg of a patient with osteomyelitis, one from a deep abscess in the groin, one from an infected leg wound and one from a large closed abscess on the buttock. In 1959, there were 2 isolations from monkeys.

S. poona. The first isolation of this serotype was in 1952, in Ontario, from a young girl recently returned from Nigeria. At the time of isolation she was suffering from abdominal pain and headache only but while in Nigeria she had had recurring bouts of diarrhea. There were no further isolations of this organism until 1961, when it was recovered from the stools of five members of the same family in British Columbia. In 1962, there were 4 more isolations, all from human sources, 1 in Manitoba and 3 in Ontario.

S. potsdam. GIBBONS & MOORE[13] reported the first isolation of this type in North America from Canadian egg-powders. This is a rare type in Canada. Four isolations were reported from human sources by CROSSLEY et al.[15] during 1945—1947 and single isolations were also reported in 1959 and 1962.

S. quiniela. A single isolate of this type has been reported. In 1959 it was isolated from a young girl who had been ill for 10 days with vomiting, fever and arthralgia.

S. reading was first reported in Canada in 1948 by CROSSLEY et al.[15] in chickens. It is an uncommon type in Canada. In 1957, it was isolated from 2 boys, members of the same family, whose grandmother had returned from a visit to Oakland, California a few days before the boys became ill. At that time there had been a sudden increase in *S. reading* infections in the U.S.A. In Canada, in each of the years following, there were one or two human cases reported. In 1962, this serotype was also isolated from poultry and a pig.

S. richmond has been isolated only once in Canada — in 1960, in Ontario, from a human source.

S. rubislaw was first isolated in Canada from a human case in 1948 in Ontario and is an uncommon type. During the ten years 1953—1962 there have been only 10 isolations of this serotype from human sources and these isolations have been distributed over 4 provinces. It is equally rarely found in animals — only two isolates being reported during the past ten years, one of these was from a rat.

S. saint paul is one of the less common serotypes (Table II). Its first appearance was in 1951 in British Columbia from a human source. It did not appear again until 1956 and was isolated in Quebec from a man with gastroenteritis who had recently returned to Canada from Peru. Another case had recently returned from Mexico and had complained of gastrointestinal trouble since his visit. During the last few years it has been reported somewhat more frequently and has also been isolated from poultry, "frozen egg", and from a cat in a familiy in which there was a human case of illness due to the same organism, and animal feed.

S. san diego. The history and behaviour of this serotype in Canada is very similar to that of *S. saint paul* (above). It was first isolated in 1950 from a man in Quebec who had recently returned from a visit to the West Indies. For the next 5 or 6 years there were one or two cases reported annually. Latterly, from 1958 (see Table II) more cases due to this type have been reported. To date it has been reported from 6 of the 10 provinces. At least two of these cases appear to have contracted the infection outside of Canada — one in Mexico and another in Europe. From the few histories available, the illnesses produced were chiefly diarrhea, but at least one showed symptoms more characteristic of "paratyphoid fever". It has also recently been reported more frequently ·from non-human sources, the chief of these, as for so many other salmonellae, is poultry but it has also been isolated from a chinchilla and from animal foods.

S. schwarzengrund is another 'rare' type in Canada. It was first isolated in 1958, in Ontario, from human sources and each year since then there has been the very occasional case. Of the 10 human isolates to-date, 7 have been from residents of Ontario, 1 from Alberta and 2 from Quebec. In 1961 it was isolated from animals in Alberta.

S. selandia. The only reported isolation of this serotype in Canada was in 1947 by GIBBONS[5] from Canadian egg powder.

S. sendai. There have been only 3 isolations of this type reported in Canada — all from human sources in British Columbia, the first in 1956 from the blood of a patient in hospital with fever and general malaise.

S. senftenberg is an uncommon type in Canada although it has been reported as early as 1945 in Ontario from a human source. Scattered over the years since then it has been the cause of an occasional case of infection. The 15 human cases, reported during 1953—62, occurred in four of the provinces. In 1958, it together with *S. typhimurium* was responsible for a large outbreak of gastroenteritis on board an ocean liner between Canada and Europe. It has been isolated from poultry, cows and monkeys and latterly there have been a number of isolations from both animal feeds and foods for human use (coconut and cake mixes) (Table II).

S. siegburg has been isolated from only 3 cases of human illness in Canada to-date. The first was in 1955 in British Columbia, the other two in Saskatchewan the following year.

S. singapore. A single isolate of this serotype has been reported in Canada. This was in 1948 from a human source in Ontario[16].

S. stanley. WYLLIE[17] reported the first isolation of this uncommon type in Canada in 1942. Two further isolations, from human sources, were made, one in 1955 from Alberta, and one in 1961 from Manitoba. It has been also isolated from laboratory monkeys, from animal foods and once from poultry.

S. sundsvall was first reported in Canada by RANTA & DOLMAN[3] from two specimens from human sources, one from British Columbia and the other from Saskatchewan and are the first reports of this serotype on the North American Continent. No other isolations have been reported in Canada.

S. takoradi. A single isolate of this type has been reported in Canada. This was from a human case in Ontario in 1961.

S. tennessee is one of the more common salmonella types in Canada. In 1947, CHASE[6] reported its isolation from poultry. Since 1950 it has been found quite frequently in man and to-date has been reported from all provinces but one, New Brunswick. In 1962 there was a marked increase in the incidence of human infections due to this type[52], and at the same time there was an exceptionally large number of isolations from cake mixes[47] and egg products[65] used for human use (Table II). Recently it has been isolated on a number of occasions from animal foods.

S. thomasville. Only two isolations of this serotype have been reported to-date. The first was in 1956 from a 3-months old baby with loose stools and intermittent fever since birth — the second was in 1961.

S. thompson (Table II) is one of the most commonly found types in Canada. GIBBONS & MOORE[13] reported the isolation of this Salmonella, in 1944, from Canadian dried egg powder, and it has been frequently found in poultry since then. Over the past ten years it has remained one of the three predominant "animal" salmonella serotypes. BIGLAND[10] reported that it was the most common salmonella serotype (exclusive of *S. pullorum*) found in poultry in Alberta in 1954 and 1956. Despite this wide distribution and common occurrence in poultry, *S. thompson* was not one of the three or four most common types causing human illness at that time. In 1958 and 1959 it actually occupied 9th. and 13th. positions, respectively, in the order of frequency of those serotypes causing salmonellosis in man. From this low in 1959, it rose suddenly to third place in 1960 and to first in importance in 1961 and 1962, when it was responsible for 27—28% of all human cases of salmonella infection. During these years, it appeared in great numbers in cake mixes, which contained egg powder. THATCHER & MONTFORD[18] reported that of 119 cake mixes examined, 51% were heavily contaminated with salmonella and in 50% of these *S. thompson* was the predominant contaminant. BUTLER & JOSEPHSON[19] reported that of 40 cases of human salmonellosis in Newfoundland in 1961, 50% were due to *S. thompson*. Of 6 different commercial brands of egg-containing cake mixes on the market, all were found contaminated with *S. thompson*, some very heavily. They point out the hazards to public health of these dried food powders. In 1962, an explosive outbreak of gastroenteritis occurred among the staff and patients of a large hospital in Edmonton, Alberta. There were 116 cases, of which 74 were bacteriologically confirmed. Of these, 59 were positive for *S. thompson*, 10 for *S. heidelberg* and 5 for both. Careful epidemiological investigation traced the source to custard. The 'custard' had been served as a pie topped with "meringue". The meringue topping had been prepared from dried 'meringue powder', which was found to be heavily contaminated with *S. thompson* and to a lesser extent with *S. heidelberg*. Bacteriological counts showed approximately 7 million salmonellae per gram of powder. Another outbreak of "food poisoning" in a highway construction camp was traced to "angel food" cake and *S. thompson* was found in large numbers in several unopened packages of cake mix in the camp kitchen.

GIBBONS & MOORE[13] described salmonellae in egg powders as early as 1944. They considerd that in the small numbers in which they found these organisms in this product, and that baking temperatures destroyed them, that the finished foods prepared with these egg powders represented, at best, a slight danger to man. The heavy contamination of egg-containing cake mixes and the general wide distribution of these preparations during the last few years have, however, clearly altered the picture and there is little doubt that these products have been largely responsible for the marked increase in *S. thompson* 'food poisoning' in man during the last 2 or 3 years. The evidence was so convincing that regulations were passed during 1962 by the Canadian Food and Drug Directorate requiring the freedom from salmonellae of cake mixes and egg-products for human use.

S. typhi (Table II). Reference has been made above to the incidence of typhoid fever in Canada. Until recent years, *S. typhi* with *S. typhimurium* were the two predominant types of salmonella in human salmonellosis. In 1958 it lost its second position to *S. heidelberg* and by 1962 it had dropped to 5th. in frequency in human salmonella infections. Since *S. typhi* is host specific for man, the control of typhoid fever is dependent on the control of the case and the carrier. With the introduction of the bacteriophage typing of *S. typhi* by CRAIGIE & YEN[20] in Canada, a most valuable tool was provided the medical officer of health in tracking down the source and following the spread of an outbreak of typhoid fever. Active programs of search for carriers by public health departments throughout the Country and prompt isolation and treatment of cases and carriers have been primarily responsible for the slow but steady control of the disease in Canada. Nevertheless, outbreaks still occur but these are getting fewer each year. Amongst the numerous outbreaks reported, special mention may be made of a large outbreak traced to contaminated well water, another to unpasteurized milk contaminated by a carrier on the producing farm, and one to home-made Italian type cheese prepared by a carrier. T.A.B. vaccination has limited value as it is fairly well restricted to the Armed Forces and very special groups. The frequency of the various phage types reported by DESRANLEAU for Canada for the years 1954—1957 inclusive[14] were E1 — 19.6%, C1 — 10.9%, A — 9.6%, D1 — 8.5%, N — 8.4% F1 — 8.1%, Group I and IV — 7.4%, E4 — 5.6%, B2 — 4.3%, F2 — 3.4%, B1 — 1.5%, M1 — 0.9%, C5, D4, T and 28 — 0.6%, C4, D2, E2, E7, E8, H, J1, 0, 29, 34 and 40 — 0.3%.

S. typhi-murium (Table II) is one of the most commonly found salmonella serotypes in Canada. For many years, 1954—1958, it was the predominant strain in both man and animals. In 1959 and 1961, it took second place in animal infections to *S. heidelberg*. In 1961 it took second place to *S. thompson* and in 1962, third place to *S. thompson* and *S. heidelberg* as the cause of human illness. Over these years, it alone has been responsible for approximately 20 — 40% of all salmonella infections. In man, the usual symptoms are diarrhea, with sometimes bloody stools, abdominal cramps, occasional vomiting and fever; the illness usually lasts about a week but there have been all degrees of severity, varying from the very mild gastrointestinal upset to severe illness lasting for weeks and very occasionally terminating fatally. The symptoms often resemble appendicitis and there have been at least two cases of appendectomies performed when the patients had perfectly normal appendices and a *S. typhimurium* gastroenteritis. The organism has also been isolated from a leg abscess, a malignant breast abscess, a knee abscess following amputation, a draining sinus on a leg, the CSF of a child with high fever and convulsions and the pleural fluid of a case of acute lung infection. Many food poisoning outbreaks have been reported but in very few has there been any epidemiological follow-up. Turkey, potato salad and salmon patties are some of the foods which have been incriminated in these outbreaks. The outbreaks have been both large, many persons becoming ill following a banquet, reception or picnic, and small family outbreaks. While one of the most common strains in chickens and turkeys, it has also been commonly found in a variety of other animals, mink, foxes, ducks, pigeons, guinea pigs, cows, rats, swine, cats, dogs, monkeys, horses, a pheasant, a canary, budgie birds, and a heron. It is curious that despite its very wide distribution in the animal population, and the fact that GIBBONS & MOORE[13] reported finding it in dried egg powder as far back as 1944, nevertheless, it remained one of the more uncommon types found in cake mixes and other egg-containing prepared food products on the Canadian market, when *S. thompson*, *S. tennessee* and *S. heidelberg* were being found very commonly and in large numbers.

S. typhi-suis would appear to be host specific for swine. In 1949, CROSSLEY et al.[21] reported the isolation of this serotype from the lung of a pig, the first report of the occurrence of this type in North and South America. It was again isolated in Ontario in 1955 and 1956 from young pigs with ulcerative enteritis.

S. urbana is an uncommon type in Canada. RANTA & DOLMAN[3] reported 2 isolations of this serotype from human sources in Quebec from the material submitted to their typing centre 1945—1947. Since then less than 20 isolations have been reported, all but one of these from human sources, scattered over the years and over 6 of the provinces. The one animal isolate was from poultry in Alberta.

S. vancouver was first reported by DOLMAN et al.[22] as a 'new' serotype in 1950. It was isolated in April 1949 from a middle aged woman in Vancouver, B.C. who developed severe gastroenteritis, shortly after hysterectomy. No further isolates have been reported in Canada to-date.

S. virchow was first isolated in Canada in Ontario from human sources in 1943[23] and not again until 1960, when another single isolate was made, also from man, in Quebec.

S. weltevreden is another very rare type in Canada. It was first isolated from 2 cases of human illness in Ontario in 1954. The only other isolation of this serotype was in 1956 from a schoolgirl with severe gastroenteritis, in British Columbia.

S. wichita has been found only once in Canada. The single isolate was from a baby girl in British Columbia with diarrhea, in 1953.

S. wien. This was first described as a "new" type — *S. montreal* — by Laidley et al.[24] — but proved to be identical to *S. wien*, which latter name was considered to have priority. The culture was isolated from a woman who had recently returned to Montreal from a visit to Italy. On the second day on the steamer out from Cannes she became ill with gastroenteritis and this salmonella was isolated from her faeces on arrival in Montreal. No further isolates of this serotype have been reported in Canada to-date.

S. worthington (Table II) until recently has been a quite uncommon type in Canada. A single isolate from human sources was first reported by Ranta & Dolman[3] from Saskatchewan in specimens submitted to their Centre 1945—1947. Since then there have been only 11 isolations of this type from man in Canada and these have been scattered over the years and 6 provinces. During 1960, 1961 and especially 1962 there have been an increasing number of isolations of this serotype from animal feeds[58] and cake mixes (Table III), but unlike the situation with *S. thompson*, this does not seem to have resulted in a marked increase in incidence of infections either in man or animals.

<h1 style="text-align:center">REFERENCES</h1>

1. Gordon, A. H. Typhoid fever from the inside. *Canad. med. Ass. J.* 1943, **48**: *358—362.*
2. Dominion Bureau of Statistics. Annual reports of notifiable diseases for the years 1929 through 1962.
3. Ranta, L. E. & Dolman, C. E. Experience with salmonella typing in Canada. *Canad. J. Publ. Health*, 1947, **38**: *286—294.*
4. Bynoe, E. T., Bailey, W. R. A. & Laidley, Rhoda. Salmonella types in Canada. *Canad. J. Publ. Health*, 1953, **44**: *137—147.*
5. Gibbons, N. E. The incidence of organisms of the salmonella group in Canadian dried egg powder. *Canad. J. Publ. Health*, 1947, **38**: *84.*
6. Chase, F. E. Salmonella studies in fowl. *Canad. J. Publ. Health*, 1947, **38**: *82—83.*
7. Division of Laboratories, Ontario Dept. Health. Ann. Report for 1944: *48.*
8. Yurack, J. A., Laidley, Rhoda M., Finlayson, Margaret & Ferguson, Mary. A new salmonella serotype. S. canada (4,12; b—1,6). *Canad. J. Publ. Health*, 1961, **52**: *72—77.*
9. Stevenson, L. Report Ont. Vet. Coll. 1938: *45.*

420

10. BIGLAND, C. H. Salmonella reservoirs in Alberta. *Canad. J. Publ. Health,* 1962, **53**: *97—104.*
11. Division of Laboratories, Ontario Dept. Health. Ann. Report for 1949.
12. Division of Laboratories, Ontario Dept. Health. Ann. Report for 1945: *48.*
13. GIBBONS, N. E. & MOORE, F. L. Dried whole egg powder. XI. Occurrence and distribution of salmonella organisms in Canadian powder. *Canad. J. Res.,* 1944, **22**: *48—57.*
14. NICOLLE, P. Rapport sur la distribution des lysotypes de S. typhi et de S. paratyphi B dans le monde. *Ann. Inst. Past.* 1962, **102**: *389—409.*
15. CROSSLEY, VERA, FERGUSON, M., IRVINE, P. & HASTINGS, M. Salmonella typing in Ontario and the use of polyvalent antisera. *Canad. J. Publ. Health,* 1948, **39**: *192—199.*
16. Division of Laboratories, Ontario Dept. Health. Ann. Report for 1948.
17. WYLLIE, J. A sporadic case of food infection due to Salmonella stanley. *Canad. J. Publ. Health,* 1942, **33**: *41.*
18. THATCHER, F. S. & MONTFORD, J. Egg-products as a source of salmonella in processed foods. *Canad. J. Publ. Health,* 1962, **53**: *61—69.*
19. BUTLER, R. W. & JOSEPHSON, J. E. Egg-containing cake mixes as a source of salmonella. *Canad. J. Publ. Health,* 1962, **53**: *478—482.*
20. CRAIGIE, J. & YEN, C. H. The demonstration of types of B. typhosus by means of preparations of Type II Vi Phage. *Canad. J. Publ. Health,* 1938, **29**: *448—463.*
21. CROSSLEY, V. M., McKAY, A., McINTOSH, R. A. & SMITH, L. T. The occurrence of Salmonella typhi suis on the North American Continent. *Canad. J. Comp. Med.* 1949, **13**: *205.*
22. DOLMAN, C. E., RANTA, L. E., HUDSON, V. E., BYNOE, E. T., BAILEY, W. R. & LAIDLEY, R. A new salmonella type. Salmonella vancouver. *Canad. J. Publ. Health,* 1950, **41**: *23—26.*
23. Division of Laboratories, Ontario Dept. Health. Ann. Report for 1943.
24. LAIDLEY, R., BAILEY, W. R. A. & BYNOE, E. T., A new Salmonella type: Salmonella montreal. *Canad. J. Publ. Health,* 1951, **42**: *99.*

EPIDEMIOLOGY OF SALMONELLOSIS IN THE UNITED STATES

BY

MILDRED M. GALTON*, Sc. M.
JAMES H. STEELE*, D.V.M., M.P.H.

&

KENNETH W. NEWELL** M.B., Ch.B., D.P.H.

Salmonellosis has been the subject of many studies throughout the world since SALMON & SMITH[1], in 1885, isolated the first member of the genus *Salmonella* from swine. Little progress was made in the study of these infections in the United States except for pullorum disease in poultry until a few years after WHITE[2, 3] and KAUFFMANN [4, 5, 6] established the present method of antigenic analysis of the *Salmonella* group and the occurrence of numerous diverse serotypes were recognized. In 1934, a National Salmonella Center was set up by Dr. P.R. EDWARDS at the Kentucky Agricultural Experiment Station, Lexington, Kentucky. A few years later (1939), a similar center was started by Dr. F. SCHIFF at Beth Israel Hospital in New York City. In 1947, Dr. EDWARDS moved the National Center to the Communicable Disease Center, Atlanta, Georgia. Much of the information on the occurrence and distribution of serotypes in this country has resulted from the work of these centers. Summary reports of the work in these laboratories have been published by EDWARDS et al.[7, 8, 9, 10], SELIGMANN et al.[11], and SAPHRA & WINTER[12]; nearly 45,000 cultures isolated from man, animals, foods and environmental sources were described. During this period, many of the State public health laboratories and some university laboratories began typing cultures they had isolated or those submitted from other laboratories in their areas. As reports of these findings began to appear, it became increasingly obvious that a variety of *Salmonella* serotypes were widely distributed in the human and animal populations. In a study of 12,331 cultures, EDWARDS et al.[8], reported that types which occurred in both man and animals were recognized in 95% of the cultures from man. These studies emphasized that members of the *Salmonella* group were widely dispersed; that they were possibly connected in a zoonotic cycle; and that further epidemiological and other studies are required to define the significance and magnitude of the salmonella disease problem in man and animals, and to develop preventive and control measures.

---

* From the Department of Health, Education, and Welfare, Public Health Service, Communicable Disease Center, Epidemiology Branch, Veterinary Public Health Section, Atlanta 22, Georgia.
** From the Division of Epidemiology, Tulane University School of Medicine, New Orleans, Louisiana.

Reporting and Current Knowledge of Prevalence

In any such program, it is desirable, first, to establish a reasonable picture of the prevalence of the disease. For many years, in the United States the collection and publication of reports of human salmonellosis was the responsibility of the National Office of Vital Statistics. In January 1961, this function was transferred to the Surveillance Section, Epidemiology Branch, Communicable Disease Center, Atlanta, Georgia. Reports sent in by State health authorities are published in the Morbidity and Mortality Weekly Report, and summarized annually.

DAUER[13] and his associates published an annual summary of disease outbreaks in *Public Health Reports*. They estimated that no more than one in 20 cases of salmonellosis were reported, and suggested several reasons why some States reported poorly or not at all. First, some health authorities failed to understand the need for investigation and reporting and felt that there were more important problems. Second, some States possessed only limited personnel to conduct epidemiological investigations. Another factor may have been poor liaison between State and local health authorities.

Over and above these factors were the differences in the basis of reporting of human salmonella incidents in different States. There was no general agreement as to what should be reported, in definitions such as what was an "incident" or a "sporadic case", or how this information should be reported or investigated.

In the United States, as in many other countries, it is usually only

Table I.

Food Poisoning in the United States: By Outbreaks 1951—1961.

Year	Salmonella	Staphylococcus	Gastroenteritis*	Total
1951	14	28	50	92
1952	31	77	50	158
1953	21	81	92	194
1954	26	100	103	229
1955	16	102	66	184
1956	23	111	88	222
1957**	30	58	135	223
1958	27	62	134	223
1959	19	89	182	290
1960	17	54	89	160
1961	20	53	25	98
Total	244	815	1004	2063

* Etiology unknown.
** Four streptococcal outbreaks.

DAUER, C. C.: Summary of Disease Outbreaks, *Public Health Rep.* **69** 538—546, 1954; **70** 536—544, 1955; **71** 797—803, 1956; **73** 681—688, 1958; **74** 715—720, 1959; **75** 1025—1030, 1960, and **76** 915—922, 1961.

the large outbreaks of salmonellosis that are reported to health agencies, and in many instances, the etiological agents are not determined adequately. In considering the reported outbreaks of salmonellosis compared to those caused by staphylococcal enterotoxin and outbreaks of gastroenteritis of undetermined etiology, the need for more complete reporting and investigation is further apparent. Nearly half of 2063 outbreaks of food "poisoning" reported from 1951 through 1961 were attributed to unknown causes, and less than 12% to salmonellosis (table I). This low figure is not surprising when it is realized that reporting of salmonellosis is not required in every State. The American Public Health Association manual on Control of Communicable Diseases, 1960 Edition, places salmonellosis in the class in which obligatory reporting of epidemics is recommended but no case reports are required. In spite of the incompleteness and inadequacy of these reports, more than a fourfold increase was noted in the number of human cases reported during the past decade, from 1733 cases in 1951 to more than 8500 cases in 1961 (table II). During this same period, there has been a steady decrease in the reported cases of typhoid fever.

Some of this increase in the number of reported human cases of salmonellosis was certainly apparent rather than real. Simultaneous to this rise, there was an increased general awareness of salmonellosis as a disease problem, an improvement in methods and facilities for

Table II.

Typhoid, Paratyphoid Fever and other Salmonelloses.
Total Reported Cases in the United States, 1951—1961.

Year	Typhoid	Paratyphoid and other Salmonelloses
1951	2128	1733
1952	2341	2596
1953	2252	3946
1954	2169	5375
1955	1704	5447
1956	1700	6704
1957	1231	6693
1958	1043	6363
1959	859	6606
1960	816	6929
1961	814	8542
Total	17,057	60,934

National Office of Vital Statistics and Communicable Disease Center, Morbidity and Mortality Weekly Report, Annual Supplements, 1951—1961.
EDWARDS, P. R.: Observations in incidence and control. *Ann. N. Y. Acad. Sci.* **70** 598—613, 1958.

the detection of salmonellae, a development of disease reporting, and a marked increase in the use of Salmonella Reference Centers. However, there are some grounds for believing that a real rise also occurred. For example, in Massachusetts[14], *Salmonella* isolations increased nearly sevenfold between 1950 and 1955, while the total specimens from which these recoveries were made, increased less than twofold.

New Developments in National Reporting and Surveillance

In May 1962, stimulated by an outbreak of *S. thompson* in Michigan[15] and a sudden increase in cases of *S. hartford* infection in the midwestern States, a limited surveillance of salmonellosis was initiated by the Communicable Disease Center. When the program began, 8 States responded to the invitation to submit monthly summaries of their isolations of salmonellae and reports of pertinent investigations; and, by November, 22 States were participating in the program. During the fall of 1962, the Associations of State and Territorial Laboratory Directors and State and Territorial Epidemiologists passed a resolution that a weekly report on *Salmonella* isolations be submitted to the Communicable Disease Center by the designated State official. In addition, all *Salmonella* isolations identified in the National Animal Disease Laboratory, U.S. Department of Agriculture, Ames, Iowa, are reported to the Communicable Disease Center. This information is summarized monthly in a Salmonella Surveillance Report and distributed to local, State, and Federal health authorities. In January 1963, reports were received from 47 State health departments and 5 additional typing centers. A total of 1111 *Salmonella* isolations were reported from human sources and 525 from animals, foods, or environmental sources. Because of the increasingly frequent identification of salmonellae in human and animal food products which cross national boundaries and which have been associated with national or regional salmonella outbreaks in animals and man, and because of an increasing awareness in the United States and Europe that *Salmonella* types new to the country have appeared in the human populations[16], the Salmonella Surveillance Report* is sent, also, to interested health authorities in other countries and to the World Health Organisation in Geneva.

Age and Sex Distribution

The reported age distribution in the United States follows the same general pattern observed in other countries in that infants and young children are most often infected. MacCready et al.[14] found

---

* Inquiries regarding reported *Salmonella* isolations in the United States should be directed to the: Salmonellae Surveillance Unit, CDC, Atlanta 22, Ga.

that in 2092 infected individuals whose ages were known, 46% were less than 10 years old, and 14% of these were under 1 year of age. Similar observations were reported by SELIGMANN et al.[11]; of 1497 cases studied, 40% were children under 10 years of age and 17% of these were infants. There was little difference in the sex of 2344 infected individuals with 1142 males and 1202 females. A few notable unexplained differences were observed with certain serotypes. For example, a 3:1 ratio in favor of males was found in S. *cholerae-suis* infections and approximately 2:1 ratio in favor of females with S. *anatum* infections.

Mortality

In general, the fatality rate is highest among salmonella-infected patients above 50 years of age, followed by infants under 1 year of age. SAPHRA[17] reported 174 (5.3%) fatalities in 3279 human infections. On the basis of age groups, the fatality rate was 15% in patients over 50 years, 5.8% in the infant group, and 2% between 1 and 50 years. Distinct differences were noted with certain serotypes. S. *cholerae-suis* had the highest rate with 21.3%. In contrast, only 5.2% of the cases with S. *typhi-murium* were fatal and 5.5% with S. *oranienburg*. Other workers have reported similar observations [8, 14].

These observations on age, sex and mortality were made on special series of isolations from persons collected for different reasons. If the infection rates in the population by serotype were known, it is possible that the proportion of those infected who became ill or died, would differ markedly from those quoted. From the existing data in the United States, it is impossible to associate the risks of infection by a serotype with the possible effects measured by the amount of human decrease.

Prevalence and Trends of Serotypes

More complete knowledge of the relative prevalence of different *Salmonella* serotypes, by time and place, is becoming of increasing importance, both from a national and international standpoint. Consideration of the reported distribution of serotypes in man and animals in this country during the nearly 30 years since serological identification procedures have been applied, suggests changing patterns. In addition, there is increasing evidence of a correlation between the types found in the human and the animal populations, as well as in their foods. S. *typhi-murium* has continued to be the most prevalent type, followed in varying orders of frequency by S. *newport*, S. *anatum*, S. *derby*, S. *oranienburg*, S. *montevideo*, and S. *muenchen*. Some 250 other serotypes have been reported. Certain of the new and/or rare types have, in recent years, gained in prevalence until they have, in fact, displaced some of the more common types. Other types, also considered rare, have been observed to rise sharply throughout the country and after 6 to 12 months, decline with almost equal suddenness. For example, *S. reading* was not detected among the 12,331 cultures identified by EDWARDS et al.[8], between 1934 and 1948, and only 24 cultures were identified during the next 8 years[10]. In contrast, 125 isolates of this serotype were identified from human cases during the next 16 months (May 1956 to September 1957). During 12 months of this period (September 1956 to September 1957) more than 300 acute sporadic cases and 3 outbreaks due to *S. reading*[18] were reported. These cases were widely distributed from Maine to California. The epidemiological picture suggested a common source of infection, but none could be identified. A few isolations of *S. reading* from poultry were identified in July and August 1956, just prior to the rise in human cases, but there was no clear evidence to incriminate poultry with the human cases, although chicken was consumed by 99% of the families. About a year after the outbreaks, *S. reading* was isolated from packages of bacon-flavored dried egg powder in New Mexico[19], but no human cases were attributed to consumption of this product. Since these episodes, *S. reading* has continued to occur only sporadically in man and poultry. SEELIGER[20] reported the isolation of *S. reading* from imported fish meal in West Germany in 1957.

In this country, a high percent of fish meal is used in poultry feeds. In 1955, *S. blockley* was isolated from fish meal in Germany[21] about the same time that a large outbreak due to this type occurred in Georgia and was traced to commercially prepared chicken salad[22]. There were approximately 300 confirmed cases and an estimated 3000 others who developed clinical symptoms. *S. blockley* was recovered from the chicken salad and from birds being processed in the plant that prepared the salad. Until this time, *S. blockley* was a relatively rare type. It was isolated first in the United States in 1955[23] and is now rather widely distributed throughout the country as a frequently encountered type in man and poultry.

The isolations of *S. infantis* have also increased markedly during the past decade. In Florida, *S. infantis* was found in poultry processing plants in 1952 and in man in 1953[24]. First seen in California in 1954, *S. infantis* continued to increase and during the first 6 months of 1960, there were 116 human infections or 48% more than in the entire year of 1959[25]. In 1958, a 1233-bed hospital in Wisconsin experienced an outbreak of *S. infantis* wound infections in 9 postoperative patients [26]. Two of these cases developed salmonella pneumonia and died. During the first 10 months of the Communicable Disease Center salmonella surveillance program (1962), *S. infantis* has been among the 7 most common serotypes reported in man and has been a frequent isolate also from poultry. *S. heidelberg* also first appeared in this country about 1953[14]. It appeared in England, Wales, and other countries about the same time[27]. Now, it is one of the most common types in both man and animals in the United States, including swine, cattle and poultry.

While some serotypes have spread rapidly throughout the country, others appear to have become more or less localized in a particular geographic area. For example, S. *miami*, first isolated from chimpanzees and man in Florida in 1942[28], became one of the most common types in the State in man, swine and fowl during the next decade. In 1944, it was the cause of an outbreak of gastroenteritis involving 60 persons in Miami[28]. This organism was isolated also from pickles served in a restaurant in which the affected individuals had eaten. More recently, it has been observed sporadically in Georgia, but only rarely in other parts of the country. In 1953, an unusual outbreak of gastroenteritis involving 17 persons occurred in Massachusetts[29]. Investigation revealed that watermelon shipped from Florida was the one food item common to all cases. S. *miami* was isolated from the patients and from samples of the watermelon.

An even more unusual situation exists in the geographical distribution of the partially host-adapted type, S. *dublin*. It has been found only rarely east of the Rocky Mountains. Most of the cultures have been recovered from cattle and foxes, although human infections have occurred and occasional isolates have been obtained from turkeys, canaries, horses, dogs, mice, chickens, rabbits, and a dove[8, 30]. In 1958, an outbreak of S. *dublin* infections occurred in Southern California[31] involving 11 confirmed human cases and 19 suspected cases. The source of infection was traced to a dairy producing certified raw milk. S. *dublin* was isolated from fecal samples from 3 cows in the herd of about 400 animals. There is no apparent explanation for the limited distribution of this serotype in a relatively confined area of the United States. Although it could be attributed, in part, to the probable limited movement of cattle from west to east, this does not account for the very low prevalence of infection in the affected area.

The 12 predominant *Salmonella* types reported for each of five varying periods[8, 10, 12, 32, 33] in man and animals between 1934 and 1963 reflects these shifts in prevalence (table III)). Similarly, localized differences in prevalence of certain types may be observed among the 12 most common types reported from widely separated geographic areas[14, 34, 35] (table IV).

It is reasonable to believe that some of these differences in the proportions of serotypes isolated in different regions reflect the different ecologies of these regions and the different sources and types of processing of foods for man and animals. Although there are numerous products which have a national distribution, many of the animal

Predominant Types of Salmonellae Isolated from Human and Animal

1934—1947[8]				1939—1955[12]		1947—1958[10]	
Human		Animal		Human		Human	
Type	No.	Type	No.	Type	No.	Type	No.
1. typhi-murium	478	typhi-murium	2860	typhi-murium	2385	typhi-murium	1479
2. newport	312	cholerae suis	970	newport	701	newport	739
3. paratyphi B	236	derby	384	montevideo	639	oranienburg	632
4. oranienburg	229	anatum	336	oranienburg	641	montevideo	554
5. montevideo	191	oranienburg	316	cholerae suis	359	muenchen	347
6. anatum	170	bareilly	312	paratyphi B	349	anatum	320
7. cholerae suis	151	bredeney	299	anatum	346	tennessee	296
8. derby	129	newport	280	muenchen	254	paratyphi B	267
9. panama	124	muenchen	270	enteritidis	240	cholerae suis	236
10. muenchen	83	meleagridis	185	tennessee	239	java	211
11. give	76	give	184	bareilly	236	reading	216
12. bareilly	63	senftenberg	163	panama	234	javiana	196
Other types	817		2713		1156		6304
Total isolations	3059		9272		7779		11797

protein foodstuffs for animals and man show a regional production and consumption pattern.

Modes of Transmission and Reservoirs of Infection

The literature abounds with reports involving our domestic mammals and fowls as a source of salmonellosis in man. The similar distribution of the most common *Salmonella* types (table III, IV) isolated from man and lower animals, ·suggests that salmonella infections in man and animals could be spread from one to the other, or derived from the same sources. This evidence by itself is inconclusive. However, a large number of studies demonstrate the continuing infection of a large number of domestic and wild animals, and it is reasonable to believe that either these infections continue, due to a natural cycle, or that there is some feedback mechanism from infected humans or animals. There has been little work to suggest that the non-typhoid salmonellae in humans are regularly perpetuated by a closed human transmission mechanism. On the other hand, paths of transmission from animals to man, with an increase in the number of organisms in transit, have been shown on a number of occasions. This incomplete and presumptive evidence leads us to believe that it is probable that the main reservoir of non-typhoid salmonellae is in animals, and that they are transmitted

III.

Sources in the United States at Various Intervals during 29 Years.

1947—1958[10]		1957—1961[32]		1962—1963[33] 10 Mos.		1962—1963[33] 7 Mos.	
Animal		Animal		Human		Animal	
Type	No.	Type	No.	Type	No.	Type	No.
typhi-murium	2390	typhi-murium	1385	typhi-murium	3222	typhi-murium	458
anatum	653	cholerae suis	378	heidelberg	574	cholerae suis	129
newport	363	heidelberg	294	newport	514	anatum	111
cholerae suis	358	anatum	291	infantis	476	heidelberg	98
enteritidis	343	enteritidis	257	enteritidis	285	derby	94
derby	308	newport	255	blockley	250	saint paul	66
heidelberg	263	san diego	229	saint paul	249	infantis	64
san diego	214	infantis	186	montevideo	173	montevideo	62
muenchen	174	chester	164	oranienburg	154	bredeney	60
montevideo	147	saint paul	136	derby	152	give	42
meleagridis	147	muenchen	114	muenchen	133	enteritidis	40
oranienburg	121	derby	113	thompson	131	oranienburg	40
	3931		1404		1641		689
	9412		5206		7954		1953

to man either directly, or indirectly in animal or animal contaminated products, incidental to the main cycle.

EDWARDS[36], stated that domestic poultry probably constitute the largest single reservoir of salmonellae among animals. More than 50% of the 12,331 cultures examined by EDWARDS et al.[8], between 1934 and 1947, were isolated from turkeys, chickens, or other domestic fowls. Although this statement covers a large time interval and a large number of isolations, it may be biased by the type of cultures sent to a reference laboratory and to the salmonella investigations which were being conducted at that time. We believe that between 1934 and 1947, more investigations were being conducted upon poultry than upon other domestic animals but that it is still probable that poultry are a major reservoir. Recently, MORAN[32] reported on the occurrence and distribution of salmonellae in animals between January 1957 and July 1961. During this 4½-year period, 6216 cultures were serotyped, and 4841 (77%) were from domestic fowls, primarily turkeys and chickens. While S. *pullorum* and S. *typhimurium* are the most common types, a greater number of *Salmonella* types have been isolated from fowl than any other species except man. The majority of these have been found, also, in humans. HINSHAW et al. [37], observed several cases of salmonella gastroenteritis, caused by contact with infected poultry, among attendants on poultry farms. Further evidence indicated that human symp-

Table IV.

Predominant Types of Salmonellae Isolated from Human and Animal Sources in 3 Widely Separated States.

	Massachusetts[14]		Florida[34]				California[35]			
	Human, 1942–1957		Human, 1942–1951		Animal, 1949–1954		Human, 1945–1950		Animal,* 1942–1950	
Type	Type	No.	Type	No.	Type	No.	Type	No.	Type	No.
1. typhi-murium	typhi-murium	983	anatum	149	anatum	1110	typhi-murium	573	typhi-murium	516
2. montevideo	montevideo	319	typhi-murium	125	derby	779	newport	127	oranienburg	151
3. newport	newport	265	oranienburg	122	newport	138	montevideo	104	bareilly	126
4. oranienburg	oranienburg	221	newport	108	bredeney	132	paratyphi B	46	anatum	81
5. paratyphi B	derby	111	derby	99	typhi-murium	121	oranienburg	35	montevideo	52
6. bareilly	montevideo	100	montevideo	77	oranienburg	113	anatum	34	newport	45
7. enteritidis	miami	83	miami	73	rubislaw	104	derby	20	derby	36
8. muenchen	paratyphi B	67	paratyphi B	35	montevideo	96	cholerae suis	16	san diego	28
9. tennessee	litchfield	57	litchfield	34	senftenberg	92	muenchen	15	bredeney	24
10. panama	muenchen	52	muenchen	34	give	82	bareilly	10	meleagridis	24
11. derby	bredeney	43	bredeney	30	manhattan	74	panama	8	cholerae suis	25
12. anatum	meleagridis	43	meleagridis	24	miami	60	saint paul	8	give	21
Other types		281		353		694		204		153
Total isolations		2625		1263		3595		1100		1282

* Data for chicken and turkey isolations tabulated by flock outbreaks of salmonellosis.

tomless excreter attendants had also transmitted salmonellae to fowl.

Considerable evidence has been accumulated concerning the presence of salmonellae in poultry meat. CHERRY et al.[38], reported the recovery of a monophasic *Salmonella* type from the skin of frozen turkeys. GALTON et al.[24], isolated *S. anatum* from material appearing to be unabsorbed egg yolk in a frozen chicken. In a survey of market meat and poultry products purchased from retail stores in Cincinnati, WILSON et al.[39] obtained salmonellae from 17% of 525 poultry specimens. Lower rates were encountered in other types of meats. SCHNEIDER & GUNDERSON[40] found 4 *Salmonella* types on the skin of 4.4% of 1014 eviscerated chickens. Most of these birds had been frozen and stored for some time. BROWNE[41] found that *S. typhi-murium* survived for at least 13 months on the skin of frozen turkeys. A large outbreak in a Federal institution in a southeastern State was due to *S. typhi-murium* in cold sliced turkey[42]. Investigation revealed that the frozen turkeys were infected with *S. typhi-murium*. While the organisms were destroyed by roasting, the cooked meat was contaminated when sliced on the same chopping block used to cut and prepare the raw turkeys for roasting. *S. typhi-murium* was isolated from 100 patients, and approximately 200 others developed clinical symptoms.

The prevalence of salmonellae in poultry processing plants was studied in Florida[24]. Salmonellae were isolated from 16% of 1244 swab samples of various materials in the plants. More than 30% of the swab samples taken on the tables on which edible viscera were wrapped, contained salmonellae, and 11% of the swabs from iced tubs containing marketable birds were positive. However, no salmonellae were recovered from 872 tissue and caecal samples cultured from 129 healthy broilers taken from the processing line just prior to evisceration. These observations and the low percentage of salmonellae obtained from cloacal swabs on 53 birds during processing (1.9%) indicated that even though the number of infected birds that come to slaughter is low, salmonellae are widely disseminated in the environment of the plant during processing. The organisms on these contaminated birds, preserved by refrigeration, would reach the kitchens and the hands of cooks and thus provide a ready source of salmonellosis in man, primarily by contamination of other foods that may be eaten uncooked.

Reports implicating raw, dried or frozen eggs as sources of salmonella outbreaks appear frequently. WATT[43] reported such an outbreak attributed to contaminated raw eggs used in mayonnaise. *S. montevideo* was isolated from 28 individuals aboard a merchant vessel docked in New Orleans. The same type, *S. montevideo*, was isolated from the meats of eggs obtained on the ship, confirming internal contamination. Ovarian transmission of *S. pullorum* was proved by RETTGER & STONEBURN[44] in 1909. A few years later, RETTGER et al.[45] suggested the possibility of infected eggs causing gastroenteritis in children. Subsequently, penetration of salmonellae through the unbroken shell of eggs was demonstrated[46, 47].

Three sharp outbreaks of salmonellosis caused by *S. typhi-murium* occurred within 2 years in an institution for mental disease in Massachusetts[48]. During an intensive epidemiologic investigation of the last of these outbreaks, involving 104 cases and 6 fatalities, eggnog was established as the vehicle of infection. *S. typhi-murium* infection was found in the flock of birds that supplied eggs for this institution, and these organisms were recovered from some of the eggs. The cultures of *S. typhi-murium* from the eggs were the same phage type found in the patients. Several other outbreaks have been attributed to the use of raw eggs in special diets or as a dietary supplement, a common practice in many hospitals.

The problem of salmonellae in powdered eggs was reviewed by SOLOWEY et al.[49]. These authors found salmonellae in 35% of 5000 samples of spray-dried whole eggs. ABRAMSON[50] described a widespread outbreak of *S. montevideo* infection in infants in the United States due to infected dried egg yolk powder that was inadequately pasteurized. McCULLOUGH & EISELE[51] were able to produce clinical salmonellosis in 32 human volunteers by experimental infection with strains of *S. meleagridis* and *S. anatum* derived from spray-dried whole eggs. Similar studies[52] with *S. newport*, *S. derby* and *S. bareilly* resulted in clinical illness in 14 subjects, and with 4 strains of *S. pullorum*[53], there were 27 cases of human illness. The effective dosage of *S. pullorum* was stated to be 10 times that of other *Salmonella* types. In spite of the apparent large numbers of *S. pullorum* organisms required to produce clinical illness, this type has been incriminated in several outbreaks of food infection[54, 55] and sporadic cases [9, 34, 56] involving approximately 450 persons in all.

Several outbreaks have been described in Morbidity and Mortality Weekly Reports involving frozen eggs. *S. pullorum* was isolated from frozen eggs used in the preparation of eggnog that was found to be the vehicle of infection in 6 persons in an Illinois institution[57]. Another outbreak in a Washington, D.C., institution was attributed to *S. tennessee* recovered from frozen egg yolk used in cake filling and from stools of the patients[58]. In Michigan, 268 persons became ill after eating coconut cream pie with meringue. *S. typhi-murium* was recovered from an unopened can of frozen egg whites from the same shipment used to make the meringue[59].

The prevalence of salmonellae in swine, a major source of meat for man, is well known. The occurrence of salmonellae in the lymph glands of normal hogs was observed in this country in 1942 by RUBIN et al.[60]. The following year, CHERRY et al.[61], recovered salmonellae from retail meat products. Data obtained in Florida[62] on the prevalence of salmonellae in hogs on farms and in abattoirs, clearly indicated an increasing proportion of infected animals during transportation and while being held for slaughter. Subsequent studies of hogs on farms, in sale barns, and in the abattoirs in the southern[63] and midwestern[64] parts of this country have shown a similar increase in infected animals as they move to the slaughter plant. Each infected animal is a possible source for the spread of salmonellae in the abattoir. In one abattoir, it was shown that by careful supervised cleansing of the dehairing machine, the number of salmonellae isolated from the sides of the carcass was reduced from 91 to 3%. After termination of the supervised cleansing procedure, the proportion of positives increased to 88%. It, therefore, appears that a build-up in the number of salmonellae occurs from the farm through the holding pens to the abattoir and processing plants. This increase must result in a greater proportion of contaminated products being offered for retail sale. It is very uncommon for this chain to start from an ill animal or animals, and the final degree of contamination depends upon a number of variables which have not yet been fully evaluated. These could well include the amount of infection upon the farm, the use of antibiotics in feed, the length and condition of transport and holding, and the method of slaughter and processing. It is probable that the place of greatest increase of contamination will vary in different areas and plants, and this will depend rather upon what is done to the animals than to a feed back of organisms from man to the product. Further studies in Florida revealed *Salmonella* contamination of fresh pork sausage ranged from 8% in samples from national producers to 58% in those from local abattoirs [65]. In addition, 12.5% of the smoked sausage samples were positive. Inadequately cooked pork products containing salmonellae frequently are cited in outbreaks of food infection. HAUSER et al.[66], in Louisiana, described an outbreak due to *S. berta* traced to consumption of contaminated sausage. It appears, therefore, that the major cause of the problem of salmonellae in retail meat products is the spread of infection among the animals just prior to slaughter and to the wide dissemination of salmonellae within the abattoir.

Until recent years, salmonellosis in cattle was not considered a serious problem in the United States. The most prevalent type was *S. typhi-murium* found primarily in cases of enteritis, usually in young animals. In 1952, 147 rectal swab cultures were taken from cattle immediately after slaughter in a Florida abattoir and salmonellae were isolated from 17 (12%) [62]. This finding was in distinct contrast to the 51% of isolations obtained from hogs after slaughter. Recent reports indicate that salmonellosis among diary and beef cattle is becoming a major problem. ELLIS[67] reported that during 1959 and 1960 in Florida, 40 *Salmonella* isolations were obtained from cattle with enteritis. Normal healthy cattle have been found to shed salmonellae for 5 months or longer. Salmonellae were recovered from beef livers which served as the basic food in a veterinary hospital where 70% of 40 dogs with diarrhea were positive for the same types. The livers were obtained locally and included those rejected for human consumption[68]. Transmission of infection to man from infected cattle has occurred primarily through consumption of raw or improperly pasteurized milk and milk products. TUCKER[69] reported an outbreak of *S. typhi-murium* involving over 300 persons in which the vehicle was cheese prepared from inadequately pasteurized milk.

In addition to our food producing animals, salmonella infections are prevalent in domestic pets that live in close association with man. WOLFF et al.[70], found 16 *Salmonella* serotypes in 18% of 100 dogs in Michigan. Rejected eggs were believed to be the source of infection in these animals. The authors suggested that dogs should be considered as a potential source of infection in man. MACKEL et al.[71], found that 15% of 1626 normal household dogs in Florida were harboring salmonellae as were 12% of 73 normal cats. Of the 66 *Salmonella* types found in man and dogs, 55 appeared in man and 53 in dogs[72]. In contrast, WATT & DeCAPITO[73] examined 1156 family dogs in Texas and recovered salmonellae from only 3.4%. Direct transmission of the infection from dogs to man, and vice versa, has been observed. However, it is reasonable to consider the possibility that man and dog could sometimes be infected from a common source.

For some years, rodents were thought to be one of the most important means of spread of salmonella infections. Surveys of rat populations reviewed by WELCH et al.[74], reported incidences of 0.7% to 13%. These authors found that salmonellae would remain viable for 148 days on rat feces. A few human infections have been reported due to contamination of food by rodents. The isolation of a variety of serotypes from rodents has dispelled the early belief that they carry only *S. typhi-murium* and *S. enteritidis*.

Salmonellae have been found, also, in snakes, lizards, tortoises [75, 76, 77] and insects such as flies and cockroaches[78]. HINSHAW & McNEIL[77] observed that since snakes and lizards frequent areas where poultry and other livestock are kept, they may form one link in the cycle for further perpetuation of *Salmonella* organisms. In cold blooded animals, it is not clear whether they are the source of infection or if they acquire the organisms through a contaminated environment. During a recent investigation of *S. hartford* infection in an infant, this type and *S. newport* were isolated from a pet turtle in the household and *S. newport* was recovered from the family dog. Whether the infant and the dog acquired their infection from the turtle or the turtle is a victim of a contaminated environment remains unknown. Although the house fly is another vector of salmonellae and has been found to be capable of depositing large numbers of these organisms on food by defecation and regurgitation, the observations of WATT & LINDSAY[79], in their fly control studies, indicate that they are of more importance in the transmission of shigellae than salmonellae.

Animal Feeds and Feed Ingredients

There is increasing evidence that widely distributed animal feeds are heavily contaminated with salmonellae and provide a means for the extensive spread of infection now apparent in domestic animals and fowls. The presence of salmonellae in prepared animal feeds was first reported by GRIFFIN[80] in New York State following unexplained explosive outbreaks of infection with *S. newport* in guinea pigs and mouse breeding colonies. Not only was *S. newport* isolated from dog feed cubes, but *S. muenster, S. minnesota, S. tennessee* and *S. senftenberg* were obtained also. About the same time, GALTON et al.[68] isolated 17 *Salmonella* types from 26.5% of dehydrated dog meal samples from 9 of 11 manufacturers. Just prior to this, salmonellae were obtained from 22% of the samples examined from hospitalized dogs and 44% of the specimens from kennel dogs. Many of the same types isolated from these dogs were recovered from the meals.

Further evidence that dogs may acquire infection through food was reported by CARAWAY et al.[81]. Salmonellae were isolated from 18 of 23 sentry dogs following varying degrees of diarrhea over a period of several months. Multiple types, including 18 different serotypes, were obtained from most of the animals. Later, salmonellae were isolated from samples taken in the plant that processed the frozen horse meat fed to these dogs. Five of the types isolated from the horse meat were identical to 5 types from the dogs[82]. During the survey in the processing plant, it was found that samples obtained from animals slaughtered in the first 3 days of the week contained salmonellae more frequently than those obtained from animals slaughtered on Thursday and Friday. The authors suggested that this might be due to spread of infection among horses kept for longer periods in the holding lots. This plant also produced boned cured horse meat for export for human consumption. Sanitary practices in the plant were observed to be strict and of high standards[83].

Recent evidence indicates that a high proportion of both domestic and imported bone meal, meat meal, fish meal, and similar protein supplements used in animal and poultry feeds are contaminated with salmonellae [84, 85, 86, 87]. Numerous serotypes have been isolated and, frequently, multiple types were obtained from the same sample. Many of these types are those found in poultry and poultry products, the meat producing animals and domestic pets, as well as those regularly obtained from man. BOYER et al.[84] reported the isolation of salmonellae from young turkeys in 2 outbreaks and young chicks in one outbreak in which the same serotypes were obtained from samples of unopened bags of feed from the same lots that had been fed these birds. POMEROY & GRADY[85] examined 980 samples of animal by-products used in feeds from 22 States and found 43 *Salmonella* serotypes in 175 (18%) samples. In Texas, WATKINS et al.[86] recovered 28 serotypes from 37 (18.5%) of 200 samples of poultry and animal by-products used in feeds.

During a study in rendering plants in Florida[63], 50% of the finished animal by-products contained salomonellae. Even after some of the suggested improvements were made in one plant, 50% of the samples of the finished product examined were contaminated. The Agricultural Research Service, U.S. Department of Agriculture, recently reported the results of a survey to gather information on the occurrence of salmonellae in animal by-products and feeds throughout the United States[88]. Questionnaires were sent to 1100 contacts, including laboratory and extension services of universities, public health and private industry laboratories, regulatory agencies, veterinary practitioner groups, rendering groups, animal by-product processors, feed manufacturing and processing groups, meat packers, fur bearing animal raisers, and biological and drug manufacturers. Although the percent of *Salmonella* isolations from animal by-products and feeds varied widely from zero to 52, the total reported from nearly 6000 samples was 12%. Egg products, poultry by-products, meat scraps, and bone meal for dog food, represented only 28% of the samples studied, but they accounted for 71% of the total *Salmonella* isolations reported.

Several recent reports have appeared concerning the presence and spread of salmonellae in such foods as soya milk[89] and dried yeast[90]. Imported coconut[91] has also been found to contain salmonellae, but human cases have not yet been traced to this source in the United States. Similarly, vegetable foods used in animal feeds have been found contaminated. GRUMBLES & FLOWERS[92] found 6 *Salmonella* serotypes in 5.14% of 135 samples of soy bean meal and cotton seed meal tested. An outbreak of acute salmonellosis in horses, including 7 fatal cases, was studied by ELLIS[67]. He isolated the same *Salmonella* serotype from the intestines and organs of several of the horses and from 1 of the bags of cotton seed meal used to feed the horses.

There has been much discussion regarding the significance of these contaminated animal feeds and their possible influence on the reported increase in prevalence of salmonellosis in our domestic animals. The evidence is mounting, however, to indicate that they may play an important role. Whether the salmonellae in these feeds produce clinical illness and mortality or a carrier state, an avenue of the transmission cycle from feed, to animals, to man, is established.

Human Carriers

In spite of the wide distribution of salmonellae in animals, animal feeds and human food products, the importance of the human carrier in the spread of salmonellosis cannot be overlooked. The role of the human carrier of S. typhi is well established. For this reason, this type is not included in the figures quoted. The definition of who is a carrier and who is a symptomless excreter in a communicable disease with a high infection rate and low attack rate is difficult. Some longitudinal studies have demonstrated continuous or intermittent excretion of one Salmonella type for long periods of time in specific individuals. However, it is believed that most people excrete salmonellae for a period as short as 2 to 3 weeks after infection. The proportions of persons with these different experiences is unknown and the usual cross sectional studies do not clarify the position. FELSENFELD & YOUNG[93] reported that human carriers were established as the cause of 26 of 56 outbreaks due to non-host-adapted salmonellae. Both infant and adult carriers of S. panama were found during and after an outbreak of 18 fatal cases in infants due to this type in Michigan[94]. In 1948, EDWARDS[8] reported that 30% of 3059 cultures were obtained from asymptomatic persons, most of whom gave no history of illness or contact with clinical cases. While MACCREADY et al.[14] reported that 21.9% of cultures isolated over a 16-year period in Massachusetts were from carriers, SAPHRA & WINTER[12] observed that 15.5% of their cultures came from healthy individuals. The incidence of different serotypes found in normal carriers varied considerably. For example, S. cholerae-suis had a low prevalence of 1.1%, while S. tennessee had a rate of 36%, and S. anatum and S. newington 25.4%. These authors estimated the carrier rate in the general population to be 0.2%. In Florida, 63% of the cultures of known origin were isolated from asymptomatic carriers and many of these were obtained from food handlers[34]. Although it is difficult to determine whether the food handler is the source of infection or has been infected from the same source as the victims, it appears that the carrier state occurs much more frequently in this group of individuals than in the general population. EDWARDS[36] suggests that the carrier state might be considered an occupational hazard, particularly among persons who handle uncooked meats and meat products. In some instances, these carrier states have cleared spontaneously, but a considerable number have been followed for several years with no change in their condition. Some authorities on salmonellosis recommend systematic and frequent stool cultures for the examination of food handlers and workers in food processing plants. In most States, however, ruling on this is the prerogative of local (city and county) health officials, and it is not required in many areas.

Prevention and Control

It has been shown that *S. pullorum* and *S. typhi-murium* can be eliminated from infected poultry flocks[95] by prompt application of testing procedures, use of disease-free replacements, and the elimination of human carriers in contact with the flocks. Now, it would be reasonable to suggest that these measures should also include the use of salmonella free feed. Much progress has been made in the reduction of *S. typhi* infections by improved sanitation, vaccination and follow-up of carriers, to prohibit them from working in food establishments. The rapid development of both human and animal food processing in this country, the widespread distribution of salmonellae in many of these foods, and the lack of bacteriological standards to assure freedom from salmonellae, has raised many problems.

The legal basis for the U.S. Food and Drug Administration[91] control of food products in interstate commerce which contain salmonellae, is set forth in Section 402(a) of the Federal Food, Drug, and Cosmetic Act. This Act defines a food as "adulterated if it bears or contains any poisonous or deleterious substance which may render it injurious to health". Salmonellae or other pathogenic organisms are not named specifically, but Section 404, Emergency Permit Control, provides: "Whenever the Secretary finds after investigation that the distribution in interstate commerce of any class of food may, by reason of contamination with microorganisms during the manufacture, processing, or packing thereof in any locality, be injurious to health, and that such injurious nature cannot be adequately determined after such articles have entered interstate commerce, he then, and in such case only, shall promulgate regulations providing for the issuance... of permits to which shall be attached such conditions governing the manufacture, processing, or packing of such class of food, for such temporary period of time, as may be necessary to protect the public health..." Animal feeds in interstate commerce are within the jurisdiction of the Act as food is defined as "articles used for food or drink for man or other animals...."

Funds and facilities have not permitted routine enforcement of this procedure. However, food samples collected during sanitary inspection of food establishments are now being tested by the Food and Drug Administration. In addition, because of reports of well-authenticated outbreaks in this country and other countries in which frozen and dried eggs have been implicated, the Food and Drug Administration is beginning a more active regulatory program to control traffic in frozen and dried products containing salmonellae.

Attempts towards prevention and control of salmonellosis should be directed at elimination of the animal cycle and the animal-vehicle-human chain of infection. In order to indicate methods of approach towards control and to determine the effect of control procedures, joint actions by the laboratory and the field workers play an important role not only in the detection of these vehicles but also in the definition of the magnitude of the problem. Consideration is being given to the development of microbiological standards for human foods. Improved procedures and practices are needed in the mass preservation and preparation of foods.

Since the salmonella infected animal provides the primary source of much of the contamination in human foods, immediate concerted attention is needed to develop methods for the production of salmonella-free animal feeds. According to MOREHOUSE & WEDMAN [88], industry groups involved in this problem are conducting an active research program to investigate problems associated with contamination of these products. They are willing to adopt necessary measures to prevent production of contaminated products as such methods are proved effective.

Training courses are being conducted with the cooperation of government agencies, to familiarize food processors with methods for the isolation of salmonellae so that they can establish a monitoring system in their plants.

Essential factors in the control of salmonellosis include close cooperation of government agencies concerned with human and animal health, and with foods, the development of a fast and widespread reporting system, as well as close collaboration between physicians, veterinarians and sanitarians who are conducting epidemiologic investigations, and the microbiologist examing the samples. In addition, training of those concerned with food microbiology, the epidemiology for control of salmonellosis, and education of food handlers and processors, should be a part of the programs of national and international health agencies.

REFERENCES

1. SALMON, D. E., & SMITH, T. 1885. Report on swine plague. U.S. Dept. of Agric., Bureau of Animal Ind. 2nd Annual Report.
2. WHITE, P. B. 1925. An investigation of the *Salmonella* group with special reference to food poisoning. Med. Res. Council. Spec. Rep. Ser. No. **91**.
3. WHITE, P. B. 1926. Further studies of the *Salmonella* group. Med. Res. Coun. Spec. Rep. Ser. No. **103**.
4. KAUFFMANN, F. 1930. Der Antigen-aufbau der Typhus-Paratyphus-Gruppe. Z. *Hyg.* **111**: *233—246*.
5. KAUFFMANN, F. 1930. Die Technik der Typenbestimmung in der Typhus-Paratyphus-Gruppe. *Zbt. Bakt. I.* Orig. **119**: *152—160*.
6. KAUFFMANN, F. 1941. Die Bakteriologie der Salmonella-gruppe. Einar Munksgaard, Kopenhagen.
7. EDWARDS, P. R., & BRUNER, D. W. 1943 The Occurrence and Distribution of *Salmonella* Types in the United States. *J. infect. Dis.* **72**: *58—67*.

440

8. EDWARDS, P. R., BRUNER, D. W. & MORAN, ALICE B. 1948. The genus *Salmonella*: Its occurrence and distribution in the United States. *Ky. Agric. Exp. Sta. Bull.* **525**.
9. EDWARDS, P. R., BRUNER, D. W. & MORAN, ALICE B. 1948. Further studies on the occurrence and distribution of *Salmonella* types in the United States. *J. infect. Dis.* **83**: *220—231*.
10. EDWARDS, P. R. 1962. Serologic examination of salmonella cultures for epidemiologic purposes. U. S. Dept. Health, Education, and Welfare, PHS, CDC, Atlanta, Ga., August., pp. *1—16*.
11. SELIGMANN, E., SAPHRA, I., & WASSERMANN, M. 1946. Salmonella infections in the U.S.A. A second series of 2000 human infections recorded by the N.Y. Salmonella Center. *J. Immun.* **54**: *69—87*.
12. SAPHRA, I., & WINTER, J. W. 1957. Clinical manifestations of salmonellosis in man. *New Engl. J. Med.* **256**: *1128—1134*.
13. DAUER, C. C. 1960. Summary of disease outbreaks and a 10-year resume. *Public Health Rep.* **76**: *915—922*.
14. MACCREADY, R. A., REARDON, J. P., & SAPHRA, I. 1957. Salmonellosis in Massachusetts. *New Engl. J. Med.* **256**: *1121—1128*.
15. FRIEDMANN, E. 1962. Surveillance of human salmonellosis. Amer. Public Health Ass. 90th Annual Meeting, Conf. of Public Health Veterinarians, Miami Beach, Fla., Oct. 17.
16. NEWELL, K. W. 1962. The value of international reporting of salmonella infections. Amer. Public Health Ass. 90th Annual Meeting, Conf. of Public Health Veterinarians, Miami Beach, Fla., Oct. 17.
17. SAPHRA, I. 1950. Fatalities in salmonella infections. *Amer. J. med. Sci.* **220**: *74—77*.
18. DRACHMAN, R. H., PETERSEN, N. J., BORING, J. R. & PAYNE, F. J. 1958. Widespread *Salmonella reading* infection of undetermined origin. *Public Health Rep.* **73**: *885—894*.
19. Morbidity and Mortality Weekly Report, National Office of Vital Statistics. 1958. Vol. 7, No. 40, Oct. 10.
20. SEELIGER, H. P. R. Cited by DRACHMAN et al., in reference 18.
21. HANDLOSER, M. 1956. Epidemiologische Beobachtungen bei einer Masseninfektion durch *Salmonella blockley*. *Arch. Hyg.* **140**: *569—580*.
22. Morbidity and Mortality Weekly Report, National Office of Vital Statistics. 1956. Vol. 5, No. 21, June 4.
23. FRIEDMAN, S., WASSERMANN, M. M., & SAPHRA, I. 1955. A new salmonella type: *Salmonella blockley*. *J. Bact.* **70**: *354—355*.
24. GALTON, MILDRED M., MACKEL, D. C., LEWIS, A. L., HAIRE, W. C., & HARDY, A. V. 1955. Salmonellosis in poultry and poultry processing plants in Florida. *Amer. J. vet. Res.* **16**: *132—137*.
25. California Epidemiological Notes, California State Department of Health. July 22, 1960.
26. CLOSE, A. S., SMITH, M. B., KOCH, MARIE L., & ELLISON, E. H. 1960. An analysis of ten cases of salmonella infection on a general surgical service. *AMA Arch. Surg.* **80**: *972—976*.
27. TAYLOR, JOAN. 1960. Salmonella in salmonellosis. *Roy. Soc. Health J.* **80**: *253—259*.
28. GALTON, MILDRED M., & HARDY, A. V. 1948. Studies of the acute diarrheal diseases. XXI. Salmonellosis in Florida. *Public Health Rep.* **63**: *847—851*.
29. GAYLER, G. E., MACCREADY, R. A., REARDON, J. P., & McKERNAN, B. F. 1955. An outbreak of salmonellosis traced to watermelon. *Public Health Rep.* **70**: *311—313*.
30. ROKEY, N. W., & ERLING, H. G. 1960. Natural Occurrences of *Salmonella dublin* in Arizona. *J. Amer. vet. med. Ass.* **136**: *381—388*.

31. Morbidity and Mortality Weekly Report, National Office of Vital Statistics. 1958. Vol. 7, No. 50, Dec. 19.
32. MORAN, ALICE B. 1961. Occurrence and distribution of salmonella in animals in the United States. Proc. 65th Annual Meeting U.S. Livestock Sanitary Ass., Minneapolis, Minn., October, pp. *441—448.*
33. Reports sent to the Communicable Disease Center from State and other United States typing centers. 1962—1963.
34. GALTON, MILDRED M., & HARDY, A. V. 1953. The distribution of salmonella infections in Florida during the past decade. *Public Health Lab.* **11**: *88—93.*
35. State of California, Dept. of Public Health, Lab. and Morbidity Reports, June 23, 1952; Dept. of Agriculture, Div. of Animal Industry, Lab. Reports. July 28, 1952.
36. EDWARDS, P. R. 1958. Salmonellosis: Observations on incidence and control. *Ann. NY Acad. Sci.* **70**: *598—613.*
37. HINSHAW, W. R., McNEIL, E., & TAYLOR, T. J. 1944. Avian Salmonellosis. Types of Salmonella isolated and their relation to public health. *Amer. J. Hyg.* **50**: *264—278.*
38. CHERRY, W. B., BARNES, L. A. & EDWARDS, P. R. 1946. Observations on a monphasic Salmonella variant. *J. Bact.* **51**: *235—243.*
39. WILSON, ELIZABETH, PAFFENBARGER, R. S., JR., FOTER, M. J. & LEWIS, K. H. 1961. Prevalence of salmonellae in meat and poultry products. *J. infect. Dis.* **109**: *166—171.*
40. SCHNEIDER, M. D., & GUNDERSON, M. F. 1949. Investigators shed more light on Salmonella problem. *U.S. Egg and Poultry Mag.* **55**: *10—11,* 22.
41. BROWNE, A. S. 1949. The public health significance of Salmonella on poultry and poultry products. Ph.D. Thesis, Univ. of Calif., Berkeley, Calif.
42. MACKEL, D. C., PAYNE, F. J., & PIRKLE, C. I. 1959. Outbreak of gastroenteritis caused by *Salmonella typhi-murium* acquired from turkeys. *Public Health Rep.* **74**: *746—748.*
43. WATT, J. 1945. An outbreak of Salmonella infection in man from infected chicken eggs. *Public Health Rep.* **60**: *835—839.*
44. RETTGER, L. F., & STONEBURN, F. H. 1909. Bacillary white diarrhea of young chicks. *Storrs (Conn.) Agric. Exp. Sta. Bull.* **60**: *29—57.*
45. RETTGER, L. F., HULL, T. G., & STURGES, W. S. 1916. Feeding experiments with *Bacterium pullorum.* The toxicity of infected eggs. *J. exp. Med.* **23**: *475—489.*
46. POMEROY, B. S., & FENSTERMACHER, R. 1941. Paratyphoid infection of turkeys. *Amer. J. vet. Res.* **2**: *285—291.*
47. HAINES, R. B., & MORAN, T. 1940. Porosity of and bacterial invasion through the shell of hen's eggs. *J. Hyg.* **40**: *453—461.*
48. PHILBROOK, F. R., MacCREADY, R. A., VAN ROEKEL, H., ANDERSON, E. S., SMYSER, C. F., SANEN, F. J., & GROTON, W. M. 1960. Salmonellosis spread by a dietary supplement of avian source. *New Engl. J. Med.* **263**: *713—718.*
49. SOLOWEY, MATILDA, McFARLAND, V. H., SPAULDING, E. H., & CHEMERDA, C. 1947. Microbiology of spray-dried whole egg. II. Incidence and types of Salmonella. *Amer. J. Public Health* **37**: *971—982.*
50. ABRAMSON, H., GREENBERG, H., PLOTKIN, S., & OLDENBUSCH, C. 1954. Food poisoning in infants caused by egg-yolk powder. *Amer. J. Dis. Child.* **87**: *1—6.*
51. McCULLOUGH, N. B., & EISELE, C. W. 1951. Experimental human salmonellosis. I. Pathogenicity of *S. meleagridis* and *S. anatum* obtained from spray-dried whole egg. *J. infect. Dis.* **88**: *278—289.*

52. McCullough, N. B., & Eisele, C. W. 1951. Experimental human salmonellosis. III. Pathogenicity of strains of *S. newport, S. derby* and *S. bareilly* obtained from spray-dried whole eggs. *J. infect. Dis.* **89**: *209—213.*

53. McCullough, N. B., & Eisele, C. W. 1951. Experimental human salmonellosis. IV. Pathogenicity of strains of *S. pullorum* obtained from spray-dried whole egg. *J. infect. Dis.* **89**: *259—265.*

54. Mitchell, R. B., Garlock, F. C., & Broh-Kahn, R. H. 1946. An outbreak of gastroenteritis presumably caused by *Salmonella pullorum*. *J. infect. Dis.* **79**: *57—62.*

55. Morbidity and Mortality Weekly Report, National Office of Vital Statistics. 1957. Vol. **6,** No. 41, Oct. 18.

56. Judefind, T. F. 1947. Report of a relatively severe and protracted diarrhea presumably due to *S. pullorum* from the ingestion of incompletely cooked eggs. *J. Bact.* **54**: *667—668.*

57. Morbidity and Mortality Weekly Report, National Office of Vital Statistics. 1956. Vol. **5,** No. 2, Jan. 20.

58. Morbidity and Mortality Weekly Report, National Office of Vital Statistics. 1957. Vol. **6,** No. 10, March 15.

59. Morbidity and Mortality Weekly Report, National Office of Vital Statistics. 1959. Vol. **7,** No. 53, Jan. 9.

60. Rubin, H. L., Scherago, M., & Weaver, R. H. 1942. The occurrence of Salmonella in the lymph glands of normal hogs. *Amer. J. Hyg.* **36**: *43—47.*

61. Cherry, W. B., Scherago, M., & Weaver, R. H. 1943. The occurrence of Salmonella in retail meat products. *Amer. J. Hyg.* **37**: *211—215.*

62. Galton, Mildred M., Smith, W. V., McElrath, H. B., & Hardy, A. V. 1954. Salmonella in swine and cattle and the environment of abattoirs. *J. infect. Dis.* **95**: *236—245.*

63. Shotts, E. B., Jr., Martin, W. T., & Galton, Mildred M. 1961. Further studies on salmonella in human and animal foods and in the environment of processing plants. Proc. 65th Annual Meeting U. S. Livestock Sanitary Ass., Minneapolis, Minn., October, pp. *309—317.*

64. Leistner, L., Johantges, J., Deibel, R. H., & Niven, C. F., Jr. 1962. The occurrence and significance of salmonellae in meat animals and animal by-product feeds. Reprinted for private circulation from Proc. 13th Research Conf., sponsored by Research Advisory Council of the American Meat Institute Foundation, University of Chicago.

65. Galton, Mildred M., Lowery, W. H., & Hardy, A. V. 1954. Salmonella in fresh and smoked pork sausage. *J. infect. Dis.* **95**: *232—235.*

66. Hauser, G. H., Treuting, W. L., & Breiffelh, L. A. 1945. An outbreak of food poisoning due to a new etiological agent — *Salmonella berta. Public Health Rep.* **60**: *1138—1142.*

67. Ellis, E. M. 1962. Salmonellosis in cattle, horses and feeds. Presented at the Midwest Interprofessional Seminar on Diseases Common to Man and Animals, Iowa State University, Ames, Iowa. Sept. 17.

68. Galton, Mildred M., Harless, M., & Hardy, A. V. 1955. Salmonella isolations from dehydrated dog meals. *J. Amer. vet. med. Ass.* **126**: *57—58.*

69. Tucker, C. B., Cameron, G. M., Henderson, M. P., & Beyer, M. R. 1946. *Salmonella typhi-murium* infection from Colby cheese. *J. Amer. med. Ass.* **131**: *1119—1120.*

70. Wolff, A. H., Henderson, N. D., & McCallum, G. L. 1948. Salmonella from dogs and the possible relationship to Salmonella in Man. *Amer. J. Public Health* **38**: *403—408.*

71. Mackel, D. C., Galton, M. M., Gray, H., & Hardy, A. V. 1952. Salmonellosis in dogs. IV. Prevalence in normal dogs and their contacts. *J. infect. Dis.* **91**: *15—18.*

72. GALTON,MILDRED M., SCATTERDAY, J. E., & HARDY, A. V. 1952. Salmonellosis in dogs. I. Bacteriological, epidemiological and clinical considerations. *J. infect. Dis.* **91**: *1—5*.
73. WATT, J., & DeCAPITO, THELMA. 1950. Frequency and distribution of salmonella types isolated from man and animals in Hidalgo County, Texas. *Amer. J. Hyg.* **51**: *343—352*.
74. WELCH, H., OSTROLENK, M., & BARTRAM, M. T. 1941. Role of rats in the spread of food poisoning bacteria of the *Salmonella* group. *Amer. J. Public Health* **31**: *332—340*.
75. McNEIL, E., & HINSHAW, W. R. 1944. Snakes, cats and flies as carriers of *Salmonella typhi-murium. Poultry Sci.* **23**: *456—457*.
76. McNEIL, ETHEL, & HINSHAW, W. R. 1946. Salmonella from galapagos turtles, a gila monster and an iguana. *Amer. J. vet. Res.* **7**: *62—63*.
77. HINSHAW, W. R., & McNEIL, ETHEL. 1948. Avian Salmonellosis: Its economic and public health significance. Report of the 8th World Poultry Congress.
78. BITTER, RUTH S., & WILLIAMS, O. B. 1949. Enteric organisms from the American cockroach. *J. infect. Dis.* **85**: *87—90*.
79. WATT, J., & LINDSAY, D. R. 1948. Diarrheal disease control studies. Effect of fly control in a high morbidity area. *Public Health Rep.* **63**: *1319—1334*.
80. GRIFFIN, C. A. 1952. A study of prepared feeds in relation to salmonella infection in laboratory animals. *J. Amer. vet. med. Ass.* **124**: *120—121*.
81. CARAWAY, C. T., SCOTT, A. E., ROBERTS, N. C., & HAUSER, G. H. 1959. Salmonellosis in sentry dogs. *J. Amer. vet. med. Ass.* **135**: *599—602*.
82. SINGER, S., & BRANDLY, P. J. 1960. Salmonellae in horse meat. *Appl. Microbiol.* **8**: *190—192*.
83. BRANDLY, P. J., U. S. Department of Agriculture, Agricultural Research Service, Meat Inspection Division, Washington 25, D.C. 1963. Personal communication.
84. BOYER, C. I., JR., NAROTSKY, S., BRUNER, D. W., & BROWN, J. A. 1962. Salmonellosis in turkeys and chickens associated with contaminated feed. *Avian Dis.* **6**: *43—50*.
85. POMEROY, B. S., & GRADY, MARGARET K. 1961. Salmonella organisms in feed ingredients. Proc. 65th Annual Meeting U. S. Livestock Sanitary Ass., Minneapolis, Minn., October, pp. *449—452*.
86. WATKINS, J. R., FLOWERS, A. I., & GRUMBLES, L. C. 1959. Salmonella organisms in animal products used in poultry feeds. *Avian Dis.* **3**: *290—301*.
87. BORING, J. R. 1958. Domestic fish meal as a source of various salmonella types. *Vet. Med.* **53**: *311*.
88. MOREHOUSE, L. G., & WEDMAN, E. E. 1961. Salmonella and other disease-producing organisms in animal by-products — A survey. *J. Amer. vet. med. Ass.* **139**: *989—995*.
89. BROWNE, A. S. Cited by P. R. EDWARDS in reference 36.
90. KUNZ, L. J., & OUCHTERLONY, O. T. G. 1955. Salmonellosis originating in a hospital. A newly recognized source of infection. *New Engl. J. Med.* **253**: *761—763*.
91. SLOCUM, G. G. 1962. Control of foodborne salmonellae under the Food, Drug, and Cosmetic Act. Presented at the Amer. Public Health Ass., 90th Annual Meeting, Conf. of Public Health Veterinarians, Miami Beach, Fla., Oct. 17.
92. GRUMBLES, L. C., & FLOWERS, A. I. 1961. Epidemiology of Paratyphoid infections in turkeys-Species encountered and possible sources of infection. *J. Amer. vet. med. Ass.* **138**: *261—262*.

93. FELSENFELD, O., & YOUNG, V. M. 1949. A study of salmonellosis in North and South America. *Amer. J. trop. Med.* **29**: *483—491.*
94. LEEDER, F. S. 1956. An epidemic of *Salmonella panama* infections in infants. *Ann. NY Acad. Sci.* **66**: *54—60.*
95. HINSHAW, W. R., & MCNEIL, ETHEL. 1951. Salmonella infection as a food industry problem. *Advances in Food Research* **3**: *209—240.*

EPIDEMIOLOGIA DE LA SALMONELOSIS EN MEXICO

POR

JORGE OLARTE & GERARDO VARELA

Hospital Infantil de México & Instituto de Salubridad y Enfermedades Tropicales, México, D.F.

(con 5 gráf.)

Importancia del problema

En el hombre

Mortalidad por Gastroenteritis y Colitis

De acuerdo con la información estadística con que contamos, la mortalidad por padecimientos diarreicos en la República Mexicana ha descendido paulatinamente en el curso de los últimos treinta años. Sin embargo, como puede verse en las gráficas 1, 2 y 3, las tasas de mortalidad por "gastroenteritis y colitis" son todavía muy elevadas, en particular en los grupos de edad que comprenden niños menores de un año, y niños de 1 a 4 años. Si tomamos en cuenta únicamente los datos relativos a los últimos diez años, nos encontramos con que durante ellos ha muerto por "gastroenteritis y colitis" un promedio anual de 69.012 personas, de las que 53.362 corresponden a niños menores de 5 años de edad. De estos últimos, 26.754 son niños de 1 a 4 años, y 26.608 menores de un año.

Estudios bacteriológicos realizados durante un largo período (1940 a 1962), los que serán analizados más adelante, indican que alrededor del diez por ciento de las enteritis graves que se presentan

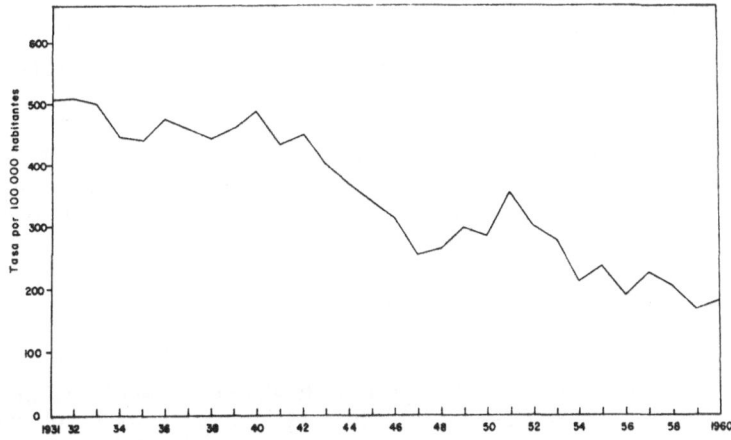

Gráfica 1. Mortalidad cruda (en todas las edades) por Gastroenteritis y Colitis en la República Mexicana 1931 a 1960.

446

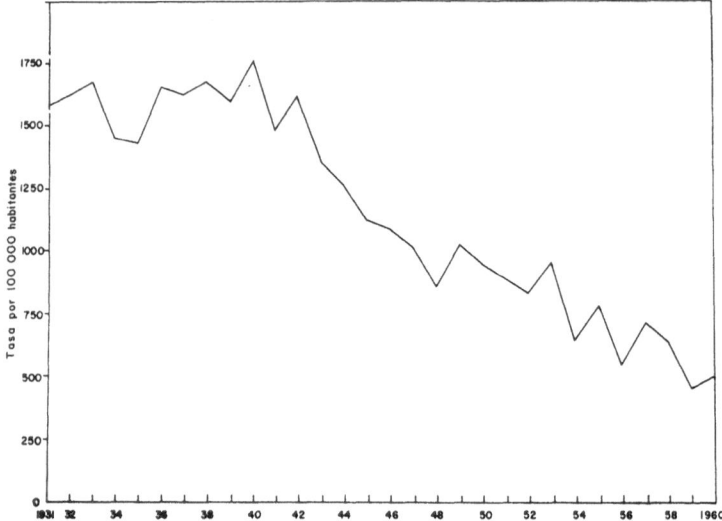

Gráfica 2. Mortalidad por Gastroenteritis y Colitis en niños de 1 a 4 años en la República Mexicana 1931 a 1960.

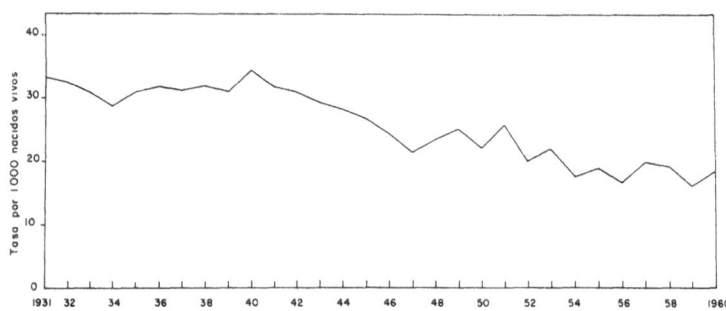

Gráfica 3. Mortalidad por Gastroenteritis y Colitis en niños menores de 1 año en la Republica Mexicana 1931 a 1960.

en el hombre en este país son causadas por *Salmonella*. Con los datos anteriores podemos estimar que la salmonelosis, sin incluir la fiebre tifoidea, ha originado en México, en el transcurso de los últimos diez años, un promedio anual de 6.901 muertes. De estas, 5.336 corresponden al grupo de niños menores de 5 años, el que comprende 2.675 niños de 1 a 4 años y de 2.661 menores de un año.

Aunque no disponemos de información adecuada que nos permita calcular las cifras de morbilidad que la salmonelosis alcanza en el país, los datos sobre mortalidad que se acaban de mencionar ponen de relieve, por sí solos, la magnitud del problema, colocándolo entre los más importantes que confronta la salubridad pública de México.

Mortalidad por Fiebre Tifoidea.

Las tasas de mortalidad por fiebre tifoidea (gráfica 4) han disminuido en los últimos once años. Sin embargo, si consideramos que esta enfermedad ha desaparecido casi por completo en la mayoría de los paises avanzados, llegamos a la conclusión de que todavía nos encontramos muy lejos de la victoria en la lucha contra este padecimiento.

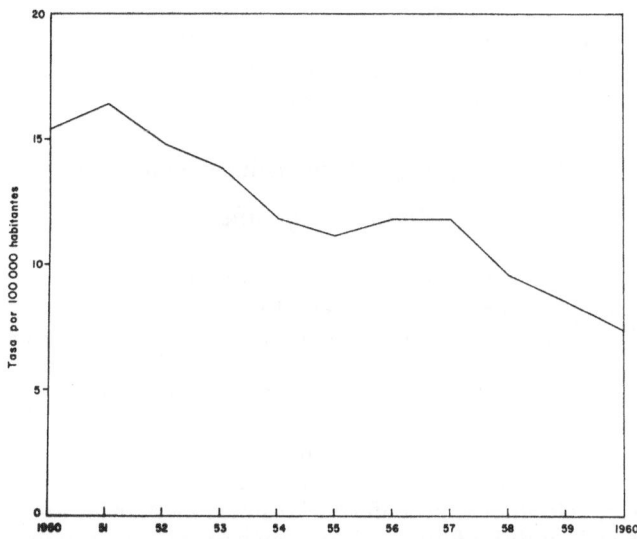

Gráfica 4. Mortalidad por Fiebre Tifoidea en la República Mexicana 1950 a 1960.

Las estadísticas del Hospital Infantil de México indican que la disminución en la morbilidad por fiebre tifoidea ha sido especialmente notable en la Ciudad de México, en los últimos ocho años, lo que puede atribuirse, sin duda alguna, a los grandes adelantos logrados en la Ciudad en este período, tanto por lo que se refiere al mejor abastecimento de agua potable, como a la construcción de una extensa red de mercados en los que se ha puesto especial atención al manejo higiénico de los alimentos.

En diferentes especies animales

Hasta donde tenemos conocimiento, no se dispone de estadísticas que indiquen la mortalidad por salmonelosis que se observa en México en los animales. No obstante, la información obtenida a través de diversos veterinarios, así como la investigación bacteriológica que hemos realizado en varias especies animales, nos permiten afirmar que el problema de la salmonelosis en los animales domésticos es en este país tan grande, o quizás aun mayor, del que se presenta en el hombre mismo. Es este, probablemente, uno de los factores económicos más importantes que dificultan el desarrollo de mejores condiciones alimenticias. Además, los animales portadores de *Salmonella* constituyen una fuente permanente de infección para el hombre.

Como se verá más adelante, exploraciones llevadas a cabo en 8 especies distintas de animales (cerdo, pollo, bovinos, caballo, gato, perro, rata, ratón), señalan que todas ellas, en mayor o menor proporción, se encuentran infectadas con los más diversos tipos de *Salmonella*. Aparentemente, los animales que sufren los mayores perjuicios desde el punto de vista de la economía, son los pollos y los cerdos, especies en las que la mortalidad de los animales jóvenes por enteritis es particularmente elevada.

Las infectiones por Salmonella en el hombre

Diarreas Infantiles

Frecuencia de la infección en niños con diarrea

Como puede observarse en el cuadro I, la frecuencia con que se encuentra *Salmonella* en niños con diarrea varía en distintos grupos estudiados. Estas diferencias pueden ser debidas a cambios en las condiciones sanitarias ocurridos en distintos años, o bien, que hayan afectado a núcleos distintos de la población estudiada. Sin embargo, hay ciertos factores que seguramente intervienen en forma directa, de los cuales mencionaremos tres. En primer lugar, el criterio médico con que se seleccionen los enfermos; es más fácil aislar los gérmenes patógenos de individuos con diarrea en su fase aguda, que de enfermos crónicos o tratados con antibióticos. En segundo lugar, el cuidado que se ponga en la toma y manejo de las muestras para cultivo; por ejemplo, en el caso particular de las salmonelas, da mejores resultados el cultivo de las evacuaciones que el de hisopos rectales; además, las muestras frescas son siempre preferibles. En tercer lugar, la técnica bacteriológica; por ejemplo, en el grupo 4, cuadro I, se obtuvo una proporción menor de aislamientos de *Salmonella* (3.5%) que en los otros grupos estudiados, lo que probablemente se debió a que, en este caso, las muestras se sembraron únicamente en placas de S S y de eosina-azul de metileno, habiéndose omitido el uso del medio de enriquecimiento de Muller Kauffmann.

Además de los datos señalados en el cuadro I, contamos con alguna información obtenida en otros Estados. VARELA & ROCH[31] cultivaron, en la Ciudad de Morelia, Estado de Michoacán, 800 muestras de materiales fecales provenientes de enfermos con diarrea, habiendo aislado *S. muenchen* 36 veces, *S. oregon* 10 veces, *S. newport* 10 veces, y *S. abony* 10 veces. VARELA & AKLE[19] encontraron 50 (11%) cepas de *Salmonella* en un grupo de 454 niños atendidos en diferentes instituciones en Ciudad Juárez, Estado de Chihuahua;

Cuadro I.

Frecuencia del Aislamiento de Salmonella en diversos grupos de niños, estudiados en diferentes años en la Ciudad de México.

Período de estudio	Grupo* y referencia	No. de niños estudiados	No. de niños con Salmonella	% de niños con Salmonella	Observaciones
Mayo de 1942 a Abril de 1943	1 (5)	1.191	89	8.1%	Casos esporádicos de diarrea**
Julio de 1950 a Julio de 1954	2 (12)	13.545	557	4.1%	Diarrea y otros padecimientos en niños hospitalizados, así como casos esporádicos
Sept. de 1953 Nov. de 1958	3 (9)	686	67	9.8%	Niños con diarrea, hospitalizados
Agos. de 1955 a Abril de 1956	4 (14)	802***	28	3.5%	Casos esporádicos de diarrea**
Marzo a Nov. de 1959	5 (9)	733	53	7.2%	Casos esporádios de diarrea**
Junio a Nov. de 1960	6 (17)	246	32	13.0%	Casos esporádicos de diarrea**

* El grupo 1 de niños fué estudiado en el Instituto de Salubridad y Enfermedades Tropicales. Los grupos 2, 3, 4, 5 y 6 fueron estudiados en el Hospital Infantil de México.

** Se designan como "casos esporádicos" aquellos que adquirieron el padecimiento en su casa. Las muestras para cultivo fueron tomadas antes de la hospitalización, cuando esto último ocurrió.

*** En todos los grupos estudiados se utilizó el medio de enriquecimiento de Kauffmann (tetrationato), con excepción de este grupo de 802 niños, lo que probablemente explique la baja proporción de salmonelas encontrada en el mismo.

predominó *S. oranienburg* (19 casos) y *S. typhi-murium* (5 casos) (cuadro II).

Frecuencia de la infección según la edad de los niños

El análisis de la frecuencia con que se encuentra *Salmonella* en niños con diarrea, en diferentes grupos de edad (cuadro III), indica que la infección es rara o presenta índices bajos en el recién nacido (0 a 5.4% en diversos grupos), aumenta durante los primeros doce meses de vida (2.2 a 12.2%), alcanzando los mayores índices en el segundo año (5.1 a 13.3%). En niños mayores de dos años la frecuencia con que se aisla *Salmonella* baja un poco (8.8% en niños de 2 a 10 años), aunque los datos de que se dispone en este grupo de edad son escasos.

Cuadro II.

Salmonelas aisladas de un grupo de 454 niños con diarrea estudiados en ciudad Juarez, estado de Chihuahua.
(VARELA, G. & AKLE, J.) (19).

Salmonella	No. de casos
S. oranienburg	19
S. typhi-murium	5
S. newport	2
S. derby	2
S. mission	2
S. menston	1
S. berta	1
S. montevideo	1
S. bonariensis	1
Otros tipos no identificados	26
Total	50 (11.0%)

Infecciones asociadas

En cierto número de niños con diarrea, en los que se encuentra *Salmonella*, es posible aislar del intestino, o bien, de otros órganos, simultáneamente más de un tipo serológico de estos gérmenes (cuadro IV). Entre más fina sea la técnica bacteriológica, y mayor el número de colonias que se aislen y se prueben con los sueros, mayores son las probabilidades de detectar estas infecciones múltiples.

Como es de esperarse, en ocasiones la infección por *Salmonella* va asociada a otras bacterias enteropatógenas, *Shigella* y *Escherichia coli*. Las combinaciones encontradas con más frecuencia en niños estudiados en la Ciudad de México pueden verse en el Cuadro IV.

En estudios recientes, RAMOS-ALVAREZ & OLARTE[17] han observado la asociación relativamente frecuente (en 12 de 26 casos de salmonelosis) de la infección por *Salmonella* con diversos virus enteropatógenos (ECHO, Adenovirus, polio, Coxsackie, etc.,) sin que se sepa por el momento la importancia que esta asociación pueda tener en la patología de la enfermedad.

Tipos serológicos más frecuentes

Como se observa en el cuadro V, en niños con diarrea estudiados en la Ciudad de México, los tipos de *Salmonella* que se encuentran con mayor frecuencia son los siguientes: *S. typhi-murium, S. derby, S. newport, S. anatum, S. oranienburg, S. infantis, S. paratyphi B, S. panama, S. newington, S. montevideo, S. cholerae-suis, S. enteritidis* y *S. poona*.

Cuadro III.

Frecuencia del Hallazgo de Salmonella en grupos de niños de diferentes edades. Enfermos con diarrea.

Grupos de Edad	Mayo de 1942 a Abril de 1943 (5)			Sept. de 1953 a Nob. de 1958 (9)			Agosto de 1955 a Abril de 1956 (4)			Marzo a Nov. de 1959 (9)		
	Casos estudiados	Positivos	%	Casos estudiados	Positivos	%	Casos estudiados	Positivos	%	Casos estudiados	Positivos	%
Menores de 1 mes				55	3	5.4	107	1	0.9	12	0	—
2 a 6 meses	373	20	5.3	242	17	7.0	277	6	2.2	215	9	4.2
7 a 12 meses				239	29	12.2	282	14	4.9	310	21	6.8
13 a 24 meses	271	36	13.3	150	18	12.0	136	7	5.1	196	23	11.7
2 a 10 años	247	22	8.8									

Fecha del estudio, referencias, no. de casos positivos y por ciento

Cuadro IV.

Frecuencia de la Asociación de Salmonella con otras bacterias enteropatógenas, encontrada en niños con diarrea en la Ciudad de México.

Asociación	Fecha del estudio, no. de niños en cada grupo y no. de casos positivos			
	(9)	(14)	(9)	(15)
	Sept. de 1953 a Nov. de 1958	Agosto de 1955 a Abril de 1956	Marzo a Nov. de 1959	Marzo de 1961 Marzo a de 1962
	En 686 niños hospitalizados	En 802 niños (casos esporádicos)	En 733 niños (casos esporádicos)	En 51 autopsias
Un solo tipo de Salmonella	51	16	42	3
Más de un tipo de Salmonella	4*	—	1**	2***
Salmonella más Shigella	4	7	6	—
Salmonella más E. coli	5	4	3	—
Salmonella Shigella y E. coli	3	1	1	—

* *S. typhi-murium, S. anatum* y *S. infantis* en un caso; *S. derby* y *S. poona* en un caso; *S. typhi-murium,* y *S. oranienburg* en un caso; *S. anatum* y *S. panama* en un caso.

** *S. typhi-murium* y *S. newport.*

*** Un caso con *S. typhi-murium, S. give* y *S. poona.* Otro caso con *S. paratyphi B, S. newport* y *S. anatum.*

Nota. Además de la asociación de Salmonella con otras bacterias patógenas, en estudios recientes se ha observado la asociación frecuente de Salmonella con virus enteropatógenos (incluyendo ECHO, Adenovirus, polio, Coxsackie y otros agentes citopatogénicos todavía no identificados) (RAMOS-ALVAREZ, M. & OLARTE, J.) (17).

Gastroenteritis en adultos

En contraste con la información que se tiene del papel que la salmonelosis juega en el niño, son más bien escasas las exploraciones realizadas en México en personas mayores. VARELA & OLARTE[24] estudiaron dos pequeños brotes de intoxicación alimenticia por *Salmonella*; en uno, que comprendió tres miembros de una familia, aislaron *S. typhi-murium*, siendo el origen de la infección una porción de queso fresco infectado con esta *Salmonella*; el otro accidente afectó a un matrimonio que ingirió salchichas enlatadas contaminadas con *S. eastbourne.*

VARELA, LAGUNA & ZOZAYA[21] aislaron *Salmonella* de un grupo de 40 personas adultas con enteritis (cuadro VI).

Cuadro V.

Tipos serológicos de Salmonella encontrados con más frecuencia en niños con diarrea en la Ciudad de México, en diferentes años.

Salmonella	Referencia, fecha del estudio y no. de cepas aisladas						Totales
	(5) Mayo de 1942 a Abril de 1943	(12) Julio de 1950 a Julio de 1954	(9) Sept. de 1953 a Nov. de 1958	(14) Agost de 1955 a Nov. de 1959	(9) Mayo a Nov. de 1959	(17) Junio a Nov. de 1960	
Grupo B:							
S. typhi-murium	18	121	23	11	19	12	204
S. derby	6	67	5	0	3	4	85
S. paratyphi B	4	22	1	0	2	1	30
Otros tipos	1	45	2	0	1	0	49
Grupo C:							
S. newport	19	53	1	1	6	4	84
S. oranienburg	0	29	7	3	4	0	43
S. infantis	0	25	3	0	0	3	31
S. montevideo	4	15	1	0	1	1	22
S. cholerae-suis	0	18	2	0	0	0	20
Otros tipos	8	21	4	0	3	1	37
Grupo D:							
S. panama	0	25	2	0	1	1	29
S. enteritides	0	14	2	0	0	1	17
Otros tipos	1	13	3	4	4	0	25
Grupo E:							
S. anatum	9	38	2	1	6	0	56
S. newington	7	15	2	0	0	1	25
Otros tipos	3	10	3	0	0	1	17
Group G:							
S. poona	6	11	0	0	0	0	17
Otros grupos	3	15	4	8	3	2	35
Totales	89	557	67	28	53	32	826

Cuadro VI.

Salmonelas aisladas de 40 personas adultas con diarrea, estudiadas en la Ciudad de México[21]. (Varela, G., Laguna, J. & Zozaya, J.)

Salmonella	No. de casos
S. anatum	5
S. typhi-murium	4
S. derby	4
S. newport	4
S. newington	3
S. thompson	2
S. anatum y	
S. manhattan	1
Otros tipos	17
Total	40

Fiebre tifoidea

La frecuencia con que se aisla *S. typhi* (10.5%) de la sangre de niños con padecimientos febriles, es bastante elevada en la Ciudad de México. (Cuadro VII).

Durante el período de 1950 a 1954, Olarte & Joachín[12] practicaron coprocultivo a un grupo de 13.545 niños con diversos padecimientos, atendidosa en el Hospital Infantil de México, habiendo aislado *S. typhi* en 215 (1.6%) casos. La proporción de niños en que se encontró esta *Salmonella* en el intestino disminuyó progresivamente en el curso de los años estudiados, de 2.6% en 1950 a 0.8% en 1954 (Cuadro VIII).

Cuadro VII.

Salmonelas aisladas de la sangre de 478 niños atendidos en el Hospital Infantil de México durante los años de 1950 a 1954 (Olarte, J. & Joachin, A.)[13], en un grupo de 4.000 niños con padecimientos febriles a los que se practicó hemocultivo.

Salmonella	No. de casos
S. typhi	421
S. paratyphi A	16
S. typhi-murium	6
S. sendai	3
S. paratyphi B	2
S. paratyphi C	2
S. panama	1
S. derby	1
Otros tipos distintos a	
S. typhi no identificados	26

Cuadro VIII.

Distribución por años de 215 casos de infección por S. typhi y de 557 casos de infección por otrasSalmonellas, encontrados en 13.545 niños atendidos en el Hospital Infantil de México, de Julio de 1950 a Julio de 1954.
(Olarte, J., & Joachin, A.)[12].

Años	No. de niños estudiados	S. Typhi		Otras Salmonellas	
		Positivos	%	Positivos	%
1950	856	22	2.6	50	5.9
1951	3.326	81	2.4	120	3.6
1952	3.944	51	1.3	255	6.4
1953	3.724	48	1.2	75	1.9
1954	1.695	13	0.8	57	4.0
Totales	13.545	215	1.6	557	4.1

En algunas poblaciones del interior, en donde se ha estudiado el problema bacteriológico de la fiebre tifoidea, se ha encontrado que es esta una enfermedad bastante común. Así, por ejemplo, Varela & Akle-Delgadillo[18], tomaron hemocultivo a un grupo de 50 sujetos febriles, y coprocultivo a un grupo de 300 personas con enteritis, atendidos en el Hospital Providencia de Ciudad Camargo, Estado de Chihuahua, México. De la sangre de estos 50 enfermos cultivaron S. typhi en 17 (34%) casos; de las materias fecales de las 300 personas estudiadas aislaron S. typhi en 31 (10.3%) casos.

Benavides, Franco & Gómez-Pagola[1] señalan que en el Hospital Infantil de México fueron atendidos 2.234 niños con fiebre tifoidea, durante los años de 1943 a 1958. El 70% de estos niños estaban en edad escolar, el 26% en edad preescolar, correspondiendo el resto a niños lactantes.

En relación con los tipos bacteriofágicos que prevalecen en México, P. Nicolle (*Ann. Inst. Pasteur*, 102: 402, 1962) estudió 186 cepas de S. typhi recolectadas por Varela et al.[23] en la Ciudad de México y por el Dr. F. Ruiz Sánchez en Guadalajara, habiendo encontrado la siguiente distribución: lisotipo A (38.2%), lisotipo E1 (26.3%), lisotipo 35 (14.5%), lisotipo 26 (6.5%), lisotipo C4 (1.6%), lisotipo 40 (1.1%), grupo I + IV y 38 (0.5%), cepas no caracterizables (10.8%). Olarte[6] recogió en el Hospital Infantil de México 8 cepas de S. typhi, aisladas durante un brote del padecimiento, habiendo pertenecido todas ellas al lisotipo 26 (Dr. W. W. Ferguson, Lansing).

Localizaciones extraintestinales

Con cierta frecuencia es posible observar, principalmente en individuos con diarrea, que las salmonelas invaden el torrente circulatorio, pudiendo establecerse, en determinadas circunstancias, prácticamente en cualquier órgano fuera del aparato digestivo. Aparentemente, estos gérmenes atraviesan la pared del intestino siguiendo la vía linfática, para establecerse luego en los ganglios linfáticos mesentéricos. Allí pueden permanecer por largos períodos, o quizás, indefinidamente en algunos casos. De los ganglios linfáticos pasan, probablemente, a la circulación sanguínea, dando origen a una bacteremia más o menos transistoria, sin que, en la mayoría de las personas, se observen fuera del intestino accidentes patológicos importantes. Sin embargo, en ciertos individuos, particularmente niños o personas con mala condiciones generales de salud, la salmonela que ha pasado a la circulación, puede invadir el parénquima de muy diversos órganos, dando origen a procesos inflamatorios localizados, algunos de ellos de tanta gravedad como meningitis, pericarditis, abscesos pulmonares, pielonefritis, etc.

A continuación se señalan las localizaciones extraintestinales de las salmonelas que con más frecuencia han sido encontradas en México.

Ganglios linfáticos mesentéricos

En un estudio de 171 individuos que fallecieron de diferentes padecimientos, en el Hospital General de México, VARELA & OLARTE[26] encontraron *Salmonella* en los ganglios linfáticos mesentéricos de 27 (15.8%) de ellos. (cuadro IX).

Cuadro IX.

Salmonelas encontradas en los ganglios linfáticos mesentéricos de 27 personas adultas (VARELA, G. & OLARTE, J.)[26].

Salmonella	No. de casos
S. oranienburg	6
S. typhi-murium	5
S. newport	4
S. muenchen	4
S. montevideo	3
S. reading	1
S. essen	1
S. chester	1
S. cholerae-suis	1
S. carrau	1

Bacteremia.

VARELA, LAGUNA & ZOZAYA[21] tomaron hemocultivo a 58 enfermos con salmonelosis intestinal, habiendo recuperado de la sangre de 19 (32.8%) personas la misma especie de *Salmonella* encontrada en el intestino. Los autores no indican los tipos serológicos aislados.

OLARTE & JOACHÍN[13], en una serie de 4.000 hemocultivos practicados a niños atendidos en el Hospital Infantil de México, durante los años de 1950 a 1954, todos ellos con procesos febriles de naturaleza aparentemente infecciosa, aislaron *Salmonella* en 478 (11.9%) casos. De estos, 421 (10.5%) correspondieron a niños con fiebre tifoidea, y 16 a paratifoidea A; los 41 niños restantes presentaron salmonelosis causada por otros tipos serológicos (cuadro VII).

OLARTE[6] estudió 80 niños con diarrea en los cuales se encontró *Salmonella* en las evacuaciones. A cada uno de ellos se tomó hemocultivo, habiendo aislado de la sangre de 12 (15%) niños el mismo tipo serológico de *Salmonella* encontrado en el intestino (cuadro X).

Cuadro X.

Resultado del hemocultivo tomado a 80 niños con Salmonelosis intestinal confirmada por coprocultivo (OLARTE, J.)[6].

No. de casos estudiados	No. de casos en los que se aisló de la sangre la misma Salmonella encontrada en el intestino		
80	12 (15%)	S. typhi-murium	6 casos
		S. newport	3 casos
		S. cholerae-suis	2 casos
		S. anatum	1 caso
		Total	12 casos

Meningitis

La meningitis es probablemente la complicación más seria que se puede presentar en la salmonelosis del niño. VARELA & OLARTE[28] señalan 2 casos en los que se aisló *S. typhi-murium* del líquido cefaloraquídeo. OLARTE[7], en un estudio que comprendió 213 niños con meningitis aguda purulenta, atendidos en el Hospital Infantil de

Cuadro XI.

Salmonelas aisladas del líquido cefalorraquídeo de 18 niños con meningitis aguda purulenta.

Salmonella	Autor y número de casos			No. total de casos
	VARELA, G. & OLARTE, J. [28]	BENAVIDES, L. et al.[1]	OLARTE, J. [7]	
S. typhi	—	8	1	9
S. typhi-murium	2	—	4	6
S. bredeney	—	—	2	2
S. saint-paul	—	—	1	1
Totales	2	8	8	18

México, encontró que las salmonelas ocupan el tercer lugar (3.8% del total de casos) en la etiología de este padecimiento, siguiendo en frecuencia a *Diplococcus pneumoniae* (26.2%) y a *Hemophilus influenzae* (14.0%). BANAVIDES, FRANCO & GÓMEZ PAGOLA[1], revelan el hallazgo de 8 casos más de meningitis producidas por *S. typhi*, en el mismo Hospital, en el período comprendido entre 1943 y 1958, en un grupo de 2.234 niños con fiebre tifoidea (Cuadro XI).

Amígdalas

VARELA & OLARTE[25] estudiaron las amígdalas extirpadas a 185 niños en la Ciudad de México. Encontraron *Salmonella* en 15 casos (8%), habiendo aislado 8 tipos diferentes (Cuadro XII).

Cuadro XII.

Salmonelas encontradas en las amígdalas extirpadas a 15 niños.
(VARELA, G. & OLARTE, J.)[25].

Salmonella	No. de casos
S. derby	3
S. panama	3
S. typhi-murium	2
S. muenchen	2
S. carrau	2
S. chester	1
S. newport	1
S. oregon	1

Osteomielitis, abscesos pleurales, infección renal

En el Hospital Infantil de México, OLARTE[6] encontró tres casos de osteomielitis por *Salmonella*, dos causados por *S. typhi* y uno por *S. paratyphi B*, así como dos niños con abscesos-pleurales de los

Cuadro XIII.

Hallazgo de Salmonella en 7 niños con osteomielitis, abscesos pleurales e infección renal[6, 11, 2].

Salmonella	Localización de la infección y número de casos		
	Osteomielitis	Abscesos pleurales	Riñón
S. typhi	2	—	—
S. paratyphi B	1	—	—
S. typhi-murium	—	1	1
S. bredeney	—	1	—
S. cholerae-suis	—	—	1
Total	3	2	2

cuales, en uno se aisló S. *typhi-murium* y en el otro S. *bredeney*. DE LA TORRE, VILLALPANDO, ESPARZA & OLARTE[2] aislaron S. *choleraesuis* del tejido renal de un niño que desarrolló trombosis del riñón izquierdo, en el curso de una enteritis causada por la misma *Salmonella*. OLARTE & GORDILLO[11] encontraron S. *typhi-murium* en la orina y en las heces de un niño con diarrea complicada con pielonefritis. (Cuadro XIII).

Placentas y vesícula biliar

VARELA & MARTÍNEZ[22], en un estudio de 81 placentas humanas, aislaron *Salmonella* en dos casos, en uno S. *anatum* y en el otro S. *oranienburg*.

En un lote de 100 vesículas biliares humanas, recogidas al practicar autopsias de personas muertas por diversas causas en el Hospital General de México, VARELA & SILICEO[32] encontraron S. *typhimurium* en un caso.

Portadores.

A pesar de que han sido realizadas en México pocas pesquisas en las que se investigue la presencia de *Salmonella* en individuos sanos, tenemos la impresión de que su existencia es en realidad limitada. Si se interroga cuidadosamente y se vigila la evolución clínica de aquellas personas aparentemente normales, en las que se encuentra *Salmonella* en el intestino, generalmente se descubre que en realidad el individuo ha sufrido ataques previos de enteritis, o bien, que en los días subsecuentes al aislamiento del germen, desarrolla el padecimiento.

En un estudio de 455 niños aparentemente sanos, de edad comprendida entre 1 y 6 años, llevado a cabo en diferentes guarderías de la Ciudad de México por OLARTE & RAMOS-ALVAREZ[15], no se encontró *Salmonella* en ningún caso.

En otro grupo de 107 niños lactantes normales, a quienes se mantuvo bajo vigilancia médica por dos semanas antes y dos semanas después de que se recogieran las muestras, RAMOS ALVAREZ & OLARTE[17] aislaron *Salmonella* en dos niños, en uno S. *typhi-murium* y en otro S. *oranienburg*.

Variaciones estacionales

Se ha observado en la Ciudad de México que durante ciertos años aumenta el número de infecciones por *Salmonella*, en tanto que en otros disminuye, presentándose este fenómeno en forma alterna de alzas y bajas (Cuadro VIII). Es importante mencionar que la infección por otras bacterias entero-patógenas, en particular el grupo *Shigella*, no presentó estas variaciones.

La frecuencia con que se aisló S. *typhi* de las evacuaciones en este mismo grupo de niños (Cuadro VIII), así como su disminución progresiva a través de los años estudiados, fue discutida al hablar de la fiebre tifoidea.

En la gráfica 5 se presenta la distribución de las infecciones por *Salmonella* (sin incluir *S. typhi*), observada en la Ciudad de México

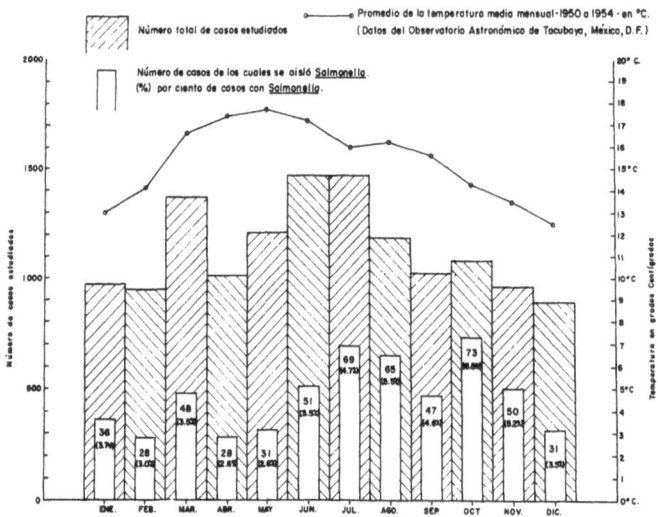

Gráfica 5. Variaciones estacionales en el aislamiento de *Salmonella*, observadas en 1 grupo de 13.545 niños estudiados en el hospital infantil de México, de Julio de 1950 a Julio de 1954. Comparación del promedio mensual de casos con *Salmonella*, con el promedio de niños estudiados y el promedio de la temperatura media mensual observada en la ciudad de México durante el mismo periodo (OLARTE, J. & JOACHIN, A.) (12).

en los distintos meses del año. Si bien es cierto que el número de casos de diarrea aumenta, en general, en los meses más calurosos, la curva de infecciones por *Salmonella*, aunque alta durante el verano, no siempre va paralela a la temperatura. Así por ejemplo, el mayor número de casos de salmonelosis observado durante el período en estudio se presentó en el mes de octubre, mismo en el que la temperatura media va en franco descenso. Resultados semejantes han sido obtenidos en otros dos grupos de enfermos con salmonelosis, estudiados en años anteriores en el Instituto de Salubridad y Enfermedades Tropicales, uno por OLARTE[5] en 1952, y otro por VARELA, LAGUNA & ZOZAYA[21] en 1947. En los tres estudios mencionados se encontró que el mayor número de infecciones por *Salmonella* se presenta, en la Ciudad de México, durante los meses de Junio, Julio y Octubre.

Las infecciones por Salmonella en los animales

Cerdo

VARELA & ZOZAYA[34] investigaron la presencia de *Salmonella* en los ganglios linfáticos mesentéricos en un lote de 209 cerdos sacrificados en el rastro de la Ciudad de México, habiendo aislado 15 cepas, entre las que predominó S. *anatum* (6 veces) y S. *cholerae-suis* (3 veces) (cuadro XIV). Dichos animales habían sido considerados normales por el servicio médico veterinario.

Cuadro XIV.

Salmonelas encontradas en los ganglios linfáticos mesentéricos en un lote de 209 cerdos aparentemente normales.
(VARELA, G. & ZOZAYA, J.)[34].

Salmonella	No. de casos
S. anatum	6
S. cholerae-suis	3
S. typhi-murium	2
S. derby	1
S. heidelberg	1
S. senftenberg	1
S. meleagridis	1
Total	15

VARELA & TÉLLEZ GIRÓN[33] estudiaron dos epizootias ocurridas en cerdos, una de disentería y otra de cólera. La primera tuvo lugar en el Estado de Jalisco y la segunda en el Estado de México. Como se indica en el Cuadro XV, en la epizootia de disentería encontraron S. *typhi-murium* y S. *cholerae-suis*; en la epizootia de cólera aislaron únicamente S. *cholerae-suis*. Los mismos autores buscaron *Salmonella* en diversos órganos en un grupo de 33 cerdos que presentaban lesiones de enteritis, sacrificados en el rastro de la Ciudad de México, habiendo encontrado 27 animales infectados con estos gérmenes. En 12 de estos cerdos aislaron un solo tipo de *Salmonella* en cada uno de ellos, en 11 cerdos dos tipos diferentes de *Salmonella* en cada animal, en otros 2 cerdos 3 tipos de *Salmonella* en cada animal, y en un cerdo 4 tipos diferentes de *Salmonella* (cuadro XVI).

Cuadro XV.

Salmonelas encontradas en 6 cerdos seleccionados de 2 epizootias, una de disentería y otra de cólera (VARELA, G. & TELLEZ GIRON, A.)[33].

Cuadro clínico	No de animales estudiados	Salmonelas aisladas
Disentería	3	S. typhi-murium 1 caso S. cholerae-suis 2 casos
Cólera	3	S. cholerae-suis en los 3 casos

462

Cuadro XVI.

Salmonelas encontradas en un lote de 39 cerdos con enteritis, provenientes de diferentes granjas, sacrificados en el rastro de la Ciudad de México (VARELA, G. & TELLEZ GIRON, A.)[33].

Salmonella aislada	No. de casos	No. de tipos serológicos encontrados en cada animal
S. derby	5	
S. cholerae-suis	4	1
S. anatum	3	
S. typhi-murium	1	
S. cholerae-suis y S. derby	2	
S. derby y S. chester	2	
S. cholerae-suis y S. typhi-murium	1	
S. cholerae-suis y S. paratyphi B	1	
S. cholerae-suis y S. anatum	1	2
S. cholerae-suis y rubislaw	1	
S. derby y S. london	1	
S. derby y S. anatum	1	
S. anatum y S. senftenberg	1	
S. anatum, S. newington y S. newport	1	3
S. anatum, S. newington y S. derby	1	
S. anatum, S. derby, S. newbrunswick y S. typhi-murium	1	4
Total de casos positivos	27	

VARELA & ZOZAYA[36] cultivaron trozos de hígado, de bazo y de intestino, provenientes de diferentes cerdos considerados sanos, sacrificados en el rastro de la Ciudad de México, habiendo aislado *Salmonella* en el 24.2% de las muestras de hígado, 9.8% en las muestras de bazo y 6.2% en las muestras de intestino. En el cuadro XVII se indica el número de muestras estudiadas así como los tipos de *Salmonella* encontrados en los diversos órganos.

Pollos

VARELA, ZOZAYA & OLARTE[37] buscaron *Salmonella* en los hisopos rectales tomados a 1.528 pollos vivos, seleccionados en el Mercado de San Juan de la Ciudad de México, habiendo encontrado estos gérmenes en 21 animales (1.3%) (cuadro XVIII).

Ganado vacuno y caballar

VARELA & ZOZAYA[36] cultivaron trozos de hígado, de bazo y de médula epinal, provenientes de diferentes bovinos considerados sanos, sacrificados en el rastro de la Ciudad de México, habiendo

Cuadro XVII.

Frecuencia del aislamiento de Salmonella de diversos órganos tomados a diferentes cerdos, aparentemente sanos, sacrificados en el rastro de la Ciudad de México. (VARELA, G. & ZOZAYA, J.)[36].

Salmonella	Organo y número de muestras estudiadas:			Total
	Hígado 198 muestras	Intestino 129 muestras	Bazo 73 muestras	
S. derby	35	3	4	42
S. newington	5	2	1	8
S. muenchen	2	0	0	2
S. newport	0	1	1	2
S. oranienburg	1	0	0	1
S. typhi-murium	1	0	0	1
S. schleissheim	0	1	0	1
S. oregon	0	1	0	1
S. thompson	1	0	0	1
S. onarimon	1	0	0	1
S. selandia	1	0	0	1
S. anatum	0	0	1	1
S. poona	1	0	0	1
Totales	48 (24.2%)	8 (6.2%)	7 (9.8%)	63

Cuadro XVIII.

Salmonelas aisladas de los hisopos rectales tomados a 1.528 pollos vivos, en el mercado de San Juan de la Ciudad de México, provenientes de diferentes granjas (VARELA, G. & ZOZAYA, J., OLARTE, J.)[37].

Salmonella aislada	No. de casos
S. typhi-murium	5
S. newington	5
S. chester	2
S. cholerae-suis	2
S. urbana	2
S. derby	1
S. montevideo	1
S. newport	1
S. oregon	1
S. anatum	1
Total	21 (1.3%)

aislado *Salmonella* en el 10.7% de las muestras de hígado, en el 10.6% de las muestras de bazo y en el 8.7% de las muestras de médula espinal. En el cuadro XIX se indica el número de muestras estudiadas así como los tipos de *Salmonella* encontrados en los diversos órganos.

Cuadro XIX.

Frecuencia del aislamiento de Salmonella de diversos órganos provenientes de diferentes animales bovinos, aparentemente sanos, sacrificados en el rastro de la Ciudad de México (VARELA, G. & ZOZAYA, J.)[36].

Salmonella	Organo y número de muestras estudiados:			Total
	Hígado 75 muestras	Bazo 246 muestras	Médula espinal 92 muestras	
S. derby	3	5	1	9
S. newport	0	7	0	7
S. newington	2	1	3	6
S. selandia	0	3	2	5
S. anatum	0	2	1	3
S. panama	2	1	0	3
S. newbrunswick	0	2	0	2
S. give	0	2	0	2
S. abony	0	1	0	1
S. oregon	0	1	0	1
S. oranienburg	0	0	1	1
S. bovis-morbificans	0	1	0	1
S. vejle	1	0	0	1
Totales	8 (10.7%)	26 (10.6%)	8 (8.7%)	42

OLARTE, ALDAMA & VARELA[8] estudiaron los hisopos rectales tomados a un lote de 350 caballos pertenecientes al establo del Colegio Militar de la Ciudad de México, habiendo encontrado *Salmonella* en 5 (1.4%) caballos (cuadro XX).

Cuadro XX.

Salmonelas aisladas de los hisopos rectales tomados a 350 caballos. (OLARTE, J., ALDAMA, A., & VARELA, G.)[8].

Salmonella	No. de casos
S. typhi-murium	2
S. derby	1
S. anatum	1
S. newington	1
Total positivos	5 (1.4%)

Ratas y ratones

En un lote de 200 ratas capturadas vivas en la Ciudad de México, OLARTE & VARELA[16] encontraron *Salmonella* en 21 (10.5%) de estos animales, habiendo predominado *S. enteritidis*. En el cuadro XXI se señalan los tipos de *Salmonella* encontrados.

Cuadro XXI.

Salmonelas aisladas de un lote de 200 ratas *(Rattus norvegicus)* capturadas vivas en la Ciudad de México (OLARTE, J., ALDAMA, A. & VARELA, G.)[8].

Salmonella	No. de casos
S. enteritidis	13
S. typhi-murium	3
S. paratyphi C	1
S. berta	2
S. paratyphi B	1
S. niloese	1
Total	21 (10.5%)

En otro grupo de 1927 ratas *(Rattus norvegicus)*, envenenadas con fluor-acetato de sodio, VARELA, OLARTE & MATA[29] cultivaron el bazo habiendo encontrado *S. pensacola* (*S. enteritidis* (?)) 33 veces y *S. newport* 1 vez.

VARELA & ZOZAYA[35] estudiaron una epizootia de enteritis en ratones criados en una granja, habiendo aislado *S. typhi-murium* del bazo de 76 (24.8%) animales, en un lote de 306 ratones estudiados. En un ratón más encontraron *S. cholerae-suis*.

Perro y gato

VARELA, PÉREZ REVELO & OLARTE[30], y OLARTE, ALDAMA & VARELA[8], encontraron *Salmonella* en el 8.5% en un lote de 235 perros. Dichos animales se encontraban en observación en el Instituto Anti-Rábico de la Ciudad de México, habiéndoseles tomado hisopos rectales (cuadro XXII).

Cuadro XXII.

Salmonelas aisladas del hisopo rectal tomado a un lote de 235 perros (VARELA, G., PEREZ REBELO, R. & OLARTE, J.)[30].

Salmonella	No. de animales
S. derby	5
S. montevideo	2
S. newsbrunswick	2
S. newington	2
S. london	2
S. give	2
S. chester	1
S. cerro	1
S. illinois	1
S. meleagridis	1
S. rostock	1
Total	20 (8.5%)

OLARTE & VARELA[16] cultivaron los hisopos rectales tomados a 80 gatos en la Ciudad de México, habiendo encontrado *Salmonella* en el 11.2% de estos animales (cuadro XXIII).

Cuadro XXIII.

Salmonelas aisladas del hisopo rectal tomado a un lote de 80 gatos, (OLARTE, J. & VARELA, G.)[16].

Salmonella	No. de animales
S. derby	3
S. typhi-murium	2
S. anatum	1
S. newington	1
S. enteritidis	1
S. montevideo	1
Total	9 (11.2%)

Distribución de los tipos serológicos de Salmonella encontrados en México

En el hombre y en los animales

Del año de 1940 al año de 1962 fueron aisladas e identificadas 2.573 cepas de *Salmonella*, habiendo sido encontrados 74 tipos serológicos diferentes (cuadro XXIV).

Transmisión

Contaminación de los alimentos

Sin duda alguna, los alimentos contaminados juegan un papel muy importante en la propagación de la salmonelosis, tanto en el hombre como en los animales. Las salmonelas no sólo son capaces de sobrevivir en la mayoría de los alimentos, sino que, generalmente, se multiplican en ellos, aumentando así su capacidad para producir la infección.

Dadas las condiciones precarias de higiene que todavía prevalecen en núcleos importantes de nuestra población, la leche mal manejada, entre otros alimentos, probablemente ofrece los mayores peligros en la transmisión de estos padecimientos, particularmente en los niños. VARELA & ZOZAYA[36] cultivaron 520 muestras de leche "certificada", recogidas en varios expendios de la Ciudad de México, habiendo encontrado *Salmonella* en 24 (4.6%) de ellas. De dos de estas muestras se aisló *S. typhi* (cuadro XXV).

OLARTE, ALDAMA & VARELA[8], señalan el hallazgo de *Salmonella*

Cuadro XXIV.

Distribución general de los tipos serológicos de Salmonella encontrados en la Ciudad de México, en el hombre y en varias especies animales. Clasificación de 2 573 cepas aisladas durante los años de 1940 a 1962.

Salmonella	No. de cepas aisladas del hombre				No. de cepas aisladas de diferentes especies animales									Leche y otros alimentos	Gran total
	Materias fecales	Sangre	Otros órganos	Total	Cerdo	Pollo	Bovinos	Caballo	Perro y gato	Rata	Moscas	Otros animales	Total		
S. paratyphi A	24	43	—	67	—	—	—	—	—	—	—	—	—	1	68
S. paratyphi B	56	4	1	61	1	—	—	—	—	—	—	—	1	5	67
S. schleissheim	3	—	—	3	—	—	1	—	—	—	—	—	1	1	5
S. abony	1	—	—	1	—	—	—	—	—	—	—	—	—	—	1
S. stanley	5	—	1	6	—	—	—	—	—	—	—	—	—	—	6
S. saint paul	2	—	—	2	—	—	—	—	—	—	—	—	—	—	2
S. reading	6	—	—	6	—	—	—	—	—	—	—	—	—	—	6
S. chester	6	—	—	6	2	2	—	—	1	—	—	—	5	3	14
S. derby	206	1	—	207	57	1	9	1	8	—	22	—	98	6	311
S. essen	1	—	—	1	—	—	—	—	—	—	—	—	—	—	1
S. typhi-murium	422	17	11	450	8	5	—	2	2	4	—	—	21	2	473
S. texas	1	—	—	1	—	—	—	—	—	—	—	—	—	—	1
S. azteca (10)	4	—	—	4	—	—	—	—	—	—	—	—	—	—	4
S. bredeney	1	—	2	3	—	—	—	—	—	—	—	—	—	—	3
S. heidelberg	—	—	—	—	1	—	—	—	—	—	—	—	1	—	1
S. san juan	1	—	—	1	—	—	—	—	—	—	—	—	—	—	1
S. cholerae-suis	30	2	—	32	18	2	—	—	—	2	—	—	22	—	54
S. paratyphi C	3	2	—	5	—	—	—	—	—	—	—	—	—	—	5
S. typhi-suis	1	—	—	1	—	—	—	—	—	—	—	—	—	—	1
S. mission	2	—	—	2	—	—	—	—	—	—	—	—	—	—	2
S. braenderup	2	—	—	2	—	—	—	—	—	—	—	—	—	—	2
S. montevideo	32	—	—	32	—	1	—	—	3	—	—	—	4	—	36
S. menston	1	—	—	1	—	—	—	—	—	—	—	—	—	—	1

Cuadro XXIV. (continuado)

Salmonella	No. de cepas aisladas del hombre				No. de cepas aisladas de diferentes especies animales									Leche y otros alimentos	Gran total
	Materias fecales	Sangre	Otros órganos	Total	Cerdo	Pollo	Bovinos	Caballo	Perro y gato	Rata	Moscas	Otros animales	Total		
S. oranienburg	68			68	1		1						2		70
S. thompson	9			9	1								1		10
S. daytona	2			2											2
S. infantis	35			35											35
S. muenchen	8			8							1		1	1	10
S. oregon	1			1	3	1	1						5	1	7
S. manhattan	2			2											2
S. newport	166	3	1	170	3	1	7			1	3		15	2	187
S. takoradi	1			1											1
S. bonariensis	2			2											2
S. blockley	1			1											1
S. litchfield	1			1											1
S. bovis morbificans	1			1	1		6				1		8		9
S. glostrup	1			1											1
S. kentucky	2			2							1		1		3
S. sendai	2	3		5											5
S. onarimon			1	1											1
S. typhi	94	576	10	680										2	682
S. eastbourne	4		1	5											5
S. berta	4			4											4
S. enteritidis	27			27					1	15			16		43
S. rostock			1	1						1			1		2
S. pensacola										10			10		10
S. panama	36	1		37			3				1		4		41
S. pullorum	3			3											3

S. kalina														
S. vejle							1						1		1
S. anatum	133	1	1	135	16	1	3	1	1	1	8		31	1	166
S. meleagridis					1				1		1		3		4
S. london	9			9	1				2				3		12
S. give	12		1	13	1		2		2		2		7		20
S. newington	46			46	10	5	6	1	3				25	5	76
S. selandia					2		5						7		7
S. cambridge														1	1
S. new brunswick	7			7	1		2		2		11		16		23
S. canoga	2			2											2
S. illinois									1				1		1
S. senftenberg	4			4	1								1	1	6
S. chittagong	1			1											1
S. aberdeen		1		1											1
S. rubislaw	2			2							1		1		3
S. poona	34	1		35	1								1		36
S. worthington	1			1				1					1		2
S. onderstepoort	1			1											1
S. carrau	1			1											1
S. sundsvall	1			1											1
S. gaminara	1			1										1	2
S. cerro										1			1		1
S. minnesota	1			1										1	2
S. urbana	3			3		2							2		5
S. alachua											1		1		1
Totales	1,533	653	32	2,218	133	21	47	5	29	33	53	—	321	34	2,573

Cuadro XXV.

Salmonelas encontradas en 520 muestras de leche "certificada".
(Varela, G. & Zozaya, J.)[36].

Salmonella	No. de muestras
S. derby	6
S. paratyphi B	5
S. chester	3
S. typhi-murium	2
S. newport	2
S. typhi	2
S. paratyphi A	1
S. muenchen	1
S. bovis-morbificans	1
S. senftenberg	1
Total	24 (4.6%)

en el 5% de 200 muestras de "carnitas" (tipo popular de carne que se consume en México), recolectadas en distintos expendios de la Ciudad de México, listas para el consumo (cuadro XXVI).

Estos son sólo ejemplos que muestran la relativa facilidad con que es posible encontrar, en nuestro medio, alimentos contaminados con *Salmonella*, poniendo de relieve el riesgo que los mismos representan para la salud pública cuando son manejados en condiciones higiénicas defectuosas.

Fecalismo humano y animal

Además del papel que el fecalismo, tanto humano como animal, juega en la contaminación de los alimentos, constituye por sí mismo uno de los factores más decisivos en la transmisión directa de la

Cuadro XXXVI.

Salmonelas aisladas de 200 muestras de carnes (denominadas "Carnitas") listas para el consumo, recolectadas en varios expendios de la Ciudad de México (Olarte, J. Aldama, A. & Varela, G.)[8].

Salmonella	No. de muestras
S. newington	5
S. meleagridis	1
S. minnesota	1
S. muenchen	1
S. gaminara	1
S. cambridge	1
Total	10 (5%)

salmonelosis de un individuo a otro. La falta de lavado de manos, y en general de buenos hábitos de aseo personal; el hacinamiento en las viviendas; la carencia de agua y de letrinas; la convivencia del hombre con ciertas especies animales como el perro, gato, pollo, y cerdo, son costumbres que prevalecen en grandes masas de nuestra población.

Moscas, pulgas, otros artrópodos

En determinadas condiciones de gran fecalismo al aire libre, en particular en áreas urbanas y suburbanas pobres, las moscas tienen cierto papel en la transmisión de la salmonelosis. Así, por ejemplo, en un exploración reciente llevada a cabo en el Instituto de Salubridad y Enfermedades Tropicales, VARELA & GREENBERG[20] pudieron demostrar un elevado índice de contaminación — en moscas y ratas capturadas vivas, en los alrededores del rastro de Tlanepantla (25.868 habitantes), población del Estado de México. De las muestras tomadas del intestino, así como de la carne del ganado bovino y porcino sacrificado en dicho rastro, cuyas condiciones de trabajo son bastantes primitivas, fué igualmente posible aislar cierto número de salmonelas (Cuadro XXVII).

VARELA & OLARTE[27] lograron demostrar experimentalmente que las pulgas (*Pulex irritans* y *Ctenocephalus canis*), cuando pican a ratones infectados con *S. enteritidis*, durante la fase septicémica del padecimiento, adquieren la infección. Estas experiencias han sido confirmadas por ESKEY, PRINCE & FULLER[3], quienes, además, obtuvieron la transmisión de la *Salmonella* de la pulga infectada al ratón, cerrando el ciclo. Estos mismos autores[4] han encontrado en la naturaleza pulgas portadoras de *Salmonella*. Aunque estos hallazgos tienen escasa importancia por lo que se refiere a la transmisión de la salmonelosis al hombre en forma directa por la pulga, juegan un papel, cuya magnitud no podemos preveer, en la propagación de la salmonelosis de roedor a roedor. Dichos animales, además de ser muy susceptibles a estas infecciones, participan en forma importante en la transmisión de la salmonelosis al hombre a través de la contaminación de los alimentos con sus deyecciones.

Otros artrópodos, que pudieran participar también en la propagación de las salmonelas, como cucarachas, piojos, garrapatas, etc., no han sido estudiados en México.

Importancia del agua

Por último, es necesario mencionar al agua como factor en la diseminación de la salmonelosis. Ante todo, queremos hacer hincapié en que el empleo liberal del agua es el arma más poderosa y eficaz con que se cuenta en la lucha contra estas infecciones. Donde falta o escasea el agua no es posible que se cumpla con las más elementales necesidades de aseo e higiene personal. Por otro lado, sabemos que el

Cuadro XXVII.

Investigación de Salmonella en las moscas y otros animales, llevada a cabo en el rastro de Tlalnepantla, D.F., durante el año de 1962 (VARELA, G., et al.)[20].

Salmonella	Animales capturados vivos en el rastro y sus alrededores		Animales sacrificados para el consumo				Total de cepas
	Moscas* (7 221 ejemplares)	Ratas** (40 ejemplares)	Ganado Intestinos (66 muestras) — Bovino	Carne (9 muestras) — Bovino	Ganado Intestinos (56 muestras) — Porcino	Carne (4 muestras) — Porcino	
Salmonella alachua	1	—	—	—	—	—	1
Salmonella anatum	8	1	3	2	1	—	15
Salmonella bovis morbificans	1	—	—	—	—	—	1
Salmonella derby	22	3	—	—	1	1	27
Salmonella give	2	—	—	—	1	—	3
Salmonella kentucky	1	—	—	—	—	—	1
Salmonella meleagridis	1	—	—	—	—	—	1
Salmonella muenchen	1	—	—	—	—	—	1
Salmonella new brunswick	11	—	—	—	—	—	11
Salmonella newport	3	—	—	—	—	—	3
Salmonella panama	1	1	—	—	—	—	2
Salmonella typhi-murium	—	9	—	—	1	—	10
Salmonella worthington	1	—	—	—	—	—	1
Total de cepas	53	14	3	2	4	1	77

* Se incluyeron 7 géneros diferentes: *Musca domestica, Lucilia, Callitroga, Phormia, Ophyra, Sepsis* y *Leptocera.*
** Ratas de la especie *Rattus norvegicus.*

agua es un medio inadecuado para la multiplicación y sobrevivencia de las salmonelas, por lo que, para que estos gérmenes se propaguen a través de ella es necesario que existan circunstancias masivas de contaminación fecal reciente. Es por esta razón que, en condiciones ordinarias, aunque el agua no posea una calidad bacteriológica elevada, si es abundante, su participación en la diseminación de la salmonelosis, con excepción, quizás, de la fiebre tifoidea, es muy limitado. El término poco afortunado de enfermedades de "origen hídrico" que algunos epidemiólogos aún insisten en aplicar a la salmonelosis, y en general, a otras infecciones intestinales, debería quedar proscrito por completo.

Nota

Los autores hacen constar su agradecimiento al Sr. Dr. Miguel Angel Bravo Becherelle, del Instituto de Salubridad y Enfermedades Tropicales, quien calculó las tasas de mortalidad basándose en datos de la Dirección Nacional de Estadística, Secretaría de Economía.

REFERENCIAS

1. Benavides, L., Franco-Gómez, A. & Gómez-Pagola, J.: Meningitis purulenta por Salmonella typhosa. Relación de un caso. *Bol. Méd. Hosp. Infant. Méx.*, **17**: *765—775*, 1960.
2. De la Torre, J. A., Villalpando, E., Esparza, H. & Olarte J.: Unilateral Renal Vein Thrombosis in an Infant with Sepsis Due to Salmonella choleraesuis. Nephrectomy with Recovery. *J. Pediat.* **52**: *206—211*, 1958.
3. Eskey, C. R., Prince, F. M. & Fuller, F. B.: Transmission of Salmonella enteritidis by the Rat Fleas Xenopsylla cheopis and Nosopsyllus fasciatus. *Pub. Health Rep.* **64**: *933—941*, 1949.
4. Eskey, C. R., Prince, F. M. & Fuller, F. B.: Double Infection of the Rat Fleas X. cheopis and N. fasciatus with Pasteurella and Salmonella. *Pub. Health Rep.*, **66**: *1318—1326*, 1951.
5. Olarte, J.: La presencia de salmonelas en las materias fecales de niños con "diarrea" (Tesis). *Ciencia (Mex.)*, **4**: *209—214*, 1943.
6. Olarte, J.: Datos no publicados.
7. Olarte, J.: Etiología de la Meningitis purulenta en los niños de la Ciudad de México. *Gaceta Méd. Méx.* **9**: *993—1002*, 1961.
8. Olarte, J., Aldama, A. & Varela, G.: Epidemiología de la diarrea y enteritis infantil en la República Mexicana y en el Distrito Federal. *Prensa Méd. Méx.* **17**: *168—176*, 1952.
9. Olarte, J., & De la Torre, J. A.: Datos no publicados.
10. Olarte, J., Edwards P. R., McWhorter, A. C. & De la Torre, J. A.: A New Salmonella Type: Salmonella Azteca. *Bol. Méd. Hosp. Infant. Méx.* **12**: *910—912*, 1956.
11. Olarte, J. & Gordillo, G. Datos no publicados.
12. Olarte J. & Joachín, A.: La presencia de Salmonella, S. typhi, Shigella y Escherichia Coli 0111: B4, 055: B5 y 026: B6 en 13.545 muestras de materias fecales estudiadas en el Hospital Infantil de la C. de México. *Bol. Méd. Hosp. Infant. Méx.* **14**: *249—255*, 1957.

474

13. OLARTE, J. & JOACHÍN, A.: Salmonella typhi and Other Organisms Encountered in Routine Blood Cultures at the Hospital Infantil of México City. *Pub. Health Lab.* **16**: *56—60*, 1958.
14. OLARTE, J., RAMOS-ALVAREZ, M. & GALINDO, E.: Aislamiento de Shigella, Salmonella y Colis Enteropatógenos de los hisopos rectales de 802 casos esporádicos de Diarrea. *Bol. Méd. Hosp. Infant. Méx.* **14**: *257—262*, 1957.
15. OLARTE, J. & RAMOS-ALVAREZ, M.: Datos no publicados.
16. OLARTE, J. & VARELA, G.: Datos no publicados.
17. RAMOS-ALVAREZ, M. & OLARTE, J.: Datos no publicados.
18. VARELA, G. & AKLE-DELGADILLO, J.: Salmonelas aisladas en el Municipio de Camargo, del Estado de Chihuahua, México. *Rev. Inst. Salubr. Enferm. Trop. (Méx.)* **14**: *117—118*, 1954.
19. VARELA, G. & AKLE-DELGADILLO, J.: Salmonelas y Shigellas aisladas en 1954 de casos de enteritis agudas, ocurridos en Ciudad Juárez, México. *Rev. Inst. Salubr. Enferm. Trop. (Méx.)* **15**, *1—-4*, 1955.
20. VARELA, G. & GREENBERG, B.: Datos no publicados.
21. VARELA, G., LAGUNA, J. & ZOZAYA, J.: Estudio de 100 casos de Salmonelosis humanas. *Rev. Inst. Salubr. Enf. Trop. (Méx.)* **8**: *15—28*, 1947.
22. VARELA, G. & MARTÍNEZ, R. A. E.: Aislamiento, en placentas humanas, de Salmonelas y otros gérmes intestinales, y posibilidades de su transmisión al recién nacido durante el parto. *Rev. Inst. Salubr. Enferm. Trop. (Méx.)* **13**: *173—176*, 1953.
23. VARELA, G., MENDOZA, P. H. & VAZQUEZ, A.: Estudio de los tipos Bacteriofágicos, fermentativos y estructura serológica de Salmonella Typhosa. Contribución a la epidemiología de la fiebre tifoidea en la Ciudad de México. *Rev. Inst. Salubr. Enferm. Trop. (Méx.)* **16**: *33—38*, 1956.
24. VARELA, G. & OLARTE, J.: Aislamiento de Salmonellas en Dos casos de Intoxicación por Alimentos. *Medicina (Méx.)* **22**: *384* 1942.
25. VARELA, G. & OLARTE, J.: Investigación de Salmonelas en las Amígdalas (Estudio de 185 amigdalotomías). *Rev. Inst. Salubr. Enferm. Trop. (Mex.)* **3**: *289—292*, 1942.
26. VARELA, G. & OLARTE, J.: Salmonella Isolated from Human Mesenteric Lymph Nodes. *Science* **99**: *407*, 1952.
27. VARELA, G. & OLARTE, J.: Transmission of Salmonella enteritidis by Pulex irritans and Ctenocephalus canis. *Science,* **104**: *105*, 1946.
28. VARELA, G. & OLARTE, J.: Classification and Distribution of 1,075 Cultures of Salmonella Isolated in the City of Mexico. *J. Lab. clin. Med.* **40**: *73—77*, 1952.
29. VARELA, G., OLARTE, J. & MATA, F.: Salmonelas en las ratas de la Ciudad de México. Estudio de 1937 Rattus norvegicus. *Rev. Inst. Salub. Enf. Trop. (Méx.)* **9**: *239—248*, 1948.
30. VARELA, G., PÉREZ REBELO, R. & OLARTE, J.: Salmonella and Shigella Organisms in the Intestinal Tract of Dogs in Mexico City. *J. Amer. vet. med. Ass.* **119**: *385—386*, 1951.
31. VARELA, G. & ROCH, E.: Salmonelosis y shigelosis en Morelia, Michoacán (República de México). *Rev. Inst. Salubr. Enferm. Trop. (Méx.)* **8**: *177—179*, 1947.
32. VARELA, G. & SILICEO, J.: Investigación de portadores de Bacterias Patógenas intestinales en vesículas biliares humanas. *Rev. Inst. Salubr. Enferm. Trop. (Méx.)* **16**: *25—28*, 1956.
33. VARELA, G. & TÉLLEZ-GIRÓN, A.: Salmonelas aisladas de un grupo de cerdos enfermos. *Rev. Inst. Salubr. Enferm. Trop. (Méx.)* **4**: *139—147*, 1943.
34. VARELA, G. & ZOZAYA, J.: Salmonelas en ganglios de porcinos. *Rev. Inst. Salubr. Enf. Trop. (Méx.).* **2**: *311—318*, 1941.

35. VARELA, G. & ZOZAYA, J.: Salmonelas aisladas en la ciudad de México. *Rev. Inst. Salubr. Enf. Trop. (Méx.)* **3**: *131—134*, 1942.
36. VARELA, G. & ZOZAYA, J.: Hallazgo de Salmonella en alimentos (Estudio de vísceras de bovinos y porcinos, y de leches). *Rev. Inst. Salubr. Enferm. Trop. (Méx).* **5**: *171—173*, 1944.
37. VARELA, G., ZOZAYA, J. & OLARTE, J.: Investigación de salmonelas en pollos normales (Estudio de 1.528 animales). *Rev. Inst. Salubr. Enferm. Trop. (Méx).* **5**: *11—14*, 1944.

SALMONELLOSIS IN SOUTH AMERICA

BY

CIRO A. PELUFFO

Instituto de Higiene, Montevideo.

An assessment of the real incidence of salmonellosis in South America either in man or animals is not possible, as there is no systematic reporting of cases, not even of food poisoning outbreaks. But the available information coming from different areas, even when patchy and incomplete, leaves no doubt that, as occurs in the rest of the world, salmonellae are widely dispersed and salmonellosis constitutes a serious public health problem.

References to the work done on the Continent are scattered in a lot of periodicals or irregularly appearing Public Health, medical or veterinary publications, many of them difficult to get or not available at all and any survey is condemned from the start to be biased according to the author's access to the sources of information. For that reason this is not intended to be a complete review of all work done in the area and less so of world literature, which has been willingly avoided; we only endeavour to give a general picture, as fair as possible under the circumstances, of the frequency and trend of Salmonella infection in man and animals, of sources of human infection and of the distribution of *Salmonella* types, except *S. typhi*, in the geographic area surveyed.

Besides South American countries, some Central American ones have been taken into consideration, particularly Costa Rica, because of the very comprehensive work done on the pathology and epidemiology of salmonellosis.

While some isolations of *Salmonella* types have been reported previous to 1936 [1-10] systematic investigations are not found until HORMAECHE and his group [11] stressed their participation in the etiology of infantile gastro-enteritis.

Most South American investigations on salmonellosis followed the Uruguayan's pattern and mainly tried to establish the incidence of *Salmonella* and other groups of enteric bacteria in childen's diarrhea and to clarify its epidemiology. That is the main reason why the South American picture will clearly differ from that of other regions.

Salmonellosis in infancy

Infantile mortality has always been one of the leading public health problems in Latin American countries, and gastrointestinal disorders rank first among its causes. Official figures are certainly below the real ones as a high number of deaths which are registered as due to other causes are primarily caused by diarrhea and enteritis. This situation led HORMAECHE and coworkers to start, in 1934, systematic studies on its etiology, which were continued through 15 years; their findings regarding the participation of *Shigella* and *Salmonella* as the cause of diarrhea in children were published in several papers [12, 13, 14, 15].

As a summary of their work in regard to salmonellae, we can say that after studying in all 5,955 children up to the age of 12, admitted to children's hospitals in Montevideo, they isolated nonhost-adapted salmonellae from 814 (13.6%) of them.

Taking into account only those cases that according to clinical features were suffering of a clear-cut enteric infection (dysenteriform, choleriform or mixed), they found salmonellae in 224 of 841 such cases (26.6%).

Letality was high amounting to 18.8% and in children under one year and with choleriform syndrome it rose to 57% [14].

The rates of infantile mortality in the rest of the continent are even higher than in Uruguay. In Brazil diarrhea is the main cause of mortality in infancy producing an estimate of 140,000 deaths annually[16]. In Venezuela, infantile mortality by enteritis in the first year of life amounts to 3,700 per 100,000 live borns and the rates in all other South American countries are over 700[17].

The pioneer work of HORMAECHE's group led others to start similar investigations in several countries. In Argentine, MONTEVERDE & DE SIMONE[18, 19], GRICHENER et al.[20]. MONTAGNA & RIMOLDI[21], VERNA [22], CATALDI[23, 24], RAMACCIOTTI et al.[25], LUBIN et al.[26]; in Brazil, DE TAUNAY et al.[27], PELUFFO et al.[28], COSTA et al.[29], MAROJA et al.[30, 31]; in Chile, VACCARO et al.[32], GARCES & PRADO[33]; in Costa Rica, ESQUIVEL[34], DE LA CRUZ[35]; in Ecuador, ALCIVAR[36]; in Venezuela, BRICEÑO IRAGORRY[37], OROPEZA et al.[38], LE MINOR et al.[39], BRICEÑO IRAGORRY et al.[40], among others have carried out systematic investigations and established the incidence of *Shigella* and *Salmonella* and in some cases enteropathogenic coli in children's diarrhea.

In Table I are summarized the results obtained in some of these investigations with regard to incidence of salmonellae. As could be expected results differ widely; besides variations in the reservoirs of infection, feeding habits and sanitary conditions that may locally influence the frequency of salmonellae and the mechanism of transmission to infants, some other factors in relation to the methodology of the investigation are of fundamental importance.

The selection of cases regarding age and clinical syndrome is a major one. We have already mentioned that HORMAECHE et al. found salmonellae in 26.6% of cases of enteritis, but only 13.6% when children with all kind of enteric disorders were studied.

Table I.

Rates of occurrence of Salmonellae in children's diarrhea.

References	Country	Cases	Positives	
			Number	%
MONTAGNA, C. P. & RIMOLDI, A. A. 1944 (21)	Argentine	100	16	16.0
CATALDI, M. S., 1944 (23), 1945 (24)	Argentine	210	59	28.1
MONTEVERDE, J. J. & DE SIMONE C., 1947 (19)	Argentine	73	20	27.39
RAMACCIOTTI, F. et al., 1962 (25)	Argentine	863	28	3.24
LUBIN, A. H. et al., 1962 (26).	Argentine	1,092	133	12.2
TAUNAY, A. de E. et al., 1945 (27)	Brazil	200	31	15.5
PELUFFO, C. A., et al., 1946 (28)	Brazil	426	53	12.44
ARAUJO COSTA, G. et al., 1957 (29)	Brazil	560	33	5.9
VACCARO, M. et al., 1943 (32)	Chile	338	28	8.28
GARCÉS, C. & PRADO, E., 1944 (33)	Chile	199	12	6.0
DE LA CRUZ, E., 1962 (35).	Costa Rica	536	48	9.0
ALCÍVAR, C., 1949 (36)	Ecuador	899	50	5.6
GARDINI TUESTA, W. E. et al., 1961 (74).	Peru	950	32	3.37
HORMAECHE, E. et al., 1936 (11), 1943 (14), 1947 (15)	Uruguay	5,955	814	13.67
BRICEÑO IRAGORRY, L., 1942 (37)	Venezuela	64	15	23.44

ALCIVAR[36] in 720 non-selected cases found a frequency of 4.86% while in 69 infants with enteritis he isolated salmonellae from 39% of them. VACCARO et al.[32] found respectively 8.3 and 13.7%.

The influence of adequate bacteriological methods and repeated examinations on the isolation of salmonellae from faeces is well known. HORMAECHE & SURRACO[41] found that using enrichment media for the isolation of salmonellae from cases of children's diarrhea the positive results are three and a half times higher than employing direct isolation. They also found that the number of positive cases is increased if several direct isolation media, two enrichment media and 1 plus 5 days incubation of tetrathionate broth are used.

All these factors may explain different results in similar regions and periods of time, but there is a trend that is clearly shown in Table I, namely the consistent reduction of incidence found in recent investigations. This has been pointed out by BRICEÑO & FOSSAERT[40] who give the following frequencies in Caracas in successive periods: 1944: 14.4%; 1954: 4.2%; 1956: 3.3%.

Many reasons could be ventured to explain this tendency that is clearly conflicting with the increasing frequency of other human salmonellosis, particularly food poisoning. But no factual evidence has been yet presented to ascertain which is the cause (or causes) to be imputed for that trend.

Table II.

Salmonella types isolated from children's diarrhea.

	Argentine (19, 22, 25, 26, 42, 43)	Brazil (27, 28, 30, 31)	Chile (32, 33, 44, 45, 46)	Costa Rica (34, 35)	Ecuador (36)	Uruguay (11, 14, 15)	Venezuela (37, 38, 39)	Total Number	Total %
S. paratyphi B	22	10	6	—	—	35	—	73	3.4
S. saint paul	—	13	—	6	2	1	3	25	1.2
S. san diego	—	1	—	1	—	15	—	17	0.8
S. derby	7	26	17	11	—	23	26	110	5.8
S. typhi-murium	175	24	100	7	4	350	22	682	31.7
S. bredeney	6	4	—	3	—	27	10	50	2.3
S. cholerae-suis	—	—	6	—	—	6	—	12	0.5
S. montevideo	3	—	—	1	—	76	7	87	4.0
S. oranienburg	—	1	1	—	—	20	11	33	1.5
S. muenchen	—	39	—	—	3	19	9	70	3.3
S. manhattan	—	1	—	12	—	—	—	13	0.6
S. newport	72	68	65	3	3	198	8	417	19.4
S. bovis-morbificans	8	4	10	—	—	2	—	24	1.1
S. enteritidis	—	5	7	—	—	1	—	13	0.6
S. panama	—	1	1	3	1	11	16	33	1.5
S. anatum	16	59	7	—	8	47	32	169	7.8
S. meleagridis	3	1	1	—	—	6	11	22	1.0
S. london	5	—	4	—	1	7	1	18	0.8
S. minnesota	16	2	3	—	1	5	2	29	1.3
Further types*	30	44	50	7	17	46	21	215	10.0
Total	366	308	282	59	50	904	184	2,153	

* Each type less than 0.5% of the total.

In Table II is given a list of the most common types isolated in several countries from "children's diarrhea" cases.

The distribution of types is not representative in all cases of their true frequency in the country, because of the small number studied, geographic limitations and/or periods of time in which they were gathered.

As has been pointed out by PELUFFO[47], COLICHON[48], BRICEÑO & FOSSAERT[40] and COSTA et al.[49], frequency greatly varies in different periods of time, in towns versus rural areas and in different regions of the same country.

There are minor differences in regard to the relative frequency of the most important types, but as a whole it shows that, as in other geographic areas and clinical conditions, among the leading types found are *S. typhi-murium*, *S. newport*, *S. derby* and *S. montevideo*.

Noteworthy are the relative incidences of *S. anatum* and *S. cholerae-suis*: this last type, common in adults and in extra-intestinal localizations in children, is very seldom found in enteric infections, while *S. anatum* shows an inverse distribution. The greater susceptibility of children to nonhost-adapted salmonellae may explain it.

S. paratyphi B occurs in children, mostly in Argentine and Uruguay, but the strains isolated belong to the so-called "animal type": dextro-tartrate positive.

When several direct and enrichment media are used and enough colonies from each plate are serologically tested, it has been found that more than one *Salmonella* type can be isolated from a single child. HORMAECHE et al.[11, 14, 15] found in 814 positive cases 57 with two types, 7 with three and 3 with more than three and up to ten types in a case with multiple examinations.

Similar results are recorded by PELUFFO et al.[28], and are not at all unusual as in other materials, mainly in animal infections, the isolation of several *Salmonella* types is a common occurrence.

Extra-intestinal localizations

They are rather frequent in infants, associated or not with enteritis. As was stressed by HORMAECHE, PELUFFO & ALEPPO in 1936[11] "Children have a sensitivity to "animal" salmonellae not suspected until now".

The widely known Kiel doctrine separated salmonellae that are primarily pathogens for man and produce in him generalized infections from those that produce a variety of diseases in animals but in man only gastro-enteritis (food poisoning) and only if the food is heavily contaminated. Investigations on infantile salmonellosis led to the conclusion that the Kiel doctrine does not apply to early childhood. In infants the border-line between enteritis and generalized infection is not well defined and pathology, prognosis and epidemiology are very different from those of the adult.

The infection starts in a slow and progressive manner, like with host-adapted salmonellae and as a rule there are premonitory symptoms; in some cases it begins with sore throat, and salmonellae can be isolated from the pharyngeal exudate[13, 50]. The disease is protracted, the evolution is longer than in adult's gastroenteritis, lasting for three to four weeks; extra-intestinal localizations are frequent and may occur, without enteric symptoms, as the only pathological manifestation of the disease.

Cases are sporadic and infection is produced after the ingestion of a small number of organisms. Frequently only infants are infected, while older children and adults in the same family are not; even one twin, breast fed, may be infected while the other remains well[11]. As a consequence of the small infecting dose, interhuman transmission occurs and limited hospital outbreaks are produced like the one observed by HORMAECHE et al.[51] in "Casa Maternal", caused by S. cerro.

All these concepts are now accepted and generally known as the "Montevideo doctrine"[52, 53, 54, 55, 56].

Tables III and IV summarize the frequency of salmonella isolations from extra-intestinal sources and types found in them. Most of these findings, with the exception of the ones from Montevideo, were fortuitous and not the result of systematic research; that explains why the number of positive isolations, specially from ear, throat and urine are very small.

In spite of numerous publications stressing the characteristics of children's salmonellosis, they are not yet widely known and very often obvious extra-enteric infections are overlooked or the diagnosis missed due to inadequate isolation methods[73]. In septicemias and meningitis we have a different picture: clinical symptoms are unmistakable and bacteriological diagnosis is always done.

It is worth to note that from the 20 cases of meningitis recorded in Brazil, 17 belonged to a serious outbreak studied by DE TAUNAY[70] and produced by S. grumpensis; most cases came from the same hospital, and S. grumpensis was also found producing gastroenteritis in children. This type was isolated for the first time in Uruguay from a guinea pig from a batch coming from Buenos Aires and as far as we know was only once more isolated in Argentine by LEIGUARDA et al.[75] from river water; it had never been previously found in Brazil.

Types isolated from extra-intestinal sources are the same as those found in the intestine but there is a clear increase of the incidence of types belonging to group C1 and specially a higher relative number of S. cholerae-suis.

Paratyphoid fever

It is difficult to assess the real incidence of the disease as few systematic investigations have been carried out expressing its

482

Table III.

Frequency of extra-intestinal localizations in children.

Country and reference	Blood	C.S.F.	Ear	Throat	Urine	Osteo-arth.	Pleuro-pulm.	Other*	Total
Uruguay (11, 14, 15, 57, 58, 59, 60)	34	3	21	22	54	1	1	1	137
Argentine (4, 61, 62, 63, 64)	5	2	—	—	—	—	2	—	9
Peru (48)	8	4	—	—	—	—	—	—	12
Venezuela (39, 65, 66)	7	3	—	1	—	1	1	—	13
Chile (32, 45, 46)	8	8	1	—	—	2	—	—	19
Brazil (67, 68, 69, 70, 71)	1	20	—	—	—	—	1	1	23
Panama (72)	3	3	—	—	—	2	—	—	8
Total	66	42	22	23	54	6	5	2	221

Source of isolation

* 1 peritonitis; 1 panophthalmia.

Table IV.

Salmonella types isolated from extra-intestinal localizations*.

	Source of isolation								Total	
	Blood	C.S.F.	Ear	Throat	Urine	Osteo-arth.	Pleuro-Pulm.	Other	Number	%
S. paratyphi A	2	4	—	—	—	—	—	—	6	2.7
S. paratyphi B	4	3	—	—	2	—	—	—	9	4.1
S. typhi-murium	34	11	16	14	31	2	—	1	109	49.3
S. cholerae suis	10	1	—	3	2	3	2	—	21	9.5
S. montevideo	3	—	1	3	4	—	1	—	12	5.5
S. oranienburg	2	1	1	—	—	—	—	—	4	1.8
S. newport	—	—	2	—	9	1	—	—	12	5.5
S. enteritidis	3	2	—	—	—	—	1	—	6	2.7
S. dublin	3	1	—	—	—	—	—	—	4	1.8
S. panama	2	1	—	—	—	—	—	—	3	1.4
S. cerro	—	—	—	2	2	—	—	—	4	1.8
S. grumpensis	—	17	—	—	—	—	—	—	17	7.7
Further types**	3	2	2	1	4	—	1	1	14	6.4
Total	66	42	22	23	54	6	5	2	221	

* Source of data: Blood: 39, 45, 46, 48, 57, 58, 59, 61, 62, 65, 66, 67, 72; C.S.F.: 14, 15, 32, 46, 48, 62, 63, 66, 68, 69, 70, 72; Ear: 14, 45; Throat: 14, 15; Urine: 14, 15, 66; Osteo-arth: 14, 15; Pleuro-pulm: 11, 62, 64, 66, 71; Periton: 60; Panoph: 71.

** Each type less than 1% of the total.

frequency in relation to typhoid fever. But it can be said that in general the rate of occurrence is low and that in several areas it does not exist at all.

S. paratyphi A: in Venezuela, BRICEÑO[65] isolated 137 *S. typhi* and 1 *S. paratyphi* A (relative frequency 0.7%) from 1,000 blood cultures; LE MINOR et al.[39] found a higher frequency, 1.4%. In Chile, SAN MARTIN & URZUA[76] isolated 48 *S. typhi* and 2 *S. paratyphi* A from an unstated number of blood cultures. In Peru, COLICHON[48] mentions the isolation of 11 strains in 4,654 examinations of faeces during a period of 6 years. In the State of São Paulo, Brazil, TAUNAY[70] isolated 27 strains from different materials, on a period of 12 years, but NORONHA[77] in the State of Minas Gerais never did, and mentions that to his knowledge no paratyphoid A fever exists in Brazil.

As far as we know no paratyphoid A fever has been recorded in Argentine, Costa Rica or Panama. In Uruguay it does not exist but we have isolated 2 strains of *S. paratyphi A* in 25 years; one from an Institution's inmate and another from a case coming from Paraguay.

S. paratyphi B: it is a common type in most South American countries but it generally produces gastro-enteritis. Strains studied in the Salmonella Centre in Montevideo, monophasic or biphasie, that came from surrounding countries, belonged to the "animal type" (dextro-tartrate positive). All strains isolated in Uruguay, even from extra-intestinal localizations in children, belonged to this same type. In Chile outbreaks of "paratyphoid B" have been reported[78] but *S. paratyphi B* has been isolated only from stools and mainly in children, which suggests that here again the "animal type" may have been involved.

S. paratyphi C: besides the reference of GIGLIOLI in British Guiana[79] to 72 isolations of *S. paratyphi C* among 1,283 blood cultures, this type has been very seldom mentioned[3] as an agent of enteric fever. In all these cases the identification of the isolated strains is an open question because the information given is not precise enough to exclude some other closely related members of Group C 1.

In Venezuela, BRICEÑO[65] isolated 5 strains belonging to Group C 1 among 1,000 blood cultures: 2 *S. cholerae-suis*, 2 *S. cholerae-suis* var. *kunzendorf* and 1 *S. oranienburg*; no *S. paratyphi C*, was found. DE SALLES GOMES & ÂRANTES[85] isolated from blood 15 strains of *S. cholerae-suis* var. *kunzendorf* in a period of 6 years.

S. enteritidis var. *chaco:* mention must be made of this type that during the Chaco war between Bolivia and Paraguay was frequently isolated from cases of clinically typical enteric fever[80].

Food Poisoning and other adult salmonellosis

There is very little information concerning acute gastro-enteritis of food origin. Outbreaks are not as common as in other geographic areas, probably because of food habits, as bulk preparation of food-stuffs is not extended and meals are not so often made in catering places where omissions in hygienic control may affect a large number of individuals. Besides, frozen or powdered eggs, an important source of infection in Europe, are not in general use in Latin America.

In the few reported outbreaks with a known source of infection, pork sausages[81], barbecued beef "con cuero"[82], minced meat[66] and bread pudding[72] were incriminated.

Salmonella types isolated from those and other outbreaks were: 3 S. *newport*,[42, 82, 139], 2 S. *oranienburg*[42, 66], 1 S. *typhi-murium*[77], 1 S. *montevideo*[72]. Double infection was found in two outbreaks: S. *typhi-murium* plus S. *panama*[81] and S. *typhi-murium* plus S. *newport*[66].

More frequently reported are sporadic salmonellae isolations from cases of chronic or acute enteritis, septicemia and localized suppurative lesions with or without a clinical syndrome of septicemia.

In Table V are given the *Salmonella* types most frequently isola-

Table V.

Salmonella types isolated from adults*.

	Faeces	Blood	Miscella-neous	Total Number	Total %
S. paratyphi A	14	—	—	14	2.9
S. paratyphi B	21	—	1	22	4.5
S. saint paul	33	—	—	33	6.8
S. san-diego	4	1	—	5	1.0
S. derby	18	—	1	19	3.9
S. typhi-murium	87	1	—	88	18.0
S. bredeney	14	—	—	14	2.9
S. cholerae-suis	5	23	5	33	6.8
S. montevideo	6	—	—	6	1.3
S. oranienburg	17	2	1	20	4.1
S. newport	55	—	1	56	11.5
S. enteritidis	6	1	—	7	1.4
S. dublin	6	—	1	7	1.4
S. panama	31	—	—	31	6.4
S. anatum	31	—	1	32	6.6
S. london	14	—	—	14	2.9
S. senftenberg	22	—	—	22	4.5
Further types**	65	—	—	65	13.3
Total	449	28	11	488	

* Source of data: Faeces: 28, 42, 48, 66, 77, 81, 83, Blood: 39, 65, 72, 81, 84, 85. Miscellaneous: 15, 32, 42, 64, 81, 84, 86.

** Each type less than 1% of the total.

ted from those cases. Again Group C 1 shows a clear predominance in extra-intestinal infection, mainly in septicemia, in which it accounts for 89% of the cases. Types isolated from enteric disorders show a similar distribution to that found in children, but with a lower frequency of S. *anatum* and a greater one of uncommon types.

Among miscellaneous infections are recorded: 2 purulent pleurisies (2 S. *cholerae-suis*), 2 suppurate gall-bladders (1 S. *newport*, 1 S. *anatum*), 2 meningitis (1 S. *saint-paul*, 1 S. *derby*), 1 liver abscess *(S. dublin)*, 1 annexitis *(S. cholerae-suis)* and one case with a lung abscess and suppurate gall-bladder *(S. oranienburg)*.

Special mention must be made of a very peculiar disease, human bartonellosis or Carrion's disease, in which salmonellae appear as secondary invaders, COLICHON & BERROCAL[48] isolated from blood in different periods of the disease, 21 Salmonella strains: 11 S. *typhimurium*, 3 S. *dublin*, 2 S. *newport*, and one each of S. *cholerae-suis*, S. *paratyphi* B and S. *panama*. In two cases they found S. *typhi* and in a further two S. *typhi* plus S. *typhi-murium*.

Human Carriers

It is becoming increasingly evident that the asymptomatic human carrier is more frequent than previously recognized. In Table VI the rates of occurrence of *Salmonella* carriers are given under three headings. Reports on asymptomatic children refer mostly to cases from hospitals; those on adults were obtained during surveys to establish the frequency of enteric bacteria among foodhandlers and the non-selected ones came from the population at large.

Again we have widely differing data which may be attributed to the selection of cases, regarding age and environment, and technical methods employed. These are important in carriers more so than in clinical cases as the number of salmonellae in normal faeces is meager, and unless enrichment and good selective isolating media are used, the chances of obtaining positive results are low. In our experience most strains isolated from carriers came from enrichment media, and often only when incubation in tetrathionate broth was extended for 5 days.

The high rate found by DE LA CRUZ[88] in Costa Rica in food handlers is striking, amounting to almost 20%, but it is in agreement with the rate of 9.2% of meat samples contaminated by salmonellae[94] and of 8% on pork sausages[95] found by the same worker, that tells of a high environmental contamination; but it is difficult to decide if food handlers are the cause or the victims of it. Unfortunately no information has been sought regarding the duration of the carrier state.

Table VII depicts the distribution of the most frequently occurring *Salmonella* types in carriers. When it is compared with the same

Table VI.

Rates of occurrence or Salmonella carriers.

Reference	Children				Adults				Non selected				Total
	(15)	(35)	(46)	Total	(39)	(87)	(88)	Total	(89)	(90)	(91)	Total	
Cases	156	187	81	424	11,203	2,012	130	13,345	116	600	454	1,170	14,934
Positives	13	17	2	32	194	154	25	373	9	4	6	19	424
%	8.33	9.9	2.4	7.31	1.73	7.65	19.23	2.79	7.61	0.67	1.32	1.62	2.83

Table VII.

Salmonella types isolated from carriers*.

	Children	Adults	Non selected	Total Number	Total %
S. paratyphi B	—	5	2	7	1.3
S. chester	—	7	—	7	1.3
S. derby	8	28	1	37	6.7
S. typhi-murium	11	33	2	46	8.3
S. bredeney	8	9	—	17	3.1
S. montevideo	9	18	—	27	4.9
S. oranienburg	2	22	—	24	4.3
S. muenchen	4	14	1	19	3.4
S. manhattan	3	7	—	10	1.8
S. newport	12	26	4	42	7.6
S. panama	2	34	—	36	6.5
S. anatum	6	73	5	84	15.2
S. meleagridis	2	9	1	12	2.2
S. london	2	12	—	14	2.5
S. borbeck	—	9	—	9	1.6
S. florida	—	22	—	22	4.0
Further types**	6	130	4	140	25.3
Total	75	458	20	553	

* Source of data: 15, 35, 39, 46, 72, 87, 88, 89, 90, 91, 92, 93.
** Each type less than 1% of the total.

in clinical conditions a few interesting facts show up, namely that *S. cholerae-suis* has never been found in 553 carriers, the change in the relative frequencies of the most common types, with *S. anatum* in the lead, and lastly the great variety of uncommon types isolated, that amount to 25% of the total.

Sources of human infections

Salmonella infections of domestic animals

Animal salmonellosis are not frequently reported and our information as a whole is rather scanty.

Host-adapted types have been reported in several areas. *S. abortus-equi* in Argentine[96], [97], Brazil[98], Peru[48] and Uruguay[99]: *S. pullorum* in Argentine[100], Brazil[101] and Uruguay[47].

Among the nonhost-specific types, *S. typhi-murium* has the leading position, not only as regards incidence but also on the range of animal species involved. It has been found in fowl[48], [81, 102, 103], dogs[48], [103, 104], cattle[81, 105], swine[103, 106], sheep[103], guinea pigs[70, 81, 103], rats[81, 103, 107], mice and rabbits [81, 103] and cat[48].

A few systematic investigations on Salmonella infections in some species have been reported.

MONTEVERDE and coworkers[104] in Argentine found salmonellae in 12.3% of 65 dogs studied (5 *S. typhi-murium*, 1 each of *S. anatum*, *S. montevideo* and a non identified type). DE LA CRUZ[108] in Costa Rica did so in 3 of 68 dogs examined (1 each *S. panama*, *S. saint-paul* and *S. derby*). Incidence of salmonellosis in dogs is much lower than that found in other areas[109], probably because dogs in Latin American countries very seldom are fed on commercial dog meal.

In cattle, ARROYO & BOLAÑOS[105] studied three outbreaks of acute enteritis of calves in Costa Rica and found *S. typhi-murium* in 2, and a non-completely identified strain belonging to group E1 in the remaining one. GALLO in Venezuela[110] found as the most common types *S. dublin*, *S. typhi-murium* and *S. paratyphi B*: GUERRERO[111] in Colombia and Ecuador isolated *S. enteritidis*; in Brazil, PENHA & D'APICE[112] and PENHA & ESQUIBEL[113] found *S. dublin*.

In hog-cholera and swine-pox QUIROGA & MONTEVERDE[106] in Argentine found salmonellae in 26 of 104 pigs studied: 21 *S. choleraesuis*, 3 *S. typhi-murium*, 2 *S. onderstepoort* and 1 each *S. paratyphi B* and *S. muenchen*.

In hens reacting to the pullorum antigen MONTEVERDE & SIMEONE [102] found nonhost-adapted types in 14: 3 *S. goettingen*, 2 each *S. brandenburg*, *S. newport*, *S. thompson* and *S. poona* and one each *S. typhi-murium*, *S. meleagridis* and *S. onderstepoort*.

Sporadic infections in several animal species have been reported and many Salmonella types, besides *S. typhi-murium*, have been isolated. In cattle: *S. newport*, *S. montevideo*, and *S. anatum*[81], *S. rubislaw*[114], *S. bredeney*[115]; in sheep: *S. newport*, *S. montevideo* and *S. senftenberg*[81]; in swine: *S. bredeney*[116]; in guinea-pigs: *S. paratyphi B*, *S. reading* and *S. bredeney*[70], *S. grumpensis*[81]; in cats; *S. anatum*[81], *S. bredeney*[115] and *S. minnesota*[108]; in monkeys: *S. bredeney* and *S. montevideo*[81].

Special mention has to be made of the isolation by MERA[117] of *S. paratyphi A* from a cat.

Salmonellae in normal animals

There is no doubt that animals are the primary hosts for nonhuman-adapted types of *Salmonella* and that food is the most common vehicle of transmission to man. Contamination of food may occur through the use of material coming from an infected animal: eggs, specially duck's eggs, are well known as source of salmonellae, and meat from diseased animals is responsible for many human outbreaks of salmonellosis. But it is increasingly evident that the main reservoir of infection is not the diseased animal but the asymptomatic carrier, much more common and of greater epidemiological significance on account of his unavoidable influence on food contamination.

The high relative frequency of children's salmonellosis in relation to adult's in the same area, found by South American workers, pointed to an extensive environmental contamination but of low density.

To investigate its sources and vehicles, the Montevideo group started work on several lines. In 1936, HORMAECHE & SALSAMENDI [118] systematically examined the lymph glands of normal pigs from the abattoir and found one ore more *Salmonella* types in 48% of pools of 20 animals. When that work was further extended[119], a similar frequency was found even when the glands of single animals were examined.

A great variety of *Salmonella* types were isolated, but very seldom *S. cholerae-suis*, a type that has been found in over 60% of diseased animals. So different a frequency is probably an expression of its high virulence for swine; it is, however, difficult to explain why this type so often found as a secondary invader in hog-cholera is not more frequently found in normal animals, as would be expected.

This work was amply confirmed in several countries and extended to other domestic animals[77, 120, 121, 122, 123, 124, 125].

In Argentine, LEIGUARDA et al.[126] systematically investigated the intestinal content of normal fish caught on the Rio de la Plata and found in pools of up to 15 individuals of the same species, close to 20% contamination by *Salmonella*. They suggest that rather than true carriers, fish are only eliminators of bacteria that witness the high *Salmonella* contamination of river's waters.

Following another line of work HORMAECHE, PELUFFO & ALEPPO, in 1942, started investigations to establish the rate of occurrence of enteric bacteria in flies. It is generally believed that flies are vectors of intestinal infections; their feeding habits and the fact that fly control brings a reduction of infantile mortality by enteric diseases [127] backs that assumption. Besides, according to the "Montevideo doctrine", children may be infected by a very small infecting dose and even when brest-fed; that also pointed to flies as possible vehicles of *Salmonella*.

They tested pools of 20 flies ground *in toto* with sterile sand and saline; an equal number of pools of *Musca domestica* and of a mixture of *Cochliomya*, *Lucilia* and *Paralucilia* were cultivated on the usual isolation and enrichment media[128].

Results were indeed rewarding as it was found that during the summer months up to 50% of pools were positives. In all 85 of 362 pools (23.48%) yielded one or more *Salmonella* types, isolating 100 strains of 15 different types; pools constituted by *Musca domestica* yielded practically the same number of positive results as those of other species. The types found were very much alike to those isolated from children's diarrhea in the same period of time and so was its distribution.

Similar results were obtained in other countries, but not all agree as regards rates of contamination; BOLAÑOS[129] in Costa Rica found the highest rate with an average of 24%, while MAROJA et al.[130] in Brazil found the lowest with 1.8%. Local conditions may differ and the high incidence found by BOLAÑOS is in agreement with those of DE LA CRUZ in normal pigs[122], in food handlers[88] and in food[94, 95] all done in the same city; technical methods are probably much to be blamed for the low frequency found by MAROJA et al. [130] as they inoculated the enrichment media with only one drop of the 10 ml of water used to wash the outside surface of flies. Besides, enrichment media were incubated for only 24 hours before plating; we have already stressed the fundamental influence on results of incubation up to 5 days when materials with a small number of salmonellae and high normal flora are investigated.

Negative results have also been recorded. COUTINHO et al.[131] tested in Brazil isolated intestine of flies, locking for parasites and enteric bacteria; they never isolated salmonellae from 185 pools with a total of 5,781 specimens. This result when confronted with those already mentioned would indicate the mechanical nature of transmission, but we think that this remains to be proved, and for that a sufficient number of parallel samples should be tested at the same time and in the same place. In our laboratory FRANCA[132] investigated the intestinal content of a small number of flies and found one *Salmonella* strain in 15 examined.

TAUNAY et al.[133] investigated the presence of pathogenic bacteria in cockroaches: from 20 pools (114 specimens) they never isolated salmonellae.

In Table VIII are summarized the rates of occurrence of salmonellae in several animal species as recorded by workers of different areas.

Salmonellae in food, water and sewage

The work done in South and Central American countries regarding food contamination is much more meager than in other areas. In Brazil, ASSUMPÇÃO[136] and in Costa Rica DE LA CRUZ[94] found salmonellae in retail meat on 15.5 and 9.2% respectively of 153 and 434 samples examined. PESTANA & RUGAI[137] in 170 samples of pork and mixed sausages found salmonellae in 3.5%; DE LA CRUZ[95] did so in 8% of 250 samples and BOLAÑOS[92] in an unstated number isolated 10 *Salmonella* strains. In several instances more than one type was isolated from a single sample.

The presence of *Salmonella* strains in fresh eggs has been reported by SIMEONE[138] in Argentine, but 38 of them, isolated from 4,206 eggs (hen, duck, goose, turkey and guinea fowl) belonged to types *S. gallinarum* or *S. pullorum* and only one did not *(S. thompson)*.

In Brazil, MAROJA & LOWERY[139] isolated *S. salford* from several turtle eggs *(Podocnemis dumeriliana)* intended for human consumption.

Table VIII.

Rates of occurrence of Salmonellae in normal animals.

	Swine						Cattle		Fish	Flies				
	(118, 119)	(119)	(77)	(120)	(121)	(122)	(124)	(125)	(126)	(128)	(129)	(130)	(134)	(135)
Total number exam.	1,703	44	50	70	100	150	600	195	?	7,240	2,000	6,600	10,080	2,860
Pools of	2/20	1	10	1	1	1	1	1	15	20	20	10	20	?
Number of pools	97	44	5	70	100	150	600	195	97	362	100	600	504	7
Positives	62	23	2	16	15	79	7	26	19	85	24	12	46	7
%	63.9	52.2	40	22.8	15	52.7	1.2	13.3	19.6	23.5	24	1.8	9.1	100
Number of strains	83	31	2	16	15	115	7	27	21	100	28	12	46	7

Table IX.

Sources of human infection.
Distribution of salmonella types*.

	Normal animals			Flies	Food	Water	Sewage	Total	
	Swine	Cattle	Fish					Number	%
S. paratyphi B	10	—	—	—	2	3	2	18	1.5
S. saint-paul	11	—	—	3	3	—	—	17	1.4
S. reading	—	—	—	2	7	—	—	9	0.8
S. chester	8	—	—	5	—	—	—	13	1.1
S. san diego	13	1	—	3	1	—	—	17	1.4
S. derby	33	—	1	1	16	16	3	70	5.9
S. california	—	—	—	—	—	8	—	8	0.7
S. typhi-murium	40	3	6	53	1	85	4	192	16.2
S. bredeney	7	—	—	3	1	11	2	24	2.0
S. montevideo	16	1	4	10	—	14	2	47	4.0
S. oranienburg	3	—	—	4	—	25	1	33	2.8
S. muenchen	3	2	—	3	—	—	—	8	0.7
S. manhattan	31	—	—	—	1	—	—	32	2.7
S. newport	27	19	1	15	10	85	7	164	13.8
S. bovis-morbificans	10	18	1	—	—	11	—	12	1.0
S. panama	10	18	—	4	9	2	—	43	3.6
S. gallinarum-pullorum	—	—	—	—	38	—	—	38	3.2
S. vejle	—	—	—	—	—	24	1	25	2.1
S. muenster	—	—	—	—	—	16	1	17	1.4
S. anatum	29	1	3	20	35	26	6	120	10.1
S. meleagridis	—	—	1	4	—	30	1	36	3.0
S. london	16	5	2	—	6	5	2	36	3.0
S. give	6	9	1	18	6	15	5	60	5.1
S. newington	2	—	—	1	8	—	—	11	0.9
S. shangani	—	—	—	—	—	24	—	24	2.0
Further types**	26	9	1	44	5	27	—	112	9.4
Total	291	68	21	193	149	427	37	1,186	

* Source of data: Swine: 77, 92, 118, 119, 120, 121, 122. Cattle: 123, 124, 125, 000. Fish: 126. Flies: 128, 129, 130, 134, 135. Food: 92, 94, 95, 136, 137, 138. Sewage: 140, 142. Water: 75, 142, 146.

** Each type less than 0.5% of the total.

Table X.
General distribution of Salmonella types.

	Argentine		Brazil		Chile		Costa Rica		Ecuador		Peru		Uruguay		Venezuela	
	Man	Other	Man	Other	Man	Other	Man	Other	Man	Other	Man	Other	Man	Other	Man	Other
S. paratyphi A	—	1	27	—	1	—	—	—	1	—	11	—	2	—	26	—
S. kisangani	1	—	—	—	—	—	—	—	—	—	5	—	—	—	—	—
S. arechavaleta	2	1	—	—	—	—	—	—	—	—	—	—	5	2	4	—
S. bispebjerg	—	128	—	—	1	—	—	—	—	—	—	1	—	—	1	—
S. abortus equi	—	6	1	—	—	—	—	—	—	1	—	—	—	18	4	1
S. paratyphi B	42	—	72	9	10	—	—	3	2	—	14	—	49	4	—	—
S. uppsala	2	—	—	—	—	—	—	—	—	—	—	—	1	—	—	—
S. abony	—	—	—	—	—	—	—	—	5	—	—	—	1	—	—	—
S. abortus-bovis	—	—	1	—	—	—	1	6	—	—	—	—	—	—	—	—
S. stanley	—	—	15	—	11	—	6	18	2	—	—	—	1	—	5	—
S. saint paul	—	—	25	5	3	—	1	5	1	1	33	—	3	—	1	—
S. reading	—	—	1	—	—	—	—	—	—	—	4	—	2	—	—	—
S. kaposvar	4	—	—	—	—	—	—	—	—	—	—	—	—	—	—	—
S. chester	—	1	3	—	—	—	7	7	—	—	5	—	—	2	1	—
S. san-diego	—	20	1	—	—	—	1	15	—	—	4	—	31	8	—	—
S. derby	8	—	147	6	17	—	14	16	—	—	2	—	64	30	80	—
S. essen	—	1	2	—	—	—	—	—	—	—	—	—	—	—	—	—
S. california	—	8	—	—	—	—	—	—	—	—	—	—	—	—	—	—
S. budapest	176	111	—	—	—	—	22	26	—	—	—	—	—	—	1	2
S. typhi-murium	6	15	216	54	103	—	10	2	4	3	74	94	710	94	83	—
S. bredeney	—	2	55	3	5	—	—	—	3	—	9	—	45	11	3	—
S. brandenburg	—	—	3	—	—	—	—	—	—	—	3	—	—	—	—	—
S. heidelberg	6	—	—	—	—	—	—	—	—	—	—	—	—	—	—	—
S. haifa	—	—	—	—	—	—	1	—	—	—	—	—	—	—	1	—
S. oslo	—	—	4	—	—	—	—	1	—	—	—	—	—	—	—	—
S. edinburg	—	—	—	—	—	—	—	3	—	—	—	—	—	—	—	—
S. paratyphi C	—	—	11	—	—	—	—	—	—	—	5	—	2	—	1	—
S. cholerae-suis	9	22	32	9	10	—	—	—	4	1	—	—	26	17	21	—
S. mission	—	—	3	—	—	—	—	—	—	—	—	—	1	—	—	—
S. amersfoort	—	—	—	—	—	—	—	—	—	1	—	—	—	3	—	—
S. lomita	—	—	1	—	—	—	—	—	—	—	—	—	—	—	—	—

S. norwich	S. montevideo	S. menston	S. oranienburg	S. thompson	S. virchow	S. infantis	S. hartford	S. bareilly	S. mikawasima	S. tennessee	S. narashino	S. belem	S. quiniela	S. muenchen	S. manhattan	S. newport	S. kottbus	S. takoradi	S. bonariensis	S. litchfield	S. manchester	S. fayed	S. bovis-morbificans	S. hidalgo	S. duesseldorf	S. glostrup	S. sanga	S. virginia	S. kentucky	S. hindmarsh	S. sendai	S. miami	S. onarimon	S. ndolo	S. eastbourne	S. israel	S. berta
11		46	2						2					19	3	20										12					1	2	1		3		
20														4	31																						1
135		40		1				2			2			69		360		1			1																
															1																						
		7	1	1		2	1			1	1	2	24		2																1	4	1				
		4								3		2		2											1												
		5	1						1			5		6	1							1															1
1		3										2	32	39			2		1			2										1					
3	1		4									3	19	19				1																	3		
			1					5								65						10			3										1		
																4																					
36	1	109	1	3								5	1	40	1	418		3	1			2									1	15		1	1		
32		26	3								2		1	99	4							13															
3							2									79						9	1				6		2								

Table X (continued)

	Argentine Man	Argentine Other	Brazil Man	Brazil Other	Chile Man	Chile Other	Costa Rica Man	Costa Rica Other	Ecuador Man	Ecuador Other	Peru Man	Peru Other	Uruguay Man	Uruguay Other	Venezuela Man	Venezuela Other
S. enteritidis	—	2	1	—	8	—	—	—	5	4	5	—	2	1	10	—
S. blegdam	—	1	13	13	—	—	—	—	1	1	6	1	2	1	3	1
S. dublin	—	—	4	—	—	—	—	—	—	—	1	—	—	—	—	—
S. rostock	—	—	—	—	—	—	—	—	—	—	—	—	—	—	—	—
S. claibornei	1	1	1	—	—	—	16	40	3	—	—	—	—	—	—	—
S. mendoza	—	3	50	—	1	—	—	—	—	—	22	—	25	1	47	—
S. panama	—	3	—	—	—	—	—	1	—	—	—	—	—	—	—	—
S. goettingen	—	—	3	—	—	—	—	—	—	—	—	—	—	—	—	—
S. dar-es-salaam	—	—	—	—	—	—	1	—	—	—	—	—	—	—	6	—
S. javiana	—	—	—	—	—	—	1	—	—	—	—	—	—	—	—	—
S. portland	—	—	—	—	—	—	—	—	—	—	—	—	—	—	—	—
S. gallinarum-pullorum	1	107	—	4	—	—	1	1	—	—	—	—	—	13	—	—
S. oxford	1	—	—	—	—	—	—	—	—	—	—	—	—	—	—	—
S. butantan	—	24	68	—	8	—	—	—	—	—	—	—	—	—	—	—
S. shangani	—	25	—	—	7	—	—	—	2	—	—	—	—	—	—	—
S. vejle	—	17	3	—	—	—	—	—	—	1	—	—	3	—	—	—
S. muenster	21	37	305	18	—	—	11	36	10	4	26	5	96	31	102	—
S. anatum	—	—	1	—	1	—	—	—	—	—	—	—	3	—	—	—
S. nyborg	—	33	—	—	—	—	—	—	1	—	—	—	—	—	—	—
S. meleagridis	5	—	1	—	—	—	—	1	1	1	—	—	12	—	21	—
S. zanzibar	—	—	—	—	—	—	—	—	—	—	—	—	—	—	—	—
S. nchanga	—	—	1	4	4	—	—	—	—	—	—	—	—	—	—	—
S. london	5	9	17	5	4	—	9	23	10	4	8	—	7	1	18	—
S. give	3	21	56	—	—	—	6	18	1	—	—	—	14	11	10	—
S. uganda	—	—	1	—	—	—	—	—	—	—	—	—	—	—	—	—
S. elisabethville	—	—	—	—	—	—	—	—	—	—	—	—	—	—	—	—
S. amager	—	—	—	—	—	—	—	—	—	—	—	—	—	—	2	—
S. clerkenwell	—	—	—	—	6	—	—	—	—	—	1	—	—	—	—	—
S. newington	—	—	2	10	—	—	—	—	—	—	1	—	3	3	3	—
S. selandia	—	3	1	—	—	—	—	—	—	—	—	—	1	—	—	—
S. new-brunswick	—	—	—	—	—	—	—	—	—	—	—	—	1	—	1	—
S. illinois	—	—	—	—	—	—	—	—	1	—	1	—	—	—	—	—
S. niloese	—	—	—	—	—	—	—	—	1	—	1	—	1	—	—	—

S. senftenberg	S. marseille	S. chingola	S. pretoria	S. abaetetuba	S. rubislaw	S. borbeck	S. poona	S. mississippi	S. atlanta	S. grumpensis	S. wichita	S. havana	S. worthington	S. del plata	S. florida	S. onderstepoort	S. caracas	S. madelia	S. brazil	S. hvittingfoss	S. gaminara	S. salford	S. cerro	S. minnesota	S. pomona	S. ballerup	S. urbana	S. morehead	S. adelaide	S. rio-grande
4					1	27						3				1								11	3					
1									1													2								
6												1			2						2		29	6		2				1
1																														
17																											2			1
															1	1										1				1
6		3		5		9	2	1		5	2	2			22	2	7						3	5		6	1			
				1																		1								1
	1								1			1																	1	
3																							3							
				1			5							1		1								2						
19	2	1		1		3				6	40	1		1			3	1						4				1		
6							2		1			1		1	3	2			3					6						
																								16						

Systematic investigations on enteric bacteria in water and sewage have been conducted by the group of the laboratory of "Obras Sanitarias de la Nación" in Buenos Aires. In 1942 MONTEVERDE & FERRAMOLA[140] examined 40 samples of sewage from the city of Buenos Aires and isolated 19 strains of 10 different *Salmonella* types; FERRAMOLA, MONTEVERDE & LEIGUARDA[141] found 10 positive samples out of 106 examined from the same city and FERRAMOLA et al.[142] did so in 13 of 24 samples of sewage from the city of Mendoza, isolating 19 strains of 12 different types.

A very thorough study on the contamination by enteric bacteria of the Rio de la Plata raw water has been made by LEIGUARDA et al.[75]. From January to December 1947 they daily gathered 6 hourly samples of water of one litre each and the 6 litres pool was concentrated by aluminium hydroxide following the technique described by PESO[143]. From 200 such samples they isolated salmonellae from 134 (67%): 60 of them yielded 2 or more types and up to 5. In all they obtained 1,022 strains belonging to 28 serological types, including a new one: *S. del-plata*[144].

With the same method FERRAMOLA et al.[142] studied the water of the Mendoza river at different points on its course and found 151 positive samples from 481 examined; they isolated in all 15 types including a new one: *S. mendoza*[145]. Looking for the source of pollution they surveyed the human and animal population at the sampling points and up current and did bacteriological examinations of their faeces. They concluded that contamination cannot be attributed to animals, but is in direct relation with the density of human population on the river's course.

PALOZZOLO et al.[146] examined 58 samples of one litre each of water of "acequias" (irrigation canals) of the City of Mendoza and found salmonellae in 48 of them (82.8%); the total number of strains isolated was 71, belonging to 12 serological types. Using methods of water analysis they established the most probable number of salmonellae in the water and found it to be generally 24/240 per 100 ml and up to 2,400 in some samples.

Table IX shows the distribution of *Salmonella* types isolated from all sources of human infection up supra analysed.

As a summary of all the information we could gather in the area surveyed, Table X shows the complete list and frequency of types isolated from man and other sources; under the first heading are included strains isolated from clinical conditions and carriers, under the second those from normal and diseased animals, food, water and sewage. Besides the work already referred to, some supplementary information has been included [147, 148, 149, 150, 151, 152].

It is worth to note that over 90% of the strains belong to only 25 types, which are all included in the first five groups of the KAUFF-

MAN-WHITE Scheme and can be serologically identified by the use of a reduced set of typing sera.

In all 133 types have been found; among them 15 were isolated for the first time in South America and most of them typed and described by the regional Salmonella Centres: *S. montevideo*[153], *S. berta*[154], *S. gaminara*[155], *S. arechavaleta*[58], *S. cerro*[51], *S. bonariensis*[156], *S. carrau*[157], *S. grumpensis*[157], *S. butantan*[158], *S. del-plata*[144] and *S. mendoza*[145]; *S. belem* and *S. brazil* were typed and described by EDWARDS & FIFE[159]; *S. maracaibo* was typed and described by LE MINOR & FOSSAERT[160]; *S. caracas*, isolated by BRICEÑO in Venezuela[161], was typed in Montevideo.

REFERENCES

1. LIGNIERES, J.: Contribución al estudio y clasificación de las salmonelosis humanas y animales. *Rev. Zootécn.*, Bs. As., **10**, *97* (1923).
2. RUIZ, F. R. & MARTINEZ, J.: Sobre un caso de septicemia por Bacillus enteritidis de Gärtner. *Rev. Med. Rosario.*, **16**, *488* (1926).
3. BACHMANN, A. & LOUREIRO, J.: Sobre la presencia de Paratifoideas C en la Argentina. *La Sem. Med.*, **39**, *1936* (1923).
4. CHIODI, E., CERVINI, P. R. & MIRAVENT, J. M.: Septicemia a B. enteritidis de Gärtner en un lactante de dos meses. *Arch. Arg. Ped.*, **3**, *527* (1923).
5. SAVINO, E., MIRAVENT, J. M. & CHIODI, E., Consideraciones sobre el trabajo de los Dres. Chiodi, E., Cervini, P. R. & Miravent, J. M. titulado "Septicemia a bacilo enteritidis de Gärtner en un lactante de dos meses". *Folia Biológica*, Bs. As., No **49/51**, *219* (1935).
6. ELKELES, G. & BARROS, E.: Infección humana aguda por Bacilo suipestifer (Salmonella cholera-suis var. Kunzendorf). *Rev. Soc. Arg. Biol.*, **11**, *158* (1935).
7. SOSA, H.: Salmonella newport en infecciones humanas. *Folia Biológica*, Bs. As., No **49/51**, *220* (1935).
8. FERRARIO, J. C.: Salmonella derby aislada de órganos de cerdo. *Rev. Soc. Arg. Biol.*, **11**, *168* (1935).
9. RISQUEZ, J. R.: Septicemias en Caracas, *Gaceta Med. Caracas*, **42**, *1* (1935).
10. SAVINO, E. & MENENDEZ, P. E.: Salmonella enteritidis (Gärtner) var. chaco n. var. agente de paratifoidea. *La Sem. Med.*, **3**, *217* (1935).
11. HORMAECHE, E., PELUFFO, C. A., & ALEPPO, P. L.: Nueva contribución al estudio etiológico de la "diarreas infantiles de verano". Las "Salmonelas" en las enterocolitis de la infancia. *Arch. Urug. Med. Cirug. y Espec.*, **9**, *113* (1936).
12. HORMAECHE, E., PELUFFO, C. A. & ALEPPO, P. L.: Zur Aetiologie der Sommerdiarrhoe bei Kindern mit besonderer Berücksichtigung der Salmonella-Infektionen. *Z. Hyg.*, **119**, *453* (1937).
13. BONABA, J., CARRAU, A., HORMAECHE, E., ZERBINO, V., ALEPPO, P. L., PELUFFO, C. A., PELUFFO, E., PRADERI, J. A., RAMON GUERRA, A. & SURRACO, N. L.: Estudios sobre la etiología infecciosa de las diarreas infantiles. Montevideo, (1940). J. García Morales, Ed.
14. HORMAECHE, E., SURRACO, N. L., PELUFFO, C. A. & ALEPPO, P. L.: Causes of infantile summer diarrhea. *Amer. J. Dis. Child.*, **66**, *539* (1943).
15. HORMAECHE, E., SURRACO, N. L., PELUFFO, C. A. & ALEPPO, P. L.: Nuevos estudios sobre las diarreas infantiles de origen infeccioso. *Anal. Inst. Hig. Montevideo.*, **1**, *33* (1947).

500

16. MARTINHO DA ROCHA, J., LUIZ DE ARAUJO MORAES, N., ARAUJO COSTA, G., COSTA, A., BUDIANSKY, E., CASTRO GARCIA, J. P. & VAZQUEZ NOBREGA, V.: Simpósio sòbre as diarréias agudas no lactante. *Bol. Inst. Puer. Univ. Brasil.*, **17,** *199* (1960).

17. VERHOESTRAETE, L. J. & PUFFER, R. R.: Las enfermedades diarréicas con especial referencia a las Américas. *Bol. Of. Sanit. Pan.*, **44,** *95* (1958).

18. MONTEVERDE, J. J. & DE SIMONE, C.: Infecciones intestinales humanas producidas por Salmonellas de origen animal. *Rev. Med. Cienc. Afines,* Bs. As., (1937).

19. MONTEVERDE, J. J. & DE SIMONE, C.: Presencia de Shigelas y Salmonelas en diarreas infantiles de verano. *Rev. Asoc. Arg. Diet.*, Bs. As., **5,** *233* (1947).

20. GRICHENER, E., MONTAUT, F. & ROMER, J.: Las Shigellosis y las Salmonellosis como causa de las Diarreas Infantiles en la ciudad de Rosario. *La Sem. Méd.*, **2,** *1356* (1949).

21. MONTAGNA, C. P. & RIMOLDI, A. A.: Importancia de la infección enteral en la etiología de los estados diarréicos agudos de la infancia. *Rev. Asoc. Arg. Diet.*, **2,** *277* (1944).

22. VERNA, L. C.: Primeras investigaciones realizadas en Tucumán sobre "Salmonelas y Shigelas" en las diarreas infantiles. *Rev. Asoc. Bioq. Arg.*, **11,** *273* (1944).

23. CATALDI, M. S.: Etiología infecciosa de las diarreas infantiles de verano. *Rev. Asoc. Arg. Diet.*, **2,** *118* (1944).

24. CATALDI, M. S.: Investigaciones bacteriológicas sobre las diarreas infantiles de verano en Buenos Aires. *Ciencia e Invest.*, *528* (1945).

25. RAMACCIOTTI, F., CURA, E., WILKOZ, T., VALLEJOS, M. E., BRIZUELA, W., MACCIONI, H., CENTENO, R., ROCA, J. A. & PUCHETA, C.: Estudio bacteriológico sobre 863 muestras de diarreas infantiles de diferentes provincias de la República Argentina durante los años 1960—61. *La Sem. Med.* **120,** *677.* (1962).

26. LUBIN, A. H., GIROLA, R. A., GRINSTEIN, S. & PIROSKY, I.: Incidencia de enterobacterias patógenas en 1092 casos de diarrea epidémica del recién nacido., Bs. As., Personal communication (1962).

27. TAUNAY, A. E., CORREA, G. A. & FLEURY, C. T.: Freqüencia de alguns agentes microbianos nas chamadas diarréias infantis em Sao Paulo. *Rev. Inst. Adolfo Lutz.*, **5,** *331* (1945).

28. PELUFFO, C. A., BIER, O., AMARAL, J. P. & BIOCCA, E.: Estudos sobre as salmoneloses em Sao Paulo. I. Incidencia dos diferentes tipos em diarréias infantis. *Mem. Inst. Butantan*, **19,** *211* (1946).

29. ARAUJO COSTA, G., COSTA, A. & BROOKING, CH.: As Shigeloses e Salmoneloses na etiologia das diarréias agudas da crianca. *Bol. Inst. Puer. Univ. Brasil*, **14,** *79* (1957).

30. MAROJA, R. C., ALMEIDA, A. J. DE, BARROS DE SOUZA, E. & DE FREITAS, E. N.: Estudios bacteriológicos de una epidemia de diarréia infantil em Fortaleza, Ceará. X Jorn. Bras. Puer. Ped., Fortaleza, Ceará (1958).

31. MAROJA, R. C., DE FREITAS, E. N. & MONTEIRO DA CRUZ, F.: Tipos de salmonella isolados na zona da Mata de Pernambuco. *Rev. Serv. Esp. Saúde Públ.*, **10,** *759* (1958).

32. VACCARO, H., PEREZ, M. & VALENZUELA, E.: Salmonelosis y Shigelosis en Chile. *Rev. Chil. Ped.*, **24,** *153* (1943).

33. GARCES, C. & PRADO, E.: Etiología de las diarreas del lactante seguida durante 1 año en el Hospital L. Calvo Mackenna. I Congr. Confd. Soc. Sudamer. Ped. Chile., *347* (1944).

34. ESQUIVEL, R.: El problem de las diarreas infantiles en Costa Rica. Contribución al estudio de su etiología. Tesis. Univ. de Costa Rica (1958).

501

35. De la Cruz, E.: Etiología de Gastroenteritis. Personal communication (1962).
36. Alcivar Ceballos, C.: Algunos datos sobre infecciones salmonellares y shigelares en niños de Guayaquil. *Rev. Ecuat. Ped.*, 1, *18* (1949).
37. Briceño Iragorry, L.: Contribución al estudio de la naturaleza infecciosa del Sindrome de "Diarrea y Enteritis" infantil. *Rev. Sanid. Asist. Soc.*, 7, *363* (1942).
38. Oropeza, P., Irazabal, P. & Briceño Iragorry, L.: La mortalidad infantil en Caracas. Las diarreas y enteritis. Su etiología. *I Jorn. Venez. Puer. Ped.*, 2, *131* (1944).
39. Le Minor, S., Fossaert, H. & Maso-Dominguez, J.: Salmonelles isolées à Maracaibo. (Venezuela). *Bull. Soc. Path. exot.*, 47, *775* (1954).
40. Briceño Iragorry, L. & Fossaert, H. C.: Estudio bacteriológico de las diarreas y enteritis infantiles en Venezuela. Mem. "I Congr. Venez. Salud Publ. y III Conf. Unid. Sanit.", *125* (1956).
41. Hormaeche, E. & Surraco, N. L.: Estudios sobre el valor de los métodos de aislamiento de salmonelas y shigelas. *Arch. Urug. Med. Cirug. y Espec.*, 28, *485* (1941).
42. Grichener, E.: Las Salmonelas en Patología Humana. *Rev. Méd. Rosario. Argentina.*, 34, *896* (1944).
43. Burgos, H. I.: Enteritis agudas por salmonelas y shigelas en la infancia. *Rev. Soc. Puer. Bs. As*, 17, *37* (1951).
44. Scroggie, A., Garces, H., Costa, A. & Agliati, J.: Salmonellosis intestinal en el lactante. *Rev. Chilena Ped.*, 24, *46* (1953).
45. Canessa, E. & Garces, C.: Distribución de especies de Shigellas y Salmonellas clasificadas en la sección de gérmenes entéricos. I Congr. Confed. Soc. Sudamer. Ped. Chile., *349* (1944).
46. Cid Rojas, L. & Gonzales Rojas, M.: Diarreas infecciosas del lactante. I Congr. Confed. Soc. Sudamer. Ped. Chile, *223* (1944).
47. Peluffo, C. A.: Distribución y frecuencia de los tipos del género salmonela aislados en el Uruguay. *Anal. Inst. Hig. Montevideo*, 4, *49* (1950).
48. Colichon, H. & Berrocal, A.: La salmonelosis en el Perú y la complicación secundaria de la Bartonelosis humana o enfermedad de Carrión. *Proc. 6th Int. Congr. Trop. Med. and Malar.*, 4, *113* (1959).
49. Araujo Costa, G., Suassuna, I. & Suassuna, Ivone R.: Tipos de salmonella e shigella ocorrentes no Rio de Janeiro. *Anais Microb.*, 5, *305* (1957).
50. Hormaeche, E., Peluffo, C. A. & Aleppo, P. L.: Las salmonelas en patología infantil. *Arch. Ped. Urug.*, 11, *8* (1940).
51. Hormaeche, E., Peluffo, C. A. & Aleppo, P. L.: S. cerro nuevo tipo de salmonela. Estudio bacteriológico y clínico. *Arch. Urug. Med. Cirug. y Espec.*, 19, *125* (1941).
52. Abramson, H., Frant, S. & Oldenbusch, C.: Salmonella infection of the new born. *Med. Clin. N. Amer.*, 23, *591* (1939).
53. Guthrie, K. J. & Montgomery.: Infections with Salmonella enteritidis in infancy with the triad of enteritis, cholecystitis and meningitis. *J. Path. Bact.*, 49, *393* (1939).
54. Bornstein, S.: The state of Salmonella problem. *J. Immunol.*, 46, *439* (1943).
55. Felsenfeld, O. & Mae Young, Viola.: The Geography of Salmonella. *Amer. J. Dig. Dis.*, 14, *47* (1947).
56. Edwards, P., Bruner, W. & Moran, Alice.: The genus Salmonella. Its occurrence and distribution in the U.S. *Ky. Agric. Exp. Stat., Bull.* 525 (1948).
57. Ramon, Guerra, A., Peluffo, E., Laguarda, M. & Aleppo, P. L.: Septicemias y bacteriemias por salmonelas en el lactante. *Arch. Ped. Urug.*, 10, *669* (1939).

502

58. Hormaeche, E. & Peluffo, C. A.: Las salmonelosis infantiles y su diagnóstico. *P. Rico J. Pub. Health Trop. Med.*, **17**, *71* (1941).
59. Peluffo, E. & Aleppo, P. L.: Las tifosis del lactante. Pediatría, p. 113, Montevideo L.I.G.U., Ed., (1951).
60. Del Campo, R. M., Aleppo, P. L.: Peritonitis por salmonelas. *Arch. Ped. Urug.*, **11**, *868* (1940).
61. Wiederhold, A. & Costa, A.: Septicemia a Salmonella bareilly. (tratada con cloromicetina). *Rev. Méd. Córdoba*, **37**, *510* (1949).
62. Ramacciotti, F., Cura, E., Roca, J. A., Santolaya, A. & Paolasso, Rosa W.: Salmonelas aisladas en la ciudad de Córdoba. Argentina. Personal communication. (1962).
63. Teobaldo, C., Actis Dato, A. & Chueca, P.: Aislamiento de un bacilo paratyphosus B de un líquido céfalo raquídeo. *Rev. Sud. Amer.*, Bs. As., **22**, *497* (1946).
64. Litmanovich, M. & Grichener, E.: Pleuresia purulenta a Salmonella kunzendorf. *La Semana Méd.*, Bs. As., **51**, *14* (1944).
65. Briceño Iragorry, L.: Sobre el papel de las Salmonelas (Grupo del Paratífico "C") en la etiologia del sindrome febril de la zona de Caracas. *Rev. San. Asist. Soc.*, **8**, *141* (1943).
66. Briceño Iragorry, L.: Contribución al estudio de las salmonelosis humanas en Venezuela. *Gaceta Méd. Caracas.*, pp. *1—84* (1948).
67. Taunay, A. de E., Alvares Correa, Gilda, & Toledo Fleury, C. Frequência de alguns agentes microbianos nas chamadas "Diarréias infantis" em São Paulo. *Rev. Inst. A. Lutz*, **5**, *331* (1945).
68. Rangel Pestana, B. & Rugai, E.: "Salmonelas isoladas de líquido céfalo raquidiano". *Anais Paul. Med. Cirug.*, **39**, *373* (1939).
69. Baracchini, O.: Salmonella typhi-murium isolada de um caso de meningite cerebrospinal. *Rev. Inst. A. Lutz*, **9**, *92* (1949).
70. Taunay, A. de E.: Personal communication (1962).
71. Taunay, A. de E. & De Brito, E. Silva, M.: Salmonelose com localizacão extra intestinal. *Rev. Inst. A. Lutz*, **3**, *244* (1943).
72. Henderson, L. L.: Salmonella infections in Panama. *Amer. J. trop. Med.*, **27**, *643* (1947).
73. Rabe, E. F.: Salmonellosis in children. *Pediatrics*, **13**, *247* (1954).
74. Gardini Tuesta, W. E. & Demarini, O. J.: Contribución al estudio de las infecciones intestinales del niño. *Bol. Ofi. Sanit. Panamer.*, **51**, *145* (1961).
75. Leiguarda, R., Peso, O. & Kempny, J.: Investigación de bacterias de los géneros salmonela y shigela en agua del Rio de la Plata. *Asoc. Interam. Ing. Sanit.*, **2**, *153* (1948).
76. San Martin, H. & Urzua, H.: Estudios sobre fiebre tifoidea: III. Epidemiología de las fiebres tifoidea y paratifoidea en Quinta Normal. *Rev. Chil. Hig. Med. Prev.*, **8**, *41* (1946).
77. Noronha Peres, J.: Investigaçõesобre o gênero Salmonella em Belo Horizonte. Tesis. Quiroz Breiner, Ed. Belo Horizonte (1948).
78. Ibañez, S., Armijo, R. & Prado, E.: Brote de paratifus B en un sanatorio. Aspecto clínico, epidemiológico y bacteriológico. *Rev. Chil. Ped.*, **20**, *227* (1949).
79. Giglioli, G. (1933). Quoted by Briceño Iragorry, L. (66).
80. Savino, E. & Menendez, P. E.: Salmonela enteritidis var. chaco. — Nuevo agente de paratifoidea. *Rev. Soc. Arg. Biol.*, **10**, *384* (1934).
81. Hormaeche, E. & Peluffo, C. A.: Las salmonelosis en el Uruguay. *Rev. Arg-Norteamer. Cienc. Méd.*, **1**, *115* (1943).
82. Ledesma, C. & Burgarelli, L.: Primer brote de Salmonella por intoxicación alimenticia causado por S. newport, en el Uruguay. *An. Cl. e Inst. Enf. Infecc.*, **2**, *617* (1942).

83. TAUNAY, A. DE E., FERNANDES PONTES, J., PRADO, E. & SALDOVAL
PEIXOTO, Ethel.: Shigueloses. — Comparação de métodos de colheita
das feces no diagnóstico bacteriológico das enterocolitis crônicas. Rev.
Inst. A. Lutz, **16,** *37* (1956).

84. MONTEIRO,A.: Dois casos de salmonelose por S. oraniemburg. *O Hospital,*
30, *259* (1946).

85. DE SALLES GOMEZ, L. & ARANTES, MARÍA.: Sôbre quinze amostras de
Salmonella cholerae suis var. Kunzendorf, isoladas de sangue humano.
Rev. Inst. A. Lutz, **10,** *89* (1950).

86. TOLEDO MELLO, J. DE: Salmonella isolada de um caso de Leptomening-
ite purulenta. *Rev. Paulista Med.,* **29,** *422* (1946).

87. ALCIVAR Z. C.: Los portadores de bacterias patógenas, un factor importante en la diseminación de enfermedades entéricas.

88. DE LA CRUZ, E.: Epidemiología de la salmonelosis en Costa Rica. III.
Salmonelas en manipuladores de carnes procesadas. *Rev. Biol. Trop.,*
7, *1* (1959).

89. MONTEVERDE, J. J. & DE SIMONE, C.: Presencia de Shigelas y Salmonelas en heces humanas normales. *Rev. Asoc. Arg. Diet.,* **7,** *28* (1949).

90. VACCARO, H., MENEGHELLO, L. & NIÑO, L.: Contribución al estudio de
portadores sanos de Salmonellas y Shigellas. *Rev. Chil. Hig. Med.
Prev.,* **4,** *335* (1942).

91. FELSENFELD, O., YOUNG, VIOLA M., TUTTEN, F. J. L. S., ARNOLD, R.
M., FERREIRA, S. & GUILBRIDE, P. D. L.: Investigation of enteric
infections in the Caribean area. Distribution of Salmonella strains in
Curacao, Jamaica and Costa Rica. *Amer. J. Dig. Dis.,* **20,** *233* (1953).

92. BOLAÑOS, R.: Sobre la distribución e identificación serológica del género
Salmonela en Costa Rica. *Rev. Biol. Trop.,* **6,** *43,* (1958).

93. HORMAECHE, E. & PELUFFO, C. A.: Unpublished data.

94. DE LA CRUZ, E., MOORE, H. A., PEREZ, F. I. & MORA, J. A.: Salmonelas
en carnes molidas destinadas al consumo. 2e. Congr. Lat. Amer.
Microb., San José (1961).

95. DE LA CRUZ, E.: Epidemiología de la Salmonelosis en Costa Rica. II.
Salmonelas en carnes procesadas. *Rev. Biol. Trop.,* **6,** *37* (1958).

96. MONTEVERDE, J. J. & GARBERS, G. V.: Infecciones debidas a Salmonella
abortivoequina. Aborto de las yeguas y piosepticemia de los potrillos.
Rev. Med. Vet., Bs. As., **31,** *138* (1949).

97. MONTEVERDE, J. J. & GARBERS, G. V.: Infecciones debidas a Salmonella abortus-equi en la Republica Argentina. Fac. Agr. Vet., Serie
Publ. No 7 (1956).

98. D'APICE, M.: Observações sobre o abôrto contagioso das eguas em
Sao Paulo. *Arch. Inst. Biol. S. Paulo,* **14,** *235* (1943).

99. DE LEON, J. P. & PELUFFO, C. A.: El aborto de la yegua "pur-sang".
Primera constatación de S. abortus equi en el Uruguay. *Bol. Dir.
Ganad., Montevideo,* **29,** *546* (1946).

100. MONTEVERDE, J. J. & SIMEONE, D. H.: Salmonelas genuinamente avia-
rias en aves "reaccionantes". *Rev. Fac. Agr. Vet., Bs. As.,* **1,** *3* (1944).

101. NOBREGA, P. (1935). Quoted by MONTEVERDE, J. J. & SIMEONE, D. H.
(100).

102. MONTEVERDE, J. J. & SIMEONE, D. H.: Salmonelas distintas de S.
pullorum y S. gallinarum en aves "reaccionantes". *Rev. Fac. Agr. Vet.,*
Bs. As., **11,** *31* (1944).

103. CASTILLO GILARDE, A.: El medio IM-SMG en el diagnóstico de la Salmonella typhi-murium en los animales. *Rev. Fac. Med. Vet., Lima,*
5, *15* (1950).

104. MONTEVERDE, J. J., SIMEONE, D. H. & KOSIK, C. V.: Microbiología
entérica en los perros. I Salmonelas. II Jorn. Vet. Plat., La Plata
(1961).

504

105. ARROYO, G. & BOLAÑOS, R.: Enteritis infecciosa de los terneros. I. Estudio de tres brotes en diferentes localidades del país. *Rev. Biol. Trop.*, **7**, *89* (1959).
106. QUIROGA, S. S. & MONTEVERDE, J. J.: Investigación de bacterias del género "Salmonella" en cerdos enfermos de peste y viruela. *Rev. Med. Vet.*, Bs. As., **27**, *226* (1945).
107. ASSUMPÇAO, L. & RIBAS, J. C.: Incidencias de bacterias do genero Salmonella em ratos da cidade de Sao Paulo. *Mem. Inst. Butantan*, **17**, *127* (1943).
108. DE LA CRUZ, E.: Salmonelas en animales domésticos. Personal communication (1962).
109. GALTON, M. M., SCATTERDEY, J. E. & HARDY, A. V. (1952). Quoted by EDWARDS, P. R.; Salmonella and Salmonellosis. *Ann. N. Y. Acad. Sci.*, **66**, *44* (1956).
110. GALLO, P.: La paratifosis de los terneros en Venezuela. *Rev. Vet. Parast.*, **1**, *34* (1939).
111. GUERRERO, R. P., Salmonellosis of Calves in Tropical Countries. *J. Amer. vet. med. Ass.*, **103**, *152* (1943).
112. PENHA, A. M. & D'APICE, M.: Observações sôbre enterite infectuosa dos becerros. Etiología e vacinação. Congr. Bras. Vet., Porto Alegre, *454* (1946).
113. PENHA, A. M. & ESQUIBEL, A.: Novas observações de salmonelose bovina en gado importado submetido a premunição contra a "Tristeza". *Bol. Soc. Paulista Med. Vet.*, **7**, *73* (1945).
114. LEITE XAVIER, V.: Ocorrência da S. rubislaw em Bovino. (Bos taurus). *Anais Escola Flum. Med. Vet.*, **1**, *1* (1958).
115. LEITE XAVIER, V.: Ocorrência de S. bredeney em gato domestico. (Primeiro registro no Brasil). *Veterinária*, **9**, *27* (1955).
116. MONTEVERDE, J. J.: Salmonella bredeney isolated in a case of septicaemic swine-pox. II. Congr. Panamer. Med. Vet., Sao Paulo, *15* (1954).
117. MERA, A. F.: Un caso de infección a paratifo A en gato. *Rev. Med. Vet.*, Bs. As., **23**, *420* (1941).
118. HORMAECHE, E. & SALSAMENDI, R.: Sobre la presencia de Salmonelas en los ganglios mesentéricos de cerdos normales. *Arch. Urug. Med. Cirug. Espec.*, **9**, *665*, (1936).
119. HORMAECHE, E. & SALSAMENDI, R.: El cerdo normal como "portador" de Salmonellas. *Arch. Urug. Med. Cirug. Espec.*, **14**, *375* (1939).
120. QUIROGA, S. S. & MONTEVERDE, J. J.: Investigación de los ganglios mesentéricos de cerdos normales. III. Jorn. Agr. Vet., Bs. As., pp. *69* (1941).
121. PESTANA, B. R. & RUGAI. E.: O porco normal como portador de Salmonella. *Rev. Inst. Adolfo Lutz.*, **3**, *232* (1943).
122. DE LA CRUZ, E.: Epidemiología de la Salmonelosis en Costa Rica. I. Salmonelosis en porcinos. *Rev. Biol. Trop.*, **6**, *27* (1958).
123. ARROYO, G.: Un aporte al estudio de los, portadores bovinos de Salmonella en Costa Rica. Tesis (1958). Quoted by BOLAÑOS, R. (92).
124. SIMEONE, D. H.: Investigación de bacterias del género Salmonella en ganglios mesentéricos de bovinos normales. *Rev. Med. Cienc. Afines.*, **9**, *9* (1947).
125. ARROYO, G. & BOLAÑOS, R.: Salmonella en bovinos adultos, aparentemente sanos, destinados al consumo. *Rev. Biol. Trop.*, **8**, *49* (1960).
126. LEIGUARDA, R. H., PALAZZOLO, ANA DE, & PESO, O. A.: Bacterias del contenido intestinal de algunos peces del Rio de la Plata. *Rev. Obras Sanit. Nación.*, Bs. As., **14**, *2* (1950).
127. GABALDON, A.: Enseñanza para la acción sanitaria de la América Latina derivadas de la Lucha Antimalárica en Venezuela. *Bol. Ofic. Sanit. Panamer.*, **38**, *259* (1955).

128. HORMAECHE, E., PELUFFO, C. A. & ALEPPO, P. L.: Investigaciones sobre la existencia de bacterias de los géneros salmonela y shigela en las moscas. *Anal. Inst. Hig. Montevideo*, **4**, *75* (1950).
129. BOLANOS, R.: Frequencia de Salmonella y Shigella en moscas domésticas en la ciudad de San José. *Rev. Biol. Trop.*, **7**, *207* (1959).
130. MAROJA, R. C., LIMA FERRO, TEREZINHA & DE FREITAS, E. N.: Enterobactérias patogênicas mecânicamente transportadas por môscas na cidade de Palmares, Pernambuco. *Rev. Serv. Esp. Saúde Públ.*, **10**, *741* (1956).
131. COUTINHO, J. O., TAUNAY, A. DE E., & PENNA, P.: Importância da Musca domestica como Vector de Agentes Patogênicos para o Homem. *Rev. Inst. Adolfo Lutz*, **17**, *5* (1957).
132. FRANCA, MARÍA.: Unpublished data.
133. TAUNAY, A. DE E., PENNA, L. & COUTINHO, J. O.: Observações sobre a transmissão de agentes patogênicos para o homem por meio de baratas. *Rev. Inst. Adolfo Lutz*, **17**, *25* (1957).
134. ALCIVAR Z., C. & CAMPOS R. F.: Las moscas como agentes vectores de enfermedades entéricas en Guayaquil. *Rev. Ecuat. Hig. Med. Trop.*, **3**, *3* (1946).
135. RAMACCIOTTI, F.: Investigación del género Salmonella en las moscas de la ciudad de Córdoba. Conferencia Clín. Chutro., Córdoba (1952).
136. ASSUMPCAO, L. DE: Pesquisa de bacterias do gênero Salmonella en carnes e seus derivados vendidos a retalho. *Arqu. Hig. Saúde Públ.*, **11**, *475* (1946).
137. PESTANA, B. R. & RUGAI, E.: Da presença de Salmonellas nas carnes preparadas. *Rev. Inst. Adolfo Lutz.*, **7**, *5* (1947).
138. SIMEONE, D. H.: Investigación de bacterias del género Salmonella en huevos de aves. *Rev. Fac. Agr. Vet.*, Bs. As., **2**, *15* (1944).
139. MAROJA, R. C., & LOWERY, W. D.: Estudos sobre diarreias agudas. III. Encontro de ovos de Tracajá (Podocnemis Dumeriliana) infectados con Salmonella salford. *Rev. Serv. Esp. Saúde Públ.*, **8**, *591* (1956).
140. MONTEVERDE, J. J. & FERRAMOLA, E.: Investigación de bacterias del género Salmonella en el líquido cloacal de la ciudad de Buenos Aires. *Anal. Soc. Cient. Argent.*, **133**, *417* (1942).
141. FERRAMOLA, R., MONTEVERDE, J. J. & LEIGUARDA, R. H.: Investigación del género Salmonella en agua y líquido cloacal. *Bol. Obras Sanit. Nación.*, Bs. As., **74**, *103* (1943).
142. FERRAMOLA, R., LEIGUARDA, R. H., ANSIAUME, E. M., PESO, O. A. & PALAZZOLO, A.: Estudio de la contaminación de aguas superficiales por bacterias del género Salmonella. *Rev. Obras Sanit. Nación.*, Bs. As., **157**, *93* (1954).
143. PESO, O. A.: Estudio de un método de concentración de muestras de agua para la investigación de bacterias. *Rev. A.N.D.A.*, Bs. As., **54**, *115* (1957).
144. LEIGUARDA, R., PESO, O. & KEMPNY, J.: S. del plata: Nuevo tipo del género Salmonella. *Anal. Soc. Cient. Argent.*, **148**, *168* (1949).
145. LEIGUARDA, R. H., PESO, O. A., PALAZZOLO, A. & ANSIAUME, E. M.: Un nuevo tipo del género salmonela: S. mendoza. *Anal. Soc. Cient. Argent.*, **152**, *230* (1951).
146. PALAZZOLO, ANA, ANSIAUME, E. M., LEIGUARDA, R. H. & PESO, O. A.: La investigación de bacterias del género salmonella en agua de acequias. Su relación con los trastornos intestinales infantiles. *Rev. Obras Sanit. Nación.*, Bs. As., **157**, *127* (1954).
147. MONTEVERDE, J. J. & DE SIMONE, C.: Presencia de shigelas y salmonelas en diarreas de verano producidas entre la población infantil de la ciudad de Buenos Aires y sus alrededores. *Rev. Asoc. Arg. Diet.*, **6**, *305* (1948).

148. MAROJA, R. C. & LOWERY, W. D.: Estudos sôbre diarréias agudas. IV. Tipos de salmonella isolados em casos de diarréia em Santarém — Pará. *Rev. Serv. Esp. Saúde Publ.*, **8**, *595* (1956).

149. MONTEVERDE, J. J. & GARBERS, G. V.: Salmonelosis de los equinos. (Infección debida a Salmonella bovis-morbificans). *Rev. Med. Vet.*, Bs. As., **38**, *1* (1956).

150. XIRINACH, HILDA: Correlación de los hallazgos parasitológicos, bacteriológicos e histológicos efectuados en 103 apéndices. Tesis. Univ. Costa Rica. (1956). Quoted by BOLAÑOS, R. (92).

151. DE LEON, J. P., EPSTEIN, B., TEDESCO, L. F. & PIÑON, J. C.: Contribución al estudio clínico y experimental con la Salmonella abortus equi. *Rev. Med. Vet.*, *Montevideo*, **25**, *849* (1949).

152. MIRANDA, H., GRADOS, O., FERNANDEZ, W., ARRASCO, M. & HIDALGO, HILDA: Estudio bacteriológico de 100 casos de enteritis ocurridos en Trujillo, Perú. *Rev. Med. Peruana* **31**, *27* (1962).

153. HORMAECHE, E. & PELUFFO, C. A.: S. montevideo. Nuevo tipo de salmonela encontrado en el Uruguay. *Arch. Urug. Med. Cirug. Espec.*, **9**, *673* (1936).

154. HORMAECHE, E., PELUFFO, C. A. & SALSAMENDI, R.: Nuevo tipo del género salmonela: "S. berta". *Arch. Urug. Med. Cirug. Espec.*, **12**, *377* (1938).

155. HORMAECHE, E. & PELUFFO, C. A.: Estudios sobre salmonela gaminara. *Arch. Urug. Med. Cirug. Espec.*, **3**, *217* (1939).

156. MONTEVERDE, J. J.: S. bonariensis. *Rev. Med. Cienc. Afines.*, Bs. As., **4**, *241* (1942).

157. HORMAECHE, E., PELUFFO, C. A. & RICAUD DE PEREYRA, V.: A new salmonella type, Salmonella carrau, with special reference to the 1,7 phases of the Kauffmann-White classification. *J. Bact.*, **47**, *323* (1944).

158. PELUFFO, C. A., BIER, O., AMARAL, J. P. & BIOCCA, E.: Estudos sôbre as salmoneloses em Sao Paulo. II. Um novo tipo de salmonela patogénica para o homem. S. butantan. *Mem. Inst. Butantan*, **19**, *217* (1946).

159. EDWARDS, P. & FIFE, MARY: Two new salmonella types: S. belem and S. brazil. *Pub. Hlth. Lab.*, **10** (1952).

160. LE MINOR, L. & FOSSAERT, H.: Un nouveau type de Salmonella isolé au Venezuela. *Ann. Inst. Pasteur*, **87**, *104* (1954).

161. BRICEÑO IRAGORRY, L.: Nueva nota sobre salmonelosis humana (Salmonella caracas). (1952). Quoted by BRICEÑO IRAGORRY, L. & FOSSAERT, H. C. (40).

SALMONELLOSES IN JAPAN

BY

HIDEO FUKUMI

Department of Bacteriology, National Institute of Health, Tokyo.

Statistical survey of Salmonella food-poisoning

It is unnecessary to draw attention to the fact that *Salmonella* widely infect the animal world including human beings, and moreover that most members of the *Salmonella* group can, and really do, infect more than one animal species. It is therefore absolutely necessary to examine a large number of species to consider the whole field of salmonelloses. Of course, human beings are our greatest concern since the primary aim of medicine is to protect human life and health.

Roughly speaking, human salmonelloses can be divided into two clinical groups, the one being enteric fever and the other gastroenteritis. The latter disease appearing, from the epidemiological point of view, as food-poisoning. Therefore, in order to obtain a complete picture of the epidemiology of salmonelloses, it may be useful to examine the statistical data of food-poisoning due to members of the *Salmonella* group as presented in the general food-poisoning statistics.

Statistical data of food-poisoning provided by the Ministry of Welfare and Public Health of Japan are based on reports from rural Health Departments; these mostly deal with outbreaks but not with sporadic cases treated by private physicians and not generally reported to Public Health Authorities. Thus there should certainly be many cases of food-poisoning that are not reported in official statistics. The details presented below should be read with this reservation in mind. Furthermore, the etiological analysis may not always be carried out efficiently because certain items of pathological material may not be fully collected or, in certain cases, there are not sufficient facilities for bacteriological or chemical examinations. For instance, an outbreak caused by a *Salmonella* may be reported as being from unknown etiology through lack of bacteriological data.

Table I shows the annual frequency of Salmonella food-poisoning giving, at the same time, details of staphylococcal food-poisoning for the period 1954 to 1960. The relative frequency of either Salmonella or staphylococcal food-poisoning compared to the total number of cases is relatively constant in each year. The relative frequency of Salmonella food-poisoning is also seen in Table II

Table I.

Annual occurrence of food-poisoning incidents in Japan.

Year	Salmonella food-poisoning			Staphylococcal foodpoisoning			Total food-poisoning		
	in-cident	patient	death	in-cident	patient	death	in-cident	patient	death
1954	42	1,161	15	28	463	2	1,354	22,528	358
1955	94	3,597	17	43	3,257	1	3,277	63,745	554
1956	50	1,411	11	36	2,170	1	1,665	28,286	271
1957	47	2,330	8	45	1,962	1	1,716	24,164	300
1958	63	2,044	7	43	987	0	1,911	31,056	332
1959	83	3,495	18	56	1,823	3	2,468	39,899	318
1960	57	1,337	9	39	1,017	1	1,877	37,253	218

where only details relevant to 1959—1960 are given. As observed, about 3% of all food-poisoning cases are reported as due to *Salmonella*, being slightly more frequent than staphylococcal food-poisoning.

Table II.

Incidents and cases of food-poisoning with their causative agents.

Variety of food poisoning with causative agents	1960		1959	
	incident	patient	incident	patient
Total number of food-poisoning cases	1,877	37,253	2,468	39,899
Bacterial food poisoning cases	279	10,220	303	11,617
Salmonella	57	1,337	83	3,495
Staphylococcal	39	1,017	56	1,823
Botulinus	1	1	3	4
Other	182	7,865	161	6,295
Chemical food poisoning cases	8	46	3	16
methanol	2	2	0	0
other	6	44	3	16
Natural toxic substances	178	583	266	1,077
botanical	69	354	125	679
zoological	109	229	141	398
Food poisoning where no causative substances were identified	1,412	26,404	1,896	27,189

The seasonal frequency (monthly) of Salmonella food-poisoning is shown in Table III from the statistics for the years 1959 and 1960. For some reason, food-poisonings are more frequent during the

Table III.

Seasonal occurrence of food-poisoning in Japan.

Month	Salmonella food poisoning				Staphylococcal foodpoisoning				Total number of food poisoning cases			
	1959		1960		1959		1960		1959		1960	
	incident	patient	incident	patient	incident	patient	incident	patient	incident	patient	incident	patient
January	0	0	0	0	4	52	1	24	36	235	21	67
February	0	0	0	0	1	130	4	158	41	318	42	560
March	1	34	1	18	5	53	3	229	42	701	30	566
April	0	0	1	22	5	117	3	92	48	710	31	197
May	7	228	1	11	8	387	6	99	70	1,999	46	2,150
June	11	280	3	24	11	172	10	210	86	1,980	69	1,432
July	18	489	8	540	25	942	13	149	339	6,005	255	6,459
August	8	309	5	75	12	185	17	354	689	10,601	321	3,865
September	10	197	10	339	12	350	18	383	712	11,535	824	15,545
October	9	1,026	4	209	14	340	3	143	313	5,049	173	4,375
November	0	0	3	314	2	60	1	2	64	515	38	805
December	0	0	1	66	0	0	0	0	30	251	27	1,232

Table IV.

Variety of food causing foodpoisoning in Japan.

Variety of food	Salmonella food poisoning				Staphylococcal food poisoning				Total food poisoning			
	1959		1960		1959		1960		1959		1960	
	incident	patient	incident	patient	incident	patient	incident	patient	incident	patient	incident	patient
Fishes and shell-fish	24	605	18	488	8	833	11	159	1,142	12,460	850	11,320
Fish and shell-fish derivates	15	579	4	92	16	291	7	89	234	2,967	177	4,317
Meats and meat products	8	335	1	3	2	15	1	8	45	3,537	24	226
Eggs and egg products	0	0	1	28	1	5	0	0	14	206	24	605
Milk and milk products	2	384	0	0	0	0	0	0	12	486	12	905
Crops and derivates products	3	556	1	3	14	313	7	154	81	2,075	66	1,968
Vegetables and vegetable products	4	375	5	125	4	143	4	355	200	2,806	129	3,472
Cakes	0	0	3	24	9	211	4	88	40	404	31	293
Others	19	225	5	86	0	0	1	18	187	7,236	173	6,063
Unknown causative food	18	436	19	488	2	12	4	146	513	7,722	391	8,084
Total	83	3,495	57	1,337	56	1,823	39	1,017	2,468	39,899	1,877	37,253

warm season, and this tendency exists both for Salmonella and staphylococcal food-poisoning.

The food-variety described as the cause of food-poisoning will be different from country to country following its food and social customs. For Japan, the details are statistically given in Table IV. The fact that among various kinds of food, fish and shell-fish and products derived from these are the most frequent cause of food-poisoning in Japan can be explained by the importance of such food in the eating-habits of the Japanese people; this influence is similar for both Salmonella and staphylococcal food-poisoning.

It is somewhat difficult to determine from the statistical data which type of *Salmonella* causes food-poisoning, because in some of the cases due to *Salmonella* there are no details concerning the sero-type of the causative agent: moreover, caution is advised in accepting the accuracy of the *Salmonella* serotyping in certain cases. Bearing these reservations in mind, the results found in the statistics are summarised in Table V. It will be noticed that *S. enteritidis* is the serotype most prevalent in food-poisoning in Japan and this fact is confirmed in a more detailed study on serotype distribution of *Salmonella* in Japan which will be described later.

Table V.

Salmonella serotypes isolated from foodpoisoning incidents in Japan. (Data reported to Ministry of Health and Welfare during the period from 1955 to 1960).

Serotypes	Incident (1955 to 1960)
S. enteritidis	88
S. typhi-murium	9
S. thompson	4
S. potsdam	3
S. senftenberg	3
S. dublin	2
S. saint-paul	2
S. oranienburg	1
S. newington	1
S. paratyphi B	1
S. give	1
S. narashino	1
S. tennessee	1
S. paratyphi C	1
S. georgia	1
S. aberdeen	1
Salmonella type not stated	112
Total	232

Salmonella types found in Japan

Salmonelloses occurring in man or animal are mainly endemic and do not spread in such a way as to become an international problem. In this respect, the serotype distribution of *Salmonella* is different from country to country, or it may be stated that there are certain characteristic features for each country or district. SAKAZAKI et al. (1959) drew up a table showing the frequency of *Salmonella* serotypes isolated from various kinds of animals, as well as from man, in Japan from 1949 to 1957, the figures given cover almost the whole of Japan thanks to the cooperation of the National Committee of Enteric Bacteria in Animals in Japan (Table VI).

The table indicates that the most common Salmonella serotypes in Japan are: *S. enteritidis, S. typhi-murium, S. senftenberg, S. new-brunswick, S. give, S. thompson, S. bareilly,* at least during the survey period. However, there appears to be a certain degree of host specificity, for instance *S. enteritidis* and *S. typhi-murium* are rather widely found in various animal species while *S. senftenberg, S. new-brunswick, S. thompson, S. give* and *S. bareilly* are more frequently isolated from chicken. On the contrary, *S. abortus-equi* and *S. gallinarum-pullorum* are more specific, the former in horses and the latter in chicken, though both are fairly common in many host animals.

Human Salmonelloses

There are two clinical forms of human salmonelloses: enteric fever and acute gastroenteritis, and some members of the *Salmonella* group have a tendency of causing enteric fever while others cause acute gastroenteritis. For instance, *S. typhi* and *S. paratyphi A* produce exclusively an enteric fever-type disease, while *S. paratyphi B* causes either form of disease. NAKAYA et al. (1952, 1954) reported the *Salmonella* types identified by the Japanese Salmonella Center during the period from 1951 to 1953 and the results are given in Table VII. Out of the 178 strains examined, 120 included are derived from human sources. Two cultures of *S. narashino* are reported as isolated from cases of enteric fever. The report of SAKAZAKI et al. (1959 — Table VI) includes 168 *Salmonella* cultures divided into 24 serotypes *S. sendai, S. narashino, S. enteritidis, S. simsbury, S. gaminara* and *S. kentucky* are described as derived from either sporadic cases or outbreaks of enteric fever. The authors also describe, in the same report, the case of a pig breeder and members of his household who developed enteric fever due to infection by *S. cholerae-suis.* Generally speaking, cases of enteric fever due to either *S. sendai* or *S. narashino* are not rare in Japan. In the same way, *S. sendai* is more or less an extreme type of *Salmonella* causing nearly exclusively an enteric fever-type disease and rarely is discovered as the causative agent in cases of acute gastroenteritis in Japan. In recent literature, cases of enteric fever-type disease are described as due to S. *bareilly*, S. *give*, S. *onarimon* and S. *curaçao* (SAGAWA et al., 1959, IMAGAWA et al., 1960, OKAJIMA et al., 1954, ISHIDA et al., 1957, MATSUBARA, 1960).

Acute gastroenteritis due to Salmonella infection can sometimes produce marked disorders of the colon, appearing as acute colitis, leading the physician to diagnose the disease as dysentery. "Dysentery-like" in Table VII refers to these cases.

The *Salmonella* serotypes most frequently isolated from cases of gastroenteric-type food-poisoning in Japan are S. *enteritidis* and S. *typhi-murium*. These two serotypes are, as already pointed out and seen in table VI, very frequent and widely distributed among warm-blooded animal species such as cattle, dogs, rats, chicken, etc. and therefore very apt to contaminate human food. Besides the above mentioned two types, SAKAZAKI et al. (1959) reported S. *paratyphi B*, S. *derby*, S. *thompson*, S. *narashino*, S. *moscow*, S. *give* and S. *new-brunswick* as causative agents for gastroenteric-type food-poisoning. Table VII adds S. *potsdam*, S. *bareilly*, S. *newport*, S. *nagoya* and S. *senftenberg* to the list. There are reports, in recent literature, that food-poisoning of the gastroenteritis-type have been caused by S. *bonariensis*, S. *london*, S. *tennessee*, S. *oranienburg*, S. *newington* and S. *hvittingfoss* (UEDA et al., 1959, 1960, BENOKI et al., 1960, SUGIHARA et al., 1955, SHIRAKI et al., 1954a).

The data presented by BENOKI et al. (1959) which show *Salmonella* serotypes isolated from cases of food-poisoning of the gastroenteric-type are also useful from this point of view (table VIII).

S. *gallinarum-pullorum* is, as can be clearly seen in table VI, a very strict host pathogen, attacking exclusively chicken. But a few cases of acute gastroenteritis in man due to this microorganism have recently been reported (MORIMURA, 1957, KUROSAKA et al. 1955, TOTANI et al., 1962). These cases are described as infected by eating eggs and one of the two patients reported by TOTANI et al. (1962) had diarrhea with blood-stained purulent stools and severe abdominal pain, and the other developed convulsion, delirium, fever, somewhat similar to "ekiri", an acute toxic form of infantile dysentery. However, generally speaking, S. *gallinarum-pullorum* appear to have a low degree of pathogenicity in human beings, because human infections due to these micro-organisms are quite rare though they are the most frequent *Salmonella* found in chicken.

Localized abscess formation is another type of salmonellosis which may appear as meningitis, osteomyelitis, cholecystitis, arthritis, etc. It is well known that cholecystitis is sometimes seen to develop after typhoid fever. SAWANISHI et al. (1960) reported cases of abscess of the epidydimites due to S. *thompson* while SAKAMOTO et al. (1956) reported that they had isolated S. *typhi-murium* from a sub-

Table VI.

Salmonella serotypes isolated in Japan (SAKAZAKI et al., 1959).

Sero-group	Serotype	cattle	horses	pigs	sheep & goats	dogs	foxes & minks	cats	rats	mice	guinea-pigs	rabbits	chickens	other fowls	eggs	man	total
A	S. paratyphi A					1			2							11	14
B	S. abortus-equi	1	105														106
	S. paratyphi B		1												6	20	27
	S. java	1				6										2	9
	S. schleissheim					1											1
	S. stanley					8										1	9
	S. schwarzengrund					1											1
	S. saint-paul					4											4
	S. reading															1	1
	S. derby	6	1	9		14										5	35
	S. essen	1														1	2
	S. typhi-murium	6	2	1	3	57			10	10	1				38	1	20	149
	S. bredeney		1			2										1	4
	S. brandenburg		1														1
C1	S. cholerae-suis			7												3	10
	S. amersfoort	1															1
	S. montevideo										1	1					2
	S. oranienburg		1														1
	S. thompson	5				38						1	18		64	5	132
	S. irumu														7		7
	S. potsdam	2				23									37		62
	S. bareilly	3				12	13						21		77	1	127
	S. mikawasima															2	2
C2	S. narashino				4	11					32						47
	S. nagoya	1	1			4					1					4	11
	S. manhattan			1													1
	S. newport	1															1
	S. tananarive					3											3
	S. glostrup										1						1
	S. virginia			1													1

	S. kentucky															2	2
D	S. sendai															4	4
	S. typhi					1										47	48
	S. eastbourne					1											1
	S. enteritidis	16	3	2	3	145	19	5	32	48	163				30	21	487
	S. blegdam					1											1
	S. moscow															3	3
	S. neasden	1															1
	S. javiana					1											1
	S. gallinarum-pullorum				1	1							418		108	1	529
E1	S. muenster			1													1
	S. meleagridis	1															1
	S. give					44							15		57		117
	S. amager					1	1										2
E2	S. newington		1			5			1				14				21
	S. newbrunswick					22							3	1	73	2	101
E4	S. senftenberg	1		5		20							49		298		373
	S. simsbury															1	1
	S. chittagong	1															1
F	S. aberdeen											1				1	2
G	S. worthington	4															4
H	S. onderstepoort															1	1
	S. horsham					1											1
I	S. hvittingfoss			1													1
	S. gaminara					1										8	9
Further groups	S. cerro						1										1
	S. weslaco	1															1
Total		53	117	28	11	429	34	5	45	58	199	3	540	39	759	168	2,481

Table VII.

Salmonella serotypes identified by Japanese Salmonella Center during the period from 1951 to 1953 (FUKUMI, 1954).

Sero-group	Serotype	human beings				geese	chick-ens	hens-eggs dying during development	rats	cattle	human food	total
		enteric fever	acute gastro enteritis	dysentery like	carrier							
B	S. paratyphi B	1		14								15
	S. typhimurium		25		2	5			1		2	35
	S. derby				1							1
C	S. potsdam		1	1	1							3
	S. bareilly		3					27				30
	S. thompson		6		1							7
	S. newport		1									1
	S. narashino	2	9								2	13
	S. bovis- mor-bificans				1							1
	S. nagoya		3		1					1	1	5
D	S. typhi	1										1
	S. enteritidis		42								2	44
	S panama			1								1
E	S. new-brunswick				2							2
	S. senftenberg		1				3	14				18
	S. anatum		1									1
Total		4	92	16	8	5	3	41	1	1	7	178

cutaneous abscess following spondylitis in a threemonth old baby who died of purulent meningitis; *S. senftenberg* was reported having been isolated at autopsy from the child's purulent lesions. It is reported that there was severe inflammation in both lateral ventricles and in the right central gyrus of the brain.

The *Salmonella* carrier-rate among healthy human beings seems to be relatively low, compared to that of *Shigella*, with a reported approximative average of 0.5% in Japan. BENOKI et al. (1956) reported that five *Salmonella* cultures were isolated from the stools of 9,903 healthy human beings, the serotypes being *S. senftenberg*, *S. paratyphi A* and *S. derby*. This corresponds to a 0.05% *Salmonella* carrier-rate in healthy people in Tokyo. Another survey was presented by HAYASHI et al. (1957) who found nine *Salmonella* carriers among 7,980 healthy people living on an island outside the port of Nagasaki (coal mining district) in 1956. The carrier-rate is here 0.12%. In the same survey, 127 dysentery carriers were found, giving a 1.6% carrier-rate, which seems to indicate that the population in question lived under fairly poor sanitary conditions. Among the *Salmonella* cultures isolated in this survey, 2 were *S. paratyphi A*, 2 *S. paratyphi B*, 3 *S. enteritidis*, and the remaining 2 were not serologically determined and, according to the report, were found to possess H antigens l, w and e, n but unknown O antigens. In Nagasaki City, YOSHIMURA et al. (1955) found 3 *Salmonella* carriers in 15,092 healthy people, a carrier-rate of 0.02%. The serotypes isolated were 2 *S. typhi* and 1 *S. enteritidis*. In the Saga Prefecture, SHIRAKI et al. (1954b) reported the isolation of 7 serotypes, 11 cultures of *Salmonella*: 5 *S. typhi*, 1 *S. enteritidis*, 1 *S. narashino*, 1 *S. gallinarum-pullorum*, 1 *S. paratyphi A*, 1 *S. derby* and 1 *S. newington*, from 10,736 healthy people; a carrier-rate of 0.11%.

In their collections of *Salmonella*, SAKAZAKI et al. (1959) found 13 serotypes which had been isolated from healthy people, namely *S. paratyphi B*, *S. typhi*, *S. typhi-murium*, *S. stanley*, *S. reading*, *S. derby*, *S. essen*, *S. bareilly*, *S. mikawasima*, *S. enteritidis*, *S. aberdeen*, *S. onderstepoort* and *S. gaminara*. FUKUMI (table VII) adds to this list the following serotypes: *S. potsdam*, *S. thompson*, *S. bovis-morbificans* and *S. new-brunswick*. Among these *Salmonella* types isolated from healthy carriers, approximately half are on the list of those causing human disease either of the gastroenteritic or of the enteric fever-type as already pointed out, namely *S. enteritidis*, *S. thompson*, *S. paratyphi B*, *S. derby*, *S. new-brunswick*, *S. potsdam* and *S. bareilly*, but the following serotypes are not on the list: *S. stanley*, *S. reading*, *S. essen*, *S. mikawasima*, *S. aberdeen*, *S. onderstepoort* and *S. bovis-morbificans*. It seems reasonable to consider that these serotypes can also produce human disease and are indeed causing it but may be too rare to be reported in scientific studies, or it may be possible that they may be of too low pathogenicity to disturb human welfare.

Animal Salmonelloses connected with human infections

In most cases, human salmonelloses are terminal infections; this means that they are caught from infected animals whereas they do not spread from man to man or from man to other animal species, except typhoid and paratyphoid fevers and some other enteric fevertype salmonelloses. Therefore, salmonelloses in animals should necessarily be taken into consideration when discussing human salmonelloses, especially from the epidemiological point of view of Salmonella food-poisoning.

Pigs

Primarily, cattle, pigs, poultry and eggs are the main sources of *Salmonella* microorganisms when taken as food. It has long been the custom to believe that pigs are heavily contaminated with a variety of *Salmonella* and that *S. cholerae-suis* is the most frequent type. However, it now appears that this is not the case, at least recently and in Japan. *S. cholerae-suis* does not appear to play an important role either in pig salmonelloses or in human infections. Table VII shows that no *S. cholerae-suis* were obtained from 110 *Salmonella* cultures derived from human materials, while table VIII shows that there was only 1 case of food-poisoning due to this serotype among 104 cases. It can also be seen from table VI that only 3 cultures were obtained out of 168 derived from a patient suffering from an enteric fever-type disease and his contacts (cited above as a pig-breeder family). Furthermore, *S. cholerae-suis* is a minor *Salmonella* as observed through cultures from material recovered from pigs, as seen in the same table. Thus, it seems that this *Salmonella* serotype does not play an important role at present in Japan, in contrast with the facts observed in European and American countries.

Furthermore, the *Salmonella* carrier-rate in pigs seems to be fairly low according to recent publications in Japan. TOKUTOMI (1957) reported 17 *Salmonella* isolates from 14 pigs among 1,082 healthy pigs in the Tokyo area; a carrier-rate of 1.02%. However, this value should be considered a little too high because an abnormally high carrier-rate in a certain district affected the percentage. Generally speaking, therefore, the *Salmonella* carrier-rate in pigs seems much lower than the figure given above. The specimens examined consisted of livers, biles, spleens, intestinal lymph nodes and stools. The serotypes isolated were 10 cultures of *S. derby*, 6 of *S. senftenberg* and 1 of *S. manhattan*. There were no *S. cholerae-suis* cultures. On the other hand, SAKAZAKI et al. (1959) stated that they

Table VIII.

Salmonella serotypes in food poisoning incidents in Tokyo (BENOKI et al., 1959).

Serogroup	Serotype	Number of incidents
B	S. paratyphi B	6
	S. stanley	1
	S. derby	1
	S. essen	1
	S. typhi-murium	17
C1	S. cholerae-suis	1
	S. paratyphi C	1
	S. thompson	5
	S. potsdam	2
	S. oranienburg	2
C2	S. narashino	5
	S. nagoya	5
D	S. sendai	1
	S. onarimon	1
	S. enteritidis	43
	S. blegdam	1
	S. moscow	1
	S. gallinarum-pullorum	1
E1	S. london	2
	S. give	2
	S. newbrunswick	1
E2	S. newington	1
E3	S. senftenberg	3
I	S. shanghai	1
	Total	104

only isolated a few *Salmonella* cultures from more than 10,000 pigs examined without the isolation of *S. cholerae-suis*. They believe that the frequency of *Salmonella* carriers among healthy pigs is less than 0.1%. ONO et al. (1958) reported that they did not isolate any *Salmonella* cultures from 137 healthy pigs. Similar results have been reported by various authors using different materials in Japan. Thus it is reasonable to conclude that the *Salmonella* carrier-rate in pigs is considerably lower than values presented by various authors in foreign countries.

The *Salmonella* serotypes isolated from pigs in Japan are: *S.*

derby, S. typhi-murium, S. cholerae-suis, S. manhattan, S. virginia, S. enteritidis, S. muenster, S. senftenberg, S. hvittingfoss, S. narashino (YAMADA et al., 1942) and *S. thompson* (SHIMIZU, unpublished).

Horses
This animal does not play an important role as it is not frequently eaten nor is it closely associated with human life, except for a particular class of people in Japan. *S. abortus-equi* is the most frequent *Salmonella* serotype in this animal species, and in a few cases *S. paratyphi B, S. derby, S. typhi-murium, S. bredeney, S. brandenburg, S. thompson, S. nagoya, S. enteritidis, S. newington* were found. The frequency of *Salmonella* in apparently healthy horses seems to be fairly low and is estimated less than 0.1%.

Cattle
Beef is one of the three most popular meat-food in Japan, the other two being pork and chicken, and so cattle is a most important source of food-poisoning in this country.

There seems to be no *Salmonella* serotypes specific to cattle comparable to *S. abortus-equi* for horses, or *S. gallinarum-pullorum* for chicken. Fairly many *Salmonella* serotypes appear to infect cattle, among which *S. enteritidis, S. typhi-murium, S. derby, S. thompson* are the most frequent. Beside these, the following are found: *S. worthington, S. bareilly, S. potsdam, S. abortus-equi, S. schleissheim, S. bredeney, S. montevideo, S. nagoya, S. newport, S. neasden, S. meleagridis, S. senftenberg, S. chittagong* and *S. weslaco.* A culture of *S. abortus-equi* in cattle was said to have been isolated from an aborting cow (KUMAGAI, 1952).

AKAZAWA et al. (1959a) reported that they had isolated *Salmonella (S. enteritidis)* from the stools of 118 Holstein cows (during dropping of young) in 3 pastures in the Tokyo-Chiba area, the carrier-rate being 0.95%. The carrier-rate determined by other workers is as follows: one isolation *(S. derby)* from 1,034 slaughtered cattle, a rate of 0.096% by NOMOTO (1952), four isolations (3 *S. typhi-murium* and 1 *S. montevideo)* from 730 cattle stools, a rate of 0.55% by the Japanese Salmonella Committee for Animals (1954).

Chicken and eggs
Chicken and eggs are one of the most popular foods in Japan and must therefore be considered as a major source of Salmonella infections in human beings. For various reasons, hen eggs seem to be fairly contaminated with a variety of *Salmonella.* For instance, there are many reports giving a high frequency of isolation of *Salmonella* from eggs not capable of embryonating or arrested during embryonation. SAKAI et al. (1956) examined eggs from two breeders in and near Hiroshima City and recovered 10 cultures (1.6%) of *S.*

gallinarum-pullorum out of 635 eggs not capable of embryonating, and 39 cultures (9.2%) of *S. give,* 26 (6.1%) of *S. senftenberg* and 8 (1.9%) of *S. gallinarum-pullorum* out of 428 hen eggs arrested during embryonation. KUWABARA et al. (1959) reported 79 isolations (12%) of *S. senftenberg* out of 717 hen eggs either unable to develop or to hatch, and during the same period, there were 193 dead chicken less than 11 days after a hatch of about 400, from which they isolated cultures of *S. senftenberg.* The findings of SATO et al. (1955) answered the question about the proportion and serotypes found in these eggs. Out of 122 cultures of *Salmonella* thus obtained, there were 37 cultures of *S. senftenberg,* 64 of *S. gallinarum-pullorum,* 7 of *S. thompson,* 12 of *S. new-brunswick* and 2 of *S. bareilly,* this in Hokkaido. For another instance, the Japanese Salmonella Committee for Animals (1954) reports isolations of 30 cultures of *S. gallinarum-pullorum,* 7 of *S. senftenberg* and 3 of *S. thompson* from 1,500 sick chicken.

The most specific and, at the same time, the most frequent *Salmonella* is *S. gallinarum-pullorum,* the next in order of frequency being *S. senftenberg* for chicken and eggs in Japan. *S. thompson, S. bareilly, S. give, S. new-brunswick* are also fairly frequent among chicken and eggs. According to table VI, *S. java, S. typhi-murium, S. irumu, S. potsdam, S. enteritidis, S. new-brunswick* were occasionally found in eggs, while *S. essen, S. montevideo, S. newington, S. new-brunswick* were found in sick chicken. *S. typhi-murium* and *S. enteritidis* are considered as playing a minor role in either chicken or eggs, which is in contrast with the fact that they both have a major role in human food-poisoning and in salmonelloses in some animal species. Particularly remarkable is the exceptional isolation of *S. typhi-murium* from chicken and eggs: only 1 culture from eggs and none from chicken out of a total of 149. Similarly for *S. enteritidis,* with 30 cultures from eggs and none from chicken out of a total of 348 cultures. Furthermore, it should be borne in mind that these two serotypes are very frequent in chicken and in eggs in foreign countries such as England and the U.S.A.

It should be noted in passing that the 38 cases of *S. typhi-murium* mentioned in table VI were derived from quails during a single outbreak (SHIMAKURA et al., 1958).

Rats

Rats are recognized as being generally fairly susceptible to Salmonella infections and develop severe typhoidal disease. They live in very close association with human life. There are, therefore, many possibilities not only for human but also domestic animal infections derived from rats carrying *Salmonella* and transferred by direct or indirect contacts. In particular, the behaviour of rats is an important factor in the contamination of food utensils with infected excreta, etc. The carrier-rate and serotype frequency of *Salmonella* in rats reported by various authors in Japan and in Korea are summarized in table IX, in which it can be seen that the most frequent *Salmo-monella* serotype is *S. enteritidis* and the second one is *S. typhi-murium*. Whether the most frequent *Salmonella* serotype in Korea is also *S. enteritidis* requires further investigations. Beside this sero-type, *S. paratyphi A*, *S. paratyphi B*, *S. paratyphi C*, *S. derby*, *S. newington* and *S. typhi* were reported as having been isolated from rats in Japan (NAKAMURA, 1953). We also isolated a culture of *S. typhi* from a rat during the investigation of an outbreak of typhoid fever in a small area in Tokyo. The phage-type of the isolate was the same as that of the epidemic strains: type B2.

Cats and dogs

The lives of cats and dogs are interrelated with those of their owners; this applies not only to domestic cats and dogs but also to wild cats and dogs which are usually found in dwelling-areas search-ing for food. Generally speaking, a great proportion of so-called wild cats and dogs, despite their name "wild", were originally kept as pets and are therefore not really wild but strays; for some reason or another, they left the household or are the offsprings of these pets unable to live without human contacts. It is therefore easy to under-stand why the *Salmonella* serotypes isolated from these animals re-semble those obtained from human beings. This is the case with dogs. It is very easy to deduce the serotype distribution of *Salmonella* in human beings in a certain area if that for dogs is already known. The question can be raised: Which of the two species, man or dog, is the donor of *Salmonella?* There will probably be frequent exchanges of *Salmonella* serotypes between man and dog.

The frequency of *Salmonella* serotypes found in dogs is seen in table X, which contains information gathered from reports by vari-ous authors obtained at different times and in different places. The extreme variety of the serotypes discovered is quite surprising, es-pecially its similarity to that of human beings. *S. typhi-murium, S. thompson, S. give, S. potsdam, S. senftenberg, S. enteritidis, S. derby, S. narashino* are the most frequent serotypes in dogs, but the follow-ing have occasionally been isolated: *S. paratyphi A, S. paratyphi B, S. stanley, S. schwarzengrund, S. bredeney, S. bareilly, S. newport, S. nagoya, S. anatum, S. amager, S. newington, S. new-brunswick, S. chittagong, S. horsham, S. gaminara, S. zanzibar, S. edinburg, S. panama, S. tananarive, S. java, S. saint-paul, S. eastbourne, S. blegdam, S. javiana, S. gallinarum-pullorum, S. typhi. Salmonella* are most frequently isolated from mesenterial lymph nodes and sometimes from liver, spleen, urine, stools, etc.

Animals of secondary importance

Sheep and goat meat is not frequently eaten in Japan. Very few

Table IX.

Salmonella serotypes and carrier-rates of rats in Japan. (Data arranged by BENOKI et al. (1959), and supplemented by the present author).

Reporters	HATTA	HAYASHI et al.	BENOKI et al.	DOI et al.	YOSHIMURA	OKUWADA et al.	OMORI et al.	OMORI et al.	Jo
Survey year	1937	1948—49	1951—52	1949—51	1955	1954	1958	1959	1954
Survey place	Tokyo	Nagasaki	Tokyo	Kago-shima	Nagasaki	Osaka	Osaka	Osaka	Pusan (Korea)
Number of rats examined	1,075	360	1,028	1,332	80	292	314	215	770
Number of rats carrying Salmonella	5	6	21	21	2	2	9	14	12
carrier-rate	0.4	2.4	2.04	1.57	2.5	0.68	2.8	6.5	1.7
S. paratyphi A			2						
S. paratyphi B			1						
S. typhi-murium		1				1	1		
S. paratyphi C							1		
S. enteritidis	4	5	16	6	2	1	7	14	
S. derby	1			15					
S. mikawasima									1
Salmonella not identified			2						11

Table X.

Salmonella serotypes and carrier-rates of dogs in Japan. (Data arranged by Benoki et al. (1959), and supplemented by the present author).

Reporters	Murase et al.	Hashizume et al.	Benoki et al.	Sakazaki et al.	Yoshimura et al.	Nagaya et al.	Nishino	Watanabe	Murata et al. Tokyo	Murata et al. Osaka
Survey year	1952	1952	1953	1954	1955	1955	1957	1957	1960	1960
Survey place	Tokyo	Osaka	Tokyo	Tokyo	Nagasaki	Sapporo	Matsu-yama	Tokyo	Tokyo	Osaka
Number of dogs investigated	534	300	204	210	200	355	274	200	200	208
Number of dogs carrying salmonella	59	3	20	54	4	11	14	29	36	48
carrier rate (%)	11	1.0	9.8	25	2.0	3.1	5.1	15	18	23
S. paratyphi A	1		2							
S. paratyphi B	7	1	1	1		6	13	4	26	9
S. typhi-murium	2		5	13		1	1	1	4	11
S. thompson	5								2	1
S. newington									1	
S. zanzibar	8			3				3	1	18
S. give	7			3					1	5
S. potsdam									1	
S. edinburg	1		1	6						
S. senftenberg	20	1	2	6	4	4		7		4
S. enteritidis	1					1		4		
S. narashino			1							
S. newport										
S. panama				2				6		
S. stanley		1		1				1		
S. bredeney								3		
S. tananarive										
S. derby	3		6	9						
S. bareilly	4									
S. anatum			2							
S. schwarzengrund				1						
S. nagoya				3						
S. amager				1						
S. newbrunswick				2						
S. chittagong				1						
S. horsham				1						
S. gaminara				1						

people eat lamb and goat meat. The carrier-rate of *Salmonella* in these animals seems to be very low and is estimated at less than 0.1%, as estimated from single reports.

Foxes and minks are mostly bred in Hokkaido for their fur. NAGAYA et al. (1956) have reported the isolation of 19 *Salmonella* carriers from 119 bred foxes (13 cultures of *S. bareilly*, 5 of *S. typhi-murium*, 3 of *S. enteritidis*, including *S. bareilly* and *S. enteritidis* also from a fox) but no isolation of *Salmonella* from 44 bred minks.

Turkeys and domestic ducks are not a major food item for the Japanese people. AKAZAWA et al. (1959b) reported the isolation of 24 *Salmonella* cultures from 100 turkeys eggs from a farm in the Tokyo prefecture in July and August 1958. They were either *S. bonn* or *S. give*. KOBORI et al. (1955) isolated a culture of *S. typhi-murium* and one of *S. bareilly* from the stools of 100 apparently healthy domestic ducks. From these and other reports, turkeys and ducks seem to be highly infected with *Salmonella*.

Quail eggs are sometimes eaten in certain restaurants, but are not popular amongst ordinary people in Japan. SHIMAKURA et al. (1958) reported the isolation of 38 cultures of *S. typhi-murium* from 55 dead 3- to 10-day-old quails out of 200 purchased on one particular occasion. It seems as though there had been an epidemic of this microorganism amongst these birds.

Whale-catching is a very important business in Japan. Recently, a food-poisoning outbreak was encountered due to whale meat contaminated with *S. enteritidis*. The whale was caught off-shore in the Wakayama Prefecture; it was sick and floating on the surface of the sea (NAKAYA, 1950). This animal species seems to be susceptible to *S. enteritidis*, becoming ill when infected.

Salmonella serotypes first isolated in Japan

Up to now, six *Salmonella* serotypes were first reported in Japan: *S. narashino, S. onarimon, S. mikawasima, S. sendai, S. nagoya* and *S. miyazaki*.

S. narashino was isolated by NAKAGURO & YAMASHITA (1938) from a patient suffering from a typhoid-like disease. This serotype is fairly common in Japan, causing either enteric fever-type illness or acute gastroenteritis as previously mentioned in the early part of this paper.

S. onarimon was reported by KISIDA (1940) as having been isolated from the stools of a healthy woman, and was also found by ANZAI & TSURUMI (1940) in the blood of a patient suffering from a typhoid-like disease. Occasionally, it is isolated in cases of food-poisoning of the acute gastroenteritis-type (see table VIII; IKEMURA & YAMA-SAKI, 1957).

S. mikawasima was isolated first by HATTA (1938) from rat faeces.

This serotype was isolated from 2 human carriers (SAKAZAKI et al., 1959) and from snakes (SAKAZAKI et al., 1960, 1961).

S. sendai was isolated from cases of enteric fever by AOKI et al. (1925). This Salmonella serotype is not rare in Japan, mostly causing enteric fever but not acute gastroenteritis.

S. nagoya was isolated by NAKAJIMA et al. (1953) from a case of actue gastroenteritis. Later, occasional isolations of this serotype have been reported not only from cases of gastroenteritis in human beings but also from domestic animals, for instance from cows, horses, sheep, dogs, etc.

S. miyazaki was isolated by FUKUDA et al. (1958) from cases of gastroenteritis and their contacts in a single family. Up to now, there have been no further reports in Japan that this serotype was isolated.

Comments

There are still many questions to be answered about the microorganisms of the Salmonella group. In this group, we have S. typhi which is strictly related to human life. It develops a typical disease known as "typhoid fever". It is directly transmitted from man to man. It survives and remains in the subject known as the "typhoid carrier". There seems to be no other reservoir for S. typhi than the human carrier. Exceptionally, there have been reports that it has been found in animals, such as rats and dogs. The same might be true for S. paratyphi A and S. sendai. There may be other similar Salmonella serotypes, but they are very few, if any.

Generally and ecologically, a Salmonella is a microorganism circulating either among animals of a single species or of several species. In such cases, animals are more or less susceptible to it and infected by contact or by food very lightly contaminated by this miroorganism. An infection cycle would be established between them. From time to time, the microorganisms leaves its own infection cycle and may infect some less susceptible animal no capable of establishing an infection cycle. Thus, the infection would end with no new victims in this or other animal groups. This is called a "terminal infection". Human food-poisoning due to Salmonella may be a typical example of this type of infection. In the same way, Salmonella serotypes happen to be isolated from an animal in which this particular serotype very rarely exists, such as stray cases, for instance S. abortus-equi in cattle, S. gallinarum-pullorum either in sheep or in dogs, S. paratyphi A in dogs or in rats.

There are reports that Salmonella has been isolated from snakes and earthworms. The rate of isolation of Salmonella from these animals is fairly high. According to SAKAZAKI et al. (1960, 1961), S. bareilly, S. mikawasima, S. ramat-gan and S. weslaco were isolated from snakes at the rate of 2:52:12:2, whereas S. potsdam, S. bonn and

S. newington were isolated from earthworms at a rate of 2:5:2. When considering the life cycle and ecology of *Salmonella*, these facts must also be taken in consideration. The high frequency with which *Salmonella* are isolated from snakes should not surprise us because rats and hen eggs, both well known reservoirs of a variety of *Salmonella* serotypes, are frequently eaten by snakes. However, there may be a specific serotype frequency of *Salmonella* isolated from snakes, or, in other words, some difference in serotype frequency between snakes and rats or hen eggs. For instance, *S. enteritidis* and *S. typhi-murium*, both frequent serotypes in rats, are not found in snakes, and there have been no reports of *S. gallinarum-pullorum* having been isolated from snakes. Rats occasionally become contaminated by the excreta of snakes. However, I do not know whether certain *Salmonella* serotypes establish an infection cycle between snakes and rats or snakes and hen.

Concerning *Salmonella* in earthworms, it is not known if there is any specific *Salmonella* serotype frequency in earthworms. Earthworms live underground and could possibly be contaminated by the excreta from various animals infected with *Salmonella*, such as rats, dogs, cattle, pigs, hen or, of course, human beings. It seems highly probably that earthworms harbour a variety of *Salmonella*. Conversely, some animals may occasionally eat earthworms, for instance pigs and hen; thus, it is possible that some sort of infection cycle occurs between these animals and earthworms.

These infection cycles of low degree added together form the higher cycle and thus, finally, the whole picture of the life of *Salmonella* is established. The characteristics of the life and epidemiology of *Salmonella* and the infections which they cause are thus explained, using as base the specific infection cycles of the various *Salmonella* found in Japan.

REFERENCES

* AKAZAWA, S. & ENOMOTO, S. (1959a): *Bull. Univ. Jap. vet. Zootechnol.*, n° **8**, *43*.
* AKAZAWA, S. & KAIZU, M. (1959b): *Bull. Univ. Jap. vet. Zootechnol.*, n° **8**, *38*.
ANZAI, H. & TSURUMI, H. (1940): *Kitazato Arch.*, **17**, *106*.
AOKI, K. & SAKAI, K. (1925): *Zbl. Bakt.*, I, Orig., **95**, *152*.
* ARAI, Y., KAWAJI, R., BENOKI, M., MATSUI, S., HAYASHI, T., KADONO, Y., HATA, K. & TAKAYAMA, Y. (1955): *J. Jap. Ass. inf. Dis.*, **29**, *237*.
* BENOKI, M., MATSUI, S. & KADONO, Y. (1956): *Bull. Tokyo Health Labor.*, n° **6**, *35*.
* BENOKI, M. & ZENYOJI, H. (1959): Bacterial Food Poisoning, Nanzando, Tokyo.
* BENOKI, M., YABUUCHI, K., TANAKA, S., UTASHIRO, Y., KO, T. & SAKAZAKI, R. (1960): *Studies of Food Hygiene*, **10**, *91*.
FUKUDA, T., SASAHARA, T., KITAO, T., TANIGUCHI, H., FUKUMI, H. & MURATA, Y. (1958): *Jap. J. med. Sci. Biol.*, **11**, *13*.

528

* FUKUMI, H. (1954): *Shindan To Chiryo*, **42**, *563*.
HATTA, S. (1938): *Jap. J. exp. Med.*, **16**: *201*.
* HAYASHI, K. & MURATA, Y. (1957): *Nagasaki med. J.*, **32**, *429*.
* IKEMURA. K. & YAMASAKI, S. (1957): *Bull. Niigata Health Labor.*, n° **20**, *20*.
* IMAGAWA, Y. (1960): *J. Jap. Ass. inf. Dis.*, **34**, *436*.
* ISHIDA, C. & MISHIMA, E. (1957): *Eisei Kensa*, **6**, *14*.
* Japanese Salmonella Committee for Animals (1954): *Jap. J. vet. Med.*, **15**, *56*.
* JO, J. (1960): *Kansai Igaku*, **19**, *2739*.
* KAWANA, T., TOCHIHARA, K., SAKAGAMI, M. & SUGIYAMA, N. (1957): *Shonika Shinryo*, **20**, *186*.
KISIDA, S. (1940): *Kitasato Arch.*, **17**, *1*.
* KOBORI, S., IWATA, T., FUJINO, K. & UEDA, S. (1955): *Tokyo J. vet. Zootechnol.*, n° **4**, *14*.
* KUMAGAI, T. (1954): *Jap. J. vet. Sci.*, **14**, *325*.
* KUROSAKA, E. & HARUHARA, K. (1955): *Shonika Shinryo*, **18**, *560*.
* KUWABARA, Y. & TABUCHI, Y. (1959): *J. Jap. Ass. Veterinarians*, **12**, *243*.
* MATSUBARA, M. (1960): *Jap. J. Bact.*, **15**, *238*.
* MORIMURA, M. (1957): *Eisei Kensa*, **6**, *39*.
* MURATA, M., MATSUMOTO, H., YAMADA, N., KOBORI, S. & SAKAZAKI, R. (1961): *Jap. J. vet. Med.*, **23** (Appendix), *398*.
NAGAYA, H. & SHIMIZU, K. (1956): *Jap. J. vet. Res.*, **4**, *75*.
* NAKAGURO, S. & YAMASHITA, K. (1938): *Gun-i Dan Zasshi*, **306**, *1241*.
NAKAJIMA, K., NAITO, S., NAKAYA, R. & FUKUMI, H. (1953): *Jap. J. med. Sci. Biol.*, **5**, *179*.
* NAKAMURA, T. (1953): *Nagasaki med. J.*, **28**, *32*.
NAKAYA, R. (1950): *Jap. med. J.*, **3**, *279*.
NAKAYA, R., NAKAYAMA, T., SAKAGUCHI, G. & FUKUMI, H. (1952): *Jap. J. med. Sci. Biol.*, **5**, *323*.
NAKAYA, R., NAKAYAMA, T., SAYAMA, E., NOJIMA, T. & FUKUMI, H. (1954): *Jap. J. med. Sci. Biol.*, **7**, *301*.
* NISHINO, H., SAKAZAKI, R. & OKADA, H. (1958): *Bull. Ehime Health Labor.*, n° **14**, *1*.
* NOMOTO, K. (1952): *Kagoshima med. J.*, **25**, *143*.
* OHMORI, G., IWAO, M., KUROSUMI, T. & IIDA, S. (1960): *Bull. Osaka City Health Labor.*, **22**, *12*.
* OKAJIMA, Y. & SATO, T. (1954): *Sogo Igaku*, **11**, *560*.
* ONO, T., OKAMOTO, T., KOSAKA, E., YOSHINAGA, S., WATANABE, J. & NISHI, T. (1958): *J. Jap. Ass. Veterinarians*, **11**, *339*.
* SAGAWA, I., YOSHIDA, S., MATSUBARA, M. & KANEDA, O. (1959): *Jap. J. Pediatr.* **63**, *2076*.
* SAKAI, M., SASAKI, S., KISHIMOTO, K. & KO, M. (1956): *Bull. Hiroshima Health Labor.*, n° **6**, *47*.
* SAKAMOTO, S., TAKAHASHI, K. & NISHIMI, Y. (1956): *Bull. Fukuoka Health Labor.*, n° **4**, *16*.
SAKAZAKI, R., NAMIOKA, S. & WATANABE, S. (1959): *Jap. J. exp. Med.*, **29**, *15*.
* SAKAZAKI, R., NOZAWA, M., MURATA, M., ISHII, F. & ISHIDA, M. (1960): *Jap. J. Bact.*, **15**, *1100*.
* SAKAZAKI, R., ISHII, F., KATO, I., TSUJIMOTO, M., MURATA, M. & TAJIMA, Y. (1961): *Jap. J. vet. Med.*, **23** (Appendix), *398*.
SATO, G., MIYAMAE, T., MIURA, SUZUKI, K. & SAKAZAKI, R. (1955): *Jap. J. vet. Res.*, **3**, *111*.
* SAWANISHI, M., WADA, K. & SUGIHARA, Y. (1960): *Eisei Kensa*, **9**, *117*.
* SHIMAKURA, S. & IWAMORI, H. (1958): *Jap. J. vet. Med.*, **20**, *263*.
* SHIRAKI, M. & FUJIMOTO, N. (1954a): *Bull. Saga Health Labor.*, n° **5**, *33*.

* SHIRAKI, M. & FUJIMOTO, N. (1954b): *Bull. Saga Health Labor.*, n° **5**, *54*.
* SUGIHARA, M., HIRAYAMA, T. & UMIHARA, M. (1955): *Jap. J. Publ. Health*, **2**, *623*.
* TOKUTOMI, T. (1957): *Bull. Nat. Inst. Publ. Health*, n° **6**, *6*.
* TOTANI, T. & NAITO, S. (1962): *Rinsho Naika Shonika*, **17**, *699*.
* UEDA, S., SASAKI, S. & KO, M. (1959): *Jap. J. Bact.*, **14**, *46*.
* UEDA, S., SASAKI, S., KO, M., TAKATA, S. & MORI, F. (1960): *Hiroshima med. J.*, *53*.
* WATANABE, S., SAWADA, A., KITAJIMA, Y., OTA, H., KOBORI, S. & SAKA-ZAKI, R. (1959): *J. Jap. Ass. Veterinarians*, **12**, *485*.
* YAMADA, T. & MATSUI, T. (1942): *Jap. J. vet. med.*, **4**, *467*.
* YOSHIMURA, K. (1955): *Nagasaki med. J.*, **30**, *1039*.

* = written in Japanese.

LES SALMONELLOSES EN INDOCHINE ET EN CHINE

PAR

L. LE MINOR

Centre des Salmonella de l'Institut Pasteur de Paris.

Les documents que nous possédons proviennent des Instituts Pasteur d'Extrême-Orient où ont été poursuivies, depuis de nombreuses années, des recherches sur les *Salmonella*, en relation avec l'Institut Pasteur de Paris. Chaque année, les Instituts Pasteur

I. Viet-Nam
A) Sud Viet-Nam

Provenances des Salmonelles isolées chez l'homme à Saïgon de 1951 à 1961 inclus.

Types	Sang	Fèces	Bile	Urines	L.C.R.	Séreuses	Suppurations diverses	Total
S. paratyphi A	154	3	1	1	—	1	2	162
S. paratyphi B	1	5	—	—	1	—	—	7
S. paratyphi B var. Java	—	1	—	—	1	1	—	3
S. typhi-murium	2	15	—	—	2	—	1	20
S. stanley	—	4	—	—	—	—	—	4
S. heidelberg	—	2	—	—	—	—	—	2
S. san diego	1	—	—	—	—	—	—	1
S. derby	1	3	—	—	—	—	1	5
S. paratyphi C	6	6	—	1	—	1	1	15
S. cholerae-suis	21	1	—	2	—	4	1	29
S. thompson	—	1	—	—	—	—	—	1
S. virchow	—	—	—	—	—	1	—	1
S. oslo	—	1	—	—	—	—	—	1
S. newport	—	5	—	—	1	—	—	6
S. bovis-morbificans	—	4	—	—	—	—	1	5
S. tananarive	—	—	—	1	—	—	—	1
S. typhi	1.537	32	5	5	7	7	9	1.602
S. enteritidis	1	—	—	1	1	1	1	5
S. dublin	2	2	—	—	—	—	—	4
S. blegdam	1	—	—	—	—	1	—	2
S. berta	—	1	—	—	—	—	—	1
S. javiana	—	2	—	—	2	1	—	5
S. anatum	—	1	—	—	—	—	—	—
S. meleagridis	—	1	—	—	—	—	—	1
S. weltevreden	—	—	—	—	—	—	1	1
S. senftenberg	—	1	—	—	—	—	—	1
Totaux	1.727	91	6	11	15	18	18	1.886

Origine zoologique des Salmonella isolées au Sud Viet-Nam de 1952 à 1961 inclus.

Types	Hommes	Singes	Porcs	Bovo-bubalin	Chiens	Rongeur	Oiseaux	Serpents terrestres	Serpents marins	Lézards
S. paratyphi A	131	—	—	—	—	—	—	—	—	—
S. paratyphi B	6	—	4	—	—	—	—	—	—	—
S. paratyphi B var. java	3	—	—	—	—	—	—	—	—	—
S. typhi-murium	24	—	—	—	—	—	1	—	—	—
S. stanley	4	—	—	—	—	—	—	—	—	—
S. heidelberg	1	—	—	—	—	1	—	—	—	—
S. san diego	1	—	—	—	—	—	—	—	2	—
S. derby	5	—	4	1	1	1	—	1	—	—
S. paratyphi C	14	—	—	3	3	—	1	—	—	—
S. cholerae-suis	25	—	110	1	1	—	—	1	—	—
S. thompson	1	—	—	—	—	—	—	—	—	—
S. virchow	1	—	—	—	—	—	—	—	—	—
S. oslo	1	—	—	—	—	—	—	—	—	—
S. newport	6	—	—	—	—	—	—	—	—	—
S. bovis-morbificans	5	—	—	—	1	—	—	—	—	—
S. tananarive	1	—	—	—	—	—	—	—	—	8
S. typhi	1503	—	—	—	—	—	—	—	—	—
S. enteritidis	5	—	—	—	—	—	—	—	—	—
S. dublin	3	1	—	—	—	—	—	—	1	—
S. blegdam	1	—	—	—	—	—	—	—	—	—
S. berta	1	—	—	—	—	—	—	—	—	—
S. javiana	5	1	—	—	—	—	—	1	—	—
S. gall. pullorum	1	—	—	—	—	—	10	—	—	8
S. give	—	1	—	—	—	—	—	—	—	—
S. uganda	—	—	—	—	—	—	—	—	—	1
S. anatum	1	7	—	—	—	—	—	—	—	—
S. meleagridis	1	—	—	—	—	—	2	—	—	—
S. lexington	—	—	—	—	—	—	—	—	—	1
S. weltevreden	1	—	—	—	—	3	—	—	—	21
S. senftenberg	2	—	—	—	—	—	—	—	—	—

d'outre-mer publient un rapport technique de leur activité, où sont entre autres citées les *Salmonella* isolées et leurs origines. Il serait ainsi possible de suivre, théoriquement, l'évolution de la fréquence des divers sérotypes. Mais ceci serait arbitraire et ne reflèterait certainement pas la réalité: la fréquence des *Salmonella* isolées varie avec le matériel étudié et aussi avec l'évolution des conditions de travail des bactériologistes dans des pays où la situation politique a subi des remous.

Nous classerons les résultats suivant les régions géographiques, en reprenant les tableaux généraux portant sur plusieurs années et comportant l'origine zoologique des isolements.

B) Centre Viet-Nam

Salmonella isolées de 1954 à 1961 dans les plateaux montagneux du Sud et au Centre Viet-Nam.

Salmonelles	Homme	Viandes et Abats de Bou-cherie	Crapaud	Serpents	Lézards
S. typhi	163				
S. paratyphi A	11				
S. stanley					2
S. san diego					19
S. dalat					1
S. derby	1	96		1	
S. budapest		1			
S. typhi-murium	4				5
S. paratyphi B var. java		1			
S. paratyphi C	1				
S. cholerae-suis	5				
S. oslo					33
S. newport					22
S. litchfield					60
S. manchester					4
S. fayed					3
S. enteritidis		1			
S. dublin	1				
S. panama	4				
S. anatum	1	34			
S. give		18			
S. meleagridis		2			
S. lexington		1	1		
S. senftenberg		7			
S. poona					43
S. hvittingfoss					26
S. welikada					32
S. bleadon					4
S. chicago					78
S. urbana					36
S. adelaïde					23
S. javiana					5
S. 9:−:1,5					1
S. vietnam					5
Total	191	161	1	1	402

Répertoire général des Salmonella isolées à Saïgon (Sud Viet-Nam) et Dalat (Centre Viet-Nam).

Salmonelles	Saigon 1952—1961	Dalat 1954—1961	Total
S. paratyphi A	131	11	142
S. paratyphi B	10	—	10
S. paratyphi B. var. java	3	1	4
S. typhi-murium	25	9	34
S. stanley	4	2	6
S. heidelberg	2	—	2
S. san diego	3	19	22
S. derby	13	98	111
S. paratyphi C	21	1	22
S. cholerae-suis	138	5	143
S. thompson	1	—	1
S. virchow	1	—	1
S. oslo	1	33	34
S. newport	6	22	28
S. bovis-morbificans	6	—	6
S. tananarive	9	—	9
S. typhi	1.503	163	1.666
S. enteritidis	5	1	6
S. dublin	5	1	6
S. blegdam	1	—	1
S. berta	1	—	1
S. javiana	7	5	12
S. gall. pullorum	19	—	19
S. give	1	18	19
S. uganda	1	—	1
S. anatum	8	35	43
S. meleagridis	3	2	5
S. lexington	1	2	3
S. weltevreden	25	—	25
S. senftenberg	2	7	9
S. budapest	—	1	1
S. litchfield	—	60	60
S. manchester	—	4	4
S. fayed	—	3	3
S. panama	—	4	4
S. poona	—	43	43
S. hvittingfoss	—	26	26
S. welikada	—	32	32
S. bleadon	—	4	4
S. chicago	—	78	78
S. urbana	—	36	36
S. adelaïde	—	23	23
S. 9:–:1,5	—	1	1
S. 6 8:eh:–.	—	1	1
S. dalat	—	1	1
S. quinhon	—	1	1
S. vietnam	—	5	5

C) Salmonella du Nord Viet-Nam (Hanoï)

Les résultats dont nous disposons sont assez anciens, car l'Institut Pasteur de Hanoï a été fermé à la suite de la guerre. Le tableau ci-dessous a été publié en 1954.

Type	Groupe	Sang	Selles	Pus	Divers	Total
S. paratyphi A	(A)	56	3		1	60
S. paratyphi B	(B)	3	29			32
S. typhi-murium			31		1	32
S. stanley			2			2
S. heidelberg			3			3
S. paratyphi C	(C)	19	1	1	1	22
S. cholerae-suis		7		1	3	11
S. newport spécifique		5	25			30
S. kottbus				1		1
S. thompson				1		1
S. oranienburg			1			1
S. bovis-morbificans		1				1
S. montevideo			1			1
S. typhi	(D)	385	34	7	5	431
S. enteritidis			1			1
S. dublin		1				1
S. blegdam		2	1	1	1	5
S. sendai		2				2
S. eastbourne			1			1
S. panama			1			1
S. anatum	(E)		3			3
S. give			1			1
S. senftenberg			1			1
S. hvittingfoss	(1)		1			1
Total		481	140	12	12	645

Lysotypes de *S. typhi* du Viet-Nam
A) Sud Viet-Nam, de 1954 à 1961

Lysotypes	Nombre	%	Lysotypes	Nombre	%
A	170	11,2	J 1	25	
B 1	4		J 3	5	
B 2	20		L 1	4	
C 3 (33, Desranleau C 2)	4		M 1	138	9,7
			M 2	3	
C 5 (Scholtens)	2		N	49	
D 1	18		O	1	
D 2	59	4,1			
D 4	10		T	36	
D 6	20		25 (Lie Kian Joe)	2	
			28 (Scholtens	1	
E 1	144	10	29 (Borman)	29	
E 2	2		37 (Nicolle et Brault)	28	
E 3 (Desranleau)	18		38 (?)	2	
E 4 (Desranleau)	3				
E 7 (Scholtens)	3		I + IV	443	31
E 9	20		Aliénosensibles	47	
E 10	6		Vi-négatifs	40	
F 1	2				
G	60	4,2			
H	2		Total	1.420	

B) Centre Viet-Nam et plateaux montagneux du Sud

Lysotypes	Nombre	%	Chimiotype	
A	37	27	35 types I,	2 types II
A —	1		I	
D 6	1		I	
E 1 (a)	4		I	
E 7	2		I	
E 10	4		I	
G	3		2 types I,	1 type II
M 1	13	9	II	
N	19	14	18 types I,	1 type II
T	2		1 type I,	1 type II
I × IV	45	33	44 types I,	1 type II
Aliénosensibles	4		I	
Total	135		116 types I, 19 types II	

II. Cambodge

Les publications sur les *Salmonella* du Cambodge sont rares. En 1958—1959, on a signalé l'isolement de

S. typhi-murium (lapins, pigeons, porcs, volailles)
S. enteritidis (souris)
S. cholerae-suis (porcs, singes)
S. bovis morbificans (cheval)
S. gallinarum-pullorum (volailles)
S. anatum (cheval)

III. Chine

L'Institut Pasteur de Chang-Haï qui, lui aussi, a été fermé du fait de la guerre, a publié en 1949 la liste des *Salmonella* isolées et identifiées depuis 1938.

Salmonelles	Eau et aliments	Rongeurs	Homme	Observation
S. paratyphi A			64	(62 malades)
S. paratyphi B			47	(45 malades)
S. typhi-murium			3	
S. derby			5	
S. paratyphi C			12	
S. cholerae-suis diphasique		1	14	(13 malades)
S. cholerae-suis kunzendorf		2	5	
S. thompson	2	5	5	
S. virchow		1		
S. potsdam		2	1	
S. newport	1		2	
S. typhi			403	(366 malades)
S. sendai			1	
S. enteritidis-Iéna			3	
S. enteritidis-Ratin		4	4	
S. blegdam		8	16	(12 malades)
S. london	1		1	
S. anatum		1	5	(4 malades)
S. aberdeen			1	
S. shanghai n.s.	1			

Commentaires

La distribution des *Salmonella* chez l'homme en Extrême-Orient diffère de celle rencontrée en Europe de l'Ouest. Plusieurs faits sont à remarquer:

1. La très grande fréquence des *S. typhi*.

2. L'abondance relative des *S. paratyphi A* que l'on ne voyait pratiquement jamais en France en temps de paix et quand les échanges de populations avec l'Afrique du Nord n'étaient qu'occasionnels.

3. La rareté de *S. paratyphi B* que l'on trouve en France au moins aussi fréquemment que *S. typhi*.

4. La fréquence de *S. paratyphi C* et *S. cholerae suis* que l'on ne trouve presque jamais chez l'homme en France.

5. Il est connu que *S. sendaï* est une *Salmonella* extrême-orientale. Mais elle est exceptionnelle. Il en est de même pour *S. blegdam*.

Des recherches intéressantes ont été faites au Viet-Nam sur les *Salmonella* dans les aliments carnés et chez les reptiles [5, 6, 7].

A Dalat (centre Viet-Nam), les *Salmonella* ont été trouvées chez 58 prélèvements sur 342 de viande et d'abats de boeuf prélevés au

Les prélèvements desquels furent isolées les *Salmonella* humaines pendant cette période sont

Salmonelles	Sang	L.C.R.	Urines	Liquide périto- néal	Pus	Matières fécales	Total
S. paratyphi A	60					4	64
S. paratyphi B	17		3		1	26	47
S. typhi-murium						3	3
S. derby						5	5
S. paratyphi C	8	1				3	12
S. cholerae-suis	10		1		1	7	19
S. thompson						5	5
S. potsdam						1	1
S. newport						2	2
S. typhi	336	5	1	8	2	51	403
S. sendai	1						1
S. enteritidis-Iéna	1				1	1	3
S. enteritidis-Ratin			2		2		4
S. blegdam	10	4			1	1	16
S. london						1	1
S. anatum	1					4	5
S. aberdeen						1	1
Total	444	10	7	8	8	115	593

marché (16,9%) et chez 81 sur 401 de viande et d'abats de porc (20,1%).

Parmi les prélèvements de muscle, 14 furent positifs sur 112 d'origine bovine (12%) et 10 sur 83 d'origine porcine (12%). Les pourcentages sont plus élevés dans les foies et la rate: 25% et 22% chez le porc, 22% et 22% chez le boeuf. Les ganglions mésentériques de boeuf contenaient dans 10% des cas de *Salmonella*, ceux de porc dans 15%.

La fréquence des *Salmonella* dans les viandes et chez les humains augmente pendant la saison sèche (novembre à mars) et cette augmentation n'est pas liée à l'eau consommée qui risquerait plus d'être contaminée à la saison des pluies.

Les viandes sont certainement une des grandes causes vraisemblablement même la principale, de la dissémination des *Salmonella* au Viet-Nam.

Les animaux à sang froid peuvent aussi souvent héberger des entérobactéries pathogènes, en particulier des *Salmonella* et *Arizona*. Les résultats suivants ont été rapportés par CHAMBON et coll.[7].

1. Serpents: *Ancistrodon rhodostoma* (crotales) 2/70 (+12 *Arizona*)
   Enhydrina shistosa   0/8 (2 *Arizona*)
   Capemis hardwickii   2/63 (+60 *Arizona*)
2. Lézards: *Hemidactylus bleker*   19/497 (+17 *Arizona*)
   Peripia peronii   14/152 (+12 *Arizona*)
3. Crapauds et escargots: 0/114 et 0/100

Ces lézards ne sont pas comestibles. Cependant les hémidactyles sont parfois absorbés vivants pour lutter contre l'asthme ou broyés dans certains médicaments antipaludéens. Ces hôtes familiers des habitations peuvent évidemment souiller les aliments de leurs déjections.

Une autre étude de MILLE et coll. [6] sur des lézards de race différente, vivant dans le voisinage des habitations (*Leilolepis bellina guttata*) a mis en évidence 242 fois les *Salmonella* sur 609 animaux (39,7%). Six animaux hébergeaient à la fois deux sérotypes différents. Au cours de cette étude fut trouvé un nouveau sérotype, *S. dalat*[8]. Le taux d'infection de ces animaux varie avec leur éloignement des villages: le taux d'infection fut de 7,8% chez 189 animaux capturés loin des villages de 50,3% chez ceux capturés dans les villages. Le rôle de ces hôtes familiers dans la dissémination des *Salmonella* au Viet-Nam peut donc n'être pas négligeable.

BIBLIOGRAPHIE

1. Rapports annuels sur le fonctionnement technique des Instituts Pasteur du Viet-Nam (1952—1961).
2. Rapports annuels sur le fonctionnement technique de l'Institut Pasteur de Chang-Haï (1938—1949).
3. Rapports annuels de l'Institut Pasteur de Hanoï (1951 à 1955).
4. Rapports annuels de l'Institut Pasteur du Cambodge 1958—1959.
5. CAPPONI, M., SUREAU, P. & LE MINOR, L. — Contribution à l'étude des Salmonelles du Centre Viet-Nam. — *Bull. Soc. Path. exot.*, 1956, **49**, *796*.
6. MILLE, R., LE MINOR, L. & CAPPONI, M. — Nouvelle contribution à l'étude des *Salmonella* du Centre et du Sud Viet-Nam. *Bull. Soc. Path. exot.*, 1958, **51**, *198*.
7. CHAMBON, L., LE MINOR, L. & MARTIN, P. — Recherche d'entérobactéries chez les animaux à sang froid au Centre et Sud Viet-Nam. *Bull. Soc. Path. exot.*, 1959, **52**, *720*.
8. LE MINOR, L., MILLE, R. & DREAN, D. — Un nouveau sérotype de Salmonella: S. dalat (4,12, 27: y: enx). — *Ann Inst. Pasteur*, 1957, **92**, *555*.
9. LE MINOR, L., EDWARDS, P. R., MILLE, R. & DREAN, D. — Un nouveau sérotype de Salmonella S. quinhon ($47:z_{44}$), — *Ann. Inst. Pasteur*, 1959, **97**, *407*.
10. LE MINOR, L., CHAMBON, L., BORIES, S., MARX, R. & CHARIE-MARSAINES. Trois nouveaux sérotypes de Salmonella observés à Dakar (S. ngor: 1, 3, 19: 1v: 1, 5 et S. santhiaba (40: $1z_{28}$: 1, 6) et à Dalat (S. vietnam 41: b) — *Bull. Soc. Path. exot.*, 1962, **55**, *213*.
11. FOURNIER, J. — S. changhaï: Rapport sur le fonctionnement technique de l'Institut Pasteur de Changhaï, 1948, p. *45*.

SALMONELLOSIS IN AUSTRALIA

BY

NANCY ATKINSON

University of Adelaide, South Australia

(with 3 figs.)

Introduction

Australia is a large country with an area of about three million square miles, roughly equivalent in size to the United States of America. Partly due to the immigration programme, the population of the Commonwealth has grown from seven and a half million in 1947 to ten and a half million in the 1961 census. Australians are mainly city dwellers, more than 56% of the entire population living in the capital cities. The 1961 figures for States and their capitals were:

New South Wales (N.S.W.)	3,900,000	Sydney	2,200,000
Victoria (Vic.)	2,930,000	Melbourne	1,900,000
Queensland (Q).	1,500,000	Brisbane	600,000
South Australia (S.A.)	900,000	Adelaide	590,000
Western Australia (W.A.)	740,000	Perth	420,000
Tasmania (Tas.)	350,000	Hobart	120,000
Australian Capital Territory (A.C.T.)		Canberra	59,000
Northern Territory (N.T.)	27,000		

The map in Fig. 1 shows the States and cities. Noticeably long distances separate adjacent capitals, the average being about five hundred miles, the longest, between Adelaide and Perth, amounting to about 1700 miles. Excellent air services provide rapid transport between States. Infectious diseases also travel easily across the continent.

Human salmonellosis has been found in every State, but official records are available only for typhoid fever, which is notifiable and endemic. The number of cases each year is small except when an epidemic occurs, such as the explosive milk-borne outbreak with over 400 cases in Moorabbin, a suburb of Melbourne, Victoria, in 1943, and the insidious, coconut-borne outbreak affecting most States (WILSON & MACKENZIE, 1955). Information on other Salmonella infections in humans is not readily accessible. In animals, Salmonella infections also occur throughout the Commonwealth, but are even more difficult to trace. The Reports of the Queensland Department of Agriculture and Stock (1950, 1951, 1952, 1958, 1959, 1960, 1961, 1962) however, stressed the importance of salmonellosis in pigs, and in the later years indicated that bovine and ovine salmonellosis were increasing in 1960—1961; one outbreak of scouring in sheep, held at a Brisbane slaughter-house, resulted in many deaths. The department's blood testing scheme and control programme for the eradication of pullorum disease in chickens greatly reduced the incidence, but *S. pullorum* was isolated occasionally.

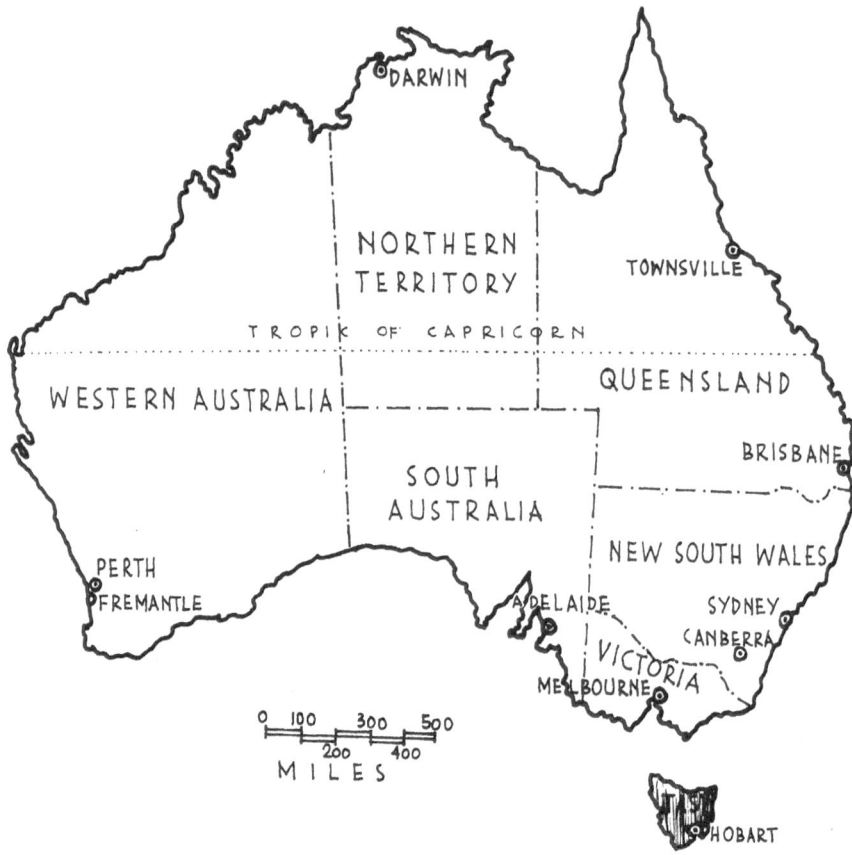

Fig. 1. Commonwealth of Australia, showing States and Capital Cities.

Although no general survey of salmonellosis in Australia has ever been published, some aspects of the problem have received atttention. SEDDON (1953) gave an account of animal salmonellosis in Australia up to 1952. My series of papers (ATKINSON et al., 1944, 1947, 1949, 1952, 1953, 1956) dealt with the occurrence of Salmonella serotypes in human, animal and other materials from all over Australia. Other publications dealt with the situation in various States. In Victoria, RUBBO (1948) and MUSHIN (1948) described a hospital outbreak of *S. derby* infection in children. Later MUSHIN

(1950) surveyed the causes of gastro-enteritis in infants, and listed the Salmonellas found in Victoria to the end of 1949. COOPER & WILSON (1957) gave figures for the serotypes in humans in Melbourne from 1950 to 1956. More recently, GRAY and colleagues (1958, 1959, 1960) investigated bovine salmonellosis in Gippsland, Victoria, and related it to the feeding of bone meal which they showed to be heavily contaminated with Salmonellas. In Queensland, MACKERRAS & MACKERRAS (1949) reviewed Salmonella infections in Brisbane, especially a severe epidemic of infantile gastro-enteritis due to *S. bovis-morbificans*. In connection with this epidemic, rats, mice and cockroaches were examined and LEE (1955) obtained an infection rate of 3% in a group of 842 urban rats. Following the discovery of Salmonellas in lizards, dying in an outbreak of enteritis in an experimental colony, LEE & MACKERRAS (1955) examined some Australian lizards and other native animals for Salmonellas, and found 62% of the lizards to be carriers with at least nine different serotypes. Meanwhile SIMMONS & SUTHERLAND (1950) and SIMMONS (1951) described their experiences with salmonellosis in domestic animals and birds in Queensland from 1946 to the end of 1950. SIMMONS et al. (1961) subsequently recorded *Salmonella* species isolated from animals and birds from 1951 to 1960. In Western Australia, KOVACS (1959) investigated the occurrence of Salmonellas in possible vehicles of infection such as coconut, egg pulp, fertilizers of animal origin and water. In South Australia, ANDERSON & WOODRUFF (1961) examined the distribution of Salmonellas in coconut and ATKINSON & EVANS (1960) found strains of Salmonella and Arizona in lizards.

Sources of the Material Presented Here

Many of the strains of Salmonella, mentioned in the foregoing references, were sent to me for serotyping. I also received many strains, not mentioned in any published work, and therefore, in gathering material for this chapter, I have drawn upon my unpublished records to supplement those in the literature. My colleagues, Dr. K. ANDERSON of Adelaide, Dr. G. N. COOPER of Melbourne, Dr. N. KOVACS of Perth and Mr. G. C. SIMMONS of Brisbane, kindly provided additional unpublished material and I wish to acknowledge with gratitude their valuable contributions. From all these sources, I collected the records of 11,273 strains of *Salmonella* isolated from human, animal and other specimens in various States, mainly over the years 1950 to 1962. More than half the strains (64%) came from humans; the remainder came from animals (16.2%) or from other materials (19.8%) such as coconut, meat meal and the like. The predominance of human strains is a reflection of the regular, routine examination of specimens from sick people through hospital or public health laboratories all over Australia.

The distribution of human, animal and other strains among the States is given in Table A.

Table A.

Distribution of 11,273 Salmonella strains

State	Number of strains from			Totals
	Human	Animal	Other	
N.S.W.	992	33+	134	1159+
Q.	1532	1372	380	3284
S.A.	1267	207	121	1595
Vic.	2921	80	99	3100
W.A.	333	154	1496	1983
A.C.T. N.T. Tas. N.G.	137	12	3	152
Totals	7182	1858	2233	11273

Queensland has the largest total and by far the largest number of animal strains, due, not to an unusually high incidence of animal salmonelloses, but to a keen interest in veterinary bacteriology, resulting in far more isolations of Salmonellas than in any other State. Victoria has the largest number of human strains, on account of special interest in public health bacteriology and salmonella serology, but the figures for animal and other strains are low. Through particular interest in public health bacteriology and especially in vehicles of infection in salmonellosis, Western Australia has the largest total for specimens other than human or animal, but figures for human or animal strains are low. Very few strains came from Canberra, A.C.T., Northern Territory, Tasmania or the New Guinea area (N.G.) so these have been recorded together.

The strains listed under "other" in Table A represented materials not coming directly from humans or animals and fell into five categories as shown in Table B. The category termed "meat or bone meal etc." includes animal feeds and fertilizers of animal origin.

Egg pulp, meat and bone meal etc., and coconut provided the largest numbers of strains, mainly isolated in Western Australia, though Queensland, through its veterinary interests, produced the greatest number of strains from meat and bone meal etc. Other States contributed relatively few strains to any category.

There is no doubt that the numbers in Tables A and B represent only a portion of the Salmonellas occurring in Australia. The human strains, though far outnumbering the animal and other strains put together, still fall far short of the numbers actually existing. They

Table B

Distribution of Salmonellas among specimens other than human or animal.

State	Number of strains from specimens of					Total
	Coconut	Egg pulp	meat or bone meal etc.	meat or food	water or moore swabs	
N.S.W.	34	27	53	2	18	134
Q.	16	6	351	6	1	380
S.A.	61	5	24	22	9	121
Vic.	?	1	94	4	–	99
W.A.	370	900	152	4	70	1496
Tas. & A.C.T.	3	–	–	–	–	3
Totals	484+	939	674	38	98	2233

mainly came from people who were sufficiently ill to call the doctor who sometimes sent them to hospital, where laboratory tests were done. But no investigations are made on the many cases of milder gastro-intestinal illness who never call the doctor. No records of such cases exist, though from popular report they are common and many may be due to Salmonella infection. Furthermore, the search for carriers has seldom been pursued; the number of carriers in our midst is unknown. From the few figures which I have been able to find (see p. 574), the carrier state seems likely to be comparatively common, especially following obvious illness. Looking at the figures for animal and "other" Salmonellas, a much more comprehensive picture would have been obtained had all the States provided animal figures comparable with Queensland's and "other" figures comparable with those from Western Australia. The most obvious gaps in our knowledge show up conspicuously in Table A. I cannot emphasize too much that the material I am presenting here represents the minimal situation.

The Kauffmann-White (KW) Groups and Serotypes of Australian Salmonellas.

The serotypes found among 11,273 strains are listed in Table I along with their KW groups.

Table I provides a catalogue of 120 Salmonella serotypes found in Australia. The total number of strains and the number in human, animal and other material are listed for each serotype and indicate its range and prevalence.

When the 11,273 strains were gathered together into their KW groups, as shown in Table II, group B included more than half

TABLE I.

Salmonella Serotypes Isolated in Australia and Territories

Serotypes in alphabetical order

Name	KW Group	Number of strains from			Total Number
		Humans	Animals	Others	
adelaide	O	194	38	27	259
altendorf	B			1	1
amager	E_1	5		15	20
anatum	E_1	102	93	68	263
angoda	N			10	10
ball	B	2			2
bareilly	C_1	13	11	32	56
berta	D_1	2		1	3
birkenhead	C_1	2	5	2	9
blegdam	D_1	40	7		47
blukwa	K			1	1
bolton	E_1		1		
bonariensis	C_2	2	7		9
bonn	C_1	1		1	2
bovis-morbificans	C_2	773	85+	49	907+
braenderup	C_1		2	8	10
brandenburg	B		1		1
bredeney	B	8	15	38	61
brisbane	M		1		1
butantan	E_1	1		7	8
cambridge	E_2	13	4	47	64
cerro	K	1			1
champaign	Q		1		1
charity	H	2		2	4
chester	B	49	21	8	78
chester or san-diego		33	11		44
chingola	F			4	4
chittagong	E_4			17	17
cholerae-suis	C_1	51	381	1	433
colorado	C_1	1			1
cubana	G_2	1	1	36	38
dahlem	Y	1			1
decatur	C_1		1		1
derby	B	254	74	82	410
derby or essen		8	2		10
dublin	D_1	1	8		9
dublin or rostock		2			2
durham	G_2	1			1
eastbourne	D_1	13		2	15
edinburg	C_1			2	2
emmastad	P	1			1
enteritidis	D_1	19	10	33	62
essen	B		1		1
ferlac	H			33	33
fremantle	T	1		1	2
give	E_1	27	14	77	118

TABLE I (continued)

Name	KW Group	Humans	Animals	Others	Total Number
glostrup	C_2	1	1		2
havana	G_2	8			8
heidelberg	B	10		1	11
hessarek	B	4	1	1	6
hidalgo	C_2		1		1
hindmarsh	C_3	1			1
hvittingfoss	I	6		7	13
irumu	C_1		2		2
jangwani	J		1		1
java	B		1	3	4
javiana	D_1	5			5
kaapstad	B			1	1
kentucky	C_3			1	1
kiel	A	3			3
kottbus	C_2	23	3		26
kotte	C_1			11	11
lexington	E_1		8	17	25
litchfield	C_2	16	1	3	20
lomita	C_1	1			1
london	E_1	5	17		22
london or give		4	2		6
manila	E_2			1	1
meleagridis	E_1	16	29	76	121
miami or sendai	D_1	3			3
mississippi	G_2	2		1	3
montevideo	C_1	4		5	9
morotai	J	2			2
moscow	D_1	3			3
muenchen	C_2	97	44	2	143
muenster	E_1			3	3
nchanga	E_1			2	2
new-brunswick	E_2	10	11	1	22
newington	E_2	40	23	18	81
newport or	C_2	340	42	15	397
kottbus		43	2		45
non-motile group B	B	16			16
norwich	C_1	2	2		4
nyborg	E_1	1		16	17
ohlstedt	E_1		1		1
onderstepoort	H	1	2		3
oranienburg	C_1	45	12	128	185
orientalis	I	4	5		9
orion	E_1	21	15	23+	59+
oslo	C_1	5	4		9
panama	D_1	1			1
paratyphi A	A	10			10
paratyphi B	B	39		39	78
paratyphi C	C_1	20	3		23
perth	P			25	25
potsdam	C_1	17	22	6+	45+

TABLE I (continued)

Name	KW Group	Number of strains from			Total Humber
		Humans	Animals	Others	
pullorum	D_1	3	1	544	548
reading	B	4	1	1	6
rough salmonellas			5		5
rubislaw	F	4	10	1	15
saarbruecken	D_1			1	1
saint-paul	B	93	15	29	137
salford	I		3		3
salinatis	B	2			2
san-diego	B	31	12	3	46
selandia	E_2			1	1
senegal	F		1		1
senftenberg	E_4	58	1	256	315
singapore	C_1	4		2	6
solna	M			13	13
stanley	B	6	1		7
taksony	E_4	1	2	17	20
tennessee	C_1			1	1
thompson	C_1	1	4	8	13
typhi-murium	B	4406	661+	225	5292+
urbana	N	3			3
vejle	E_1	9	8	3	20
vejle or anatum		14	9		23
victoria	D_1	5			5
virginia	C	1			1
wandsbek	L			8	8
wandsworth	Q	5			5
warragul	H	2			2
waycross	S	32	9	17	58
weltevreden	E_1			6	6
worthington	G_2	16	22	47	85
zanzibar	E_1		3	2	5
untyped		29	40	37	106
Totals		7182	1858	2233	11273

Number of different serotypes 120.

(55.7%) and with groups C (21.3%), D (6,3%) and E (11.1%) covered 94.4% of the strains.

Among the serotypes in Table I are some which were described
for the first time in Australia. They are listed in Table III with
details of their origin.

With the exception of S. *morotai*, the serotypes in Table III were
isolated in Australia. S. *morotai* was isolated during World War II
from military personnel in the town of Morotai, near the Equator,
on the northern tip of the Moluccas Islands to the north-west of Dar-
win; two strains were sent to me for serotyping. S. *adelaide*, which I

TABLE II

Distribution of Australian Salmonellas among the Kauffmann-White groups.

K-W group	Number of serotypes	Number of strains			Total
		Human	Animal	Other	
A	2	13	—	—	13
B	19	4965	817	432	6214
C_1	20	167	448	207	822
C_2	8	1295	186	69	1550
C_3	3	2	—	1	3
D_1	12	97	26	581	704
E_1	16	205	200	315	720
E_2	5	63	38	68	169
E_4	3	59	3	290	352
F	3	4	11	5	20
G_2	5	28	23	84	135
H	4	5	2	35	42
I	3	10	8	7	25
J	2	2	1	—	3
K	2	1	—	1	2
L	1	—	—	8	8
M	2	—	1	13	14
N	2	3	—	10	13
O	1	194	38	27	259
P	2	1	—	25	26
Q	2	5	1	—	6
S	1	32	9	17	58
T	1	1	—	1	2
Y	1	1	—	—	1
Totals: 23	120	7153	1812	2196	11161

first described in South Australia (ATKINSON, 1943), has proved to be an endemic Australian type (ATKINSON et al., 1952). *S. perth* was isolated from Ceylon coconut and is not an Australian type. The other serotypes in Table III are rare. *S. brisbane* and *S. hindmarsh* have not been found since the original isolations.

In contrast to such countries as Africa, Australia has provided few new serotypes and only a small portion of the serotypes listed by KAUFFMANN (1961). The predominance of serotypes of KW groups B, C, D and E, recognized elsewhere as the major groups, was to be expected.

Human Salmonellosis

The KW groups and serotypes

The 7153 human strains of Salmonella recorded in Table I have been put together into their KW groups in Table IV.

TABLE III

Salmonella Serotypes First Described in Australia

Serotype	KW group	Antigenic formula	Source of first strain
adelaide	O	35; fg; —	faeces of a fatal case of gastro-enteritis (ATKINSON, 1943).
atherton (now known as waycross)	S	41; z_4 z_{23} —	faeces of 26 cases of mild gastro-enteritis in a military hospital on the Atherton Tableland, Q. (FERRIS, HERTZBERG & ATKINSON, 1945; ATKINSON, WOODROOFE & MACBETH, 1947; ATKINSON, WOODROOFE & MACBETH, 1950a).
ball	B	1, 4, 12; y; enx	faeces of adult with gastro-enteritis (ATKINSON, WOODROOFE & MACBETH, 1950b).
brisbane	M	28; z; enz_{15}	intestine of sheep which also gave *S. adelaide* from the liver, *S. london* from the spleen and *S. typhi-murium* from the kidney (ATKINSON, unpublished).
fremantle	T	42; (f) gt; —	faeces of an adult with gastro-enteritis (ATKINSON, unpublished).
hindmarsh	C_3	(8); r; 1, 5	intestine of a fatal case of gastro-enteritis (ATKINSON, unpublished)
morotai	J	17; lv; 1, 2	faeces of two adults in an outbreak of food-poisoning in an army unit at Morotai (ATKINSON, WOODROOFE & MACBETH, 1950b).
perth	P	38; y; enx	desiccated coconut from Ceylon (ATKINSON, unpublished, KOVACS, 1959).
victoria	D_1	1, 9, 12; lw; 1, 5	faeces of case of gastro-enteritis (COOPER & WILSON, 1957).
warragul	H	1, 6, 14, 25; gm;—	faeces of a case of gastro-enteritis (COOPER & WILSON, 1957).

Group B occupies a major part (69.3%) and with group C (20.5%) covers nearly 90% of human strains. With the addition of groups D (1.4%), E (4.6%) and O (2.7%), 98.5% of the human strains are included. The remaining groups A, F, G, H, I, J, K, N, P, Q, S, T and Y represent only 1.5% of human strains; strains of groups L and M were not encountered.

The number of different serotypes and the commonest serotypes are listed for each KW group in Table IV. *S. typhi-murium* was the commonest serotype, not only in group B, but in the whole Table, and accounted for 4406, or well over half, of the human strains. *S. bovis-morbificans*, the commonest serotype in group C, had the next highest total of 773 strains, followed by another group C sero-

TABLE IV

KW Groups and Commonest Serotypes of Human Salmonellas

KW grp	Number of		Commonest Serotypes		
	Sero-types	Strains	Name	Number	Percentage KW grp
A	2	13	paratyphi A	10	77%
B	14	4965	chester (or san-diego)	113	2.3%
			derby (or essen)	262	5.3%
			paratyphi B	39	0.8%
			saint-paul	93	1.9%
			typhi-murium	4406	88.8%
			remainder		0.9%
C_1	14	167	bareilly	13	7.8%
			cholerae-suis	51	30.5%
			oranienburg	45	27.0%
			paratyphi C	20	12.0%
			potsdam	17	10.2%
			remainder		12.5%
C_2	7	1295	bovis-morbificans	773	55.8%
			muenchen	97	7.5%
			newport (or kottbus)	406	31.4%
			remainder		5.3%
D_1	11	97	blegdam	40	41.4%
			eastbourne	13	13.4%
			enteritidis	19	19.6%
			remainder		25.6%
E_1	9	205	anatum	102	50.0%
			give	27	13.2%
			meleagridis	16	7.8%
			orion	21	10.3%
			remainder		18.7%
E_2	3	63	cambridge	13	20.7%
			new-brunswick	10	15.7%
			newington	40	63.6%
E_4	2	59	senftenberg	58	98.5%
			remainder		0.5%
Further Groups			Total 287		% of Further Grps.
G_2	5	28	worthington	16	5.6%
O	1	194	adelaide	194	68.0%
S	1	32	waycross	32	11.3%
			remainder		15.1%

type *S. newport* (or *kottbus*), with 406 strains. Other serotypes contributing noteworthy totals are *S. derby* (262 strains) in group B, *S. adelaide* (194 strains) in group O and *S. anatum* (102 strains) in group E.

In group C, the main contribution came from C_2 (1295 strains) representing 88.5% of the group C total of 1464. Group C_1 provided only 167 strains or 11.4% but they consisted of 14 different serotypes. Group C_3 was rare with only 2 strains.

In group E, the total number of strains was 327 of which 205 or 62.2% came from E_1, 93 or 19.3% from group E_2, and 59 or 18.1% from E_4. There were no E_3 strains. S. *anatum* was the commonest serotype followed by S. *newington* and S. *senftenberg*.

Groups A and D, though contributing little to the total numbers, contained important human pathogens. In group A S. *paratyphi A*, of which 10 strains are recorded in Table IV, occasionally caused serious illness in Australia, though the infection was usually contracted abroad. In group D, represented only by D_1, S. *blegdam* which had the largest group total of 40 strains, came from cases of serious and sometimes fatal infection in Australian soldiers and others in the New Guinea area in World War II (see p. 555). No cases were found in Australia. S. *typhi*, also causing severe illness, belongs to group D_1 but is not listed. The further groups from F on, apart from O, contributed little to human salmonellosis.

Time and place distribution

Although the commonest serotypes, determined by numbers, can be seen in Table IV, a better assessment of their significance can be made from Fig. 2, showing the distribution of all the human serotypes through the years from 1950 to 1962 in the four States, S.A., Victoria, N.S.W., and Queensland, for which good totals for human strains were available.

The years are marked at the top and bottom of Fig. 2 and correspond to the columns. The serotypes are listed at the left in alphabetical order in their KW groups and correspond to the rows. Each square corresponds to one serotype and one year and is divided into quarters to represent the four States, the upper left for S.A., the upper right for Victoria, the lower left for N.S.W. and the lower right for Queensland. When a serotype has been found, in any number whatever, in a particular State and year, the appropriate quarter is filled in. Thus at the top of Fig. 2 part 1, in group A, S. *kiel* is a newcomer appearing only in 1962 in the two States N.S.W. and Queensland. In the second row, S. *paratyphi A*, by contrast, was found prior to 1950 in S.A. and Victoria, but subsequently reappeared only in N.S.W. in 1953 and 1955.

In group B, the completely black bar for S. *typhi-murium* in the last row is outstanding and represents regular occurrence in all four States in every year. S. *derby*, in the fourth row, is also continually present through the years and except for 1953, 1954, 1956, 1957 and 1962, it was found in all four States; it is obviously an endemic type. In the third row above S. *derby*, is S. *chester*, more important in the later years showing a continual occurrence from 1957 to 1962, often

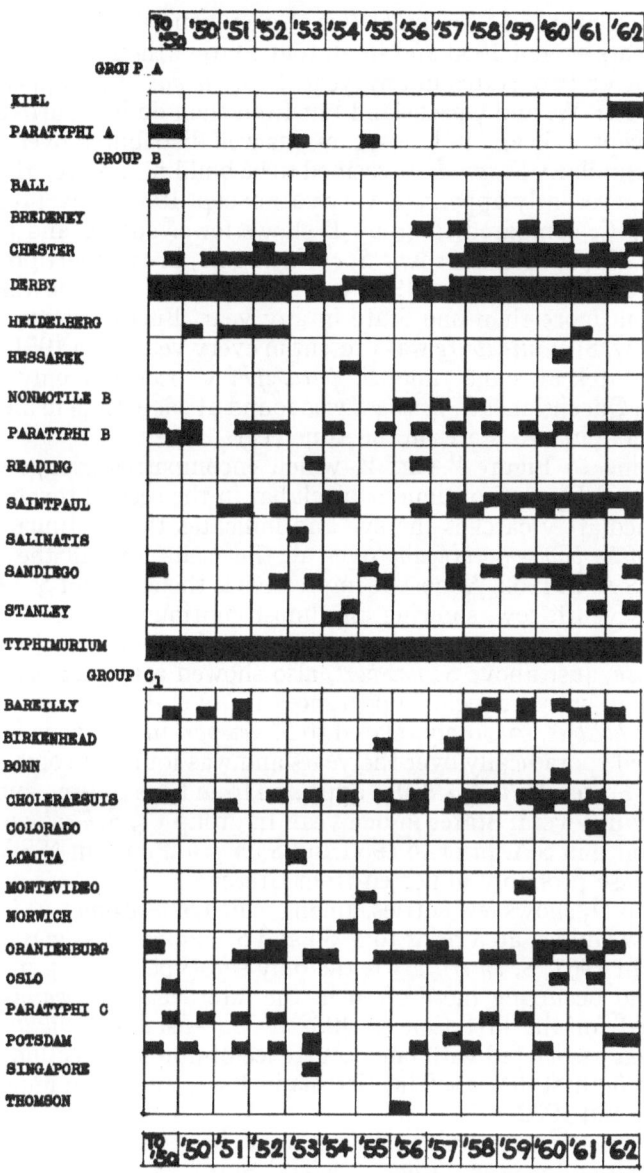

Fig. 2. Part 1. Occurrence of Human Salmonellas in four States from 1950 to 1962.

in the four States. *S. paratyphi B* is also endemic and has been found in all years except 1953 and in all four States, though not in more than one or two States in any year. *S. saint-paul* first appeared in 1951 in N.S.W. and Queensland but from then on it occurred every year except 1955 and in later years was well distributed over several States. Similarly *S. san-diego* seemed to be building up over the years and more recently appeared every year except 1957 in several States.

Group C_1 compared to group B shows fewer entries and no continuous black bars. However *S. cholerae-suis*, in the fourth row, appeared every year except 1950 and in all States at some time, but seldom in more than one State in any year. Further down, in the ninth row, *S. oranienburg* was present in every year from 1951 on and in all States at some time. *S. paratyphi* C occurred only in five years in Queensland. *S. potsdam* was found at some time in all States but seldom in two States in the same year.

Turning to Figure 2 part 2, which encompasses groups C_2, C_3, D_1, E_1 and E_2, the continuous black bar in the second row of group C_2 immediately catches the eye and indicates the continual occurrence of *S. bovis-morbificans* over all the years and all the States, except S.A. in 1962. Almost as impressive is the entry for *S. newport*, in the seventh row, showing an almost continuous black bar indicating occurrence in every year in two, three or four States. *S. muenchen*, just above *S. newport*, also showed a regular occurrence each year with frequent appearances in several States in a single year. *S. kottbus*, so closely related to *S. newport* in antigenic formula, appeared sporadically over the years and was found at some time in all four States. *S. litchfield* also appeared from time to time and occasionally in several States in one year. In group C_3, *S. hindmarsh* was found once in S.A. prior to 1950 and *S. virginia* once in N.S.W. in a single case probably imported from Greece.

Group D_1 shows few entries, among which *S. blegdam* was found in the New Guinea area prior to 1950 and *S. miami* (or *sendai*) in Tasmania in 1961. *S. enteritidis* is the only serotype showing many entries and occurring more often in the later years. *S. panama* was recorded for the first time in 1962 in Queensland. *S. pullorum* in 1961 was associated with one occurrence of human infection in S.A. *S. victoria*, first described in Victoria in 1954, was again found there in 1961 and 1962.

The most important human pathogen in group E_1 is *S. anatum*, occurring every year, often in several States. *S. give* has increased in incidence over the years from 1955 onwards and was often found in two States in one year. In group E_2, *S. cambridge* occurred sporadically in various States, *S. new-brunswick* was found mainly prior to 1950, and *S. newington* seemed to be appearing more frequently in recent years when it occurred from 1957 to 1961, in Queensland and Victoria.

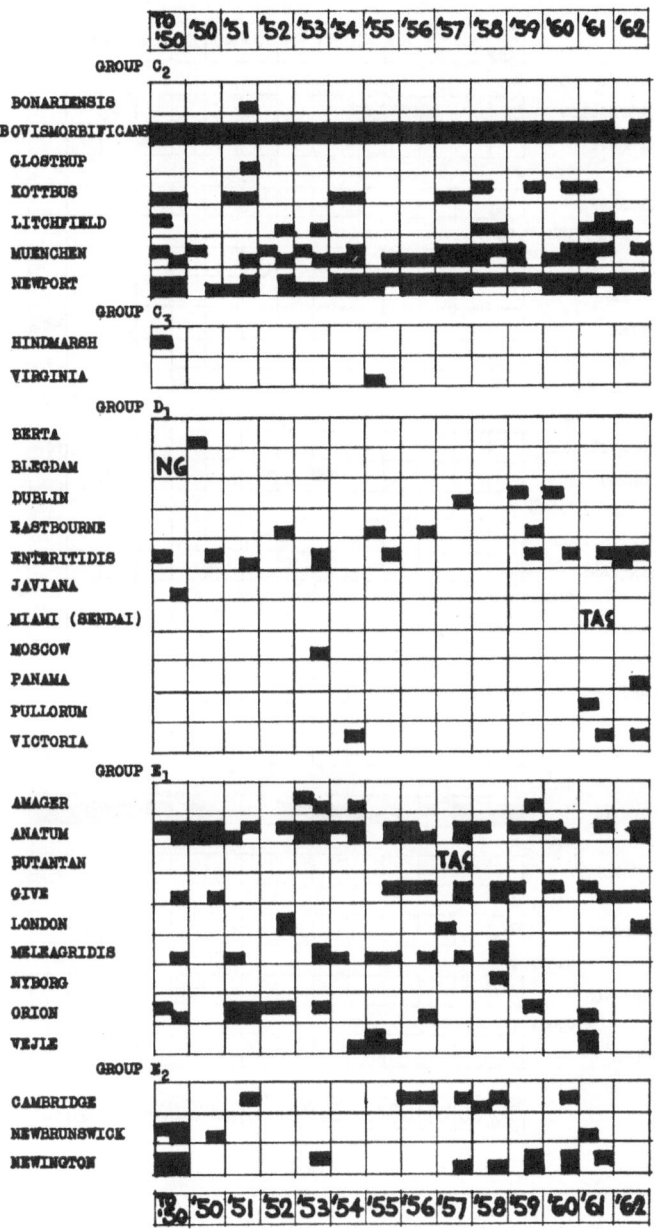

Fig. 2. Part 2.

554

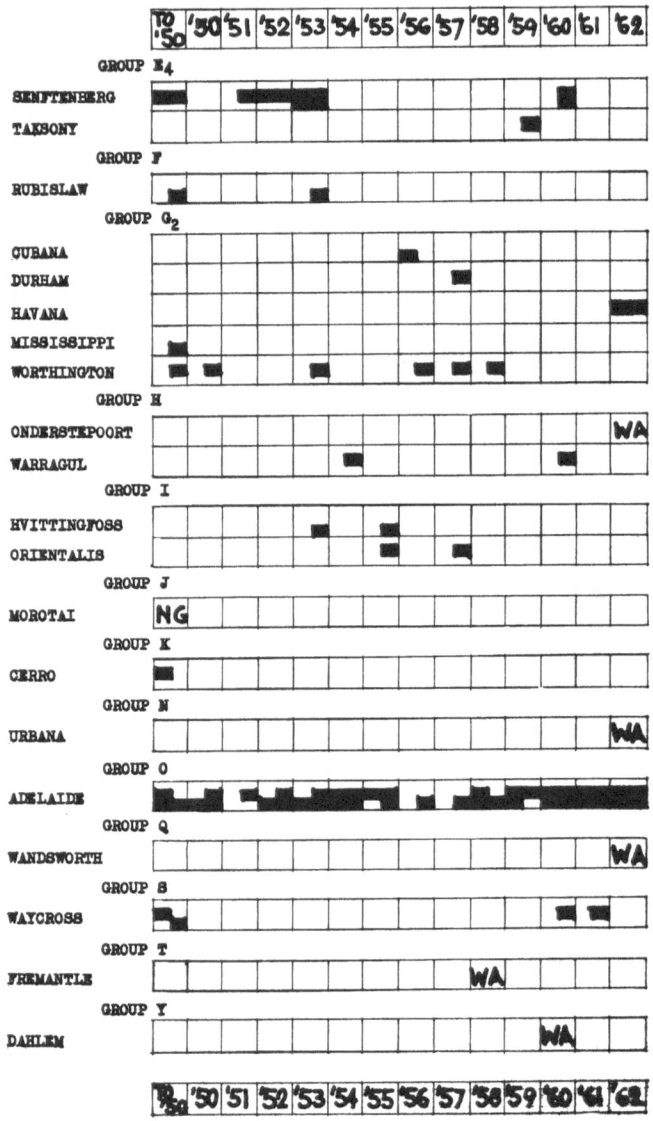

Fig. 2. Part 3.

In Fig. 2 part 3, covering group E_4 and the further groups from F on, only one serotype *S. adelaide* in group O shows a continuous, black bar indicating regular occurrence and importance in human infections; it is endemic (ATKINSON et al., 1952). The widespread occurrence of *S. senftenberg* in 1952 and 1953 corresponds with its first isolations from coconut (see p. 582); it was recorded from humans again in 1960 when it was also found in coconut. In group G_2, *S. worthington* was found in several years in Victoria, *S. havana* was found there as well as in S.A. in 1962. In the remaining groups very few entries occur and they are apparently of little significance in human disease. Some rare serotypes were found only in Western Australia (marked W.A.).

In summary, the KW groups important here in human disease were A, B, C_1, C_2, D_1, E_1 and O. However, their importance hinged entirely around a few serotypes, other members of the groups being rarely or sporadically found. The accent is thus placed on the individual serotype as opposed to the KW group. The main serotypes in human infections are (in order of appearance in Fig. 2) *S. paratyphi A*, *S. chester* (and related *san-diego*), *S. derby*, *S. paratyphi B*, *S. saint-paul*, *S. typhi-murium*, *S. cholerae-suis*, *S. oranienburg*, *S. bovis-morbificans*, *S. muenchen*, *S. newport* (and related *kottbus*), *S. enteritidis*, *S. anatum* and *S. adelaide*.

Human Outbreaks of Salmonellosis

The epidemiology of human salmonellosis in Australia is not well understood. Few outbreaks have been studied and individual cases have seldom been followed up. The majority of outbreaks occurred in hospitals or institutions, but family infections were occasionally recorded. Adults as well as children were often involved and the severity of the illness varied greatly. Rarely was the reservoir or the vehicle of infection discovered. Family outbreaks, in which child or adult carriers were probably the reservoir, occurred with *S. adelaide* and *S. javiana* (MACKERRAS & MACKERRAS, 1949; ATKINSON et al., 1952). Outbreaks, in which a particular food was mentioned, occurred with *S. bovis-morbificans* (corned beef), *S. cholerae-suis* (brawn and Spanish cream), *S. newport* (or *kottbus*) (tongue), *S. paratyphi B* (tinned salmon, ATKINSON & WOODROOFE, 1944), *S. reading* traced to sausage made with raw pork (COOPER & WILSON, 1957), *S. typhi-murium* traced to Spanish cream (ATKINSON & WOODROOFE, 1944), or after eating roast duck meat or boiled fish, *S. give* and *S. typhi-murium* traced to mussels from a beach where sea water contained *S. berta* (KOVACS, 1955), *S. orion*, *S. potsdam*, *S. senftenberg* and *S. typhi-murium* traced to eating contaminated coconut (WISON & MACKENZIE, 1955). Outbreaks of severe and fatal disease occurred with *S. blegdam*, *S. bovis-morbificans*, *S. derby*, *S. paratyphi B*, and *S. typhi-murium*. The *S. blegdam* outbreak, involving salmonella fever or septicaemic conditions, occurred in the New Guinea area in Australian soldiers and natives during World War II (FENNER & JACKSON, 1946; COBLEY & WILSON, 1946; ATKINSON, 1946; JONES & FENNER, 1947; ATKINSON et al. 1949). Recovered cases often remained urinary or faecal carriers for long periods and FENNER & JACKSON (1946) suggested that carriers were important as a reservoir of infection.

A major outbreak of gastro-enteritis in young children, due to S. *bovis-morbificans*, occurred in Brisbane, Queensland. It was extensively investigated by MACKERRAS & MACKERRAS (1949) who showed that the organism was widespread in the environment of cases and occurred in the hospital in ward nail brushes, sinks and boilers; mice and cockroaches found harbouring the Salmonella, were considered to be innocent victims of the environment and not original carriers, though they probably played some part in spreading the infection. Human carriers seemed important and the reservoir of infection was thought to be a housing camp in which living conditions were not hygienic. Another major outbreak in young children in hospital was due to S. *derby* and occurred in Melbourne, Victoria. RUBBO (1948) and MUSHIN (1948) described the outbreak in detail and related it to contamination of the environment from cases and carriers. S. *derby* was found in ward dust and towels, in hospital mice which were probably innocent victims again, but helped to spread the organism. The original reservoir of the infection was obscure. Other serious outbreaks in young children were caused by S. *derby*, S. *paratyphi* B or S. *typhi-murium* (ATKINSON et al., 1953) but their sources were unknown.

S. *kottbus* and S. *muenchen* illustrate how Salmonellas may enter maternity hospitals with expectant mothers, who are carriers or develop diarrhoea, and infect their own and other babies (MACKERRAS & MACKERRAS, 1949). I have a number of records suggesting that diarrhoea, due to a Salmonella, is not uncommon after confinement and could be an activation of latent infection, which may be much more frequent than we suspect. Certainly, a number of carriers of Salmonellas were found among women attending a post-natal clinic in Sydney, N.S.W. (see p. 574).

S. *mississippi*, S. *waycross* (or *atherton*) and S. *new-brunswick*, all rare types, produced outbreaks of gastro-enteritis in military hospitals in the Northern Territory and North Queensland during World War II (ATKINSON et al., 1947, 1950; FERRIS et al., 1945). The origins of these outbreaks were not known. Less severe outbreaks provided strains of S. *anatum* (MUSHIN, 1950; ATKINSON et al., 1947), S. *san-diego* or S. *typhi-murium* (ATKINSON et al., 1953). Two hospital outbreaks involved more than one serotype. The first occurred in new-born babies suffering from very mild illness, from which S. *typhi-murium* was isolated in some cases, and an unusual organism, a non-motile group B Salmonella, was isolated from others. The second outbreak concerned cross-infection from children to nurses, but when the strains were serotyped the Salmonellas from the nurses were different from those of the children whom they blamed for infecting them.

S. typhi-murium was involved more often than any other serotype in both severe and mild outbreaks. Obviously much detailed work remains to be done before a clear epidemiological picture of human salmonellosis in Australia can emerge. However, the human carrier probably plays an essential role as a reservoir of infection and there is no doubt of the danger of hospital cross-infection following the admission of open cases of gastro-enteritis.

Severe or Fatal Infections in Humans.

Salmonella infections in man may be manifested in a variety of forms, some of which are particularly severe.

The commonest serotype, *S. typhi-murium*, was responsible for many severe infections and deaths. Others similarly involved were *S. bovis-morbificans*, *S. derby*, *S. cholerae-suis* and *S. paratyphi B*. Some rare serotypes such as *S. hindmarsh*, *S. salinatis*, *S. vejle* and *S. virginia* were occasionally found in serious illnesses. The range of severe diseases represented here covered acute gastro-enteritis, middle ear infection, meningitis, salmonella fever, bacterial endocarditis, cholecystitis, osteomyelitis, encephalitis, wound infection, abscesses and suppurative arthritis. Although this is only a small sample of the severe cases, it suffices to accent the potential dangers of Salmonella infections in man.

Animal Salmonellosis

The KW groups of animal strains

The distribution of 1812 animal strains of Salmonella in the KW groups is shown in Tables V and VI.

Compared with the human strains, the animal strains contained a smaller number in group B (45%) and larger numbers in groups C (35%) and E (13.4%), suggesting that group B might be more important to humans, and groups C and E more important to animals. Group D (excluding *S. pullorum*, for which suitable figures are not available though it still occurs in chickens) represented only 1.4% of the animal strains. In the further groups from F on, group O, with its single serotype *S. adelaide*, formed 2.1% of the animal total compared with 2.7% of the human total. Group G provided 1.3% of the animal total. The remainder of the further groups comprised only 1.8%.

The KW groups and serotypes from various kinds of animals.

Tables V and VI show the different kinds of animals from which strains of Salmonella were isolated.

TABLE

Distribution of Serotypes & KW Groups of

KW Group	Serotype	Domestic animals						Number of Birds		
		cat-tle	sheep	pigs	hors-es	dogs	cats	chick-en	duck	tur-key
B 12 sero-types	brandenburg			1						
	bredeney	6		3				4		
	chester	4	1	2		2	1	4		
	chester or san diego	6	1			1		2		
741 strains	derby	14	3	14		2	2	27	2	
	derby or essen	1		1						
	essen			1						
	hessarek		1							
	java			1						
	saint-paul	3	4	3				3		
	san-diego	3	2	3				4		
	stanley									
	typhi-murium	203	99+ (& 2 goats)	63	13	10	1	57	59+	2
C$_1$	bareilly			1				7		
	birkenhead	2								
11 sero-types	braenderup			1				1		
392	cholerae-suis	5		322			1	2		
strains	decatur									
	irumu	1		1						
	norwich	1								
	oranienburg	3		2				2		
	oslo							4		
	paratyphi C									
	potsdam	7	2	3				3		
	thompson	1								
C$_2$ 8 sero-types. 178 strains	bonariensis	1						4		
	bovis-morbificans	18	13+	13	1	1	1	10	1	
	glostrup							1		
	hidalgo									
	kottbus	1			1			1		
	litchfield	1								
	muenchen	16	1	4	2			12		
	newport	13	10	3	9		1	15		
D$_1$ 5 sero-types	blegdam									
	dublin	4								
	eastbourne									
	enteritidis	3								
21 strains	pullorum							unknown number		

V

Salmonellas Found in Various Animals in Australia

Strains found in									
Birds	Lab. Stocks			Native animals		Wild animals			Totals
miscellaneous	G' pigs	mice	rab-bits	Liz-ards	miscellaneous	rats	mice	miscellaneous	
									1
									13
				6		1			21
									10
	3			1			3		71
									2
									1
									1
									1
									13
									12
			1						1
7 pigeons 2 magpies 8 canaries 2 cage birds 2 pheasants	42	3	5		3 kangaroos	9		{ 1 gnu 1 cockroach	594
									8
1 bird				1	1 kangaroo				5
									2
	1								331
								1 flies	1
									2
								1 bandicoot	2
						3			10
									4
						2			2
				6					21
						3 (N.G.)			4
				2					7
1 canary	2	2		2		3	7	3 cockroaches	75
	1								1
									1
									3
				4					1
									39
									51
	1	6							7
									4
	1								1
1 cygnet	2	2							8
									1+

KW Group	Serotype	Domestic animals						Birds		
		cattle	sheep	pigs	horses	dogs	cats	chicken	duck	turkey
E₁ 10 sero-types 194 strains	anatum	27	5	7	2	4	2	35	6	
	bolton									
	give	1		1			1	8	1	1
	lexington			1				6		
	london	7	1	1				7	1	
	meleagridis	5	1		1	1		15		
	ohlstedt				1					
	orion	2		1				10	1	1
	vejle	7	1 goat							
	vejle or anatum	8						1		
	zanzibar	1						1	1	
E₂ 3 sero-types 38 strains	cambridge	2					1	3		
	new-brunswick	6						5		
	newington	8		4	1			6		3
E₄ 2 sero-types 3 strains	senftenberg	1								
	taksony									
F 2 sero-types 11 strains	rubislaw	1								
	senegal									
G₂ 2 sero-types 23 strains	cubana									
	worthington			2				20		
H 2 strains 1 serotype	onderstepoort									
I 2 sero-types 6 strains	orientalis	1								
	salford		1					2		
M 1 strain 1 serotype	brisbane			1						
O 1 sero-type 36 strains	adelaide	6	7		1		2	5		
Q 1 strain 1 serotype	champaign							1		
S 1 sero-type 9 strains	waycross	3						4		

Strains found in									Totals
Birds	Lab. Stocks			Native animals		Wild animals			
miscellaneous	G' pigs	mice	rab-bits	Liz-ards	miscellaneous	rats	mice	miscellaneous	
1 cygnet					1 wallaby	2			92
									1
						1			14
					1 bandicoot				8
									17
					2 kangaroos	1			26
									1
									15
									8
									9
									3
									6
									11
								1 elephant	23
									1
									2
				9					10
					1 bandicoot				1
									1
									22
2 birds									2
1 bird							1		3
									3
									1
				14		1			36
									1
				2					9

TABLE VI

Distribution of Animal Salmonellas in the KW Groups

Animal	Number of strains in KW groups														
	B	C_1	C_2	D_1	E_1	E_2	E_4	F	G_2	H	I	M	O	Q	S
Domestic															
cattle	240	20	50	7	58	16	1	1			1		6		3
pigs	92	330	20		12	4			2						
sheep	111	2	24+		7						1	1	7		
horses	13		13		4	1				1			1		
dogs	15		1	1	5										
cats	4	1	2		3									2	
goats	2				1										
Birds															
chickens	101	19	43		83	12	2		20		2		5	1	4
ducks	61		1		10										
turkeys	2				2	3									
pigeons	7														
magpies	2														
canaries	8		1												
cagebirds	2														
pheasants	2														
wild birds		1								2	1				
cygnets				1	1										
Natives															
lizards*	7	7	8						9				14		2
kangaroos	3	1			2										
wallabies					1										
bandicoots		1			1			1							
Miscell.															
rats*	10	8	3		4								1		
mice	3		7									1			
gnu	1														
elephant							1								
cockroaches	1		3												
flies		1													
Lab. Stock															
guinea pigs	45	1	3	4											
mice	3		2	8											
rabbits	6														

* Besides the Salmonellas, lizards provided 16 Arizona strains & rats provided 1 Arizona strain.

Table VI shows the number of strains occurring in the KW groups, of which B, C_1, C_2, and E_1 were found in a wide range of animals, E_1 and O in a smaller range and the other groups in only a few kinds of animals. Group A was not represented but strains of the Arizona group, though not specially sought, were found in lizards and in a rat (ATKINSON & EVANS, 1960).

Cattle and chickens provided strains covering 11 different KW groups; pigs, sheep, horses, cats, lizards, and rats produced strains

TABLE VII

KW Groups and Commonest Serotypes of Animal Salmonellas

KW grp	Number of		Commonest Serotypes		%age of KW grps.	Occurrence
	sero-types	strains	name	number		
B	11	741	chester or san-diego	43	5.8%	wide range of animals.
			derby	71	9.6%	wide range of animals.
			typhi-murium	594	80.2%	very wide range of animals.
			remainder		4.4%	
C₁	12	392	cholerae-suis	331	84.7%	mainly pigs
			oranienburg	10	2.6%	cattle, pigs, chickens, rats.
			potsdam	21	5.4%	domestic animals, chickens, lizards.
			remainder		7.3%	
C₂	8	181	bovis-morbificans	78	43.1%	wide range of animals.
			muenchen	39	21.5%	domestic animals, chickens, lizards.
			newport	51	28.1%	domestic animals, chickens
			remainder		7.3%	
D₁	5	21	enteritidis (including bleg-dam & dublin)	19	91 %	cattle, guinea pigs, lab. mice.
			pullorum	unknown No. —		mainly in chickens
			remainder		9 %	
E₁	10	194	anatum	92	47.5%	wide range of animals.
			give	14	7.2%	wide range of animals.
			london	17	8.8%	domestic animals & birds.
			meleagridis	26	13.4%	wide range of animals.
			orion	15	7.8%	domestic animals
			remainder		15.3%	
E₂	3	38	new-brunswick	11	29 %	cattle, chickens
			newington	23	60.7%	wide range of animals.
			remainder		10.3%	
Further Groups Total 89						
F	2	11	rubislaw	10	11.2%	Cattle, lizards
G₂	2	23	worthington	22	24.7%	pigs, chickens
O	1	36	adelaide	36	40.5%	wide range of animals.
S	1	9	waycross	9	8.0%	cattle, chickens, lizards.

covering five to seven groups, but strains from the remaining animals belonged to relatively few groups.

In Table V, the serotypes are listed under their KW groups with the number of strains found for each kind of animal. There were more than 60 different serotypes, the commonest of which are given in Table VII, with their group percentages and their animal range.

The commonest serotype was *S. typhi-murium* with a total of 594 strains; together with *S. chester* (or *san-diego*) and *S. derby*, all found in a wide range of animals, it accounted for over 80% of group B. The commonest serotype in group C_1 was *S. cholerae-suis* (84.7%) coming almost entirely from pigs; *S. oranienburg* and *S. potsdam* occurred in much smaller numbers but in a variety of animals. *S. bovis-morbificans* formed over 40% of the strains in group C_2, to which *S. muenchen* and *S. newport* each contributed over 20%; all three occurred in a wide range of animals. Group D_1 was mainly represented by various *enteritidis* types found in cattle and laboratory stocks of mice and guinea pigs; *S. pullorum*, essentially a chicken type, was recorded once from a dog. In group E_1, the commonest serotypes, *S. anatum*, *S. give*, *S. london*, *S. meleagridis* and *S. orion*, occurred in many different kinds of animals. Group E_2 contained only 38 strains, but over 60% were *S. newington* and nearly 30% *S. new-brunswick*. There were just three strains in group E_4, two of which were *S. taksony* and the other *S. senftenberg*. These two types could have arisen, in some animals, from feeds containing contaminated meat or bone meal etc. (see p. 581). The main serotypes in the further groups from F on, were *S. rubislaw* (group F) in cattle and lizards, *S. worthington* (group G_2) in pigs and chickens, *S. adelaide* (group O) in a wide variety of animals, and *S. waycross* (group S) in cattle, chickens and lizards.

Animal outbreaks

The epidemiology of animal salmonellosis in Australia is even more obscure than that of human salmonellosis. Numerous outbreaks occur but few have been studied or even recorded. References to some of the earliest recorded outbreaks can be found in the papers by SEDDON (1953) and ATKINSON & WOODROOFE (1944). Table VIII lists some outbreaks which I have collected from the more recent literature and from my unpublished records.

For each kind of animal, Table VIII shows the total number of outbreaks and the number involving one or more serotypes. An impressive number (45 out of 255) gave two to six different serotypes and occurred generally in cattle, chickens or sheep. These figures stress the likelihood of mixed infections in animals and the need for careful bacteriological examinations in any animal outbreak. The greatest number of outbreaks were recorded for chickens (and these do not include *S. pullorum*). Cattle and sheep provided the next larg-

TABLE VIII

Outbreaks of animal Salmonellosis in Australia

Animal	Number of outbreaks involving:					
	Total	1 sero-type	2 sero-types	3 sero-types	4 sero-types	6 sero-types
Canaries	4	4				
Cattle	39	22	13	3	1	1
Chickens	83	67	12	3	1	–
Dogs	2	1	1	–	–	–
Ducks	31	30	1	–	–	–
Goats	1	1	–	–	–	–
Guinea pigs	18	17	1·	–	–	–
Horses	8	8	–	–	–	–
Kangaroos	1	1	–	–	–	–
Lizards	1	–	1	–	–	–
Mice	7	7	–	–	–	–
Pigeons	1	1	–	–	–	–
Pigs	13	12	–	1	–	–
Rabbits	1	1	–	–	–	–
Sheep	43	38	3	1	1	–
Turkeys	1	–	1	–	–	–
Totals	255	210	33	8	3	1

est numbers of outbreaks. In pigs, the outbreaks covered only those due to Salmonellas other than *S. cholerae-suis*, which is well known to cause many outbreaks, for which no reliable numbers were available. In other animals relatively few outbreaks have been investigated.

The following series of Tables gives details of individual outbreaks in various kinds of animals.

TABLE IX

Outbreaks of Salmonellosis in sheep in Australia

Serotype	Number & source of strains	Number and description of outbreaks
38 outbreaks with 1	serotype	
bovis-morbificans	4 meat, internal organs	1 outbreak at abattoirs, S.A. 1 outbreak in rams, many of which died after travelling from S.A. to W.A.
chester or san-diego	1 ovine joints	infected shearing wounds & arthritis, Q.
meleagridis	1 —	1 outbreak of abortion in ewes, S.A.
newport		3 outbreaks, 1 in Q. with sheep sick & dying after staging during a journey, probably ate a poison plant; 2 in N.S.W.

TABLE IX (continued)

Serotype	Number & source of strains	Number and description of outbreaks
	3 intestine	where lambing ewes were dying in 2 days with watery scours and sheep in a drought feeding experiment were dying with salmonellosis.
potsdam	3 M.L.G. intestine, lung, heart blood	2 outbreaks in Q. in experimental sheep, neonatal fatalities in a lambing trial and sheep dying of salmonellosis during depletion of vitamins.
salford	1 —	rams dying after transportation from N.S.W. to Q.
typhi-murium	50 internal organs heart blood	28 outbreaks, including 7 in sheep after travelling (1 in W.A., 1 in S.A., 5 in Q.) 4 considered to be due to contaminated water (1 occurring throughout a large sheep growing district in S.A., and dam water and magpies were shown to contain S. typhi-murium, the other 3 were in W.A., and 1 sample of water contained S. typhi-murium), 2 in experimental sheep (1 in Q. in copper metabolism test, 1 in N.S.W. in drought feeding trials), 15 in paddocked sheep (1 in Vic., 1 in N.S.W., 2 in W.A. in ewes showing septic metritis vaginal discharges and abortion, and 11 in sheep suffering heavy losses from haemorrhagic enteritis, colitis, scours and wool plucks in W.A., S.A. and Q.).

3 outbreaks with 2 serotypes

anatum and	1 —	
typhi-murium	1 —	Sheep dying after trucking, Q.
derby and	1 heart blood	Many rams sick and some dying after a
typhi-murium	1 spleen	train journey, Q.
muenchen & rough	1 intestine	1 outbreak, Q.
Salmonella	1 M.L.N.	

1 outbreak with 3 serotypes

adelaide	3 faeces, intestine liver	
bovis-morbificans and	2 faeces, intestine	Sheep dying after trucking, Q.
san-diego	1 intestine	

1 outbreak with 4 serotypes

adelaide	1 liver	
brisbane	1 intestine	
london and	1 spleen	Salmonellosis in sheep, Q.
typhi-murium	3 kidney, spleen & liver	

Outbreaks in sheep are listed in Table IX. The majority (28) were caused by *S. typhi-murium* which occurred in all States and was found also in association with other serotypes. Many outbreaks followed a long interstate journey by rail, and some appeared during experimental procedures; in either case, resultant debility might have resulted in activation of latent infection. Other outbreaks involved lambing ewes and caused abortion, metritis and vaginal discharges; RAC & WALL (1952) in South Australia, described abortion in sheep due to *S. meleagridis*. Some outbreaks occurred in various States in paddocked sheep, and in South Australia, WATTS & WALL (1952) related a widespread, summer outbreak to water from dams which had fallen very low and were shown to contain *S. typhi-murium*. Suggested sources of pollution were wild birds, such as magpies, from which *S. typhi-murium* was isolated. The sources of infection in the other outbreaks were not known.

Outbreaks in cattle are shown in Table X. *S. typhi-murium* occurred in various States and in the majority of outbreaks. At least two followed a long journey, and in one outbreak in calves, in Queensland, whey from a cheese factory was suspected as the vehicle of infection but bacteriological proof was lacking (SIMMONS & SUTHERLAND, 1950). In an outbreak in Gippsland, Victoria, involving at least six serotypes, GRAY et al. (1958, 1959, 1960) related the infection to the feeding of cow meal containing many different types of Salmonella.

TABLE X.

Outbreaks of Salmonellosis in Cattle in Australia

Serotype	No. and source of strains	Number and description of outbreaks
anatum	1 intestine	1 outbreak with deaths from enteritis
bovis-morbificans	2 liver, M.L.N.	1 outbreak, cows dying.
typhi-murium	35 faeces, internal organs, M.L.N.	19 outbreaks (16 in Qld., 2 N.S.W., 1 in Vic.) cows, heifers or calves sick and dying with diarrhoea or enteritis; in one outbreak in calves whey from a cheese factory seemed to be the cause but no Salmonellas were detected in it; one outbreak was in new born calves and cows after trucking.

13 outbreaks with 2 serotypes

adelaide and	1 faeces	1 outbreak of scouring in calves, Q.
derby	1 faeces	
anatum and	1 lung	1 outbreak in cows, Q.
bredeney	1 lung	
anatum and	2 uterus, faeces	1 outbreak in cows, Q.
newport	1 intestine	

TABLE X (continued).

Serotype	No. and source of strains	Number and description of outbreaks
bovis-morbificans and typhi-murium	3 calf and cow liver 2 cow kidney, small intestine	2 outbreaks of salmonellosis, one in calves, the other in cows, Q.
bovis-morbificans and derby	1 bullock heart 2 bullock lung, intestine	about 100 head of cattle died at abattoirs in very hot weather after travelling several hundred miles by rail, Q.
derby or essen and typhi-murium	1 M.L.N. 1 M.L.N.	salmonellosis in a herd, Q.
muenchen and typhi-murium	4 kidney, intestine, abomasom 3 liver, faeces, abomasom	3 outbreaks, cows dying with salmonellosis, Q.
newington and typhi-murium	1 calf intestine 1 calf heart blood	1 outbreak of salmonellosis in calves, Q.
newport and typhi-murium	2 bull liver, calf intestine 2 bull intestine, calf heart blood	2 outbreaks of salmonellosis, one in bulls, the other in calves, Q.

3 outbreaks with 3 serotypes

adelaide anatum and derby anatum,	1 calf faeces 1 calf faeces 1 calf faeces 2 internal organs	1 outbreak of scouring in twin calves dying with salmonellosis, Q.
muenchen and typhi-murium bredeney, san-diego and typhi-murium	1 internal organ 1 intestine 1 cow faeces 1 cow faeces 1 cow faeces	cows with scouring and reduced milk yield in a herd, Q.

1 outbreak with 4 serotypes

anatum, london, muenchen and typhi-murium	3 M.L.N., lung, intestine 1 faeces 2 spleen, intestine 2 M.L.N. intestine	bulls sick or dying with diarrhoea, Q.

1 outbreak with 6 serotypes

cambridge chester, newport, san-diego, senftenberg & typhi-murium	1 faeces 1 liver 3 faeces, liver 1 faeces 1 faeces 3 faeces, intestine	1 outbreak on several farms in one district, connected with cow meal and bone meal containing numerous Salmonellas, Vic.

Mixed infections with several serotypes also occurred in Queensland but the sources of infection were not known.

Outbreaks of salmonellosis in birds are described in Table XI, which begins with S. *typhi-murium* septicaemia in canaries, and goes on to chickens, in which many different serotypes were found. In 66 outbreaks, involving one serotype, S. *typhi-murium* was found less frequently than in sheep or cattle, but occurred in all, except four, of the 16 outbreaks involving more than one serotype. Several of the rarer serotypes, such as S. *bareilly*, S. *oslo*, S. *lexington*, S. *worthington* and S. *waycross*, were found in chickens. No source of infection was indicated in any of these outbreaks.

In marked contrast to chickens, ducks showed a small variety of serotypes and were mainly infected with S. *typhi-murium*, which also caused an outbreak in pigeons. One outbreak in turkeys provided S. *give* and S. *newington*. Sources of infection were unknown.

In table XII are recorded two outbreaks in dogs, one of which involved S. *typhi-murium* and S. *anatum* and was very severe, killing all the puppies. S. *meleagridis*, S. *newport* and S. *typhi-murium* were responsible for outbreaks in foals in various States. Healthy foals were shown to be carriers in the S. *newport* outbreak (COTTEW & FRANCIS, 1954), and the S. *newington* outbreak was connected with travel. In pigs, S. *typhi-murium* was associated with seven outbreaks and S. *give*, S. *potsdam*, S. *newington*, S. *newport* and S. *bredeney* occurred in others. Goats also suffered from S. *typhi-murium* infection. Except for the Victorian outbreak in foals following the feeding of bone meal, there is no suggestion of the source of infection, but animal carriers no doubt play a part as reservoirs of various Sal-

TABLE XI

Outbreaks of Salmonellosis in Birds in Australia

Serotype	No. and Source of strains	Number and description of outbreaks
Canaries typhi-murium	6 heart blood, liver	4 outbreaks, 2 in N.S.W. and 2 in Q., numerous young birds died with septicaemia.
Chickens		
66 outbreaks with one serotype (excluding S. pullorum)		
adelaide	3 liver, lung, intestine	1 outbreak in chickens, Q.
anatum	20 liver, lung, intestine	9 outbreaks with medium to heavy losses especially in young chickens, Q.
bareilly	1 liver	1 outbreak with numerous losses, Q.
bonariensis	4 liver	1 outbreak in chickens, Q.
bovis-morbificans	2 lungs	deaths in chickens with congested lungs, Q.

TABLE XI (continued)

Serotype	No. and Source of strains	Number and description of outbreaks
bredeney	2 chickens	1 outbreak, Q.
cambridge	2 chickens	1 outbreak, numerous losses, Q.
chester	2 chickens	1 outbreak, numerous losses, Q.
derby	19 lung, liver, intestine	12 outbreaks with medium to heavy losses especially in young chickens, Q.
give	6 lung, liver, intestine	4 outbreaks with medium to severe losses in young chickens, Q.
lexington	3 chickens	1 outbreak, Q.
london	7 liver, intestine	3 outbreaks with numerous losses in chickens, Q.
meleagridis	1 intestine	1 outbreak with heavy losses in chickens, Q.
muenchen	8 lung, intestine	5 outbreaks with numerous losses in young chickens, Q.
newington	4 chickens, liver	3 outbreaks of salmonellosis in chickens. Q.
newport	5 lung, liver, intestine	1 outbreak with heavy losses of chickens within 14 days of hatching, Q.
oranienburg	1 chicken	1 outbreak with heavy losses, Q.
orion	2 lung, intestine	1 outbreak, Q.
oslo	4 lung, liver, intestine	4 outbreaks with numerous losses in young chickens, Q.
rough Salmonella	3 chickens	1 outbreak in chickens, Q.
san-diego	3 chickens	1 outbreak with numerous losses, Q.
typhi-murium	17 liver, lung, intestine	8 outbreaks, numerous losses in chickens, some with diarrhoea, Q.
waycross	3 liver, intestine	1 outbreak, Q.
worthington	10 live, lung, intestine	3 outbreaks with many chickens sick and dying, Q.

12 outbreaks with 2 serotypes

adelaide and	2 chickens	1 outbreak, Q.
typhi-murium	3 chickens	
anatum and	1 liver	1 outbreak with heavy losses in chickens, Q.
give	1 liver	
anatum and	2 chickens	1 outbreak with heavy losses in chickens, Q.
muenchen	1 chicken	
anatum and	2 chickens	1 outbreak with 7% mortality, Q.
bareilly	1 chicken	
anatum and	2 liver, lung	1 outbreak with losses in day-old brooding chickens, Q.
bovis-morbificans	3 intestine	
bovis-morbificans &	2 lung, liver	1 outbreak of salmonellosis, Q.
typhi-murium	1 lung	
kottbus and	1 chicken	1 outbreak, Q.
typhi-murium	1 chicken	
meleagridis and	3 liver, lung, intestine	2 outbreaks in young chickens, Q.
typhi-murium	3 liver, lung, intestine	
meleagridis and	6 lung, intestine	2 outbreaks in chickens, Q.
worthington	4 lung	

TABLE XI (continued)

Serotype	No. and Source of strains	Number and description of outbreaks
orion and typhi-murium	1 intestine 3 lung, liver	1 outbreak in week-old chickens, Q.

3 outbreaks with 3 serotypes

bareilly, bovis-morbificans & typhi-murium	4 chickens 2 chickens 3 chickens	1 outbreak in brooding chickens, similar to outbreak of previous year with 4 serotypes (see below) Q.
champaign, chester and meleagridis	1 chicken 2 chickens 2 chickens	1 outbreak with numerous losses, Q.
new-brunswick, newport and typhi-murium	1 chicken 1 chicken 3 chickens	1 outbreak with heavy losses in two day old chickens, Q.

1 outbreak with 4 serotypes

glostrup, meleagridis, typhi-murium & worthington	1 chicken 3 chicken 1 chicken 2 chickens	1 outbreak in brooding chickens, Q.

Ducks

30 outbreaks with one serotype

anatum	2 intestine	2 outbreaks of enteritis in ducks and ducklings, many died, Q.
bovis-morbificans	1 heart blood	1 outbreak with 50 out of 50 dying, Q.
derby	3 lung, liver	1 outbreak with severe losses, Q.
london	1 duckling	1 outbreak of salmonellosis in 3 days old ducklings with numerous deaths, Q.
typhi-murium	42 liver, lung intestine	25 outbreaks with medium to severe losses in ducklings, often with diarrhoea, Q.

1 outbreak with 2 serotypes

give and typhi-murium	1 liver 1 liver	1 outbreak with numerous losses, Q.

Pigeons

typhi-murium	4 joints, liver	1 outbreak, N.S.W.

Turkeys

give and newington	1 intestine 1 intestine	1 outbreak with losses of turkey poults, Q.

TABLE XII

Outbreaks of Salmonellosis in dogs, goats, horses and pigs in Australia

Serotype	Number and source of strains	Number and description of outbreaks
Dogs		
derby	2 faeces	1 outbreak of enteritis and coccidiosis in a kennel, Q.
anatum and typhi-murium	3 faeces, intestine 4 heart blood, liver	1 outbreak in puppies in a kennel of Afghan hounds; 12 puppies died within 24 hours of having "fits" or diarrhoea with blood; treatment with chloromycetin had no apparent effect on course of disease nor was it prophylactic; adult dogs were not affected, Q.
Goats		
typhi-murium	2 faeces	Several goats sick in a herd, Q.
Horses		
meleagridis	1 intestine	Foals dying in one stud which had an outbreak of S. newport infection the following year, Q. (see below).
newington	2 intestine	Horses dying after trucking, severe inflammation of the colon, Q.
newport	8 faeces, liver, lung, lymph node, intestine	Foals dying in one stud which had S. meleagridis infection the previous year, some foals were sick with diarrhoea and other healthy foals also gave S. newport from their faeces, Q.
typhi-murium	12 blood, faeces, kidney, spleen, lung, liver, bowel abscess	4 outbreaks in Vic. and one outbreak in W.A. involving mortalities in foals with diarrhoea or intense colitis; one Vic. outbreak followed the feeding of bone meal.
Pigs		
(S. cholerae-suis outbreaks are not included)		
give	1 —	1 outbreak in which many died and many were sick, Q.
newington	1 lung	deaths and illness in a herd, Q.
newport	1 liver	severe losses in suckling pigs with necrotic enteritis, septicaemia and pneumonia, Q.
typhi-murium	10 M.L.N., spleen, heart blood, liver, bronchial L.N., kidney	7 outbreaks with medium to severe losses, many sick, some with septicaemia, others with enteritis, Q.
bredeney	2 kidney, intestine	1 outbreak with deaths, Q.
cholerae-suis & potsdam	1 liver 1 liver	

monellas. In view of the occurrence of Salmonellas in pet foods, especially kangaroo or horse meat and dog biscuits (see p. 582), I consider these to be an important reservoir for domestic pets, such as dogs and cats.

Table XIII records outbreaks in laboratory stocks, in which 13 out of 17 outbreaks in guinea pigs, and three out of eight outbreaks in mice, were due to *S. typhi-murium*, but *S. bovis-morbificans* and some *enteritidis* types also infected them. The sources of infection

TABLE XIII

Outbreaks in Laboratory Stock Animals

Serotype	No. and Source of strains	Number and description of outbreaks
Guinea pigs		
blegdam	1 heart blood	1 outbreak, Vic.
bovis-morbificans	1 liver	1 outbreak of salmonellosis, Q.
derby	3 heart blood, liver	deaths after experimental inoculation, probably activation of latent infection, Q.
rough Salmonella (—; gm;—)	1 —	epidemic, Vic.
typhi-murium	33 liver, spleen, heart blood	13 outbreaks, often with diarrhoea, apparently healthy animals were often carriers harbouring the organism in internal organs, some animals died only after experimental manipulation, probably through inactivation of latent infection; 3 of the outbreaks were in Vic., 5 in S.A., 3 in Q., 2 in W.A., and 1 N.S.W.
eastbourne and hidalgo	1 heart blood 1 —	1 outbreak of diarrhoea, Q.
Mice		
blegdam	5 heart blood	1 outbreak, Vic., associated with blegdam outbreak above in guinea pigs
bovis-morbificans	2 internal organs	baby mice dying after experimental inoculation, mother mice were healthy carriers, Q.
enteritidis	5 heart blood	2 outbreaks, Vic. and W.A.
rough Salmonella (—; gm;—)		1 outbreak, Vic., associated with the rough Salmonella outbreak above in guinea pigs.
typhi-murium	4 liver	3 outbreaks, 2 in Vic., and 1 in Q.
Rabbits		
typhi-murium	2 —	1 epidemic in Vic.

were not known but healthy carriers of S. *typhi-murium* or S. *bovis-morbificans* were found in some stocks. I believe that feeding stuffs, such as mouse biscuits and the like, could be vehicles of infection.

Few outbreaks have been recorded in native Australian animals. In Table XIV, an outbreak in kangaroos was associated with S. *meleagridis*, and one outbreak in lizards in captivity produced S. *birkenhead* and S. *chester*, though a number of healthy lizards were shown to harbour numerous serotypes (LEE & MACKERRAS, 1955). Apparently, these animals form a reservoir of Salmonellas.

TABLE XIV

Outbreaks in Native Animals

Serotype	No. and Source of strains	Number and description of outbreaks
Kangaroos meleagridis	2 intestinal contents	1 outbreak in a small zoo, 4 animals died, P.M. showed haemorrhagic enteritis, Q.
Lizards birkenhead and chester	1 liver 3 large intestine	1 outbreak of enteritis in lizards captured and kept in captivity for experiments, Q.

Carriers in Human and Animal Salmonellosis

The human or animal carrier is probably the most important reservoir of infection.

Infants, children and adults may all become carriers of one or more of a variety of serotypes. Infants frequently became carriers following infection with S. *typhi-murium*, S. *bovis-morbificans*, or S. *derby* and often remained carriers for months after clinical recovery. Examination of women in a post-natal clinic in Sydney, N.S.W., revealed carriers of S. *anatum*, S. *bovis-morbificans*, S. *heidelberg*, S. *kottbus*, S. *meleagridis*, S. *senftenberg*, and S. *vejle*. Casual occurrence of Salmonellas in connection with typhoid carriers has been found: routine checks showed S. *newington* in one carrier and S. *derby* in another. The work of WILSON & MACKENZIE (1955), on typhoid contracted from coconut, showed that people frequently became carriers of the Salmonellas contaminating coconut and some also developed typhoid fever. I have been struck by the number of carriers, in my records, among children sent to hospital as feeding problems. A Salmonella was often isolated during routine stool examinations and I get the impression that many of these were, in fact, cases of mild salmonellosis.

Not many animals have been examined for the carrier state. Those for which results are available have usually been selected on account of association with outbreaks, such as the mice, cock-roaches and rats in connection with the Brisbane *S. bovis-morbificans* outbreak and the mice in the Melbourne *S. derby* outbreak. *S. typhi-murium* carriers were not recorded among the wild mice examined, nor among the lizards, though they occurred in other animals. Rats or lizards were carriers of *S. adelaide, S. bovis-morbificans, S. meleag-ridis, S. muenchen, S. oranienburg, S. paratyphi C* or *S. rubislaw.*

Salmonellas from Materials Other Than Human or Animal

The KW groups and serotypes found.

The materials dealt with here could be vehicles or reservoirs of human or animal salmonellosis. Many more varieties of materials, not so far examined, may eventually be incriminated, but their discovery depends upon more extensive enquiries into the food and surroundings of known cases of salmonellosis.

The 2233 strains of Salmonella from specimens other than human or animal, were recorded in the literature already quoted, in my unpublished records, or in those kindly supplied by my colleagues.

Compared with the figures for humans and for animals, there is a marked change in distribution of the KW groups. Group B has become much less significant (19.7%), group C has also diminished (12.6%), but groups D (26.5%) and E (30.7%) have greatly increased. Among the further groups from F on, G, H, O and P all contained many strains.

The distribution of the KW groups and their serotypes among the various kinds of material examined, is shown in Tables XV and XVI.

The large variety of serotypes can be seen in Table XV where they are listed in alphabetical order under their KW groups.

Table XVI shows that the KW groups B, C_1, C_2, E_1 and E_4 were represented in all categories of materials. Groups F, M, N, P and S were found only in coconut, in which groups D_1 and O failed to appear though they occurred elsewhere. The commonest serotypes and the materials containing them are given in Table XVII.

In group B, 13 different serotypes were found. Occurring in many materials were *S. typhi-murium*, the commonest type, forming 52.1% of the group B total, *S. derby* (19% of group B) and *S. saint-paul* (6.8%). *S. paratyphi B* (9.1%) occurred in coconut, dog biscuits and sewer swabs and *S. bredeney* in meat or bone meal etc. In group C_1, 13 different serotypes occurred, but *S. oranienburg* was dominant (61.6% of the group total) followed by *S. bareilly* (15.4% of the group), both found in many different materials. Group C_2 had only four serotypes, of which *S. bovis-morbificans*, from a wide variety of

TABLE XV

Distribution of Serotypes and KW Groups of Salmonellas Found in Specimens Other than Human or Animal

KW Group	Serotype	Number of Strains Found in					Totals
		Coco-nut	Egg pulp	Meat or bone meal etc.	Meat or food	Water, Moore Swabs etc.	
B	altendorf					1	1
13	bredeney			38			38
sero-	chester	1	2	5			8
types	derby		41	39	1	1	82
432	heidelberg				1		1
strains	hessarek		1				1
	java	3					3
	kaapstad					1	1
	paratyphi B	34		1		4	39
	reading		1				1
	saint-paul			22	3	4	29
	san-diego			2	1		3
	typhi-murium	17	169	10	12	17	225
C₁	bareilly	20	7	3		2	32
13	birkenhead					2	2
sero-	bonn				1		1
types	braenderup	1		7			8
207	cholerae-suis				1		1
strains	edinburg	2					2
	kotte	11					11
	montevideo		1	4			5
	oranienburg	4	67	56		1	128
	potsdam	?	3	1	2		6+
	singapore			2			2
	tennessee	1					1
	thompson	8					8
C₂	bovis-morbificans	1	40	3	3	2	49
4	litchfield	1				2	3
serotypes	muenchen		1	1			2
69 strains	newport	7	1	7			15
C₃1 sero-type strain	kentucky			1			1
D₁5 sero-	berta					1	1
types	eastbourne				1	1	2
581 strains	enteritidis			33			33
	pullorum		544				544
	saarbruecken			1			1
E₁	amager	13		2			15
	anatum		5	59	4		68
13 sero-types	butantan	7					7
	give		4	72	1		77
315	lexington			17			17
strains	meleagridis		11	52		13	76
	muenster	2	1				3
	nchanga	2					2
	nyborg	7	5	3		1	16
	orion	?	2	20		1	23+

TABLE XV (continued)

KW Group	Serotype	Number of Strains Found in					
		Coco-nut	Egg pulp	Meat or bone meal etc.	Meat or food	Water Moore Swabs etc.	Totals
	vejle		2	1			3
	weltevreden	6					6
	zanzibar	1		1			2
E₂ 5 sero- types 68 strains	cambridge			44		3	47
	manila			1			1
	new-brunswick			1			1
	newington		4	14			18
	selandia			1			1
E₄ 3 sero- types 290 strains	chittagong	17					17
	senftenberg	201	4	48	1	2	256
	taksony			17			17
F₂ sero- types, 5 strains	chingola	4					4
	rubislaw	1					1
G₂ 3 sero- types 84 strains	cubana	3	2	31			36
	mississippi	1					1
	worthington		2	45			47
H 2 sero- types 35 strains	charity					2	2
	ferlac	33					33
I 1 sero- type 7 strains	hvittingfoss	6			1		7
K 1 sero- type, 1 strain	blukwa					1	1
L 1 sero- type, 8 strains	wandsbek					8	8
M 1 sero- type, 13 strains	solna	13					13
N 1 sero- type, 10 strains	angoda	10					10
O 1 sero- type 27 strains	adelaide		14	7	2	4	27
P 1 sero- type, 25 strains	perth	25					25
S 1 sero- type, 17 strains	waycross	17					17
T 1 sero- type, 1 strain	fremantle		1				1
3	untyped	4	4	2	3	24	37

TABLE XVI

KW Groups of Salmonellas from Specimens Other than Human of Animal

Specimen	Number of Strains in KW Groups																				untyped	Total
	B	C_1	C_2	C_3	D_1	E_1	E_2	E_4	F	G_2	H	I	K	L	M	N	O	P	S	T		
coconut	55	47+	9			42	4	218	5	4	33	6			13	10	14	25	17	1	4	488+
eggpulp	214	78	42		544	98		4		4											4	1007
meat or bone-meal etc.	117	73	11	1	34	156	61	65		76		1					7				2	603
meat or food	18	4	3		1	4		1									2				3	37
water, moore swabs etc.	18	5	4		2	15	3	2			2		1	8			4				24	98
Totals	432	207	69	1	581	315	68	290	5	84	35	7	1	8	13	10	27	25	17	1	37	2233

TABLE XVII

KW Groups and Commonest Serotypes of Salmonellas from Specimens Other than Human and Animal

KW grp	No. of sero-types	stains	Commonest Serotypes		dosage of DW gp	Occurrence
			name	No.		
B	13	432	bredeney	38	8.8%	meat or bone meal, etc.
			derby	82	19 %	egg pulp, meal or bone meal etc., meat, liver, ocean water
			paratyphi B	39	9.1%	coconut, meat & bone meal etc., sewer swabs
			saint-paul	29	6.8%	meat & bone meal, etc., sewer swabs, river water, pork, kangaroo meat for pets.
			typhi-murium	225	52.1%	coconut, egg pulp, meal or bone meal, etc. horse meat for pets, dam water, pork, lamb, veal, sausage, roast duck meat, flummery, span-ish cream, ice cream
			remainder		4.2%	
C$_1$	13	207	bareilly	32	15.4%	abattoirs effluent, coconut,
			oranienburg	128	61.6%	egg pulp, meat or bone meal etc., coconut, egg pulp, meat and bone meal, river and ocean water.
			remainder		23.0%	
C$_2$	4	69	bovis-morbificans	49	71.3%	coconut, egg pulp, meat and bone meal, etc. mutton, sew-er swabs.
			newport	15	21.8%	coconut, egg pulp, meat & bone meal, etc.
			remainder		6.9%	
D$_1$	5	581	enteritidis	33	5.7	meat & bone meal, etc. egg
			pullorum	544	94.0%	pulp
			remainder		0.3%	
E$_1$	13	315	anatum	68	21.6%	egg pulp, meat or meal etc., metwurst sausage, ham, pork
			give	77	24.5%	egg pulp, meat or bone meal etc., mussels.
			lexington	17	5.3%	meat or bone meal etc. egg
			meleagridis	76	24.2%	pulp, meat & bone meal etc., sewer swabs, river water.
			nyborg	16	5.0%	coconut, egg pulp, meat or bone meal, etc. sewer swabs.

TABLE XVII (continued)

KW grp	No. of sero- types	stains	Commonest Serotypes name	No.	dosage of DW gp.	Occurrence
			orion	23	7.3%	coconut, egg pulp, meat or bone meal, etc., sewer swabs
			remainder		9.1%	
E_2	5	68	cambridge	47	69.2%	meat or bone meal, etc. oyster beds, river and ocean water.
			newington	18	26.6%	egg pulp, meat & bone meal etc.
			remainder		4.2%	
E_4	3	290	chittagong	17	5.9%	coconut
			senftenberg	256	88.2%	coconut, egg pulp, meat and bone meal, etc.
			taksony	17	5.9%	meat and bone meal, etc.

Further Groups

KW grp	No. of sero- types	stains	Commonest Serotypes name	No.	dosage of DW gp.	Occurrence
G_2	3	84	cubana	36	16.0%	coconut, egg pulp, meat and bone meal, etc.
			worthington	47	20.2%	egg pulp, meat and bone meal etc.
H	2	35	ferlac	33	14.2%	coconut
I	1	7	hvittingfoss	7	3.0%	coconut, milk mixture
M	1	13	solna	13	5.6%	coconut
N	1	10	angoda	10	4.3%	coconut
O	1	27	adelaide	27	11.6%	egg pulp, meat and bone meal etc. kangaroo meat for pets, abattoirs effluent.
P	1	25	perth	25	10.8%	coconut
S	1	17	waycross	17	7.2%	coconut
			remainder		7.0%	

specimens, formed 71.3% of the group total, and *S. newport*, from coconut, egg pulp and the meat meal category, formed 21.8%. Group D_1, with five serotypes, was much overemphasized by the inclusion of 544 strains of *S. pullorum* from egg pulp, forming 94% of the group total; I included them to accent the extent to which this type may be present in egg pulp, even though the flocks producing the eggs may be tested and not obviously suffering from pullorum disease (KOVACS, 1959). *S. enteritidis*, in the meat meal category, was next with 5.7% of the group total.

In group E_1, in which 13 serotypes were found, the commonest were *S. anatum* (21.6% of the group), *S. give* (24.5%), *S. lexington* (5.3%), *S. meleagridis* (24.2%), *S. nyborg* (5%) and *S. orion* (7.3%). They all occurred in many different materials. Group E_2 had five serotypes of which the commonest were *S. cambridge* (69.2% of the group) found in the meat meal category and in water, and *S.*

newington in meat and bone meal etc. and egg pulp. Unlike human and animal specimens, these "other" specimens contained numerous group E_4 strains, consisting of three serotypes, *S. senftenberg* from coconut, egg pulp and meat and bone meal etc., forming 88.2% of the group, *S. chittagong* (5.9%) from coconut and *S. taksony* (5.9%) from meat and bone meal etc.

In further contrast to the human and animal specimens, the 'other' specimens contained a considerable number of strains of serotypes in the further groups from F on. Group G_2 had three serotypes, of which *S. cubana*, from coconut, egg pulp and meat meal category, formed 16% of the total number of strains in the further groups, and *S. worthington*, from egg pulp and the meat meal category, formed 20.2% of this total. In group O, *S. adelaide* (11.7% of the total) came from all kinds of material except coconut. In group I, *S. hvittingfoss* was found in coconut and milk mixture from a milk bar. Unusual serotypes from groups F, H, M, N, P and S, found in coconut, were *S. chingola*, *S. ferlac*, *S. solna*, *S. angoda*, *S. perth*, and *S. waycross*. Groups K, L and T had unusual serotypes of which *S. blukwa* and *S. wandsbek* came from water, and *S. fremantle* from egg pulp. Compared with human and animal specimens, the 'other' specimens contained a much larger number of rare or exotic serotypes. The small number of strains of *S. typhi-murium* isolated from coconut or meat meal and bone meal etc. was remarkable in view of its abundance in human and animal material. Such a common type could be expected to appear frequently, especially in the meat meal category, prepared from materials of animal origin. The reasons for its absence are obscure, but it is possible that conditions produced by processing may become unfavourable for its survival or the laboratory techniques which succeed in detecting it on other occasions, fail to detect it in these materials (see coconut inhibition of selenite medium, later in this section). *S. cholerae-suis*, frequently found in pigs, presents a somewhat similar paradox, occurring once in pork and not at all in any 'other' material. What happens to these two common serotypes in processed material of animal origin? Maybe they have become too well adapted to survival in the living animal, wheras other serotypes such as *S. senftenberg*, found frequently in coconut, meat meal etc., but only occasionally in humans or animals, have become better adapted to survival in the environment than in the living animal.

Extent of Contamination of 'other' Materials.

Organic fertilizer, meat meal etc., pet food and coconut, with their impressive loads of Salmonellas, must be regarded as dangerous sources of human or animal salmonellosis. The extent and frequency of contamination of these materials is indicated by the findings in various States in recent years. Of samples of fertilizers of animal origin, meat and bone meal etc. from several manufacturers in Victoria, 91% contained Salmonellas, often as many as five different serotypes in a single sample (GRAY, HARLEY & NOBLE, 1960). In Queensland, at least 23 out of 53 samples from numerous sources contained from one to four Salmonella serotypes (SIMMONS, 1962). KOVACS (1959) in Perth, W.A., isolated Salmonellas from almost 50% of 94 samples of which many had two serotypes and one had 13 serotypes.

Seeking sources of contamination in a bone meal factory, GRAY, HARLEY & NOBLE (1960) found that certain bones arrived at the factory heavily contaminated, 13 out of 17 samples containing Salmonellas. Furthermore, on the drying grids, which were well seeded with Salmonellas, massive multiplication took place, and the final milling ensured a uniform distribution throughout the product. Bone meal seemed to be a splendid medium for preserving Salmonellas, which were isolated from samples up to three years old. In South Australia, investigation of a meat meal factory showed the ground in the loading area to be contaminated with at least three different Salmonellas (Institute of Medical and Veterinary Science, 1960-61). As a likely source of Salmonellas in meat meal, organic fertilizers etc. was the animal supplying the raw material, KOVACS (1959) examined 200 mesenteric lymph glands from cattle and pigs from a slaughter-house in Perth, W.A., and found 30.5% to contain Salmonellas, sometimes as many as five serotypes in one gland. These results were considerd to confirm a heavy Salmonella infection in cattle and pigs.

Pet foods, of which few samples were tested, frequently harboured Salmonellas and constituted an obvious hazard to dogs and cats; human food could also become contaminated from them by contact or handling in the store or home. In Victoria, GRAY, HARLEY & NOBLE (1960) found that four out of four samples of horse meat for pets contained Salmonellas; S. paratyphi B was isolated from dog biscuits and S. saint-paul from kangaroo meat in Adelaide (Institute of Med. and Vet. Science, 1960—61); in Perth, W.A., frozen kangaroo meat for pets was heavily contaminated, three out of four samples providing two Salmonella serotypes and numerous E. coli and enterococci (Public Health Lab., Perth, 1960).

Coconut presents a somewhat similar picture to bone meal. The manufacturing process is much the same, and massive contamination again occurs on the drying racks. Like bone meal, coconut is also a good medium for preserving Salmonellas. ANDERSON & WOODRUFF (1961) recovered S. thompson from artificially contaminated coconut after four weeks and ANDERSON (1962) isolated it from the same coconut after at least eighteen months storage in a closed jar unprotected from light. I believe that Salmonella serotypes might vary in their ability to survive in coconut or bone meal etc., S. typhimurium being less favoured than S. senftenberg or some of the exotic types. Investigations of the survival times of various common and rare serotypes in coconut, bone and meat meal etc. could provide valuable information.

The extent to which coconut samples were contaminated with Salmonellas can be seen from the figures from KOVACS and the Public Health Laboratory, Perth, W.A., where in 1959 nine out of 35 samples contained Salmonellas, and in 1960 29 out of 286 samples gave Salmonellas. In 1961 a higher incidence was found using a newer method of isolation, which overcame the inhibition observed in selenite medium with coconut. Of 1282 samples examined, 272 or 21% contained Salmonellas. Thus the extent of contamination indicated by the older method represented only a small percentage of the actual contamination. In other States at least seven serotypes were isolated from coconut during 1960, 1961 and 1962. There seems little doubt that organic fertilizers, bone and meat meal etc., pet foods and coconut are still being sold with a dangerous Salmonella content.

Bacteriological screening of 'other' materials.

ANDERSON & WOODRUFF (1961) pointed out the difficulties which faced public health authorities in deciding whether a shipment of coconut was safe for human consumption. They showed that Salmonellas could be unevenly distributed, not only in a shipment, but in a single bag of coconut. Selecting a 100 lb. bag from which the routine test on 20 g in selenite medium had given *S. kotte*, they took samples from 29 different parts of the bag. Of the 29 resulting cultures of 20 g each (equivalent to 580 g of coconut), only one gave a growth of *S. kotte*. (Referring back to the findings of KOVACS on coconut inhibition of selenite medium, this result could be partly due to the use of selenite medium.) The number of isolations increased when the sample size was increased. Four 20 g samples of Philippine coconut failed to yield Salmonellas, but in two out of four 180 g samples *S. cubana* was detected. ANDERSON & WOODRUFF emphasized that, while existing bacteriological screening methods for coconut remained so unsatisfactory, the continued importation of coconut contaminated with Salmonellas presented a real public health problem. Some contaminated batches seemed certain to escape detection and would be passed as fit for human consumption.

The problem of screening meat and bone meal etc. for Salmonellas is even more acute, as shown by the results of KOVACS (1957) in Perth, W.A., and SIMMONS (1962) in Queensland. Their investigations showed how deceptive the accepted tests can be. Using 10 g amounts of blood and bone or meat meal, known to contain Salmonellas, KOVACS compared selenite medium with tetrathionate broth. Five tubes of each medium received 10 g each of the specimen. In one meat meal sample no Salmonellas were detected in the selenite medium but two serotypes, S. *san-diego* and S. *nyborg*, were obtained from four of the tetrathionate tubes. From one sample of blood and bone, two of the selenite and one of the tetrathionate tubes produced S. *derby*, and from another sample, S. *oranienburg* was isolated from three of the selenite and four of the tetrathionate tubes. Plating from selenite broth was better on SS agar than on bismuth sulphite agar. These results suggested uneven distribution of Salmonellas in the original samples and a variation in the ability of selenite and tetrathionate broths to pick them up. SIMMONS obtained even more striking irregularities in isolation of Salmonellas from meat and bone meal products in Queensland. No Salmonellas were obtained by direct plating on selective media, but 12 samples yielded Salmonellas when 1 g was cultivated in tetrathionate broth before plating on selective media (SS agar and brilliant green agar), and 21 samples were positive when a 10 g amount was used. Strangely enough, only nine samples yielded Salmonellas in both 1 g and 10 g tests, so that there were 3 samples which were positive in 1 g and negative in 10 g amounts. Salmonellas were isolated from 46 SS and 34 brilliant green agar plates, but of all the pairs of plates corresponding to single samples, only 28 pairs both gave Salmonellas; in the remaining pairs of plates, one or other was positive but not both. Tetrathionate broth, after incubation for 24 hours, gave fewer positive plates than after incubation for 48 hours, but here also there was variation in the samples found positive and the serotypes isolated after the two different incubation periods. In some samples, Salmonella serotypes, found in the 1 g tests, were not detected in the 10 g tests. The picking of six colonies enhanced the possibility of finding more than one serotype. Sampling the bags from the manufacturer was also a problem and a certain number of bags was sampled taking material from the top of one bag, the middle of the next, the bottom of the next and so on. Though only a small number of investigations have been quoted here they are sufficient to indicate the complexity of the problems involved in the bacteriological screening and control of coconut and meat or bone meal and the like.

There seems no good reason at present to introduce *E. coli* or enterococcus counts for these materials. Convincing figures would have to be produced on their correlation with Salmonella content before they could even be considered to replace the present methods of testing for the Salmonellas themselves.

Bacteriophage Grouping of Australian Salmonellas

To assist in the epidemiological investigation of salmonellosis in Australia, I have been working on bacteriophage grouping schemes for two of our important serotypes, S. *bovis-morbificans* and S. *adelaide*, and also for S. *waycross* (ATKINSON, 1957). Five phage groups were proposed for *S. bovis-morbificans* (ATKINSON, 1956 a and b) and evidence for their epidemiological significance was presented from results of testing a large batch of strains of mixed origin, including some from Brisbane at the time of the 1947 epidemic. Most of these Brisbane strains proved to belong to phage group 6, while unrelated strains belonged to various phage groups (ATKINSON & BULLAS, 1956 a and b). With *S. adelaide* (ATKINSON, 1955), six phage groups were established but their epidemiological value was hard to assess as the strains available for testing were of heterogenous origin (ATKINSON & KLAUSS, 1954, 1955). The phage grouping scheme for *S. waycross* (ATKINSON, et al. 1952) appeared to be epidemiologically useful as the 29 strains from the Atherton outbreak all formed one group (called Atherton group) and the other human strains made a second group. Two lizard strains, isolated much later by LEE & MACKERRAS (1955) also belonged to the Atherton phage group, suggesting that lizards may have been connected with the earlier Atherton epidemic. The bacteriophages associated with the grouping schemes have been described (ATKINSON & GEYTENBEEK, 1953; ATKINSON & CARTER, 1953; ATKINSON & BULLAS, 1956 c_1, d and e; ATKINSON & BULLAS, 1957). Further development of these schemes is required to include new groups to take care of those strains which at present fit into none of the existing groups.

Analysis of the Collected Results

The general relationship of the KW groups and their serotypes to the specimens from human, animal and other sources can be seen in Fig. 3.

KW groups B, C_1, C_2, E_1 and O occurred frequently in all three sections. Groups D_1 and E_2 were less frequently found and covered a smaller range of categories. Group E_4 and the further groups from F on (except O) occurred mainly in 'other' material, especially in coconut. Fig. 3 emphasizes the earlier suggestion that the occurrence of each KW group depended upon a few serotypes, the others contributing little. Many of the common serotypes were widely distributed. *S. chester*, *S. derby* and *S. typhi-murium* in group B, *S. potsdam* in C_1, *S. bovis-morbificans*, *S. newport* and *S. muenchen* in C_2, *S. anatum*, *S. give* and *S. meleagridis* in E_1 and *S. adelaide* in O, were found in all or most of the listed categories. These serotypes seemed to be well adapted to living in humans, numerous kinds of animals and in 'other' materials (except for the peculiar scarcity of *S. typhi-murium* in coconut and meat and bone meal etc.). It is worthy of note that *S. typhi-murium*, in spite of its name, was found comparatively rarely in mice or rats and very frequently in man and his domestic animals. We must therefore resist the temptation to refer *S. typhi-murium* contaminations to a rodent reservoir and rather look

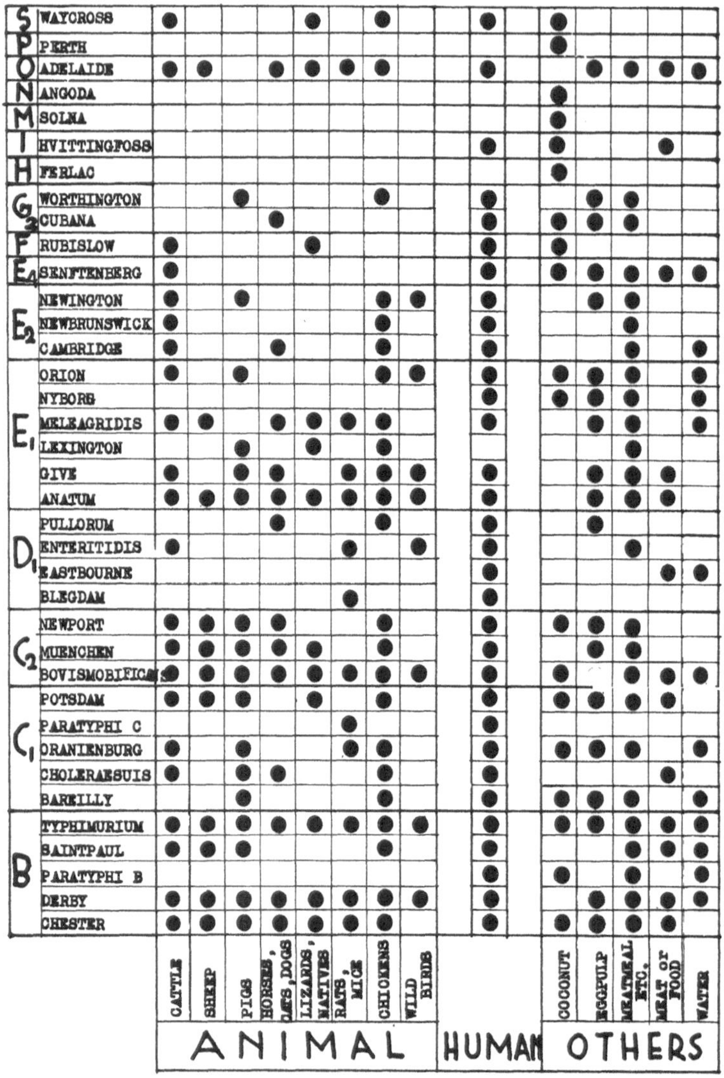

Group	Serotype	CATTLE	SHEEP	PIGS	HORSES, DOGS	LIZARDS, NATIVES	RATS, MICE	CHICKENS	WILD BIRDS	HUMAN	COCONUT	EGGPULP	MEATMEAL ETC.	MEAT OR FOOD	WATER
S	WAYCROSS	●			●		●			●	●				
P	PERTH									●					
O	ADELAIDE	●	●		●	●	●	●		●		●	●	●	●
N	ANGODA									●					
M	SOLNA									●					
T	HVITTINGFOSS									●	●			●	
H	FERLAC									●					
G₂	WORTHINGTON			●				●		●		●	●		
G₂	CUBANA				●					●	●	●	●		
F	RUBISLOW	●				●				●	●				
E₄	SENFTENBERG	●								●	●	●	●	●	●
E₂	NEWINGTON	●		●			●	●		●		●	●		
E₂	NEWBRUNSWICK	●					●			●			●		
E₂	CAMBRIDGE	●			●		●			●			●		●
E₁	ORION	●		●			●	●		●	●	●	●		●
E₁	NYBORG									●	●	●	●		●
E₁	MELEAGRIDIS	●	●		●	●	●			●		●	●		●
E₁	LEXINGTON			●		●	●			●			●		
E₁	GIVE	●		●	●		●	●	●	●		●	●	●	
E₁	ANATUM	●	●	●	●	●	●	●	●	●		●	●	●	
D₁	PULLORUM			●			●			●		●			
D₁	ENTERITIDIS	●				●		●		●		●			
D₁	EASTBOURNE									●				●	●
D₁	BLEGDAM					●				●					
C₂	NEWPORT	●	●	●	●			●		●	●	●	●		
C₂	MUENCHEN	●	●	●	●	●		●		●		●	●		
C₂	BOVISMOBIFICANS	●	●	●	●	●	●	●	●	●		●		●	●
C₁	POTSDAM	●	●	●		●				●	●	●	●	●	
C₁	PARATYPHI C						●			●					
C₁	ORANIENBURG	●		●			●	●		●	●	●	●		●
C₁	CHOLERAESUIS	●		●	●		●			●				●	
C₁	BAREILLY			●			●			●		●	●	●	●
B	TYPHIMURIUM	●	●	●	●	●	●	●	●	●		●	●	●	●
B	SAINTPAUL	●	●	●				●		●			●	●	●
B	PARATYPHI B									●		●			●
B	DERBY	●	●	●	●	●	●	●		●		●	●	●	●
B	CHESTER	●	●	●	●	●	●	●		●		●	●	●	

ANIMAL — HUMAN — OTHERS

Fig. 3. Occurrence of Commonest Serotypes among Human, Animal & Other Materials

to humans or other animals for the source of this common organism.

Serotypes which were more limited in distribution, were *S. paratyphi A* (not shown in Fig. 3) limited to man, *S. paratyphi B* in humans and several categories of 'other' materials, *S. paratyphi C* in man and a few rats, *S. cholerae-suis* found mainly in pigs, sometimes in man and not in any category of 'other' specimens except once in pork. Group D_1 strains seemed the most specialized and included the human pathogens *S. typhi* and *S. blegdam* and the fowl pathogen *S. pullorum*, found almost exclusively in chickens and egg pulp. On the whole, group D_1 strains were rare in 'other' materials. *S. nyborg* (in group E_1) occurred in most 'other' categories and occasionally in man. The three serotypes in group E_2 appeared chiefly in cattle and chickens, egg pulp, and meat and bone meal etc., and sporadically in humans. *S. senftenberg* in group E_4 was an environmental type found in all categories of 'other' materials and only rarely in cattle and man. *S. rubislaw* in group F came mainly from lizards, but sometimes was found in cattle, man and coconut. *S. worthington* (group G_2) occurred in chickens and pigs, egg pulp, and meat and bone meal etc., and occasionally in humans. *S. ferlac* (group H), *S. hvittingfoss* (group I), *S. solna* (group M), *S. angoda* (group N) and *S. perth* (group P) appeared in little else besides coconut, which also supplied several rare serotypes in group C_1, *S. edinburg*, *S. kotte*, and *S. tennessee*, and in group E_1 *S. butantan*, *S. nchanga* and *S. weltevreden*, not listed in Fig. 3.

The overall outcome seems to be the designation of certain serotypes as:

1. widespread dangerous pathogens passing between man, animals, and environment with considerable efficiency, and typified by *S. typhi-murium*, *S. derby*, *S. bovis-morbificans*, *S. anatum* and *S. adelaide*;

2. dangerous pathogens, mainly of humans, rarely in animals or 'other' materials, typified by *S. paratyphi A*, *S. typhi*, *S. blegdam* and *S. paratyphi B* (which had the greatest occurrence in 'other' materials);

3. dangerous pathogens of certain animals, typified by *S. cholerae-suis* in pigs and infrequently in man but not in 'other' materials, and *S. pullorum* in chickens and egg pulp but very seldom in anything else;

4. mainly environmental and rarely infecting man or animals, typified by *S. senftenberg* and the various serotypes found in coconut.

Nevertheless the largest range of serotypes occurred in humans, only five types in Fig. 3 not being recorded for man. This suggests that all Salmonellas are capable, on occasion, of infecting humans, so all should be regarded as potentially dangerous. Much more detailed work is required on the factors influencing infectivity and environmental resistance of selected serotypes. There seems little doubt that characteristics beyond those embodied in the Kauffmann-White scheme are involved in these special properties of individual serotypes.

From the figures presented here, which represent only a fraction of the existing Salmonellas in Australia, it seems safe to say that, over the years, salmonellosis has been increasing rather than diminishing, and the situation in 1962 was certainly no better, and probably worse, than it was years ago. If we wish to reduce this disease, much more careful work is required in tracing outbreaks and following up individual cases. Only by painstaking laboratory testing, in conjunction with enlightened public health policies, can we hope to understand the mode of spread and ultimately defeat this ever present disease.

REFERENCES

ANDERSON, K. & WOODRUFF, P. (1961): *Med. J. Austral.*, 1, 856.
ANDERSON, K. (1962): unpublished.
ATKINSON, N. (1943): *Austral. J. exp. Biol. med. Sci.*, 21, 171.
ATKINSON, N. (1946): *Med. J. Austral.*, 1, 326.
ATKINSON, N. (1955): *Austral. J. exp. Biol. med. Sci.*, 33, 371.
ATKINSON, N. (1956a): *Ibid.*, 34, 231.
ATKINSON, N. (1956b): *Ibid.*, 34, 361.
ATKINSON, N. (1956c): *Ibid.*, 34, 369.
ATKINSON, N. (1957): *Ibid.*, 35, 1.
ATKINSON, N. & BULLAS, L. R. (1956a): *Ibid.*, 34, 225.
ATKINSON, N. & BULLAS, L. R. (1956b): *Ibid.*, 34, 349.
ATKINSON, N. & BULLAS, L. R. (1956c): *Ibid.*, 34, 27.
ATKINSON, N. & BULLAS, L. R. (1956d): *Ibid.*, 34, 445.
ATKINSON, N. & BULLAS, L. R. (1956e): *Ibid.*, 34, 455.
ATKINSON, N. & BULLAS, L. R. (1957): *Ibid.*, 35, 193.
ATKINSON, N. & CARTER, M. C. (1953): *Ibid.*, 31, 591.
ATKINSON, N., CARTER, M. C., WOLLASTON, J. M. & WALL, M. (1953): *Ibid.*, 31, 465.
ATKINSON, N. & EVANS, E. (1960): unpublished.
ATKINSON, N. & GEYTENBEEK, H. (1953): *Austral. J. exp. Biol. med. Sci.*, 31, 441.
ATKINSON, N., GEYTENBEEK, H. SWANN, M. C. & WOLLASTON, J. M. (1952): *Ibid.*, 30, 333.
ATKINSON, N., GEYTENBEEK, H. & WOODROOFE, G. M. (1952): *Austral. J. exp. Biol. med. Sci.*, 30, 177.
ATKINSON, N. & KLAUSS, C. (1954): *Ibid.*, 32, 221.
ATKINSON, N. & KLAUSS, C. (1955): *Ibid.*, 33, 375.
ATKINSON, N. & WOODROOFE, G. M. (1944): *Ibid.*, 22, 51.
ATKINSON, N., WOODROOFE, G. M., CHIBNALL, H. & MANDER, S. (1949): *Ibid.*, 27, 597.
ATKINSON, N., WOODROOFE, G. M. & CULVER, D. E. (1952): *Ibid.*, 30, 73.
ATKINSON, N., WOODROOFE, G. M. & MACBETH, A. M. (1944): *Ibid.*, 22, 301.
ATKINSON, N., WOODROOFE, G. M. & MACBETH, A. M. (1947): *Ibid.*, 25, 25.
ATKINSON, N., WOODROOFE, G. M. & MACBETH, A. M. (1949): *Ibid.*, 27, 375.
ATKINSON, N., WOODROOFE, G. M. & MACBETH, A. M. (1950a): *Ibid.*, 28, 367.
ATKINSON, N., WOODROOFE, G. M. & MACBETH, A. M. (1950b): *Ibid.*, 28, 377.
COBLEY, J. F. & WILSON, T. E. (1946): *Med. J. Austral.*, 1, 439.
COOPER, G. N. & WILSON, M. M. (1957): *Ibid.*, 1, 829.
COTTEW, G. S. & FRANCIS, J. (1954): *Austral. vet. J.*, 30, 301.
FENNER, F. & JACKSON, A. V. (1946): *Med. J. Austral.*, 1, 313.
FERRIS, A. A., HERTZBERG, R. & ATKINSON, N. (1945): *Ibid.*, 2, 368.
GRAY, D. F. (1960): *Vic. vet. Proc.* 1959–60, 31.
GRAY, D. F., HARLEY, O. C. & NOBLE, J. L. (1960): *Austral. vet. J.*, June, 246.
GRAY, D. F., LEWIS, P. F. & GORRIE, C. J. R. (1958): *Ibid.*, Nov. 345.

589

Institute of med. and vet. Science, Adelaide, S.A. Ann. Rep., 1960–61, 1961–62.

JONES, H. I. & FENNER, F. (1947): *Med. J. Austral.*, **2,** *356.*

KAUFFMANN, F. (1961): Die Bakteriologie der Salmonella-Species, Munksgaard, Copenhagen.

KOVACS, N. (1957): *Rep. pub. Health Lab. W.A.*

KOVACS, N. (1959): *Med. J. Austral.*, **1,** *557.*

LEE, P. E. (1955): *Austral. J. exp. Biol. med. Sci.*, **33,** *113.*

LEE, P. E. & MACKERRAS, I. M. (1955): *Ibid.*, **33,** *117.*

MACKERRAS, M. F. & MACKERRAS, I. M. (1949): *Ibid.*, **27,** *163.*

MUSHIN, R. (1948): *J. Hyg., Camb.*, **46,** *151.*

MUSHIN, R. (1950), *Austral. J. exp. Biol. med. Sci.*, **28,** *493.*

Queensland Dept. of Agriculture and Stock, Annual Reports, 1950, 1951, 1952, 1955, 1958, 1959, 1960, 1961, 1962.

RAC, M. & WALL, M. (1952): *Austral. vet. J.*, **28,** *173.*

RUBBO, S. D. (1948): *J. Hyg. Camb.*, **46,** *158.*

SEDDON, H. R. (1953): C'wealth of Australia, Dept. of Health Service Pub. No. 10, part 5, vol. 2.

SIMMONS, G. C. (1962): unpublished.

SIMMONS, G. C. (1951): *Austral. vet. J.*, **27,** *296.*

SIMMONS, G. C., CONNOLE, N. D. & ELDER, J. K. (1962): in press.

SIMMONS, G. C. & SUTHERLAND, A. K. (1950): *Austral. vet. J.*, **26,** *57.*

WATTS, P. S. & WALL, M. (1952): *Ibid.*, **28,** *165.*

WILSON, M. M. AND MACKENZIE, E. F. (1955): *J. appl. Bact.*, **18,** *510.*

POSTFACE

Au moment où il élaborait le projet de ce livre, l'éditeur se trouvait devant une double alternative: ou bien laisser aux auteurs une liberté totale non seulement pour la conception générale de leurs articles mais également pour tous les détails — la nomenclature y compris — et cela même si personnellement il ne partageait pas l'opinion de l'un ou de l'autre, ou bien suggérer aux collaborateurs, avec une persuasion plus ou moins prononcée, un schéma à suivre et certaines règles à observer.

L'option pour la deuxième alternative aurait sans doute contribué à donner à ce livre une plus grande homogénéité. Par contre, elle aurait empêché les auteurs à developper librement des jugements personnels et à exprimer certaines convictions intimes.

Le choix ne fut pas facile, car il comportait des risques dont les conséquences ne pouvaient être prévues dans leur ensemble. La possibilité de prévenir les aléas dans un cas comme dans l'autre par une attitude de liberté mitigée ou de pression tempérée ne parut pas digne d'être retenue.

C'est finalement la première solution qui fut choisie, ce qui explique pourquoi ce livre exprime par endroit des opinions divergentes et aussi pourquoi certains problèmes — surtout d'ordre épidémiologique — ont été envisagés selon des points de vue très différents. Nous croyons qu'il convient en fin de compte de s'en féliciter.

Nous sommes en tout cas pénétrés de la conviction que ce livre comble une lacune.

LIST OF AUTHORS
LISTE DES AUTEURS
AUTOREN-LISTE

1. E. S. ANDERSON, M.D.: Director, Enteric Reference Laboratory, Central Public Health Laboratory, Colindale Avenue, London N.W. 9
2. NANCY ATKINSON, D.Sc.: Reader in Microbiology, Department of Bacteriology, University of Adelaide, Box 498D, G.P.O., Adelaide, South Australia.
3. H. D. BREDE, Prof., Dr. med.: Professeur de Microbiologie médicale et Chef du Laboratoire de Microbiologie de la Faculté de Médecine de l'Université de Stellenbosch, P.O. Box 53, Bellville/C.P., Afrique du Sud.
4. E. T. BYNOE, Ph.D.: Director, Laboratory of Hygiene, Department of National Health and Welfare, Ottawa, 4, (Ontario) Canada.
5. H. FEY, Prof., Dr.med.vet.: Direktor des Veterinär-Bakteriologischen Institutes der Universität Bern, Engehaldenstrasse 6, Bern, Schweiz.
6. H. FUKUMI, M.D.: Chief, Department of Bacteriology, Chief, Japanese National Salmonella Centre, National Institute of Health, 284 - Chojamaru, Kamiosaki, Shinagawa-ku, Tokyo, Japan.
7. MILDRED M. GALTON, Sc.M.: Chief, Veterinary Public Health Laboratory, Communicable Disease Centre, Atlanta 22, Georgia, U.S.A.
8. CH. B. GERICHTER, M.D. (Collaborator of Dr. W. SILBERSTEIN).
9. G. GHYSELS, Dr.méd.: Attaché à l'Institut d'Hygiène et d'Epidémiologie, 14 - rue Juliette Wytsman, Bruxelles 5, Belgique.
10. T. IINO, M.D.: Head, Laboratory of Microbial Genetics, National Institute of Genetics, Misima, Japan.
11. F. KAUFFMANN, Prof., Dr.med.: Leiter der Internationalen Salmonella-Centrale, Statens Seruminstitut, 80 - Amager Boulevard, Kopenhagen S, Dänemark.
12. K. LACHOWICZ, Prof., M.D.: Chief, Department of Bacteriology, Laboratory for Enteric Infections, State Institute of Hygiene, 24 - Chocimska, Warsaw, Poland.
13. J. LEDERBERG, Prof., M.D.: Professor of Genetics, Department of Genetics, Stanford University Medical School, Palo Alto, California, U.S.A.
14. L. LE MINOR, Prof.agr., Dr.méd.: Chef de Laboratoire à l'Institut Pasteur de Paris, 25 - rue du Docteur Roux, Paris (XV), France.
15. A. E. MAYA, Dr.med.: Ehemahliger wissenschaftlicher Mitarbeiter am Hygiene-Institut der Universität Bonn, 4 - Am Feldschlössel, 759 Achern, Deutschland.
16. K. W. NEWELL, M.B., Ch.B., D.P.H.: Professor of Epidemiology, Tulane University School of Medicine, New Orleans, Louisiana, U.S.A.
17. P. NICOLLE, Dr.méd.: Secrétaire du Comité International de la Lysotypie Entérique, Chef du Service des Bactériophages de l'Institut Pasteur de Paris, 25 - rue du Docteur Roux, Paris (XV), France.
18. J. OLARTE, Dr.med., D.Sc.: Jefe del Laboratorio de Bacteriología intestinal, Hospital Infantil de México, 162 - Calle Dr. Marquez, México 7, D.F., México.
19. E. VAN OYE, Prof., Dr.méd.: Chargé de cours à l'Université Libre de Bruxelles, Chef du Centre National Belge des Salmonella et des Shigella, 14 - rue Juliette Wytsman, Bruxelles 5, Belgique.
20. C. A. PELUFFO, M.D.: Professor of Bacteriology, Head Department of Bacteriology, Instituto de Higiene, 3051 - av. Alfredo Navarro, Montevideo, Uruguay.
21. M. RAYNAUD, Prof.agr., Dr.Sci., Dr.méd.: Chef de Service à l'Institut Pasteur de Paris, 25 - rue du Docteur Roux, Paris (XV), France.

22. H. P. R. SEELIGER, Prof.,Dr.med.: Leiter der Salmonella-Zentrale am Hygiene-Institut der Universität Bonn, Bonn/Venusberg, Deutschland.
23. F. N. SICKENGA, M.D.: Secretary to the Health Council of the Netherlands, 8 - Dr. Kuyperstraat, 's-Gravenhage, Holland.
24. W. SILBERSTEIN, Prof.,M.D.: Director, Government Central Laboratories, Ministry of Health, Jaffa Road, Jerusalem, Israël.
25. ANNE-MARIE STAUB, Dr.Sci.: Chef de Laboratoire à l'Institut Pasteur de Paris, 25 - rue du Docteur Roux, Paris (XV), France.
26. J. H. STEELE, D.V.M., M.P.H.: Chief, Veterinary Public Health Section, Communicable Disease Centre, Atlanta 22, Georgia, U.S.A.
27. G. VARELA, M.D., M.P.H.: Director del Instituto de Salubridad y Enfermedades Tropicales, 470 - Carpio, México 17, D.F., México.
28. P. VASSILIADIS, Prof. agr.,Dr.méd.: Chef du Laboratoire de Bactériologie de l'Hôpital Général "Reine Frederica" du Pirée, 119 - avenue de la Reine Sophie, Athènes 602, Grèce.
29. J. A. YURACK, Ph.D.: Bacteriologist-in-Charge, National Salmonella Reference Centre, Ottawa 4, (Ontario) Canada.

SUBJECT INDEX